READING THE ROCKS

SIR WILLIAM E. LOGAN
1798–1875

READING THE ROCKS

MORRIS ZASLOW □ THE STORY OF THE GEOLOGICAL SURVEY OF CANADA 1842-1972

Published by

The Macmillan Company of Canada Limited,
Toronto, Ontario, in association with the
Department of Energy, Mines and Resources
and Information Canada

Ottawa, Canada
1975

ISBN 0-7705-1303-4

Published May, 1975, by
The Macmillan Company of Canada Limited,
Toronto, Ontario

CONTENTS

PREFACE

Like other students of Canadian history, I have long known and admired the work of such celebrated explorers as Robert Bell, G. M. Dawson, A. P. Low, J. B. Tyrrell, and Charles Camsell. These encounters whetted my curiosity about the institution that these men had served with distinction, so I willingly accepted a commission from the Geological Survey of Canada to prepare a history of its eventful 130 years of service to the nation. It soon became apparent that the project presented greater challenges and even more opportunities than I had at first envisaged. The travels and explorations generally associated with the Survey comprise only a small part—the visible tip of an iceberg as it were—of its total achievements. Since the Geological Survey is a long-lived branch of the national government, its history must also be concerned with how the Survey functioned so successfully in conjunction with a variety of administrative and political superiors, with the mining and other industries, with the academic and scientific communities, and with five generations of the Canadian public. It is also the history of how a never-ending mission was conducted during many decades of changing methods that reflected the continuing advances in mining, transportation and laboratory technologies, and drew on constantly-evolving bodies of scientific knowledge. But more important still, it is the history of the Survey's staff, the hundreds of men and women who laboured in different capacities to win a broader, deeper understanding of Canada's vast territory.

While my training as a historian may have equipped me to deal with some aspects of the subject, for others, such as evaluating scientific work or tracing the transformation of the youthful science of 1842 into the highly-specialized group of geosciences of today, it has been necessary to seek assistance from persons with special technical knowledge or experience. In attempting to cope with these topics, I was pleasantly surprised to find that some common ground exists between the approach of a geologist and that of a historian. Since both are concerned with change in relation to time, they follow similar premises with regard to causality and accept similar standards of accuracy, objectivity, and critical evaluation of sources. Moreover, because this history traces a science from its early beginnings, I was able to see new concepts and technical innovations in their proper order of development, and to align my thinking to the increasing sophistication of the modern geosciences. (My experience, indeed, leads me to wonder whether geologists might not also benefit from the experience of studying the history of their science as an aid in developing concepts that may be valid for solving current problems, and perhaps in relieving the present tendencies to over-specialization.)

Nevertheless, a non-geologist would find it utterly impossible to investigate the scientific or technical dimensions of the Survey's history without assistance from a great many specialists in various fields. Almost every page of this work reflects my immense debt to the gentle, patient, good-humoured scholarly H. M. A. Rice, whose death in 1970 was a shock to his many friends. J. F. Henderson, whose intimate experience with developments at the Survey over the past 35 years gave him a special insight into the latest period, provided valuable aid with the final part of the study. Present and former members of the Survey generously supplied me with personal or other information and helped clear up innumerable points of fact or interpretation. Of special value were the contributions from many retired staff whose experience went back to the beginning of this century (M. E. Wilson, Eugene Poitevin, M. Y. Williams, and B. R. MacKay), or to the interwar years (F. J. Alcock, J. F. Walker, R. T. D. Wickenden, H. S. Bostock, George Hanson, Diamond and Mrs. Jenness, Mrs. F. B. Forsey, C. O. Hage, J. C. Sproule, J. F. Caley, A. H. Lang, C. H. Stockwell, P. S. Warren, S. C. Ells, and T. E. Warren—the last two, former employees of the Mines Branch). The present staff provided a quantity of information on current and recent developments (including several papers outlining the background of various units) and, as mentioned earlier, were always available with answers to specific questions to help me with difficult topics. Thanks are also due to a number of other organizations for help with the research and illustrations. These include the Public Archives, the National Library, and the National Museum, all in Ottawa; the New Brunswick Museum in Saint John; the Redpath Library, McGill University Archives, McCord Museum, and Montreal City Archives, in Montreal; the University of Toronto Library and the Metropolitan Toronto Public Library; the Saskatchewan Research Council; the British Columbia Department of Mines and Petroleum Resources; and the Departments of Geology at the Universities of Saskatchewan, British Columbia, and Western Ontario.

Several Survey officers also read and commented on parts of the various drafts of text. This burden, however, fell mainly on H. M. A. Rice, J. F. Henderson, and the members of the History Committee: A. H. Lang and G. B. Leech, its successive chairmen, S. C. Robinson, Edward Hall, E. I. K. Pollitt, Peter Harker, and R. G. Blackadar. The History Committee also facilitated my interviewing retired staff as well as present employees stationed outside Ottawa. In addition, Harker and Blackadar were instrumental in enlisting the support of various units of the Geological Information Processing Division, while Pollitt and Hall also attended to the financial and administrative aspects.

The many other persons at the Survey who helped in preparing this study included the personnel of the library, headed by Mrs. D. M. Sutherland; of the Branch Registry, headed by R. D. Robillard; of the Financial Services Section, headed by Mrs. M. I. Going; and of the Personnel Section, headed by K. J. Fracke; also G. J. Charlebois of the Purchasing and Supplies Section; L. J. Lajoie, who attended to the space requirements; and J. L. L. Touchette, who supplied copies of publications and maps. The illustrations side owes a great deal to the efforts of the Photographic Services and Cartography Sections, especially D. C. Beckstead, the photographic librarian, and Mary Raddatz of Cartography Unit 'B', who drew the many new maps; also to Marguerite Rafuse and Dorothy Whyte of the editorial staff of the department, who in addition to editing the text and seeing the book into final form, also attended to the layout.

Besides giving me access to many of its records and making its facilities and personnel available, the Survey also provided a series of university students who were employed as research assistants in the summers from 1965 to 1968 inclusive: T. B. Smyth, James Bradburn, G. R. I. MacPherson, W. E. Eagan, G. A. Gartrell, R. J. Montague, and Douglas Wyatt. Their work was most helpful during the early stages of the project. Two of the several typists employed at various times, Mrs. Irene Baker and Miss Pauline Laurin, typed most of the manuscript. The Survey also reimbursed most of my travelling and living expenses and paid me for some of the time spent on the project, notably during the year 1968 - 69 when I was on sabbatical leave from my university post and could turn my full attention to research and writing. I can hardly give enough credit to the Survey administration for these kinds of assistance and for the interest shown in the progress of the work—especially since the support was without any strings attached and I received *carte blanche* to write this book in my own way. Apart from matter quoted from Survey publications, the Survey bears no responsibility for the contents, interpretations, or points of view expressed in this book, which are the responsibility of the author.

As with many large scholarly undertakings, the preparation of this study entailed many long separations from my family, all the more because the bulk of the research and writing had to be carried on in Ottawa. I have very special reason, therefore, for recording my appreciation of my wife's patient understanding and encouragement during the trying months and years needed to bring this study to completion, as well as my thanks for her assistance with the manuscript, the proofs, and the index.

MORRIS ZASLOW
Professor of History
University of Western Ontario
London, Canada

PART ONE

The Formative Years 1842–1881

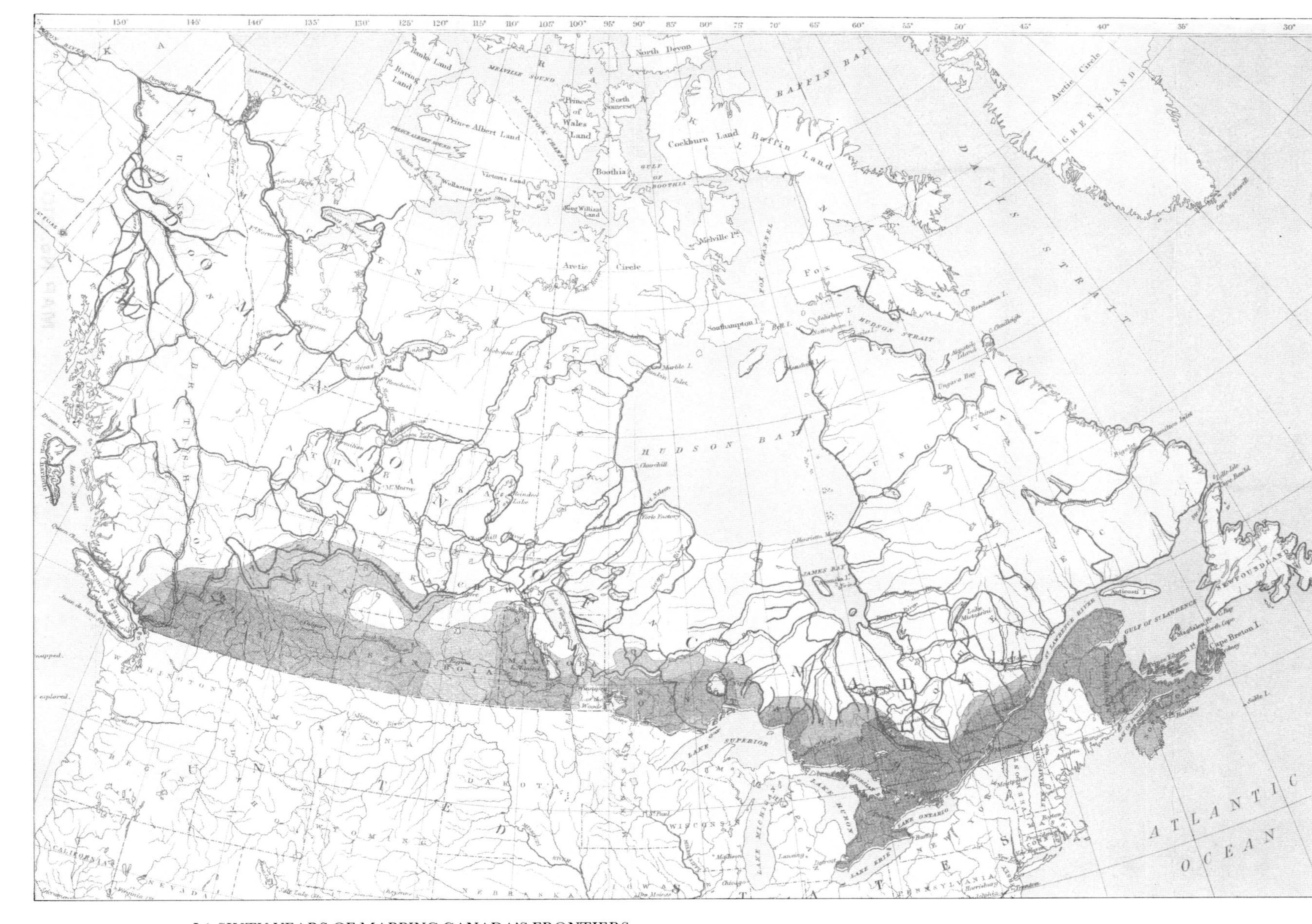

I-1 SIXTY YEARS OF MAPPING CANADA'S FRONTIERS
This map was published in 1902 by Robert Bell, the Survey's chief, to celebrate the diamond jubilee of the Geological Survey of Canada. The districts and routes surveyed since 1842 (shown in red in the original) demonstrate how much the work of mapping depended on the network of waterways that crisscrossed the unsettled parts of the country. The map shows the major role the Survey played in opening the Canadian west and north. Scale: approx. 1 inch = 400 miles.

1

Founding the Geological Survey of Canada

WHEN THE GEOLOGICAL SURVEY OF CANADA was founded in 1842 no one could have imagined that such a tentative creation would still be flourishing 130 years later as one of the most illustrious institutions of its kind anywhere in the world. Even less could one have envisaged its present scale and complexity, the magnificent part it has played in the development of Canada, its pioneering role as the first scientific arm of the Canadian government, or its contributions to expanding mankind's knowledge in the field of the earth sciences. Indeed, along the way from its humble beginnings as a two-man operation to its present full-time staff of over 500, its achievements have been quite out of proportion either to the size of the work force, or to the public funds spent in sustaining its efforts. In its first 60 years, while engaged in mapping the geology of the subcontinent that is Canada, the Survey began the systematic mapping of the vastnesses of the Canadian Shield and the rugged Cordilleran country, made the first accurate maps of most of Canada's northern rivers, and was the first arm of the federal government to make an appearance in many districts.

For when the Survey was begun, the geographical entity extending over half a continent and known as the Dominion of Canada was still some distance in the future. What was then called Canada—the government that established the Geological Survey and to whose territory its activities were confined for 25 years—was the Province of Canada, roughly those parts of present-day Quebec and Ontario lying within the St. Lawrence-Great Lakes drainage basin. Newfoundland, Nova Scotia, Prince Edward Island, and New Brunswick were British colonies too, each with its own government and more or less with the same boundaries as today. The remainder of present-day Canada—in which the Survey was to do its major work after 1870—was under the nominal control of the Hudson's Bay Company as its charter territory of Rupert's Land, as was the North-Western Territory, which the company held from the British Crown under 21-year licence. By 1870 the western part of the last region would constitute another British Crown colony, British Columbia.

The vast extent of country over which the Survey ranged made its men the foremost authorities on the unsettled parts of Canada—their soils, forests, water-powers, wildlife, minerals, climate, navigation and communications, native peoples, and the need for, and problems of governing these areas. Their advice was invaluable in drawing attention to Canada's resources and opportunities, in guiding the Dominion government to exercise its sovereignty and to determine its policies of resource management. Through the reports, articles, and lectures of its personnel, but especially through its exhibits at the great international fairs, the Survey played an important role in making an interested world aware of Canada's potentialities.

The men of the Survey not only mapped the geology; they collected materials from every phase of life they encountered—samples of rocks and minerals, plants, animals, fishes, birds, Indian and Eskimo artifacts, even languages and legends. Through the work of exploring, mapping, and collecting information and specimens from across Canada, the Survey developed its museum at very little cost to the public into a comprehensive collection of all aspects of the Canadian scene, of natural and human history as well as of rocks and minerals. Soon it was being referred to as a national museum, which in time it was to become, after more than 80 years as the Survey's museum—the National Museum of Canada.

But after all, the purpose of the Survey's founding and the justification for its continuance was its role of assisting the progress of Canadian mining. How well has this responsibility been discharged? Through their maps and reports its geologists have guided generations of prospectors to the districts where particular kinds of mineral bodies are likeliest to occur, enabling them to concentrate their efforts on the most promising areas and ignore the much larger areas of barren rocks. Even more, by studying different types of rocks and learning the histories of their deposition, succession or intrusion, geologists have been able, with increasing success, to point out the clues that could reveal possible orebodies, to pinpoint with increasing accuracy those areas in which mineable deposits are most likely to be found, and to advise mine operators on the possible behaviour of mineral-bearing formations at depth.

In carrying out these functions the Survey has had to tread a careful path between activities that are of immediate and direct benefit for the mining industry and those that are significant at longer range, for the second category often produced misunderstandings that brought down criticisms and attacks on the Survey, its directors, and its work. Investigations of remote, inaccessible districts, and general researches into the qualities, characteristics, and histories of rock formations especially have given rise to misunderstanding, for who can know—least of all, the public—where the limit of practicality lies in such kinds of work? Such activities may appear to have no relation to current mining needs, yet, as we shall see, seemingly unimportant, esoteric researches have often borne tremendous economic consequences. Throughout its history the Survey has faced constant pressure to be as "practical" as possible, and its programs have been compromises between short-range and longer-term work, the proportions varying from period to period according to current economic, political, and industrial conditions. The trend, however, in recent years has been for the Survey to devote a somewhat greater part of its effort to general scientific studies.

Beyond question, the Survey has been a significant force in the Canadian mining industry's rise to its present prominence, and it has assisted the economic growth of the nation in a major way. The accurate, knowledgeable, disinterested advice of its staff, the published reports and maps of the geology of districts, or on particular minerals, have contributed inestimably to the opening of major mining regions and to the development of important mines. The Survey's researches into the genesis of orebodies and the other phenomena have helped mine operators to solve their most serious geological problems. By training many scientists, engineers, and managers in the industry, it has raised the standards of exploration for minerals, of mining practice, and of mineral resources administration in all parts of Canada, and more recently, in other countries as well.

The desire to serve the mining industry led the Survey into many areas of endeavour not commonly associated with geological work—collecting statistics of mineral production, investigating particular mineral industries and their markets, studying production and refining processes, and the like. These activities, which were undertaken from the early days, eventually developed into a Section of Mines, and when that proved insufficient to meet the needs of the growing mining industry, the section became an important constituent of the present-day Mines Branch.

In addition to the National Museum and the Mines Branch, the Survey can claim a part in shaping another agency of government. The work of surveying the geology of large parts of Canada compelled the Survey to organize a staff of topographers to relieve geologists of some of the time-consuming work of mapping the terrain, and thus enable them to concentrate on geological mapping. The Topographical Division was a highly efficient, professional organization that became a major component of the present-day Surveys and Mapping Branch. In more recent years the Survey has evolved other units and undertaken new activities that have been handed over, together with some staff, to other agencies of government.

The Geological Survey of Canada thus has played an important and unusual role as an institution, an arm of the national government. It has followed a strange administrative progression, starting out as a Crown corporation, operating under a short-term (two-year) contract. Brought to Ottawa and placed directly under government after 35 years of independent operations, it became for a time an autonomous government department, then one of two branches of a Department of Mines, and subsequently a branch or division within three successive multi-branched departments. At its commencement in 1842 it was a novelty—an agency financed entirely by government yet remaining outside its immediate control, to which the usual roles of political patronage appointments and of political direction of its work did not apply. The Survey went on to establish new standards of scientific and technical skills in the government service, and it was a strong force for the professionalization of the public service; indeed its members played a major part in setting up the professional association of civil servants. As an early example of the involvement of government in scientific work, the Survey established many trends, too, in working out such practical problems as recruitment and classification of technical staff, students' summer programs, and the like.

1. Keweenaw Peninsula	11. Furnace Falls	21. Gaspé Peninsula	31. New York
2. Chicago	12. Georgian Bay	22. Joggins	32. Wilkes-Barre
3. Detroit	13. Bytown	23. Parrsboro	33. Maunch Chunk
4. Olinda	14. Montreal	24. Cumberland Co.	34. Pottsville
5. Normanville	15. Laprairie	25. Albion Mines	35. Philadelphia
6. Niagara	16. Richelieu River	26. Pictou	36. Washington
7. Toronto (York)	17. St. Johns	27. Sydney	37. Pittsburgh
8. Kingston	18. St. Maurice River	28. Boston	
9. Blairton	19. Quebec	29. Springfield	
10. Madoc	20. Saguenay River	30. Albany

I-2 EASTERN NORTH AMERICA IN 1840

The Survey has also played a powerful role in promoting science in Canada. It has stood in the forefront of several advancing scientific fields, contributing to the fund of world scientific knowledge. Its work has made Canada known as an advanced progressive country. Indeed, the first occasion that such recognition was accorded Canada or Canadians was as early as 1856 when the Survey's first director, William Edmond Logan, was honoured with a knighthood. Within Canada the Survey has always maintained a close liaison with universities, the scientific and professional societies, and the scholarly journals. Its employment of generations of young scientists as field assistants, and more recently as researchers, has helped train many future leaders in the professions and sciences, who spread the knowledge and standards they imbibed while working for the Survey, around the Canadian community. As a result, the level of scientific training was raised, and the Survey was enabled to draw on better-qualified staff who could provide more advanced, more demanding types of service. Through this interrelationship the Survey was also able to draw on new reservoirs of talent to maintain its forward momentum, for an agency can be no better or stronger than the men who serve it. It is propelled forward as an institution by a changing group of people, who come to the task equipped with the latest scientific knowledge to illuminate their own work and to update the operations. This process is repeated every generation and every decade in the men recruited by the Survey. Each generation has contributed its talents and devotion to renew the vigour of a historic organization and enable it to face the challenges and the opportunities of a new time. Today, and for the last two decades, it has been confronted by both in abundance.

* * *

IN 1840 THE WESTERN WORLD was astir with excitement at the vistas of unparalleled material progress that lay before it. Distances overland were being overcome by railways, while the dreaded crossing of the Atlantic was made comfortable, speedy, and safe by steamships that brought remote empty, or backward continents within the ambit of European civilization and exploitation. The recently invented telegraph was capable of carrying news from afar virtually with the speed of light. Just as significant was the transformation of industry. The Industrial Revolution was in full swing, and hand methods of manufacture were giving way to factories which mass-produced cheaper, better goods than anything previously available. Minerals played a vital part in the transformation of the western world, and one in particular, the fossil fuel coal, was the driving force behind the movement—coal to heat homes and buildings, drive engines, smelt iron ore, or to produce coke, illuminating gas, and other by-products. Coal released industries from their dependence on swift-flowing rivers and allowed them to migrate to hitherto green, peaceful countrysides. Chimneys belching smoke began darkening the skies and surfaces of the English Midlands, while other "black countries" began to take form in south Wales, the Tyneside, the Clydeside of Scotland, the Rhineland and Silesia in Prussia, and New England and Pennsylvania in the United States.

Iron, another mineral, was the second fundamental ingredient for the new age. The manufacture of iron, tied to dwindling supplies of charcoal, had been a limiting factor in industrial expansion. Now that coke could be used to make iron and steel, the world output of pig iron soared, almost trebling in the 1830's alone. Iron and steel found new uses in bridges and buildings, railway locomotives and every type of engine, rails, and water pipes—even ships' hulls. Steam engines, applied particularly for de-watering mine shafts, made it possible to work mines more than 1,200 feet deep and win increasing tonnages of copper, tin, lead, and zinc metals, for which many new uses were being found. Men turned also to the earth for the greater quantities of building stone, clay, sand, and gravel needed in construction, for gypsum and phosphates for fertilizers, and for precious metals and gems for currency and luxurious living. Minerals were the key to modernity and progress; hence countries began looking to their subterranean resources for these all-important gateways to wealth and national power.

Britain, the workshop of the world and the first nation to attain this plateau of industrial achievement, pointed the way for other advancing nations. By the end of the 1830's, British steamships sailed virtually every sea, and England was crisscrossed by a network of railways connecting manufacturing centres, seaports, and mining districts. To provide the raw materials for these industries, the British mining industry expanded to unprecedented levels. Coalfields and iron mines were opened in the Midlands, the North, and in Wales, while tin, lead, and copper were won from ancient mines in the Midlands and the southwest. At the same time, the busy factories

I-3 THE ADVANCE OF INDUSTRIALIZATION: THE UNITED STATES IN 1841
During a number of trips in 1840-41 around Montreal, the northern United States, and the Maritime colonies, Logan set down his impressions in a bound notebook preserved in the Metropolitan Toronto Library. Foreshadowing modern-day concerns is the note he penned on August 22, 1841, beneath this sketch of the Lehigh River below Maunch Chunk (now Jim Thorpe), Pennsylvania: "Through the valley of the Lehigh run a turnpike a river a towing path a canal & a rail way filling up the whole of the flat ground."

and increasing population of Britain afforded a large market for raw materials from abroad—principally foodstuffs and timber, but also iron, lead, and copper ores to augment Britain's now-inadequate mineral resources.

Besides stimulating the industries of other lands through providing an enormous market for their primary products, Britain also had the surplus capital and the technology to help them establish modern industries like her own. The English railway builders, for example, built railways all over the world, often arranging the loans to finance construction. In this way the Industrial Revolution spread quickly to other favourable regions—particularly western Europe and eastern North America. In the United States conditions were exactly right for this economic transformation. The people were vigorous, alert, and progressive, while the natural resources of their land put those of Britain far in the shade. After the end of the Napoleonic Wars (and its own War of 1812 with Britain) the United States began to expand rapidly westward, aided by public investment in roads and canals which uncovered new resources for exploitation, and enlarged the home market for domestic manufactures. Steamboats began to ply the coasts, the mid-continental rivers, and the Great Lakes, and quickly the United States followed Britain in developing a network of railways. The iron industry, based upon numerous, widely-scattered, small, bog iron deposits, had expanded west to Pittsburgh as early as 1810. Copper was mined in small amounts in Vermont, New Jersey, Virginia, and Tennessee, and lead in Missouri and Wisconsin—this last from the days of the French Empire in America.

I-4 FIRST STEAM VESSEL TO PLY THE ST. LAWRENCE RIVER
Though the 81-foot *Accommodation* was rather slow, the service was well patronized; passengers during the War of 1812 included troops en route to the fighting in the west.

The real clue to future advances, however, was the adoption of coke for smelting iron ore, which freed the iron and steel industry from its dependence on wood-charcoal. The introduction of the new smelting processes in the years after 1837 at once led to the feverish exploitation of the magnificent deposits of anthracite in Pennsylvania and other states. The production of metal trebled in the United States during the 1840's, and the country advanced by 1860 to stand second only to Britain as an industrial power. William E. Logan, an expert on the Welsh coal deposits who visited Pennsylvania at the height of the boom in 1841, saw this as an inevitable development: "The coalfields of Britain are a trifle to those of America, & when it is recollected that with the coal is associated various veins & seams of Iron ore & huge beds of limestone it would seem as if providence intended that Virginia & Pennsylvania & Ohio should become the workshops of the world. The wealth that those three states will in the course of a quarter or half a century arrive at is incalculable."[1]

The British American colonies shared a little of this advance, as far as they could. The Province of Canada was emerging from her political and economic difficulties of 1837-38, thanks to a bountiful flow of British capital and British immigration, and to the political union of the colonies of Upper and Lower Canada in 1841 into a single government with a population that was to reach 1,800,000 within a decade. The province still was in a pre-industrial stage of development, producing raw materials—notably timber and wheat—for the British market, with the assistance of favourable imperial preferential tariffs. Its system of communications, principally the St. Lawrence waterway, was oriented towards this goal. Recognizing the importance of transportation to Canada's economy, Canadians came early to adopt the new media of communications. Barely eight years after the invention of the first steamboat in Britain, John Molson's *Accommodation* was carrying freight and passengers from Montreal to Quebec (1809), and beginning in 1816, the Great Lakes were served by Canadian steamboats. During the 1820's schemes were also propounded for railways that would link Lake Ontario with Lake Huron, Niagara with Detroit, and the Atlantic coast with the St. Lawrence River. A first passenger railway was in operation by 1836—the 16-mile Champlain and St. Lawrence, between Laprairie and St. Johns, which cut off a difficult 100-mile passage by rivers between the Richelieu River points and Montreal. At the same time tracks were laid down the Niagara Escarpment for a portage railway, but the road could not be operated until iron rails and more powerful engines became available. In Nova Scotia, too, a short six-mile railway between Albion Mines and the coast was in operation by 1838 to facilitate the movement of coal to shipboard.

Canada had neither the markets nor the political climate to embark on an industrializing course like the United States, though she too had long experience with small bog iron manufactories that went back to the St. Maurice forges, opened early in the eighteenth century to help New France achieve a measure of self-sufficiency. Other early iron works in Upper Canada included those at Furnace Falls (Lyndhurst) in 1800; at Normanville, Norfolk County, perhaps the longest-lived (1820-37), best managed (by Joseph Van Norman) and largest (750 tons per annum) of all; and at Blairton, Madoc, and Olinda. These depended on charcoal for smelting, but the main limiting factor was lack of markets. So small was the domestic need that most of their production was exported in the form of pig iron to the United States. Unfortunately, Canada seemed deficient in coal as well, which would be essential if the colony, like the neighbouring states, was to advance into steel manufacturing. There were rumours, of course, of other minerals, notably of copper at various places along the north coast of Lake Superior, arising from an exploration of Alexander Henry the elder in the 1770's. A more promising possible site for an iron and steel industry than Canada was the sister colony of Nova Scotia, which possessed large, excellent coalfields in Pictou, Cumberland, and at Sydney, as well as local supplies of iron ore, and of limestone, essential for fluxing. Nova Scotia's coal exports increased dramatically during the twenties, thirties, and forties when the province emerged as an important coal producer—the source of much of the fuel shipped down the Atlantic coast to American users. Additionally, the colony was an important exporter of other minerals, notably of gypsum, used for soil dressing. Canada, Nova Scotia, and the other colonies too, were aware of the important part that mineral resources would play if they were to fulfil their destinies in the world of the nineteenth century, and each colony was conscious of the need to develop its own mines and factories.

To discover deposits of mineable minerals a great deal of effort was being expended in many countries, most of it in a very 'hit and miss' fashion, with little knowledge of, or attention to, the principles of mineral occurrence. Coal was sought in every conceivable geological environment, prospectors were constantly misled by accidental showings, and investors by exaggerated or falsified evidence. Speculators and the public at large seemed perpetually willing to risk their money on the most fantastic schemes. Even in such coal-rich lands as England or Pennsylvania, large sums were wasted with little commensurate return through entrepreneurs not knowing where to search. Most indeed knew so little about mines and ores that they could scarcely recognize an outcrop of coal or of possible ore even when standing on it, or tell in which direction to seek its extension. However, a practical science of ore-finding and mine development was evolving, based on observations on the occurrences of minerals and experience of the varied structures of the mineral deposits encountered during mining. Seventeenth and early eighteenth century writers on coal mining already referred to outcrops, dips, rises, strikes, dykes, and faults; one (Strachey, 1719) even sketched a complete succession of coal measures broken by a fault and picked up again in the next valley.

I-5 NOVA SCOTIA'S FIRST STEAM LOCOMOTIVE
Both the locomotive "Samson" and the passenger coach "Nova Scotia Pioneer" were built in England in 1838 for use on the short Albion Mines railway line.

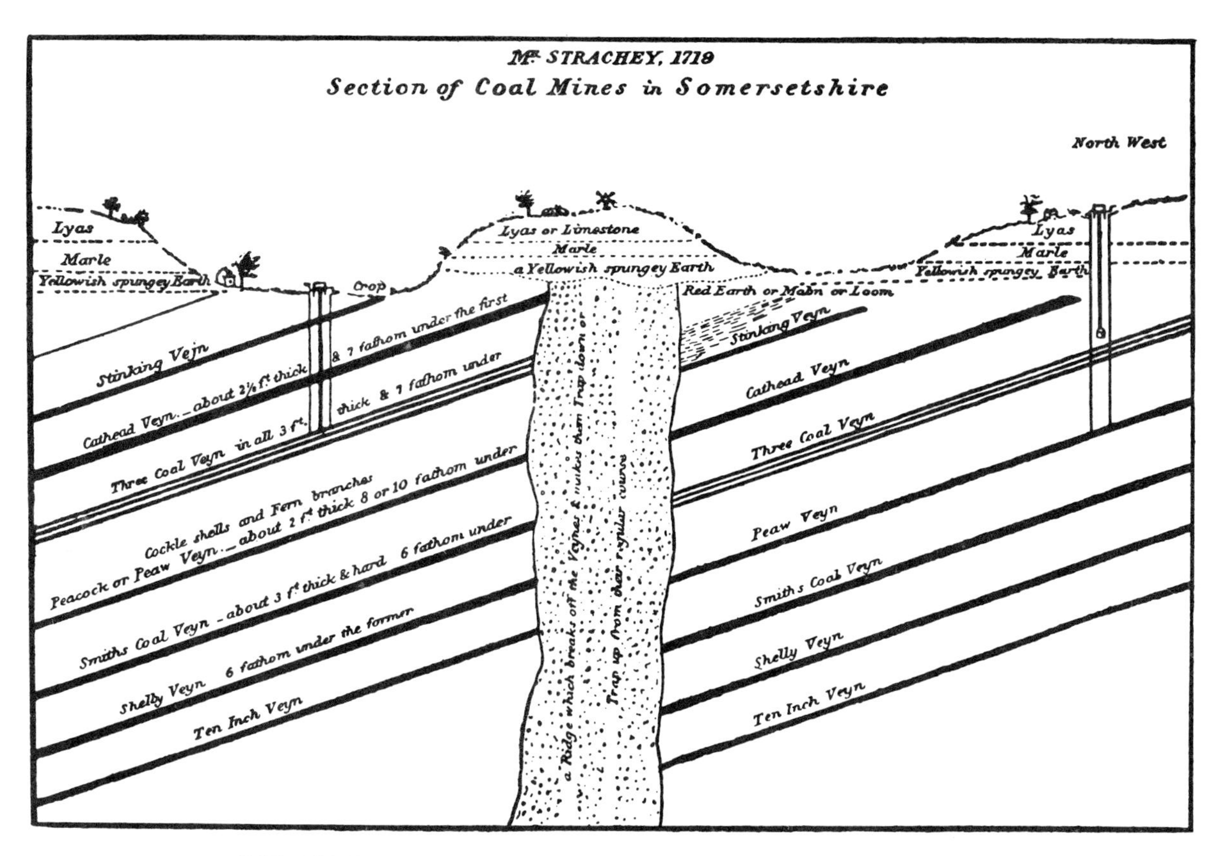

I-6 EARLY DIAGRAMMATIC REPRESENTATION OF BRITISH GEOLOGY
Prepared by John Strachey for a paper on the Somersetshire coal district.

The possibility of increasing the rate of mineral discovery by mastering the principles under which certain types of rocks are formed and where they might be found powerfully stimulated the developing science of geology. Though speculations about the nature and origin of the earth's surface are surely as old as man himself, it was not until the early nineteenth century that scientific study began to achieve meaningful, coherent results. The science of geology (the name, in the sense of 'the study of the earth,' apparently was coined by a Norwegian author in 1657 and was first used in an English work in 1661) was beginning to develop principles and theories, based on the observation of phenomena around the world. Characteristically, geology was the hobby of comfortably-situated gentlemen or professional men, especially physicians and clergymen, but students were also beginning to form associations to exchange information and ideas, disseminate information, and provide forums for discussion. A leader almost from its establishment in 1807, was the Geological Society of London, formed for the purpose of communicating "the result of their observations, and of examining how far the opinions maintained by the writers on geology were in conformity with the facts presented by nature."[2] The members of such associations described the geology of particular areas, and their maps of the strata often were invaluable to prospectors and others. The value of geological researches and maps for mineral exploration was quickly recognized, and governments began sponsoring geological surveys to ensure that the fullest use was made of the natural resources of their lands. By the 1820's some of the more progressive states of the American union were inaugurating geological surveys with a view to attracting settlers to their vacant lands, encouraging investment, and enhancing the general prosperity of their regions.

Geological mapping was carried on increasingly by the regular map-making agencies of governments, or by units organized specially for the purpose. As early as 1780 a geological map of France in sixteen sheets, made by Messieurs Guettard and Monnet, was published by royal order. In Britain, in 1794, the Board of Agriculture began issuing a series of county maps that described the soils and rock exposures, and between 1811 and 1817 it sponsored the three volumes of the first official geological memoir in Britain, *General View of the Agriculture and Minerals of Derbyshire,* by John Farey. About the same time

I-7 SIR H. T. DE LA BECHE

work was begun on the geological map of Scotland by an employee of the Board of Ordnance, and on that of Ireland by the Inspector of Mines. Somewhat later, the Ordnance Trigonometric Survey, which had been founded in 1791 to prepare topographical maps for military purposes, under the command of Captain T. F. Colby, began to compile data for historical, statistical, and geological as well as topographical purposes. Following the publication in 1815 of the great geological map of England by the travelling engineer William Smith, the Ordnance Survey became interested in geological mapping to its scale of one inch to the mile (1 inch = 1 mile).

In 1832 the Board of Ordnance was approached by Henry De la Beche, a member and officer of the Geological Society of London and author of a recently published *Manual of Geology* (1831), to be allowed to use the Ordnance maps of Devonshire as the basis for a detailed geological map of the county. The upshot was that De la Beche was appointed geological colourer to the Ordnance Survey and was granted £300 to complete the geology for the map of Devonshire, which task he finished in 1834. In the spring of 1835 Colonel Colby consulted a number of leading geologists to elicit their views on combining geological investigations with the work of topographical surveys. He was advised that such a survey would be invaluable to industry and would promote geological science, also that it could be performed economically by the topographical surveyors under the direction of an experienced geologist. A position was established, and De la Beche was appointed "director of the ordnance geological survey," with an appropriation of £1,000 a year, a salary of £500, and instructions to "commence the Geological Survey of Cornwall without delay."[3] De la Beche completed the maps of Cornwall and Wales, published a *Report on the Geology of Cornwall, Devon and West Somerset* (1839), was given charge of a mining record office (1839), opened a Museum of Practical Geology (1841), and finally, in 1845, was appointed director general of the Geological Survey of Great Britain and Ireland, at the head of an establishment that included regional directors for England and Ireland.

In making geological surveys a state activity, the British were only following some of the more progressive, ambitious member states of the United States. For interest in the new science of geology, though perhaps less well informed and more pragmatic in its emphasis, was almost as keen in the United States as in Britain. Both of the pioneer learned societies—the American Philosophical Society of Philadelphia and the American Academy of Science of Boston—published scientific works from 1771 onward and included geology as part of their study of natural history. They were joined in 1799 by the American Mineralogical Society and, rather later, by a first, short-lived American Geological Society (1819-30).

Following Benjamin Silliman's pioneer course in chemistry and mineralogy at Yale in 1806, the teaching of geology and mineralogy spread very rapidly in American universities. Silliman founded the *American Journal of Science* in 1818, and directed it himself until 1838, when he handed it over to his son. The magazine was by no means exclusively devoted to geology; its original title page read "American Journal of Science, more especially of Mineralogy, Geology and Other Branches of Natural History, including also Agricultural and the Ornamental as well as useful Arts." Perhaps this catholicity of interests accounted for its success; within four years it was self-sustaining. The *Journal* had a great influence upon the development of geology, however, since for years it was one of the most important organs for the promulgation of scientific information, used by Americans and Canadians alike. Many early contributions by officers of the Geological Survey of Canada appeared in its pages. Of equal importance for the progress of geology in America was Silliman's role as a teacher. Trained as a lawyer and completely unacquainted with the science when he was appointed in 1802, he not only mastered the field but trained pupils who became professors of geology, directors of state surveys, and the future leaders of American geological science in their turn. The elder Benjamin Silliman has truly been described as the father of geological science in the United States.

I-8 BENJAMIN SILLIMAN
A graduate of Yale University, he became professor of chemistry and natural history at the age of 23, later enlarging his fields of instruction to include geology, mineralogy and pharmacy.

Parallel with Silliman's work at Yale and his missionary labours on behalf of science, went the steady onward progression of geological knowledge in the United States. At first serious scientists had to cope with a host of charlatans, pseudo-scientists, Biblical exegesists, and well-meaning but misguided commentators and practitioners. Still, the most advanced geological works written in Europe were quickly received and studied in the United States, and America frequently was visited by leading foreign geologists. Then too, from the early nineteenth century there began to appear a succession of geological textbooks, written and produced in the United States, describing American conditions. These helped to elevate the science of geology, as practised, above the level of the mere cataloguing of data accompanied by very elementary theories of causation, occurrence, and structure. Gradually, American students began to recognize the principles of sedimentary deposition, to collect and study fossils, and to apply the new definitions of geological periods and categories where they seemed applicable. The geology of the United States itself became increasingly studied, notably that of the eastern seaboard states, the Appalachian mountain chain, and the coal basins of Pennsylvania, Ohio, and Virginia. Official expeditions to the western territories began to include men trained in geology, or professional geologists, with the result that by the mid-1840's considerable good work had been done on the geology of such areas as the Ozarks, the Mississippi basin, and the Rocky Mountains. In most instances this was the work of geologists in the employ of the United States Congress, which restricted its role to the territories, geological information being regarded in the same category as the other surveys of the western territories—that is, as assisting Congress to dispose of the public domain.

Similar economic motives impelled many states to undertake geological surveys in the years after 1823, when one of Silliman's disciples, Denison Olmsted, persuaded North Carolina to sponsor a geological survey as a means of attracting capital and settlers. Olmsted, subsidized by a $500 grant, produced for the state two reports that amounted to little more than catalogues of current developments, after which the survey was discontinued. Most of the state surveys were interested largely in economic goals—to evaluate mineral resources and potentialities, assist existing and prospective mining industries, furnish propaganda to attract settlers, and help with the sale of state lands. Some states appointed men who had completed successful surveys elsewhere, others entrusted the survey to a professor of geology at one of the state universities. No doubt such an appointment secured more scholarly and conscientious work than was normal for the pittances most states were prepared to pay.

Most state surveys, like that of North Carolina, were of brief duration. The small area of some, Delaware or South Carolina for example, meant that the state could be covered quickly. Besides, most legislatures had very limited understanding of what was needed, and they obtained surveys that consisted of little more than a few rapid traverses of the territory, with slight emphasis on describing outcrops and their qualities. The histories of some state surveys show that they suffered from indifferent or ignorant legislators who could not sufficiently appreciate the importance of the work to pay for its proper conduct or for its continuance. Others were ruined by excessive political interference and patronage-mongering.

Still other surveys were terminated because of the financial embarrassment of states caught up in the Panic of 1837 and the ensuing depression. At least one state survey was brought down by the rivalries of the geologists within the survey itself. A few years constituted the lifespan of most state surveys; Delaware's lasted two years, Ohio's and New Hampshire's three, and Tennessee's and New Jersey's four years. The survey of even an important mining state like Pennsylvania had to be terminated after six years, in 1842, because of serious financial difficulties of the state treasury. Michigan, despite the outstanding work of the dynamic Douglass Houghton and the development of a very important copper mining industry there partly as a result of his work, could not be persuaded to continue its support beyond the four years originally contracted for. Even New York's geological survey, which was the most successful of all in a scientific sense, establishing the standards and classifications for much of the United States and Canada, for all practical purposes was discontinued after seven years (1836-43), though the paleontological work of Dr. James Hall was permitted to continue until his death in 1898.

In comparison with these, the establishment in the Province of Canada of a continuing survey that has carried on without interruption for more than 130 years—the first 50 years under only two directors—ranks as a major achievement. The experience of the brief, quickly-terminated state surveys for many years reminded the first director of the Canadian survey of the transitoriness of his profession and the waywardness of provincial politicians. It made him ever conscious both of the need to remain on cordial terms with his governmental sponsor and to gain for his survey a most favourable public image. His successors have ignored this public relations function only at their peril. It has not been enough for the Geological Survey of Canada to do good work consistently; it must also be seen to be doing good work.

Despite its troubled institutional history, the Geological Survey of New York was particularly significant for the early development of the Canadian survey. Thanks to the enlightened public men who controlled the state at the time, New York secured well-trained practising geologists to staff the organization, and began the work by dividing the state into four regions, each under the control of a geologist. In seven years they defined the principal rock formations of the state and gave them names, some of which are still in use. Moreover, they established type sections with which similar occurrences elsewhere in the United States and Canada could be compared. Two of the original survey parties worked along the borders fronting on Canada, Ebenezer Emmons along the St. Lawrence and James Hall along the Niagara frontier.

I-9 DOUGLASS HOUGHTON
Houghton went to Michigan Territory as a youth of 20 to practise medicine and lecture on scientific subjects. Through his urgings, the legislature of the newly-created state set up a geological survey under his direction. Despite the drastic reduction of his financial support, he continued his work, extending it to the Upper Peninsula. He was drowned in Lake Superior in 1845 at the age of 36.

Both were keen observers who worked out stratigraphic successions that extended north beyond their districts into Canada, so that when the Canadian survey was begun, it could apply these New York definitions and names to their Canadian extensions.

Hall, a paleontologist, was fortunate in being able to study a series of relatively undisturbed fossiliferous rocks, and as these spanned many different geological periods, he could describe in detail the succession of strata and their characteristic fossils. It was the high cost of publishing his expensive reports, illustrated with drawings of fossils which seemed of no practical value to the layman, that led to the curtailment of the survey's operations.

I-10 DENISON OLMSTED

I-11 JAMES HALL

I-12 EBENEZER EMMONS

Ironically, though, Hall himself was permitted to continue in the work while it was his colleagues who were dispensed with. During his long, productive life he published in his *Palaeontology of New York* alone, fifteen volumes, comprising 4,539 pages and 1,081 full-page plates of fossils, besides numerous articles and other works.[4] As the leading paleontologist in the United States, and one who was in close contact with the types of fossils likely to occur in Canada, it was natural that he should often be resorted to for information and advice in the early years before the Canadian survey secured a paleontologist of its own.

* * *

THE BRITISH POSSESSIONS IN NORTH AMERICA, like the neighbouring United States, had also received a measure of geological study by the end of the 1830's, principally by British-trained members of the armed services or persons attached to parties employed on official government business. Such a man was Dr. John J. Bigsby, a British army surgeon employed on the boundary survey between the Great Lakes and the Lake of the Woods under the joint commission set up after the War of 1812. Besides his delightful book, *By Shoe and Canoe,* he read papers on the geology of the Montreal district and of the country between Lake Huron and Lake Winnipeg, which were published in American and British journals as well as in the *Transactions of the Quebec Literary and Historical Society,* which apart from the newspapers was practically the sole medium in Canada at that time for publishing such scientific work. Lieutenant F. H. Baddeley, R.E., explored the Saguenay and Gaspé regions and later was the geologist to an expedition that examined the country between Georgian Bay and Ottawa River. Captain H.W. Bayfield, R.N., who was engaged in the official survey of the upper Great Lakes and the Gulf of St. Lawrence, published an article on the geology of the Lake Superior district. Other British army officers of the period who contributed to the unfolding of the geology of Canada were Lieutenant F. L. Ingall, who examined the country between the St. Maurice and the Ottawa, and Captain R. H. Bonnycastle, R.E., who reported on geological and mineralogical aspects of the Kingston district where he was stationed.

Special mention must also be made of the geological work of the Arctic expeditions of the period following the end of the Napoleonic Wars, most of them by parties of such naval officers as Captain John Franklin, Lieutenant Edward Parry, Captain John Ross, Captain G. F. Lyon, and Lieutenant George Back. These expeditions examined the resources and natural phenomena of the districts they traversed, and usually included persons trained in natural sciences, sometimes in geology specifically. Often, too, they brought back rock samples to be examined by experts like Professor Robert Jameson of the University of Edinburgh, who published a report on the geology of the areas covered in Captain Parry's second and third voyages. The result was the securing of considerable information—for the time—about the geology of the Arctic islands, the Mackenzie River district, and the western margin of the Precambrian Shield. Since the overland expeditions often used the trade routes of the Hudson's Bay Company, the first districts of the present-day provinces of Saskatchewan and Alberta to be examined were the more forbidding northern parts rather than the southern prairies and parkland that were to attract most future settlers. Thus credit for "the first substantial geological account of Western Interior Canada"[5] goes to Dr. John Richardson, who accompanied Franklin on his expeditions across the continent from York Factory down the Mackenzie River in 1819. That expedition took him also into the Barren Lands and along the Arctic coast of the continent before the party returned in 1822. As the naturalist to the expedition, Richardson published notes on the fish, botany, aurorae, as well as "geognostical observations" which describe the geology not merely of Churchill, Athabasca, and Mackenzie Rivers in some detail, but also of the country east and north of Great Slave and Great Bear Lakes. He returned along much the same route in 1825, and again a third time some 20 years later when he was engaged in the search for the missing Franklin expedition, so his three reports afford an interesting comparison of the advances in geological science over the 30-year interval.

The Atlantic colonies also were subjects of early geological study, partly because of the example of the states to the south, but mainly because of their indicated mineral wealth. In Nova Scotia, for example, the commencement of the Cape Breton coal mines by the General Mining Association awakened an interest in the geology and minerals of the province, and within two years two articles were published on the geology of Nova Scotia. The real pioneer of geology in the Maritimes, however, was the self-taught geologist, Dr. Abraham Gesner, who, like so many other medical men, developed an interest in natural history and made geology his hobby. His practice at Parrsboro was within easy reach of the geologically interesting and economically significant Cumberland and Joggins coal measures. In 1836, he published an outstanding book for its day, *Remarks on the Geology and Mineralogy of Nova Scotia,* which included a geological map of the interior of the peninsula identifying the main formations and deposits of iron ore and of coal.

As a result of this book, Gesner was approached to undertake the geological survey of New Brunswick. The legislature, egged on by the provincial university, was desirous of having the mineral potential of the province explored, and Gesner, in the years between 1838 and 1843, swept through southern New Brunswick, writing enthusiastic and eulogistic reports of the mineral wealth of the colony. When these touched off a mining boom in which many investors lost money, Gesner was unfairly blamed for the outcome. Coupled with the colony's shortage of funds, it led to discontinuance of the survey. To add insult to injury, Gesner himself was caught up by the mining excitement and invested unwisely in a de-

I-13 SIR JOHN RICHARDSON

I-14 ABRAHAM GESNER

posit of albertite (a solid hydrocarbon mixture), then lost a lawsuit over his right to mine the property in question. In order to pay his debts, he had to dispose of his extensive natural history collections, which became the nucleus of the New Brunswick Natural History Museum in Saint John. However, the wheel of fortune took a favourable turn for him. He immigrated to the United States where he developed and patented a process for manufacturing kerosene that brought him a modest fortune, so that he was able to return to Halifax, where he died in 1864.

The last years of Gesner's work in New Brunswick coincided with a brief attempt by the government of Newfoundland to have a geological survey made, the upshot of which was the securing from England of J. B. Jukes, a future leader of British geology. Jukes soon fell into disfavour with the colony's legislature, which tried to reduce his grant after one year and discontinued the survey at the end of the second year. There was a further dispute over the publication of Jukes' report, *Excursions in and about Newfoundland, during the years 1839 and 1840,* which, however, appeared in London in 1842. Thus the experience of the Atlantic colonies paralleled that of the American states as regards the troubled, brief careers of their geological enterprises. The single important contribution was the work of Gesner, which may be credited with bringing the Joggins coal beds to the attention of other leading geologists. As a result, first William Logan, then Charles Lyell, were attracted there. Lyell's interest, in turn, stimulated that of a brilliant young scholar from Pictou, J. W. Dawson, who was destined to play a leading role for half a century in Canadian education and science.

I-15 J. B. JUKES
After leaving Newfoundland, Jukes surveyed the northeastern coast of Australia, worked in North Wales, directed the Irish geological survey from 1850 to 1869, and published widely-used manuals of geology.

* * *

MEANWHILE, THE INCREASING TURNING to geological explorations on the part of the neighbouring states was observed with interest in the colony of Upper Canada. As early as January 18, 1832, Lieutenant Governor Sir John Colborne recommended for the favourable consideration of the Legislative Assembly a petition from Dr. John Rae, a peregrinating economist and pioneer social scientist who sojourned in Canada between 1821 and 1848. Rae appealed for the support of the colony to prepare a comprehensive study of Upper Canada, including "the leading features in the Geological Structure of the Country, from whence the nature and peculiarities of the soil in the different sections of it may, with most certainty, be deduced."[6] In the following session the Attorney General, H. J. Boulton, introduced another petition, this time from the York Literary and Philosophical Society, requesting a vote of money for "the appointment of persons duly qualified to investigate thoroughly and scientifically, the Geology, Mineralogy and general Natural History of the Province."[7] Both petitions were referred to the committee of supply, but neither received any financial assistance from the hard-pressed legislature.

Three years later, at the session of February 1836, the Reformer-dominated House of Assembly, with its inclination to look to the United States for ideas and examples, approved a motion from William Lyon Mackenzie, seconded by Charles Durand:

> Ordered—That Messrs. Dunlop, Gibson, Charles Duncombe, and MacIntosh, be a committee to consider and report a plan for the Geological survey of this province; that they have power to send for witnesses and report by bill or otherwise.[8]

The report was received on March 1, ordered to be printed, and its recommendations were referred to the committee of supply. Brief and to the point, the report referred to the known salt beds, and hinted at the possibilities of copper on Lake Huron, and at hoped-for coal deposits ("we believe every indication exists of their presence in several districts"). Developed mines, it said, would furnish much-needed local markets for farm produce, while coal would displace inconvenient wood as a domestic fuel, as a source of power for steam engines, and aid "in the manufacture of our various minerals." Since industrialization was out of the question for a new country whose prosperity consequently was tied to the progress of agriculture and mining, proper information about its natural resources was essential to good government. The report concluded by "strongly" urging that the survey be made "so that a report of our natural wealth and resources may be laid on the table at the commencement of the next Session of Parliament."[9] That the committee expected a useful report from only a few months' work indicates a grasp of the need for serious investigation that was little if any better than that of the legislators of North Carolina or Michigan.

The session was halted by the dissolution of the legislature by Lieutenant Governor Sir Francis Bond Head, and in the general election that followed, the Reformers were badly beaten. However, during the next session (November 1836-March 1837), Dr. Dunlop returned to the fray with a successful motion:

> Resolved—That an humble Address be presented to His Excellency the Lieutenant Governor, praying that he will inform this House, whether there are means in his power to effect a Geological survey of this Province.[10]

Dunlop also gave notice of another motion for the setting aside of Crown lands to pay for the cost of making "a correct Geological survey of this Province."[11] He did not even introduce this second motion, and though the first was passed by the House of Assembly, no action appears to have been taken to implement it.

For the next five years the project marked time. In the interval, rebellions broke out and were suppressed in each of the Canadas, and after a period of rule by appointed councils, the reunification of Upper and Lower Canada was enacted by the Imperial Parliament. The new government was inaugurated under the skillful guidance of Lord Sydenham, under whom a thoroughgoing reorganization of government departments and activities was begun. A spirit of progressive, efficient, up-to-date administration permeated the government, an atmosphere in which a geological survey seemed an entirely desirable and appropriate activity. It was not long before the government received requests to undertake this function.

This time the prime mover was the Natural History Society of Montreal, founded in 1827. Under the presidency of Dr. A. F. Holmes, a founding member of the society and professor of materia medica and chemistry at McGill College, the society was strongly oriented towards geological and mineralogical studies and collections. On July 5, 1841, the first session of the Parliament of the united province entertained petitions from the Montreal society, supported by the Literary and Historical Society of Quebec, and presented by Benjamin Holmes and Henry Black, the members for Montreal and Quebec respectively, to have a systematic geological survey of the province carried out. After being referred to a five-man committee the proposal was adopted by the government, whose leader, S. B. Harrison, on September 10, 1841, himself moved the necessary financial provision:

> *Resolved*—That a sum not exceeding one thousand five hundred pounds, sterling, be granted to Her Majesty to defray the probable expense in causing a Geological Survey of the Province to be made.[12]

The item, the 81st resolution of the committee of supply, like the others, carried on division. At last the geological survey was a reality, and the search for a suitable geologist to undertake it could be commenced.

The news of the forthcoming establishment of the new survey was received with considerable interest in scientific circles. One man who gave the matter attentive thought was an expatriate Montrealer who was visiting the city from Britain that summer, as he had also done the previous year. William Edmond Logan was a 43-year-old bachelor in comfortable circumstances who had become interested in geology some ten years previously. He was on the point of embarking on a long-anticipated tour of the coalfields of Pennsylvania and Nova Scotia when he heard the news of the projected geological survey of Canada. While he proceeded along his way, visiting the rich new anthracite mines of Pottsville, Maunch Chunk, and Wilkes-Barre, he turned over in his mind the possibility of applying for the position. He wrote his brother James in Montreal on September 6, 1841, that "I have almost made up my mind, if I can make the necessary arrangements in business matters to offer myself as a candidate to undertake the survey of Canada & if I once begin it will not be my fault if it does not go ahead."[13] At Halifax, after viewing the Nova Scotia coalfields, which he did not feel possessed as good coal as those of Pennsylvania, he wrote again to James, bidding farewell to America "perhaps only for a short time," and closed his letter with the offhand remark: "By the by have the £1500 for geological investigation really been voted."[14] Back in England, Logan began his campaign to be the man chosen to make the geological survey of Canada.

The appointment was in the hands of the Governor General of Canada, Sir Charles Bagot, who succeeded Lord Sydenham on the latter's sudden, untimely death. Bagot was besieged by requests on Logan's behalf from Montreal citizens who were friends of Logan from his youth in that city. The son of a Scotsman who had come to Montreal around 1784 and became a prosperous baker and property owner, William Edmond Logan was born in Montreal in 1798, and educated at Alexander Skakel's school prior to being sent in 1814 to Edinburgh with his older brother Hart for further education. Their parents also retired to Scotland in 1815, but the family still maintained its business interests in Montreal, managed by the oldest son, James, who was William's adviser, agent, and loyal supporter, and an uncomplaining benefactor to the Geological Survey down to his death in 1865. No doubt James managed William's suit, using his influence with such notable local families as the Molsons and Chamberlins. The *Montreal Gazette* supported Logan, "knowing the high qualifications of MR. LOGAN for this important duty," and added its opinion "that, from many considerations, nothing could prove more gratifying to scientific men throughout the Province than his appointment."[15] That there was an organized campaign on Logan's behalf is evident from a letter in The Logan Papers from Dr. A. F. Holmes in which he, writing on January 13, 1842, referred to James' attempt to recruit him to support the suit of his brother, then explained, regretfully, that he did not feel able to support the application because of his grave doubts concerning Logan's qualifications in chemistry, mineralogy, and "conchology." However, despite Holmes' demurrer, the campaign attained its objective, for the Governor General by February 18, 1842, had practically decided in Logan's favour. As Bagot wrote his superior, the Secretary of State for the Colonies, Lord Stanley,

> To obtain the services of a Gentleman whose only object would be the pecuniary remuneration attached to the duty and who would not enter on it with the zeal and enthusiasm which generally distinguishes scientific enquiries would of course be unsatisfactory and I should therefore be anxious, if possible, to select a person who is either a native of this Country or who feels a strong interest in it. The only individual answering this description who has been mentioned to me is Mr. W. E. Logan who is at present in England, but who would come out if appointed to this service . . . . If he has been correctly described to me, he would certainly be the fittest person on every ground to be employed in this Survey.[16]

In Britain, Logan had the support of the leading figures in British geology, who were glad to provide the necessary testimonials. Bagot's letter to Stanley had been in the nature of a request that he verify Logan's standing with such agencies as the Geological Society of London or with the professors of geology at the universities; and this the Colonial Secretary proceeded to do. De la Beche—whose support was beyond price because of his important position as director of the Ordnance geological survey—was unstinting in his praises:

> I have examined several portions of country with Mr. W. E. Logan, and can safely affirm that no one can be more careful, able, or desirous of attaining the truth . . . . I would further observe that Mr. Logan is highly qualified as a miner and metallurgist to point out the applications of geology to the useful purposes of life, an object of the highest importance in a country like Canada, the mineral wealth of which is now so little known.
>
> I should anticipate the best results, both to the science of geology and its application, from the employment of Mr. Logan on the Geological Survey of Canada.[17]

The Reverend William Buckland, professor of geology at Oxford University, had previously suggested Logan's name to Sir Robert Peel, the Prime Minister, as the man best qualified to survey the Crown-owned coalfields of Nova Scotia. Asked now for his views regarding Logan's suitability to undertake the geological survey of Canada, he not only praised Logan's scientific attainments, but went on to write a third letter for Sir Charles Bagot in which he commented on Logan's personal qualities:

> . . . he is not only enthusiastically devoted to and highly qualified for field-work in geology, but he is also a man of modest and gentlemanly demeanour, and of high principle, and good conduct and right feeling, with whom it is pleasing to have intercourse, and in whom it is quite safe for persons in authority to place confidence.[18]

Their praises were deserved, for notwithstanding Holmes' doubts, Logan's qualifications were unimpeachable. His education in Edinburgh had exposed him to some of Scotland's ablest teachers, and his classmates were among the intellectual elite. He may have studied some geology there, although he seems to have preferred the mathematics and languages. He entered the University of Edinburgh in 1816 and studied logic, chemistry, and mathematics, but left after one year without proceeding into medicine, his intended program. From his school years he acquired habits of strict application and self-discipline, and an earnest approach to life which did not completely efface a lighter, even exuberant, touch about his personal relationships. From Edinburgh he went to London to work at his uncle Hart Logan's

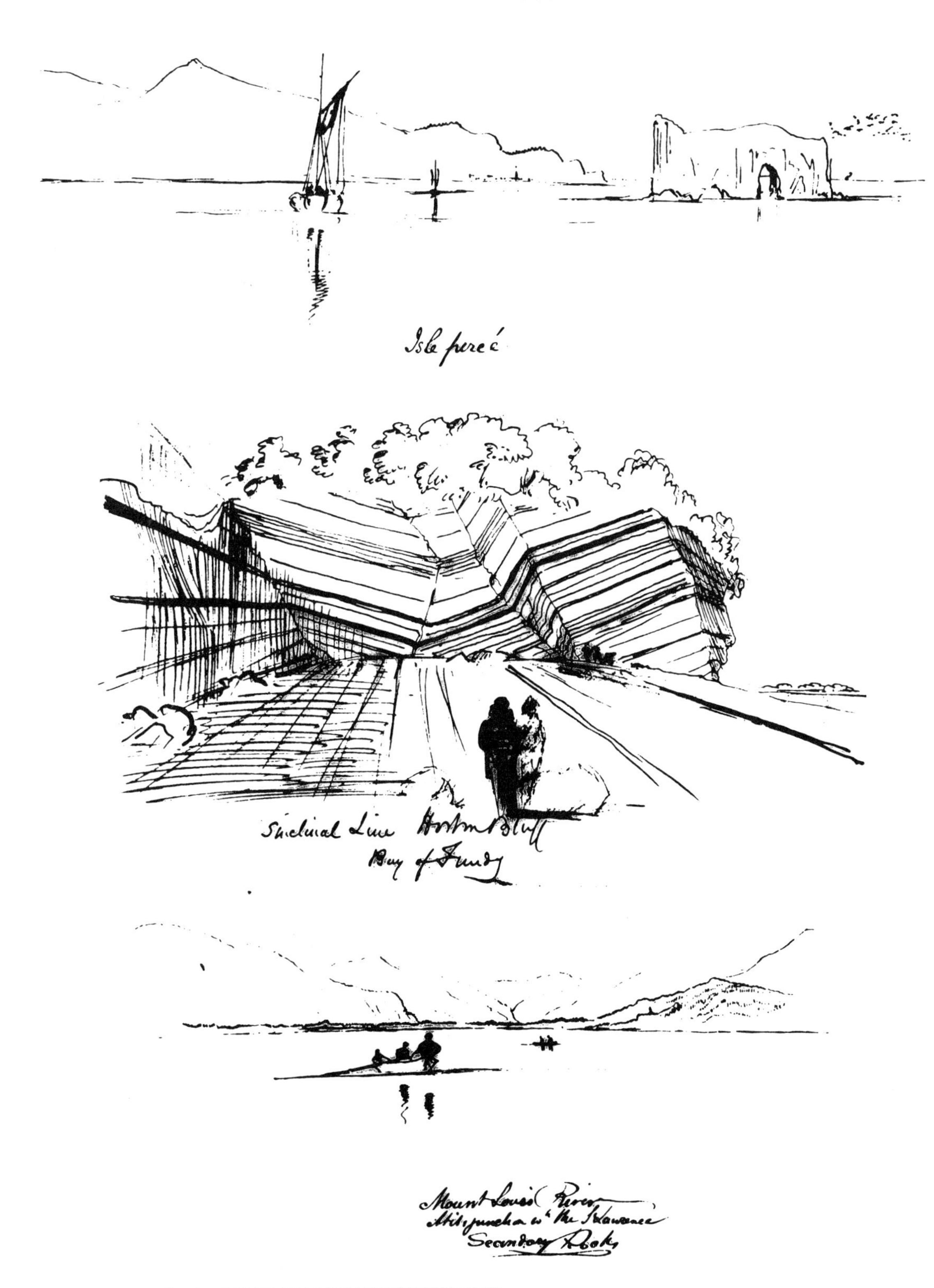

I-16 MORE PAGES FROM LOGAN'S NOTEBOOK
Isle Percé and Mount Louis River were drawn on August 17, 1840, during a voyage from Pictou to Quebec on the steamer "Unicorn." The strata at Horton Bluff, Nova Scotia, were sketched on September 16, 1841, on a trip from Boston via Windsor and Halifax to Pictou.

counting house, where he acquired a thorough grounding in bookkeeping and accounting. At this time his inclinations lay in artistic directions, to judge from his interests in sketching, playing the flute, studying French, Italian and Spanish, as well as the sightseeing tour he took to Italy in 1829. The London phase of his life may be credited with developing his orderly, businesslike habits, his strict scrupulousness in business dealings, the careful accuracy of his measurements and mapping so characteristic of his geological work, and the skill in drawing that make his notebook sketches such a joy to behold.

In 1831 his uncle sent him to Swansea, Wales, as his local representative and to manage the Forest Copper Works, which Hart Logan controlled. In addition to learning the business thoroughly (including the processing of the ores and the mining of coal), Logan became an enthusiastic amateur geologist, studying the stratigraphy of the coal measures, examining coal mines, collecting fossils, and preparing a geological map of the district. In the decade after 1831 Logan participated in the activities of the Swansea Institute, was a valued member of the prestigious Geological Society of London, demonstrated his talents as an uncommonly gifted geological mapper, and contributed a notable paper that broke new scientific ground in adding to the understanding of the process by which coal beds are formed.

Yet another source of support came to Logan through a rather fortuitous family connection. His brother Hart practised law in Edinburgh in partnership with a certain Anthony Murray, who was related to Sir George Murray, a former Colonial Secretary and currently chief of the Board of Ordnance, which was the sponsor of the geological mapping project in Britain. On November 26, 1841, Sir George Murray was advising Anthony Murray that he would recommend Logan to Sir Charles Bagot for the position if Logan would supply suitable testimonials, while Anthony Murray, in turn, arranged an appointment for Logan to meet Sir George Murray at the Ordnance office in London, some time in December.

Coincidentally, Sir George Murray was being besieged to use his good offices from another quarter. Anthony Murray's brother, Alexander, was also in London seeking a position appropriate to his talents and needs. He had been educated at the Royal Naval College, Portsmouth, and had served some years in the Royal Navy, but left it in 1834, when he was a lieutenant, as so many others had done, because of the slowness of promotion. He immigrated to Upper Canada and bought a farm near Woodstock shortly before the outbreak of the Rebellion of 1837, in which he took part in the naval detachment that operated at Chippawa against the rebels. By the summer of 1841 he was back in Britain, attempting through his kinsman Sir George Murray to obtain appointments with Lord Haddington and Sir J. Cockburn respecting his re-entry into the Navy. The two suppliants, Logan and Murray, undoubtedly knew each other's situation—their brothers were law partners after all—and they must have discussed the possibility of linking their fortunes not long after Logan's return from America. Before Logan saw Sir George Murray he received a letter from Alexander Murray (London, December 2, 1841) in which Murray referred to their conversations in the following language:

> As Sir J. Cockburn could give no encouragement to Sir Geo. Murray as to my prospects in the Navy, I may say I conceive it is almost useless, & do not intend prosecuting either it or the memorial [to Lord Haddington] until I see what I may expect from other quarters . . . . I carefully abstained from mentioning that I had any prospect through you, but told him that in consequence of a letter from my brother, I should defer presenting my memorial for the present. Should you have any thing to suggest to me, or should any advancement in your own case transpire, be so kind as to drop me a line to say so . . . .[19]

In this fashion the Canadian Survey gained the services of an able, trustworthy assistant, and Logan a lifelong colleague and friend.

Meanwhile, the appointment campaign was making progress. In February 1842, as we have seen, Governor General Bagot wrote to the Colonial Secretary inquiring about Logan's qualifications, and authorized Stanley to offer the position to Logan if his references were satisfactory. On April 2, Stanley's under-secretary, G. W. Hope, M.P., invited the four leading British geologists nominated by Logan—the president of the Geological Society of London, the director of the Ordnance Geological Survey, and the professors of geology at Oxford and Cambridge—to submit their references. The testimonials, which were "most satisfactory," were received by Hope a few days later; on April 9, acting on the authority of Bagot's dispatch, Hope made Logan an official offer of the appointment, which Logan accepted formally by letter on April 14. The news of the appointment was reported with pleasure by the *Montreal Gazette* on May 16, and was a feature of the celebrations when Sir Charles Bagot visited Montreal on June 22, 1842, to be received by the Natural History Society, of which he became patron. Following his reply to the address, Bagot informed the society that he had recently appointed Logan provincial geologist, and explained: "Previous to his appointment of Mr. Logan he had considered it proper to refer to England for an account of his qualifications, and the result was that a mass of testimonials was sent out . . . affording ample testimony that to no one could the important office be more appropriately entrusted than to that gentleman."[20]

I-17 SIR GEORGE MURRAY
After an important military career (some of it in Canada), General Murray held a variety of military and administrative positions in the British government. His family, the Murrays of Ochtertyre, was distantly related to the Murrays of Dollerie, to which Alexander Murray belonged, both families being small landowners near Crieff, Perthshire. A much closer tie arose, however, through the marriage of the general's only daughter to Alexander's brother Anthony, the Edinburgh lawyer.

Logan had asked for two or three months' delay before assuming the appointment to enable him to clear up several professional engagements, which Stanley granted in view of Bagot's failure to specify the immediate commencement of the survey. In fact, Logan did not reach Kingston, currently the seat of the government, until August or early September 1842. During this time Murray was occupying himself with studies of land surveying and geology, as he had been doing ever since "it appeared probable that he would be Employed on the Survey of Canada."[21] By July he wrote Logan that though he felt he had mastered the principles of land surveying, he still remained completely ignorant of the practical side of geology, "In other words, I cannot discriminate one rock from another; so that I have I suppose every thing to learn yet."[22] This was remedied by Murray's going to Swansea to join Logan for a period of practical work, accompanying Logan on his geological excursions prior to his return to Canada. No doubt Murray's training continued during the ensuing year, for there is mention of his serving some time as a volunteer with De la Beche in the Ordnance geological survey, both in the field and at the Museum of Practical Geology.

Logan arrived in Kingston in the autumn of 1842 to find the government in the midst of a political crisis and not prepared to deal with his position. So, while he waited, he made a round of visits to the iron mines at Marmora, the Silurian limestone beds at Brockville, and to Perth to meet and confer with Dr. James Wilson, the local amateur naturalist and geologist. Then followed his interview with the government and the discussion of the terms of employment. He was to be allowed an assistant of his own choosing, be permitted to return to England once more and have his passage paid back to Canada the following year, and be at liberty to use the information he submitted in his reports to the Governor General so long as he avoided direct copying. Before departing for England to spend the winter of 1842-43, Logan completed his studies of the existing information on the geology and topography of the province and submitted a 5,000-word "Preliminary Report" (which was printed in 1845 as part of his first *Report of Progress for the Year 1843).* In England Logan dealt with his father's estate and other business matters, conferred with De la Beche regarding the task he had undertaken and the possibilities of collaboration between their two surveys, underwent treatment by a Dr. Robert Dick to be cured of an irritating skin disorder, upset stomach and nosebleeds, and presumably attended to the final instruction of his assistant. For Murray was now judged sufficiently well trained to be sent immediately on his own to examine the country between Lakes Huron and Erie, while Logan worked at the opposite end of the province, investigating the south coast of the Gulf of St. Lawrence between Pictou and Gaspé. With nothing more substantial than the £1,500 appropriation and the appointment of Logan to make a geological survey of the Province of Canada, the Geological Survey of Canada was launched on its long, distinguished career.

II-1 GEOLOGICAL FORCES AT WORK
These folded sedimentary layers exposed near Sullivan River in southeastern British Columbia dramatically illustrate the awesome forces constantly at work reshaping the features of the earth. One of the Geological Survey of Canada's tasks is to examine, study, and report on the effects and the causes of such forces.

2

The Science and the Man

WHAT WAS THE STATE OF GEOLOGICAL KNOWLEDGE when the government of the Province of Canada appointed Logan to make the geological survey of Canada? After some 50 years of rather confused investigation, study and discussion, geology was emerging as a distinct science based on a body of clear, well-understood principles founded on observation and reasoning. The underlying assumptions of this new science were that the earth had evolved over an enormously long period of time; that its development resulted from the operation of regular, ascertainable, natural forces; and that many of these same physical, chemical, and biological processes are still going on around us. Consequently, from careful observation of the phenomena of the present-day world the scholar can elucidate the principles according to which the earth developed to its present situation, and which explain how it might be expected to develop in the future.

This doctrine that "the present is the key to the past," so upsetting to the centuries-old religious beliefs in Creation and a 5,000-year-old universe, had within the previous decade received its most articulate (though by no means original) exposition in the work of a brilliant young lawyer, traveller, and scholar, Charles Lyell. In the first chapter of the third volume of his *Principles of Geology,* published in 1833, the central thesis is stated in the following key paragraphs:

> All naturalists, who have carefully examined the arrangement of the mineral masses composing the earth's crust, and who have studied their internal structure and fossil contents, have recognized therein the signs of a great succession of former changes; and the causes of these changes have been the object of anxious inquiry. As the first theorists possessed but a scanty acquaintance with the present economy of the animate and inanimate world, and the vicissitudes to which these are subject, we find them in the situation of novices, who attempt to read a history written in a foreign language, doubting about the meaning of the most ordinary terms . . . .[1]
>
> They imagined themselves sufficiently acquainted with the mutations now in progress in the animate and inanimate world, to entitle them at once to affirm, whether the solution of certain problems in geology could ever be derived from the observation of the actual economy of nature, and having decided that they could not, they felt themselves at liberty to indulge their imaginations, in guessing at what *might be,* rather than inquiring *what is*; in other words, they employed themselves in conjecturing what might have been the course of nature at a remote period, rather than in the investigation of what was the course of nature in their own times . . . .[2]
>
> . . . [This] produced a state of mind unfavourable in the highest possible degree to the candid reception of the evidence of these minute, but incessant mutations, which every part of the earth's surface is undergoing, and by which the conditions of its living inhabitants is continually made to vary . . . . Geology, it was affirmed, could never rise to the rank of an exact science,—the greater number of phenomena must forever remain inexplicable, or only be partially elucidated by ingenious conjectures. Even the mystery which invested the subject was said to constitute one of its principal charms, affording as it did, full scope to the fancy to indulge in a boundless field of speculation.[3]
>
> . . . We shall adopt a different course, restricting ourselves to the known or possible operations of existing causes; feeling assured that we have not yet exhausted the resources which the study of the present course of nature may provide, and therefore that we are not authorized, in the infancy of our science, to recur to extraordinary agents. We shall adhere to this plan . . . because history informs us that this method has always put geologists on the road that leads to truth,—suggesting views which, although imperfect at first, have been found capable of improvement, until at last adopted by universal consent. On the other hand, the opposite method, that of speculating on a former distinct state of things, has led invariably to a multitude of contradictory systems, which have been overthrown one after the other,—which have been found incapable of modification,—and which are often required to be precisely reversed.[4]

Lyell's views, known as *uniformitarianism,* were certainly not original to him, but as a biographer has said of him, "He found uniformitarianism an hypothesis, and he left it a theory."[5] His idea of continuous world development and the evidence that geologists amassed concerning the changing forms of life were among the strongest influences upon Charles Darwin, whose theories of evolution afterwards burst on the world with *The Origin of Species.*

Lyell was not an explorer or discoverer, but an open-minded seeker after knowledge, an observer with an uncanny skill for seeing the inter-relationships among natural phenomena, detecting their significances, and deducing from them the clearest, most logical conclusions. The germ of the idea of uniformitarianism had been disseminated by an earlier generation of British geologists, in particular James Hutton (1726-97), a well-to-do, well-educated Edinburgh Scot who had concluded that all the characteristic features of the landscape could have been produced by ordinary natural processes in everyday operation, provided one were prepared to postulate for the process an unlimited amount of time.

Since the evidence seemed incontrovertible, the contradiction must arise from man's current ideas of time, derived from religion. At the urging of friends Hutton set down his ideas in a paper for the Royal Society which was read in 1785 and published three years later. The paper was so poorly written that it received little attention at the time, but in 1802 when John Playfair presented his dead friend's ideas in an extremely readable book, *Illustrations of the Huttonian Theory,* the scientific world began to take notice of the idea that the past can be understood by studying the present; and geology, as we know it today, can be said to have made its start, even though the theory still was far from having wide acceptance.

The painstaking work of William Smith (1769-1839), a self-taught English civil engineer, furnished a very convincing early demonstration of how uniformitarian principles operated in nature, even though Smith himself did not enter into the realm of theoretical speculations but rested content with careful, accurate observations and conclusions. Despite his lack of formal training, Smith was an exceptionally acute observer, with a keen, logical mind. He was in charge of the excavation of many English canals, in the digging of which the nature of the rock strata to be excavated is of immediate practical importance. Smith's canals were dug in relatively flat lying sedimentary beds and he soon noticed that one particular type of bed was always succeeded by another type of bed, and that by observing exposed rocks it was possible to predict the sequence of the underlying rocks. Smith also collected fossils from various strata and in the process perceived that those from one type of bed differed from those in the beds above and below. He went on to recognize that certain fossils (guide fossils) were characteristic of a certain stratum and served to identify that stratum wherever it might be found, even though its appearance had changed. Similarly, beds of like appearance but of different geological ages could be distinguished from one another and could be related to their proper geological settings on the basis of the fossils they contained. The practical engineer Smith did not indulge in sweeping generalizations, but he published a detailed description of each geological formation and of its contained fossils, as well as a careful map showing the surface exposures of each. The publication of Smith's great map had an almost immediate impact. Some of his formations (groups of strata that together form a logical unit) were found to occur on continental Europe as well, while others filled gaps in the succession there. The method of *correlating* remote groups of rocks on the basis of their contained fossils had been discovered, the groundwork was laid for building a geological chronology or time scale, and historical geology was born.

Notwithstanding the practical, everyday uses that William Smith made of fossils, the science of paleontology (the study of ancient life) could scarcely be said to have begun. What exactly were fossils? Why were the fossils from a certain type of bed always the same, and different from those in beds above or below? Was there any system linking the fossils of similar forms occurring in these different beds? Could they be described and named so that the results of work in one area could be published for the benefit of workers in other areas? These questions had to be answered before the study of fossils could become a proper science.

Men have always been interested and intrigued by fossils which are mentioned in writings dating back to classical times. Even though some of the earliest of these suggest that fossils are the remains of once-living things, most people well into the eighteenth century regarded them as curious sports of nature. It was mainly due to the efforts of the Frenchmen J. B. Lamarck (1744-1829) and G. Cuvier (1769-1832) that the study of fossils was put on a proper footing, and the science of paleontology was born. From 1793 onward Lamarck, the founder of invertebrate paleontology, was enunciating radical principles that are little different from those held today. Half a century before Darwin, he boldly advocated evolution, insisting that fossil shells must be compared with modern shells and that the differences between them were due to changes that had taken place over countless generations

II-2 SIR CHARLES LYELL

II-3 JAMES HUTTON

II-4 WILLIAM SMITH

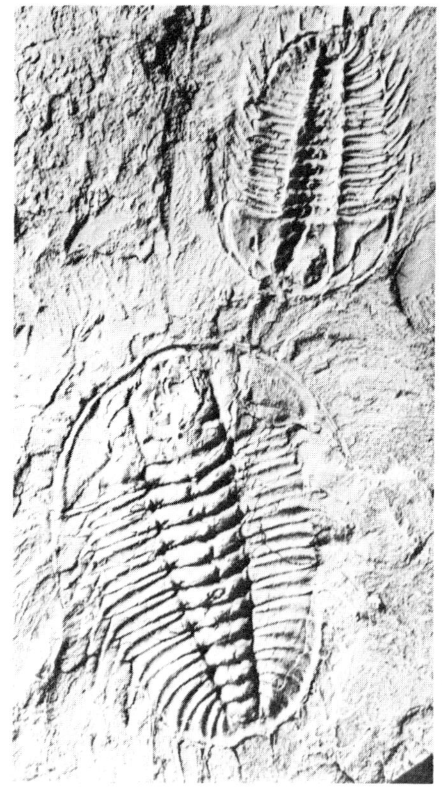

II-5 CAMBRIAN TRILOBITES
The imprints of two of these large and impressive fossils and several smaller ones are shown on this slab from Yoho National Park, British Columbia.

in the creatures themselves. Furthermore, the differences between related shells in older and younger beds in a sequence are perfectly understandable as being inherited modifications caused by gradually changing conditions. Lamarck's approach, then, was essentially that of a stratigraphic paleontologist.

Cuvier's work, on the other hand, lay in the study of the actual morphology of these ancient creatures. He was a specialist in vertebrates, and in the Tertiary rocks of the Paris basin he found plenty of vertebrate fossils to study. While making detailed anatomical descriptions of these ancient animals he was struck by both the similarities and differences between these creatures and their modern counterparts. He realized that some of the Tertiary animals were quite unlike anything living today, although the anatomical structures clearly foreshadowed those of living beasts and in some instances even explained hitherto inexplicable structures. Cuvier never was able to accept Lamarck's bold assumption that younger species gradually developed from older, and suggested instead that each was a separate creation, wiped out at the appropriate time and replaced by another one, modified presumably in the light of an earlier experience. This hypothesis had the merit of being theologically acceptable. Cuvier should be credited at least with persuading his contemporaries to see that every anatomical part of a fossil has its counterpart in a modern creature.

Cuvier and Lamarck were only two of the foremost contributors to the science of paleontology, which by the time of the inauguration of the geological survey in Canada in 1842 was a well-established branch of geology and the essential guide for comparative dating of rocks. It was not until the theory of evolution had been accepted by the scientific world, later in the century, that paleontology could be studied on its own terms, free from preconceptions, as being the development of life forms in relation to changing environments over vast stretches of time. In the meantime, much excellent work was being done in collecting, figuring, and describing fossils from all parts of the world, and in establishing a worldwide succession of major rock units with their typical fossils so that geologists everywhere might fix the appropriate stratigraphic position for any fossiliferous stratum studied. During the early days of the Canadian Survey it made some notable contributions to world science through its paleontological work, exchanging information with fellow specialists in Europe and the United States.

But how did the earth's rocks originate and how did they develop their present features? The occurrences of fossils in flat-lying sediments seemed to corroborate a theory of the origin of rocks postulated by Abraham Gottlob Werner (1750-1817) and promulgated by him at the school of mines in Freiberg, Saxony, and by his enthusiastic students in many lands. Werner taught that all rocks were

II-6 J. B. P. A. DE M. LAMARCK

II-7 BARON G. L. C. F. D. CUVIER

II-8 A. G. WERNER

precipitated out of water. He arranged the earth's crust into a succession of formations which he believed were originally developed through deposition at the bottom of an ancient, earth-girdling sea. These were simply ranked in five orders of superposition as indicative of their relative ages: Primary rocks, Transition rocks, Secondary rocks, Tertiary rocks, and Volcanic rocks. Arguing from observations over a very limited area, and on the indisputable fact that the secondary rocks had formed in water, Werner concluded that all rocks had this origin. He assumed the existence of a universal ocean out of which the Primary crystalline rocks were precipitated and the later groups were formed, partly by further chemical precipitation and partly by mechanical accumulation, after dry land had emerged here and there.

Against the supporters of Neptunism—as the followers of Werner were called for obvious reasons—were ranged the upholders of an opposite view, that the main origin of rocks was as matter emitted by innumerable volcanoes. This theory of Vulcanism was the work mainly of the Frenchmen N. Desmarest and Élie de Beaumont, derived from the study of the volcanic massif of the Auvergne and of the mountains of Europe. Nowadays we know each group was partly right, that rocks originate in both fashions, and are in general divisible into three broad groups—sedimentary, metamorphic, and igneous rocks—of which the sedimentary rocks are almost always of Neptunian origin, igneous rocks uphold the Vulcanist interpretation, while metamorphic rocks share both origins.

Vulcanists and Neptunists agreed that the early history of the earth was a succession of profound catastrophes. Obsessed with their supposed prime-cause of rock formation, they assumed that the age of great changes was past and failed to use the present world as their laboratory, feeling that what occurred now was "the mere faded relics of the power with which geological changes were formerly affected."[6] The Vulcanists believed the earth had developed according to a succession of far-spread catastrophic events, each of which weathered into different sorts of sediments. The Neptunists, to explain the frequent departure from horizontally-lying strata—such as would be the unfailing result if deposition by water were the only cause of rock formation—were forced to assume periodic cataclysmic revolutions, culminating with the great flood recounted in the Old Testament. So also assumed the paleontologist Cuvier, who was hard put to explain the sudden disappearances of the extinct life forms whose remains he was studying except on the basis of catastrophic revolutions. Writing in 1826, he instanced the disappearance of the hairy mammoth, whose ice-embedded carcasses and ivory were causing much wonderment at this time. Asserting that the animals must have been exterminated in an instant by a force that brought glacial conditions to their Siberian habitat, he went on to argue that

> This change was sudden, instantaneous not gradual, and that which is so clearly the case in this last catastrophe is not less true of those which preceded it. The dislocation and overturning of the older strata show without doubt that the causes which brought them into the position which they now occupy, were sudden and violent . . . . Life upon earth in those times was often overtaken by these frightful occurrences. Living things without number were swept out of existence by catastrophies . . . . The evidences of those great and terrible events are everywhere to be clearly seen by anyone who knows how to read the record of the rocks.[7]

Catastrophism, besides being a convenient explanation, was also a most acceptable theory to reconcile the stubborn geological evidence with the accounts of the Creation and the early history of the earth as set forth in Genesis, as well as with the theological concepts of an ever-present God regulating the affairs of the universe, and, providentially providing man with a central and unique place in it. Many important geologists of the nineteenth century, including Gesner, Sedgwick, Buckland, and J. W. Dawson, accepted some variant catastrophic interpretation as the only way of reconciling their sincere religious convictions with their integrity as scientists. For how else except through sudden overwhelming catastrophic events could the enormous variety of familiar rocks have been produced in the limited time allotted by the Book of Genesis? Thus Abraham Gesner, in his early, able, *Remarks on the Geology and Mineralogy of Nova Scotia* (1836) was at pains periodically through his text to reaffirm his faith in the Biblical account even while he accepted the evidence of his eyes:

> Might not many of the changes which have taken place upon the earth, have been produced between the period when the globe was first created, and the Noachian deluge? And might not many of those effects, the causes of which are now almost inexplicable, have been produced at that momentous period, when the "windows of heaven were open," and "the fountains of the great deep broken up?" From what we have endeavoured to examine, and our feeble penetration into these dark problems, we are compelled to believe, that in no way can these phenomena be so satisfactorily accounted for and explained, as by admitting the brief account of the creation of the world, in the first Chapter of Genesis; and that there is no necessity for making the world appear older than its date given by Moses. Fortunately however, diversified as the opinions of modern Geologists may be, there are few who do not add much testimony to corroborate the statements of that inspired historian.[8]

Because uniformitarianism seemed to run counter to prevailing religious dogmas and concepts, it, like evolutionism later, faced the full blast of religious prejudice. It had also to overcome the stubbornly-held dogmas of the Catastrophists who could not conceive of the earth as having been shaped by slow, gradual, almost imperceptible, processes at work in the present-day world. Still, so cogent and so sustained by innumerable details was Lyell's reasoning that it was ultimately triumphant. The mystery of the mammoths, for example, was explained along gradualist lines—that the species was not overwhelmed in an instant but gradually declined to eventual extinction as the climate grew increasingly more difficult for survival even with the adaptations developed by the species; that over the prolonged period of this decline, individuals and small groups perished under circumstances that led to their bodies becoming encased in ice and so preserved for thousands of years. This might happen, for example, through animals becoming overwhelmed in avalanches, falling into crevasses, or being drowned in floods and transported north by Siberian rivers to a land of ice and permanently-frozen ground. By such means, given a sufficiently-long period of time, thousands of skeletons and corpses of mammoths might eventually accumulate to perplex men who came on the scene many centuries later. How much more satisfying to the understanding was such an explanation that accounted plausibly for all the facts and did not attempt to answer one question by posing another totally unanswerable one, like Cuvier's hypothesis of the instantaneous, catastrophic cold. Small wonder that Lyell's *Principles* should prove such a epochal work. Even more than the multitude of facts Lyell marshalled, and the daring hypotheses that he postulated, was the attitude of mind he fostered—that geological and biological problems should be approached free from preconceptions and examined by the light of everyday reason. And Lyell practised what he preached. When Darwin enunciated his theory of evolution, Lyell at first was cautious about accepting it until he had investigated the subject thoroughly for himself. Then, once convinced of its correctness, he battled valiantly for its acceptance, and not only incorporated it into his own writings, but thought through the ramifications for his field of earth science.

* * *

THE TRIUMPH OF UNIFORMITARIAN ideas meant that by the time the Geological Survey of Canada was founded geologists had accepted the principles that rocks are formed by processes analogous to those that are visible today; that in an undisturbed sequence a given stratum is older than all the strata above it and younger than all those below it; that there is a progressive change in the fossils from older beds to those in younger beds; and that beds far removed from each other may be correlated by comparing their contained fossils. Researches had indicated that strata and groups of strata could be correlated from country to country, and even from continent to continent. But local descriptive terms like "upper shale beds" were completely unsatisfactory to show these correlations or associations. A uniform system of nomenclature was needed to indicate rocks that contained the same kinds of fossils and therefore were of the same age, even when they were entirely dissimilar in outward appearance.

The stratigraphies of many local areas in Britain and on the continent had already been worked out, while Smith and others had prepared maps that attempted to correlate rocks over wider areas. What was needed was a grand succession, or timetable that included the whole range of fossiliferous strata, from ancient (Primordial or Axoic) rocks that showed no traces of former life, up to the present era. With this it would be possible to correlate all these local successions. Early in the nineteenth century, accordingly, work began in Britain and Europe on such a universally-applicable geological timetable. One of the most vexing problems concerned a group of rocks widely developed in Wales and the west of England, known as the Old Greywacke Series. The difficult task of subdividing this series was undertaken by two famous geologists, Roderick Murchison and the Reverend Adam Sedgwick, professor of geology at Cambridge University. Murchison worked down from the top of the series and Sedgwick upward in the more disturbed rocks to the north. Murchison recognized a great series of fossiliferous formations not previously described, to which he gave the name Silurian after the Welsh tribe that had once lived in the region. Sedgwick, on the other hand, recognized that part of the Old Greywacke Series was older than the rocks Murchison was studying, and to these he gave the name Cambrian, after the Roman name for the mountains in which these rocks were best exposed. Incidentally, some very old, badly deformed rocks known to lie below Sedgwick's Cambrian came to be known as Pre-Cambrian at this time; and the term Pre-Cambrian (now 'Precambrian') has since attained worldwide usage, despite attempts to substitute a more distinctive name.

So far all was well, but as Sedgwick worked up through his succession and Murchison down through his, the fossils collected by each were compared, and it was discovered that Sedgwick's Upper Cambrian was identical

II-9 REV. ADAM SEDGWICK

II-10 SIR R. I. MURCHISON

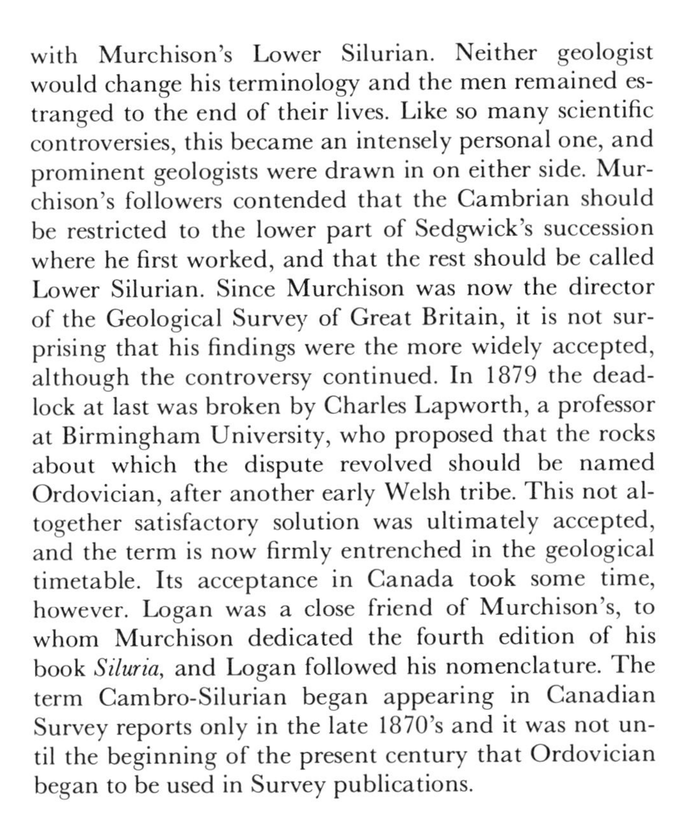

with Murchison's Lower Silurian. Neither geologist would change his terminology and the men remained estranged to the end of their lives. Like so many scientific controversies, this became an intensely personal one, and prominent geologists were drawn in on either side. Murchison's followers contended that the Cambrian should be restricted to the lower part of Sedgwick's succession where he first worked, and that the rest should be called Lower Silurian. Since Murchison was now the director of the Geological Survey of Great Britain, it is not surprising that his findings were the more widely accepted, although the controversy continued. In 1879 the deadlock at last was broken by Charles Lapworth, a professor at Birmingham University, who proposed that the rocks about which the dispute revolved should be named Ordovician, after another early Welsh tribe. This not altogether satisfactory solution was ultimately accepted, and the term is now firmly entrenched in the geological timetable. Its acceptance in Canada took some time, however. Logan was a close friend of Murchison's, to whom Murchison dedicated the fourth edition of his book *Siluria,* and Logan followed his nomenclature. The term Cambro-Silurian began appearing in Canadian Survey reports only in the late 1870's and it was not until the beginning of the present century that Ordovician began to be used in Survey publications.

In southern England, in the counties of Devon and Cornwall, a great series of strata had also been assigned to the Greywacke Series. In 1836 Sedgwick and Murchison had left Wales to study this series. So disturbed, twisted and altered were these rocks that both men were sure they were far older than Murchison's Silurian in nearby Wales, and at first assigned them to Sedgwick's Cambrian. They were, however, fossiliferous; and by a study of the fossils alone the paleontologist William Lonsdale discovered that the series was neither Cambrian nor Silurian but younger than either, although it was older than the series of which the English coal measures were a part. Further field work confirmed this, and in 1837 Sedgwick and Murchison proposed the term Devonian for these rocks.

The younger series to which the coal measures belong had already been studied in considerable detail because of its economic importance, and in 1822 the English geologist W. D. Conybeare had proposed the term Carboniferous for it. In much of the world the Carboniferous is not easily subdivided, although in places, upper and lower parts have been designated. In the great coal basins of the eastern United States, however, these subdivisions are clear enough that early American geologists were able to regard them as separate units, and in the late nineteenth century Alexander Winchell and H. S. Williams, respectively, proposed the terms Mississippian for the lower unit and Pennsylvanian for the upper. All three terms are still used; Carboniferous therefore can refer to both units taken together, or to a situation where only one may be present. Murchison was also responsible for identifying and naming the next great series above the Carboniferous. While studying the geology of the Ural Mountains on a commission from the Russian government, he found a great series of rocks overlying the coal measures which he was able to correlate with an English counterpart. The development of these younger rocks in the government of Perm was so excellent that in 1841 Murchison proposed the name Permian for the system.

The subdivision of the lower part of the stratigraphic succession composing the earth's crust in general had now been completed. At the bottom was a complex group of altered, unfossiliferous rocks known by the over-all term Precambrian. The subdivision of the Precambrian was to come much later and to no small extent through the labours of Canadian geologists–as will appear in later chapters. Above the Precambrian the succession eventually established was Cambrian, Ordovician, Silurian, Devonian, Carboniferous, and Permian, each being regarded as a 'period' of time during which a 'system' of rocks was formed. The time span in years was not then known, but the age of each system relative to the others was firmly established, and its recognition all over the world was made possible by means of its contained fossils. Hence the whole succession from the bottom of the Cambrian to the top of the Permian was named Paleozoic ("ancient life").

Similarly the succession of systems above this was called Mesozoic or "middle life," the period in which life had evolved to the point that fossils of animals with back bones were found in the strata. The earliest known Mesozoic group was the New Red Sandstone formation of England, so called to differentiate it from the Old Red Sandstone, which is Devonian. When the New Red Sandstone was traced to the continent a clearly recognizable three-fold subdivision soon became evident there, particularly in Germany, which led F. von Alberti in 1834 to propose the over-all name Triassic to describe the series. Long before this, in 1795 Alexander von Humboldt had proposed the term Jurassic for a series of rocks excellently exposed in the Jura Mountains of France and Switzerland. A study of the fossils soon established the identity of these rocks with strata in England that lie above the New Red Sandstone, so the Jurassic became the second Mesozoic period. A system of rocks above the Jurassic is well developed in southern England and neighbouring France. In both places it is characterized by thick beds of chalk, made famous by the white cliffs of Dover, so the name Cretaceous (from the Latin *creta* chalk) was proposed for this system in 1822. The fact that in the vast majority of Cretaceous rocks around the world chalk does not occur, demonstrates that geological periods are not associated with nor can they be identified by particular kinds of rock or minerals. These three systems together comprise the Mesozoic Series.

The characteristic life of the younger series above the Mesozoic was modern in aspect, that is, many of the forms are identical with forms existing today or were evidently closely related. The series was therefore known as the Cenozoic ("recent life"), and when Lyell attempted its subdivision in 1833, he did so on the basis of the percentages of fossil species with living counterparts. His earliest units he named Eocene ("dawn of recent"), Miocene ("middle recent"), and Pliocene ("more recent"), to which in 1839 he added a fourth unit, Pleistocene ("most recent"), above the Pliocene. While later collecting has thrown doubt on the accuracy of Lyell's percentages, his names still stand. After further stratigraphic studies the name Oligocene ("few recent") was given to a group of beds formerly considered the top of the Eocene and the bottom of the Miocene; and Paleocene for another group recognized in 1874 to lie below the Eocene just above the Cretaceous. With these, the subdivision of the Cenozoic was complete.

One last step remained. It was clear that the whole Cenozoic extending from the Paleocene to the top of the Pliocene was probably of no longer duration than one of the Paleozoic systems, so it seemed desirable to coin a term that embraced all five series. As they were roughly equivalent to the consolidated rocks that in Werner's system were called Tertiary, that term was selected to designate the whole Cenozoic up to the Pliocene. This left the Pleistocene, much of it scarcely consolidated into rock, which is overlain by material deposited by processes that have only just stopped or are still going on. This recent material was appropriately named Recent, and the Pleistocene and Recent together were grouped as the second and last major period of the Cenozoic, under the name Quaternary.

The task of subdividing, classifying, and naming the stratigraphic sequences of the earth's crust was largely complete in outline by the middle of the nineteenth century, as a result of work done mainly in the 1830's and 1840's. Much work remained, to refine the definition of the established terms, and to define both physically and faunally the top and bottom of each period and system. As originally defined, each of these boundaries was a major stratigraphic discontinuity representing a dramatic change in existing conditions, known as a geological revolution. These revolutions were once believed to be worldwide but extended investigation has not borne this out. In many parts of the world, strata of a particular system have been found that are older or younger than those at the type section, and in some places one system grades imperceptibly into the next without any evidence of an intervening revolution. In such places the boundary between the systems had to be made on faunal grounds, giving rise to much discussion and some controversy. Although the task has been completed for most systems, the work still goes on; the attention of most stratigraphers and stratigraphic paleontologists has turned to the finer and finer subdivision of the systems, permitting more precise field mapping, better informed generalizations, and more accurate basic information that can be brought to bear on economically-important problems like the occurrence of oil-bearing strata.

These periods of geological history are represented in today's earth by the consolidated sediments, derived from material worn or dissolved from existing rocks, carried away and accumulated in water, mostly the sea, or occasionally, on land. They include the numerous sedimentary rocks, such as sandstones, limestones, and shales, most of which were formed in the same way as silt settles out of dirty water in a glass. These were deposited as layers, became hardened into rock, and each layer became a "stratum." One of the first great principles arising from the theory of uniformitarianism is that each layer so formed must be younger than all the layers below it, and older than all the layers above it. It is this rule of superposition that is at the core of all stratigraphic work. William Smith's great contribution was to show that a stratum could be correlated with others of similar age anywhere else in the world by matching their contained fossils.

The reconstruction of a succession of the rocks forming the earth's crust would appear to be simply a question of identifying which bed lay above which, and listing their contained fossils. Unfortunately, this is far from a simple task. Sedimentary rocks are laid down in their regular sequence, but everywhere there are vast gaps in this record when no rocks were formed, or where great thicknesses have been removed by the erosive action of wind, water, and ice. Not all sedimentary rocks are fossiliferous, and even if they originally were, the fossils may have subsequently been destroyed. Indeed, to find well-preserved fossils is the exception rather than the rule. Finally, the earth's crust is solid only in relation to our time-span. In geological time, a matter of billions of years, the land rose and fell, seas came and went, and the sequence was interrupted again and again. The solid crust has been folded, twisted and broken, squeezed up into high mountain ranges, then worn down to the roots, once again to form level plains. The original flat-lying, buried strata were thrown high into great folds and vast fragments, marked by major dislocations along the fractures (faults). In each fragment the succession of strata is always upward from oldest to youngest, but in some the strata may have become overturned, with the result that the older rocks are now above the younger ones; or because of folding, the same strata may be repeated; or the strata in one fragment may be very different in age from those in the next fragment which has been thrown up

GEOLOGICAL TIME-SCALE

ERA	PERIOD	EPOCH	MILLIONS OF YEARS BEFORE NOW[1]
	QUATERNARY	Recent	
		Pleistocene	1.5-2
		Pliocene	7
		Miocene	26
CENOZOIC	TERTIARY	Oligocene	38
		Eocene	54
		Paleocene	65
	CRETACEOUS		136
MESOZOIC	JURASSIC		190-195
	TRIASSIC		225
	PERMIAN		280
		Pennsylvanian	325
	CARBONIFEROUS		
		Mississippian	345
PALEOZOIC	DEVONIAN		395
	SILURIAN		430-440
	ORDOVICIAN		500
	CAMBRIAN		570
P R E C A M B R I A N			

[1] Based on "Standard for Geological Time" used in Geotectonic Correlation Charts 2, 3, and 4, in *Geology and Economic Minerals of Canada*, edited by R. J. W. Douglas (GSC, Economic Geology Report No. 1, fifth edition, 1970) Maps and Charts volume. Dates obtained by measuring the amount of decay of certain radioactive minerals, giving under suitable conditions the length of elapsed time since the mineral formed. The method has only recently been developed into a powerful geological tool. See Chapters 21 and 22, below.

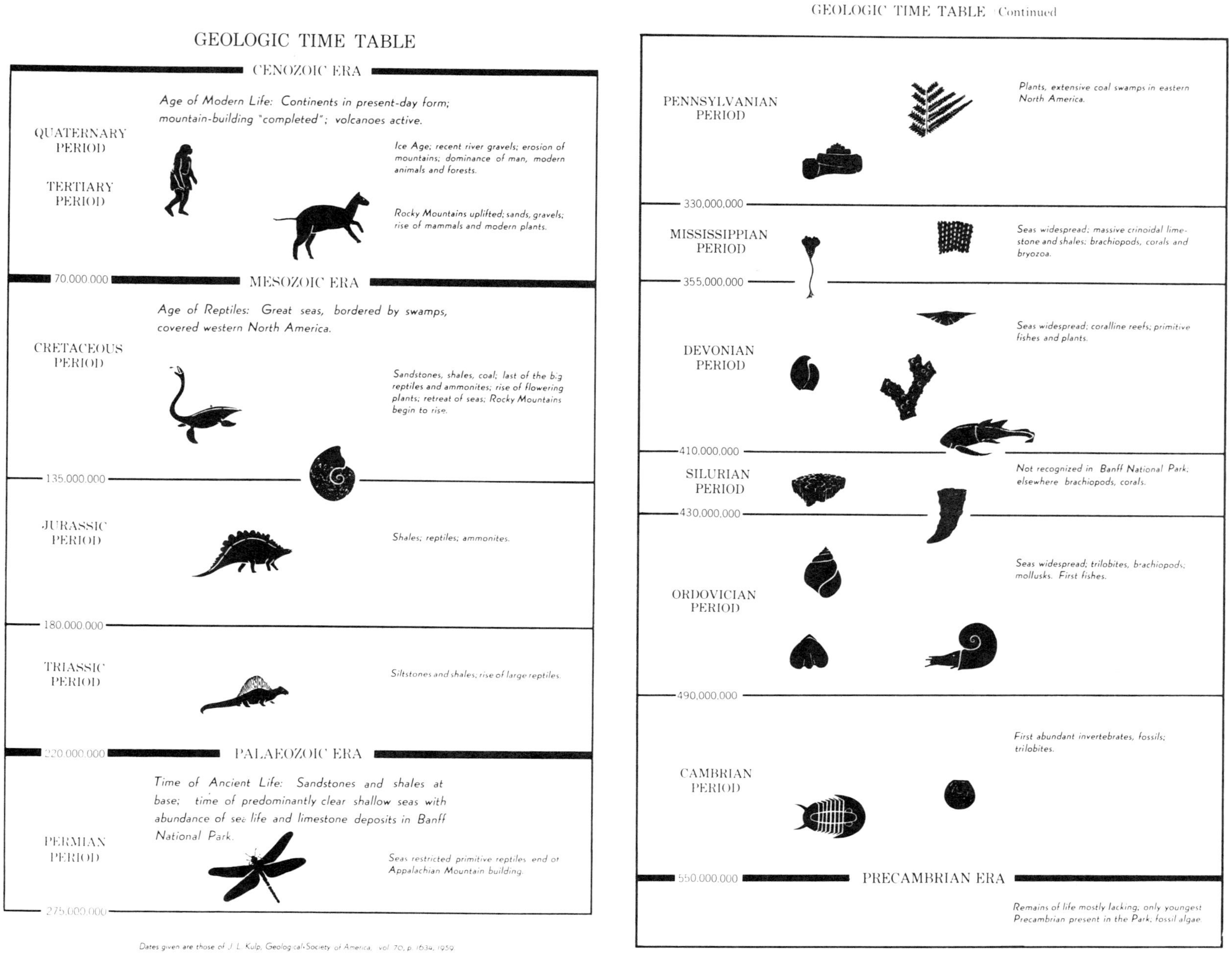

GEOLOGIC TIME TABLE

CENOZOIC ERA

Age of Modern Life: Continents in present-day form; mountain-building "completed"; volcanoes active.

QUATERNARY PERIOD	*Ice Age; recent river gravels; erosion of mountains; dominance of man, modern animals and forests.*
TERTIARY PERIOD	*Rocky Mountains uplifted; sands, gravels; rise of mammals and modern plants.*

70,000,000 — MESOZOIC ERA

Age of Reptiles: Great seas, bordered by swamps, covered western North America.

CRETACEOUS PERIOD	*Sandstones, shales, coal; last of the big reptiles and ammonites; rise of flowering plants; retreat of seas; Rocky Mountains begin to rise.*
135,000,000	
JURASSIC PERIOD	*Shales; reptiles; ammonites.*
180,000,000	
TRIASSIC PERIOD	*Siltstones and shales; rise of large reptiles.*

220,000,000 — PALAEOZOIC ERA

Time of Ancient Life: Sandstones and shales at base; time of predominantly clear shallow seas with abundance of sea life and limestone deposits in Banff National Park.

PERMIAN PERIOD	*Seas restricted primitive reptiles end of Appalachian Mountain building.*
275,000,000	

Dates given are those of J. L. Kulp, Geological Society of America, vol. 70, p. 1634, 1959

GEOLOGIC TIME TABLE (Continued)

PENNSYLVANIAN PERIOD	*Plants, extensive coal swamps in eastern North America.*
330,000,000	
MISSISSIPPIAN PERIOD	*Seas widespread; massive crinoidal limestone and shales; brachiopods, corals and bryozoa.*
355,000,000	
DEVONIAN PERIOD	*Seas widespread; coralline reefs; primitive fishes and plants.*
410,000,000	
SILURIAN PERIOD	*Not recognized in Banff National Park; elsewhere brachiopods, corals.*
430,000,000	
ORDOVICIAN PERIOD	*Seas widespread; trilobites, brachiopods, mollusks. First fishes.*
490,000,000	
CAMBRIAN PERIOD	*First abundant invertebrates, fossils; trilobites.*

550,000,000 — PRECAMBRIAN ERA

Remains of life mostly lacking; only youngest Precambrian present in the Park; fossil algae

II-11 ILLUSTRATED GEOLOGICAL TIME TABLE
Prepared by Helen Belyea for a guide to Banff National Park, 1960. Dates for the various periods vary slightly from those presented in the text of this chapter.

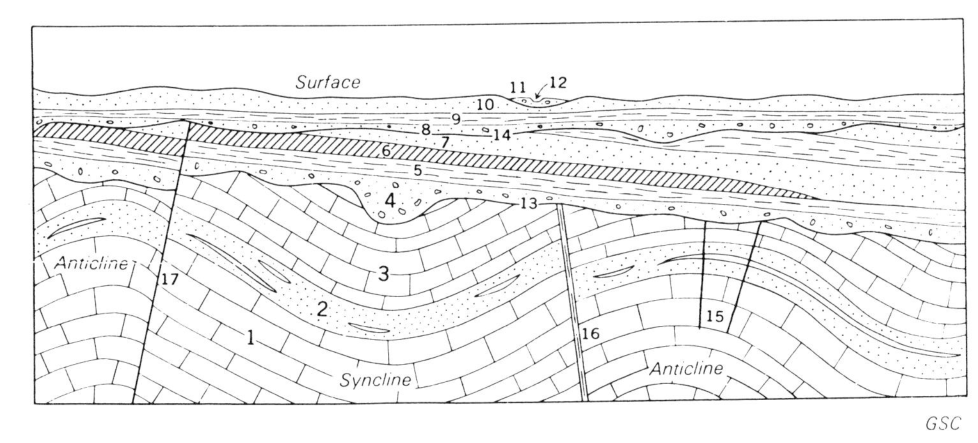

II-12 MODERN DIAGRAMMATIC REPRESENTATION OF GEOLOGICAL SUCCESSIONS

An ancient series (1-3) consists of three conformable formations comprising limestone (1), sandstone with lenses of shale (2), and younger limestone (3). These beds were folded into anticlines and synclines, after which the upper ones were eroded to form the ancient land surface (13). This surface was submerged and on it was deposited another conformable series (4-7), comprising conglomerate and sandstone (4), shale (5), lava flows (6), and sandstone (7). These formations and those of the underlying, folded series were uplifted and tilted. Erosion then produced another land surface (14). This was submerged and a third series (8-10) was deposited, comprising conglomerate and sandstone (8), shale (9), and sandstone (10). The beds of this series remain in their original horizontal position. A modern stream (12) has cut a valley in formation (10) and has deposited sand and gravel (11). The lines (15) represent fractures older than surface (13). A fault, (16), which has formed a wide shear zone, is also older than surface (13). A later fault, (17) is younger than surface (13) but older than surface (14).

alongside it. The task of the field geologist is to piece together a continuous story out of this seemingly jumbled, confused record.

Sedimentary strata comprise only one major group of rocks. Igneous rocks, which provided the basis for the theory of Vulcanism, originated deep within the earth's crust as hot molten masses called magmas. In all periods magmas have worked their way upward through the earth's crust until cool enough to crystallize as igneous rock. Most magmas crystallize while still deep in the crust, and the resulting igneous rocks become visible only when the rocks above them have been weathered away. Commonly, they have ascended along faults and fractures in the overlying rocks to form distinctive veins and stringers (sometimes depositing associated metallic minerals), or interacting with the containing rocks among which they pass. Only occasionally do they emerge directly to the surface, and then in spectacular fashion, by passing right through the crust to pour out of a volcano as lava.

Metamorphic rocks, the third main type, present an even more difficult problem. They originate from either sedimentary or igneous rocks that have been so deeply buried within the earth's crust that heat, pressure, and hot gases or liquids have completely altered (metamorphosed) their original nature. In some metamorphic rocks the original sedimentary character may still be sufficiently recognizable to permit the rules of stratigraphy to be applied, but only with great care, for its sedimentary origin may be clear in one place, but quite uncertain in another.

Far less was known in the 1840's, or for long afterwards, about the properties and methods of analysis of igneous and metamorphic rocks than about sedimentary successions. For one thing, igneous rocks do not fall into regularly characterized periods, but are the results of geological events that have occurred in all ages. Their significance lies in the processes by which they have been formed, and since each case is unique, the resulting rocks and structures possess individual characteristics depending on the physical and chemical conditions under which the geological events took place—conditions that can now be estimated only very roughly because they occurred at great depths and many millions of years ago. Hence there are many imponderables about these structures.

From the standpoint of mapping, the relations of igneous and metamorphic rocks to sedimentary strata (and sometimes to other igneous rocks) have to be ascertained by a procedure comparable to the rule of superposition. If a magma or 'pluton' has cut through certain rocks, it is younger than the youngest of these; if strata overlie its *weathered* surface, they are younger than the plutonic mass. The relationships between igneous bodies in an area may be recognized by chemical and mineralogical similarities, and by the nature of their contacts with one another; but dating could only be done relative to the sedimentary host rocks until radioactive age-determination methods became perfected in recent times. Dating such rocks has both scientific and economic significance, for working out the relations can provide clues where to look for other products of the geological event, which may include mineable ore deposits.

The heart of geological work is field work, accuracy in recognizing and identifying rocks in the field, measuring their positions, and working out their relations with other rocks. The field methods in use by Logan's day were the basis of those in use today, though modern geology has the advantage of nearly a century and a half of experience and much more sophisticated equipment. Good field mapping is far from easy, aside from the physical problems of working in remote districts and under difficult conditions. The geologist must be well trained, keep his wits about him at all times, and never take anything for granted. A great deal of training, experience, and natural ability are required in order to perceive and visualize the three-dimensional relationships of the rocks which may occur on vast scales, or present very imperfect, fragmentary, outcrops with great gaps in the record. Uncertainties are inevitable from the nature and history of the rocks—the great disturbance, distortion and deformation of sedimentary strata, as well as those inherent in most igneous and metamorphic rocks. The science is always evolving through the accumulation of new knowledge based on field observation and laboratory analysis, which may offer new interpretations and insights leading to better understanding of the rocks and their field relations. The geologist requires special talent to be able to pick out the key elements in a situation and arrive at the most reasonable and accurate interpretation afforded by developments of the science to date.

* * *

GEOLOGISTS ARE NOT EMPLOYED BY GOVERNMENTS for the primary purpose of contributing to the increase of theoretical geological knowledge; geological surveys, in Canada and elsewhere, were established with the objective of advancing the mining economy of the province. Geology began with, and developed from, efforts to understand the principles behind the occurrence of minerals so as to assist the development of mineral resources. By the time the Geological Survey of Canada was founded, much practical information had been accumulated on the nature of many mineral deposits; and ideas had been formulated on the types, ages, and structures of the rocks and formations in which they occurred. Geologists mapping the country attempted to indicate where conditions were favourable for the presence of valuable minerals. They studied deposits under development, or minerals of economic interest encountered in their investigations, to find out more about how and where these minerals were most likely to be found.

Economic minerals produced by the normal processes of sedimentation, like limestone, gypsum, oil, and gas, were formed usually at the bottoms of shallow seas, or else in marshy terrain, like coal. Since they are parts of a stratigraphic succession, overlaid by many subsequent strata, searches for these minerals could be localized in the formations in which they were known to occur. Thus coal, the chief *desideratum* of early nineteenth century industrial civilization, was known to occur in certain parts of the Carboniferous, so searches were directed towards finding whether or not these particular Carboniferous formations were present in the Province of Canada. When it was learned that in western America coals also occurred in later periods, investigations for coal concentrated on the appropriate parts of the Mesozoic and Tertiary successions. Since coal is an easily recognizable solid and surface showings could be followed in depth, coal beds were quite easy to locate, though in mining coal many structural and economic problems had to be overcome. Far more difficult minerals to find were petroleum and natural gas, which presented problems because they migrated from their points of origin. Geologists had first to discover the strata in which the minerals originated, their derivation, and in what sorts of structures they tended to localize. Eventually it was found that much petroleum and natural gas originated in the Devonian period from the alteration of marine organisms,

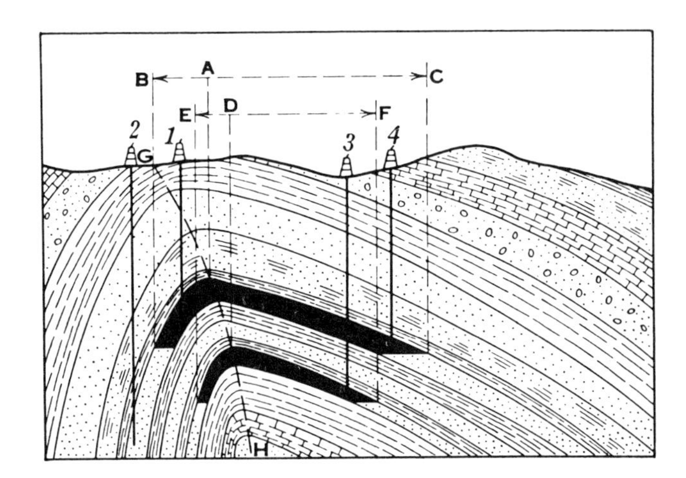

II-13 PETROLEUM OCCURRENCE AND OIL WELL DRILLING
Two oil-bearing strata are enclosed within the anticline, the axis of which is indicated by the line G-H. Wells 1 and 4 would tap the upper stratum within the productive limits B-C. Well 3 would also penetrate the lower stratum within its productive limits E-F and produce from both levels, while Well 2 would be a "dry hole."

and being lighter than water they migrated to permeable higher strata and tended to become concentrated in the anticlinal arches beneath impermeable strata. Consequently, geologists were able to direct searchers to the places where favourable conditions prevailed, greatly reducing the areas to be investigated, though the chances of discovering a mineable deposit were still extremely small.

More difficult to localize are the metallic minerals which, as miners have known for centuries, commonly occur in veins in sedimentary rocks, usually in conjunction with igneous and metamorphic rocks like granite. Nothing was known then about the manner of their formation—that some were actual components of magmatic intrusions, parts of the mineral content of the magma which separated out from the mass, according to the conditions of cooling, in large enough bulk to produce a massive orebody. More often, however, they are carried by water or other materials into the cracks or fissures of host rocks, to crystallize out, in conjunction with the other intruded material (usually quartz), in concentrations, which if sufficiently high, could result in economic ore. Early in the history of the Survey its members were examining major faults and fractures as the likeliest places, structurally, for such minerals to concentrate. When they learned to differentiate between the types of Precambrian rocks, they felt they could geographically delineate a particular period (Huronian) in which the metallic minerals were most likely to occur.

These metallic minerals, present in so many places in the Canadian Shield and Cordilleran regions of Canada, have been a major subject of inquiry on the part of the Canadian Survey as well as of the mining industry of Canada. Here especially, the geologist of the Survey has concentrated on learning more and more exactly the

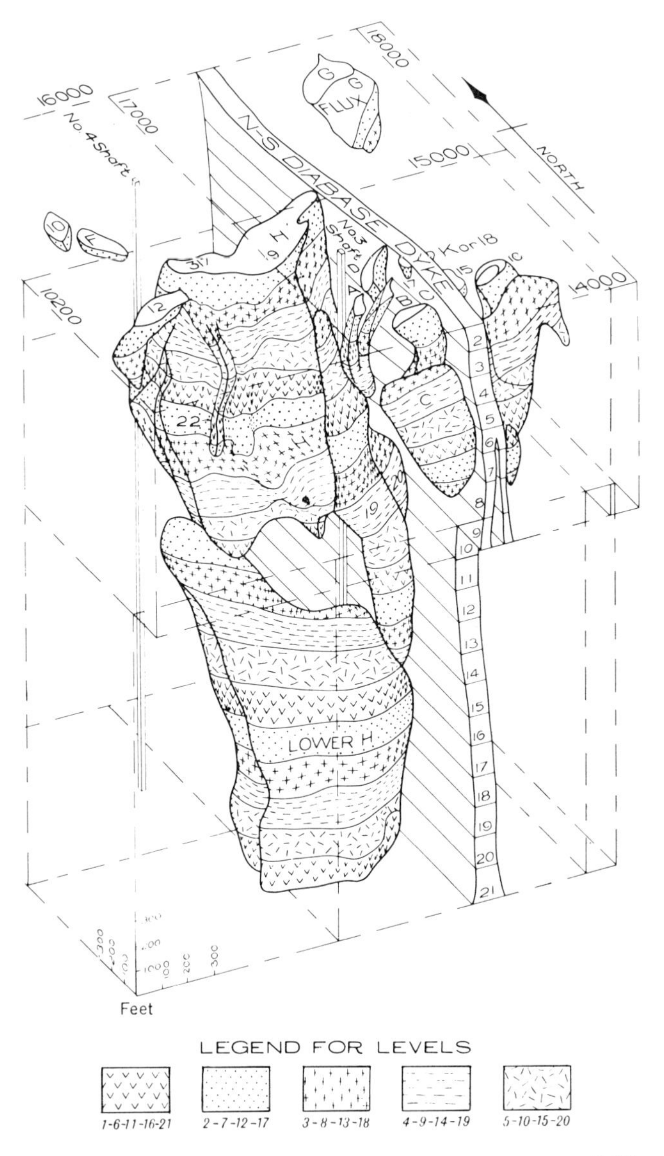

II-14 A MAJOR ORE DEPOSIT
This isometric projection (three dimensional) diagram of a group of contiguous and related orebodies was made possible by extensive test drilling and production analysis incidental to developing and operating one of Canada's largest base metal mines.

types of rocks in which these minerals occur, and on localizing the search for economically-workable deposits. He did not attempt to discover these deposits himself, except accidentally, though the geologist's role in this regard has been commonly misconceived. For the task of mapping the geology of a region allowed geologists, as they proceeded about their work, time only to indicate structures that were favourable for the occurrence of these minerals, and to note traces of valuable minerals. Discovery and development of an orebody required infinitely more time, expense, and risk than the Survey could afford.

The task of finding a workable deposit, even in a known favourable area, entails hunting for hundreds of 'showings' spread over a wide area; stripping away the overburden, digging and trenching to get an idea of their extent and trends; searching for an 'occurrence' that might warrant further efforts to assess its size, content, grade and shape by additional work including diamond drilling; and sinking shafts and tunnelling that could raise the 'occurrence' to the status of a 'prospect.' Hundreds of occurrences subjected to such efforts might yield only a single prospect; and even then, dozens of prospects are abandoned as too small or too low grade to justify the heavy capital outlays needed to develop them. Even developed mines still are not assured of automatic success, but must depend on the accuracy of estimates of ore reserves, as well as on such external factors as improved transportation, changing mining technology, and favourable market conditions. The appropriations provided the Survey could never have paid for more than one or two experiments of this sort a year—and the work of mapping the geology of Canada would have had to go by the board. The staff and money of the Survey unquestionably were far better employed over the years in exploring and mapping the geology of Canada.

In the nineteenth century ignorant optimism on the score of mining success was the rule rather than the exception, and the magnitude of the odds against developing a showing into a mine was completely hidden to miners or public. Mining men blissfully proceeded to install mining plants without adequate assurance that the outlays were justified and that the expenses would be recovered. Investors were even more ignorant, and often they were misled by highly coloured, optimistic propaganda, or deceived by activities undertaken deliberately to contribute to the excitement—a practice still not completely eradicated from the mining world, despite the efforts of regulatory bodies. At such times the geologists served a negative as well as a positive role—not only showing where the best chances for finding mineral deposits were, but also guiding efforts and investment away from places where success was highly unlikely. They exposed inherently impossible claims or deliberate falsehoods, and they criticized extravagant or unrealistic efforts by mining developers—often to be answered by considerable attacks and abuse. But their work contributed to the eventual emergence of new professional standards of prospecting and mining development, and they helped place the industry on a firmer footing.

The inauguration of the Geological Survey of Canada came at the conjunction of two important trends—at the height of a period of intellectual ferment that was turning geology into a more exact science, operating under a system of strict working rules and directed towards expanding man's intellectual horizons; and at a time of increasing concentration on ways of winning greater quantities of minerals from the earth to meet the insatiable needs of the expanding industrial revolution. Both trends enjoined the same task on the new Geological Survey of Canada—of carefully mapping and studying sedimentary beds, contorted and metamorphosed strata, igneous and volcanic rocks, for the dual purposes of assisting the finding and exploitation of mineral deposits, and shedding light upon particular geological phenomena and fundamental processes.

The undertaking of a geological survey of Canada in the 1840's imparted a great responsibility to those men who conducted it, but also offered unparalleled opportunities to explore the large territory comprised in the Province of Canada, a territory greater than the British Isles, than most of the major countries of Europe, or than any single state of the United States. Within this area, extending from the Strait of Belle Isle past the head of Lake Superior, would certainly be found important mineral wealth to enrich the people and government of Canada. As well, the operations of many geological principles still unrecognized by man might be discovered there, which might revolutionize important aspects of geological science and facilitate the finding of mineral deposits. In addition, on the margins of Canada and beyond its borders stretching far to the north and west lay a vast expanse of more than one million square miles of Precambrian rocks, possibly the largest group of such rocks exposed anywhere in the world, in which were buried great treasures of metallic riches. Perhaps somewhere in that vast region, too, might be found the answers to the mysteries of the origin of the earth's crust, or of life on earth, the precursors of the trilobites and brachiopods that flourished so abundantly in the next, or Cambrian period, the earliest Paleozoic era.

To conduct a survey that promised such great opportunities required a man of sterling qualifications and of the highest integrity, and familiar with the most modern methods. These the Canadian government secured when it chose William Edmond Logan for the task, for he could meet the greatest geologists of the day on terms of equality and friendship. He had proved his ability beyond question in the most advanced and critical arena; as had already been indicated by his admission to membership in the highly-influential Geological Society of London.

That Logan had come to the attention of the most eminent British geologists of the time was a result, in the first instance, of his outstanding ability as a stratigrapher and mapmaker. Perhaps as early as 1831, certainly by 1834, like many other amateurs in the British Isles, Logan began mapping the rock formations and coal measures of the district in which he was living, the South Wales (Glamorganshire) coal basin, recording the geological field data on the recently-published 1 inch = 1 mile Ordnance Survey maps of the region. By 1837 his work was so far advanced that he could present a carefully-executed geological map at the Liverpool meeting of the British Association for the Advancement of Science, supplemented by a paper "On that part of the South Welsh Coal Basin which lies between the Vale of Neath and Carmarthen Bay. In explanation of a geological map of the district, laid down by the author on the sheets of the Ordnance Survey."[9]

The Transactions of the Association contain a summary of this accompanying statement in which Logan analyzed the results shown on the map and drew conclusions regarding the location and quality of the coal deposits, both known and undiscovered. When De la Beche, after finishing his work in Devon and Cornwall, turned next to South Wales, Logan made his work available to the Ordnance Survey. For his part, De la Beche considered Logan's map "beautifully executed" and decided that "The work on this district being of an order so greatly superior to that usual with geologists . . . we shall adopt it for that part of the country to which it relates."[10] In fact, Logan's work was taken over intact, and eventually his name appeared on the published sheets for the area: "Geologically surveyed by W. E. Logan and Sir Henry De la Beche."[11] Logan was also credited with an important innovation in geological cartography, the introduction of cross-sections, representing hypothetical vertical slices of the earth's crust to depict the land surface and succession of strata beneath it, a practice the Geological Survey of Great Britain adopted in 1844.[12] That Logan considered stratigraphy his *forte* is shown by the letter to his brother James from Pennsylvania in September 1841. During his visit to the coalfield at Wilkes-Barre he was shown some work executed by members of the state geological survey. He commented "It was very well done but I could beat it hollow," then went on to observe:

> In this department [mapping] as also in what may be termed geology proper, such as following out stratification however contourted, detecting dislocations, anticlinal & sinclinal lines & in short in all kinds of geological mapping I am not afraid of competition with any one I know of. De la Beche once told me he considered me the best geological mapper in Britain.[13]

In having his work accepted for the official Geological Survey of Great Britain, Logan had passed a most exacting test as a field geologist.

He also played a part while in Wales in the formation of the Swansea Philosophical and Literary Institute in 1835, becoming its honorary secretary and curator for geology, setting up the collections, labelling and displaying specimens and performing other duties—experience that would stand him in good stead in attending to this important side of the Geological Survey of Canada's work and would bring his greatest public triumph. His generous nature was also displayed in his patronage of the institute's work; he was a generous contributor to the fund to erect a handsome new building for the Royal Institute of South Wales (as the Swansea Institute had become in 1839), as well as donating meteorological instruments, and books for the library. Time and again, to assist the successful operation of his Canadian Survey, he

II-15 THE ROYAL INSTITUTE OF SOUTH WALES, SWANSEA

would have to do as much and more. Indeed, his altruism, his complete devotion to the interests of the Survey, and his undeviating belief in the importance of the institution and the work it was doing, were basic factors in the success the Geological Survey of Canada enjoyed under his direction.

But Logan's principal claim to be ranked among the pioneers of geological science, that which brought him to the attention of Lyell, was his proposing a satisfactory solution to the problem of the origin of coal by rational, uniformitarian principles. This he presented in his paper "On the Characters of the Beds of Clay immediately below the Coal-Seams of South Wales, and on the occurrence of Boulders of Coal in the Pennant Grit of that district," read to the London Geological Society on February 26, 1840, and published in its *Transactions*, with a summary in its *Proceedings*.[14] Logan reported that in examining nearly one hundred coal seams in South Wales he had invariably found them underlain by a bed of greyish, sandy clay, in which were immense numbers of interwoven roots and stems of *Stigmaria Ficoides*. From this he concluded:

> . . . when it is considered, that in so wide a district of country abounding in coal, there is not a seam which is not immediately underlaid by a bed wholly monopolized by these peculiar vegetable organic remains, it is impossible to avoid the inference, that some essential and necessary connection exists between the production of the one and the existence of the other. To account for the unfailing combination by drift, seems an unsatisfactory hypothesis; but whatever may be the mutual dependence of the phaenomena, they give us reasonable grounds to suppose that in the *Stigmaria Ficoides* we have the plant to which the earth is mainly indebted for those vast stores of fossil fuel which are now so indispensable to the comfort and prosperity of its inhabitants.[15]

He had discovered, by careful field investigation and logical deduction, that most coal was produced in place by the plants growing in vast numbers in great peat swamps, a hypothesis confirmed by continuing examination of coal-bearing seams in other parts of the world, by microscopic studies of the coal itself, and by frequent discoveries of trunks and branches of the fossilized plant within the coal seams.

In undertaking the new survey, then, Logan brought to the work a thorough training in the latest geological principles, a reputation as a careful, conscientious observer and maker of maps, and a man who had contributed, and would contribute, to the furtherance of the science. The succeeding years showed him to be a most able, painstakingly cautious worker. As he told a correspondent in 1845 (echoing a view De la Beche had expressed to him earlier, in a letter of April 25, 1843), he was prepared to issue reports of progress, "merely for the purpose of indicating that due diligence is exercised in the prosecution of the task committed to me. Confident opinions stated upon partial data & general conclusions drawn from limited premises are in Geology of all things to be avoided."[16] Still later, he wrote to an associate that "I cannot express myself with so much confidence as you do in regard to what science may achieve in endeavouring to pry into the bowels of the earth. I rarely ever venture to use the word *certainty*, and seldom go beyond *probability*."[17] His biographer, B. J. Harrington, said of Logan that "As a close observer and careful delineator of facts he excelled, but his mind was not of the speculative type, and he rarely indulged in the flights of fancy so common among geologists."[18] Logan, indeed, preferred to leave to others the opportunities of deriving theoretical speculations from the evidence accumulated by the Survey, in particular his staff associates Billings and Hunt, or his friend, Sir William Dawson.

II-16 A *SIGILLARIA* STUMP
Many Carboniferous plants from whose roots and trunks coal is formed have survived as fossil stumps in which the woody matter has been replaced by clay or shaly material. This specimen was photographed in 1922 near Coal Mine Point, Cape Breton, Nova Scotia.

In view of the newness of the Canadian Survey and its precarious position, it was all to the good that early in its history it should achieve a reputation for careful, sound, accurate work of a high order of competence. In this fashion it largely avoided criticisms of hostile geologists, and publicly-aired internal disputes, and was able to proceed steadily with productive work that commended it to the government and the public, especially searching for, and reporting on, economic minerals. And, in due course, the Survey began to make its own contributions to the advancement of the science. Among the earliest of these, that immediately attracted the attention of a wide public because they were so pertinent to the great intellectual controversies of the age, were the finding of tracks of creatures in early Cambrian rocks and the reported discovery of life in Precambrian formations, the so-called *Eozoon Canadense.* By comparison, far less notice was taken of Logan's more fruitful and permanent contribution of dividing these oldest rocks, the Fundamental Gneiss, into two great groups, Laurentian and Huronian—the first major victory in the long battle to unravel the mysteries of the Precambrian.

III-1 THE SURVEY'S MONTREAL HOME, 1852-81
After being housed in three separate locations in its first nine years, the Survey remained in this building for almost 30 years until the move to Ottawa. Erected in 1814 and the one-time home of Peter McGill, the building was acquired by the government of the Province of Canada for the offices of the Crown Lands Department while the capital was at Montreal. After the Survey left, the building was occupied by various Quebec government agencies such as the Council of Agriculture and the Department of Health, then by business firms. It was demolished only in 1963, when the city block on which it stood, along with two others, was razed to the ground to make way for a giant multistory complex of provincial courthouse and government offices.

3

Early Trials and Triumphs

When Logan set out on what was to be his life's work, he brought to the task a mature character which had been developing over his forty-four years. For as long as he remained provincial geologist and director—and in some ways down to the present—the Survey he founded would be marked indelibly with his personality. Throughout Logan's directorship the Geological Survey of Canada remained a small organization and the close-knit empire of a single man. He was well-to-do, had not sought the position for material ends, and acknowledged favours to no one for his appointment. He brought to the work a complete dedication to the pursuit of scientific truth but at the same time a respect for practical endeavours, a good business head, an ability to deal with all sorts of people, and a fair capacity for managing employees. He would have to use these abilities to the full to establish, preserve, and expand the geological survey.

Logan began with a definite strategy—so to utilize the two-year commission "to make a geological survey" as to secure its extension on a continuing basis. For this he must demonstrate the value of the survey to the government and the public, and keep his expenses as low as possible. Logan realized that the practical aspect must override the scientific. To advance the search for new mineral deposits and evaluate those already known was always an important part of his program. A search for mineral resources could produce important geological discoveries, while an investigation carried out for scientific ends could yield valuable practical results. So Logan proceeded about his work by careful, accurate, imaginative collection of basic data, including data on mineral resources that might become the foundations for new industries.

The Province of Canada consisted only of the southern parts of present-day Ontario and Quebec, but with the conditions of travel then existing, this was a vast area, whose geology was virtually unknown. First Logan familiarized himself with contiguous areas, particularly New York, where the geology was the best known and most carefully studied in America. From the start he formed a close friendship with Professor James Hall of Albany based on mutual respect and common interest, which, despite Hall's notoriously jealous, choleric nature, lasted through Logan's lifetime. Logan concluded that the province fell into three major geological units, a conclusion that subsequent study has extended and refined. By the end of the first season's work by Murray and Logan in 1843, the limits of all three were well defined: an Eastern Division of folded Paleozoic rocks covering most of the Eastern Townships and the Gaspé peninsula; a Western Division of flat-lying Paleozoic rocks extending west from Montreal to Detroit River, but split in two parts by a band of older gneisses and schists that came to be known as the Frontenac Axis; and a Northern Division comprised of these last-named rocks, the Primitive or Metamorphic (now Precambrian) rocks of the Canadian Shield, occupying all the known region to the north. Separating the Eastern and Western Divisions was a major structural discontinuity, extending from the northeastern corner of Lake Champlain to Quebec City and beyond. Destined to involve Logan in much effort and thought before he could work out its implications to his satisfaction, this remarkable tectonic feature is known to this day as Logan's Line.

Logan was conscious always of the need to justify his operation by immediate economic benefits. Nothing would

III-2 THE EARLY YEARS: WORK IN CENTRAL CANADA, 1842-55

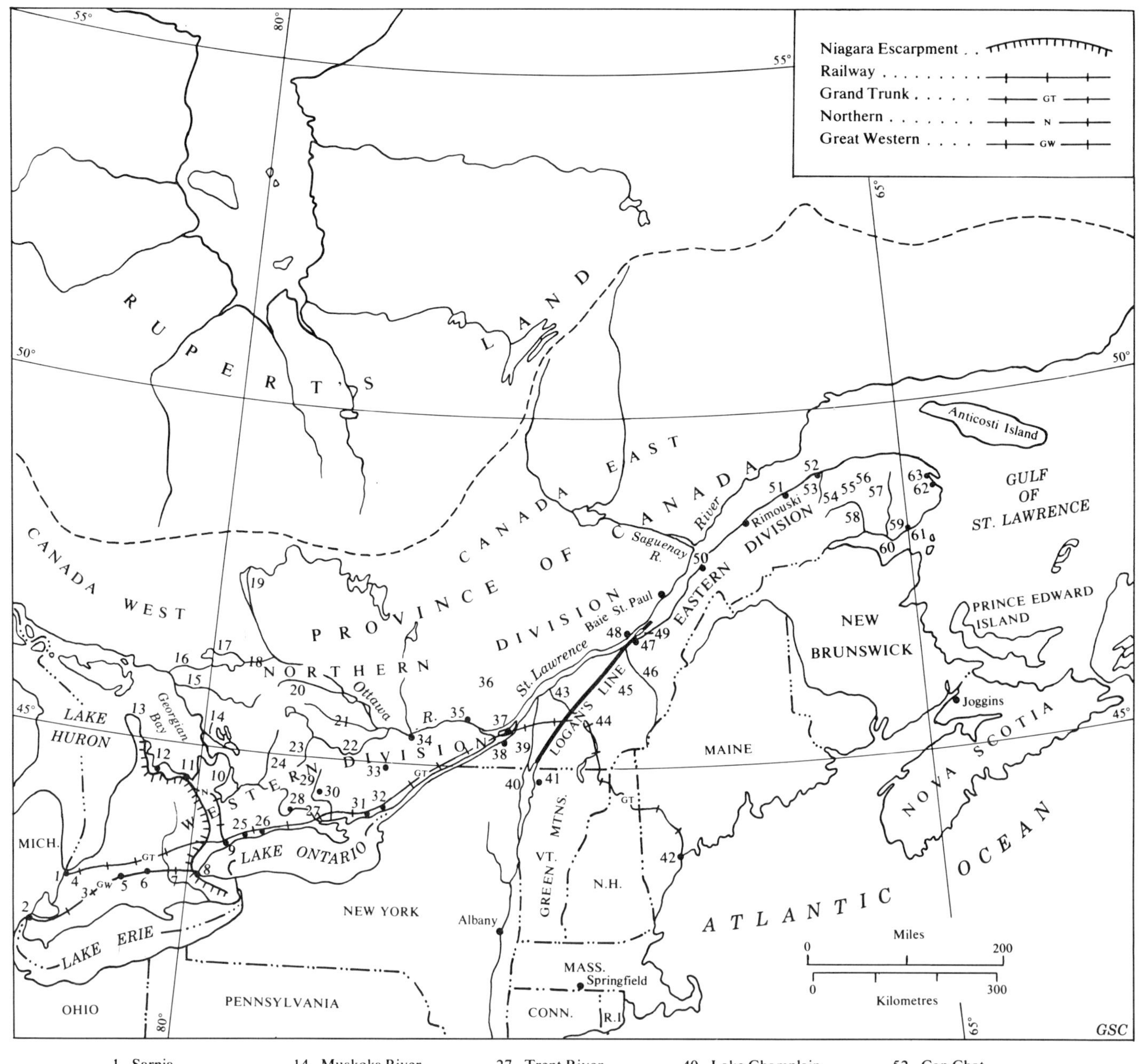

so commend the survey to his political superiors as the discovery of valuable mineral resources, especially coal. Consequently, the first concern was to determine whether the coal-bearing strata of the United States or New Brunswick extended north or west into the sedimentary basins. This was certainly possible, inasmuch as Paleozoic rocks of the Carboniferous succession, near the top of which coal occurred in the United States, were known to be present in Canada.

In his first season Murray made a reconnaissance across the Paleozoic strata of the Western Division, using as his reference the succession worked out in New York. In the United States, the Carboniferous formations occurred near the top of a great saucer-shaped basin of sedimentary rocks nearly 30,000 feet thick, near whose base was a persistent formation abounding in characteristic fossils, the Trenton limestone. This formation could easily be traced in Canada, and from it the uncomplicated, undulating succession of superimposed strata could be worked out. This Murray proceeded to do by examining a stretch of country cutting right through the heart of the Ontario peninsula, bounded by Whitby and Grand River on the south, and by Lake Simcoe and Georgian Bay on the north. He soon discovered, unfortunately, that the upper half of the New York and Pennsylvania succession was completely missing from that part of Canada, the uppermost Paleozoic formation in Canada being at least 4,000 or 5,000 feet below the Carboniferous. Hence, as Logan observed, "we are therefore not warranted reasonably to anticipate the occurrence of any part of those true (coal) measures in the district in question."[1]

Logan undertook a similar task for the Eastern Division, tackling the more complicated folded and faulted rocks without the benefit of an established succession in adjacent areas to guide him. He began by compiling a detailed section of the tilted Carboniferous rocks exposed in a 30-mile stretch of coast at Joggins, Nova Scotia. This section, included in Logan's first report, dated April 28, 1844, has never been bettered in a century and a quarter. And no wonder! Logan catalogued 76 coal beds and 90 underclays, totalling 14,570 feet, comprised of 1,460 separate strata ranging in thickness from 132 feet to one-eighth of an inch. Next he followed the exposures along the east coast of New Brunswick, paying special attention to the south shore of Chaleur Bay, before crossing to the north side, the Gaspé peninsula of Canada. But again, the rocks on the Canadian side of the bay proved to be older than those of the Atlantic colonies' coal basins. Logan observed of the uppermost bed, that "It is probably an inferior member of the carboniferous series, but it seems to be too low down to contain any of the profitable beds of coal."[2] This tentative conclusion was confirmed the following year when he returned to Gaspé, this time with Murray, to undertake a traverse across the peninsula (1844). Entering from Cap Chat on the St. Lawrence, they ascended Cap Chat River to its source in the picturesque Shickshock Mountains, climbed the peak that now bears Logan's name, then descended by way of Cascapedia River to its mouth in Chaleur Bay. Then, while Logan completed his examination of the coast east to Port Daniel, Murray surveyed Bonaventure River for 50 miles. The next year, 1845, Murray returned to Gaspé for another traverse, this time in an east-west direction from Matane to Douglastown.

On all three explorations of Gaspé a substantial part of the effort was spent surveying the topography of the unmapped interior—a foretaste of what was to become a major preoccupation in years to come, since no accurate maps existed for most of the territory over which Logan and his successors were to operate; and accurate base maps were essential if the stratigraphic successions were to be correctly recorded and interpreted. For his first exploration of Gaspé in 1843, Logan began with Captain Bayfield's charts of the coast and the prominent physical features inland. For the coast, while the canoeman-guide, an Indian named John Basque, paddled along the shore, Logan filled in the details of the topography by pacing out the distances, taking his bearings by compass, and noting the geology as he went along. On the river traverses, canoes were used for transportation where possible. Bearings were measured by compass or theodolite, and distances by a Rochon micrometer, a telescope with internally mounted cross-hairs by which intercepts could be measured on a rod held at a distance by an assistant, and the distance calculated. On smooth water, changes in elevation were estimated by measuring the rate of the water flow and computing the fall of a stream in inches per mile, waterfalls or rapids were measured by instrument levelling, and absolute elevations by barometers. These rather crude methods were supplemented from time to time by astronomical observations. By such means the surveyor could travel considerable distances daily, produce maps accurate enough for the reconnaissance type of geology being undertaken, and still have some time left for geological observations.

Only those familiar with this type of work will realize its arduous nature—the long hours of paddling and scrambling over rocks or through dense bush along the shore, portaging around falls and rapids, enduring swarms of pestering insects, and collecting specimens. "No quarryman ever worked harder. I began at 6 a.m., and did not leave off until seven in the evening,"[3] wrote Logan during his survey of the Gaspé coast. Finally, the geologist finished his day, that started at dawn, by devoting one hour, or two, or more, to completing his field notes after sundown, sometimes toiling in a smoky or leaky tent. Still, Logan's field notebooks, written in his small, neat hand and inked over, enlivened by skillful, graceful and ingratiating sketches, give little outward sign of the difficulties under which many of the pages were written.

In 1845, while Murray carried on in Gaspé, Logan embarked on an initial reconnaissance of the third division, an enormous area almost three times the size of the two that had so far engaged their attention. Little was known of these ancient rocks that were the dominant geological feature of such a large part of Canada and of Rupert's Land beyond. They were almost without counterpart in Europe, except perhaps among the iron-bearing gneisses of Scandinavia. Logan proposed to attack the work by compiling a section along Ottawa River, which had practical as well as scientific advantages to recommend it. The region had been mapped only to Deep River; Logan, assisted by J. McNaughtan, provincial land surveyor, surveyed the river all the way to the head of Lake Timiskaming and also west from Ottawa via Mattawa River to Lake Nipissing. He noted far more than the geology, describing waterways, travel routes and conditions, timber, settlement and agriculture, and collecting meteorological records.

As he ascended Ottawa River Logan found himself in the midst of great uninterrupted stretches of Precambrian rocks, with here and there, patches of fossiliferous formations. Following Lyell, he classed the predominant type of rock, fine-grained, highly crystalline gneisses, as Metamorphic Series, "possessing an aspect inducing a theoretic belief that they may be ancient sedimentary formations in an altered condition,"[4] he shrewdly added, anticipating the long debate that would ensue on the origins and nature of the Precambrian. His acute eye detected more: south of Mattawa River these were interbedded with layers of another "primitive" rock, a usually coarse-grained, whitish crystalline limestone, which "constitutes so marked a character that it appears to me expedient to consider the mass to which they belong as a separate group of metamorphic strata."[5]

Also of significance, though he would only become aware of this when he had worked further in the Precambrian north of the Great Lakes, was a succession he encountered about Lake Timiskaming. There the gneiss was overlain by greenish chloritic slates and conglomerates with embedded fragments of the gneiss, followed by generally fine grained, pale sea-green sandstones, then by limestone beds containing fossils he identified as Middle Silurian. For the present, Logan was content simply to record the observation but, from his knowledge of the general structure of the country, he anticipated encountering the slates somewhere to the west, along the north shore of Lake Huron. Already Logan was embarked along the course of analyzing and subdividing the Precambrian, a path that led directly to the Canadian Survey's most vexing scientific problems and to its greatest scientific achievements. Logan also observed and commented upon the numerous veins and dykes, some containing mineable quantities of metallic minerals, a type of occurrence which was to become the basis of the great hard-rock mining industry of twentieth-century Canada. He also was much perplexed by the evidences of glaciation. Such evidence as polished, scraped rock surfaces, parallel scratches and grooves, and moraines, pointed to the conclusion that the Ottawa valley "may have been the seat of an ancient glacier."[6] But Logan could not quite reconcile the scale of these glacial effects with the level character of the valley or the glacial markings (striae) pointing in more than one direction. Here again, Logan was opening up another important scientific area for future investigation by his survey.

In keeping with the need to justify his survey in practical terms, Logan in his reports carefully enumerated mineral occurrences that he found or that had been brought to his notice, and suggested which of these might be worth attempting to develop. To take an early example, in his third report, covering his exploration of the Ottawa valley, he closed with a 12-page section headed "Substances Capable of Economic Development" which listed and described occurrences of a wide variety of materials, ranging from metallic minerals to medicinal springs. Murray, who that summer worked in the interior of the Gaspé peninsula, was sufficiently impressed by the disturbed limestone formation about Mount Albert to speculate that it might be "the equivalent of the lead-bearing limestone of Wisconsin" and to recommend the district as "worthy of research."[7] Today the large copper mines near Murdochville, less than 30 miles from Mount Albert, confirm Murray's early prediction. Again Logan, not content with describing a limestone occurring at Rama and Madoc, sent samples to England to be tested for lithographic purposes, and quoted the favourable report he received as indicating the feasibility of establishing a thriving export market. Unfortunately, the main economic conclusion of the first two years' work was a negative one, sorely felt by Logan because of his roots in Welsh mining—that the coal-bearing formations did not extend into Canada, and further expenditures on searches for coalfields would be fruitless. In time it would be claimed as an important early achievement of the Geological Survey that it had checked the expenditure of thousands of pounds of public and private money in

III-3 SKETCHES FROM LOGAN'S SURVEY OF GASPÉ, 1844
From Field Notebooks Nos. 1965 and 1966. *Upper left,* Gros Maule; *upper right,* Great Fox Valley; *centre left,* labradorite boulder above Tourette; *centre right,* Cape Chatte; *lower left,* Metis Falls; *lower right,* sandstone pillar near Marsouin River.

Canada—unlike Britain or the United States where large sums were spent in fruitless searches for coal in places where its occurrence was geologically impossible.

Still, notwithstanding Logan's conclusion, reports of coal discoveries continued to excite the public and take up his time. When the reported coal was not an outright fraud "planted" by interested parties—like those he unmasked at Baie St. Paul and Bowmanville—it was found to be pieces of black shale that sometimes were rich enough in bitumen to yield a flame when ignited. Murray and Logan encountered many such bitumen-impregnated shale beds, occasional cavities of solidified bitumen, and slow-flowing liquid bitumen. Logan's description of one such occurrence in the Gaspé has been cited as being possibly the earliest reference to the occurrence of petroleum in an anticlinal structure. The beds occurred in formations overlying the Trenton limestone, well below the Carboniferous coal measures. But reported coal occurrences from Quebec City, Levis, and other places continued to annoy and disturb Logan until, almost 20 years later, the nature of the bituminous formations began to be properly understood.

The practical side of the Survey's operations accentuated the importance of making chemical analyses of the ores and minerals being collected in the field, so Logan determined that the next addition to his staff should be a chemist. Early in 1844 he had arranged to hire, at a salary of £300 per annum, Count E.S. de Rottermond, a young Pole who had recently arrived in Canada bearing excellent testimonials from L'École Polytechnique in Paris. His beginning was not very auspicious. Taken on the exploration of the Gaspé peninsula with Logan and Murray in 1844, he found "roughing it" very onerous and soon was sent back to Montreal. But Logan was very partial to his young protégé and even after a year still was making allowances for him:

> De Rottermond has not done so much, but he has been in love, and is to get married on the 15th inst. into a highly respectable family which has some French political influence. He is a young man, of gentlemanly manners, and, I think, of some energy, though he was completely knocked up in the forest last year, to which I carried him at his own earnest desire, just to show him there was no romance in the matter . . . I fancy he will do, though, perhaps, he will require some management.[8]

Logan, in fact, stood as his chemist's parent at de Rottermond's wedding in May 1845, to a daughter of P. D. Debartzch, a powerful seigneur and member of the Executive Council (the appointive upper house of the Canadian legislature). By his marriage de Rottermond also became a brother-in-law to L. T. Drummond, a member of the Baldwin-LaFontaine Reformer administration of the 1840's and fifties.

Unfortunately, de Rottermond's and Logan's ideas of work differed radically. De Rottermond remained behind in Montreal during the summer of 1845 to analyze some 44 rock samples. When Logan returned in the autumn from his exploration of the Ottawa valley, he found the samples still sitting on the window-ledge where he had left them in the spring. Incensed, Logan demanded some action on the analyses, but a month passed and still nothing had been done. De Rottermond could not, or would not, produce a daily report of his work during the previous months. He left off his work on December 20 to attend to family matters, and early in January 1846, announced he was going into a family business. Logan never did get the analyses or any account of de Rottermond's activities during the spring and summer of 1845.

But Logan was still not quit of his chemist, who, relying on his newly-acquired family connections, started a cam-

III-4 E. S. DE ROTTERMOND
The records of the Notman Photographic Archives show that this photograph of de Rottermond was made in 1867, probably from a daguerrotype.

III-5 T. STERRY HUNT

paign on his own behalf. He approached the Provincial Secretary, to whose department the Survey was attached, about his final salary, and, much to Logan's disgust, received payment for the first quarter of 1846. Logan had to content himself with a stated intent never to employ de Rottermond again in any capacity whatsoever. De Rottermond counterattacked by charging that Logan had not given him proper equipment or assistance, and had asked him to undertake unreasonable projects like analyzing lithographic stone. He accused Logan of "wishing to interfere with a branch of the sciences which he knows nothing about, and which was only given to him to assist him in researches which might be useful to the Country, and not to be subject to his whims."[9] Egotistically, de Rottermond asked to be appointed head of an independent scientific division devoted to chemical studies, with Logan heading a separate, co-equal geological division. On the face of it, this was an absurd proposal. The government, which barely tolerated one scientific organization, was not likely to authorize a second, and de Rottermond never secured his separate command. The otherwise unedifying episode did have its bright side—Logan was put on his guard against any future challenge to his control over the Survey, and insisted that it function as an integrated whole to attack the problems of Canadian geology on all fronts. He also determined to brook no further political interference, and, despite the patronage-hungry mood of that period, Logan thereafter appointed the personnel and managed the work of the Survey entirely to his own satisfaction.

A sequel to the episode was the publication in the *Journals* of the Legislative Assembly of Logan's correspondence with Provincial Secretary Daly and de Rottermond, an indication of the legislature's interest in the affair. From the correspondence de Rottermond emerged as a lazy, high-strung, self-centred, unreliable employee, and Logan as a patient, tolerant, too-trusting employer. The letters exonerated Logan, except, perhaps, for having hired de Rottermond in the first place, and being too kind or timid to dismiss him outright for dereliction of duty. De Rottermond continued to flit in and out of the public eye as a self-styled authority on all sorts of matters scientific, until his declamations and posturings exhausted everyone's patience. He died in Montreux, Switzerland, in 1858 at the age of 45, a completely discredited figure.

Fortunately for Logan and his work, the position of chemist was eventually filled by an extremely capable individual, Thomas Sterry Hunt, who was to add great lustre to the Survey's record of achievement. Logan had turned to the United States for de Rottermond's successor, and first hired Denison Olmsted, junior, the son of the Yale professor who had surveyed North Carolina in the 1820's. Olmsted took sick and was forced to retire soon after his appointment, whereupon Logan again canvassed the geological surveys of the various states. It was on this second round that he secured the services of Hunt, a twenty-year-old chemist with the Vermont Geological Survey whom Professor Silliman highly recommended. Hunt was appointed in December 1846 and took up his duties in Montreal early in 1847.

* * *

By the time of the de Rottermond trouble, Logan had successfully weathered his first legislative hurdle, the two-year term for which the survey had been instituted. Despite his desire "to bring the investigation to a conclusion in as short a time as a due regard to geological truth and the applications of the science will permit,"[10] Logan knew from the outset that a proper geological survey was an infinite undertaking, and that it was essential that the survey be made a permanent, continuous institution like that of Britain. As he later explained, "In a geological survey of a new country, at first you obtain only a general sketch, as it were, of the subject, which you must fill up afterwards by degrees, and the more you enter into detail the greater will be your results. What you first point out will furnish the means of farther discoveries."[11]

At the close of the first year Logan interviewed the Governor General and government leaders to urge that the survey be put on a more satisfactory and definite footing. The head of the government, William H. Draper, was completely in sympathy with the request, but because of the unsettled political situation, he told Logan, the main task would be to persuade the Assembly. "Unless it can in some way be indicated that value will be returned to the country for the expenditure, it is in vain to expect that the Legislature will support the Survey for the sake of science—in which opinion I thoroughly agree with him," wrote Logan.[12] One immediate result of this advice was to persuade Logan to institute a museum to place some tangible results of his survey before the public and the legislators. Already a large collection had been accumulated—70 large boxes of specimens, which brother James stored in his warehouse on St. Gabriel Street, were brought back by Murray and Logan from their first season's work alone. Following his interview with Draper, Logan rented a house at 40 St. James Street and ordered display cases. He took Murray into his plans:

> I must get a house or a set of rooms for our collection. Managing this, we must put our economic specimens conspicuously forward; and it appears to me that in the exhibition of these, large masses will make a greater impression on the mind than small specimens . . . . This induces me to say that I should like you to send to Montreal, as soon as can be done by water communication in the spring, a thundering piece of gypsum. Let it be as white as possible.[13]

Another action was his placing before the legislature a report of his activities, much as he disliked the American practice of annual reports and favoured delaying them until the completion of the field work and the maturing of thought. On April 21, 1844, he addressed his first *Report of Progress for the Year 1843* to Governor General Metcalfe for submission to the legislature.

When the legislature met early in 1845 in the new provincial capital of Montreal, Logan was able to point to considerable achievements in return for the £1,500 grant (in fact, by then he had overspent it by a considerable amount from his own pocket, anticipating its renewal). He and Murray had devoted two long field seasons apiece exploring the central part of Upper Canada and the Gaspé peninsula and studying their results, a chemist had been hired, an office was opened, a museum was in operation, and a first report was before the legislature. Furthermore, Logan had carefully tended his political fences by writing and talking to legislators of all parties, and performing small favours on their behalf. The Assembly quickly passed the bill Logan had helped draft to provide for the future needs of the Survey. Assented to on March 17, 1845 (8 Vict., Chap. 16), it still did not create a perpetual Survey, but spoke instead of the expediency "that the said Survey should be continued to a completion," and authorized the government to "employ a suitable number of competent persons . . . to make an accurate and complete Geological Survey of this Province." It also ordered the Survey to make a collection of its maps, diagrams, drawings, and specimens to serve as "a Provincial collection," and to furnish a report of progress "on or before the first day of May in each year." A sum of £2,000 per annum now was provided "to defray the expenses of the said Survey, or any arrears of expenditure already incurred" for a five-year period during which an annual financial statement would be submitted to the legislature.

Thus Logan's original brief assignment had become an institution, his survey had become the Geological Survey of Canada, with a statutory foundation and at least five more years of guaranteed existence. The additional funds enabled Logan to recover the £800 he had advanced and left something over to expand the operation. Evidences of the new-found prosperity were the increase in Murray's salary to £300, the formal appointment of de Rottermond who speedily revealed his true colours that summer, and the moving of the office and museum to a more suitable location in the building of the Natural History Society of Montreal, 10 Little St. James Street, which Dr. A. F. Holmes offered at the moderate rental of £120 per annum, the society to retain the third floor for its own displays. Here the two allied organizations co-existed, to their mutual benefit, until 1852, when the government (which by then had abandoned Montreal in favour of alternating terms in Toronto and Quebec City) moved the Survey into a building that it owned in Montreal. This was the former residence of Hon. Peter McGill, at 5 St. Gabriel Street (later renumbered 58, then 76), which had housed the Crown Lands office.

III-6 THE ST. GABRIEL STREET NEIGHBORHOOD, MONTREAL
At 76 St. Gabriel Street the Survey occupied one of the choicest locations in Old Montreal facing the Champ de Mars and directly across the street from the historic St. Gabriel Street Church, the first Protestant church in Montreal. In 1874 the building was extended southwest and connected with properties owned by Logan on St. James (formerly Little St. James) Street, where the growing number of staff were given their offices. The area bounded by Notre Dame, St. Gabriel, Craig, and St. Lambert (now St. Laurent) Streets is now a single provincial government complex; St. Gabriel terminates at Notre Dame, apart from a pedestrian way to Craig Street.

That Logan was personally elated by the action of the legislature is shown from his reaction to a flattering offer he received through De la Beche shortly after the passage of the act. Though he might have had a salary of £1,200 a year for directing a survey of the coalfields of India on behalf of the East India Company, Logan declined the offer, largely because the legislature had been so generous in meeting his every request, even to increasing the grant:

> Perhaps the Canadians are leaning on me for the Survey, and might think it not very handsome if I were to leave the country before the expiration of the five years. I am persuaded, though I say it, who should not say it, they will not find any one to take the trouble I do. It has been hinted to me that in continuing the Survey the Government have been in some degree influenced by the circumstance of finding a person who is a Canadian by birth considered competent to do the work.

In thanking De la Beche, Logan added that the offer would do good—he could use it, if necessary, to impress "my Canadian friends that geological investigations are something thought of in other parts of the world."[14]

The five-year term was a serious inconvenience and annoyance, as Logan was to find. Early in 1850 he had to travel to Toronto to lobby for a renewal of the statute, "lest being out of sight I should get out of mind."[15] Indeed, the unanimity of 1845 had disappeared, and there was every prospect that the Survey would be caught up in the political feuding. The government was in a state of crisis, and Logan had to spend precious spring months waiting for the legislation, while Murray began casting about for another position—just in case. A few legislators, led by William Hamilton Merritt, were critical of the modest expenses incurred, but they were more than offset by staunch supporters, like Sir Allan MacNab. The attitude of the members might be characterized as a dash of criticism, a sprinkling of enthusiasm, and a cupful of indifference. The brief debate, however, saw many of the indifferent members brought around by the argument that it was not a good thing to leave the task half done. The new act (13-14 Vict., Chap. 12), merely renewed and continued its predecessor. Assented to on July 24, 1850, it provided that the act of 1845 "is hereby revived, and shall continue during five years from the passage of this Act and then until the end of the then next Session of the Provincial Parliament," and extended all its provisions, including the £2,000 annual grant, for the new period.

Some of the opposition to the Survey may have been inspired by Logan's unavoidable involvement, in the line of duty, on the side of the entrepreneurial groups developing the mineral resources of the province—the new aristocracy of wealth that was coming under attack from the radical political fringe, the Clear Grit and Rouge parties. Indeed, when Logan went to Toronto he half-anticipated such a challenge, needlessly, as it proved: "There may be some of this among the *clear Grits* as they are called."[16] An example of the Survey's involuntary identification with business was its role in the Lake Superior mining boom that began in 1845 following the rise of a large, profitable copper mining industry on the rich Keweenaw peninsula of upper Michigan. This had been launched by the discoveries of Douglass Houghton, the state geologist of Michigan, during his survey of the region—itself a powerful testimony to the practical value of the governmental geological explorations.

The government of Canada received thirty applications for mining locations along the north shore of Lake Superior in 1845, but specific regulations to deal with mining land grants were lacking. Logan, when asked for recommendations, suggested that licences should be granted only to bona fide operators and for 21-year terms, also that all claims should be surveyed before being awarded. This the government proceeded to do. Under the regulations promulgated on May 9, 1846, a claimant was to be allowed only one licence, for a single location, five miles by two miles, the long dimension to follow the direction of the vein as determined by the provincial geologist, who would also settle disputes on the spot, after which the boundaries of the location would be laid out by a provincial land surveyor. The government, however, did not follow Logan's advice about leasing the lands; instead it provided for their outright sale at a price amounting to £1,280 per ten-square-mile lot. This effectively prevented lots being taken up by individual businessmen, with the result that wealthy, politically-powerful groups, the Montreal Mining Company in particular, purchased the most promising locations, then left them undeveloped while they waited for buyers. For decades this policy—for which Logan was in no way to blame—was denounced as retarding the development of the mineral resources north of Lakes Superior and Huron.

The government party headed by Logan that left for Lake Superior in May 1846 had a definite duty of taking charge of the mining boom on the spot. Accompanying Logan were Murray, John McNaughtan, the provincial land surveyor who had been Logan's colleague the previous summer in the survey of the Ottawa valley, and an assistant, James Richardson, a farmer and schoolmaster from Beauharnois, near Montreal, who was to make a long, useful career in the Survey. While McNaughtan surveyed the claims, mostly around Fort William, Michipicoten, and Mamainse, Logan examined the rock formations about these regions, and Murray made traverses inland up Kaministikwia and Michipicoten Rivers.

III-7 THE EARLY YEARS: WORK IN THE CANADIAN SHIELD, 1845-55

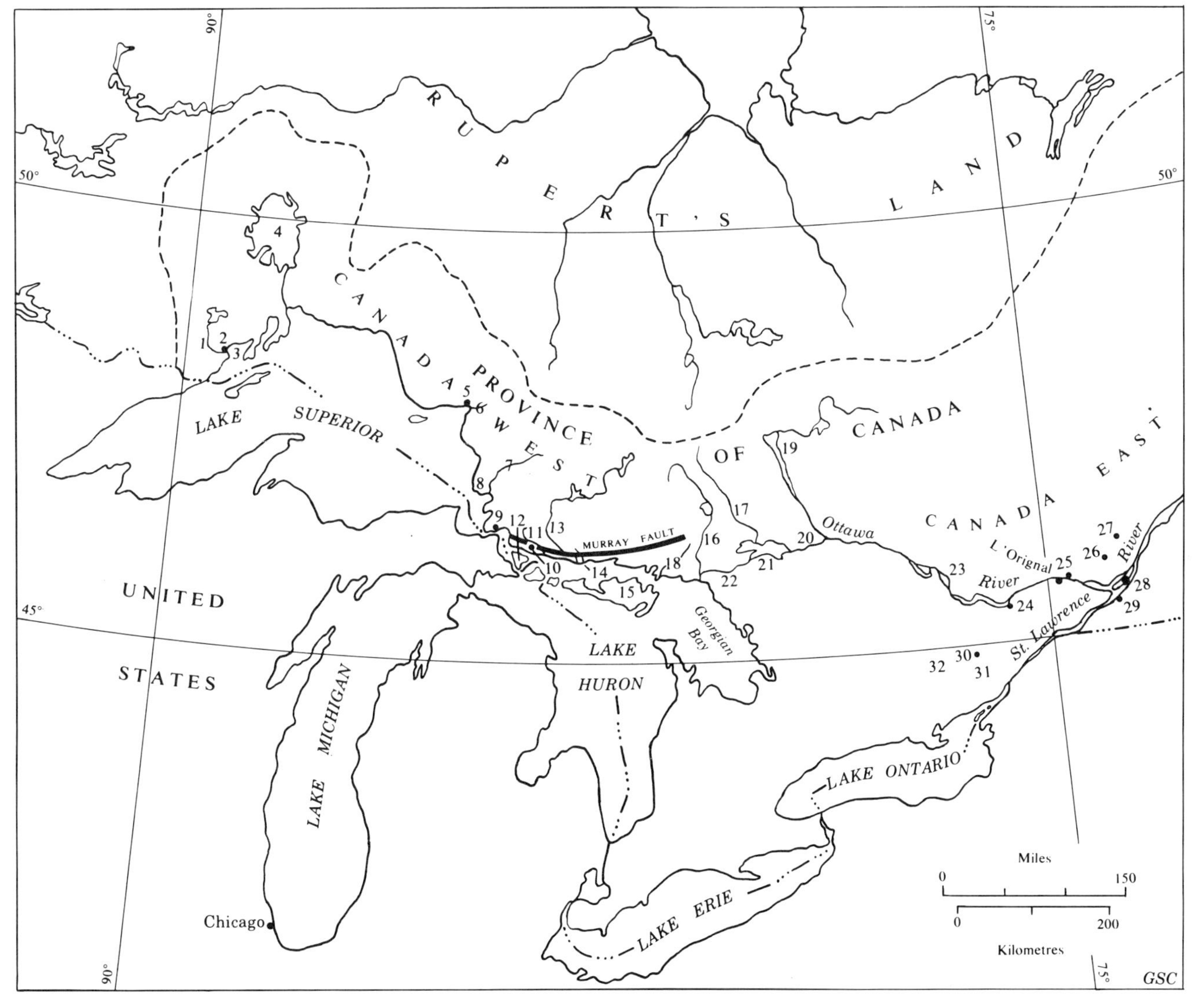

1. Kaministikwia River
2. Ft. William
3. Thunder Bay
4. Lake Nipigon
5. Michipicoten
6. Michipicoten River
7. Batchawana River
8. Mamainse
9. Sault Ste. Marie
10. Bruce Mines
11. Thessalon River
12. St. Joseph Island
13. Mississagi River
14. Blind River
15. Manitoulin Island
16. Wanapitei River
17. Sturgeon River
18. Whitefish River
19. Lake Timiskaming
20. Mattawa River
21. Lake Nipissing
22. French River
23. Grand Calumet Island
24. Ottawa
25. Grenville
26. St. Jérôme
27. Rawdon
28. Montreal
29. Beauharnois
30. Perth
31. Burgess Tp.
32. Tudor Tp.

Logan found the region a most complex one, in which because of intense volcanic activity, only some of the uppermost formations were identifiable, tentatively, as Lower Silurian sandstones and limestones. At the base of his section was gneiss and granite, overlain by beds of slate and conglomerate resembling the rocks he had observed on Lake Timiskaming, intruded by many dykes and sills, in which occurred veins bearing copper, lead, zinc, and silver values. In his report dated May 1, 1847, the section on "Economic Application of Materials" drew largely upon Logan's knowledge of the economics of the industry from his years with the Forest Copper Works in Wales. The report pointed to the uncertain quantities and grades of the metallic minerals in the narrow, irregular veins, and added that the lack of coal to refine the ores locally, and the high costs of transportation, machinery, and labour (particularly before Lake Superior was made properly accessible by the first Sault Ste. Marie canal in 1855), would render all but the richest deposits uneconomic. Logan's report did much to damp down the excitement, and he was blamed for the abrupt demise of the rush.

The Survey's concern with the Precambrian formations of the northwestern corner of Canada continued two years longer, because of the shift of the mining excitement to the north shore of Lake Huron. Twenty-two locations were taken up there in 1847, by the Montreal Mining Company and others, and copper began to be mined at Bruce Mines. In 1847 and 1848 Murray examined the coast and major islands between Sault Ste. Marie and French River, and traversed the principal southward-flowing rivers as far east as Lake Nipissing, topographically mapping a large part of the distance. The government in 1848 requested Logan to visit Bruce Mines to look at the claims for locations and at the developmental work under way. He proceeded there in July, with some trepidation lest his presence amid promoters and brokers should be used to further the activities of unscrupulous speculators. He examined the coast between Bruce Mines and Sault Ste. Marie, paying particular attention to the lodes and workings on the Montreal Mining Company's Cuthbertson Location, gave the operators the benefit of his stratigraphic knowledge and mining experience, took samples of the ore extracted and in place, and before the year closed presented the government a clear, factual account of the situation and prospects of the new mine field. His private opinion he confined to letters from the field to his brother James. The mine, he felt, was a risky venture because of extravagant management rather than the quality of the ore, and he advised his brother to reduce his holdings of Montreal Mining Company stock. Subsequently, under different management, the mine became a significant producer and a pioneer in opening up mining interest and activity in the region.

III-8 JAMES RICHARDSON IN 1865

Following the Lake Superior and Lake Huron mining booms, the Survey returned to its systematic study of the geology of the province. Murray resumed his more detailed examination of the Ontario peninsula, which he had not wholly abandoned in 1847 and 1848 in favour of the Precambrian work. He traced the entire coast of Lake Huron in Canada, including the whole of Georgian Bay in 1849, and in 1850 and 1855 completed his examination of Logan's Western Division, tracing the geographical distribution of the principal Paleozoic formations within the region, and assessing the potential economic minerals—mostly structural materials, gypsum and marl, some bog iron and peat, and petroleum. The petroleum was reported from oil springs in Mosa and Enniskillen townships, which he examined for himself.

Murray also traced the southern boundary of the Canadian Shield in 1851 and 1852 by examining the triangle of country lying between Gananoque, Bytown, and Ottawa and St. Lawrence Rivers, and the geology between Kingston and Lake Simcoe along the rivers and lakes of the Moira and Trent waterways. In the next three seasons he worked north in this part of the Canadian Shield, examining the country between Georgian Bay and Ottawa River, and between the Trent watershed and Lake Nipissing. This work included traverses of Muskoka, Petawawa, Bonnechere, Madawaska, York,

Gull, Magnetawan, French, and other rivers of the beautiful resort country of lakes and rivers of our day, which was then in its natural, unsullied, state. Most of this country was unmapped, or inadequately mapped, and Murray's were the first accurate maps and descriptions of large parts of the region; he also measured the heights of the principal rapids and falls, and the altitudes of the major lakes. He encountered mainly Precambrian gneisses and crystalline limestones, with fewer and fewer patches of Paleozoic rocks the farther north he proceeded.

These were, perhaps, Murray's happiest years with the Survey, for in 1856 he became partially incapacitated by a stroke, and family problems also began to weigh down upon him. By now his knowledge of geology was sufficient to permit him to make astute observations and correlations, and he had mastered the techniques of bush travelling and topographical surveying. He must have cut a fascinating figure as he wandered about the country, expostulating at every obstacle in the best sailor's vernacular. An ardent sportsman, he loved dogs, hunting and fishing, and his crack marksmanship kept his crews in fresh meat. His addiction to a cold bath every morning, even in winter, made him something of an oddity, for "Cleanliness was a sort of hobby with him."[17] Murray's ability to lead his own field party virtually doubled the amount of field work the Survey could do each season, but his continued residence in Woodstock, more than 400 miles from Montreal, reduced his effectiveness. Since he usually went directly to the field from Woodstock and returned thither at the end of the season, the work at times was affected by delays in his getting to the field; by equipment left behind at Woodstock, not forwarded from Montreal, or temporarily lost in transit between the two places; by misunderstandings over plans and instructions; by difficulties about expenses and letters of credit, and the like. The arrangement was that Murray should come to the office in the winter to complete his reports, but often these, or the maps, had to be finished at home, occasioning other inconveniences. Because of his domestic arrangements, Murray was in no position to assist Logan with his increasing administrative burdens or assume management of the Survey in Logan's unavoidable absences; instead, these responsibilities devolved upon Hunt.

Logan, as far as time allowed, continued field work, mainly close to Montreal—the Eastern Townships, St. Lawrence valley, and the lower Ottawa River. The intensely contorted and complicated metamorphosed formations of the Eastern Townships were his main preoccupation during the five seasons from 1847 to 1851. His objective was to trace the extension into Canada of the New England mountain formations and to study the effects of the Appalachian mountain-building epoch upon the Paleozoic sedimentary rocks; also, to find out as much as he could about the potential economic deposits of this highly-mineralized region. In 1847 he studied between Montreal and the southeastern corner of the province, including the formations underlying the valleys and ranges paralleling its eastern boundary. In subsequent seasons, accompanied by Hunt, Richardson, or Murray, he gradually worked out the geology from the southern border northeast to the base of the Gaspé peninsula. Among the many interesting formations he studied were two roughly parallel belts of serpentine rocks extending from the border northeast to culminate in a large body about Black Lake. He noted the presence of asbestos, which was then of no economic value; rather, it was regarded as a defect that kept the serpentine from being used as a decorative stone. What interested Logan at the time was its association with chromic iron (chromite), a rather rare mineral used in dyes for textile printing.

III-9 ALEXANDER MURRAY

He traced and attempted to explain the occurrences of alluvial gold along Chaudière River and its tributaries, noting that gold was also present in upper terraces of former streams. However, he cautioned against over-optimism, or growth of gold rush mentality, warning that the gold was widely distributed and not likely to yield easy wealth. He also examined more closely the occurrences of copper he had first observed in 1840, and described their geological settings, outlining and calling attention to what was soon to become an important mining belt and Canada's first major mining development. These occurrences were to draw Logan and his staff back to the region again and again. Indeed, the descriptions and locations he published of three separate copper occurrences were responsible for attracting private groups to prospect in the area in the first place. While none of the described locations proved mineable, prospectors continuing farther along the trends Logan had indicated soon found the important deposits that became the basis for mines, including the Acton, believed to have been the largest copper mine in the world at one time, and the Eustis, which was operated for 74 years.

In and after 1851 Logan centred his attention mainly on the region west of Montreal, where, after tracing the Paleozoic formations along the north sides of the St. Lawrence and Ottawa Rivers between Quebec City and Grenville on the Ottawa, he came to grips in earnest with the great problem of the Precambrian. In the Grenville district he encountered a group of formations that gave him an indication of the complexity of the Precambrian era—the now-familiar gneisses and crystalline limestones with their very distinctive characteristics, occurring separately or interbedded; a confusing sequence of early intrusive rocks; greenstone dykes cut by an intrusive syenite which in turn was cut by feldspar porphyry—all of them of pre-Paleozoic age. Then in 1853 he found in Morin and adjacent townships the remarkable anorthosite bodies, which he noted were composed of labradorite and andesine.

For this wide assortment of altered and intrusive rocks, the then-current general term "Metamorphic Series," no longer seemed appropriate. So in his *Report of Progress* for 1852-53, Logan introduced a more specific name, Laurentian Series, for these early Canadian rock formations:

> The name which has been given in previous Reports to the rocks underlying the fossiliferous formations in this part of Canada is the Metamorphic series, but inasmuch as this is applicable to any series of rocks in an altered condition, and might occasion confusion, it has been considered expedient to apply to them for the future, the more distinctive appellation of the Laurentian series, a name founded on that given by Mr. Garneau to the chain of hills (Laurentides) which they compose.[18]

The problem of differentiating among several important, widely-occurring groups of formations of pre-Paleozoic age still remained. Separable immediately from the Laurentian gneisses, however, were the chloritic slates and conglomerates of the Lake Timiskaming, Lake Superior, and Lake Huron districts, proved beyond question to be distinct and younger formations by their superposition over the gneisses as well as by the gneissic debris they contained. They were also known to be widespread and of great economic importance for their contained valuable metallic minerals. Accordingly, in the handbook prepared by Logan and Hunt for the Paris Universal Exposition of 1855, the *Esquisse Géologique du Canada*, this younger, ore-bearing formation was named the "Huronian formation." In the accompanying map they were assigned, though no fossils had ever been found in them, to the Cambrian System.

For Hunt his first years with the Survey were an apprenticeship, a prelude to the international recognition that came to him in 1855. He arrived at the Survey in 1847, little more than a youth, at an impressionable age, and he found in Logan, almost 30 years his senior, an ideal chief, colleague, patron, mentor, and friend. Hunt had youthful enthusiasm, self-confidence, imagination, brilliance, and capacity for hard work to contribute greatly to the progress of the Survey. He accompanied Logan or Murray to the field, to Bruce Mines, Baie St. Paul, the Ottawa valley, and the Eastern Townships. He made numerous trips on his own to examine mineral deposits on the spot, to collect specimens for the great exhibitions, and for his own mineralogical and chemical studies. He collected spring waters in one hundred pound glass bottles, samples of soil and peat from all corners of the province (for a study that Logan initiated in 1849, perhaps with a view to wooing the agrarian interest in the legislature), anorthosite, serpentine and other intrusive rocks, quartz crystals, bitumen and mineral pitch, and ores of iron, lead, copper, and gold.

Besides assessing the economic importance of specific Canadian mineral resources, Hunt's work also advanced geological science. He discovered two new minerals, which he named loganite and wilsonite. He proposed an organic origin for the nodules of phosphate of lime found in Lower Silurian rocks. His investigation of the serpentine and associated rocks of the Eastern Townships blossomed into a broader study of the effects of local metamorphism upon certain minerals. From his analyses of spring waters he began to evolve a theory of the solution and deposition of minerals. His studies of the Grenville feldspar rocks contributed to the fuller understanding of the Morin and other anorthosites. Thus Hunt added an important dimension to the work of the Survey that enabled Logan to study and interpret the complicated formations of the Eastern Townships and the Grenville district with increasing confidence.

* * *

FROM 1850 ONWARD, LOGAN'S TIME BEGAN to be occupied increasingly with administrative and public duties that took him away for long intervals from his favourite field work of tramping about unexplored areas, studying rock formations and mineral occurrences. Much as Logan disliked it, these involvements marked a new stage in the progress of the Survey—the recognition of the great achievements of its early years. They signified, simultaneously, the public's growing awareness of the Survey's important role in discovering and making known the mineral resources of Canada; Logan's important place in the practice and theory of geological science; and Canada's place in the world as a present and future source of natural resources and a rapidly developing centre of scientific and intellectual achievement. When honours were showered upon Logan, he never ceased to protest that they were not so much a tribute to his work as to Canada and Canadian achievements. In truth, all three were linked. In honouring Logan, the Survey was being honoured, and Canada was being placed in a favourable light before the world. The tributes to Logan brought greater support, strength, and influence to the Survey, and lifted it beyond the reach of those who might wish to stand in its way.

The first opportunity to take advantage of this recognition followed shortly after the act of 1850 renewed the Survey for another five-year term. Canada had been invited to participate in the Exhibition of the Industry of all Nations, under the patronage of Prince Albert, the Prince Consort, and held in 1851 in the Crystal Palace in London, which was England's monument to vigorous scientific and industrial technology. In such an industrial exhibition, a well-organized collection of mineral samples was bound to attract considerable attention and comment, and bring favourable publicity and trade opportunities for Canada. The United States showed little interest in the exhibition, and the individual states were too preoccupied with their own problems to send good mineral displays. Consequently it was to Logan and Canada that the world turned to learn about the main geological features of North America.

In the preparations for the London exhibition, which took much of the summer of 1850, Logan received enthusiastic public support. The business community of Montreal was anxious to affirm its loyalty to Queen and Empire and make amends for the disgraceful behaviour of some of its members during 1849—which included pelting the Governor General, burning the Parliament Buildings, and signing a manifesto urging annexation to the United States. They willingly contributed rock and mineral samples to Logan and his staff, who also scoured the country for suitable specimens, and the whole collection was displayed in Montreal before being despatched to London. Logan was kept busy organizing the large display, as well as with correspondence and arrangements; for besides managing his own exhibit, he was a member of the jury that evaluated the exhibits in the mineralogical and metallurgical division.

Pride in Canada's displays, of which the mineral section was the most successful, inspired outbursts of patriotic fervour and optimistic speculations about Canada's dazzling future. The press gave full and favourable publicity to the Canadian exhibits in London, crediting Logan for the success of the mineral display. The exhibit was viewed by the most influential personages of England and Europe, headed by Queen Victoria; and following the close of the exhibition Logan received a formal letter of thanks and a medal for his valuable services as juror from the president of the exhibition commission, Prince Albert.

More important were the laudatory tributes of the experts in London, who admired Logan's systematic arrangement of the exhibit and the high standard of the specimens selected. The specimens were arranged according to their uses, into categories of metals and ores, metals requiring peculiar chemical treatment, mineral or stone paints, minerals applicable to fine arts, materials applicable to jewelry, minerals applicable to glassmaking, building materials, and miscellaneous materials. This eventually became the standard format for mineral exhibits. One Canadian correspondent reported to his editor that

> On Minerals, one of the principal Geologists was kind enough to say to me one day, that Mr. Logan's collection is the most complete and interesting from any country in the Exhibition, and most of the leading journals extol it highly. The jurors regret, that as Mr. Logan was himself a juror, it is against regulation to award him a medal; but they make highly complimentary mention of his collection.[19]

From January to August 1851, Logan was outside Canada on the business of the exhibition. Besides the welcome opportunity to visit his brother and sister, his stay in England meant a chance to renew professional and scientific connections that had been somewhat eroded by distance and time, as well as by his increasingly close ties with American geologists. Through Sir Roderick I. Murchison, on March 6 Logan was elected a fellow of the Royal Society. On April 30 he attended the meeting of the Geological Society of London to give an address "On the Occurrence of a Track and Foot-Prints of an Animal in the Potsdam Sandstone of Lower Canada." These fossil tracks, discovered near Beauharnois earlier that summer, were apparently made by a large crustacean-like creature with limbs or claws and a body-case or tail that

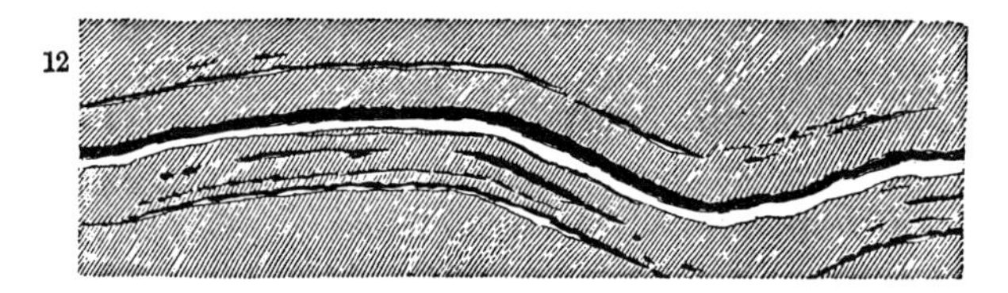

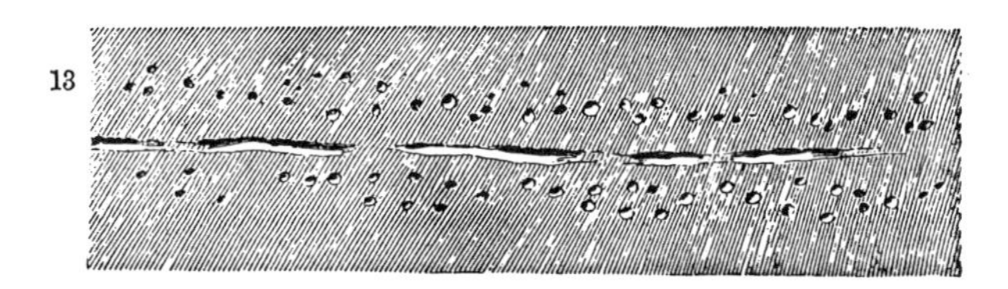

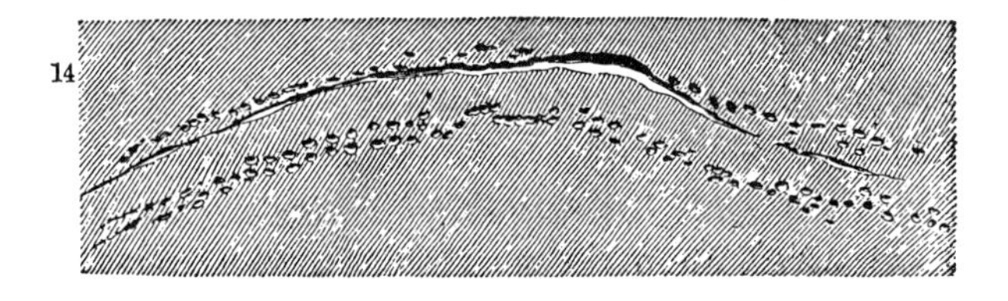

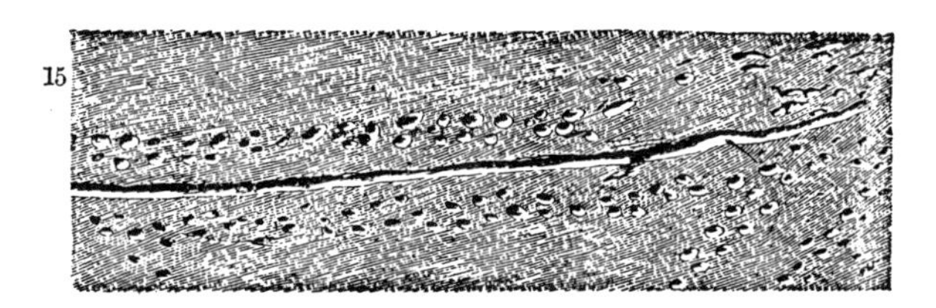

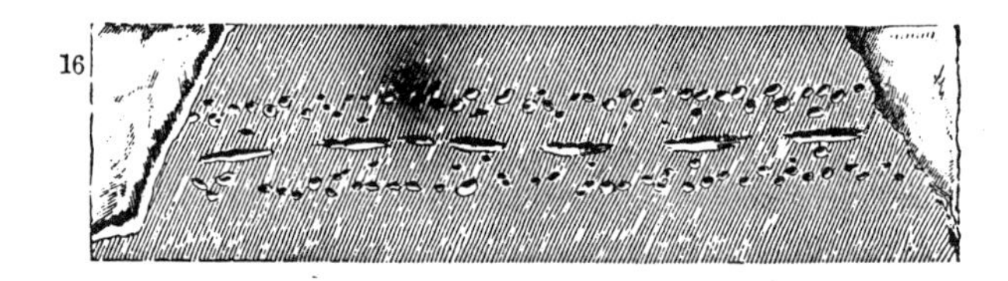

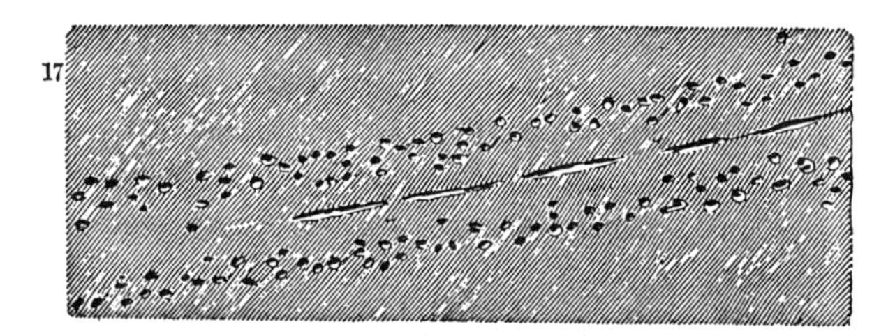

III-10 TRACKS FROM BEAUHARNOIS
Scale 1/10.
12—*Protichnites lineatus* (Owen)
13—*P. alternans* (Owen)
14—*P. multinotatus* (Owen)
15—*Protichnites septemnotatus*
16—*P. octonotatus* (Owen)
17—*P. octonotatus* (Owen)
As published in *The Geology of Canada,* 1863.

dragged in the sand of the beach or shallow sea bottom. This attracted much attention in England, because it seemed to provide conclusive evidence that primitive life forms appeared much earlier and evolved much more rapidly than had hitherto been suspected. Logan was followed by Richard Owen, professor of anatomy of the College of Surgeons, who concluded that the tracks were apparently made by a tortoise-like, air-breathing quadruped. In July, Logan presented another paper before the British Association for the Advancement of Science meeting at Ipswich, in which he outlined the principal results of his surveys to date—the boundaries and general characteristics of the three great geological provinces of Canada, and the attempts at correlating the lower, unfossiliferous formations of the Lake Superior, Lake Huron, and Lake Timiskaming districts.

When Logan returned to England in the early months of 1852 to wind up the affairs of the Canadian exhibit, he brought along many more casts of tracks, plus some of the original rocks with footprints, the result of a further search for specimens by Richardson. These were displayed in the museum of the Geological Society, to which Logan presented a fuller, well-illustrated paper on the Potsdam sandstone deposit and the footprints. Professor Owen, on the basis of this added evidence and further reflection, decided that the prints must have been made by a crustacean, a crab-like creature.

Logan was able to devote most of the next two years to the Survey, particularly to his own work in the Grenville district. By 1854, however, he had to turn to two more public engagements—the preparations for a second great exhibition, the Universal Exposition to be held in Paris in the following year; and the political ordeal represented by the appointment on September 24, 1854, of a select committee of the Legislative Assembly on the Geological Survey.

Logan's written presentation to the select committee listed both the Survey's triumphs and its problems. Among the former were the reconnaissance mapping of the sedimentary rock systems of the southwest, the more detailed study of areas of special interest like Lake Superior's north shore and the Eastern Townships, the topographical mapping of several districts, the discovery of new fossils, the extension of the knowledge of Canada's mineral resources—all this at a cost of below £20,000 in dozen years. The problems included the country and climate in which the Survey had to operate, the absence of topographical maps which doubled the amount of time the geologists had to spend in the field, the insufficiency of scientific staff in the Montreal office, and the inability to publicize the results of its work adequately. As for this last, too few copies of the reports were available to the public, while the lack of money to illustrate the texts with maps and plates seriously restricted their effectiveness.

A number of experts were called to assist the committee evaluate the work of the Survey. Dr. James Hall paid high tribute to its work and expressed amazement at the low cost to the public. The New York survey, dealing with a much smaller area, had submitted an annual bill of $20,000, exclusive of publications. Hall also praised Logan's policy of a unified organization as avoiding the errors into which New York had fallen. He followed his praise with several practical suggestions—the annual appropriation should be increased from £2,000 to £5,000; and to meet the existing demand past reports should be republished, as well as a large-scale geological map of the province, and lists setting out the economic minerals characteristic of each geological formation. He also indicated that a large sum could be usefully spent enlarging and equipping the Survey's museum.

The select committee received suggestions from a number of other authorities, including Professor E. J. Chapman of the University of Toronto, Alexander J. Russell, draftsman with the Department of Crown Lands, the Reverend Andrew Bell of L'Orignal, an amateur geologist, and the inevitable Count de Rottermond. Chapman, who echoed Hall's views on publication, suggested the Survey should be allowed £2,500 to prepare and publish a general textbook on Canadian geology, a task Chapman later carried out as a private venture. Russell praised Logan's work as a surveyor and outlined the contributions of the Survey to the development of the province. It had helped open unused land to development; the information it supplied on agricultural, timber, mineral, and water-power resources helped the government plan new surveys, and assisted farmers, miners, and lumbermen to order their operations; and it did most valuable work publicizing Canada abroad by bringing her resources and opportunities to the attention of influential European manufacturers and capitalists.

Logan proved a most persuasive witness during his appearance before the committee. First he outlined in detail the regions studied by the Survey, then described the financial situation, noting how he had kept the Survey alive through periods of difficulty by loans and gifts of books and instruments from his own means. To operate the Survey efficiently, he felt he needed two topographical surveyors, "young men, who can stand hard work, and would consider reputation a part of their reward,"[20] a museum assistant, two or three additional explorers, an occasional accountant, and above all, an intelligent office worker to assume many of the routine duties in the office, for so many people were now coming for advice that dealing with them took up half the director's time. In response to questions, Logan gave his own evaluation of the Survey's accomplishments. Those he mentioned on the practical side included locating good building stone and many other minerals, and proving that coal did not exist anywhere in the province; among the scientific accomplishments, the recognition of the Laurentian Series of rocks, mapping the boundary between them and the Silurian strata, the descriptions of various types of rocks, formations and structures, and the new minerals identified by Hunt.

Following these and other testimonies, the select committee, headed by the able John Langton (who became auditor general the next year), tendered its report. Its recommendations were quite unprecedented for a governmental investigative body, since they did nothing but praise the Survey's work to date—"In no part of the world has there been a more valuable contribution to geological science for such a small outlay."[21] It suggested that more money should be made available to the Survey to enable it to expand its work in every way. The substance of past annual reports should be republished, fuller reports should henceforth be issued and printed in larger editions to allow for wider distribution to universities, learned societies, legislators, and others. More money should also be granted to establish a large, well-staffed museum and to hire more staff—an assistant to supervise the museum, office assistants to relieve the director for closer supervision of the field work, plus more topographers and field workers. It recommended dividing the Survey into geology, mineralogy, chemistry, paleontology, and mining sections, all under the immediate supervision of the director. To facilitate the collection of data, all deputy provincial surveyors (responsible for topographical surveys) should be trained in the rudiments of geology, local centres should be established from which surveys could be made, railways should be required to furnish plans and sections of their surveys, and ordinary citizens should be encouraged to help by being supplied with standard lists of questions, along with short instructions on how to proceed and what to look for when investigating geological phenomena and collecting unusual specimens. Finally, the select committee recommended that the annual appropriation of the Geological Survey should be trebled—from £2,000 to £6,000 per annum. Was there ever another such governmental investigation?

While the hearings proceeded, so did the preparations for the Survey's participation at the Universal Exposition in Paris the next year. In view of his success at the Crystal Palace Exhibition, Logan was appointed one of two special commissioners in charge of the Canadian exhibit, his colleague being Dr. J. C. Taché, M.L.A. for Rimouski, medical doctor and journalist. Assembling and organizing the specimens consumed nearly the whole energies of the Survey during 1854, for virtually all the mineral specimens were collected by its staff. In addition, Logan and Hunt prepared a handbook containing an essay on Canadian geology and also a catalogue of Canadian minerals as a guide to the minerals exhibit. Carrying both Logan's and Hunt's names as co-authors, this *Esquisse Géologique du Canada* was sumptuously printed in Paris in 1855 by Hector Bossange et fils, and included a small, finely-detailed geological map of Canada and adjoining parts of the Atlantic colonies and the United States. Preparing the map was another of the preliminary tasks that engaged the Survey, for this was the first instalment of the project that the select committee had recommended to the Survey—a detailed geological map of the province to accompany the republished reports of the Survey. The Legislative Assembly, with unaccustomed promptitude, had already voted £2,000 for the map; and Logan, thus encouraged, prepared a base map on a 1:1,600,000 scale (approximately 1 inch = 25 miles) from existing topographical survey maps of the province, which he took to Paris to put on display and serve as the model for the small map in the *Esquisse*.

By May 1855, Logan and Hunt (who was appointed an associate juror of the exposition) were in Paris preparing their own display and overseeing the entire Canadian department. Logan was kept busy from morning till late at night advising individual exhibitors on display techniques and showing them how a good exhibit was organized and presented. He also had burdensome social responsibilities of attending functions he did not enjoy, and participating in all official deputations, like the one that waited on Napoleon III shortly after an unsuccessful attempt on the emperor's life.

To a degree, Logan's heavy burdens were occasioned by the relative inactivity of his fellow commissioner. Dr. Taché left most of the work on the exhibits to Logan while he spent his time at the unofficial functions and in preparing an essay describing Canada's historical development and economic potentialities. *Esquisse sur le Canada* took only third prize in the Canadian section of the essay competition but nevertheless was the one the Canadian government published for the exposition. Another of the prize entries, interestingly enough, was that of Elkanah Billings of Bytown, who before long joined the Survey as its first paleontologist. Taché's conduct came under attack in Canada from a segment of the English-language press, mainly as a result of reports sent back from Paris to the *Montreal Gazette* by Alfred Perry, himself one of the hardest-working members of the Canadian delegation. One of these drew the following contrast between the two commissioners:

> Mr. Taché has exhibited, throughout, the most deplorable want of activity, of business habits and knowledge, and of urbanity, even while he has displayed an excess of vanity, and a determination that, while he does nothing for himself, nobody else shall get the credit of doing anything.
>
> Mr. Logan, it is proper to say, has throughout manifested the greatest activity, and exerted himself to the utmost . . . . Mr. Logan is too quiet-tempered a man to deal with Mr. Taché. That is his only fault: and that is to accuse him of being too much of a gentleman for the exigencies of his situation.[22]

It was Logan, however, who carried away the praises and honours for Canada's participation at the Paris Exposition to the point where the overshadowing of his fellow commissioner became something of an embarrassment and a source of complaints of discrimination on the part of Dr. Taché's sensitive compatriots. Logan's outstandingly fine exhibit, organized along the same lines as the one in London but containing even better samples and specimens, was admired by all. *The Times*, of September 7, 1855, in an article on "The Paris Universal Exhibition" paid a graceful tribute to the Canadian exhibit, and singled out William Logan in particular for its praises:

> It was hardly to be expected that those provinces, not yet emerged from the first labours of settlement should, nevertheless, in a rough way have taken count of their mineral resources. Yet such is the case. In this Exhibition the Canadian Commissioner, Mr. Logan, himself the surveyor of the geological structure of the colony, and a man of rare scientific attainments, has arranged a magnificent collection of all that in this field of industry the provinces may be expected to yield. Here are fine building stones, and slate and marbles, masses of phospate of lime imbedded in calcareous rock, mica and whetstones, and sandstone so pure as to be considered well fitted for use in glass manufacture. Here, also, is a good display of copper ore, rich enough to promise fairly for the future, and great blocks of magnetic iron, containing 65 per cent of the metal, being in itself a natural loadstone, and extracted from a bed, 500 feet thick. Unfortunately, there is no coal. The American fields terminate just beyond the verge of the southern frontier, and the great source of wealth is withheld from the colonists—who shall say for what wise purpose? Perhaps, to stimulate their industry in clearing away those interminable forests interposed between western civilization and the Rocky Mountains. Certainly we may hope to enable Canada to compete with Sweden in supplying our iron trade with an abundance of the finest quality of iron smelted with wood charcoal. Like Australia, Canada has her goldfields, and Mr. Logan exhibits numerous specimens of nuggets collected there, but, with exemplary patriotism, he expresses, in showing them, his hope that these fields may remain unprospected by the digger, and that the sturdy industry of the colony may escape that source of demoralization.

In all, Logan spent more than two months in the active preparation of his mineral exhibit. When the jurors came to review the mineral displays, they awarded the first prize to the exhibit of the Geological Survey of Canada.

This acclaim contrasted with another poor showing by the Americans. As in London, the state surveys did not make impressive displays, and the Canadian Survey was left with the opportunity of representing the entire continent. Logan, in fact, took over 7,000 square feet of space not used by the Americans. Yet he remained on such good terms with the American officials that they directed visitors to the Canadian exhibit as giving a more complete, accurate picture of North American conditions. The friendly co-operation between the two delegations was important, ultimately, because Americans were prominent on many of the juries. However, some Canadians could not refrain from exulting over the well-patronized Canadian exhibit and sneering at the Americans' efforts. A Montreal observer reported that "One gentleman . . . says he has just passed their division in the main building. He found in one court, about 18 feet by 29, three American Commissioners, two of whom had their feet on the counter, and all were whittling."[23]

As usual, Logan took the opportunity to establish firm contacts with French geologists. At the height of the preparations for the exhibition, on May 7, he found time to attend a meeting of the Société Géologique de France to present a "Note sur le terrain silurien du Canada." He told the society of the latest find by the Survey in 'Silurian' beds at Levis, of fossil graptolites, the first of their kind discovered in perfect condition, according to Professor Hall. He also arranged for Hunt to be elected to the Société Géologique and to give papers to leading French societies, four different articles on chemical geology being published by Hunt in the French periodicals during 1855. Both Logan and Hunt returned from France carrying the decorations of newly-appointed members of the Legion of Honour.

A still greater honour awaited Logan when he returned to England. Queen Victoria had visited the Canadian displays and expressed warm approval of Logan's work; now it was indicated to Logan that Her Majesty wished to bestow a knighthood upon him, "in consequence of your having been twice selected by the Province of Canada to represent first in this country, and afterwards in Paris, the productions and industrial resources of the colony at the respective Exhibitions in London & Paris."[24] Apparently Lyon Playfair, chemist to the Geological Survey, professor in the new School of Mines, a particular friend of Logan's and an intimate of the Prince Consort, had convinced Prince Albert and Lord Palmerston, the Prime Minister, of Logan's worthiness for a knighthood. A letter to Palmerston from Lord Elgin, the recently-retired Governor General of Canada and warm admirer of Logan's work, completed the process. Further in the background lay suggestions and appeals from Canada. On September 6, 1855, the *Canadian Statesman* of Bowmanville, mindful perhaps of Logan's part in exposing a coal hoax in that locality, closed a report on the mineral exhibit with the statement:

> . . . we cannot leave this subject without suggesting that no man in Canada deserves a Knighthood more than Mr. Logan, nor can our good and gracious Queen honor us more than by conferring one on him, let him then be knighted, and let his coat of arms be henceforth the "Maple Leaf."

The Queen knighted Logan at Windsor Castle on January 29, 1856, and this was followed shortly afterwards by the Geological Society of London bestowing on Sir William the Wollaston Palladium Medal, perhaps the highest recognition a geologist of that day could earn. Because Logan had been absent in Europe for almost a year, it was imperative he should return to Canada in time to plan the field work for the coming season. So he missed the formal presentation at the annual meeting of

Confidential

M. H. 10th Nov/55

My dear Sir William!

Let me congratulate you; the affair is settled & the Government have determined to recommend you to the Queen for Knighthood. Keep your own Counsel yet, but you will soon have official intelligence

Yours truly
Lyon Playfair

III-11 LYON PLAYFAIR'S NOTE TO LOGAN

III-12 THE LOGAN MEDALS
Most of these are on display in Logan Hall at the Geological Survey. They include: International Exhibition, London, 1851—Nos. 11, 16, 29, 36; Paris Exposition, 1855, Grand Gold Medal of Honor—No. 27, with silver and bronze copies—Nos. 25, 26, 28, 31, 37; Exhibition at London, 1862—Nos. 9, 19; International Exhibition at Paris, 1867—Nos. 3, 13, 15; Chevalier of the Legion of Honour, 1855—Badge Nos. 1, 4, 7, 33, 35, and miniatures—Nos. 2, 6, 8, 14, 20, 23, 24, 30; the Wollaston Medal of the Geological Society of London, 1856—No. 32; the Royal Medal, Royal Society of London, 1867—No. 22; Order of the Tower and Sword of Portugal, 1855—No. 17; medal commemorating the visit of Prince Edward to Canada, 1860—No. 18; copy of the Logan Gold Medal endowed to McGill University as a student award—No. 12.

the society—no doubt to his immense personal relief. Instead, Murchison, who had just succeeded to the direction of the British Geological Survey, accepted the medal on Logan's behalf. The president, W. J. Hamilton, stressed Logan's contributions to geological knowledge and his labours in Canada, in particular the admirable large geological map of Canada displayed in Paris, where all were impressed "with its execution, and the clear idea it conveyed of the geological structure of the country."[25] In his reply, Murchison reviewed Logan's connection with the society and British geology, then took the occasion to stress that Logan's work sustained his own definitions of both the Upper and Lower Silurian.

Logan's reception when he returned to Canada proved the old adage that 'nothing succeeds like success.' When he landed, laden with medals and bearing a "Sir" before his name as signs of the high regard in which he was held in the leading centres of world civilization, Canadians vied with one another to hail his achievements. Small-town newspapers proudly acclaimed the native son who had voyaged across the Atlantic to impress a continent and win the approbation of his sovereign. Toronto, currently the capital of the province, was the first city to honour him formally. The Canadian Institute, mindful that Logan had been its president from 1850 to 1852, arranged to have his portrait painted by Berthon.[26] The presentation, on April 5, 1856, was an unnerving experience. In his speech of acceptance the new knight emphasized that his honours held a practical value for the progress of Canadian science:

> . . . I am proud to think that it was, perhaps, more because I was a Canadian in whom the inhabitants of the Province had reposed some trust, that the honour which has been conferred upon me by Her Majesty was so easily obtained. That I am proud of the honours which have been bestowed upon me by the Emperor of France, in respect to my geological labours, and also by my brother geologists in England, there can be no doubt. But I have striven for these honours because I considered that they would tend to promote the confidence which the inhabitants of the Province have reposed in me, in my endeavours to develope the truth in regard to the mineral resources of the Province;[27]

Another ordeal, a complimentary dinner of the city of Toronto, followed on April 12. Not to be outdone, Logan's native city followed suit. He was the guest of honour at a soirée given by the Natural History Society of Montreal on May 20, and received an Honorary Doctor of Laws from McGill University on May 25. The citizens of Montreal, led by William Dawson (the Nova Scotia geologist and educator, who had just become principal of McGill University), organized a drive to purchase for their distinguished fellow citizen who had brought such distinction to their city, a large, suitably-inscribed silver fountain. The money was quickly raised, but because of delays in commissioning and manufacturing the fountain, which was "engraved with designs illustrating the palaeontology of the Carboniferous era" resting on a pedestal of ebony,[28] the presentation did not take place until March 1859. Sir William Logan's medals now repose in the Survey's exhibition hall, thanks to the generosity of the late Dr. J. B. Tyrrell, but the silver fountain has vanished without trace. Can any reader furnish a clue to its present whereabouts, or to its fate?

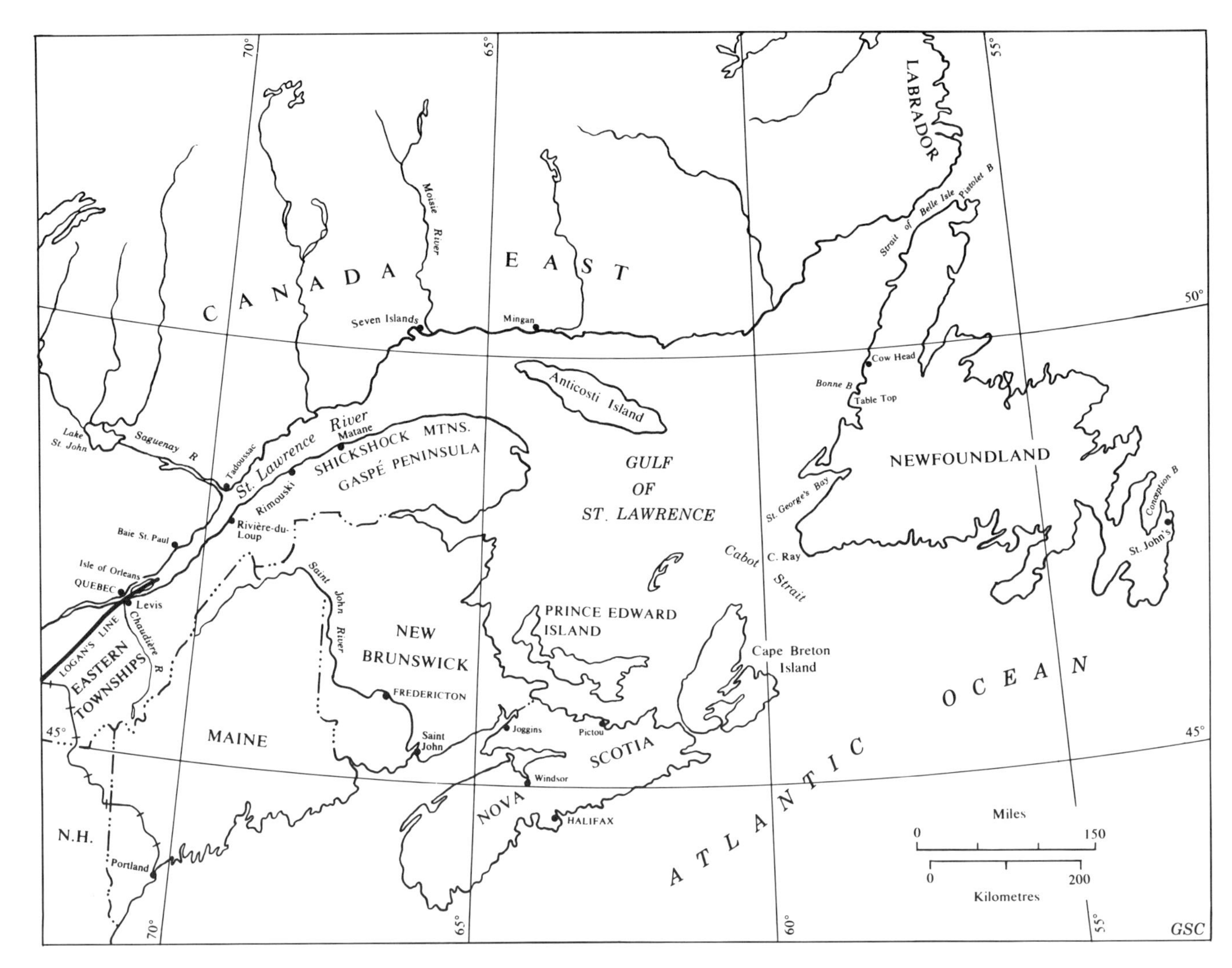

IV-1 ATLANTIC BRITISH NORTH AMERICA, 1850–67

4

Geology of Canada, 1863

DESPITE THE PRAISES THAT THE SURVEY had received at the hands of the select committee in the autumn of 1854, and the unprecedented honours showered on its director in the autumn and winter of 1855-56, the organization was on the brink of extinction when Logan hastened back to Canada in February 1856. The previous Survey act was assented to on July 24, 1850, so the Survey was due to expire as a publicly-supported agency unless a new act continuing its life was passed by July 24, 1855, or at the very latest, at "the end of the then next Session of the Provincial Parliament." Nothing had been done to renew the Survey act during the session in the spring of 1855, the last to be held, for the time being, in Quebec City; and nothing was done, in Logan's absence overseas, by the new session that began in Toronto in October 1855. In the early months of 1856, accordingly, Logan was compelled to deal with a shaky coalition government of Liberal-Conservatives under Sir Allan N. MacNab and George E. Cartier that was disintegrating in the face of internal differences over the separate schools question, discontent over the ineffectual leadership of MacNab, and the vigorous assaults of a powerful opposition, notably the Clear Grits, led by George Brown.

As soon as he arrived in Toronto, Logan began lobbying with all his friends in the legislature, with the new Governor General, Sir Edmund Head, with the scientists at the university, with the Anglican Bishop of Toronto—with anyone who might have some influence. The coming man in the government was the Attorney General for Canada West, John A. Macdonald, so it was to Macdonald that Logan turned to get his bill through the legislature. In March Macdonald promised to take the matter in hand; in April the government agreed to sponsor the bill; by mid-April its details were settled; and finally, at the end of April 1856 the Survey bill reached the floor of the Assembly. Only one voice was raised—that of the aging William Lyon Mackenzie, the former rebel—who wanted the government to assure him that the increased grant to the Survey did not conceal the surreptitious intrusion of political patronage into its operations. Macdonald replied that the entire grant would be at Sir William Logan's disposal, and there would be no interference with his power to appoint his assistants, which assurance satisfied Mackenzie. Since Mackenzie was a long-time supporter of the Survey, the question may well have been intended to help Logan. Certainly, it was all to the good that the government had been forced to affirm publicly its policy of upholding the full autonomy of the Survey under Logan's leadership. Mackenzie's acceptance of the statement also silenced other opposition spokesmen, since he was the leading critic on the public accounts. The Toronto *Leader* interpreted the exchange as reflecting credit on both sides and bringing about the best state of affairs as far as the public interest was concerned. The bill passed, and Logan was at last free to proceed to Montreal, to the reception that awaited him there. Soon after he left Toronto the long-delayed political storm broke in the shape of a revolt of the government members that forced MacNab's resignation on May 24, and elevated Macdonald to the leadership of the Canada West (Upper Canada) section of the dual ministry.

The Survey had been saved. The new act, which was assented to on May 16, 1856 (19-20 Vict., Chap. 13), was strongly influenced by the recommendations of the select committee. The grant was increased, though not quite so much as had been suggested—to £5,000, instead of

£6,000 per annum—and Logan was permitted to use it to pay "any arrears of expenditure already incurred," as well as for current expenses. (In 1859 the amount was altered to $20,000 per annum, an equivalent sum, when the province made the change to the dollar standard.) The act of 1856, unlike its predecessors, made specific reference to the museum function of the Survey, by empowering it "to establish a Geological Museum at some convenient place, which shall be open at all seasonable hours to the public." Intending provincial land surveyors were to be examined on the rudiments of geology, and railway and canal companies were to supply the Survey with copies of all their plans and sections. An important new provision related to the Survey's mapping function was the authority to establish permanent bench marks as points of reference and to fix their latitudes, longitudes, and altitudes. As previously, an annual report of progress was to be submitted, but the reporting time was moved ahead to March 1 from May 1. Another unfortunate relic of past acts, that had already caused so much trouble and would cause more in years to come, was the limited term, now stated as being for a period "not exceeding five years from the passing of this Act."

The increased Survey grant enabled Logan to move forward with a long-desired expansion of staff and operations. Four new members were added to the full-time staff during the next three years, more than doubling the number of its effectives. Foremost among them was Elkanah Billings, a member of a pioneer Ottawa family, who had been trained as a lawyer and had edited the Ottawa *Citizen.* He developed an interest in geology and natural history and undertook a systematic study of the fossils of the Ottawa region; to popularize the study of natural history, in 1856 he founded the *Canadian Naturalist,* which periodical he continued to edit for some years. Logan had long felt the need for a paleontologist to increase the Survey's effectiveness and free it from its dependence on foreign scientists. Finding it impossible to entice Hall to Canada, Logan offered the position to Billings, who joined the Survey as its first paleontologist and remained until his death in 1876.

At the outset Billings relied heavily on the published descriptions of species by the New York, British, and other surveys, but before long he felt confident enough to embark on the description of some of the many new species of fossils continually being added to the collection. He was an able, observant paleontologist, and something of a genius at identifying and classifying fossils and recognizing their basic characteristics. Soon his descriptions were being published in the international journals, where they attracted wide and favourable comment. In 1858 he travelled to England to establish personal contacts with the leading British and European specialists. For 20 years Billings devoted himself largely to invertebrate paleontology to become a recognized authority in the field and give the Survey a high reputation in this aspect. Besides his work on Paleozoic fossils he made occasional forays into entomology and other branches of natural history.

Now that the Survey was able to contemplate publishing maps, the need for a staff draftsman was obvious. Logan found his chief draftsman in Robert Barlow, a former sapper and miner who had been in charge of a corps of surveyors in the Royal Engineers. Extremely competent, he continued with the Survey until his death in 1881, to be succeeded by his son, Scott Barlow, who had entered the office as his assistant. Between them, the two Barlows controlled the Survey's drafting operations until 1894, while a younger son of Robert, Alfred E. Barlow, became a noted geologist with the Survey.

IV-2 SIR JOHN A. MACDONALD
Government Leader, Province of Canada, 1856-62, 1864-67, and Prime Minister, Dominion of Canada, 1867-73, 1878-91

IV-3 ELKANAH BILLINGS

IV-4 SCOTT BARLOW

IV-5 HORACE SMITH

The expansion of the exploration side of the Survey was accomplished by promoting James Richardson to lead independent field operations, beginning in 1856. Richardson had originally been hired to do camp work, but possessing some native ability and no small amount of curiosity, he soon took an interest in geology, learned the rudiments of reconnaissance exploration under Logan's direction, and became "a most valuable and indefatigable explorer, capable of mapping accurately his survey of rivers, lines of traverse through the forest, and outcrops of rocks."[1] An important addition to the Survey's summertime operations in the field was the hiring of a youthful prodigy, Robert Bell, the son of the Reverend Andrew Bell, the Presbyterian minister and amateur geologist who had testified on the Survey's behalf in 1854. The son displayed an early interest in natural history, geography, and surveying, and had already come to Logan's notice. When his father died, young Robert became something of a protégé of Logan's, and in 1857, as a mere schoolboy of 15, he worked as a summer assistant. His first report was published that year, and he headed his own party in 1859, though in other seasons he still assisted Richardson and Murray. During these years he was enrolled as a student at McGill University, from which he graduated as a civil engineer in 1861. Between 1863 and 1869 he taught chemistry at Queen's University and conducted surveys for Logan during some of the summers. His real geological contributions can be said to have begun only in 1869, the year Logan retired, when Bell joined the staff full-time and made ready to begin his memorable expeditions into the north and west. But from the very first season, his keen powers of observation and his broad interests were in evidence, as witness the outstanding botanical collection and extensive descriptions of the fauna and flora of the Saguenay region he made during the summer of 1857.

To assist with the work at St. Gabriel Street, particularly in connection with the museum and paleontology, Logan secured two more employees, both from England, Thomas C. Weston, lapidary, and Horace S. Smith, artist. Logan had sought to hire Weston's father, a lapidary in Birmingham, to help with the display of the museum specimens, but he was unwilling to give up his business and sent his son instead. Youthful and only partly trained, T. C. Weston arrived in Montreal and immediately came under his employer's wing. Logan's counsel, his occasional financial assistance, and his friendship, turned the young immigrant into a trustworthy field companion, an expert preparator of fossils, a very useful museum assistant, and the Survey's first librarian. Though Weston's time was taken up mostly with the museum prior to 1865, he also did considerable field work, collecting fossils in the neighbourhood of Quebec City, Anticosti, Newfoundland, Joggins, and after Logan's retirement, in far-off southern Alberta. His autobiography, *Reminiscences among the Rocks* (Toronto, 1899), is a highly informative, delightful account of life and work in the early days of the Geological Survey of Canada. Horace Smith was lured to Canada by Billings during the latter's visit to Britain and continued with the Survey until his death in 1871. He was a quietly competent employee, who reached his peak as illustrator with his sketches for the *Geology of Canada,* published in 1863. Finally, rounding out the list of the Survey's staff though he was certainly no newcomer, was Michael O'Farrell,

IV-6 SIR WILLIAM LOGAN IN A FORMAL POSE AND IN HIS WORKROOM

the handyman, janitor, and caretaker of the successive Survey premises from 1842 until his superannuation in 1889. The arrival on the scene of the Barlows and Smith was immediately reflected in an improvement in the Survey's publications, which now began to feature maps, sections, and drawings of fossils.

The expansion of the staff in the late fifties at last produced a small but well-rounded organization—one that conducted explorations, made maps, issued publications, operated a museum, and one in which the results of field investigations were bolstered and checked by the findings of chemistry, mineralogy, and paleontology. Formal scientific training was not a feature of the employees' backgrounds, except for Hunt (and even his was limited). The men were gifted amateurs, who largely learned on the job. The fact that they had all been selected and trained by Logan went far to make the Survey an integrated, cohesive team. Logan's role as their employer, patron, and teacher gave him a natural ascendancy over every one of them, and created a bond of personal allegiance between each man and the director. Hence the Survey was doubly strong—from the standpoint of the talents of its individual members, and from its ability to function as a team deriving the maximum advantage from the particular skills of each man. Small as it was, the Survey was an efficient organization with a fine esprit de corps, an extension of the personality and character of the director.

The character with which Logan stamped the Survey was a reflection of his own—unostentatious, industrious, painstaking, practical, and devoted. Despite his considerable wealth and fame, Logan remained a modest, self-effacing, basically shy person, of extremely simple tastes, who often was mistaken on the roads for a common labourer. He lived in a single room "that served for office and mapping room, reception-room, bed-room, wardrobe,"[2] sparsely and plainly furnished, with walls decorated with work clothes, instruments and collecting basket hanging from pegs or nails, and the floor lined with pairs and pairs of old boots. The Survey was Logan's whole life, to which he devoted all his waking hours, energies and talents; and he expected the same sort of high-minded dedication from his men. In the field he set an example for his men to emulate. Short, slight, wiry, he bore up under every hardship, worked incredibly long hours, checked his data with infinite care and thoroughness, and extracted information from every possible source; again, he demanded the same accurate, careful, neat work from employees. He never cared for writing–and perhaps had little time for it beyond the preparation of his official reports. He published a few articles and addressed learned societies when he considered he had significant new information to impart, or felt the public image of the Survey would be enhanced thereby. He was the perfect leader in that he never was resentful nor jealous of his staff or of their accomplishments. Instead he encouraged and made opportunities for them because he realized that the more they made names for themselves, the more they enhanced the reputation of the Survey. So he was anxious that his men should receive due recognition in the scientific world for their discoveries and their theories. He himself was most circumspect about giving them credit, particularly in the annual reports.

In his dealings with his staff, Logan was scrupulous and thorough, but also sympathetic and humane. Despite occasional outbursts of temper, Sir William was a fair-minded, mild-mannered, fatherly counsellor and friend. He often helped out members of the staff with gifts and loans, besides frequently keeping the Survey going from his private means pending reimbursement from the government treasury. At the office he attempted to review the work of every employee each day. He tried to make the Survey a pleasant, good-humoured place in which to work, and despite his advancing years, could always understand and make allowances for his young colleagues' high spirits. Where he was a strict taskmaster was in demanding from them conscientious, thorough efforts, and the careful handling of Survey funds. It was not merely

TABLE OF THE PROBABLE EQUIVALENTS AMONG THE PALÆOZOIC ROCKS OF GREAT BRITAIN AND NORTH AMERICA.

	I. Great Britain.	II. Western Canada.	III. Eastern Canada.	IV. New York.	V. Pennsylvania.		VI. Tennessee.
Carboniferous.	Carboniferous series.	....	.... Bonaventure formation.	....	XIII. XII. XI. X.	Seral. Umbral. Vespertine.	X. Coal measures. IX. Mountain limestone. VIII. Silicious group.
Devonian.	Upper Devonian. Middle Devonian. Lower Devonian.	.... Chemung and Portage group. Hamilton formation. Corniferous formation. Oriskany formation.	Gaspé sandstones, and Famine River limestones.	Catskill group. Chemung group. Portage group. Genessee slates. Hamilton group. Upper Helderberg gr.: Onondaga and Corniferous limestones. Schoharie grit. Cauda-galli grit. Oriskany sandstone.	IX. VIII. VII.	Ponent. Vergent. Cadent. Post-meridional. Meridional.	VII. Black shales. VI. Dyestone and grey limestone group.
Upper Silurian.	Ludlow group.	.... Water limestone. Onondaga formation.	Limestones of Gaspé and the Bay of Chaleurs.	Lower Helderberg gr.: Upper Pentamerus, Encrinal, Delthyris, Pentamerus, and Tentaculite limestones. Water-lime group. Onondaga salt group.	VI.	Pre-meridional. Scalent.	
Middle Silurian.	Wenlock limestones. Upper Llandovery rocks. Lower Llandovery rocks.	Guelph formation. Anticosti group: Niagara formation. Clinton formation. Medina formation.	Limestones of the Chatte River.	.... Niagara limestone. Clinton group, Medina sandstone. Oneida conglomerate.	V. IV.	Surgent. Levant.	
Lower Silurian.	Caradoc or Bala group. Upper Llandeilo rocks. Lower Llandeilo rocks. Lingula flags.	Hudson River gr.: Hudson River formation, and Utica formation. Trenton group: Trenton formation. Black River and Birdseye formation. Chazy formation. Calciferous formation. Potsdam sandstone.	.... Hudson River and Trenton groups probably wanting. .... Quebec group. Potsdam group.	Hudson River gr.: Hudson Riv'r shales, or Loraine shales. Utica slate. Trenton, Black River, and Birdseye limestones. Chazy limestone. Calciferous sandstone. Potsdam sandstone.	III. II. I.	Matinal. Auroral. Primal.	V. Central limestones and shales; including the Stones River and Nashville sub-groups. .... IV. Magnesian limestones III. Chilhowee sandstones
							II. Ocoee conglomerates I. Mica slate group.

IV-7 This table, which appeared in 1863 in *Geology of Canada*, reflects Logan's agreement with Murchison's interpretation of Silurian. Today his Lower Silurian is identified as Ordovician and the Upper Silurian is added to Devonian, leaving his Middle Silurian to represent modern-day Silurian. It was natural to correlate the formations of Western Canada (now Ontario) and Eastern Canada (now Quebec) with New York, but the other columns were probably intended mainly for the benefit of future readers.

that he was intensely scrupulous with the government's money, but, as he wrote one of his men: "Be as economical as you can. The more economical you are, the better I shall be able to pay you."[3]

The addition of Billings and Weston to the staff made it possible to set the museum in order and make it ready for visits from the public. There was reason for haste—the American Association for the Advancement of Science was to hold its annual meeting in Montreal in August 1857. The display cases and furnishings were ready in the summer of 1856, the building was extensively remodelled, and Logan and Billings spent most of the following year at work on the museum, which was in operation by the time of the meetings. The first floor was devoted to economic geology, specimens of rocks and minerals "as can be applied to the useful purposes for life,"[4] prepared so as to illustrate those uses, and arranged in the classes or categories Logan had found so effective in exhibition displays. A collection of minerals and rocks, particularly metamorphic and igneous rocks, as well as the Survey's geological maps, sections, models, also were situated on the first floor. The second and third floors housed the fossils that Billings had classified and arranged into species, and that Weston had prepared for display.

Billings also was responsible for describing the new forms collected or received; by 1863 he had published descriptions of no fewer than 526 new species, mostly brought in by explorers for the Survey, particularly when they examined new formations. The survey of Anticosti by Richardson in 1856 brought the museum an extensive and most important collection—excellent sets of specimens from a series of perfectly conformable beds, correlative to some that were poor in fossils in the standard New York region. As Billings reported, "The fossils of the middle portion of the rocks of Anticosti fill this blank exactly, and furnish us with materials for connecting the Hudson River group with the Clinton, by beds of passage containing some of the characteristic fossils of both formations, associated with many new species which do not occur in either."[5] He made many field trips to examine the distribution of formations and to collect specimens, often at Logan's request. In 1857, for instance, he visited Ottawa and Bonnechere Rivers, "investigating some points bearing upon the grouping of the organic remains in the Black River and Trenton limestone;"[6] then he proceeded to southern Hastings County, Niagara, and the Grand River district to collect specimens. After his return from England he was often in the Eastern Townships helping to elucidate the difficult stratigraphic problems.

Billings was not the only recorder of the new fossils being discovered in Canada. Logan, implementing the select committee's suggestion that the Survey publish illustrative material, had already arranged for a series of volumes on Canadian fossils. Since each volume was supposed to include ten plates of illustrations, plus the descriptions in the associated text, it was known as a decade. Logan had appointed J. W. Salter, paleontologist to the Geological Survey of the United Kingdom, to prepare the first decade and Professor Hall a second, before Billings could undertake responsibility for further volumes. Salter's *Figures and Descriptions of Canadian Organic Remains, Decade I,* was published in 1859, but Hall's did not appear until 1865, though as early as April 1855 he had submitted a report to Logan, "Descriptions of Canadian Graptolites," on the subject of his Decade II. The third (1858) and fourth (1859) decades, which incorporated articles by J. W. Salter, T. Rupert Jones, and J. V. Thompson, were chiefly the work of Billings, who also published eight other volumes for the Survey, besides contributing to Canadian, American, and European periodicals.

* * *

NATURALLY, THE FIELD OPERATIONS remained the basic activity of the Survey. The outlying regions of Canada were examined and reported on by Murray, Richardson, and Bell with a view to tracing the trends of the major rock groups to the Strait of Belle Isle and to Lake Superior. Regions like the Ontario peninsula were re-examined to study in detail the boundaries and characteristics of the various formations. In a few areas special studies were being undertaken to solve the mysteries of such complicated formations as the Grenville rocks or the Quebec Group. After 1856, however, the emphasis seemed to change markedly. There was less stress on pure exploration and more on the examination of specific problems. The work in the field seemed designed to fill gaps in the geological map and the comprehensive account of Canadian geology that Logan was preparing.

For every season between 1856 and 1863, except for 1859, Murray worked in the territory north of Lakes Huron and Superior, continuing the survey of the Precambrian formations (differentiated now between Laurentian and Huronian, the latter to be recognized before long as favourable ground for occurrences of metals). In 1856 he operated north of Lake Nipissing and French River, surveying and mapping French, Sturgeon, Wanapitei, and Whitefish Rivers, examining the geology and supplementing the base lines that the land surveyor A. P. Salter was running through the country. Murray observed and traced the appearance of the Huronian above the Laurentian as he moved west and north from French River, and also noted that the Huronian lay unconformably below the lowest fossiliferous strata where these were found. In the course of this exploration, Salter informed Murray of a major magnetic deflection he had encountered along his meridian line north from Whitefish Lake. Proceeding to the place, Murray reported "an immense mass of magnetic trap,"[7] and took away a number of specimens which Hunt found to contain titaniferous iron, magnetic iron pyrite with traces of nickel, and non-magnetic iron pyrite with 2-3 per cent copper and 1 per cent nickel. According to the Ontario Nickel Commission (1917) "Murray walked over and examined the long gossan-stained ridge, at the foot of which in later years the greatest nickel mine in the world was discovered"[8]—the Creighton mine.

After 1857 Murray worked farther west, between Blind River on Lake Huron and Batchawana River on Lake Superior, mapping the Huronian and Lower Silurian (now Ordovician) formations around Sault Ste. Marie, St. Joseph and Manitoulin Islands. Bell also led a party into the region in 1859 and assisted Murray in 1860, his work being mainly in the Paleozoic of Manitoulin Island. The end-product of Murray's geographical investigations in this district was represented by a map (1 inch = 8 miles) in the Atlas of the *Geology of Canada* of the distribution of the Huronian rocks between Batchawana and Mississagi Rivers, and the adjacent Lower and Middle Silurian formations. A second purpose of Murray's work was to examine the copper-bearing rocks. For the benefit of the readers of the report Logan explained the reasoning behind the investigation. The veins in cracks or fissures in the rocks in which metallic ores occur are likeliest to be found along the axes of the anticlinal and synclinal folds where the rocks are most strongly flexed. Hence Murray was tracing and mapping the undulations of the strata, using as a marker bed a broad band of limestone occurring near the middle of the succession and sharply differentiated from the rocks on either side of it. Murray's main find, in 1858, was the great fault, afterwards known as the Murray Fault, which runs a few miles inland and parallels the north shore of Lake Huron. Observing it first at Thessalon River, he traced it past Bruce Mines to Blind River, commenting that according to theory, it should be associated with mineralized zones. Murray was the first of many members of the Survey, notably W. H. Collins, to study this area; the reports of their operations encouraged prospectors for uranium to concentrate on the district, and the result has been the great modern uranium mining industry at Elliot Lake, in the Blind River district.

At the same time "James Richardson, Explorer" was filling in other gaps in the geological map of Canada, notably along the North Shore of the Gulf of St. Lawrence, and Gaspé. In 1856 Richardson visited the Mingan Islands and made a circuit of the coast of Anticosti. He found a very uniform succession of fossiliferous strata on Anticosti which he separated into six divisions without attempting to correlate them with mainland formations, but which Billings assigned to the Lower and Middle Silurian. Richardson also reported fully on the geography and resources of the little-known island because of his belief that it offered "about a million of acres of good land" of which "not a yard of the soil has been turned up by a permanent settler"[9] to a Canada that was then desperately seeking *lebensraum* for the inhabitants of its overcrowded farmlands.

In 1857 and 1858 Richardson's explorations were mainly on the south shore, that of 1857 being an examination of the mountainous northeast tip of the Gaspé peninsula, and the following year, of the rivers and mountains between Matane and Rivière du Loup. He was assisted by Scott Barlow, who mapped the country, and by Robert Bell, who helped with the surveying and studied the flora and fauna of the region, on which he reported separately. Richardson was also the first member of the Survey to make an extensive examination of the geology of the North Shore of the Gulf of St. Lawrence. In 1857, accompanied by Barlow and Bell, he crossed from Gaspé to Saguenay River and went up to Lake St. John on a reconnaissance in which he noted the Laurentian gneisses that encircle the Royaume du Saguenay, the labradorite outcrop at the outlet of Lake St. John, and the Lower Silurian rocks along the south shore of the lake. Possibly he accompanied Logan on an examination of the coast around Mingan Harbour in June 1859; certainly he returned to the North Shore to survey the series of harbours from Mingan east to the Strait of Belle Isle during the summer of 1860. This trip is of special interest as being apparently the first occasion on which photographs were used by the Survey to show geology. At any rate, the glass plate negatives Richardson brought back from the North Shore are almost the earliest photographs in the Survey's unique, priceless photographic collection. In 1861 and 1862 Richardson returned to complete his survey through the Strait of Belle Isle and to explore the coast of Newfoundland from Pistolet Bay on the north to Bonne Bay at the base of the northern peninsula. He made a number of traverses inland to the Laurentian spine of the peninsula, and made the first reports on the interesting geological formations at Table Top, Bonne Bay, and the spectacular breccia of Cow Head. Richardson prepared a large-scale map of the northwestern coast of Newfoundland, on three large sheets. His field notes contain many comments on the fisheries and settlements of the North Shore and Newfoundland coast, but failure to publish reports of the Survey's activities during the years after 1858 means that Richardson's important, interesting work in the North Shore region is generally overlooked.

Part of the field work of Murray, Richardson, and Bell during the years 1859 and 1863, inclusive, was not exploring new country but collecting or checking data for the *Geology of Canada.* They re-examined certain localities to clear up specific stratigraphic problems, collect fossils, and secure samples of minerals for description. Some of this was done on the way to or from their main work. For instance, the Bruce Peninsula or the Niagara Escarpment could be studied on the way to the north shore of Lake Huron. On the other hand, some entailed special trips, like Murray's to a series of points along Ottawa River, between Montreal and Niagara, and the Owen Sound and Manitoulin Island districts in 1859 and 1861. No doubt these investigations supplied some of the data on the Quaternary geology of the Ontario peninsula, and helped locate and describe the Erie clay, and the overlying Saugeen clays and Artemisia gravels discussed in the 1863 report and mapped in the Atlas. Richardson also spent a large part of these years, when not engaged in his work at the Strait of Belle Isle, checking on various details in the Eastern Townships.

Logan could not devote so much time to field work after his return from Europe in 1856 as he would have liked. Until late 1857 he was fairly taken up with the affairs of the museum and the forthcoming meetings of the American Assocation for the Advancement of Science. By 1859 he was again involved in a succession of urgent, time-consuming problems—dealings with the government in 1859 over a temporary reduction in the money available to the Survey, and troubles about getting the Geological Survey Act revived and extended when it expired in 1861 (probably the most difficult of his many dealings with the governments of the Province of Canada); the ambitious publications program; and the responsibility for Canadian participation at still another international exhibition—to be held in London in 1862. For several years his field work appears to have been limited to intervals of freedom from these several commitments, and the short periods he could spare from the office. In 1857 he was able to do a little work in the Grenville at the end of the season; in 1858 he devoted a full season; in 1859 he spent only short periods in the area, about Mingan Harbour on the North Shore, and in Vermont. He seems to have been desk-bound in 1860 and free to do geology for no more than brief intervals, often in the immediate vicinity of Montreal. In 1861 he devoted at least four

IV-8 RICHARDSON'S SURVEY PARTY ON THE LABRADOR COAST, 1862

IV-9 BROOM POINT, NEWFOUNDLAND
This striking geological feature, on the west coast of Newfoundland, is formed of brecciated material similar to that of nearby Cow Head. In 1767 the towering vertical layers suggested the name to navigator-cartographer James Cook.

IV-10 JAMES LOWE

months to an extensive examination of the Eastern Townships, and in 1862, after his return from England, he spent parts of the autumn on field work that ranged from St. Albans, Vermont, to Levis. In sum, during the years between his return from England until the publication of the *Geology of Canada,* Logan was able to work extensively only in the seasons of 1858, 1861, and possibly, 1862. He concentrated chiefly on two main areas: Grenville and the Eastern Townships, where he was concerned more with solving geological problems than with mapping—though Logan, the stratigrapher, relied on accurate field observations to solve or verify every geological problem.

In the work in the Grenville in 1858 Logan was assisted by James Lowe, a local farmer who possessed "much aptitude in geological field-work,"[10] and by the Englishman, W. M. S. D'Urban, who made a study of the natural history of the area. Logan had set forth his thoughts on the Grenville region in two papers before the Montreal meeting of the American Association for the Advancement of Science, on August 13 and 14, 1857. In these he described how the name "Huronian" had been established for the younger, distinctive succession of slates and conglomerates of the Precambrian as represented at Lake Timiskaming and north of Lake Huron; and the name "Laurentian" was given the still-undifferentiated gneisses and crystalline limestones; also how it seemed desirable to map the limestone beds, which he had traced for a distance of eighty miles, and which, according to reports, extended all the way to Labrador and to Georgian Bay. In 1858 the main effort was to trace out the interrelationships between these limestone beds with the Laurentian gneisses and with the intrusive and volcanic formations, notably the rocks exposed at Morin, St. Jérôme, and Rawdon. The problem moved no nearer solution in ensuing years. The statement on the Laurentian System in the *Geology of Canada* merely described the various types of rocks without attempting to establish any new formations or systems, explaining that,

> To determine the superposition of the various members of such an ancient series of rocks is a task which has never yet been accomplished in geology, and the difficulties attending it arise from the absence of fossils to characterise its different members. Bands of the crystalline limestone are easily distinguished from bands of the gneiss, but it is scarcely possible to know from local inspection whether any mass of limestone in one part is equivalent to a certain mass in another . . . the only reliable mode of pursuing the investigation and working out the physical structure is patiently and continuously to follow the outcrop of each important mass in all its windings as far as it can be traced, until it becomes covered up by superior unconformable strata, is cut off by a great dislocation, or disappears by thinning out.[11]

SECTION FROM MONTMORENCI TO ORLEANS ISLAND.

Montmorenci. North Channel. Orleans Island.

Horizontal and vertical scale, one inch to a mile.

g, Laurentian gneiss.
t, Trenton limestone.
u, Utica and Hudson River formations.
q, Quebec group.
F, Fault.
O, Overlap.
S, Level of the sea.

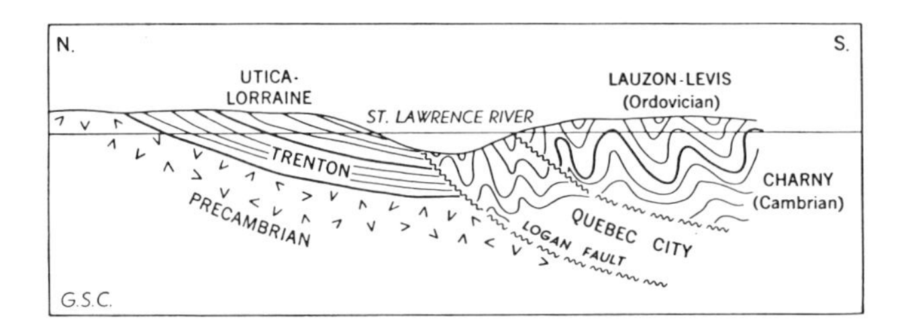

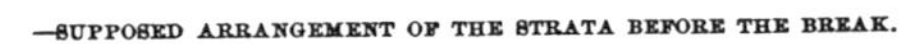

—SUPPOSED ARRANGEMENT OF THE STRATA BEFORE THE BREAK.

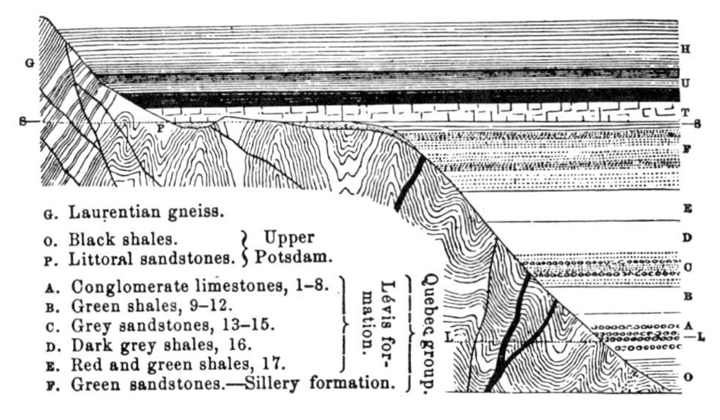

T. Trenton group of limestones.
U. Utica shales.
H. Hudson River sandstones and shales.
L. L. Sea level at the commencement of the Quebec period.
S. S. Sea level at the close of the Potsdam, and also at the beginning of the Trenton period.

IV-11 LOGAN'S LINE
Following his discovery of this great fault, Logan revised his interpretation of the geology of the district around Quebec City. His "Section from Montmorenci to Orleans Island (*upper*) from *Geology of Canada* still remained acceptable a century later, as the diagram (*centre*) from *Geology and Economic Minerals of Canada,* fourth edition, 1963, indicated. Logan's explanation of the event postulated a lateral force (about whose origin he refused to speculate) that jammed the Lower Silurian succession against the slanted front of the harder Laurentian gneiss, causing lower members of the Quebec Group to slide upward along the margin of the Laurentian and overlap the younger strata. This explanation was illustrated by the sketch (*lower*), also from *Geology of Canada.*

South of the St. Lawrence, in the meantime, the region between Montreal and Chaudière River was beginning to emerge as Canada's main mining centre, even before the stimulus of the high copper prices of the American Civil War years carried its mines into all-out production. The new found prosperity gave point to Logan's careful tracing of the bands in which the copper showings occurred and to his careful listing of their locations in his report. The work formed part of the broader problem of correlating the strata of the Eastern and Western Divisions on either side of the great fault now known as Logan's Line—the solution for which depended on the correct interpretation of the Quebec Group. Other problems included the nature and origin of certain rocks, particularly the serpentines, and of deposits of economic minerals, notably gold and copper.

The rocks of the Quebec Group were the real problem—a very thick assemblage that contained many important copper deposits and serpentine bodies. They overlay considerably altered and metamorphosed Utica and Hudson River shales which were the uppermost of the Lower Silurian formations identified in New York; but they did not correspond to the formations that succeeded the Hudson River shales either in New York or in the Western Division. The discovery of graptolites in 1854 at Levis did not upset Logan's interpretation of the age of the Quebec Group, for the place of graptolites in the paleontological sequence still had not been definitely established. But fossils Billings subsequently collected at Levis created doubts that set Logan to reconsider the horizon of the Quebec Group. The great European paleontologist, Joachim Barrande, commented that the orthoceratites and trilobites reported from the Quebec Group at Levis appeared related to members of his Primordial Zone, and that attributing them to a formation younger than the Hudson River shales seemed in error. Billings, in a letter of July 12, 1860, to Barrande, concurred that the fossils were indeed earlier than Hudson River shales. Logan, writing to Barrande on December 31, 1860, confirmed that Billings had identified a number of the 137 species as belonging to two earlier New York formations, Calciferous and Chazy. Logan had therefore returned to Levis, checked the position of the fossil beds, examined the stratigraphy of the Quebec Group, and now had an explanation for the mystery. His conclusion, which he reported in the letter to Barrande, was:

> From the physical structure alone, no person would suspect the break that must exist in the neighbourhood of Quebec, and without the evidence of the fossils, every one would be authorized to deny it .... Since there must be a break, it will not be very difficult to point out its course and its character. The whole Quebec group ... appears to be a great development of strata about the horizon of the Chazy and Calciferous, and it is brought to the surface by an overturn anticlinal fold with a crack and a great dislocation running along the summit, by which the Quebec group is

brought to overlap the Hudson River formation. Sometimes it may overlie the overturned Utica formation; and in Vermont, points of the overturned Trenton appear occasionally to emerge from beneath the overlap.[12]

This letter is one of the earliest published recognitions of thrust faults anywhere in the world, according to a modern authority, and a first step in unravelling the complicated tectonic history of the Atlantic region of Canada.[13]

Logan proceeded to work out the geological history of the Eastern Townships in the light of this revised interpretation. He traced the overturn southwest into Vermont along the east side of Lake Champlain, and curving northeast to past Quebec City, across the north end of the Island of Orleans, under the river and Gulf of St. Lawrence, and emerging again near the tip of Gaspé. The Paleozoic succession north of this line was checked as far as possible. On the Island of Orleans and Gaspé the fault was skirted on the north by Utica or Hudson River rocks, but the long North Shore of the St. Lawrence provided few other places to view the Paleozoic succession. At Bonne Bay and elsewhere along the west coast of Newfoundland, exposures were good. There the Lower Silurian formations apparently fitted into the New York sequence, with Potsdam, the lowest Paleozoic formation followed by rocks analogous to the Quebec Group. In a paper to the Natural History Society of Montreal in May 1861 (published in the *Canadian Naturalist* and *American Journal of Science* and repeated in the *Geology of Canada,* chapter 11) Logan expanded this explanation of the formation and overturn of the Quebec Group.

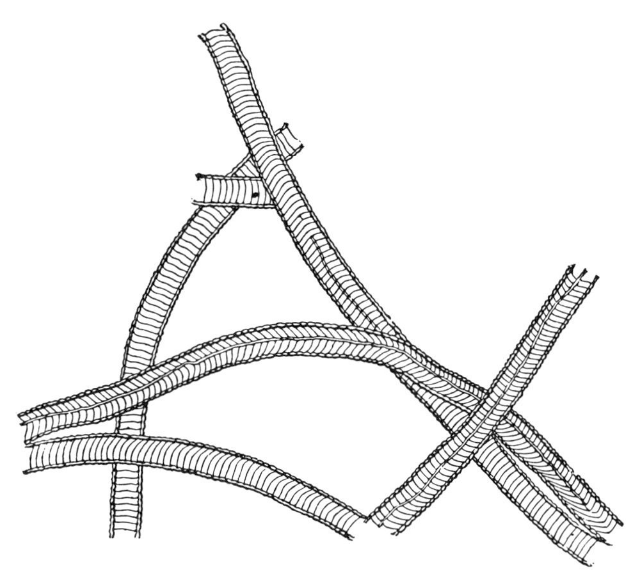

IV-12 THE PERTH FOSSIL TRACKS
Tracks of *Climactichnites Wilsoni.* Published in *Geology of Canada.*

He also attempted to relate the Quebec Group to the crystalline strata on the south and east of them that formed the ranges of the Appalachian mountain system. The geology of this region had given rise to much controversy and was the subject of one of the bitterest of geological disputes in the United States, the Taconic controversy. Emmons and others regarded the Green Mountains as Primary (i.e., Precambrian), while other geologists insisted that they were altered Paleozoic rocks of Lower Silurian age. Logan, who followed this last interpretation, regarded them as altered Quebec Group, and therefore, put them in earlier Paleozoic time than others had done, thus adopting a halfway position between the two factions. He also made an attempt to subdivide the Quebec Group. By 1863 he suggested that the Quebec Group consisted of two main formations—of which the lower, or Sillery Formation, included the conglomerate limestones, slates, sandstones, shales, and the mineral occurrences that made the district such an important mining centre; while the upper, or Levis Formation, contained green sandstones. The actual relations and succession of rocks that constitute the Quebec Group was to be a most difficult, highly controversial question for future investigators.

Logan also became involved in these years with two more episodes in which fossils played a central role. The first, a repetition of the Beauharnois find of the early fifties, excited little or no controversy. This was the discovery, near Perth, by Logan's old friend Dr. James Wilson, of another example of fossil tracks, also in Potsdam sandstone. There were no fewer than six separate tracks, the longest about 13 feet, clearly impressed in what was once soft sand or mud, and presumably made by a single creature. In appearance these tracks resembled the impression that would be made in mud by a rope 6¾ inches thick, consisting of two bounding ridges, a series of curving, convex, transverse depressions between the two ridges, and, a little faintly, a third ridge in the middle of the track undulating between the two bordering ridges. Since the tracks resembled those of a modern-day mollusc, Logan assumed they had been made by a species of gigantic mollusc, which in 1860 he named *Climactichnites Wilsoni.* Richardson, who went in winter to collect the material, took up a 76-square-foot piece, an upper stratum containing the mould of the impressions, since the prints themselves were on a very thin, friable sheet of rock that could not be moved.

The other episode was destined to become one of the major scientific controversies to engage the Survey's attention. This was the discovery of apparent evidences of life in the Laurentian rocks (the crystalline limestones), a discovery which, if true, pushed back the origins of life on earth much further than anyone had been able to prove to that time. The affair was set in motion with the arrival in October 1858 of a piece of crystalline limestone collected that summer at Grand Calumet on Ottawa River by John McMullen, an assistant Logan had employed to examine the Laurentian rocks on the south side of the Ottawa while he himself worked in the Grenville district. The rock was a pyroxenite scored on the surface by a series of roughly parallel concentric grooves, which were filled with crystalline carbonate of lime. What caused the excitement was that these markings were identical with those on another rock sample collected in Burgess Township some years before by Dr. Wilson of Perth. In that specimen, however, the ridges were loganite and the filling, dolomite.

Two identical forms, each composed of different mineral constituents, created a real puzzle, because they seemed to refute the possibility that the form might have developed from mechanical or chemical causes. Logan was led to the thought that the configuration of the corrugations or laminations was evidence of a fossil origin, the end-result of the alteration of the remains of a coral-like creature replaced selectively by other minerals that could survive the metamorphism and recrystallization to which the rocks were subsequently subjected. He may have been encouraged in this view by his colleague Hunt, who was arguing strenuously the notion that some form of organic life had existed in Laurentian time. In a paper before the Geological Society of London in 1859,

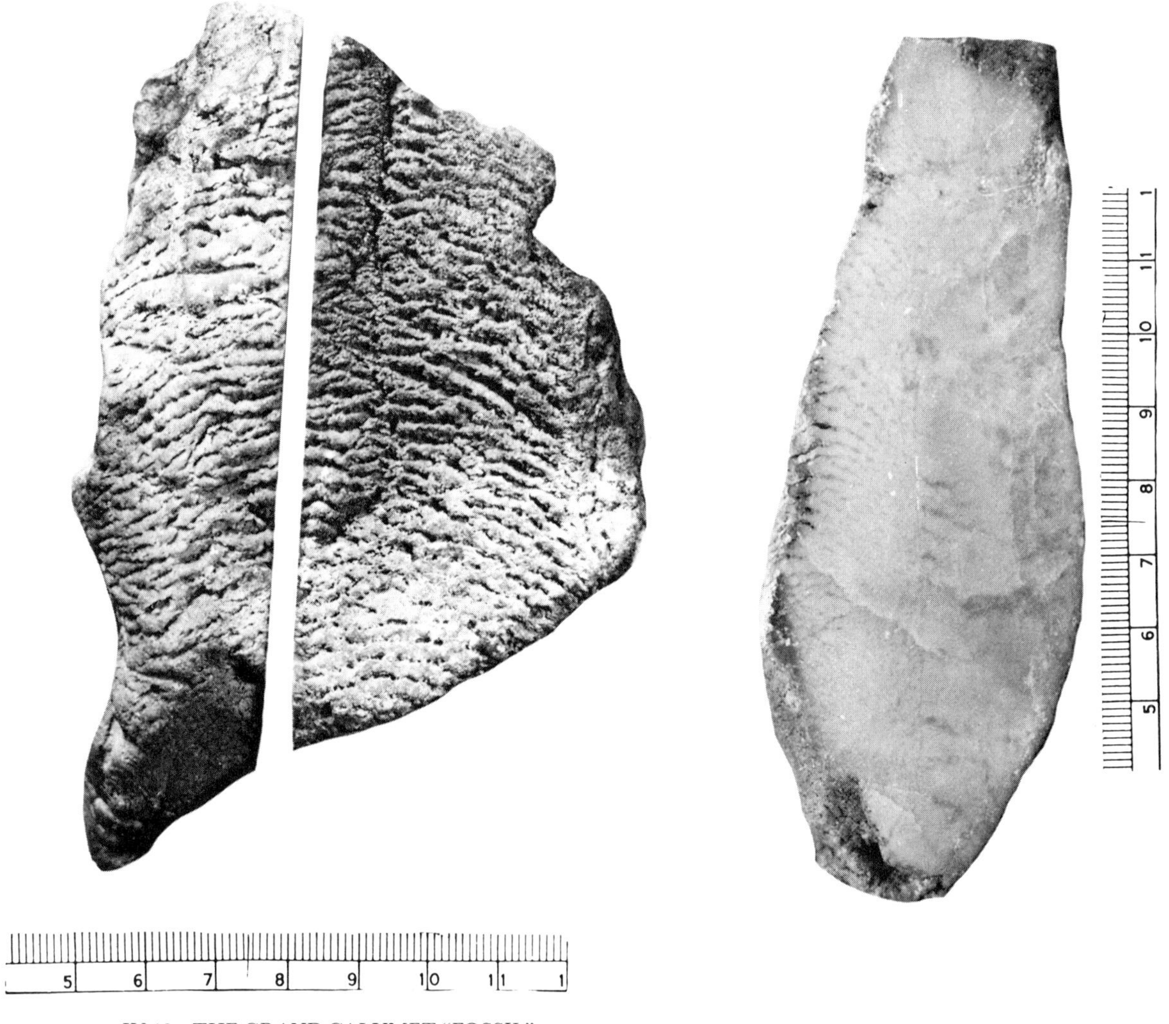

IV-13 THE GRAND CALUMET "FOSSIL"
Photographed from above (*left*), and in cross-section at the point of slicing (*right*) to show the depth of the corrugations.

Hunt pointed to the presence in the Laurentian of beds of iron oxide apparently reduced from sediments, metallic sulphides, and graphite, as "evidence of the existence of organic life at the time of the deposition of these old crystalline rocks."[14] Finally, it was argued that the form of the supposed fossils resembled a known Lower Silurian fossil *Stromatopora rugosa.* Logan was sufficiently convinced to display the specimens at the August 1859 meeting of the American Association for the Advancement of Science in Springfield, Mass., and to present his conclusion that they were of fossil origin. He also took the specimens to England when he went there in 1862. But because of the inability of paleontologists to detect any positive signs of organic structure under a microscope, Logan made few converts at this time apart from his friend Professor Ramsay. The smouldering ember was fanned into a blaze, however, with the discovery of still more specimens in 1864 that under the microscope showed apparently organic structural characteristics. With these the *Eozoon Canadense* boom was under way, to reach its height in the decade after 1866, under the doughty championship of Sir William Dawson.

* * *

HUNT AFTER 1855 ENTERED UPON A NEW PHASE of work that marked him as a most unusual scientist and gave him the reputation of the foremost chemical geologist of America of the time. He carried out the routine analyses required for purposes of the Survey without even a laboratory assistant, did a certain amount of field work, prepared his own reports, helped Logan with the editorial tasks, and took charge of the office during Logan's absences. In his capacity as chemist and mineralogist, he drew attention to new methods of ferrous metallurgy, discussed the possible uses of peat, proposed the use of sesquioxide of chromium for printing banknotes (a proposal adopted by the United States government for its "greenback" currency, but from which Hunt received no financial reward). Like Logan, he too was a bachelor at that time, for he married late in life after he had left the Survey. He lived in a small study and smaller bedroom that opened off his laboratory and was free to devote as much time as he could wish to science. He was far more outgoing than his superior, however, for he enjoyed social gatherings and was a brilliant conversationalist. From 1856 to 1863 he went to Quebec each spring to teach chemistry at Laval University, lecturing in French "which language he spoke with fluency and elegance."[15] In 1862, perhaps to be closer to the laboratory, he accepted an appointment to the chair of applied chemistry and mineralogy at McGill.

In addition to his labours in the laboratory, Hunt after 1855 began increasingly to use his chemical skills to solve theoretical problems, to explain how things *could* have happened (which often was transformed in his mind into how they *did* happen). He became a kind of chemist-uniformitarian, confident that almost all scientific problems could be explained in terms of chemistry. From limited theories intended to cover only a specific phenomenon or answer a specific question, his theories became more sweeping in range, until finally they coalesced into a complete scientific system that attempted to cover the whole of earth history, expressed in his "Report on the Chemistry of the Earth," prepared in 1870.

Such courage or foolhardiness aroused criticism, and even in Canada the recognition of Hunt's extreme brilliance and versatility was coupled with an awareness of his tendency to jump to conclusions, disregarding facts that did not support them, and to flit from subject to subject without working any of them out to the full. His theories injected him into the Taconic controversy and a host of other scientific debates, as did his frequent efforts to uphold his prior claims to have originated certain ideas or theories. His broad speculations ranged across many sciences, even in the space of a brief lecture. He once replied to a critic that

> In the lecture in question, I endeavored to bring together the results of modern investigations in physics, chemistry, mathematics and astronomy, and to construct from them a scheme which should explain the development of our globe from a supposedly intensely heated vaporous condition down to the present order of things. I could not pretend to discuss from their various standpoints, all the conclusions arrived at by different investigators, inasmuch as, even had my attainments permitted, the limits of an hour's lecture would have proved far too short.[16]

Specialists in other fields resented his meddling in their fields without proper credentials, and complained bitterly that his conclusions carried findings based on the study of chemistry beyond their proper sphere and failed to take into account those of other sciences crucially involved in the problems. Unquestionably, geochemistry had a large, important part to play a century ago in improving the understanding of the processes by which the earth develops. But Hunt attempted to use its findings (more precisely *his* findings) to explain almost every physical process.

Hunt carried the Survey well into the realm of speculation and theorizing, into which it perhaps was bound to move now that the initial reconnaissance phase was nearing completion and the prime task was to concentrate on the geology of those parts of the country that were of economic or scientific interest. It may well be that the presence of Hunt and Billings as colleagues relaxed Logan's caution slightly and made him readier to express conclusions that reflected their ideas as well as his own. While Logan's work in his own field of stratigraphy was almost perfection, when he went outside it, as he had to do in order to deal effectively with other aspects of Canadian geology, he became dependent on the views of other specialists—Dawson and Hall as well as Hunt and Billings—with their limitations and inadequacies. Equally, Logan was a moderating influence on Hunt, forcing him to check his ideas against actual conditions and to square his theories with reality. Hunt could trade ideas freely with men of differing points of view, and from this new ideas ware generated. But this interplay of minds makes it difficult to know where a new idea originated. Hunt found the Survey a most congenial environment where he was able to do his most effective work, develop his most brilliant theories, achieve recognition, and attain the summit of his reputation.

So prolific was he as a writer and thinker that by 1862, when he was barely 35, he had already published 71 papers. His studies led him, naturally, to investigate how certain minerals could have been formed by chemical action. As early as 1853, having burned and analyzed lingulae shells, he speculated that beds of phosphate rock might have been formed from the remains of just such fossil creatures. His work on the chemistry of mineral springs led him to a prolonged study of the origin of dolomite and magnesite, and to the conclusion that certain rocks that occur in association could have been produced by the discharge of mineral springs with a carbonate base into confined shallow saline seas. He described the chemical composition and crystalline structures of the various mineral components of anorthosites, and classed the rocks as metamorphic, while his studies of the chemistry and mineralogy of serpentine rocks led him to regard these as being of sedimentary origin.

Hunt has been credited also with a role in promulgating the anticlinal theory of petroleum accumulation. The sudden rise of a petroleum industry in southwestern Canada in 1857 and in Pennsylvania in 1859 led Hunt as well as many others to study the chemistry and geology of petroleum. His major paper on the subject, published in the August 1861 issue of the *Canadian Naturalist,* was delivered as a paper to the Natural History Society of Montreal on or about February 28, 1861, since it was reported in the *Montreal Gazette* of March 1. Charles Robb, a mining engineer who later worked for the Survey, also read a paper on the same subject on February 2 to the Canadian Institute in Toronto which was published in the *Canadian Journal of Science* in its July 1861 issue. Furthermore, a Professor E. B. Andrews had an article on the location of oil and gas springs in Ohio in the December 1861 number of the *American Journal of Science.* Robb's paper touched on the main points of Hunt's, but without the same clarity on some essentials. Hunt's analyses of petroleum confirmed his and Robb's view that it originated almost entirely from marine fossils of the Upper Silurian and Devonian periods. Using Logan's analysis of the stratigraphy of the region (as Robb had also done), Hunt pointed out that the place to search for oil was at the top of the major anticline, or of subordinate folds in the strata, since petroleum, being lighter than water, had a tendency to migrate upward to the highest point in the bed of origination or of any higher beds which it penetrated, until eventually some of it would reach the atmosphere through fissures in the strata. (The important idea, that the petroleum occurrence depended on the structure possessing an impervious stratum as a cap to trap the mineral, apparently originated in 1865 with Alexander Winchell.) Hunt argued that petroleum permeated the strata through which it migrated, and that it did not originate in the bituminous shales that were commonly considered its source—but rather that those shales were made bituminous by being permeated with petroleum from below.

The most controversial of Hunt's theories was one that arose from his studies of the chemistry of crystalline and altered rocks—that it might be possible through chemical analysis to identify the geological ages of metamorphosed sediments since "the crystalline minerals which are formed, being definite in their composition, and varying with the chemical constitution of the sediments, may perhaps to a certain extent, become to the geologist what organic remains are in the unaltered rocks, a guide to the geological age and succession."[17] Of all his theories, this was perhaps the one that was most bitterly attacked, though Hunt's defenders have pointed out that "he nowhere pretended (as has sometimes been unjustly said of him) that we were yet able to interpret aright all of the phenomena they presented."[18]

The most persistent and grandiose of Hunt's theorizing related to the history of the earth. One important element of this he enunciated in his report for 1856 and in a paper in 1857, then expanded and refined it during the next twenty years—his theory of metamorphism and vulcanism. Faced with such facts as metamorphosed strata occurring in the midst of unaltered formations, and igneous rocks that are chemically akin to recognized Precambrian and Paleozoic sedimentary rocks, he took up a little-known theory that had been first suggested by Sir John Herschel in 1836, and proceeded to test it, develop it, and work out its implications. Herschel's theory was that sediments were altered by a sort of auto-metamorphism—as they accumulated, their mass increased to a point where the weight pressing on the lower strata raised the temperature (as did heat from beneath), altered them, and made them plastic. Hunt's main contribution was to study how water in the rocks lowered the temperatures at which they became plastic, and the role such water would play in chemical reactions within the rocks or in volcanic eruptions. As early as his 1856 report he concluded "that rocks covered by 10,000 feet or more of sediment, and permeated by alkaline waters, are in the conditions required for their alteration, and that elevation and denudation would exhibit these lower strata to us in the state of metamorphic crystalline rocks."[19] His paper of 1858 "On the Theory of Igneous Rocks and Volcanoes" added this summary:

> We conceive that the earth's solid crust of anhydrous and primitive igneous rock is everywhere deeply concealed beneath its own ruins, which form a great mass of sedimentary strata permeated by water. As heat from beneath invades these sediments, it produces in them that change which constitutes normal metamorphism. These rocks at a sufficient depth are necessarily in a state of igneo-aqueous fusion, and then in the event of fracture of the overlying strata, may rise among them, taking the form of eruptive rocks. Where the nature of the sediments is such as to generate great amounts of elastic fluids by their fusion, earthquakes and volcanic eruptions may result, and these, other things being equal, will be most likely to occur under the more recent formations.[20]

Any discussion of the work and thought of the officers of the Survey in the Logan years must necessarily refer to the many articles written by Survey personnel in addition to their contributions to the reports. These papers are all the more valuable for tracing the thoughts of the geologists because the official reports of necessity tended to be more factual than interpretive. They also give an indication as to the activities of the Survey in the years between 1858 and 1866 when there were no annual reports. But reports or not, papers continued to appear each year, usually first in the form of an address to some organization, perhaps to be printed in the local newspaper, then published in a scientific journal and perhaps reprinted subsequently in two or three others. Hunt, the most prolific of the group, furnishes a good illustration of the number of papers and the outlets. The full list of his publications, compiled by Dr. F. D. Adams, gives 59 different pieces (many of them subsequently reprinted) between the years 1855 and 1863, exclusive of his chapters in Survey reports, an average of better than six articles a year over the nine-year period. These articles first appeared in ten different journals in five countries, the commonest outlet being the *American Journal of Science*, with 21 pieces. His Canadian publications numbered 20 papers, five in the *Canadian Journal of Science*, the organ of the Canadian Institute in Toronto, and the remainder in the *Canadian Naturalist*, by now the journal of the Natural History Society of Montreal. Overseas, his work appeared in the publications of the Royal Society, British Association for the Advancement of Science, Geological Society of London, and in the *Philosophical Magazine* in Britain, besides which nine papers appeared in French journals and one in a German publication.[21] Billings also was a frequent contributor to European, American, and Canadian journals, but Logan, Richardson, Bell, and Murray made fewer appearances in print.

Logan encouraged the preparation, presentation, and publication of these articles because they brought the Survey personnel and their work to the notice of the scholarly world, and afforded an opportunity to present points of view, conclusions, and arguments, as well as put on record facts of scientific importance that perhaps were not "practical" enough to warrant extensive treatment in reports which were scrutinized by politicians and businessmen. Logan, Hunt, and Billings were honoured members of Canadian, American, and European scientific societies and were the mainstays of the Natural History Society of Montreal, which relied mainly on the Geological Survey and McGill University to furnish the nucleus of its speakers, officers, and members. For example, eight of the thirty-five papers to the society in 1861-62 were by Survey personnel—one apiece by Logan and Bell, two by Billings, and four by Hunt. Hunt and Billings also were members of the seven-man board that edited the *Canadian Naturalist.*

Among the strongest, most effective supporters in Canada of the Survey and its work were the two scientific societies in Montreal and Toronto, and the province's universities. Faced with similar problems of trying to secure support from an indifferent public and government, and of sharing a rather limited amount of special talent available to do scientific work in Canada, the groups were allied in sustaining one another's efforts. The Survey's publications received favourable and sympathetic reviews in the scientific journals, while university personnel were among the strongest supporters Logan could muster when it came time to do battle with the reluctant legislature. The Survey and the universities looked to one another for employment opportunities and to make the most effective use of the available talent. Principal Dawson of McGill checked fossils, and Robert Bell when he was at Queen's did field surveys, while Hunt, as we have seen, taught at Laval and McGill, and Billings prepared specimens for McGill. The Survey also furnished university museums and laboratories with mineral and fossil specimens.

The Survey owed a very great deal to university staff members for their independent geological and mineralogical work and writing, for they promoted the cause of Canadian geology and added considerably to the geological knowledge the Survey was accumulating. The *Geology of Canada* was not merely being polite when it listed Principal Dawson at McGill, Professors E. J. Chapman and Henry Croft at the University of Toronto and Professor Henry Y. Hind at Trinity College, Toronto, as important contributors to the work. Logan supported Hind's application to serve as geologist on the expedition the Canadian government got up hastily in 1857 to examine the country between Lake Superior and Red River, and even invited Hind to work for him during the summer if the Red River project fell through. Hind went to Rupert's Land twice, in 1857 and 1858, and then in 1861 made an interesting exploration on the Labrador plateau by way of Moisie River. The geological findings of these expeditions were of some value to the Survey, and this was only proper, for Hind made considerable use of the Survey's work, notably Richardson's surveys of Anticosti and the North Shore, and copied profusely from the latter's unpublished reports.

IV-14 J. W. DAWSON, 1863
Principal, McGill University, 1855-93

How much he needed the support of the academic and scientific community, as well as any other friends, was brought home to Logan in 1859. The government, apparently with little warning or explanation, simply cut the grant to the Survey in half, to $10,000. As Logan wrote to his brother Edmond on April 18, 1859:

> I am sorry to say that the Government have found themselves under the necessity of reducing the grant to the Geological Survey to one-half the usual amount, making it £ 2,500 instead of £ 5,000. The sum will be no more than sufficient to pay the salaries of the staff and keep up the museum. It will allow nothing for exploration; so that we have been instructed to stay at home and work out a condensation of all our reports, and thus show the present condition of Canadian geology.[22]

The principal economy effected by Logan was to forego publication of a report for 1859, and instead, concentrate on preparing the *Geology of Canada* and geological map, with the field work being directed to that end. The reduction in funds also made it less likely for Logan to employ outside staff, like university personnel, though he did not incline very strongly towards this in any event, since he preferred persons whose work he could supervise in the analytical and reporting stages after the summer months had passed.

The work on the *Geology of Canada* was interrupted for Logan by another call, on September 26, 1861, to serve as chairman of a four-man commission in charge of Canadian participation at the International Exhibition to be held in London during the summer of 1862. Dr. J. C. Taché, interestingly, was one of the division heads in charge of the exhibits derived from the forests and waters of Canada East. The notice was short, and the preparations had to be rushed; in fact, the Canadians barely took up their space in time to prevent its being reassigned. The whole affair was of a piece, plagued by a series of squabbles. Logan protested insufficient space had been allotted to the Canadian displays, and there were peevish arguments about a timber display in the Canadian exhibit that towered so high as to overshadow an adjoining display of stained glass. Logan's touchiness seemed to reflect his failing health and increasing years. Even the disposal of the collection after the exhibition created a good deal of confusion and bad feeling, in which Logan fortunately was not involved.

* * *

THE PLAN TO PREPARE A GRAND SUMMATION of what was known about the geology of Canada had arisen from a suggestion by the select committee of 1854 that the Survey's early reports should be reprinted in view of their scarcity and the valuable information they contained. An unusually amenable legislature had appropriated $8,000 (£ 2,000) in late 1854 for the publication of a geological map and condensed report, so work could begin at once. Of the two, the projected map was to take much longer and was not to be finished and published until 1866, three years after the completion of the comprehensive account. Yet it was the first of the two to be started. A preliminary map was prepared in 1855 for display at the Universal Exposition in Paris and for publication on a greatly-reduced scale in the *Esquisse Géologique du Canada.* Although praises were lavished on it, Logan was painfully aware of the map's shortcomings, particularly on the topographical side. He discovered discrepancies in the base maps of parts of Canada, and very large errors when it came to correlating the maps of the contiguous colonies, states, and territories. For example, on two supposedly authoritative maps there was a discrepancy in the position of Lake Michigan that amounted to almost a whole degree (55′, or over 40 miles) in the longitude of Chicago. To clear up these difficulties, Logan proposed to establish a few key points astronomically by means of the telegraph, a power thoughtfully included in the Survey act of 1856. Lieutenant E. D. Ashe, R.N., of the Quebec Observatory, in 1857 determined the positions of Quebec, Montreal, Ottawa, Kingston, Toronto, Collingwood, Windsor, and Chicago, the last with the co-operation of the United States authorities. By the end of 1857 Logan reported that he now had a satisfactory topographical map on which to place the geology, and that the map was almost ready for the engraver. However, there was no urgency to push the map ahead of the account it was to accompany, and besides, the longer it was held back, the more up-to-date the topographical and geological information that could be incorporated in it.

Furthermore, printing the map would be expensive, and the Survey once more was feeling a financial pinch, the result of reduced and uncertain government grants. After 1857 when the government failed to balance the budget every effort was made to reduce expenditures. The Survey's grant being one item under its control, Logan experienced rough handling even from his political friends, the Macdonald-Cartier government. The 50 per cent cut from the grant in 1859 was restored in 1860. However, the Survey act's time for renewal, in May 1861, once again coincided with another period of internal political trouble compounded by the outbreak of the American Civil War and the still-unbalanced government budget. Instead of renewing the act, or even continuing the $20,000 a year grant, the government made the Survey a grant for 1861 of $9,000 additional money, and left Logan to pay for its operation from the $8,000 reserved for the special report and maps, plus a $3,000 reserve Logan had saved from past appropriations. In the following year the government again voted $20,000 for the Survey, but did not replace the publication money, so the situation was not improved materially.

Then, on May 26, 1862, with the *Geology of Canada* nearing completion and Logan in Europe at the International Exhibition, the government changed hands, and an economy-minded Reformer coalition attained office under John Sandfield Macdonald and Louis V. Sicotte. A year later on May 15, 1863, Parliament was dissolved without supply being voted for the year. The government won a narrow election victory during the summer, and discussion of the financial position after six successive deficits was resumed in the autumn. Logan's position had grown intolerable. By the end of 1863 he had to report that the Survey's means were exhausted. He himself advanced the salaries of the staff during 1863 in view of the non-passage of the appropriation when Parliament had been dissolved in May, and paid about $3,000 besides for the cost of the type for the *Geology of Canada,* so he was now the Survey's creditor to the tune of some $10,000 in all. Ahead he saw another $1,200 to be paid for printing the *Geology,* $1,500 for the engraver's bill on Professor Hall's decade on Canadian fossils, and the still larger sum that would be required to engrave and print the geological map of Canada—not to speak of the continuing costs of operating the Survey and museum. Indeed, in this letter of January 9, 1864, he even suggested to the current Minister of Finance Luther H. Holton, that unless the Survey were put on a more satisfactory financial footing and saved from having to make an annual struggle for its money, it should be reduced to a caretaker status like that of New York, with only the museum and paleontologist kept in service to maintain the investment already made, while Logan continued exploration as a private citizen from his own means, assisted by "two of the less expensive explorers" paid by the government.[23]

Such was the difficult background against which the *Geology of Canada* was prepared and published. The bulk of the work appears to have been done during 1861 and 1862 to judge from the text, though Logan and the others, of course, had been collecting data and checking specific points in connection with the book for years. The work aimed at incorporating all information secured by the Survey down to the end of 1862, and Logan's preface was dated May 1863. It would seem that chapters 1 to 16—on the geology of Canada treated by separate chapters on the major systems, series, formations, and groups recognized in Canada (i.e., from Laurentian to Lower Carboniferous in the Bonaventure formation of Gaspé)—were prepared by Logan in 1861 before he departed for the exhibition. At any rate, these chapters included findings down to the end of 1861, while additions and revisions resulting from the work of 1862 were presented in the supplementary chapter 22. The five chapters, 17 to 21, dealing with petrology and mineralogy—the work of Hunt and amounting to almost half the book—were prepared to the end of 1862 in the original text. So presumably, was the final half of the supplementary chapter, devoted to "the Drift," that is, surficial geology.

As published by Dawson Brothers of Montreal, the *Report of Progress from Its Commencement to 1863; Illustrated by 498 Wood Cuts in the Text and Accompanied by an Atlas of Maps and Sections* (short title, *Geology of Canada*), was a massive tome of 983 pages. It was more than a reprint of the thirteen separate reports issued to that time, though it incorporated all, or most, of the geological material of those reports and in places quoted extensively, sometimes for several pages consecutively, from those reports, or from articles published in scientific journals by Survey personnel. All the contemporary and background material in the reports—descriptions of routes traversed, conditions and occurrences in the field, the order and progression in which the work was done, the evolution of concepts and conclusions—was excised, apart from chapter 1 and Logan's preface. These presented only a brief outline of the geography of Canada, and a short recapitulation of the work done by various persons, in the form of acknowledgments and tributes. Instead, the *Geology of Canada* offered the reader the purely geological, petrological, and mineralogical findings of the Survey (and of some 50 other persons, where they were relevant), set out as never before, on a province-wide basis, gathering up all the pertinent information on the Laurentian System, or Quebec Group, or Gaspé Series, in a single chapter that described typical successions, formations, associated minerals and the forms the associations took, fossils, peculiar, unusual or noteworthy rock occurrences—and indicated the geographical distribution of the groups and their various formations. Similarly, in the chapter on economic geology, each mineral was discussed separately in an article that gave distribution, occurrences, natural forms, mining and metallurgy, prices and markets. The result was a competent, thorough, inventory or catalogue of all that was known about Canadian geology, and a most valuable compendium for university teachers and students, mining men and engineers, and amateur geologists.

The *Geology of Canada* was more than a simple catalogue and review of what had been published since 1844, or of the work since 1858. True, many of the latest ideas and interpretations that had come to the fore as a result of the work and thought of recent years had been incorporated—Logan's view of the Quebec Group or the supposed Laurentian fossils, or Hunt's anticlinal theory of petroleum accumulation and his theories of metamorphism and eruptive rocks, for example. Through collecting and preparing the material in topical form, the Survey was moving from description to a new stage of generalizing and theorizing about Canada's geological formations and their character. By discussing a system such as the Laurentian Series in brief space, or describing what "usually," "sometimes," "occasionally," or "in general" was the case, a considerable step was being taken towards generalizing. Bringing together scattered instances of the same or related phenomena revealed their common elements, compelled guesses as to structural or petrological trends, or pointed up important differences that required explaining.

The *Geology of Canada* was received enthusiastically. The Toronto *Leader* of May 6, 1864, compared Logan's sense of accomplishment to that of Gibbon when he had completed his great history of the Roman Empire, and expressed the hope that this *magnum opus* would by no means be Logan's last. Principal Dawson, writing in the *Canadian Naturalist,* gave a characteristically weighty, censorious verdict:

> The value of this work to Canada can scarcely be over-estimated. It must be regarded as of vast importance, whether we consider readers abroad or at home, whether we consider scientific objects purely or those which are practical. Its mechanical execution is an evidence of the progress of the arts among us. Its publication to the world is proof of the interest taken in science in this country, and of the enlightened patronage afforded by the Government to such investigations, and at the same time, of the immense value of our mineral resources, as well as of the extent to which they have already been made available. It gives for the first time to geologists abroad the means of making themselves thoroughly acquainted with the geology of this country; and it thus places Canada on a level with those older countries whose structure has been explored, and the knowledge of it made the common property of the world. In some departments of geology, it even makes Canadian rock-formations rank as types to which those of other countries will be referred. This is especially the case with regard to those oldest of known rocks, the Laurentian series, whose intricacies have for the first time been unravelled by the Canadian survey, their mineral character explained, and the earliest known traces of animal life obtained from them; so that the term *Laurentian* is applied as the general designation for the most ancient formations of Europe as well as of America. To the people of Canada the publication of this Report must mark an era both in science and practical mining. Any one desirous of studying geology, has here to aid him a detailed account of the structure of his own country; an advantage not hitherto enjoyed by our self-taught geologists, and one which in a reading country like this, must bear good fruit. The practical man has all that is known of what our country produces in every description of mineral wealth; and has thus a reliable guide to mining enterprise, and a protection against imposture. Even in the case of new discoveries of useful

> minerals which may be made, or may be claimed to be made, after the publication of this Report, it gives the means of testing their probable nature and value, as compared with those previously known.
>
> No one, in short, need henceforth have any excuse for professing ignorance of the labors of the Geological Survey, or for representing it as a useless expenditure of the public money. Persons not interested in science or in practical mining might heretofore have been excused for not having read the annual reports of progress, with their dry details and want of suitable illustrations; but after the publication of this attractive volume, such want of knowledge can no longer be tolerated; and it is to be hoped that no public speaker or writer will venture so to proclaim his own ignorance as to pretend that Canadian Geology is one of those little matters which have, in the midst of more important affairs, escaped his attention, or to underrate the labors of those who have devoted themselves to this great work.[24]

With the *Geology of Canada* Logan could be said to have completed the original commission with which he had been charged—to make the geological survey of Canada. On the material side the reports and the *Geology of Canada* more than fulfilled any reasonable expectations that its original sponsors could have had. The Survey had disproved the occurrence of coal in Canada and explained where to look for petroleum; it had shown the types of formations in which metallic minerals occur. It had devoted a great deal of attention to the locations and uses of many deposits of minerals, whether they occurred in veins, beds, or alluvial gravels. The reports had discussed the uses for particular kinds of stone, gypsum, and marl beds, and had pointed out the locations of exploitable deposits. They had related soil qualities to the rocks from which the soils were derived, investigated peat and manures, analyzed mineral springs, studied and reported on mining methods, metallurgies, and potential markets for various minerals.

Fulfilling the task begun in 1843 had taken not two years, but twenty, and had left its mark on the health of Logan and Murray. It had necessitated the creation of a full-fledged organization, with specialists in charge of field surveys and mapping, chemical and mineralogical studies, paleontology, museum, as well as such craftsmen as the draftsman, artist, and museum assistant. In those twenty years the Province of Canada had become one of the better known parts of the world in a geological sense thanks to the efforts of the Survey; and many of the facts learned and theories developed were important contributions towards the advancement of the science. During the period the three leading figures of the Survey—Logan, Hunt, and Billings—had gained personal honours and international recognition as scientists of world stature, and brought great credit to Canada as a modern, progressive land with a great future. As Hon. G. W. Allan had said in 1859, "It may safely be asserted that the geological survey has done more for the reputation of Canada among intelligent and scientific men abroad and in England, than anything else connected with the country."[25] The appearance of the volume elicited the following tribute to the Survey and to Logan from the *Saturday Review* of Britain:

> The style in which this work has been got up, the precision of the drawing, and the accuracy of the woodcuts, may almost challenge comparison with the execution of similar productions on this side of the Atlantic. There has been a steady persistence in the conduct of this remarkable Survey, honourable alike to the successive Governments that have encouraged it, and to the officers who have carried the work. No other Colonial Survey has ever yet assumed the same truly national character; and the day may come—if ever the "Imperial Colony" shall claim and obtain independence—when the scientific public of a great nation, looking back upon the earlier dawnings of science in their land, shall regard the name of Logan, native born, with the same affectionate interest with which English geologists now regard the names of our great geological map-makers, William Smith and De la Beche.[26]

V-1 SIR WILLIAM LOGAN IN LATER LIFE

5

From Provincial to Dominion Survey

WITH THE COMPLETION OF the *Geology of Canada,* the Survey seemed to cross a watershed and enter upon a period of marking time. Logan was frequently absent on business, and his guiding hand was missing for months at a time. Perhaps the organization had already grown to the point where it needed a head who could devote his main energies to administration and supervision, forswearing field work or research. Then, again, a turning point had arrived as regards the staff. Most of the stalwart pioneers—Murray, Billings, Richardson, Robert Barlow, and especially Logan himself who was over 65—were growing old and past their peak of effectiveness. With the publication of the 1863 report, the Survey appeared to have virtually achieved its primary goal of exploring and mapping the province, and locating and identifying its characteristic rock formations. Now it could turn more fully in other directions, towards greater emphases on mining and economic geology, and special studies of structural and petrological problems.

Then along came Confederation, to inaugurate a new phase in the history of the Survey. Two million square miles of territory was joined to Canada, increasing its area tenfold, and giving Logan's Geological Survey of Canada an entire subcontinent in which to operate. Logan had long dreamed of some day being called to commence the investigations in the western reaches of Rupert's Land, and even before the work was properly commenced in Canada he had worked in the Maritimes and maintained his ties there ever since. Expansion into the almost unknown, unmapped regions to the west and north confirmed the Survey in its earlier role of exploring, mapping, elucidating, and recording the geology, this time for a vastly larger region of some four per cent of the land surface of the earth.

Coincidentally, with the expanded area of responsibility went considerable reorganization. A larger, younger, more highly specialized and academically-trained staff had to be secured, headed by a new director who would replace the hitherto virtually irreplaceable Logan. Administrative reorganization also was required, including a new relationship with the Dominion government and civil service, and a modified internal structure to permit the Survey to discharge its added responsibilities as effectively as possible. By 1881, when these major changes had been completed, the Survey was a far different institution than the one that had examined the St. Lawrence valley from Newfoundland to Thunder Bay, from Lake Champlain to Michipicoten, under Logan's inspired direction and sealed its work with the *Geology of Canada.*

That great task completed, Logan followed the publication and favourable reception accorded to the *Geology of Canada* by renewing his campaign to secure the revival of the five-year financial clause of the act of 1856 that had expired in 1861. Since then, he had been compelled to go cap-in-hand each year for the funds with which to carry on, and to use his own money to overcome delays in the grants. It says much for Logan's persistence and his great prestige that despite the kaleidoscopic political changes of these years, he continued to secure from successive governments approximately $20,000 each year for the Survey.

Still, the uncertainties interfered with Logan's ability to operate on the scale or along the lines he desired, or to secure and retain valuable assistants. In 1864 Murray was assigned to duty in Newfoundland, and began to draw his pay from the government of Newfoundland, saving the Survey his salary of $1,600. In the end, however, Murray remained with the Newfoundland survey, made a new career for himself there, and was permanently lost to the Geological Survey of Canada. Robert Bell, too, took up an academic post at Queen's University in November 1863, which made him available only in summers and then only when he was not improving his academic qualifications, as he was in 1864 at the University of Edinburgh.

Logan began his dealings with Luther H. Holton, the Montreal businessman and well-disposed friend of the Survey, who held the portfolio of Minister of Finance in the Sandfield Macdonald-A. A. Dorion Liberal administration, formed in May 1863. To Holton, Logan addressed a lengthy appeal for a larger grant to help cover the outlays on the publications and to reimburse him for the money he had advanced to keep the Survey going through 1863. Even more important, in Logan's opinion, was the passage of legislation to give the Survey an assured income for a term of years. As he informed Holton,

> The professional responsibilities and difficulties of conducting a geological Survey over so large an area as Canada presents are quite sufficient, without being complicated with those of finance. But the minds of those charged with the investigation can never be free from pecuniary anxieties while the support given to the Survey lasts only from year to year, and has to be struggled for every session of Parliament. No good plan of investigation can be formed under such an arrangement, and the officers and explorers on whom I depend, uncertain of continued employment, may be tempted away from me. The peculiar training which they require would make it very difficult to replace them, and though, as you will understand from a reference to the preface of the late Report, some of them are not highly paid men, they are all from their practice very valuable.
>
> Whatever the grant may be, it should be secured for a term of years by an act . . . .[1]

The Parliament of the united Province of Canada was approaching chaos. The general election of the summer of 1863 had produced two almost evenly-divided rival blocs of Liberals and Conservatives. The Sandfield Macdonald-A. A. Dorion government struggled on until March 21, 1864, then resigned, to be followed by a Liberal-Conservative administration headed by the elderly Colonel E. P. Taché but dominated by John A. Macdonald, George E. Cartier, and Alexander T. Galt—all friendly to the Survey. This, too, was a very shaky government, kept alive mainly by the continuing discussions of a legislative committee that included the main leaders of both sides and was presided over by the Liberal George Brown, which was studying the reform of the Canadian constitution. Through the spring, while the committee held its sittings, the Taché-Macdonald government staggered on. On June 14 the committee brought forward its report, in which the majority favoured a change of constitution to a federal union embracing either the two sections of Canada, or all of British North America; and that same day the government was defeated in the House on a vote of non-confidence on a different matter. During the next four days (June 15-18) the government resigned and inter-party negotiations brought about the formation of a new administration embracing members of both blocs to deal with the problem of constitutional reform. Thus was formed the Great Coalition that was to be the active proponent in effecting Confederation in the next three years.

Despite its weakness, the Taché-Macdonald government of March-June 1864, took up the project of a new Survey act, and the legislation was put in the hands of the Minister of Finance, Alexander T. Galt. Galt introduced the resolution that the House go into committee to consider a new Survey act on May 6, and the resolution was before committee of the whole during the sitting of June 3. Considerable opposition was expressed against the proposal to make the grant for five years from January 1, 1864, and the resolution carried only after division. When it came before the House on Wednesday, June 8, it faced a formal amendment by the leader of the previous government, John Sandfield Macdonald, to change the appropriation to a sum to be voted annually by Parliament. The amendment was defeated by 42 to 71 on a

V-2 L. H. HOLTON
A Montreal businessman and a founder of the Grand Trunk Railway of Canada, Holton was a firm supporter of the Survey in a party whose leadership was generally indifferent or hostile to it.

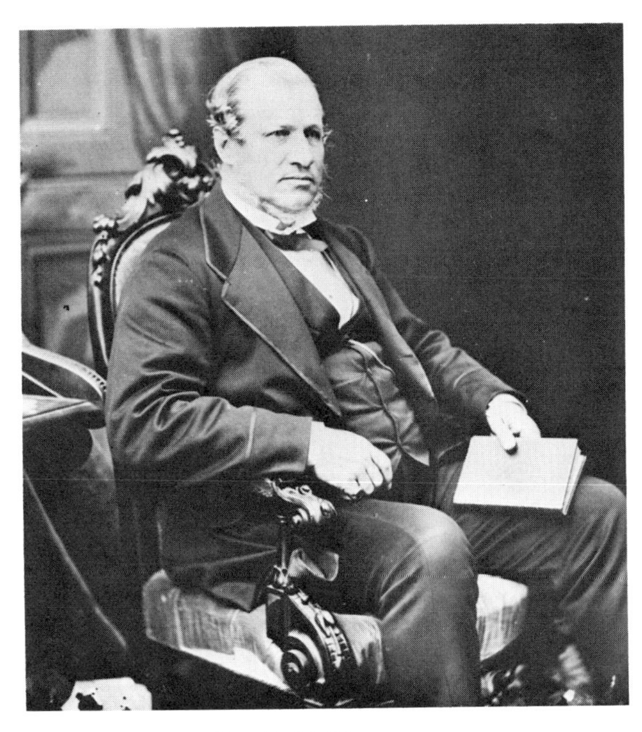

V-3 SIR ALEXANDER T. GALT
Minister of Finance, 1858-62, 1864-68

recorded vote, a very good majority considering the near-deadlocked situation in the House; and the resolution carried on division. Galt proceeded with his bill, which passed its third reading on June 25, after the formation of the new coalition government, and it was given assent on June 30, to become Chapter 8, 27-8 Victoria, of the Statutes of Canada.

The debate on the resolution on June 8, 1864, is well worth examining, since it revolved around the identical issue that today—over a century later—once more is under serious consideration: the desirability or otherwise of term or cyclical budgeting, particularly for the research and scientific agencies of government. The debate, which lasted for two hours, was reported by the Toronto *Globe* of June 9, 1864, as follows:

> Mr. GALT moved concurrence in the resolution relating to the geological survey of the Province.
>
> Mr. J. S. MACDONALD said that he was not opposed to the grant for the geological survey, but he wished it voted annually. He had not had time to devote to geological studies, but he understood from those who had knowledge of those things that there were complaints that the geological researches and reports did not afford that satisfaction which people expected from the appropriations made. Last year the appropriation was reduced to $15,000, but this Government had increased it again to $20,000 and he objected to a Bill which would vote *en bloc* for five years $100,000. He thought that what was necessary should be voted annually, and that there should be yearly reports of the progress made.
>
> Mr. HOLTON thought that there were advantages in having provision made for the survey for a certain term of years. He considered, however, that a great deficit in the present system was that the survey was not attached to some government department, the Crown Lands for instance. He hoped that the Minister of Finance would embody this idea in his Bill. He believed that the survey had been of immense value to the country. It had contributed to the development of our mineral resources, and he thought it also something to be proud of, that in a young country like this, and by a native of this country, so much had been added to the body of scientific knowledge.
>
> Mr. GALT thought Mr. Holton's suggestion about the survey being attached to some department a valuable one. He was glad also to hear him speak in such terms of the importance of the survey. If one thing more than another had tended to raise the character of Canada with respect to its resources and in a scientific point of view, it was its geological survey, and at this moment more than ever it was our duty to support this survey. At this time, through discoveries of gold, copper and other minerals, this subject was becoming more than one of ordinary importance, and he thought that the staff employed in the survey should be placed in so efficient a footing that the best possible information regarding our resources would be communicated. He would be happy to see the Lower Provinces acting with Canada in a joint geological survey. As regarded making provision for five years, so large a work must be undertaken according to a general plan—the isolated portions of which, unless completed, would be comparatively valueless.
>
> Mr. BROWN—Why not on the same principle vote the supplies for the militia for five years?
>
> Mr. GALT said that the two cases were entirely different. The services of competent scientific men would not be obtained unless there was some guarantee of their being required for some time. The permanence of the survey, or at least its continuance for a certain fixed period, was necessary to its efficiency.
>
> Mr. A. MACKENZIE said he was sure that the House was willing to treat Sir Wm. Logan and his associates with all reasonable liberality, but it was a principle to which the British people had adhered very closely, that they would only vote public money from year to year. He read, as an illustration, in the reply by the Canadian Government in 1863 to the Duke of Newcastle, when he asked a vote for the militia for five years.
>
> Mr. BROWN heartily endorsed all the eulogies passed upon Sir William Logan, and all that had been said with regard to the value of the survey, and would be prepared to vote any reasonable sum, even should it exceed $20,000, that was shewn to be necessary in order to carry on the survey with efficiency. At the same time he failed to see that any sufficient argument had been adduced in support of the resolution. The statute passed in 1856 set apart $20,000 a year for five years, and since the expiry of that period the money had been voted from year to year without any question. The Act itself, so far as regarding the permanency of the survey, had not expired, but merely the provision which voted the money for five years in advance. He agreed with the members for Cornwall and Lambton, that it was a most salutary practice to have the public expenditures voted from year to year. It would be better, also, for the good working of the system, that Parliament should be at liberty to vote each year what the varying exigencies of the service might require. Circumstances might arise under which Sir William Logan might find more than $20,000 necessary. It might be found that our Chaudiere gold fields were really valuable, and geological explorations in that district might become necessary, which a $20,000 grant would not admit of. Or suppose we annexed the western territory, the $20,000 would not be sufficient. At all events, the House should not depart from the principle that it was bound to exercise, from year to year, a control over the whole expenditure of our public departments.

Mr. CAMERON said that there had been no difficulty while the Act sought to be revived was in existence, but since the expiry of the Act difficulties about continuing the survey had been raised every time the vote came up.

Mr. McDOUGALL spoke in support of the idea which had been thrown out, that the survey should be attached to some department, and here his testimony to the great value of the survey generally, and particularly of the maps prepared under its auspices, a matter which had come especially in his cognizance while in the Crown Lands Department.

Mr. DUNKIN objected to the survey being attached to a Government Department, as the result would be that the popular pressure on the departments in the direction of economy might lead to the survey being discontinued or curtailed in its efficiency.

Messrs. SHANLY, McGEE and MORRIS spoke in support of Mr. Galt's proposition.

Dr. PARKER opposed it. He held that future Parliaments should be competent to decide, from year to year, what was required for the support of the survey, and that this Parliament had no right to bind its successors for five years to come.

After the recess. MR. BROWN on a matter of privilege . . .

The act, as passed, reflected some of the points raised during the discussions of June 8. Short and direct, it renewed the act of 1856 in its entirety and made the $20,000 a year appropriation payable for a term of five years beginning January 1, 1864, and "until the end of the then next Session of the Provincial Parliament" in 1869. The act settled the matter of the status of the Survey in the absence of financial provision for its operation, by declaring its existence to be permanent, and subject only to suspension by express act of Parliament. This was done by declaring that "all the provisions of the said Act [of 1856], shall apply to the said appropriation as hereby continued; the said Act, with the exception of the provision limiting the duration of the said appropriation, having been and being hereby declared to be permanent." Thus the Survey finally lost its temporary, *ad hoc*, character, and became an institution rather than a process—"a Survey" rather than "a survey." The act also, for the first time, met the suggestions of attaching the Survey to some regular department of government by making provision for this to be done, while at the same time preserving the Survey's special status and separate position in any such change, as well as its cherished autonomy as regards appointments:

II. The Governor in Council may attach the said Geological Survey to any Department of the Civil Service, which he may think most expedient, as a Branch of such Department, of which the Director of the said Survey, and his Assistants shall thereupon become Officers while employed on the said Survey; but such Assistants shall continue to be nominated by the said Director subject to the approval of the Governor.

V-4 JOHN SANDFIELD MACDONALD
Government Leader, Province of Canada, 1862-64; Premier of Ontario, 1867-71

This done, Logan threw himself into the various enterprises associated with the operations of the Survey, among them two vexing matters that took him to Europe no less than three times, for 15 months in all, during the next three years. These were the completion of the Atlas that was to accompany the *Geology of Canada* and other coloured maps, also the presentation and defence of the much-disputed Laurentian fossils discovered by the Survey. This last was destined to give rise to an extremely awkward scientific controversy, though fortunately the Survey had only a limited role in the protracted dispute. The paternity of the Survey was acknowledged by Principal Dawson in his presidential address to the annual meeting of the Natural History Society of Montreal on May 18, 1864, when he proclaimed that "The discovery of this remarkable fossil, to be known as the *Eozoön Canadense*, will be one of the brightest gems in the scientific crown of the Geological Survey of Canada."[2] However, the Survey was fortunate in being able to turn the matter over to Dawson and the noted British authority, W. B. Carpenter, on whom fell the burden of defending *Eozoön* for years to come—in Dawson's case, till his death in 1899.

It will be recalled that Logan, in 1858, had first taken up the idea that evidences of organic life were present in the oldest Laurentian rocks of Canada, and had publicly announced his discovery before the American Association for the Advancement of Science meeting in Springfield in the following year. He had taken specimens with him to England but had not encountered too great success, the verdict generally being one of "not proven."

In the *Geology of Canada* Logan contented himself with describing the "supposed fossils" or "certain forms strongly resembling fossils," in his chapter on the Laurentian System.[3] In the autumn of 1863, however, Logan detected new specimens of the supposed fossil among certain rock samples returned from the Grenville district by James Lowe, which appeared clear enough to permit positive identification. Billings suggested (wisely it proved) that they be referred to Principal Dawson, "our most practised observer of natural-history objects with the microscope"–and a noted controversialist.[4] Examining the alternate layers of serpentine and calcite, Dawson believed he detected an organic shape, concluding that the layers of serpentine represented the interspaces of the fossil, not the skeleton. He announced that the evidence pointed to the former presence of a coral-like creature; which Logan, in turn, took up in his *Report of Progress for 1863-66:*

> This according to him belongs to a Foraminifer growing in large sessile patches after the manner of *Polytrena* and *Carpenteria,* but of much larger dimensions, and presenting at the same time in its minute structure a close resemblance to other foraminiferal forms, such as *Calcarina* and *Nummulina.* To the fossil Dr. Dawson has given the name of *Eozoon Canadense.*[5]

Hunt studied its mineralogical character and prepared an explanation of the supposed process of chemical replacement. When Logan went to England in the spring of 1864 he took specimens of the new finds which he referred to Professor Rupert Jones, whose support helped convince Sir Charles Lyell. Lyell then singled out the find in his presidential address to the British Association for the Advancement of Science in September 1864 as one of the greatest geological discoveries of his time.[6] When Professor Carpenter returned to London, he too was persuaded by these specimens, and a group of papers was prepared, together with plates illustrating the fossil, for the *Journal* of the Geological Society of London. The articles on *Eozoon* by Logan, Dawson, Carpenter, and Hunt appeared in its February 1865 number and were subsequently reprinted in the *Canadian Naturalist.* Dr. Carpenter presented still another paper on the subject before the British Association meeting the following year in Birmingham.

By now, however, *Eozoon* was coming under attack from mineralogists. Two Irish scholars, Professors William King and T. H. Rowney, in a paper "On the So-called

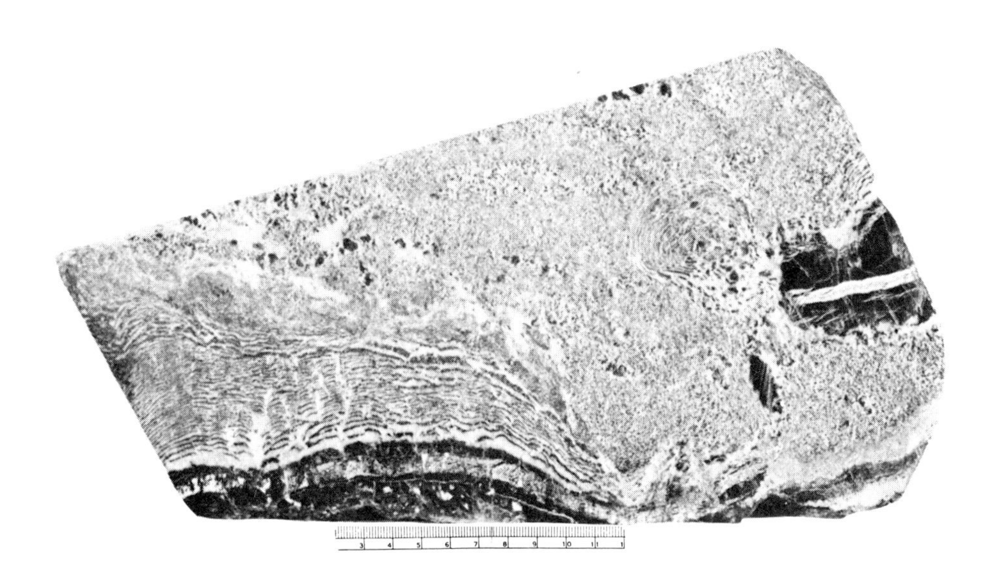

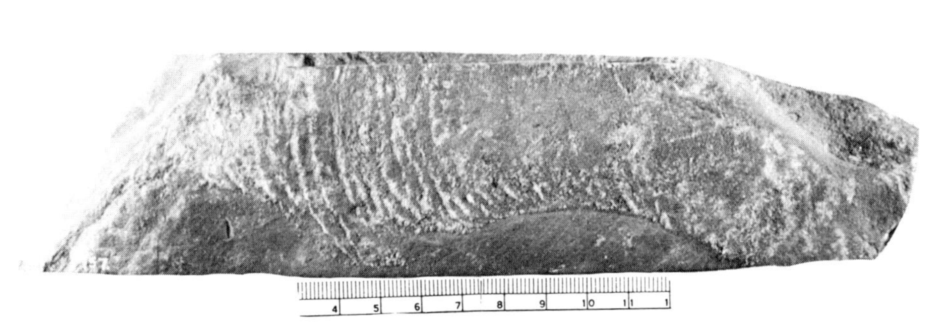

V-5 TWO SPECIMENS OF *EOZOON CANADENSE*

Upper–The Côte-St-Pierre specimen collected by Lowe near Grenville, Quebec, in or after 1863, which Dawson studied under the microscope, concluded was of organic origin, and named *Eozoon Canadense,* "the dawn animal of Canada." *Lower*–The Tudor specimen collected by Vennor in 1866 near Millbridge, Ontario, which Logan, Dawson, and Carpenter described in 1867.

'Eozoonal Rock'," argued that the formation was of mineral origin, an example of chemical metamorphism. They pointed to the fact that it was always found in metamorphosed rocks, that the structure or form bore analogies to other types of crystalline formations, and concluded that *Eozoon* was "solely and purely of crystalline formation."[7] The eozoonists soon found allies among other paleontologists in Europe, while other mineralogists raced to join the onslaught on *Eozoon*.

In the summer of 1866, more samples were discovered by H. G. Vennor during his examination of the Hastings district in the Township of Tudor, which, together with others discovered around Madoc by Dr. Dawson, were paraded by the eozoonists as cutting the ground from under Professors King and Rowney. For in these samples, which came from comparatively unaltered limestone, the forms were etched entirely in limestone. Logan announced this newest find in a paper he read before the Geological Society of London in May 8, 1867 "On New Specimens of Eozoön," which was published in that society's quarterly journal for August 1867. It was followed there by an article by Dawson and notes by Carpenter responding to his critics. In his article, Dawson interpreted the peculiar characteristics of the Tudor specimen as those of "a young individual, broken from its attachment, and embedded in a sandy calcareous mud."[8] Fortunately, the Survey's involvement appears to have ended with Logan's second paper on *Eozoon*.

For the controversy grew increasingly unpleasant, involving personal abuse, and was kept alive by exchanges of papers in the learned quarterly journals and by animated correspondence in the monthly publications. Mainly it was conducted between micro-paleontologists and mineralogists, neither of whom paid any attention to the arguments or evidence adduced by the other side. Dawson in 1875 published his writings on the subject as a book, *The Dawn of Life on Earth* (or *Life's Dawn on Earth* in a British edition). In 1878, the German zoologist Karl Möbius, a specialist in Foraminifera, made a thorough examination of all the available specimens from the paleontological standpoint and concluded they failed to show the regularity in shape and outline that a true fossil foraminifer would show. He concluded, therefore, that the *Eozoon* must be of mineralogical or mechanical origin. Möbius' opinion seemed to convince the vast majority of scientists on the matter, though Dawson never abandoned his belief.

The prolonged, bitter dispute reflected the excessive specialization of scientists, the unwillingness of some to examine a subject from all sides, and the pioneer state of the science of micro-petrology; for microscopic studies were still only in their infancy. Furthermore, *Eozoon* was a product of the intellectual environment of the time. The new doctrines of evolution were making their way into the public mind, and the evolutionists desperately desired to push back the origin of life to the earliest possible time. Ironically, although Dawson was not one of the evolutionists, these gladly seized upon *Eozoon* as important evidence in their cause, notably Darwin in the fourth edition of his *Origin of Species*. For Dawson and the other defenders of *Eozoon* were correct in their contention that the origin of life on earth lay somewhere in Precambrian time, and that other discoveries eventually would corroborate their claim. Modern geologists, including scientists of the Geological Survey of Canada, have indeed discovered life-forms in Precambrian rocks, but *Eozoön Canadense* is not among them.

* * *

LOGAN'S TWO VISITS TO ENGLAND in connection with *Eozoon*, as well as a third one in 1868, were devoted mainly to the preparation of coloured maps for the Survey, since these could not yet be lithographed in Canada to Logan's exacting standards. Most of the maps and sections were drawn by the Barlows by 1863, and most of these were engraved in Montreal. For the printing and lithographing, however, Logan selected the firm of E. Stanford of London, and he spent most of the remaining months of 1864 after his arrival in August preparing the maps for publication. First he took the geological map of Canada to Mr. Bone, the colourer of the British Survey maps, in order to have the same colours applied to his map as were used in the British maps for the corresponding geological periods and groups. During November and December he was hard pressed transferring twenty-nine different colours to base maps for printing the 1 inch=125 miles map. When he received the first trial copy of the general map in March 1865 he found the colours satisfactory, but the registration was defective, and he anticipated many trials would be required before the map would be ready.

The general map, and the others as well, were finished by the end of 1865, with the Atlas appearing some time in 1866. The completed Atlas contained six coloured maps: the Grenville district; northwest of Lake Huron; east of Lake Champlain; the Quebec Group; surficial deposits from Gaspé to Lake Superior; and especially the beautiful 18 x 8¼-inch "Geological Map of Canada." In addition, it presented fifteen geological cross-sections associated with these and other parts of the geology of Canada. Incidentally, the foreword to the Atlas announced the inauguration of a program to publish geological maps of the Province of Canada on a scale of 1 inch = 4 miles—a program that is still in progress. That of the Eastern Townships region was prepared and partly engraved, while a second was in preparation for the area from Montreal west to about Bowmanville. The map of the Eastern Townships was to have a tragic sequel.

Logan returned to London early in 1867 and again in 1868 in connection with other maps. The first was a geological map to be exhibited at the Canadian exhibit in Paris, on which Logan was said to have seriously impaired his eyesight. The second visit in 1868 was to oversee the publication of the large-scale geological map of Canada of which the general map in the Atlas is a small prototype. Though this beautiful map bears the date 1866 it was actually not published until 1869. As late as November 19, 1869, Logan was writing Stanford's for the still undelivered fifty copies, closing, pathetically, with: "I shall certainly get into disgrace with the government and the Canadian public on account of these maps in regard to which I have made so many promises."[9]

V-6 H. G. VENNOR
From a Notman photograph of 1862, when Vennor was 22. In addition to his work for the Survey, Vennor wrote on meteorology and in 1876 also published a noteworthy ornithological book *Our Birds of Prey.*

The work of the Survey in the years after 1863 was interrupted a good deal by these absences of Logan in England, by his summer-long vigil in Montreal in 1865 over his dying oldest brother James, and by the necessity of preparing another exhibit of Canadian minerals, this time for the Paris Industrial Exhibition of 1867. These preparations were a considerable drain upon the time of the entire staff during 1866, while Richardson and especially Hunt spent most of 1867 manning the exhibit. Murray's departure for Newfoundland further diminished the number of full-time staff capable of operating field parties. Indeed, there remained only Richardson; and he went into the field each year to examine the formations along the south shore of the St. Lawrence, with a shift in 1869 to the North Shore between Tadoussac and Sept-Iles. Robert Bell was a seasoned field man, but he was now employed at Queen's University and could only explore during three of the summers—in 1865 and 1866 when he studied the Paleozoic formations of Manitoulin and the other islands of Lake Huron, and in 1869 when he examined the Upper Copperbearing Series at Thunder Bay and the geology of the Lake Nipigon district.

Richardson, Bell, and James Lowe continued tracing the Laurentian formations north of Ottawa River in 1865 when Logan was out of the country. In addition, two mining engineers were engaged primarily to investigate metallic mineral occurrences—A. Michel, a gold miner, who reported on the distribution of lode and placer gold deposits in the Eastern Townships in conjunction with a project he was doing for a private company; and Thomas Macfarlane, a British mining engineer who had managed a copper mine in Norway before coming to Canada. In the single season of 1865 Macfarlane investigated the northeastern coast of Lake Superior from Sault Ste. Marie to Michipicoten and the Hastings district of eastern Canada West. He was followed in the Hastings area by Henry G. Vennor, who worked there for the first three of his fifteen years with the Survey. It fell to Vennor during his first season's work to identify a flaky, shiny, yellow, malleable metal being mined near Madoc as gold—the first recorded gold discovery in the Precambrian of Ontario—thereby contributing to the short-lived Hastings gold rush of 1866.

The uneven course of operations in these years is reflected by the production of only two volumes of reports to cover Logan's last six years as director, one for 1863 to 1866 containing six field reports, the other for the years 1866 to 1869 with four field reports on the former Province of Canada (plus five others representing the work in 1868 in the Maritimes, including four on the Pictou and Cumberland coalfields, one by Logan). Neither volume contained a separate paleontological report, but Billings'

V-7 HUNTINGDON COPPER MINE, 1867
One of several copper mines opened in the Eastern Townships of Quebec in the 1860's, situated in Lot 8, Range VIII, of Bolton Township, Brome County, and described in Richardson's comprehensive report of 1866.

first volume on Paleozoic fossils appeared in 1865. No doubt Weston was doing a good deal of the work of this section now, as he gained experience and Billings was frequently "ill." Much in evidence in both reports, however, were Hunt's contributions. Hunt was tirelessly studying, experimenting, and travelling about to secure data and check his conclusions in the field, as well as managing the Survey in Logan's absences—all without help, until he acquired an assistant, Gordon Broome, in 1869.

His reports reflect his energy and versatility; they ranged from investigations of mineral occurrences, to mineralogy and metallurgy, chemical analyses, extraction and refining processes, production data, marketing problems of petroleum, salt, iron, gold, and peat, and general investigations of the porosity of rocks and the possible sedimentary origin of some granites.

These two reports show a very considerable shift in the orientation of the work, a tendency to accept the *Geology of Canada* as an authority and to feel that the major geological formations had received their definitive statement therein; and in the subsequent work to place more emphasis on aspects with strong economic and 'practical' overtones. Comparatively little reconnaisance work was done, apart from Macfarlane's exploration of part of the coast of Lake Superior (1865), Bell's of Lake Nipigon (1869), and Richardson's of a part of the North Shore (1869). The concern was to search out metalliferous formations and mark their limits, and help to identify and find ore deposits. Macfarlane's study of the Hastings district (1865) was occasioned by a petition to the government to offer a large land grant for the construction of a railway into the district, and was devoted to investigating iron and other metallic mineral prospects there. Michel's was intended as a guide to prospectors, to show the types of rock in which gold was likely to occur. Similarly, two of the four reports on Nova Scotia analyzed and evaluated the coals. Richardson's careful survey of the Eastern Townships traced the boundaries of the metal-bearing formation, and included an appendix on

"Copper Distribution in Eastern Canada" giving a lot-by-lot list of properties known to possess showings of copper. Hunt's reports abundantly illustrate this concern for practical results, being designed specifically to assist mining operations in every possible way.

This increased concentration of the Survey upon 'practical,' economic aspects of the work was a consequence among other things, of the fact that for the first time considerable mining activity was under way in various districts of the province. This was partly a result of the Survey's labours since 1843, partly a reflection of the generally improved economic situation, and partly the effect of the new economic climate that followed the dismantling of the imperial trading system and the opening of Canada more readily to American investments and markets. The coming into existence of a mining industry added a powerful material interest to supplement the intellectual and academic community whose backing had sustained the Survey in its difficult early years. Besides moral support, the mineral industry could also assist the Survey's investigations, for the underground openings and the discovered deposits provided information about the character of the strata, the mode of occurrence, and the genesis of mineral deposits. This applied particularly to the many borings for petroleum and salt in the flat-lying strata of Canada West; also, to some extent, to the copper mines and gold workings in the Eastern Townships, the graphite, phosphate, and other mines in the Ottawa district, the iron and gold mines in the Hastings district. The co-operation of the mining industry was secured through emphasizing those aspects of the work that were relevant to its problems, particularly finding, extracting, processing and marketing metallic and industrial materials—in short, precisely the kind of work Hunt was doing so effectively.

New geological 'truths' were not overlooked, though promulgating them seemed somewhat inhibited by the recent publication of the *Geology* and, perhaps, a feeling that this should be left to Logan himself. At any rate, it was Logan who introduced a major reinterpretation of the Quebec Group by reclassifying it into three instead of two formations. Between the Sillery and Levis, he established the Lauzon formation (taken from the upper part of the Levis), which was said to be particularly rich in metallic minerals. Again, it was Logan who hinted at a tentative division of the Lower Laurentian to include a Hastings Series, comprising formations of calc-schists, dolomites, and micaceous slates, and carrying gold, copper, iron, and other metals as well as new specimens of *Eozoon*. In a footnote to Macfarlane's report in 1865, he suggested that "These Hastings rocks may be a higher portion of the Lower Laurentian series than we have met with elsewhere."[10]

Bell got into difficulties over a statement in his 1869 report that the Upper Copperbearing rocks along Thunder Bay probably were Triassic or Permian, rather than Lower Silurian as tentatively stated in the *Geology*. The grounds Bell cited were their position in relation to other formations, and because they differed from Lower Silurian rocks in any contiguous part of the continent and "bear a strong resemblance to the rocks of Permian or Triassic age in Nova Scotia."[11] This report was received after Logan's retirement and was printed when he was out of the country. But when Logan returned in 1870 he prepared a "Postscript" which was published in the *Report of Progress for 1866-69* where Bell's own report appeared. After disputing minor statements respecting the area of Lake Nipigon and its elevation above Lake Superior, Logan went on to attack Bell's statement of the age of the Upper Copperbearing Series, pointing out that the succession all along the northern and southern shores of Lake Superior indicated the series was of Lower Silurian age or earlier, and deploring that Bell had not paid sufficient attention to the structural evidence but based his opinion "on lithological grounds alone." He concluded his observations with a cutting reminder, "But it is not the duty of the Geological Survey to predict what the age of the northern trappean rocks may be, but to investigate the evidence carefully and state it impartially."[12]

Logan's attitude was that he was personally responsible for whatever appeared in the official reports. Had he seen the report earlier, he almost certainly would never have allowed it to be published until he was completely satisfied with its contents. This particular incident helps explain why Logan had such strong reservations about using part-time employees in any but a subordinate capacity, especially individuals who could not confer with him frequently so that he could check their data, descriptions, and conclusions. It also brings up the never-ending problem of publications policy, as between the rights of the author and those of the Survey.

For his part, Hunt carried on as before, to arrive at, and publish, conclusions based upon his chemical and mineralogical studies. In these years he supplemented them by a considerable number of field investigations to see how the problems actually manifested themselves, and how they might best be met. His work tended to be related to practical problems of Canadian mining, though simultaneously he was writing and publishing widely on theoretical subjects and perfecting his grand design of earth history. His economic studies led to theories about how and where the minerals occurred, as when Hunt speculated that graphite was formed in Laurentian rocks by the alteration of organic matter and was subsequently transformed into veins and lodes in a crystalline form, probably in an aqueous solution at temperatures "not far below a red heat."[13] A small fraction of his two reports was devoted to such general scientific problems as the porosity of various kinds of rock under varying temperatures, or the speculation that granitic rocks could be produced from solutions draining into fissures as well as intruded from below in the form of dykes. When it came to questions of stratigraphic relationships and succession, however, Hunt endorsed the views of his superior, for example, on the Quebec Group, about which he had serious private misgivings.

The natal day of the new Dominion does not appear to have been marked in any special way by the Survey. Logan had been close to the centre of things, for when he was in London in 1864-65 he breakfasted at the Westminster Palace Hotel with Galt and Cartier on April 27, 1865, and he met them and their confrères Brown and Macdonald on other occasions as well. No doubt he succeeded in upholding the interests of the Survey at that crucial time. He was again in London from early in 1867 until May, so it may be assumed that he also witnessed, or followed, some part of the final stages—the last-minute negotiations, and the passage of the British North America Act through Parliament in February-March 1867. But when Logan returned to Montreal early in June 1867, Hunt and Richardson were still in Paris at the Exhibition, and the office was in quite a disordered state, to judge from Logan's extant letters to certain recalcitrant employees. On July 6 Scott Barlow was told that his unwarranted absences from the office must cease, that he would be fined two days' pay for each day he was absent, and that if he did not reform his conduct he could quit the Survey altogether. A young field assistant, who complained about Richardson's unfair reports of his behaviour, was told that Logan knew Richardson too well to believe such an accusation, that "the trouble & expensive loss of time which your shameful neglect of your work has put me to and the many blunders I find in what you have done, are not likely to keep me in good humour in regard to what concerned you,"[14] and that he should be more temperate in his habits. Similar advice, perhaps, may have been forthcoming to Billings, whose attendance at the office left much to be desired.

During the winter of 1867-68, Logan proceeded with the important arrangements for the continuance of the Survey in the expanded Dominion. The Survey act, having

V-8 CANADA'S FIRST OILFIELD AT PETROLIA, ONTARIO
Bitumen and mineral oil were encountered by early settlers, and in 1857 or 1858 J. M. Williams drilled the first successful oil well. By 1862 an oil industry was springing up in Lambton County that reached its peak of production in 1894. As early as 1849 the Survey studied the local and regional geology and tested samples of the bituminous materials. By 1860 Logan and Hunt had made considerable progress in working out the geological and physical controls governing petroleum and natural gas occurrence in southwestern Ontario—an early Survey contribution to the growth of Canada's fossil fuel industries.

been passed in 1864, had some time to run; but some changes would have to be made before then, since a larger appropriation was required if the operations were to extend to the new provinces to the east. Accordingly, on February 14, 1868, Logan wrote to the Minister of Finance, Sir John Rose, suggesting the need for revising the act since "From the conversation with which you lately favoured me, it appears to be the intention of the Government to extend the Geological Survey over the whole Dominion, and to provide by an act of Parliament for its continuance for five years."[15] In a subsequent letter of February 29, he outlined his proposed staff arrangements and drew up a budget of $30,000 per annum, seventy per cent of which was for staff, the remainder chiefly for explorations and printing. The cabinet by an order in council of March 10, 1868, agreed to put a new bill before Parliament. Later Logan asked the government to make the grant for five years from the date of passage of the act and to the end of the next following session of Parliament. The act, Chapter 47, 31 Victoria, assented to on May 22, 1868, did not follow this suggestion, but it extended a grant of $30,000 per annum back to July 1, 1867, and for the five years from that date, thus placing the Survey on the same fiscal year basis as other branches of the government. In addition, the valuable privilege of carrying unspent funds over from one year to the next was renewed. In effect, the Survey continued to be what it had been in the past, in Logan's words, "a sort of contract by which certain work has been undertaken for a certain lump sum per annum, whether the hands be many or few. We have to account for every penny that is spent, however, & shew that it has really been disbursed for geological purposes."[16]

There was no statement in the act specifically proclaiming the extension of operations over the whole of Canada, except as might be inferred from its continuing to call the Survey "the Geological Survey of Canada," describing the museum collection as "a collection for the whole Dominion of Canada," and reaffirming its authority to establish reference points and markers "in Canada" and to receive copies of survey plans and sections of all federally-incorporated railway and canal companies. The positive statement, that the Survey shall "take charge of and conduct the Geological Survey of the several Provinces and Territories of the Dominion," first appeared in the act of 1877.

One change that did follow shortly after Confederation was in the arrangement whereby the Geological Survey reported to Parliament and government. Before 1867, Sir William Logan had reported to the Provincial Secretary. By an order in council of December 6, 1869, it was transferred to the jurisdiction of the Secretary of State for the Provinces, the incumbent being the former tribune of Nova Scotia, Joseph Howe. The rationale, apparently, was that the Survey "extends over all the Provinces of Canada, and its operations may cause correspondence with the local Governments."[17] Howe was a very helpful, sympathetic minister, but soon after he retired in 1873 the department was dissolved. Thereupon the Geological Survey, along with Indian Affairs and several other branches, fell under the scrutiny of the gigantic Department of the Interior, newly established for the principal purpose of administering the western and northern territories of Canada. Here it remained until 1890, when it became elevated to the status of a department, and its director to the rank of deputy minister.

* * *

EXPANDING THE SURVEY TO THE Maritime provinces renewed Logan's association with the geological structures of the Maritime coalfields, one of his early interests and the starting point for his survey of Canada's geology in 1843. The formations of the region related easily to those of Gaspé and the south shore of the St. Lawrence, and included the continuation of the Quebec Group, which was such a difficult geological problem. Perhaps the most interesting aspect of the new territory from the geological standpoint was that it afforded an opportunity to deal with the full range of Carboniferous and early Mesozoic formations, whose absence from the record in the old Province of Canada accounted for the lack of coal there.

Because of the scientific work in progress or already completed in the Maritimes, Logan was entering a region whose geology was well known, at least in outline, and one in which mining held a considerable economic importance. Able, devoted work had been done by a number of residents, some of whom had achieved considerable professional stature. From among the best of these Logan hoped to find the men he required, for he was astute enough to know that bringing in outsiders to the exclusion of the local men would certainly be resented. Besides, he had a lifelong respect for amateurs who were well versed in local situations—after all, he had been one himself.

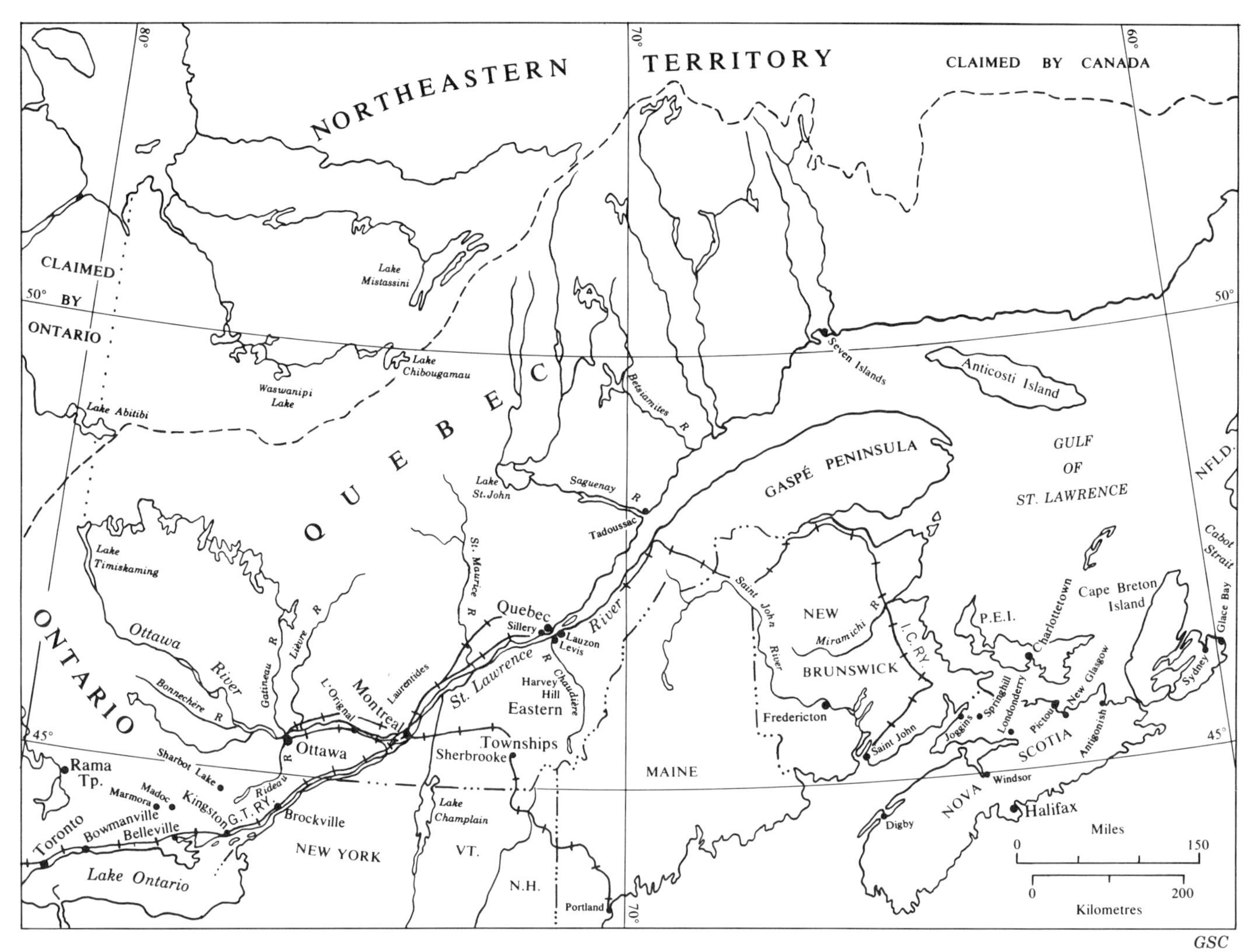

V-9 EASTERN CANADA IN THE 1870's

Nova Scotia was a celebrated locale for geological studies during the previous quarter century, beginning with Charles Lyell's visit in 1842 to the Carboniferous formations at South Joggins, undertaken as a result of the recommendation of Logan who had been there the previous year. Logan, in fact, had discovered in Lower Carboniferous rocks at Horton Bluff, fossil tracks which were the first evidences of animal life at that early time. For assistance Lyell drew upon Dr. Gesner; William Brown, a mine manager of the General Mining Association that controlled the coalfields of Nova Scotia; and a young native of Pictou, John William Dawson, a graduate of Pictou Academy who had just returned from a term of study at Edinburgh University. Lyell's infectious enthusiasm inspired in Dawson a lifelong interest in geology and paleontology that was to supplement his more important public career of educational administrator, first as superintendent of education for Nova Scotia, and later as principal of McGill University. Dawson collaborated with Lyell on a later visit to Joggins in 1852 and on an article describing two fossil animal skeletons from the Carboniferous, the first to be recovered. In 1855 Dawson published the first edition of his comprehensive *Acadian Geology*, the major work on the geology of Nova Scotia and parts of New Brunswick to that date. His departure for Montreal in 1856 left a great void, but important work continued to be done by such amateur collectors as W. B. Webster and the Reverend David Honeyman, a Presbyterian minister who had come out from Scotland to practise his ministry but soon abandoned it in favour of a geological career. That geology and natural history were popular studies was indicated by the formation of a Nova Scotia Institute of Natural Science in 1862 as an offshoot from the Literary and Scientific Society of Halifax.

V-10 THREE OUTSTANDING NEW BRUNSWICK GEOLOGISTS
C. F. Hartt L. W. Bailey G. F. Matthew

The provincial legislature, which had been disinclined to assist geological surveys because of a feeling that owners of land should be expected to discover the value of their own property, was growing increasingly aware of the need to advertise the resources of the province in order to attract settlement and investment. In 1862 the province had presented a highly-regarded exhibit at the London Exhibition, principally through the efforts of Honeyman as general secretary, and of Professor Henry How of King's College, Windsor, who collected the minerals and compiled the catalogue. As a result, the legislature established a provincial museum in 1865, set aside a large room for it in the new Province Building, and began developing a collection under Honeyman as curator. Hence, by Confederation the basic task of outlining the major geological features of the province had been completed and published in a highly-regarded work, a useful cadre of experts on the geology of the province had been formed with whom Logan could work, and the province and public were educated to the value of geological work.

In New Brunswick, also, geological studies had made notable advances after the termination of Gesner's short, controversial tenure as geological surveyor for the province and the publication of his four reports (1839–42). Dr. James Robb, chemist and naturalist at King's College, Fredericton (the future University of New Brunswick), crowned several years of investigations in 1849 with a good geological map of the province prepared as part of a report on the soils and agricultural potentialities of New Brunswick, a map Logan could use for the main source on New Brunswick in his "General Map of Canada." Robb, who died in 1861, was succeeded at the university by Loring Woart Bailey, a graduate of Harvard and Brown universities, who came to Fredericton in 1860 as professor of chemistry and natural science and was to dominate geological studies in the province for 46 years. He found two very able young associates in C. F. Hartt and G. F. Matthew, the latter a Saint John customs house officer who continued his geological work to his death in 1923. Hartt, a graduate of Acadia and Harvard, worked in New Brunswick during the summers until 1865, when ill health forced him to move to a warmer climate. An able paleontologist, he made notable collections of fossils from the various formations of southern New Brunswick; and in 1863, through the study of certain fossils recovered from the Saint John slates which he identified as being of lowest Cambrian age, he was able to make a tentative correlation of the sequences there with those of eastern America and Europe. Afterwards Hartt spent some time in Brazil, published an important *Geology and Physical Geography of Brazil* (1870), was a professor of geology at Cornell University, and died in 1878.

New Brunswick, then, had a group of very able people who had done useful work; it also had very nearly established a provincial geological survey in the years immediately before Confederation. In 1864 Bailey, Hartt, and Matthew presented their usual report on their summer's work to Lieutenant Governor A. H. Gordon, which he in due course presented to the legislature. A group of members, led by Peter Mitchell, opposed the acceptance of this voluntary report, because Mitchell hoped to appoint Professor Henry Y. Hind as provincial geologist and feared the position would be offered to Bailey instead.

The following year the government appointed Bailey and his associates to survey the southern part of the province, while Hind, with a $500 grant, did similar work in the more remote, wilder northern section. Neither Lieutenant Governor Gordon nor Premier Tilley wanted to establish a provincial survey at this late date since they preferred allowing geological surveys to be undertaken by the new confederate government. Hind further reduced his chances of appointment by scurrilous attacks on his supposed rival Bailey. Frustrated in his hopes, he soon moved to Nova Scotia to a teaching post at King's College, while New Brunswick settled down to await the coming of the Geological Survey of Canada.

Logan took up the matter of extending the Survey to the Maritimes in his letter to Sir John Rose of February 29, 1868, in which he named six New Brunswick and Nova Scotian geologists whom he wished to employ, two as full-time and four as part-time staff. Of these, he proposed to pay Charles Robb, an experienced mining engineer and brother to the late Professor Robb, the highest sum, $1,600 per annum; and a lesser sum to Henry Poole, Jr., the son and assistant to the manager of the Glace Bay colliery who had studied at the School of Mines in London. The four temporary employees—Professor Bailey and G. F. Matthew for New Brunswick and Professor How and D. Honeyman for Nova Scotia—were to be paid $500 each. As soon as he had received the approval of the government, in March 1868 Logan wrote to each man, inviting him to work for the Geological Survey of Canada. All except Poole accepted his offers, so Logan was able to begin with the staff he wanted, apart from the single substitution of Edward Hartley for Poole as junior full-time officer. Apparently a preliminary meeting was held in Portland, Maine, in May 1868, with Bailey, Matthew, How, and Honeyman when Logan passed that way on his return from England. Final plans for field work for the summer of 1868 were arranged at this meeting, and the Survey's work in the Maritimes was properly launched. The Survey's role, in fact, had begun the year previous, as Logan had assured Rose in February 1868, when Rose expressed anxiety that the Survey should lose no time assuming its place in the Maritimes. Logan informed him at that time, no doubt with a sense of pleasure, that he had already "shown the flag," and practically on the morrow of Confederation at that: "But I may say that we have already made a commencement, for Dr. Hunt with the aid of Mr. Michel effected a preliminary examination of the Nova Scotia gold mines in the month of November last."[18]

V-11 SIR JOHN ROSE
Minister of Finance, 1867-69

Logan, the Canadian patriot, could feel pleasure and pride over his part in extending the benefits of Confederation to other parts of British North America. But while Confederation was adding new ground for the Survey, the important bridgehead Logan had secured in Newfoundland as a preliminary step towards intercolonial co-operation and unity was being severed. In 1863 members of the Newfoundland government, interested in publicizing the mineral resources of the island, had approached Logan for assistance in establishing a geological survey of their own. An arrangement was reached whereby Murray would undertake the field survey in 1864 and 1865, with Billings and Hunt assisting in the analyses of specimens and Logan acting as honorary director of the operation. Murray was rather anxious for an independent command outside Canada now that his wife had died and his children were grown. The only drawback was the pay, which was £600 per annum including expenses, and would leave him less than he was currently receiving from the Canadian Survey.

Murray's principal work in the two years was on the well-exposed east coast of the island to the base of the northern peninsula. He submitted his report to Logan, who added his comments, discussed its implications, and

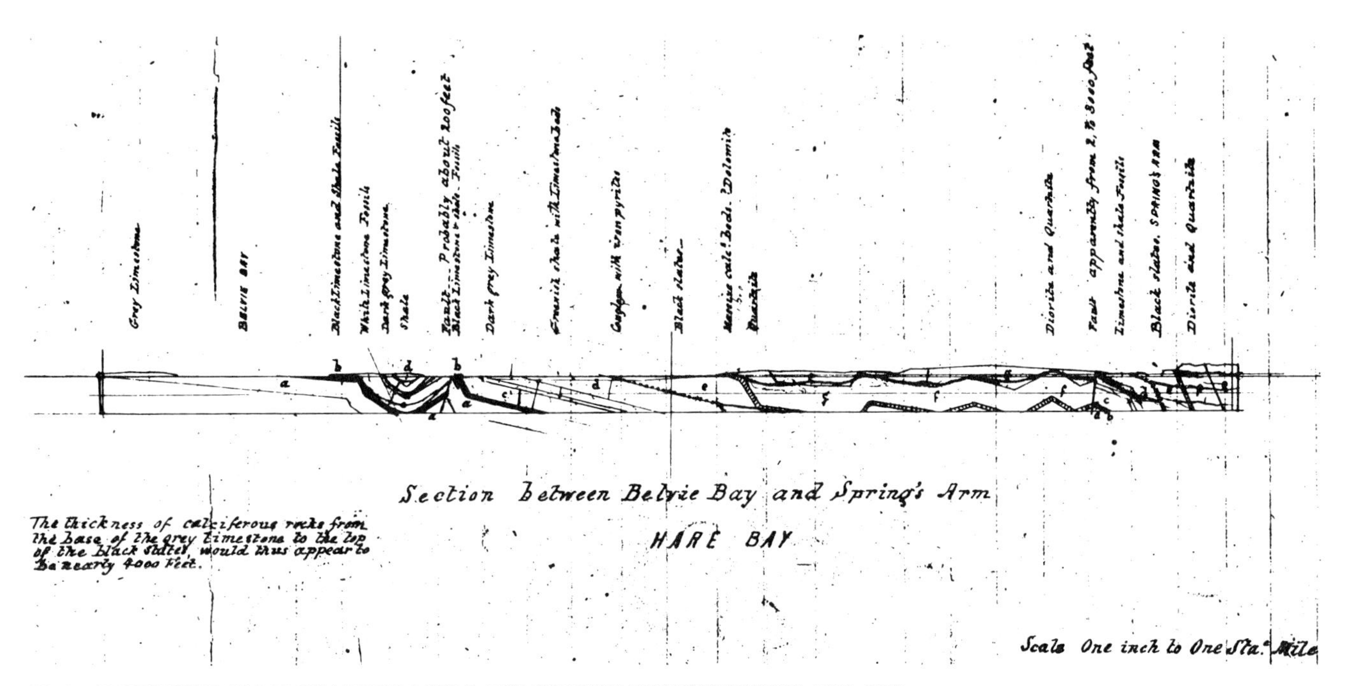

V-12 SECTION FROM ALEXANDER MURRAY'S SURVEY OF NEWFOUNDLAND, 1865
Belvie (Belvy) Bay, now Ariege Bay, is a southwestern arm of Hare Bay, the northernmost major indentation along the east coast of the Great Northern Peninsula that terminates in Cape Bauld. Murray's neatly-written report is in the Survey's library.

presented it officially to the government of Newfoundland. In it Logan referred to the possibilities, if the survey were continued, of discovering coal beds similar to those of Cape Breton, and deposits of metals like those of the Quebec Group. The Newfoundland government, highly encouraged, extended Murray's appointment to 1866 and subsequent years. Unfortunately, that summer while exploring the Carboniferous formations along the southwest coast from Cape Ray to St. George's Bay for possible coal occurrences, Murray suffered a painful accident, tearing an achilles' tendon while jumping from one rock to another. Efforts were made to secure a pension for him, but neither the Canadian nor the Newfoundland government would accept the responsibility, and no compensation was forthcoming. The Newfoundland government, however, retained his services until his retirement some 15 years later, so Murray was not too badly off.

Assisted now by the able James P. Howley, Murray continued to explore and map the coasts of Newfoundland and the major rivers of the interior. He also arranged a collection of economic minerals for the Paris Exhibition of 1867, organized a provincial geological museum, published a succession of annual reports (as well as a complete reprint of the series in 1881), and prepared a geological map of the colony which he had printed in London in 1873. Though Newfoundland went its separate way politically and the Dominion government in 1870 actually refused permission for Survey officers to visit the island in any official capacity, relations between the two Surveys remained extremely close right down to Murray's retirement. Murray referred his mineral samples and fossils for analyses to Hunt and Billings or their successors, had his maps prepared by the Barlows, and visited Montreal to discuss geological and administrative problems with Logan, Bell, and Richardson (and was visited by Logan, in turn, in 1871). Occasionally, notwithstanding the Canadian government's edict, a member of the staff would accompany Murray to the field, as Weston did for two months in 1874 when he received leave from Logan to seek out fossils for Murray in the Cambrian formations around Conception Bay. Altogether, Murray's years in Newfoundland were apparently happy ones; he secured there the independence, recognition, and prestige he craved. He married again, raised a second family, and lived in St. John's as a local notability, aide-de-camp to three governors, and the recipient of a C.M.G. He retired with a full pension to the ancestral Scottish home at Crieff, Perthshire, in 1883, where he died in the following year.

* * *

ANOTHER IMPORTANT QUESTION that Logan raised in February 1868 in his correspondence with Rose was the matter of his retirement. He was approaching the age of 70, his health was failing, and his temper was not always the best. He had lost his brother James in 1865 after a lengthy illness, and he now had far greater business responsibilities to oversee. Their father had divided his considerable property in Scotland and Canada among his four sons, and since none of the brothers married, the property of each reverted to the surviving brothers. James, who received the Canadian possessions, had built them into a sizable fortune during his lifetime of trading and investment, so that Sir William was now quite a man of property. Probably the most valuable of these family possessions was the 210-acre tract of real estate on the eastern outskirts of Montreal known as the Logan Farm, which the father had acquired in 1788 and 1811 by purchasing two adjoining properties. As Montreal grew, after 1840, the southern part of the property was subdivided into building lots and became part of the expanding city. A further £27,596 was realized from the sale in 1846 of just under 50 acres in the centre of the estate for use as an ordnance ground. The northern part of the farm, beyond the city limits, however, retained its rural character for another 30 years. Here the imposing stone house Rockfield (also shown on the map as "Rocklands"), with its gardens and farm plots, continued to provide pleasant pastoral surroundings for Logan's guests and for Logan himself, especially after his retirement.

As early as November 29, 1867, Logan advised his friend Professor A. Ramsay of his intention to retire, and in his letter of February 14 he explained to Rose that

> The extension proposed will undoubtedly greatly [increase] the labour of directing the survey, while at the same time my increasing age (now closely verging on 70 years) makes me feel that I am not so well fitted for work as I used to be. This has led me seriously to consider whether it would not be prudent on my part to retire, and if I could see how those at present associated with me in the investigation could successfully manage it without me, I would not hesitate to withdraw, notwithstanding that scientific friends on both sides of the Atlantic have strongly urged me still to persevere.[19]

His projected budget, drawn up on February 29, listed among the officers an "Assistant director gradually to take the place of Sir W. E. Logan 2000," together with an explanation that the deputy "is gradually in case of need to assume my duties. It appears to me only prudent to make this arrangement; but who the deputy is to be, or when [,] I shall have to look for him on the other side of the Atlantic."[20] In May he confessed to his friend Principal Dawson "This extended Survey is too big for me, at my time of life." That comment had been inspired by the news that Mr. Poole had rejected his offer and Logan therefore would have to look for "a cleverish youngster" and then train him in the field.[21]

Who, indeed, was this deputy to be, who should be groomed as Logan's successor? Within the Survey only one choice seemed possible—Dr. Hunt—since Billings was obviously unsuitable and Murray was happily ensconced in Newfoundland with his own little command. Hunt had been playing an increasing part in the work and administration of the Survey; as he later told a House of Commons committee,

> I occupied myself with a great many cognate questions, applying chemistry and mineralogy to the geological questions raised, and also doing a great deal of field work. I may say that during the last two or three years of Sir William Logan's administration, he was absent from the country a great deal, and practically the whole duty of directing the Survey devolved upon me. I had power of attorney to receive and pay all moneys for two years, and also organized all the parties and employed the assistants.[22]

Logan admitted as much in his letter of February 14, 1868, when he argued that Hunt's salary of $1,600 should be greatly increased: "His professional reputation is spread over both Europe and America, and his name carries authority wherever it is known. The survey owes much of its reputation to his labors. Its efficiency would be much crippled without him, and yet I am in constant fear of losing his services."[23] Logan obtained the increase, and also made it retroactive to July 1, 1867, the start of the Survey's latest financial period.

Yet Logan apparently never seriously considered Hunt as his successor. In fact his proposal to raise Hunt's salary to the amount of $2,800—just below Logan's own $3,000—was with the view of retaining his services (though Logan had told Rose that Hunt could easily have secured a $5,000 salary elsewhere) in order to give him a new position of mining inspector. As such, Hunt would deal with explorations for economic minerals and the problems of mining development, for which he was so eminently well qualified, as his recent reports demonstrated. Even Hunt's pupil, friend, and lifelong associate, James D. Douglas, the future guiding genius of the great Phelps-Dodge Arizona copper firm, suggests that Hunt was temperamentally unsuited for the position:

> That he was not appointed head was due to Sir William's knowledge of his lack of tact and of his irritability under business worries. His fitness to direct the technical operations of the Survey was beyond question .... But no one knew better than Sir William how intolerant of contradiction he was, and how impolitic were his comments on political leaders upon whose good will the votes for the support of the Survey depended. His decision, therefore, not to recommend Hunt as business director was an act of kindness to his old colleague.[24]

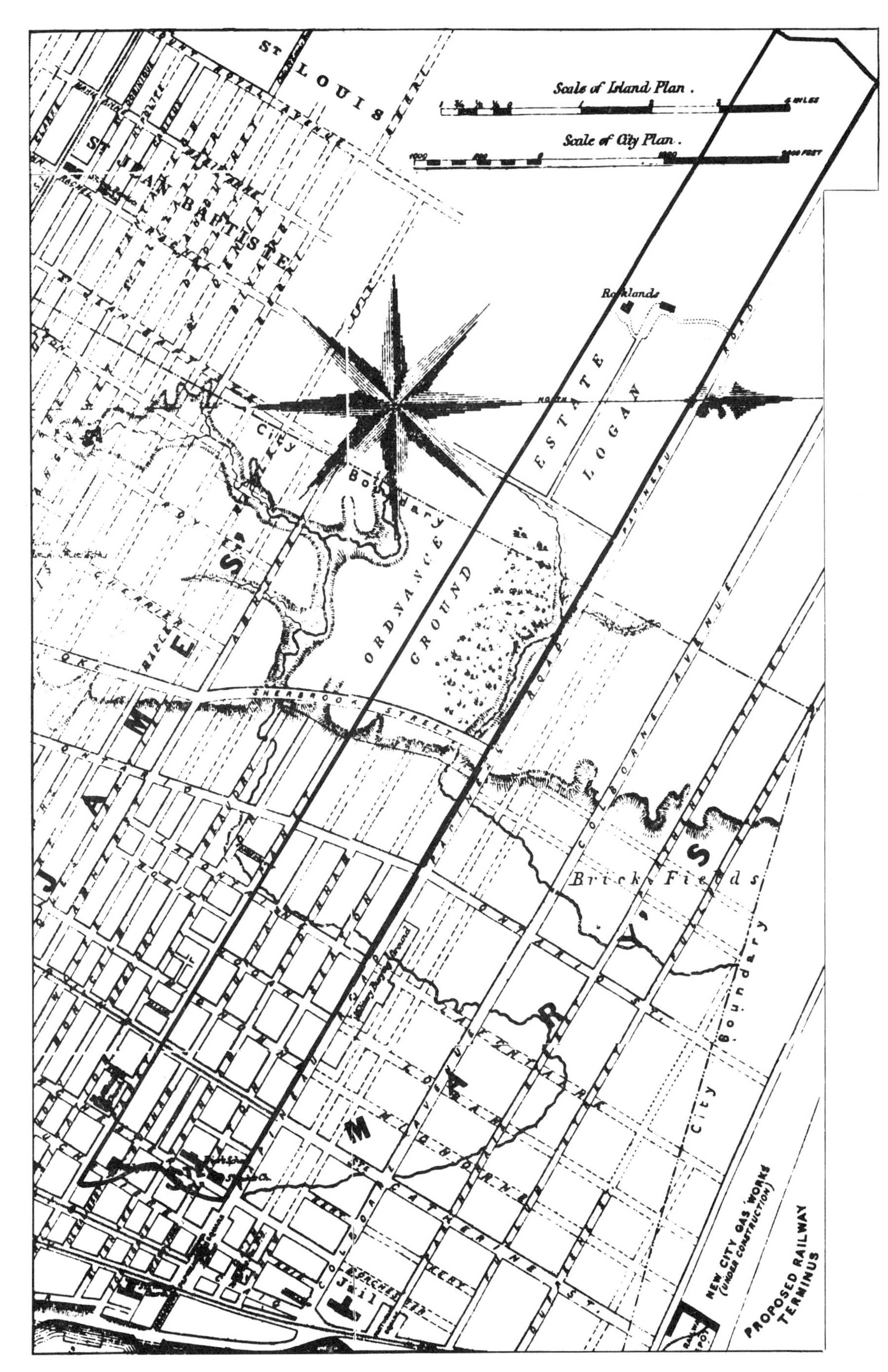

V-13 THE LOGAN FARM, 1872

The farm, whose original dimensions are shown in outline on the Johnston Map of Montreal of 1872, was about 5 arpents wide. From its southern limit, the Little River (Rivière St. Martin) that crossed this part of Montreal between Dorchester and Craig Streets, it extended northeasterly for some 54 arpents (10,000 feet) through the city, across the village of St. Jean Baptiste, and into Côte St. Louis, the next village. The Ordnance Ground eventually was transferred to the city, and was converted into a public park known since 1901 as Lafontaine Park. The considerable tract beyond the city acquired little or no settlement during Logan's lifetime, apart from the buildings and staff of Rockfield. The detailed Hopkins maps of the Island and City of Montreal (1879), however, depict the northern remnant of the Logan Farm as completely subdivided into streets and lots, though its buildings were almost the only ones shown in that part of the villages' plans.

Indeed, Logan's February 14, 1868, letter to Rose indicates that he was thinking of bringing in someone from outside the Survey, someone from "the other side of the Atlantic." Already he had written to his friends in England about his needs, and no doubt the field was canvassed thoroughly on Logan's visit during the spring of 1868. He learned that the director of the Geological Survey of Victoria in Australia, Alfred R. C. Selwyn, had the requisite scientific and technical skills, and administrative experience, and that he was looking for a new position. Soon they were in correspondence, and by the end of 1868, Logan was able to write Rose of Selwyn's willingness to accept the post, and to ask for permission to make a formal offer: "I have long personally known Mr. Selwyn and can bear testimony to his high qualifications to fulfill the duties of the post in question with credit to himself & to the dominion. He was for a long time on the geological staff of the United Kingdom & was selected from it to conduct the survey of Victoria which has now been under his charge for upward of sixteen years."[25] Selwyn arrived in Canada on October 1, 1869, and Logan began training his intended successor in the intricacies of the organization he had founded. On November 30, 1869, Sir William Logan served his last official day as director of the Geological Survey of Canada, though until his death he continued to take temporary charge of the Survey almost annually and to carry on geological work on his own account. He had contributed 27 years to the establishment and the shaping of an organization which he hoped would combine the highest ideals of scientific inquiry with disinterested service to the government and the people of Canada. The growth of international respect for its work, the continuing support of the Canadian government, and the wide popular acclaim that was accorded him upon his retirement, were measures of how abundantly he had succeeded in these objectives.

The man on whom fell the burden of keeping the Survey at the same high level of accomplishment in a period of new, heavy responsibilities, had a scholarly background very like Logan's. Alfred Selwyn was born in Kilmington, Somerset, on 28 July 1824, the son of the Reverend Townshend Selwyn. He had been educated privately, first by tutors at home and then for a few years in Switzerland, where he developed a lifelong interest in geology and mountain climbing. Although without formal training as a geologist, he advanced rapidly in his chosen profession. He began his career in 1845 by being appointed to the Geological Survey of Great Britain as an assistant to Logan's friend, Professor Ramsay, who characterized Selwyn's work as "the perfection of beauty."[26] Resigning from the British survey in 1852, he was appointed on the recommendation of his British superiors, to direct the new geological survey of the colony of Victoria in Australia. Victoria was then at the height of a feverish prosperity, in the midst of the gold rush, and Selwyn enhanced his reputation by his studies of the structure of the goldfields. He also organized and served as a commissioner on several of the great exhibitions of the 1860's, where he probably made Logan's acquaintance.

V-14 SIR ANDREW C. RAMSAY
A long-time officer of the Geological Survey of Great Britain and its director general from 1871 to 1881, he was also professor of geology in the University of London.

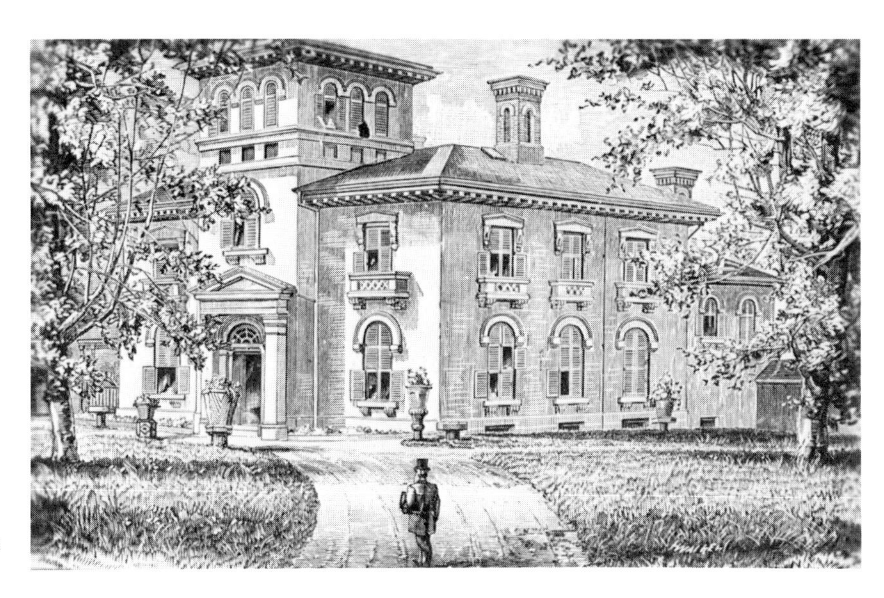

V-15 ROCKFIELD, LOGAN'S FARM, MONTREAL
The last Canadian residence of Sir William Logan.

As director of the Victoria survey he also encountered some of the difficulties and criticisms that were later to plague him in Canada. He built up a very able organization and trained many future leading Australian geologists. But his seeming tardiness in producing results and apparent emphasis on the scientific rather than the practical side of work got him into difficulties with pragmatic legislators in Victoria (as later in Canada) who were more interested in immediate results to justify the expenditures than in developing sound scientific concepts for use in the future. In 1867 the Victoria legislature debated the continuance of the appropriations for the survey, and the opinion was expressed that "all Selwyn had accomplished was to produce highly coloured maps, which might have been done at ½ the cost."[27] A writer calculated that at the rate the survey was producing the maps, it would have taken 400 years to map the colony completely. The high quality of the work, and the obvious lack of adequate staff, did not influence legislators who were interested only in practical results for their money. In 1869 the legislature abolished the survey, and freed Selwyn to assume the post of director of the Geological Survey of Canada about to be vacated by the retirement of Sir William Logan.

The new director was a capable, energetic geologist, with an established reputation as a stratigrapher, an interest he had in common with Logan, which he was to pursue in his years with the Canadian Survey. He was also in the Logan tradition in being an exacting scientist, careful that nothing should be published that was not completely accurate and beyond possible misinterpretation. He was scrupulously honest, jealous of the honour of the Survey, personally industrious and conscientious, quick and alert, prepared to travel in the worst terrains and do what he regarded as his duty, whatever the obstacles. These were important qualities, and perhaps they account for his having been chosen as that great man's successor.

As time was to reveal, however, Selwyn lacked other of Logan's strongest qualities. Though a sound enough geologist, he was without the instinctive, almost magical talent for interpreting geological evidence correctly that was so strongly in evidence in Logan and several gifted later officers of the Survey. Morever, he most notably was lacking in Logan's immense tact for dealing with people, his highly developed political sense, and his talent for public relations. Instead, Selwyn's sometimes rash, offhand verbal statements and disputatious nature embroiled him in many controversies and earned him many enemies. These weaknesses were to bring frequent difficulties and frustrations for Selwyn and for the Survey during the long years of his directorship.

After retiring from the Survey, Logan spent a number of his winters in Wales and his summers in Canada, serving as acting director in Selwyn's absences and carrying on investigations in the field, particularly in the Eastern Townships. At first his work was impelled by a desire to enjoy the sublime mystery of the Great Fault and to work out the intricate geology for his own satisfaction. But increasingly it became an obsession with him that he had to defend his scientific reputation against an unwarranted slur upon his ability. That came, of all people, from Hunt, who as early as 1870 began criticizing Logan's map of the Eastern Townships and the "errors in the interpretation of the rocks and with regard to their distribution" of the Quebec Group in particular.[28] Principal Dawson tried to prevail on Hunt to moderate his criticisms, but Hunt insisted on airing his views in the paper he presented at the meeting of the American Association for the Advancement of Science in August 1871, "Geognosy of the Appalachian System, and the Origin of Crystalline Rocks." His biographer Douglas explained the action as an outgrowth of Hunt's fearlessness, stubbornness, as well as his conviction of his infallibility: "Whether the subject was important enough to warrant the vehemence with which he maintained his ground and sustained his arguments is quite another question; but to Hunt truth was truth, and facts, as he saw them, were facts, and the relative proportion and importance of the truth and the facts had no weight with him."[29]

Logan was bitterly hurt by this attack on his professional skill and complained that "My present investigations have been undertaken with much inconvenience to myself, in consequence of some of my work having been (needlessly, as I am persuaded) called in question."[30] He felt Hunt was overstepping his proper fields of chemistry and mineralogy and meddling in stratigraphic and paleontological problems about which he knew little. About the same time on December 26, 1871, he wrote Professor Bailey in a rather similar vein: "Dr. Hunt is an excellent lithologist, and it was to aid you in this branch that he was sent to New Brunswick. But when he says that that which gives an upper-silurian aspect to the rocks over a very extensive area is merely accidental, he appears to me to be travelling out of his beat . . . . We are depending on you and Matthew to enlighten us."[31]

For his own part, Logan continued to work in the Eastern Townships each summer, usually assisted in the collection of fossils by Weston and in the measurements by Arthur Webster, who was delegated to this duty by Selwyn. By the end of 1874, he was contemplating testing the succession in a selected site by diamond drill to prove the matter once and for all. On August 17, 1874, he wrote Richardson of his plans: "I have almost made up my mind to purchase a diamond drill, & use it in the Eastern Townships next season. By going down through some of the isolated patches of serpentine South east of Harvey Hill & getting into the black slate below I shall be able to convince every one of the sequence."[32] He went back to Wales for the winter and it was while he was arranging to return in 1875 with a diamond drill to bore a 950-foot hole at a cost to himself of about $8,000 that his health broke down completely. After some months of suffering he died on June 22, 1875, at the home of his sister, Elizabeth Gower, at Castle Malgwyn, Cardiganshire. He was commemorated in Canada with a marble bust by the London sculptor, Marshall Wood, which Murray arranged for and which was ready in 1880; by a Logan Gold Medal for geology and the Logan Chair of Geology, both at McGill University, which he helped endow; also by the fine sympathetic biography, *Life of Sir William E. Logan, Kt., LL.D., F.R.S., F.G.S.&c.,* written by B. J. Harrington with the assistance of Logan's devoted friends, notably Murray, and published in 1883. But his true monument is the Geological Survey of Canada, which he founded, nursed, and reared to healthy vigorous maturity in the face of tremendous odds.

A decade later Hunt still was unrepentant. As he told the House of Commons committee in 1884, "I say it with all respect, that Sir William was wrong in the position which he took in regard to these rocks, and that I was right, and subsequent investigations, not only in North America, but in the Alps and other parts of the world, have shown that Sir William was wrong in his views."[33] He admitted that Logan worked for a long time, "even while his health was failing, and when he has told me that he had never passed a night without severe pain and sleeplessness." But all Logan was trying to do, according to Hunt, "was endeavouring to establish a theory in which he was undoubtedly in error."[34]

By then Hunt had departed from the Survey, having tendered his resignation which took effect on June 30, 1872, three years before Logan's death. He complained Selwyn had been brought in over his head and, jealous of Hunt's superior attainments and wider public reputation, had put obstacles and annoyances in Hunt's way; while Selwyn retorted, perhaps accurately, that Hunt had resigned in a temper with Logan for not having named him as his successor.

Hunt went to a professorship at the Massachusetts Institute of Technology, but "teaching was not congenial to him"[35] and he left that position in 1878. He missed the Survey and his Canadian years a great deal, for they had been his most successful period, and what followed was rather anticlimactic. He frequently returned to Canada and at every opportunity he visited the Survey. He was deeply involved in scientific and learned societies (including the Royal Society of Canada and the International Geological Congress, of both of which he was a "founding father"), and he published often on a great variety of chemical, mineralogical, and geological subjects. He became increasingly regarded in the United States as a brilliant eccentric, who sometimes produced important insights that pointed to great truths. He died in New York in 1892, already a half-forgotten figure, outmoded by a science that no longer appreciated either his flights of fancy or his sparks of true genius. Today his reputation, however, is once more on the rise, and he is being increasingly recognized as an important pioneer of geochemistry.

Though the Survey would never again see the likes of Logan, or of Hunt, and instead appeared to be entering on an age of lesser men, its days of greatness as an organization still were ahead of it. Under Selwyn a whole team of younger men—Dawson, Bell, McConnell, Tyrrell, Low, Ells, Lawson, Adams, Barlow, and Macoun—would carry the Survey forward to new heights and give it far greater breadth in the gigantic theatre afforded by the transcontinental Dominion.

VI-1 U.S. GUNBOAT *SARANAC* TAKING ON COAL AT NANAIMO, B.C., 1875
Upper—James Richardson's photograph of June 15, 1875. *Lower*—Engraved copy prepared by Burland, Desbarats and Company, Montreal, for the *Report of Progress,* 1876-77.

6

The Survey from Sea to Sea

ON THE VERY DAY SELWYN assumed his new position as director of the Survey in Montreal, Canada was to have taken possession of Rupert's Land, the vast, empty proprietary domain of the Hudson's Bay Company that stretched from the borders of Canada all the way to the Rockies on the west and to the Arctic shores on the north. The Métis uprising in the Red River Settlement unexpectedly delayed the transfer until the following summer, so that it was not until July 15, 1870, that Rupert's Land was annexed to the Dominion in the form of the small province of Manitoba and the enormous North-West Territories. A year later, on July 20, 1871, following negotiations between the Canadian government and representatives of the Crown colony of British Columbia, that colony also entered Confederation. Canada had reached the shores of the Pacific and achieved transcontinental dimensions. And all this new territory meant additional, urgent work for the Geological Survey of Canada. The Dominion government that controlled the land and the resources of the North-West Territories and even of the province of Manitoba, expected the Survey to proceed at once to examine, map, and report on the new domain. As for British Columbia, geological surveys were expressly set forth as one of the duties of the Dominion government in the Terms of Union, and the new province was insistent on the speedy implementation of this anticipated benefit from Confederation.

Clearly it would be the new director's first task to familiarize himself with this vast country as quickly as possible. In his first season, in 1870, while Canada was still in the process of acquiring the North-West, Selwyn began his initiation, in what was also a new country to him, by visiting the Eastern Townships, New Brunswick, and Nova Scotia. Though the ostensible purpose was to examine the gold occurrences in the light of his experience in the goldfields of Australia, he also gained first-hand knowledge of some of the rock formations and of conditions of travel. The following year, 1871, he made a remarkable exploratory trip across the Dominion's newest province, British Columbia, and in 1872 he travelled from Lake Superior to Lake Winnipeg by canoe. Now he had some idea of the structural and stratigraphic complexities of British Columbia and of the Precambrian terrain of Ontario and the North-West Territories. Finally, in 1873, he travelled by Red River cart and river boat from Winnipeg to the Rocky Mountains and back to familiarize himself with the western prairies. Thus, by the end of his fourth season as director, Selwyn had at least a broad idea of the conditions in the major geological provinces for which the Geological Survey was responsible.

Perhaps the most pressing task facing the Survey was the exploration of the geology of the North-West, with its rugged mountains, endless forests, turbulent rivers on the one hand, and the vast stretches of flat-lying strata that formed the prairies on the other. This was almost a *terra incognita* so far as previous work was concerned. In the Maritimes, the Survey had been able to move forward from reasonably good foundations of previous studies, with men knowledgeable in the geology of their particular areas. The North-West was different. There were virtually no local experts to be recruited for the task, and precious little earlier geological work on which to rely. True, there had been a few increments to the geological knowledge of the region after 1840 in the form of better travel accounts, while extensive work was being carried on in the adjoining territories and states of the United States. Just as geological studies in New York afforded a valuable base from which to launch the study of the geology of Canada, so, too, the work in the western United States provided a good general idea of what might be encountered in the regions north of the 49th parallel.

Mainly, though, what was known in 1870 about the geology of the Canadian west resulted from the work of two exploring expeditions launched in 1857 by the British and Canadian governments. The Canadian was the more limited, being restricted to the two summers of 1857 and 1858, the first season of which was spent mainly in trying to locate an all-British communication route between Lake Superior and Red River; while in the second the party ranged over the eastern half of the prairies to about 106°W longitude. The naturalist and geologist in both years was Professor Henry Y. Hind, then of Trinity College, Toronto. The fossils he collected

VI-2 RECONNAISSANCE SURVEYS IN WESTERN CANADA IN THE 1870's
Base map—Department of the Interior, *Map of the Dominion of Canada*, 1880.

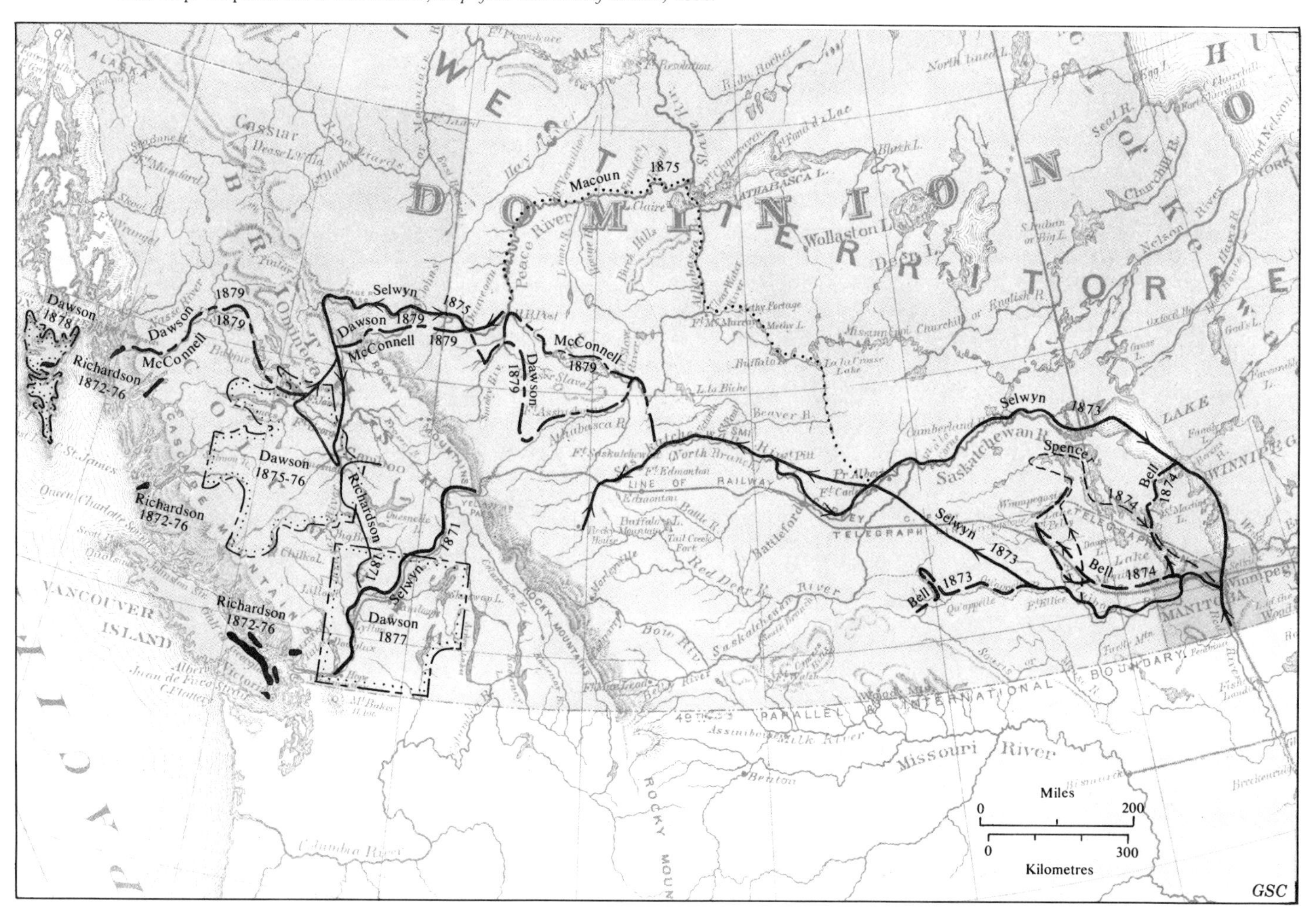

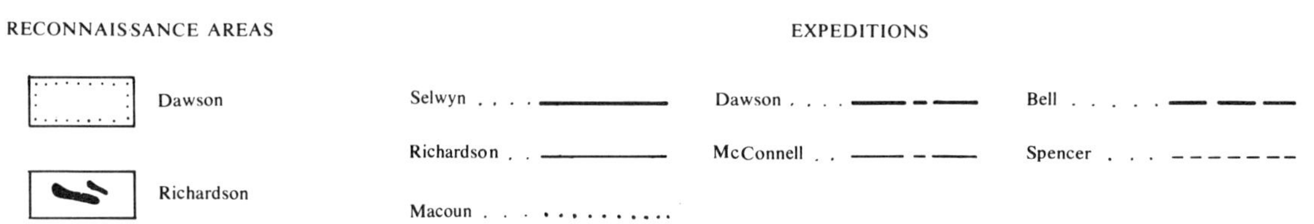

were examined by Billings, and referred to the American experts F. B. Meek and F. V. Hayden, who identified them as being of Jurassic and Cretaceous ages (periods with which Billings had no experience), comparable with those of Nebraska Territory. Similarly, the coals Hind discovered were of Cretaceous or Tertiary age, younger and of lower grade than the coals of the Carboniferous. He surmised that the rock types of the Prairie region corresponded with those observed in the western United States, except for their lying more to the west than their American equivalents, to conform with the northwesterly trend of the Laurentian gneisses in their sweep from Lake Winnipeg towards the Arctic Ocean. He identified a sequence of younger formations as he travelled west from Red River—first a belt of Silurian rocks west of Lake Winnipeg, then Devonian strata in which lie the basins of Lake Manitoba and Lake Winnipegosis, then rocks of the Cretaceous period displayed on the second prairie level, and finally on the Grand Coteau in the southwest, rocks that reached into the Tertiary. The principal geological problem, aside from the practical difficulties of finding outcrops since so much of the country was overlaid by great thicknesses of drift, was the absence of Carboniferous rocks anywhere between the Devonian and Cretaceous exposures. Hind speculated that these might occur as a narrow band masked by drift on the flanks of Riding Mountain. Following his return, he published a geological map of the region as far west as Saskatchewan River.

VI-3 H. Y. HIND

The British expedition, headed by Captain John Palliser, and with James Hector, M.D. as naturalist and geologist, was far more ambitious than its Canadian counterpart. It spent almost three years (1857-60) in the country, ranged as far west as Vancouver Island, and covered the prairie region and the Rocky Mountains between the 49th and 54th parallels very comprehensively. Hector's geological report was not prepared till after he returned to England and was published in 1863, so that his conclusions had the benefit of Hind's findings. Hector was in a better position than Hind to delineate the three great prairie levels, or steppes, and the sequence from the Laurentian and Huronian west of Lake Superior to the Tertiary of the Coteau and Cypress Hills country. He concluded that the formations in the Rocky Mountains and the valleys to the west of them were of Carboniferous or Devonian age. He also made some study of the geology of Vancouver Island, especially the newly-opened coal mines at Nanaimo, which he regarded as a good commercial lignite coal, possibly of Cretaceous age.

Both Hector and Hind wondered about the origin of the thick overlying drift on the prairies. Both accepted the prevailing theory that floating icebergs were responsible for depositing it in place, then attempted to account for the moraines and similar formations as produced by erosion. Hind, more than Hector, remained puzzled by some of the phenomena—the glacial striae in the country west of Lake Superior, the extremely high elevations at which some of the rock debris was found, and the angles at which it sometimes lay. He concluded that while the agent for transporting the drift "is almost universally acknowledged to have been water and floating or moving ice,"[1] some of the occurrences indicated the action of glaciers or stranded ice; and that the resulting formations were the product, first, of glacial action, then of icebergs.

A pressing reason for the Geological Survey to turn immediately to the investigation of western Canada was the government's plan to have a railway built across the continent to hasten the integration of the region into the rest of the Dominion. Engineering parties set to work locating prospective routes across British Columbia almost as soon as that province had joined Confederation. But the final selection had to be based on both the engineering aspects and the economic potential of the regions opened up. And no group in Canada was so well qualified as the officers of the Survey to judge the mineral potential of the North-West, or to appraise the whole complex of natural resources of the undeveloped regions. Hence Selwyn was instructed to undertake the first Survey reconnaissance of British Columbia in person, to acquire "such a personal and practical knowledge . . . as will enable you hereafter to successfully direct Geological expeditions in that portion of the Dominion."[2] He was to range as broadly as possible, to pay attention "to the various subjects in addition to those appertaining especially to geology, to which your attention should be directed, such as the nature of the soil, the vegetation, the quality and kind of timber, the distribution of plants and animals, the character of the climate, etc., etc., on all of which interesting and valuable information may no doubt be gathered."[3]

This relationship with the Canadian Pacific Railway surveys greatly affected the Geological Survey during the 1870's, for its operations often were directed into regions and along lines that were of particular value in planning and executing the great railway enterprise. The two organizations helped each other largely informally, the Survey furnishing information on various travel routes and on the minerals and other resources contiguous to them, while the railway survey teams provided material assistance to geological explorers in the field as well as some of the first reliable topographical maps. The assistance of the railway survey crews was assured in 1871 when Sandford Fleming, the engineer in chief, instructed the surveyors in British Columbia to aid Selwyn "and cooperate with him in his explorations as far as is convenient for them to do so without retarding the work they are specially engaged in."[4] Consequently the geologists were able to use the surveyors' camps and trails, to augment their supplies at their camps and to receive advice and aid that must have been invaluable to Selwyn in his first brush with the British Columbia wilderness.

On June 26, 1871, Selwyn, accompanied by Richardson, left Montreal for the long journey to British Columbia. The only feasible route as yet entailed a two-week trip by rail to San Francisco, thence by coastal steamer to Victoria. Selwyn's difficulties still were not ended when he arrived in Victoria, for it took him from July 10 to July 26 to reach Yale, the head of steamboat navigation on the Fraser and the departure point for the interior. Then followed a journey by wagon road from Yale to Kamloops Lake, and by boat to Kamloops, at the mouth of North Thompson River. This part of the journey was easy, except that transportation was at a premium and Selwyn's impatience drove him to travel much of it on foot, with a few Indian packers carrying essential supplies and the bulk of the baggage to follow. Each packer carried 100 pounds as much as 27 miles in a single day for his $1.00 a day.

At Kamloops the party divided, with Richardson making a side trip up the Cariboo Road to Quesnel, then returning to start work on the Vancouver Island coalfields, a project that kept him occupied off and on until 1876. Selwyn started up the North Thompson with a string of packhorses, over a reasonably good trail at first. But beyond the valley of the North Thompson the trail ended and as the party began to cross the divide on the way to the Yellowhead Pass, the tangled forests and rushing rivers slowed it almost to a standstill. At one point the going was so rough that they only were able to make headway at the rate of four miles in five days of work, "the whole of it being through dense forests, alternating with boggy creeks and steep sideling hills."[5] The party faced the usual problems of mountain travel—hunting for strayed horses, fording streams, corduroying muskeg, and searching for horse feed. A lesser spirit might have quailed before the difficulties of the task. And these were not the only hazards. On October 1 Selwyn returned from a solitary side trip no horse could negotiate, to find that his hungry animal had apparently eaten the records of two days' observations. Who can fail to admire the scientific spirit that rises above such a disaster to test his surmise by watching the horse greedily devour blank pages from the same notebook?

The original plans had called for the expedition to pass over the Yellowhead or Leather Pass and proceed to Jasper House, but the lateness of the season and the shortage of provisions precluded this plan. Selwyn examined the neighbourhood of Leather Pass and Tête Jaune Cache, then returned to the North Thompson, where he had four canoes built, and the entire party floated down to Kamloops in less than half the time it had taken to ascend the same river.

Throughout the hardships and frustrations, Selwyn never lost sight of his main objective as a geologist: to record geological data, to generalize, and to observe *everything*. Even yet he had not fully grasped the immensity of the Dominion, for he did not perceive the impossibility of correlating formations between eastern and western Canada, and he compared the rocks round Victoria with the Quebec Group. Nevertheless he returned from British Columbia with an eight-fold classification of the rocks, which, although later work forced its almost complete abandonment, served in the meantime as an invaluable guide to systematic observations.

In a more general vein, he also returned to Montreal a convinced booster for British Columbia's potentialities.

VI-4 MOUNT SELWYN
Photographed by Selwyn on September 22, 1875, on his return journey up the Peace River, and engraved for the 1875-76 *Report of Progress.*

He summed up his view of the province with "There can scarcely be a doubt in the mind of any one who has visited the country, that a bright and prosperous future is in store for the Alpine Province of the great Dominion; only to be realised, however, when the iron road shall have brought her into closer communication with her elder sisters in the east."[6] Richardson, the farmer, too, was impressed by the agricultural and other resources of the areas through which he passed that summer. He even brought back to Montreal several samples of hops grown on Vancouver Island which William Dow and Co. found to be far superior to anything grown locally in Quebec. He was astounded also by the rich fish resources of the region. He noted that at Qualicum River, near Nanaimo, the bottom was literally covered with fish of excellent size and an Indian had only to cast his spear at random in the water to get a large fish.

Accompanying Selwyn on his expedition were two photographers, supplied by the Notman firm of Montreal, who secured some creditable photographs of the unknown country through which they went, notwithstanding the frequent rains and snows that obscured the light, and the difficulties inherent in their bulky, awkward equipment. Fleming had agreed to pay half the costs of the photographic team out of Canadian Pacific funds, but Selwyn would pay only half their expenses while their regular employer paid their salaries. Later, in 1873, it was suggested that the Survey should establish its own photographic department to carry on the work more efficiently. But Selwyn, although he was satisfied with the photographs taken in 1871, did not feel justified in making such an expensive addition. Survey parties, however, continued to take photographs themselves in the course of their work.

In 1875, Selwyn returned to British Columbia, again in conjunction with the Pacific Railway project, this time to examine the possibilities of a northern route via the Peace River district. Accompanying him were Arthur Webster of the Survey, and the botanist John Macoun of Belleville who had been through the country in 1872 with Charles Horetzky of the Pacific Railway survey staff. They travelled from Victoria to Quesnel by steamers and the Cariboo Road, and from Quesnel by pack train and canoe across the height of land to reach the navigable headwaters of Peace River. For an experiment, Selwyn took along a collapsible canvas boat, which despite some imperfections in design convinced him that "Every exploring party travelling in such country as British Columbia should have one of these boats."[7] Where Finlay and Parsnip Rivers unite to form the Peace, he and Macoun spent July 11 climbing a mountain—Mount Selwyn—and found their 5½ hours of continuous toil "amply repaid by the magnificent scene around us."[8] The party descended the Peace as far as

VI-5 B. J. HARRINGTON

"The Forks" (the present town of Peace River, Alberta), where Selwyn turned around and returned to Victoria by the same route he had come. Macoun left Selwyn's party at Fort St. John, and with a single companion proceeded on an arduous, adventurous journey by canoe down Peace River 700 miles to Fort Chipewyan.

At Fort Chipewyan, Macoun had an experience that quite erased the discomforts he had endured. Fort Chipewyan was a major transfer point along the Hudson's Bay Company's supply route to the far North-West, and there Macoun met the officers in charge of the districts beyond. They were able to inform him in gratifying detail about the conditions for raising field crops and vegetables at places like Fort Simpson, Fort Liard, and Fort Yukon, so that Macoun "obtained more accurate knowledge of the vast interior than had been obtained by any former explorer."[9] After a brief rest, Macoun left Fort Chipewyan to make his way via Methye Portage southeast to Saskatchewan River, Lake Winnipeg, and Fort Garry. With him he brought copious notes on the floras along the route. Though he had no training in geology, his notebooks also contain many observations on the rocks along the route, including mention of the now-famous Athabasca River oil sands.

Selwyn's immediate reaction to his trip was to become a supporter of the Peace River Pass as a better route for the Pacific railway than the one through the Yellowhead Pass farther south favoured by Fleming. He also made many useful and shrewd observations on the geology of the country he traversed, even though, strictly speaking, his journey had been one of reconnaissance, designed to expose problems rather than to solve them.

In the meantime, Richardson went out every year between 1871 and 1875 to British Columbia by the United States route for a five or six months' season in the field, departing usually from Montreal in May and leaving Victoria around the end of November. His main work was examining and measuring the coal-bearing formations of Vancouver Island and checking on the reported occurrence on the Queen Charlotte Islands. There was coal there, true enough, but very uneven in quality and uncertain in quantity. Hunt's successor as Survey mineralogist, B. J. Harrington, rated the samples he received as anthracite, however, and also classed the Vancouver Island coals as bituminous, rather than lignite. Part of Richardson's time was spent aiding the Canadian Pacific Railway surveys; in this connection he devoted the 1874 season to a general reconnaissance of several coastal fiords as far north as Wrangell, Alaska. Meanwhile, the fossils he was bringing back were helping to fix a Jurassic or Cretaceous age for the metamorphosed rocks of the coastal regions in which he worked. All the Queen Charlotte fossils were of new species, as were many of the others, so the museum soon had the largest collection anywhere of North Pacific coast fossils. He also brought back specimens of the flora and fauna of the region, among which were some previously unrecorded species.

Selwyn's visits to British Columbia had impressed him with the fact that the work was sufficiently important and complex to require the presence there, full-time, of a thoroughly-trained geologist capable of dealing with the intricate stratigraphic and structural problems of the province. Richardson, elderly and self-trained, obviously was not the man for this difficult task, particularly since he had no experience with volcanic rocks which comprised a large part of the formations. Furthermore, there were the difficulties of travelling back and forth between Montreal and Victoria before the completion of the Canadian Pacific Railway in 1885-86, as well as the wishes of the inhabitants to have permanently in their midst a geologist whom they could consult. It was highly desirable, too, for the Survey to be represented on the west coast by an officer capable of operating on a year-round basis and without supervision.

VI-6 PLAINS INDIAN CAMP, INTERNATIONAL BOUNDARY VICINITY, 1872
One of the many photographs taken by the Western Boundary Commission, which surveyed the International Boundary from Lake of the Woods to the Rockies in 1872-75. G. M. Dawson was the commission's geologist.

VI-7 HOUSES AND CARVED POSTS, SKEDANS VILLAGE
Engraved from G. M. Dawson's photograph of July 1878, taken during his study of the Queen Charlotte Islands that resulted in his fine report on the Haida Indians of that region.

Such a man Selwyn found in the person of George Mercer Dawson, one of the most remarkable Canadian scientists of the century, and one of the few geologists of the Survey whose stature approaches that of Logan. The son of Principal Dawson, he was born in Pictou in 1849 and came to Montreal as a child when his father went to McGill University. He was educated by tutors, attended McGill for a time and then went to England to study at the Royal School of Mines, from which he graduated with almost every honour that institution could bestow. Selwyn was prepared to hire him to serve in British Columbia in 1873, but Dawson, with Selwyn's approval, accepted an appointment instead as geologist and botanist to Her Majesty's North American Boundary Commission under Major D. R. Cameron. Dawson produced a first-rate report of his work with the commission, an extremely detailed, accurate, critical account of the geology and botany of the area contiguous to the International Boundary, from Lake of the Woods to the Rockies. This met his stated purpose, "to make the forty-ninth parallel a geological base line with which future investigation may be connected."[10] That task completed, he accepted an appointment on the staff of the Geological Survey, effective July 1, 1875, to work initially in British Columbia.

At first glance, the most remarkable thing about G. M. Dawson was his physical appearance. A congenital condition, or illness or accident in infancy, left him deformed, a hunchback who grew no taller than a boy of 12 or 13, and with such weak lungs that his life was in danger from even a common cold. Yet he refused to accept an office appointment and insisted on remaining a field geologist and explorer in British Columbia almost until 1895 when he became director of the Survey. Far from allowing his physical handicap to restrict his activities, he sought out the most arduous exploring assignments and made a name for himself as a swift, untiring traveller and an intrepid climber who wore out more robust colleagues. His work was always careful and precise. The acid test of any geologist is how well his work will stand up when it is re-examined later on the ground by a geologist equipped with new tools and informed by new scientific knowledge. Dawson meets this test better than any man of his generation. Those who have followed his paths and examined his notes and observations continually stand amazed at how rarely he reached an incorrect conclusion, how precise and acute were his powers of observation, literally overlooking nothing; also how much territory the little man covered in a single day, week or season. His powers of observation were matched by a keen intelligence; his generalizations showed the rare clarity, imagination, and originality of his mind. His ability to form sound and lasting general conclusions from a few reconnaissance observations and distant scannings was unique.

He was also extremely versatile. He wrote scholarly articles in several fields of geology, on anthropology, botany, zoology, and history, and compiled several handbooks on Canada of an encyclopedic nature. He was an excellent public lecturer and teacher, with a facility for imparting complicated technical material in lucid and arresting form enlivened by examples from everyday experience, that was reminiscent of the great scientific popularizer of that day, Thomas Huxley. He was an above-average sketcher and water-colourist, and probably the best of the several poets in the Survey's history. He was also a diplomat, who served Canada on various missions relating to the Alaska Boundary and Bering Sea fur seal problems.

The years 1875 to 1877 were spent reconnoitering in south-central British Columbia, including the Okanagan country. At first Dawson followed closely the basic classification devised by Selwyn, but his report of 1877 displayed his ability to observe accurately and generalize with clarity and imagination. His originality made the report a milestone in the geological study of the province. He enunciated the necessity of setting up local geological units, leaving correlation from one area to another until sufficient data could be secured; and he completely abandoned any attempt to correlate from one side of the continent to the other. He concluded that no rocks in central British Columbia, except perhaps some gneisses, are older than Carboniferous; that the Triassic was a period of intense vulcanism throughout much of the Pacific coast; that sediments and volcanics may grade into granite, and hence that some granite is of metamorphic origin although other granites are magmatic and melt their way into place. These observations were revolutionary in their day and were not generally accepted for another 50 years. He made pertinent remarks on the origin of some of the major river valleys, described and gave names to some of the principal rock units that are still in use, and suggested the preglacial origin of some placer deposits. In that season of 1877 he also covered 18,000 square miles of mountainous country without the help of a serviceable map. Indeed he had to make one as he went along.

In the next two seasons, Dawson transferred his attention to northern British Columbia. In 1878 he examined, in more detail than Richardson could, the coal deposits of Queen Charlotte Islands. His geological conclusions, as usual, were accurate and penetrating, but while he studied the geology he also found the time to compile a comprehensive report on the customs and life of the Haida Indians, including a vocabulary of the Haida with related vocabularies from other Northwest Pacific tribes

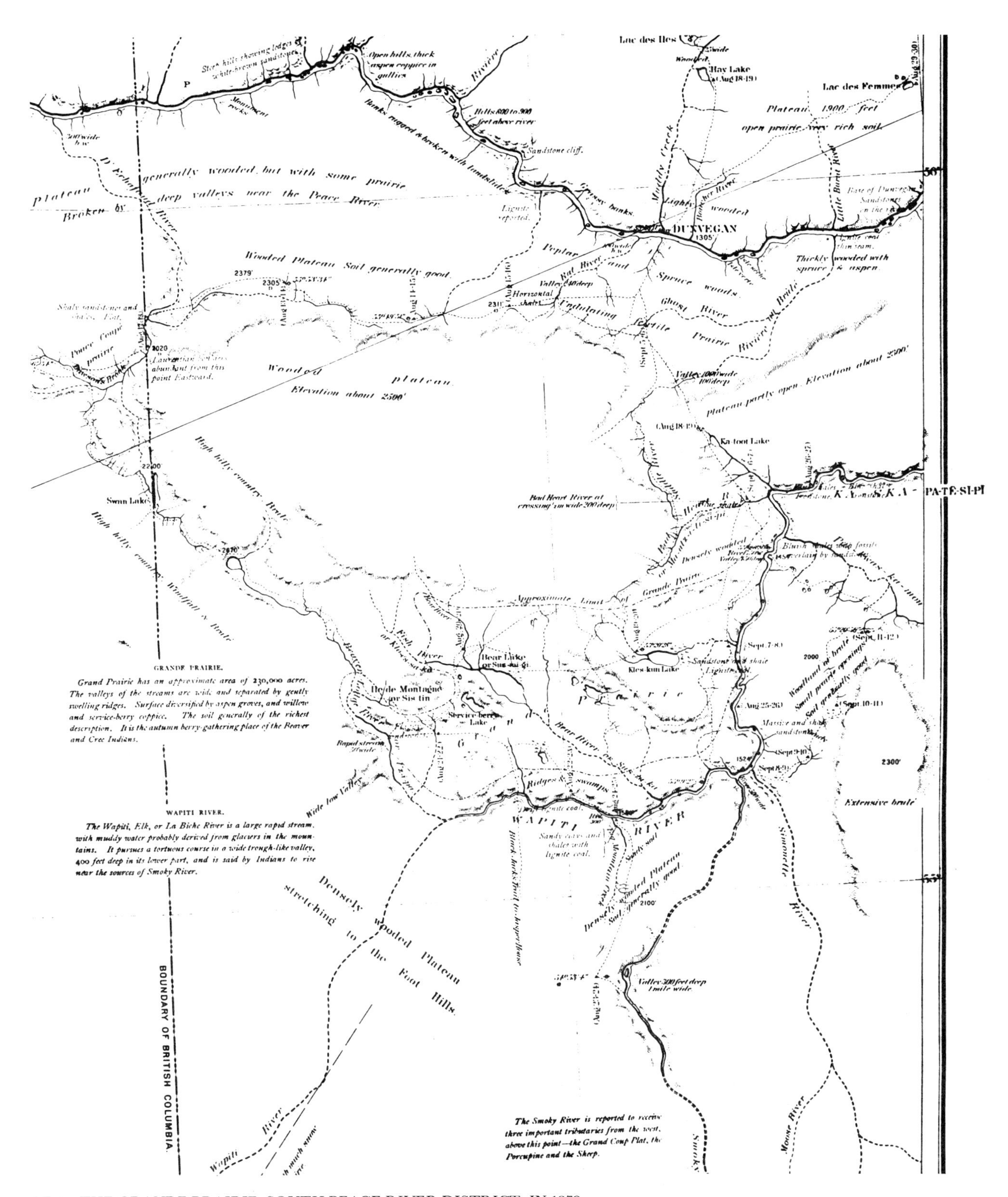

VI-8 THE GRANDE PRAIRIE, SOUTH PEACE RIVER DISTRICT, IN 1879
Detail from G. M. Dawson's map of the territory between the Pacific Ocean and Fort Edmonton, which he and his party explored in the summer and autumn of 1879. His description is prophetic: "The soil of Grande Prairie is almost everywhere exceedingly fertile, and is often for miles together of deep rich loam which it would be impossible to surpass in excellence."

for comparison. This remains a basic source work on the anthropology of this important people. In 1879 he made a tremendous seven months' reconnaissance from Port Simpson up Skeena River, through the Rockies by the Pine River Pass, and on to Edmonton, the purpose being to secure "all possible information as to the physical features and economic importance of the country, for the purpose of determining to what extent it offered advantages for the passage of the line of the Canadian Pacific Railway."[11] In this railway location survey Dawson was covering some of the same ground as Selwyn traversed in 1875. Again, as was his custom, he found time for his many side interests in the natural and human history of the region traversed; he even compiled a rudimentary Cree vocabulary. A memorable feature of this trip was that it marked the first association with Dawson of the young geologist, R. G. McConnell, a recent graduate of McGill. The expedition was not all work and no play. On July 1, McConnell recorded in his field notebook, the party ran up the Union Jack at remote Fort St. James in honour of the day, then devoted the afternoon to horse races.

Much of Dawson's reconnaissance work in British Columbia during the 1870's was directed towards the search for coal, that foremost necessity for the transcontinental railway. From his surveys in south-central British Columbia, Vancouver and Queen Charlotte Islands, and the upper Skeena and Peace River basins, a fairly clear picture of the various coal measures of the province had emerged. The Nanaimo and Comox coalfields on Vancouver Island were parts of a large continuous basin of Cretaceous strata; the Queen Charlotte Islands measures were part of a much smaller basin of older, Jurassic or early Cretaceous rocks; the scattered seams in central British Columbia were deposited in isolated lake basins, none very large, and all of younger, Tertiary age; while the Peace River coals were of two ages, like those of the prairies: one of Middle or Lower Cretaceous; the other Tertiary, presumably extensions of coal-bearing horizons of the Plains region farther south and east.

By the end of British Columbia's first decade as a Canadian province, the general outlines of its geology were fairly well understood, as a result of the reconnaissance surveys of Selwyn and Dawson and a few specialized studies like Richardson's of the Vancouver Island coal basin. Some idea had been gained of the types of rocks and structures to be expected, and the work had demonstrated the complexities of the Paleozoic and Mesozoic successions, as well as the intensity and long duration of the mountain-building forces to which the formations had been subjected.

Following his first trip to British Columbia in 1871, Selwyn had spent the next two summers getting acquainted with the newly-acquired North-West Territories. In 1872, accompanied as far as Lake of the Woods by Robert Bell, he made a reconnaissance along the proposed railway route from Lake Superior to Fort Garry. The next year he was in the field from July to October, covering some 2,350 miles between Fort Garry, Fort Edmonton and Rocky Mountain House, returning by way of North Saskatchewan River and Lake Winnipeg. His trip added to his repertoire of methods of travel. In British Columbia he had employed pack train and canoe (including the collapsible canvas boat). Here, on the prairie he used Red River carts, buckboards, canoes, and Hudson's Bay Company batteaux. For measuring distances, on the way out he used an odometer (a wheel with a revolution counter) attached to the cart wheel; and he records that the party averaged 26.4 miles a day crossing the prâiries to Edmonton.

The trip was a frustrating experience to a geologist, for days might pass without a single outcrop to examine. Selwyn noted important coal seams west of Edmonton, and that the chances for oil and salt on the prairies were excellent. His purpose, as on his trips to British Columbia, was not so much to examine the geology *per se*, as to observe what geological problems required solution, and how these ought to be attacked. He also brought back other impressions—the great potentialities of "these magnificent prairie lands" as illustrated by the crops at Rocky Mountain House, the problems the Plains Indians faced with the rapid disappearance of the buffalo, the discomfort of the insect life that assailed his camps in "myriads," "swarms," and "clouds," and the autumnal cold as his party floated down-river in an open canoe.

Bell also spent the summers of 1873 and 1874 in the unfamiliar surroundings of the flat prairies, frustrated by "the great depth of the superficial deposits which almost cover up and conceal the fundamental rocks, together with the nearly horizontal attitude of the latter, and their poverty of fossils."[12] His work in 1873 took him to the southern part of the district, between Qu'Appelle River and the International Boundary. He came back very much alarmed at the unrest which prevailed among the Indians of the region, and suggested that government activities should be kept away from the area until agreements had been made with the Indians. He also reported that the situation would get completely beyond control unless examples were made of the more intransigent Indians or Métis. Selwyn, whose travels had been in a much less threatening part of the prairies, considered these observations unduly alarmist; all that was needed, Selwyn suggested, was to "deprive them of whiskey and take care that they are not permitted to starve from lack

VI-9 G. M. DAWSON'S PARTY, FORT McLEOD, B.C., JULY 1879

VI-10 GROUP OF BLOOD INDIANS OUTSIDE FORT WHOOP-UP, ALBERTA, IN 1881

of a little timely assistance when the Buffalo have failed them, and I am convinced that they would give no cause of alarm."[13] However, he did instruct Bell not to try to reclaim his horses if the Indians should steal them, and to be particularly tactful and careful around the Indians.

Bell returned to the prairies in 1874, but this time he worked in the area to the north of Selwyn's track, between Lake Winnipeg and Athabasca River. From these three traverses, as well as one carried out in 1874 by Bell's assistant J. W. Spencer, a picture of the overlapping sequence of beds began to emerge. In 1874, Bell also was involved in another undertaking on behalf of the Survey. Geologists working in the prairies were faced with the difficulty of reaching satisfactory conclusions where the bedrock was almost entirely covered with drift. Without outcrops to observe, the geologist was lost, so borings seemed the answer, particularly since chances were good that coal, oil, salt, or artesian water might be discovered in the process. The Pacific Railway survey was sufficiently interested in ascertaining the mineral resources along its proposed route to pay half the cost of securing a drill. In 1873 the government advanced $6,000 to the Survey to purchase a steam-driven diamond drill, and the rig, under the direction of Joseph Ward, an engineer from Petrolia, Ont., was sent to the North-West late that summer. The following year borings were made with the Survey's apparatus at Rat Creek and Fort Ellice, by Alexander Macdonald, Ward's assistant; and a crew was sent west by the Fairbank Company of Petrolia to drill at Fort Pelly at a point indicated by Bell. Unfortunately, Bell did not wait for the arrival of the driller but left instructions with the trader at Fort Pelly for Ward to drill at Swan River Crossing, then returned to his own work. The upshot was that Ward drilled the wrong site, for there were two Swan River Crossings. A lengthy correspondence ensued, until Selwyn was glad to drop the whole matter.

Selwyn's report for 1875 referred to the limited success of the activity but added, hopefully, that "so far as they have gone, they lead to the conclusion that no difficulty will be found in obtaining a good supply of water on any part of the western plains at a moderate depth beneath the surface."[14] That year R. W. Ells, who had worked with a diamond drill in New Brunswick, was sent to take charge of the operation. The heavy equipment was moved with difficulty from Fort Ellice to the proposed drill site at the elbow of the North Saskatchewan, and then, when the Indians made menacing gestures, into Fort Carlton itself, where a depth of 175 feet was reached by the end of the season. In his report, Ells outlined the difficulties of moving bulky, heavy equipment across unsettled country, and of having to depend upon unreliable local freighters and unskilled local labour. He observed that half the working season was lost travelling to and from the scene of operations and suggested that a small party would have to be wintered in the North-West for effective results. But the Survey saw little reason for further work on the prairies until the early eighties, when it returned in force to play its part in the opening of the southern plains by the Canadian Pacific Railway.

* * *

FOR ALL THE EMPHASIS upon expanding operations to the western plains and British Columbia, the geological needs of the eastern partners in Confederation could not be neglected. Like the British Columbians, the inhabitants of Nova Scotia and New Brunswick also expected that their provinces would receive their due share of work as part of the Confederation bargain. Besides, in view of the known important resources of coal, gold, iron, and gypsum, the region demanded attention in its own right. Logan had sent Hunt and Michel to Nova Scotia in the fall of 1867 to make a brief general examination of the goldfields and had followed that by a special report on their findings. Then for the summer of 1868 he had deployed a force of seven geologists (including himself) in the two provinces by way of impressing the Maritimers with the value of the service they could now expect from the Geological Survey of Canada.

In New Brunswick the work was principally scientific in tone, Charles Robb being employed to trace the outlines of the Carboniferous basins of central New Brunswick and to correlate the underlying strata with those of Gaspé, while the summer-vacationing geologists, Professor Bailey and G. F. Matthew, were assigned work close to home in southern New Brunswick. There they had at hand an extremely complicated succession of rocks ranging from Precambrian to Carboniferous, and Logan hoped they might work out a table of the succession that would help with identifying the formations farther afield. In Nova Scotia, however, the emphasis was upon economic work that would commend the Survey to the mining interests, and through them, to the provincial government and public. Honeyman was employed in Antigonish and Pictou Counties examining Silurian and Carboniferous formations with particular reference to

VI-11 COAL SLIDE NEAR JOGGINS, N.S., 1879
Photographed by T. C. Weston.

the gypsum deposits near Antigonish; while How investigated iron, copper, and lead ores in Digby County, at the opposite end of the peninsula.

The main effort was devoted to the Pictou coalfield, where Logan and young Edward Hartley conducted separate operations mapping the badly-disturbed coal formations. Their work resulted in a large scale (1 inch = 1 mile) coloured map of the Pictou coalfield, and four reports that amounted to almost 200 pages of text. Together these presented a thorough, detailed description of the geology and mines, analyses of the coals and iron ores of Pictou County, and, as an afterthought, coals from the newly-discovered field at Springhill.

This pattern was continued in subsequent years, as far as staff permitted. Selwyn visited the Maritimes in 1870 to meet the geologists of the region and to examine the goldfields of Nova Scotia with a view to applying his Australian experience there. Robb continued his traverses of the Carboniferous of New Brunswick, notably along the upper Saint John and Miramichi Rivers, but by 1872 he had transferred to the Cape Breton coalfield. Bailey and Matthew continued to work, as much as their regular occupations permitted, on the complex rocks that flanked the Carboniferous basin on the south and southwest. Selwyn respected their abilities and their efforts, but, as he once wrote, the arrangement was "in every respect unsatisfactory" since "no man can serve two masters."[15] The situation improved considerably in 1872, when the Survey obtained the services of R. W. Ells, a recent graduate of McGill and recipient of the Logan Gold Medal in geology, and placed him at work in New Brunswick with Bailey and Matthew. From the outset, though, Ells worked on his own, as when the provincial government requested his services to examine the borings made by a diamond drill it had loaned to a private group to aid them in prospecting for coal. Still the work advanced year by year, and by the end of the decade enough progress had been made to permit the publication of the first three sheets of the geological map of the province, as well as a summary by Bailey of the work of previous years in the *Report of Progress for 1878-79,* his "Report on the Geology of Southern New Brunswick, embracing the Counties of Charlotte, Sunbury, Queen's, Kings, St. John and Albert."

Nova Scotia was considerably better off, since a substantial share of the Survey's manpower was devoted to investigating the geology of this mineral-rich province. Indeed, in his summary report in 1874-75 the director felt obliged to defend himself against charges of unduly favouring the province with: "It may, perhaps, be thought that an undue share of attention was being devoted to investigations in Nova Scotia, to the neglect of that of other extended and less known portions of the Dominion."[16] For the work Selwyn relied on the Survey's own staff rather than on the part-time geologists Logan had employed in 1868, and that continued to be employed in New Brunswick. Young Hartley was making good progress on the coalfields at Springhill and on Cape Breton, when he took sick and died in November 1870, at Pictou. He left a few hurriedly worked-out suggestions for expanding production and improving health and safety standards. His assistant Scott Barlow continued the work in the Springhill coal district for a number of seasons, and published reports on it in 1873-74 and 1875-76. In 1872 the director himself visited the iron mines at Londonderry, at the request of Sir Hugh Allan, the recently selected builder of the transcontinental railway, to report on their extent and productive capacity. As Selwyn wrote, "The establishment of works such as are proposed would be of such vast importance not only as regards the development of the coal and iron resources of Nova Scotia but to the Dominion generally that I think my time could scarcely be more profitably employed."[17]

The situation apparently changed for the better in 1873 when both Walter McOuat and Charles Robb were sent to the Cumberland County and Cape Breton coalfields respectively. Their reports were the first real geological studies of the strata enclosing the coal basins, without which efficient exploration and exploitation of the deposits were impossible. Robb made a start at determining the position of the coal measures in the stratigraphic column and published a table in 1874 tentatively correlating the coal seams of the Sydney field. However, McOuat left Nova Scotia after the single season of 1873 to work on the geology of the Eastern Townships, while Robb resigned in 1875 to return to the mining industry. Their place was taken by Hugh Fletcher, a graduate in 1870 of the University of Toronto with first class honours in the natural sciences, who made the geology of Nova Scotia his life's work, filing his first report in 1875 and his last in 1907.

Although Selwyn may have felt the need to defend himself against the charge of devoting an undue proportion of the work to Nova Scotia, certain parties were convinced that the truth lay in the opposite direction, as witness two articles in the Halifax *Morning Chronicle* in December 1870. These were probably the work of the Nova Scotian geologists H. Y. Hind and D. Honeyman, the former of whom had been refused work because of his connections with mining interests and his lack of professional scruples, while the latter had worked for the Survey in 1868 but had produced a report that was too poor to publish. The articles blasted the Survey for neglecting to recognize the work of the Nova Scotian geologists and failing to promote them in favour of "some favorite son or nephew of some nabob in authority," at Ottawa.[18] They complained, too, that by establishing its museum at Montreal, the Survey deprived Nova Scotia of the benefits of that institution, and that the Survey reports contained glaring inaccuracies through failing to consult the work of the local geologists.

When queried about the accusations by E. A. Meredith, the Under-Secretary of State for the Provinces, Selwyn's indignant retort was that the accusations "deserve only silent contempt."[19] He contended the Survey was already doing more than it could afford in Nova Scotia, and that most of its officers working in the Maritimes were natives of that region, not political appointees brought in from outside. He probably got to the heart of the matter when he wrote to H. Y. Hind that "It is thro'out so ignorantly untrue that it would be only a waste of time to reply to it.—It seems to be rather an attack on the Dominion Government than on the Geological Survey the latter being used only as a convenient peg on which to hang the abuse of the Government."[20] This opinion was corroborated when the newspaper went on to state that the management of the Survey afforded a good illustration of how the Maritimes were going to be treated by the Dominion government.

Closer to home were two areas that received the attention of the Geological Survey in the 1870's. In both, rich mineral deposits were already under development, and they were known to possess others; and each was characterized by some of the most difficult, complicated geological structures anywhere in Canada. These were the Eastern Townships region of Quebec, and the southern margin of the Precambrian Shield in the vicinity of Ottawa River. Sir William Logan's continuing interest in the Eastern Townships was reflected in his work on the stratigraphic succession of the Quebec Group after his retirement. The Survey provided assistance but refrained from involving itself in the controversies over the Quebec Group, notably with Hunt, who was proposing a stratigraphic succession in which the sequence of rock units was arranged according to similarities in mineral composition and degree of metamorphism rather than their positions and structural relationships. After Logan's death McOuat worked there during the season of 1875 before his own death in November of that year. Then Selwyn studied the Quebec Group in 1876 and 1877, and published his own conclusions in the *Report of Progress for 1877-78*. He attacked Hunt's theories ("Unfortunately in Canadian geology, hitherto the stratigraphy has been made subordinate to mineralogy and palaeontology"[21]), but he also expressed reservations about Logan's classification. In place of Logan's three divisions of Sillery, Lauzon, and Levis, he proposed to divide the rocks into three different groups—a Lower Silurian group, a "Vol-

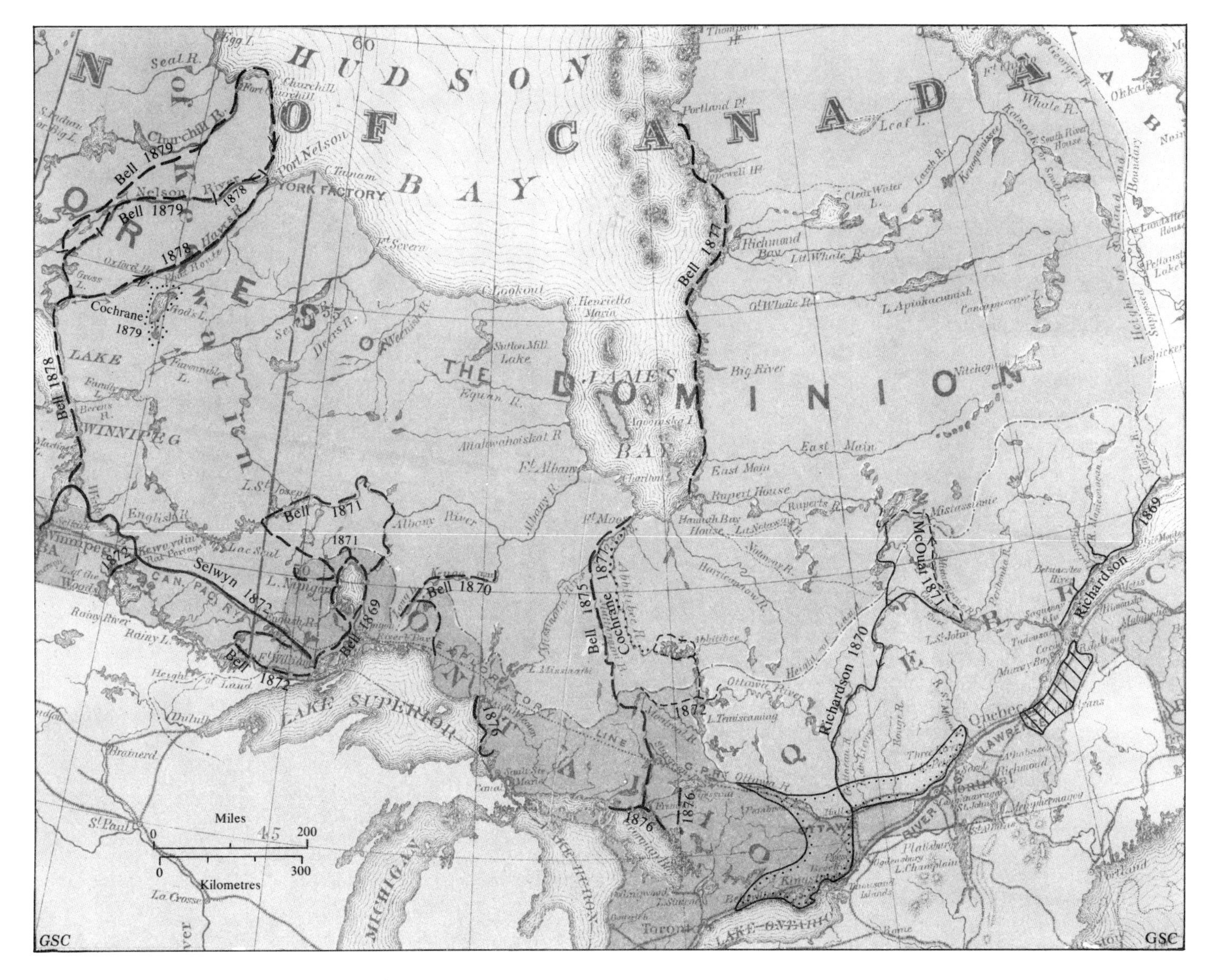

EXPEDITIONS

Bell . . . . . . — — — —

Cochrane . . . .

Richardson . . . ————————

Selwyn . . . . ————————

McOuat . . . . – – – – – – – – –

RECONNAISSANCE AREAS

Vennor 1866-81 (including Ord and McConnell 1878-79)

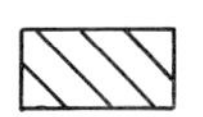

Richardson 1867-68

VI-12 RECONNAISSANCE SURVEYS IN CENTRAL CANADA, 1867-80
Base map—Department of the Interior, *Map of the Dominion of Canada,* 1880

canic group, probably Lower Cambrian," and a "Crystalline Schist group (Huronian?)."

This brought down abuse on Selwyn from all sides, but he continued in his course, and proceeded to a drastic, long-overdue redefinition of the nomenclature of the lower half of the Paleozoic. In 1879 he and Webster reexamined the country between the Chaudière and the Vermont border, and then he joined his opposite number for New Hampshire, Professor Hitchcock, to review the points at issue. Next he redefined the Paleozoic succession for Canada on a basis more consistent with international practice. Henceforth the term Silurian would include only the former Upper and Middle Silurian, while Lower Silurian was to be divided into Cambro-Silurian and Cambrian. Selwyn explained his action as follows:

> It having, in many cases, been found impossible to identify, trace out and define the limits of the above-named subdivisions, it has become necessary to adopt the more comprehensive nomenclature, and at the same time to define its precise signification as now used, and it is proposed henceforth to use it in the reports and maps of the Canadian Geological Survey.[22]

Though this still did not solve the disputes over the precise divisions of the Quebec Group, or their distribution, it had the effect of changing the problem. The rocks would now be looked at in a different context—their equivalence with the members of the established international succession—in which sense they would be redefined, renamed, and given different boundaries. The re-examination of the Quebec Group was carried out mainly by Ells, who studied its northeastern extension in northern New Brunswick and Gaspé in the early 1880's and gradually worked his way south from there. But much would remain to be done over many years before the problem of the Quebec Group could be satisfactorily resolved.

In the southern margin of the Precambrian, skirting the Paleozoic strata of Ontario and Quebec, was a region known to contain valuable mineral deposits, notably gold, lead, graphite, magnetic iron, mica, and apatite (a phosphate mineral). The work started by Macfarlane in 1865 in the north of Hastings County, was continued in 1866 and subsequent years down to 1881 by H. G. Vennor. Vennor began in the Marmora-Madoc district but gradually worked his way east to the Rideau Lakes and north to Bonnechere River, where he connected with Murray's work. He crossed over to Quebec in 1876, working east past Logan's Grenville district to the Laurentides north of Montreal. He was assisted in the years after 1876 by L. R. Ord, who, in partnership with R. G. McConnell in 1880, carried the work farther east towards the St. Maurice. Vennor's principal geological task was to trace the Precambrian formations, a work similar to that carried on by Logan's assistant Lowe until 1870 in the country north of the Ottawa. As Vennor gained experience, he attempted to work out the relationship among the different Laurentian formations he encountered—crystalline limestones, gneisses, the Hastings Series, the Grenville Series, and the anorthosite of the Laurentides. On the basis of Vennor's report in 1880, Selwyn was confirmed in his opinion that the anorthosite "or Norian rock of Hunt" was an intrusive rock. These extremely complex rocks would require the work of far more skillful and experienced geologists before they would begin to give up their secrets.

In addition, Vennor studied the occurrences of mineable minerals and mining activities throughout the territory, noting good prospects for possible development, and discussing mining methods, equipment, operations, marketing possibilities (for example, those resulting from the opening of the railway from Kingston to Sharbot Lake) or economic difficulties (the impact of the depression of the 1870's). Besides these highly informative mining studies, he also was required to exercise the art of land

VI-13 R. G. McCONNELL

surveyor to a surprising degree. He found the existing land surveys so unsatisfactory that for a time he followed the method of completely resurveying whole townships, then going back over the country with his own topographical map to draw in the geology.

From the Ottawa and the upper Great Lakes, work continued in earnest on Canada's largest geological province, the Precambrian Shield. In the east, Richardson in 1870 and McOuat in 1871 and 1872 made remarkable canoe trips north from the edge of the Shield in the direction of Lake Mistassini. Richardson returned to the Ottawa via Gatineau River, but McOuat investigated more thoroughly the belt of sedimentary and volcanic rocks Richardson had discovered, and found that they stretched from Lake Mistassini to Lake Timiskaming, where they joined Logan's Huronian rocks. On the west, as well, Robert Bell continued mapping the Precambrian formations north of Lakes Huron and Superior, carrying on the work that he had begun in 1869 in the Lake Nipigon and Thunder Bay areas. In 1870 he explored between Michipicoten River and Lake Nipigon, covering 1,080 miles in track surveys. In 1871 he com-

VI-14 HAULING A YORK BOAT OVER THE ROBINSON PORTAGE
Engraved from a photograph by Robert Bell, taken in 1878 during his reconnaissance of the region between Lake Winnipeg and Hudson Bay. The 1,315-yard portage between Echimamish and Franklin's Rivers was used by travellers in crossing from the Nelson to the Hayes watersheds.

pleted the mapping of Lake Nipigon and proceeded north to Albany River, returning to Lake Superior via Kenogami and Pic Rivers. In 1872 his work was farther west, between the head of Lake Superior and Lake Winnipeg. Then, following two seasons on the prairies, in 1875 Bell ranged northward from Lake Huron to James Bay. These journeys, in which the geologists made their own maps, were conducted principally in canoes along the ubiquitous waterways, using the magnetic compass to measure directions, and estimating distances by elapsed travelling time, or occasionally, by Rochon micrometer. From time to time astronomical observations were taken in order to control and correct the ground measurements.

Since Bell was the practitioner *par excellence* of his method of reconnaissance survey, his exposition of the advantages of this technique of exploration, as set forth in the *Report of Progress for 1866-69,* is of interest:

> Throughout the season, I worked upon the principle which we have always pursued upon the Geological Survey, in exploring or surveying in a new region namely, that of following, as much as possible, the water-courses, instead of cutting "exploratory lines" in the woods. The following may be mentioned amongst the many advantages of this system: (1) We avoid the expense of cutting the lines, which would add but little to our knowledge of the natural features of the country, and would soon be obliterated. (2) The clear space afforded by the surface of the water serves better than an artificial opening in the woods for measurements by the micrometer, which may be made as accurately as by the chain. It also admits of triangulation, which is impossible in the forest. (3) A greater distance may be surveyed per day. (4) A smaller party can do the work. (5) The canoes or boats by which the survey is carried on, also serve to convey, at the same time, the supplies of the party, and allow of a considerable quantity being taken, thus enabling the work to go on continuously for a greater distance, or a longer time, without reference to the base of operations, than where everything has to be carried on men's backs. (6) The riverbeds and lakeshores afford many more exposures of the rocks than are to be met with in the woods, where they are covered by earth or moss. (7) The same measurements, which serve to determine correctly the distribution of the rock-formations, also enables us to lay down the topographical features of the country, and thus we obtain a knowledge of its geography simultaneously with that of its geology.[23]

Much of the value of these reconnaissance journeys lay

in the maps the geologists produced, which were usually the first accurate ones of the routes they traversed. Another was the information about the features and resources of little-known sections of Canada. Just as the Survey had advised the Canadian Pacific survey on the potentialities of various railway routes, so it also informed other agencies of government about travel routes, forests, water-powers, wildlife, native populations, climate, and engineering problems. In his traverse of the Albany in 1871, for example, Bell delved into the agricultural possibilities of the area by examining the previous 40 years' records of the Hudson's Bay Company's establishment at Martin's Falls; and he reported that the Albany was navigable for shallow draft boats from that point all the way to James Bay. Besides, these visits carried into the regions the first affirmation of Canadian authority, and they afforded welcome companionship to lonely wilderness settlements, as witness the many correspondents Bell secured among fur traders and others as a result of his journeys.

In the geological sense, these reconnaissances by members of the Survey could do little more than catalogue the kinds of rocks that they encountered along a narrow strip of country adjoining their traverse. Scores of miles of uncharted country generally lay between two track surveys, an area that for many years was filled in on the topographical and geological maps largely by guesswork. The officers lacked the time during these surveys to collect enough data, nor did they have the specialized training and the knowledge of the basic principles to generalize correctly from the information they did secure. The elucidation of the difficult Precambrian problems would have to await careful systematic work later, after the basic principles began to be worked out and the men with specialized training were available.

* * *

SUCH WERE THE PROBLEMS AND CHALLENGES that faced the Geological Survey in the new Dominion of Canada under the directorship of Alfred Selwyn. The Pacific coast, that sea of mountains, the endless prairie, the vast lake-studded, rocky Precambrian Shield, the complex formations of Nova Scotia, New Brunswick, Gaspé, and the Eastern Townships—any one of these could have fully occupied a staff the size of Selwyn's for several lifetimes. How could the Survey, given its present personnel and means, best be deployed to meet and accomplish its gigantic burdens? This was the central problem that Selwyn had to face while maintaining the Survey's position in relation to the government and people of Canada.

For the Survey in these years underwent major changes in its administrative structure, in its relations with the government, even in the nature of its field activities. These were reflected in the Survey act of 1872, and particularly that of 1877. With the current act under which the Survey had operated since 1867 due to expire on July 1, 1872, Selwyn requested a moderate increase in the annual appropriation from $30,000 to $40,000 in the forthcoming renewal of the act, so as to augment the field staff with men possessing "the training and acquirements absolutely requisite for the creditable performance of the work."[24] Somewhat more diffidently, he pleaded for an increase in his own salary to the amount he had been paid in Victoria and had been given the impression he would receive in Canada, instead of which he had been kept at Logan's modest level of $3,000. The cabinet more than met his request—as of July 1, 1873, the Survey's appropriation would become $45,000 a year, and Selwyn's own salary $4,000 per annum.

When the bill embodying the increase to $45,000 was presented to Parliament, it touched off a debate reminiscent of that of 1864. The Leader of the Opposition, Alexander Mackenzie, again urged instead, an annual grant coming yearly before Parliament. Selwyn afterwards wrote to Mackenzie to inform him of the Survey's view on this and other matters:

> The five years' clause in the Act is not as supposed an expression of want of confidence in the action of the Parliament, but for the purpose of avoiding the annual delay which would otherwise necessarily arise, and which, owing to the period at which the supplies are generally voted, would often prevent operations being commenced till long after the time when, to secure a good season's work all arrangements should be completed, and the explorers be already in the field. It also enables the Director to secure and retain the services of a far higher class of men, than he would otherwise be able to do, and to plan his operations in advance, and in the way best calculated to secure valuable results.[25]

An interesting constitutional point was raised during the debate by David Mills, the member for Bothwell, Ontario—that since the provinces had charge of mineral resources within their borders, geological surveys were really a provincial responsibility. The question of jurisdiction was of considerable importance in the United States, but Joseph Howe did not take up this constitutional challenge. Instead, he merely observed that the present system was the best one, since the Dominion was in a better position than were the several provinces to hire first-class scientists.

The point raised by Mills respecting the status of the Geological Survey of Canada deserves a brief comment. In the founding provinces, there was at the very least an understanding regarding federal authority in this field, because Sir William Logan, who assuredly discussed the matter with the principals involved, encountered no opposition whatever in extending the Canadian Survey to the new provinces, hiring the leading geologists, and commencing work in the Maritime provinces. In the provinces of Manitoba, British Columbia, Prince Edward Island, and Newfoundland, a federal government obligation to conduct geological surveys was expressly written into the arrangements under which they entered Confederation.[26] The Dominion government's responsibility for managing the natural resources of federally-controlled territories entails the authority to have them surveyed, geologically and otherwise, as help with their proper management. Moreover, the various acts passed by Parliament have assigned to the Survey functions that extend over the whole of Canadian territory, either implicitly (for example, by proclaiming its museum "a collection for the whole Dominion of Canada" in the acts of 1868 or 1872), or explicitly (as in the act of 1877, to "take charge of and conduct the Geological Survey of the several Provinces and Territories of the Dominion"). Hence it appears that a federally-sponsored geological survey *must* be carried on because of express commitments and understandings made with several provinces of the Dominion; and that it *might* be carried on because of the Dominion's continuing responsibility for federally-administered territories and their resources, and because of its own legislation.

At the same time, provinces have never been prevented from operating their own geological survey organizations. Some, notably Ontario, were doing so before the end of the last century. Interestingly, as early as February 14, 1868, in one of his letters to Sir John Rose, Logan had anticipated this development and proposed a solution: "The present constitutional arrangements place the mineral lands within the control of the local governments, and it might properly belong to them to undertake the more detailed examination and description of mines and useful minerals, leaving to the general survey the investigation more immediately belonging to physical structure."[27] This, obviously, would leave the federal Survey unhampered in the task of studying basic geological structures that override provincial boundaries, for the better elucidation of the geologies of the several provinces. But he went on to suggest that his Survey might be expected, all the same, to function in both the general and the local aspects of the work.

When the Survey act next came up for renewal in 1877, a depression was in full swing, and a Liberal government was in office, with Alexander Mackenzie as Prime Minister, and David Mills was the newly-appointed Minister of the Interior—the department that now had oversight of the Survey. No move was made to disband the Survey in the name of provincial autonomy, but the new act (39-40 Vict., Chap. 9) reflected the stands of the Liberal party during the debates of 1864 and 1872 by introducing the first major alterations in the structure of the Geological Survey since its inception. Henceforth the Survey would be a regular branch of government, enjoying permanent status and no longer requiring periodic reaffirmations of its existence. Its 35-year career as a virtually private organization operating under contract for the government was ended. In future it would be a "branch of the Department of the Interior known as the Geological Survey Branch," operating in "the several Provinces and Territories of the Dominion." Its appropriations would now be annual grants, subject every year to parliamentary vote and audit. Ended, then, were the grants for five-year periods capable of being transferred from

VI-15 JOSEPH HOWE
Secretary of State for the Provinces, 1869-73

year to year as required, which Logan had used so effectively because he had the wherewithal to act as its private banker, and had been permitted to do so because of his known great integrity and his reputation for managing the Survey in the public interest.

The new act placed the staff of the Survey on the same footing as the other employees of the Department of the Interior by empowering the minister to direct them "to perform any duty in or with respect to any other branch." It also extended an important benefit to the staff by bringing all the permanent employees within the provisions of the Civil Service Act for superannuation and other benefits. The lack of civil service status had certainly been a serious problem for most of the members, since only the director, chemist, and paleontologist had been ranked as eligible for civil service benefits. Selwyn in December 1875 had tried in vain to have the remaining fifteen members of his staff brought on the civil service list. The absence of such protection was on everyone's mind, for three deaths from illness had occurred in the early seventies; those of Edward Hartley in November 1870, Horace Smith in June 1871, and Walter McOuat in November 1875. The last case was particularly distressing. McOuat, who was only 33, left a wife and three small children, and the Survey could only provide a gratuity equal to two months' salary—the princely sum of $183.32. The lack of provision for superannuation or security for one's family must have been a considerable worry; under the new act, employees became eligible for superannuation and could go about their work more confidently, and therefore more efficiently. In 1880 two employees became the first beneficiaries of the new superannuation provisions—Robert Barlow, aged 67, and a most unwilling James Richardson, aged 70.

Though the Liberals were the party of economy, they made no effort to reduce the annual appropriation allotted to the Survey. However, while they left the appropriation at its current level, they increased the duties they expected the Survey to perform with the existing staff and budget. Section 3 of the act added to its functions the important new duties "to study and report on the fauna and flora of the Dominion," and "to continue to collect the necessary materials for a Canadian museum of natural history, mineralogy and geology."

For a whole generation the organization carried a new official title of "The Geological and Natural History Survey of Canada," and this was reflected by a significant change of emphasis and approach to the work. Selwyn had resisted an earlier effort to make the Survey sponsor botanical explorations, arguing they bore no relationship to geological work and would require special parties to conduct the field work, as well as a special supervisory staff in the office. Now he had no option. He had to issue instructions to field geologists to collect natural history specimens "when the doing so will not interfere with the main objects of the exploration. Even when collections of these cannot be preserved, useful and interesting observations on occurrence and distribution of species can be given."[28] Geologists, complying, began gathering specimens of the fauna and flora, observed forests and other vegetation, collected artifacts and vocabularies, and made anthropological notes on native societies.

Some of this already was being done by the Survey, but the new legislation encouraged the naturalist proclivities of the officers and directed them away from their main tasks of geological reconnaissance, observation, and reporting. Still, the nation was ultimately the gainer. To this provision of the 1877 act, Canada owes some of the most valuable natural and human history collections in the present-day National Museum of Canada. The new field of study also led to new staff being secured who were qualified to deal with the new sciences. Dr. John Macoun, whom Selwyn had employed in conjunction with his own surveys and would gladly have hired permanently, became the first of these in 1882. Eventually a considerable staff qualified in sciences other than geology would be employed by the Survey. However, inasmuch as no additional funds were provided for this work, the immediate effect was to encumber the Survey with new responsibilities without supplying the means for properly discharging the old ones.

The act of 1877 also attacked the last vestige of the Survey's special status—its being located in Montreal, away from the seat of government. This had occasioned some minor difficulties. Selwyn had had to go to Ottawa periodically for discussions with E. A. Meredith, the deputy minister of the Department of the Interior, or with the minister. Officers occasionally had to go to Ottawa to testify, as Bell did in 1870 before a Commons committee on immigration and agriculture. But against this there were the many advantages of living in Canada's largest, wealthiest, most advanced city, rather than in the backward backwoods capital. Now, under the act, the government was empowered to "direct the removal of the museum, and the officers and others connected with the Geological Survey Branch of the Department of the Interior, to the City of Ottawa." Such a move, it was felt, would bring the Survey into closer relations with the cabinet and Parliament, with other branches of government interested in exploration, surveying and mapping, and more directly under the scrutiny of Parliament. For the present this remained only an idea, but within a few years it was destined to be implemented.

Not everything about the Survey's situation in Montreal was satisfactory. Its accommodations in the St. Gabriel Street building had become quite inadequate by 1869 for all the services the organization was expected to perform, and as well, to house the museum's rapidly growing collections. In 1873 Selwyn proposed having the building enlarged. The Dominion Architect looked over the premises and the Secretary of State for the Provinces approved the $6,000 estimate for the needed addition—only for the project to be dropped because of the developing economic depression. As so often in the past, Sir William Logan again came to the rescue. He constructed a building directly behind the Survey's quarters and offered to rent the space the Survey needed for $1,000 a year, a mere four per cent return on his investment. All that was needed was to cut a door between the two buildings, and the addition would provide offices (including Selwyn's), drafting rooms, and a good deal of storage space, leaving the old building for laboratories and a considerably enlarged museum. The rental arrangement was authorized in February 1874, the Survey's space was very nearly doubled, and its needs were adequately served for the remainder of its stay in Montreal. Here was another example of the benefits the Survey continued to derive from Logan's active and continuing interest. It was almost his last service, for he left Canada that autumn and did not return.

Although the Survey was expected to deal with an area the size of Europe and collect and report on natural history as well as geology, its staff was increased only slightly during the seventies, from about sixteen full-time professional staff in 1870 to twenty-one a decade later. There were other changes in personnel, of course, for officers were lost through death, resignation or retirement, including Hartley, McOuat, and Robb from the field staff, who were succeeded by Ells, Fletcher, Dawson, McConnell, Wallace Broad, A. S. Cochrane, and W. H. Smith. In 1872 Hunt's large, important role was partly filled through a half-time arrangement with B. J. Harrington, a graduate of Yale who lectured in chemistry and metallurgy at McGill University.

Later, when the university was able to employ Harrington full-time, he resigned from the Survey, to be succeeded by Frank D. Adams. On the chemical side, Hunt was replaced by a young chemist from Australia, G. C. Hoffmann. Selwyn had known of his work, and when Hoffmann came to America but failed to find a suitable position, Selwyn appointed him to the Survey in 1872. While Harrington and later Adams continued Hunt's petrological and mineralogical work, Hoffmann confined himself to chemical analyses. Following the death of Billings, the Survey's paleontologist, in 1876, Selwyn appointed J. F. Whiteaves, the English-born recording secretary and curator of the museum of the Natural History Society of Montreal, who had been working for some years for the Survey as Billings' part-time assistant. Whiteaves proved a valuable addition to the Survey staff not only as paleontologist but also as the effective museum director. He prepared separate annual reports, a noteworthy series of reports on Paleozoic and Mesozoic fossils, and a series, *Contributions to Canadian Palaeontology,* in which the guide fossils for various rock units were carefully described, many for the first time. Equally important, was his aid in identifying the field men's fossil collections from their widely-scattered areas of operations, and helping the officers with their correlations by establishing the positions of their fossils in the stratigraphic column. Other additions to the museum staff in the seventies were A. S. Foord and C. W. Willimott.

A measure of the growing specialization, as well as recognition of the contributions of the officers in the various fields, was the establishment in 1877 of four positions of assistant director, awarded to the field geologists Dawson and Bell, and to Harrington and Whiteaves of the inside staff. Designed to introduce a more formal chain of command and to prepare the organization for a time when it would acquire a staff commensurate with its enlarged responsibilities, the move reflected the Survey's role as a national institution soon to be based in the national capital.

As in Logan's day, the Survey under Selwyn continued to function as one of the principal agencies for publicizing Canada's natural resources at home and abroad. It assisted government departments with the decennial censuses, lent unpublished maps to the compilers of the Dominion Atlas, and supplied information and advice on remote sections of the country to several departments. The public also was better informed about the geography of Canada through the numerous newspaper articles, lectures and speeches given by the geologist-explorers, particularly Robert Bell, as well as by their publications in Survey reports and scholarly journals. An interesting step to bring geology to the people was the preparation of kits of mineral and rock samples for distribution to schools, colleges, and other educational institutions. Weston and the new museum assistant Willimott began assembling the kits as a sideline, but the demand developed so quickly that the work eventually took most of Willimott's time. From a first kit of 277 specimens, supplied in 1875 to the school trustees of Elora, Ontario, the number quickly rose to 39 in just two years. This service, now numbering thousands of sets per annum, continues.

Of all its services in the information field, perhaps the most effective, though it required a great deal of effort, was the Survey's role as exhibitor at important fairs and exhibitions, most notably the Philadelphia International Exhibition of 1876 and the Paris Exhibition of 1878. These exhibitions publicized the Survey's work and brought the mineral resources of Canada to the attention of large and influential audiences. They afforded lucky officers welcome relief from the routines of office or field work, and brought them into contact with new commodities, processes, and equipment on display by other exhibitors, and with scholars and administrators from all over the world. Their preparation and presentation was a considerable drain on the energies of the Survey however. Selwyn spent two months (March and April) in Paris setting up the exhibit, then the remainder of the year managing it or serving as president of the jury charged with judging the maps, geographical, and cosmographical apparatus (Class XVI). In all, he was absent from Montreal between February 5 and December 30 in 1878. The field parties were put to considerable effort and expense collecting the desired specimens, in preparing the catalogues, and setting up and disposing of the exhibits. Sometimes the participation was also a sizable drain on the Survey's limited funds.

Still, the results were worthwhile. They publicized Canada's mineral resources and brought them to the attention of interested capitalists and settlers; and they contributed to a sense of national pride. One Canadian visitor to Philadelphia commented that "Her department looked more like one of the old and fully developed nations, rather than of a comparatively small and young colony,"[29] while the Montreal *Gazette* of July 22, 1876, observed of the Philadelphia exhibit that "Canadian visitors cannot but be highly pleased with the ostensible rank which the country takes among the great nations; her products, natural and manufactured, are so many and varied as to astonish even a well-informed Canadian himself. This is no flattery, no puff, and will be acknowledged on all hands."

The permanent record of the Survey's achievement, covering the entire range of its activities, in contrast to the temporary brilliance of participation in great exhibitions, appeared in the annual reports of progress. These reflected Selwyn's guiding hand in their style and scholarly excellence, for Selwyn put a good deal of effort into them, including much rewriting of badly-prepared manuscripts. A master of clear English expression himself, he often was faced with reports that contained errors, or were so badly written as to be misleading, and then had to decide whether to let them go forward after editorial revision, or to withhold them from publication pending satisfactory rewriting. There were frustrating delays because some officers were slow in preparing their reports and maps, and sometimes returned to the field before the previous year's reports were completed. Then there were the slow printing processes, the technical language that occasioned delays while proofs were being checked, and the director's own absences. These all conspired to make regular and punctual issue of the reports highly problematical. Still, though they did not always appear as speedily as some people wished, nor include all the work they hoped to see, the reports did appear in regular sequence during the seventies, a volume a year, and ten for the decade from 1870-71 to 1879-80, a regularity not matched since Logan's earliest years.

Of all the connections binding the Geological Survey to Montreal, none was closer than that with McGill University, with its endowed Logan Chair of Geology, and the Montreal Natural History Society. Particularly intimate were its relations with Principal Dawson, the moving spirit behind the university and the society, who probably exercised a greater influence on the conduct of the Survey and on its work than anyone except the director himself. Recognized as one of the leading Canadian geologists of the period, he was often used as a consultant. His son, too, was playing an increasingly important and respected part in the affairs of the Survey as an assistant director by this time. The father commanded great influence in the highest scholarly, political and financial circles, which he used to advance the work of the Survey and the careers of people in whom he had a special interest, like his son George and the chemist Harrington, his future son-in-law. Without Dawson's powerful support for the Survey during these years, it might not have achieved so much.

The close connection between the Survey and the scientific community of Montreal inspired a most vociferous opposition to the removal of the Survey to Ottawa, a matter that had been argued almost from the beginning of the decade. The proposal presented a frightening prospect, especially to people in Montreal interested in science. So when the Geological Survey bill of 1877, with its provision for the possible removal of the Survey to Ottawa came before Parliament, the Montreal M.P.'s were vocal in their opposition. One argued that the usefulness of the museum would be destroyed if it were removed to Ottawa, since the Members of Parliament would take no interest in it and scientists would have no access to it; another, that the removal could only be regarded as an insult to the city of Montreal.

Selwyn was not pleased at the prospect of moving the headquarters, for he foresaw it as bringing the Survey under the direct scrutiny of Parliament and of the minister, and allowing the Members of Parliament better facilities for interfering with its operations. Accordingly, he found strong arguments against leaving Montreal—that removal of the facilities would weaken McGill and other educational institutions in Montreal, while the Survey would be separated from the greatest centre for mining capital in the country. He also mentioned the loss of the valuable assistance of Principal Dawson and McGill personnel, and the possible breakage of valuable specimens during the move. In short, the Survey's "removal to Ottawa could not fail to operate in every respect most prejudicially."[30]

But despite the objections, the government in 1878 went ahead with acquiring facilities for the Survey in Ottawa. The building finally selected and ultimately purchased, at a cost of $20,000, was the Clarendon Hotel on the corner of Sussex and George Streets, a former luxury hotel that had fallen on difficult times but recently had been the locale of the first exhibition of the Canadian Society of Art under the patronage of the Marquess of Lorne, the Governor General. The implementation of the government decision brought out a concerted opposition. The Montreal Natural History Society in 1881 officially registered its opposition to the move, to which Principal Dawson added a personal letter to the Prime Minister, Sir John A. Macdonald, reciting his objections. The Montreal Board of Trade also sent a delegation to Ottawa to protest early in 1881, although Selwyn had warned it that the government's decision was past changing.

Indeed, there remained nothing for the Survey to do except move, and move it did in April and May of 1881, without serious incident or dislocation. The greatest problem was one that had not been considered at all: the matter of the Survey's right to many of Sir William Logan's effects. Now that the Survey was in the process of moving, it was discovered that many of the instruments and specimens in the museum belonged not to the Survey but to the estate of Sir William. The Survey had to pay nearly $8,000 for equipment it had been using, in order to avoid having to replace it at even greater cost.

By the summer of 1881 the move was completed, at a cost to the government of a little over $10,000, excluding the payment to the Logan estate. The dire predictions of the effect of the removal to Ottawa were not realized. In Montreal, Sir William Dawson secured a large gift from the Redpath family to rebuild McGill's museum. As for the Survey, in its new location in tiny Ottawa it seemed more popular than ever. Selwyn noted that the visitors to the museum in its first full year in Ottawa had numbered 9,549, whereas the previous high in Montreal had been 1,652. Evidently the accessibility to interested parties had not been affected so adversely as had been predicted. With the Geological Survey now transferred to the nation's capital Selwyn could carry forward with the work begun in the seventies. The range of activities and area of operations had been enlarged. Now, using the opportunities his new proximity to the centre of political authority afforded him, it was for Selwyn to do justice to the immense tasks that lay ahead.

PART TWO

Growing Pains

1881–1907

VII-1 HEADQUARTERS OF THE GEOLOGICAL SURVEY, 1881-1910
547 Sussex Street, Ottawa

7

Washing Dirty Linen

THE YEARS THAT FOLLOWED THE MOVE to Ottawa in 1881 brought serious tests and trials to the Survey and its director. In Montreal the Survey had enjoyed a sheltered existence near the campus of McGill University, and its relationship with the government that paid its expenses was like that of a modern-day Crown corporation, that left it free to follow its own programs with little or no interference. The move to Ottawa was ordered by the government of the day in part to remove the Survey from this protected environment and bring it more directly under federal control, to do the government's bidding in return for continued financial assistance.

The first few years in Ottawa were necessarily a time in which a new relationship was being worked out between the Survey and the Dominion government. Did the federal government intend to use the Survey as an agency for the rapid exploration and mapping of the vast unknown territories of Canada? Would it pursue a 'hands off' policy, leaving the Survey free to direct its own program? Would it be generous with funds, or would it, in Sir William Dawson's words, fall "into the popular mistake of limiting our scientific expenditure by a narrow and slavish utilitarianism which defeats its own end?"[1] And what of the Geological and Natural History Survey of Canada? Would it have to compromise its programs, and if so, how far and in what directions? Or, could it actually improve its position and gain stronger support from a government in the midst of which it now found itself?

By the early eighties the Survey was well on the way to implementing an employment policy that Selwyn had enunciated as early as 1871, to upgrade the scientific attainments of his staff and "to give the preference to young men who have received an education specially fitting them for the work, and who wish to make it their profession."[2] With few exceptions, his appointments by the eighties were drawn from the ranks of the university graduating classes, an indication of the way Canadian universities had expanded their programs in geology and the other sciences, as well as a testimony to the great advances in geological science since the middle of the century. Then, geology had not been a very advanced science, beyond the abilities of men like Murray, Richardson, Bell—or Logan or Selwyn for that matter—to master through private study and practical experience. Increasingly, however, geology was becoming a complex science based on a very considerable accumulation of ascertained facts derived from all parts of the world, and an ever-enlarging body of hypotheses and principles. Far more training was now required for a geologist to make himself competent. The same held true for paleontology, mineralogy, chemistry, and the other sciences with which the Survey had now to concern itself.

Moreover, the kind of work the Survey was expected to perform was also beginning to change. In Logan's day field work had consisted mainly of geological reconnaissance and mapping, and not especially of interpretation. There was still vast scope in Canada's unexplored northland for that traditional form of activity, but there was need, too, for concentrated work on difficult formations or on important highly mineralized districts that only specially-trained personnel could perform. More than formerly, the continuing success of the Survey depended on the abilities and training of the young men secured by Selwyn. That he obtained so many highly talented new staff members was a tribute to his judgment and to the high reputation of the Survey, as well as to the quality of instruction available in the Canadian universities. Thanks to the pool of talent becoming available, Selwyn was able to raise the qualifications demanded of new members of his staff, and to elevate the work to professional status. Eventually, in 1890, he would place in the Survey act a daring specification that new employees on the scientific staff had to be university graduates, or their equivalent.

During the seventies Selwyn had hired a number of young, university-trained scientists, chiefly from among the graduates of McGill. After 1880, McGill's pre-eminence became less pronounced, however, even though it contributed A. E. Barlow, a younger son and a brother of the Survey's draftsmen and a Logan gold medallist, added to the field staff in 1879; Albert Peter Low, a graduate of McGill with first class honours in geology, and Henry M. Ami, a Dawson gold medallist, both appointed in 1882, Low as a field explorer and Ami as assistant paleontologist. University of Toronto graduates included Joseph B. Tyrrell in 1881, and Andrew C. Lawson in 1884. Tyrrell was something of an exception, an arts graduate with an avocation for the natural sciences. But his career demonstrated that self-training could still suffice, for he was to become a major name in the Survey, though, significantly, mainly in reconnaissance mapping rather than solving complex geological problems. Lawson, by contrast, was a gold medallist in geology, and thanks to leaves of absence to attend Johns Hopkins University during the winter months, he became in 1888 the first officer of the Canadian Survey to hold a Ph.D. in geology—a harbinger of the day when the Ph.D. would become the accepted standard.

VII-2 J. B. TYRRELL

Other universities that contributed officers in this period were Royal Military College, Kingston, one of whose early graduates, Lawrence Lambe, joined the Survey in 1885 as assistant paleontologist. William McInnes and E. Rodolphe Faribault, hired in 1882 and 1884, were graduates of the University of New Brunswick and of Laval University, respectively. Both were students of professors who worked for the Survey in summers, McInnes of L. W. Bailey, and Faribault of the Abbé J. C. K. Laflamme.

Another trend evidenced by Selwyn's appointments was the selection of men qualified in specialties other than geology and its laboratory ancillaries. The selection of Donaldson B. Dowling was significant in this respect, for Dowling, a gold medallist at McGill, was a civil engineer, and his appointment in 1885 implemented Selwyn's plan of employing topographers to accompany the geologists in the field and assume the main burden of preparing the base maps, freeing the geologists to concentrate on the geology. In 1883, too, Selwyn made a start towards developing a mining section by hiring two mining geologists, Eugene Coste and Elfric D. Ingall, graduates of the École des Mines in Paris and of the Royal School of Mines in London. The year before, he had also made a beginning in the natural history field by appointing Professor John Macoun of Albert College, Belleville, Ontario, and in 1883 he added a taxidermist to the staff to cope with the zoological specimens for the museum. The newcomers also reflected the slow turning of French Canada away from the traditional professions and academic interests towards the new field of science and technology. Ami, Faribault, Coste, and Laflamme—the last-named being a one-time pupil of Hunt's who worked many summers for the Survey in Quebec and who has been called the father of geology in French Canada—were the first group of French origin to join the scientific staff in Selwyn's day. Others included N. J. Giroux, a field geologist, and the draftsmen C. O. Senécal and O. E. Prud'homme.

What lay behind this remarkable expansion of personnel was a sudden improvement in the Survey's financial position. In July 1883 the permanent officers' salaries, hitherto carried on the Survey's budget, were transferred to the Civil List, which now covered the salaries of 31 employees, including the director; 4 chief clerks (the assistant directors); 5 first, 5 second, and 8 third class clerks; the librarian, resident caretaker, and messenger, etc. This increased the funds available to the Survey by about $30,000 a year, for the annual appropriation remained at $60,000 as before. The extra funds went to increase the number of field parties, to expand the office facilities and ancillary services in Ottawa, and also to hire additional staff as temporary clerks who were paid out of the annual appropriation; thus establishing two different classes of employees within the Survey. The terms of the Civil Service Acts of 1882 and 1885 rated the professional and technical officers as clerks, and the assistant directors as chief clerks, with salaries attached to these grades that were wholly inadequate for the highly-trained specialists the Survey required. Furthermore, they provided that each employee entering a particular class, had to be paid the minimum salary for the grade. Selwyn was thus placed in a dilemma; he could not hire a qualified young university graduate at a third-class clerk's salary of $400, while to appoint him immediately as a second-class clerk at $1,100 (on the limited establishment for the class) would be unfair to more senior men. Consequently, to overcome the difficulty, Selwyn paid some of his new officers from the annual budget, though as temporary staff they would also be excluded from the benefits of civil service status. As late as 1890 Barlow, Dowling, Lambe, and McInnes (of the newcomers cited above) were still being paid as temporaries out of the annual appropriation. The system was unwieldy, unfair, and bound to breed discontent and jealousies among the "temporary" employees.

None of the salaries could be termed liberal, and most of the staff could have earned better incomes outside the Geological Survey; especially the field geologists, who with a few seasons' experience could command good salaries in the universities or elsewhere. Selwyn found himself—like succeeding directors in certain other periods—training men, then losing them to more lucrative employments just when they became particularly useful to the work of the Survey. Arthur Foord left in 1883 to become curator of the York Museum, Harrington and Adams to teach chemistry at McGill, A. C. Lawson in 1890 to teach at Berkeley University, and J. W. Spencer to join the Geological Survey of Georgia, then the University of Missouri. These men received far better salaries in their new positions, often more than double those paid in Canada. A government not fully convinced of the necessity of paying adequate salaries to retain qualified professional staff, hampered the effectiveness of the Geological Survey.

The financial relief of the transfer of the senior officers' salaries to the Civil List did not last long. Salaries kept rising, as did publication expenses and the costs of the supporting services, and before long Selwyn was forced to major economies. In 1885 he notified 12 employees on the temporary list, including Macoun, Ingall, and A. E. Barlow, that their services would have to be dispensed with after April 30, 1885. The threatened men (as was perhaps intended) turned to the members of Parliament, who gave way and voted supplementary funds, increasing the Survey's money to $78,557 from the previous $60,015. An effort to cut back the appropriation for the fiscal year 1886-87 was countered by the threat that the Survey would have to curtail its activities, and the increase remained. By the late eighties and nineties, however, Selwyn was again almost continuously in difficulty trying to make ends meet. More than once after the field parties had been assigned, the prospect arose that field operations would have to be suspended until after July 1, the start of the new fiscal year.

The Survey's coming into the public service in 1877 was accompanied by close oversight of its finances by the Auditor General, J. L. McDougall. From the various statements and printed exchanges of correspondence with that official, it would appear that Selwyn was not a careful or systematic administrator, or any stickler for routine. He did little about frequent complaints over the tardiness of officers in presenting their field accounts and settling their spring-time advances, beyond warning the delinquents and posting their names. He ignored "channels" when purchasing furnishings, stationery, books for the library, having repair work done, or arranging for the printing of the publications. He was criticized for not observing the proper forms in the purchase of the instruments, etc., from the Logan estate, or of collections for

VII-3 JOHN MACOUN

the museum. The Auditor General protested that the Survey never supplied him with an inventory of its stores and equipment. He complained about the payment of some temporary staff on the 15th of the month, several temporary employees being paid by a single cheque drawn on the director, and payments made without proper certification as to work done and verified attendance at work.

Since many of these complaints resulted in refunds and pay deductions, it would appear that Selwyn's oversight sometimes was at fault. From time to time the Auditor General administered a reprimand: "I observe that you are still in the habit of paying various tradesmen for painting and repairs,"[3] or in more censorious vein, "Extra work in keeping accounts is not a drawback serious enough to overbalance the advantage of having the expenditure as much as possible under the control of one responsible head. I hope you will see your way clear to conforming with the practice of other Departments."[4] Selwyn, for his part, explained, justified, argued, stood on his dignity, and occasionally counter-attacked: "It appears to me that the queries in this letter are somewhat out of place. I fail to see why a calculation need be entered into as to what 'would have been'."[5] That riposte was inspired by a long correspondence about a horse Coste had brought back from the field, boarded all winter in Ottawa, and then had to sell in the spring for little more than its board bill.

The Auditor General's scrutiny of the field officers' accounts helps to explain their dissatisfaction with the working conditions. At various times McDougall disallowed expenditures on cigarettes, tobacco, and $2.00 worth of wine; R. G. McConnell was compelled to refund $88.00 he had received in payment for clothing destroyed by a prairie fire during an expedition, the grounds cited being that the government did not assume responsibility for the property of employees on its service, since—another 'would have been'—"There is no greater reason for responsibility in the North-West than in any other place: the clothing, although destroyed by a prairie fire, might have been destroyed in Ottawa if he had been here."[6] McDougall complained that officers who did not live in Ottawa should not be allowed living expenses when they came there to report, and he watched the living expense claims like a hawk.

Besides the low salaries, tight controls, and slow promotions, some men leveled serious complaints against the director and his management. Some of these were justified, though others were unfair as well as unkind. Selwyn's instinct was profoundly conservative—to protect the existing order and distrust innovations. Time and again he defended his administrative policies by claiming they were originally those of Logan—as though that was the ultimate, irrefutable argument in their defence. Members of the staff found him aloof, overbearing, lacking in tact, stubborn, autocratic, and unwilling to explain his policies. They grumbled that he did not always select the right men for various duties, or train men to do the work he wanted; that he failed to give clear, explicit instructions or guidance to his staff, then criticized the men for not doing what he wanted. Some said he could not be trusted, that he changed his mind, that he forgot understandings or commitments, then blamed the men unfairly when they acted as he had originally ordered. Those who felt the lash of his criticism complained that he gave the best assignments and the highest pay to certain favourites, allowed them to do what they wished, and published everything they wrote. The educated young men found grounds for complaint in Selwyn's editorial hand. They were annoyed when he interfered with their reports and charged that he rewrote them to suit his own purposes, held back their publications, and did not publish their maps. Frustrated, some took their complaints to the newspapers, while others pulled whatever political strings they could.

Unfortunately, Selwyn was no Logan in his relations with the public, though he did appear to enjoy the firm, unshakeable support of the government, particularly of the Prime Minister, Sir John A. Macdonald. Sir William had been beyond compare, with his advantages of social position, wealth, the knighthood that placed him above criticism, and the good press he always cultivated. In his day, too, Canada's Survey had stood out because it had few rivals with which to compare it. Now there were excellent surveys in several American states, in most countries of Europe, in the major British colonies and other lands. The Geological Survey of Canada, overwhelmed by its many tasks in a territory as big as Europe, no longer commanded the same prestige as in Logan's time. Selwyn seemed to lack any instinct for public relations, or for cultivating the press or Parliament. He seemed almost impervious to public opinion and did practically nothing to educate the public or the government to the Survey's role or needs.

The restlessness inside the Survey was paralleled by a measure of public questioning and criticism of the policies of its director. Canada, like other countries, was experiencing a severe depression in the years after 1873. The mining industry, too, felt the pinch; entrepreneurs suffered from a lack of investment to enable them to develop their mineral holdings. The public and government, sympathetic with the struggling mining industry, saw in the Geological Survey a major instrument that might better serve the national interest. The Survey, they said, should do more to encourage, rather than dis-

VII-4 PART OF THE MINERALS AND ROCKS COLLECTION
First Floor, Geological Survey Museum

courage, investment and development; it should undertake studies of immediate practical benefit to the mining industry, like detailed mapping of mining districts, examining mineral deposits, testing rock samples, collecting mining statistics, and reporting on mining methods in operation elsewhere. The approaches and achievements of the United States Geological Survey and of various state surveys in these respects were held up as something for the Canadian Survey to emulate.

Instead, these critics charged, Selwyn was turning the Survey into a purely scientific organization, rather than one that fostered the development of the mining industry to the utmost degree, as in the good old days of Logan. Selwyn was accused of refusing to conduct special surveys to aid mining development unless they had a bearing on some question of scientific interest. Some of the reports on current mining activities were needlessly discouraging, it was alleged, and important mining areas were ignored for years on end while surveys were undertaken of remote sections of the country unlikely to be developed for years to come. The director apparently sat on many reports, so that valuable work of the staff, such as that on the important phosphate mining district along the Ottawa River, was withheld from the public. Altogether, the criticisms went, the Survey did not seem to be achieving very worthwhile results in terms of the money allotted to it, and seemed to spend an inordinate proportion of its funds on museum and library acquisitions or other purposes unrelated to helping mining development. Several anonymous articles in the press, apparently written by persons in intimate contact with the work and hinting at trouble within the organization, added to the uneasiness about its management and purposes.

Early in 1884, Robert N. Hall, M.P. for Sherbrooke and a considerable capitalist in that industrialized region, rose in the House of Commons to suggest that the Survey was "not keeping pace with the geological progress of the country;"[7] and the time had possibly come for a second committee, like the one 30 years before that had helped materially in promoting its work. His motion received unanimous consent on February 25, 1884, and the committee was appointed with terms of reference empowering it to inquire into a matter of increasing public concern—the relevance of the Survey's program of activities to the country's present needs. The committee was asked specifically to "obtain information as to the methods adopted by the Geological Surveys of this and other countries in the prosecution of their work, with a view to ascertaining if additional technical and statistical records of mining and metallurgical development in the Dominion should not be procured and preserved."[8]

* * *

In the context of the tensions and frustrations that had been developing within the Survey, this innocent-sounding inquiry into its affairs was like breaking open a hornet's nest. Perhaps that was what some of the promoters of the inquiry had intended—to turn it into an occasion to air all their complaints and pull down the director. Selwyn believed so, and was partly correct in interpreting the hearings as being motivated by personal animosities, as "simply an organised conspiracy the outcome of the jealousy of some of those who have had to leave the Survey."[9]

Most of his critics, however, were still in the employ of the Survey. The leader of this 'organised conspiracy' was Dr. Robert Bell, who harboured the feeling that his unique talents were not sufficiently appreciated, and that he was being held back from receiving his just dues by the ill-will of the director and his partiality for G. M. Dawson. Perhaps this went back to 1871 when Bell's request to take over the work in British Columbia was denied. Richardson had been sent instead, to be followed four years later by Dawson, a complete newcomer, for whom the position was kept open, Bell believed, through the backstairs influence of the Dawsons in the cosy environs of the McGill campus. Bell was even more put out by the appointment in 1877 of the four assistant directors, when in view of his seniority, he felt, he should have been appointed the one senior, or chief, assistant director. The subsequent appointment of Dawson as chief assistant director, and therefore presumably the heir-apparent to the directorship, was the last straw. Selwyn attributed much of the discord within the Survey to Bell's jealousy of Dawson, while Bell told the committee that he regarded his treatment by Selwyn as "amounting, I may say, almost to persecution; I know it amounts to such in my own case."[10] His testimony was one long indictment of Selwyn's capabilities and management, and a thinly-veiled hint that the Survey could do with a new director—preferably a Canadian, with wide experience of the country and its people.

Bell's campaign began early in 1881, shortly after the move to Ottawa brought the Survey again into close proximity to the seat of political power. In April 1881, he wrote to Colonel J.S. Dennis, deputy minister of the Interior, and to a private M.P., for support in his campaign for advancement. In April 1883, he wrote a letter published in the Toronto *Mail* under the pseudonym of "Bystander" in which Selwyn was accused of incompetence and of plagiarizing the work of his subordinates. That attack was viewed so seriously that Selwyn communicated with the Prime Minister about it, and Bell was severely reprimanded by that august personage. Next, Bell enlisted a group of seven M.P.'s and drafted a letter protesting his treatment, in particular the promotion of Dawson over his head, for them to send to the Prime Minister. He also used his influence on other members and undoubtedly made some impact by these political intrigues, though at least one friend warned him not to overdo it lest the resentments become directed against himself rather than Selwyn. This was what eventually happened. After the committee had reported, another friend noted that there had developed "a prejudice against Bell in Ottawa,"[11] because of his constantly irritating tactics.

Behind Bell was a group of younger men who either did not like the work rules or had been censured (or dismissed) for breaking them. Henry G. Vennor, for example, had been dismissed for using his position to traffic in phosphate lands in the Ottawa valley. Hugh Fletcher had been suspended for two weeks without pay for writing to the newspapers. A. S. Cochrane, a young protégé of Bell's who referred habitually to Selwyn as "The Prince of Liars" or "The Thing," was annoyed at being criticized for coming to the office seven minutes late in the mornings, for deficiencies in a map for which he considered Coste's assistant was to blame, and especially for his own lack of promotion and low salary in comparison with Coste. (Coste, on his first season in the field in 1883, was paid more than Cochrane on his seventh. Of course Cochrane had little schooling and was employed mainly on topographical work, while Coste was a mining engineering graduate, which accounted for the discrepancy.) Vennor, J. F. Torrance, and Wallace Broad also had been in trouble with Selwyn over tardiness in presenting their reports, or his refusal to publish them without substantial revisions. Bell failed to recruit the Abbé Laflamme, who wrote in response to Bell's letter that while he was not entirely satisfied with conditions, Selwyn had always dealt fairly with him.

Bell had two allies outside the Survey in the congenial work of trying to get rid of Selwyn, though each probably was aiming at the same objective for himself—the directorship. Dr. T. Sterry Hunt had failed to find a remunerative or satisfying career in the United States since leaving the Survey, and had grown increasingly bitter. Perhaps he hoped to regain some of his one-time magic by returning to the scene of his early triumphs; perhaps he wanted the prize that had eluded him in 1869 when Logan chose Selwyn for his successor. Still further in the background and largely playing his own game since he had a direct line of contact with Mr. Hall, the committee chairman, was Thomas Macfarlane, the mining engineer who had been employed briefly in 1865-66, then had made his reputation for his role in discovering the remarkable Silver Islet mine in Lake Supe-

VII-5 SIR WILLIAM DAWSON

VII-6 THOMAS MACFARLANE

rior. Afterwards he had worked as a mining engineer for American interests in Canada, the United States, and South America, but found the remuneration inadequate and the living uncertain. To Hall he suggested that the committee should recommend five-year terms for the director, and that the appointment should be offered to Dr. Hunt, then to himself as second possibility if Hunt declined or was considered unacceptable. To his wife he confided that if any change was brought about in the Survey then "I may have my chance."[12]

The select committee was a fairly representative group of M.P.'s, seven Conservatives to four Liberals, mostly from constituencies in which the development of mining was an important concern. Next to Hall the member who carried the weight of the questioning, was S. J. Dawson, M.P. for Algoma, the renowned surveyor, member of the exploring expeditions of 1857 and 1858, and founder of the Dawson route from the lakehead to Manitoba. E. C. Baker, M.P. for Victoria, British Columbia, was another frequent questioner. At the other extreme, Wilfrid Laurier does not appear to have asked a single question or given any sign of his presence beyond his name on the list of members.

As might be expected, much of the hearings, which occupied eighteen sittings and lasted from March 6 to April 3, 1884, was spent in bitter personal wrangling, as the director was plied with leading questions about the staff members and their work, and they retaliated in kind to his critical comments. First to appear before the committee, for four sittings, was Selwyn, who provided the information regarding the history and structure of the Survey. When asked whether it was as efficient as in times past, he trotted out his familiar complaints about lack of money or staff and the vast areas the Survey was expected to deal with topographically as well as geologically. He threw back the allegation that its reputation had deteriorated since Logan's day, and combatively suggested that if improvements were expected, a larger staff must be provided, salaries must be increased to attract first-class men, geologists must be relieved of all topographical mapping, the director must be given full authority to dismiss incompetent or disloyal employees, and the Survey must be completely relieved from having to assist private mining ventures. At first he evaded questions about the efficiency of the Survey being affected by personal feuds, and attributed internal difficulties to inadequate salaries.

A favourite topic of questioning was the *Report of Progress for 1880-81-82,* the slender 211-page volume published just prior to the committee's meeting. As Selwyn explained, it by no means represented all the work accomplished during the years 1880 to 1882. The committee persisted in asking Selwyn whether he felt that this was an adequate output for those years, or why so much of the work performed in those years had not been included. It was in vain that Selwyn made a special appearance before the committee to reiterate the point, and called on the committee to examine the fatter earlier volumes as well, observing that in Logan's 20 years the Survey had published only 2,200 pages of reports, as against the 3,800 pages in 10 years under his directorship. But all to no avail. The committee had satisfied itself that the volume was an insufficient return for the government's money and no amount of explaining could efface the impression, as the final report revealed: "Judged even as to quantity, it is a meagre result of two years time, for a staff of about thirty highly educated geologists, chemists, palaeontologists and botanists, maintained at a cost to the public, during those two years, of over $110,000."[13]

Woven into the testimony of many of the members, or ex-members, of the Survey was a formidable assortment of accusations against the director, against his competence as a geologist, his holding up reports and maps, and his treatment of his men. It was also asserted that he had made costly mistakes with certain programs (such as his well-drilling program on the prairies), and that he mishandled the funds. One emotion-laden charge was that the Englishman, Selwyn, did not understand how to handle Canadians and that some of his difficulties arose through his failure to treat his staff with sufficient consideration or respect. No doubt Selwyn did have an Old World conception of himself as the chief and of the officers as his underlings that grated on the nerves of some of the men. When Selwyn let slip an impression that he did not regard young Canadians as amenable to discipline, he quickly found himself in difficulties. A newspaper picked up this slight upon the national honour and claimed that Selwyn wished to reduce Canadians to slavish submission. Here was a subject that the committee loved to pursue, and the aspirants to high place were given ample opportunities to testify that they, unlike the present director, considered Canadians made good members of parties, excellent officers, distinguished scientists, and were entirely competent to hold any position of trust or authority within the Survey.

The debate about the operation of the Survey centred mainly on the lack of harmony within the organization, and the causes. The thesis expressed in varying degrees by Selwyn's critics was that the Survey had been one big happy family under Logan, and since then a serious breakdown of morale had arisen because of Selwyn's shortcomings, until by 1884 virtually the whole staff had been alienated. Selwyn argued that the internal difficulties were the inevitable results of the increased staff and specialization of functions, and of the wholly inadequate salaries. As he saw it, the only discord lay between himself and Dr. Bell, and any extension arose from Bell's intrigues: "I believe Dr. Bell is always scheming. Sir William Logan himself said so, years ago."[14] G. M. Dawson, who followed Selwyn, largely supported his superior, maintaining that the Survey's reputation stood as high as it ever had done. As for any lack of harmony, "I fear there are some gentlemen on the staff with whom it would be very difficult to be harmonious, and that they might object to something, whatever was done."[15] But mainly the discontent arose from the insufficient salaries. He hinted that the move to Ottawa tempted "the least useful members of the staff to work other methods of interest than those of doing their work"[16] and pleaded that the director should be trusted to conduct the management of this—or any other scientific service—free from interference. In a series of extremely polished replies to questions, he revealed his talents for clear diplomatic exposition, and stepped down at the end of his session before the committee without alienating anybody, having kept clear of the bitter aspersions that characterized the evidence of most of the other staff members.

Selwyn found other defenders of his administration in the persons of Scott Barlow, the topographer-turned-geologist, J. F. Whiteaves, the paleontologist, John Macoun, the botanist, G. C. Hoffmann, the chemist, and John Marshall, the accountant. This curious array of supporters drew comments. The Montreal *Herald* waxed sarcastic about the efforts of non-geologists to defend Selwyn's geological reputation, and remarked, not quite accurately, on the director's failure to produce active field geologists to testify on his behalf.

Behind, and completely overshadowed most of the time by these personal feuds, was considerable information regarding the Survey's roles since its inception, and advice as to the desirable courses of action in the future. In his submission Macfarlane stated that the Survey's work was spread too thinly, that it paid too little attention to mining needs and too much to matters of theoretical geology. This, he claimed, was a complete reversal from Logan's attitude, for "his [Logan's] opinion seemed to be that the chief work of the Survey was to assist in discovering and developing the mineral resources of the country, by all the scientific means at its command, and in a subsidiary manner only collecting and studying scientific data."[17] Macfarlane suggested that the Survey should concentrate on mineral-rich districts, making surveys, publishing maps, and describing typical mines for the benefit of potential investors; do assays and analyses for prospectors and mine owners; and collect mining statistics. It should employ many more mining specialists—

chemists, assayers, mineralogists, petrographers, mining engineers, metallurgists, and statistical clerks. Professor Bailey's evidence also emphasized the need for mining engineers and compilers of mining statistics.

Robert Bell's testimony, stripped of personalities and criticisms, also contained a plea for the development of the mining statistics function. He agreed with Macfarlane that more thorough geological work should be carried out, particularly in areas that held promise of mineral wealth. In keeping with his own broad-ranging interests, he stressed the importance of initial exploration and mapping. He felt that the number of field officers should be increased, and explorations "of our great unknown territories" that embraced all aspects of natural history besides geology should be continued. He urged that for reasons of economy parties should do the topographical as well as the geological mapping on these exploratory journeys, and "if the Geological Survey is also to be a Natural History Survey" it should include specialists in all types of natural history in addition to botany.[18]

Hunt also stressed the practical bent of the Survey in Logan's time—his concern to study mineral locations and mining processes, and Hunt's own efforts to analyze soils, building stones, phosphates, brines, oil shales, and other commercial minerals on behalf of the mining industry. In his opinion, the Survey had taken a wrong turn under Selwyn in abandoning its policy of looking into "every chemical, mineralogical or metallurgical question, that could apply to the mineral resources of the country,"[19] and, instead, concentrating too much on explorations, field work, mapping and topography, and especially, natural history. Professor E. J. Chapman of Toronto, another contemporary of Logan's, agreed with Hunt in feeling that the Survey "was clearly instituted to convey to the Canadian people practical information respecting the mineral resources of the country." Since Logan's retirement there had been a tendency to "ignore the primary object for which the Survey was instituted."[20] He felt the officers were too concerned with establishing their own scientific reputations, and that as a result "the present reports seem rather to be addressed to the Geological Society of London or the geological section of our Canadian Royal Society, instead of to the people at large."[21]

From the welter of testimony, the committee emerged with a report that in large measure reaffirmed those preconceptions that had led to its summoning. The terms of reference had been designed to criticize the Survey for not being sufficiently helpful to the mining industry and so the committee dutifully reported. The primary purpose, they said, was "to obtain and disseminate, as speedily and extensively as possible, practical information as to the economic mineral resources of the country, and scientific investigations should be treated as of only secondary importance, except where necessary in procuring practical results."[22] The report noted "the serious

VII-7 J. F. WHITEAVES

VII-8 G. C. HOFFMANN

lack of attention to the mining industries of the country in actual operation,"[23] and commented favourably on the interest the Geological Survey of the United States took in these matters. It expressed surprise that the Survey did not show any interest in "the great geological and scientific facts demonstrated by the opening and continued prosecution of these mining industries," or feel the necessity of preserving records of them "for the proof or refutation of existing theories, and more especially for the guidance of future explorers in similar fields."[24] Finally, it concluded this section of its report with a recommendation to appoint a mining engineer with rank of assistant director to report on mining and metallurgical developments and gather mining statistics.

The report criticized the current activities, including delays in publication and too much work not published at all. It claimed that field parties spent too much time on non-geological matters, and that the reports contained entirely too much trivia as details of scenery or anecdotes of Indians. It reported that "the administration of the Department under its present management is unsatisfactory;" it recommended that the operations should follow a more systematic program, and that "the field operations should be confined to subjects more closely allied, practically and scientifically, to a Geological Survey."[25] Finally, the committee discussed the disaffection within the Survey, as instanced by the resignations and dismissals of members of the staff, averaging one per year:

> Your Committee also feel obliged to report that the relations between the Director and some of his staff have been, and are, of such an unpleasant character as to have greatly impaired the usefulness of the Survey. By some of the witnesses this difficulty is thought to result from insufficient salaries; by others, to defects in the temper and tact of the Director; and by others still to jealousy and insubordination on the part of the members of the staff. Your Committee have not felt it to be within their province to investigate and decide as to a defect of internal administration of this kind, but its existence and the unfortunate effects resulting from it are too apparent to be wholly ignored.[26]

This seemed a curious conclusion, but even more curious was what did *not* happen as a result of the committee report. It had called for further inquiry into the internal workings of the Survey; Bell demanded an investigation to clear his reputation from the slights cast upon him; the director immediately requested his minister to appoint a board to clear the Survey of the aspersions cast upon it in the committee room. But the government was set against a repeat performance; and there was to be no rematch between Selwyn and Bell. No one left as a result of the committee or its report, and Selwyn continued as director. Macfarlane found solace and security in an appointment as analyst with the Customs Department, while Hunt returned to his rather aimless career in the United States. As for Selwyn, Bell, and Dawson, they continued in their accustomed places pursuing their same activities for eleven years, and Dawson and Bell for seven years after that. The mind fairly boggles at the tensions and strains that must have permeated the decrepit building on Sussex Street in the long years that followed the airing of so much dirty linen before the House of Commons select committee hearing of 1884. The remainder of the staff seem to have grown resigned to the inevitable, if one may judge from the testimony presented before a select committee appointed in 1892 to inquire into the civil service. Selwyn restated his position to this committee, while the delegates from the staff—Fletcher and W. H. C. Smith—reiterated the familiar complaints of low pay, slow promotion, insufficient support staff, and interference by the director with the work of the officers.

* * *

THE REPORT OF PROGRESS FOR 1882-83-84, which appeared in 1885, may be said to be Selwyn's reply to the strictures of the committee. Its appearance carried a reproach, for in contrast to its slender predecessor which had aroused so much ire this new volume was a robust 784-page compilation. Comprising fifteen reports in addition to Selwyn's, it reported mainly on work done in 1883 (the season *before* the 1884 committee met), with one report based on work done between 1880 and 1882 and another on work in the period 1879 to 1883. The volume also included six plates of sections and thirty-seven maps. Selwyn's own report refuted some of the aspersions of the committee. A letter from Major J. W. Powell, director of the United States Geological Survey, lauding the ethnological work of the Survey was printed. Readers were reminded that the committee had not mentioned the move from Montreal in 1881 nor the necessity of completely recataloguing the library. Selwyn also pointed out that the Survey had never been so fortunate as to possess thirty highly-trained scientists on its staff, as the committee had alleged. As a final stroke, he reviewed the many activities that were never dealt with in the reports—educational collections, advice to practical miners, including assays and analyses, preparing

VII-9 EUGÈNE COSTE

VII-10 E. D. INGALL

VII-11 H. P. H. BRUMELL

and superintending displays at international exhibitions, papers and reports on botany, paleontology and ethnology published in the *Transactions* of the Royal Society of Canada or other journals. Then he closed with a final dig: "It will thus be seen that the last annual volume, dated 1880-81-82, does not 'profess to give the useful work of the survey for two years,' neither do any of the 10 volumes which have preceded it."[27]

The report did reveal a considerable change in emphasis, for which the committee's comments may be given some credit. Besides the reports of exploration of the Athabasca River basin and the coasts of Labrador, Hudson Strait, and Hudson Bay, there were also J. F. Torrance's long-awaited report on the apatite mines of the Ottawa valley, Coste's on the gold mines of the Lake of the Woods district (a critical account of misdirected efforts and dubious promotions), and Willimott's on the mines he had inspected in Ontario, Quebec, and Nova Scotia during 1883. Nor could there be doubt of the immediate practicality of the 'scientific' report by Dawson and McConnell on the extensive coal-bearing strata of southwestern Alberta, through which the Canadian Pacific Railway had just built, plus Hoffmann's analyses of coals from nearly 40 separate localities in western Canada.

The volume also indicated an effort to meet the complaints over the tardiness in the publication of Survey results. Besides catching up with the 'backlog' of unpublished reports, the 1882-83-84 volume introduced the practice of printing the separate reports as received and processed. This encouraged officers to prepare their reports as quickly as possible, and prevented the tardiness of one man—or one report—from delaying the publication of the work of others. Beginning in 1886 the collected reports began appearing more regularly in an annual volume known as the *Annual Report (New Series),* Volume I of which covered the work of the years 1884 and 1885, while subsequent volumes covered each of the succeeding years. The director's report, Part A of each volume, presented a comprehensive review of the work during the year covered.

Maps were a bigger problem, especially now that the Survey was beginning to issue areal maps on the 1 inch = 4 mile scale (or in some cases the more detailed 1 inch = 1 mile scale). Besides the great labour and expense they occasioned in the field, they presented further difficulties in publication. The chief draftsman, Scott Barlow, complained regularly over the shortage of experienced draftsmen, the great accumulation of unpublished material, and the lack of money to pay for engraving finished maps. At the end of 1894 his successor, James White, listed 81 maps, of which only 35 were "ready for engraver" or "in hands of engraver," and only 10 were actually published that year. Delays in map production were, and are, a continuing frustration, holding back completion of many studies, because the map is always *a* major end-product, and sometimes *the* major end-product.

The committee findings also influenced the sort of activity the Survey undertook, and intensified its efforts (never wholly neglected but perhaps insufficiently publicized), to assist the mining industry. Henceforth Selwyn took care to underline the Survey's part in aiding favourable mining developments. In one report he claimed that a certain gas discovery in southern Ontario was the result of a recommendation in a previous report that deep-well drilling should be extended north and east of producing fields. In another he pointed to the Survey's associating the Huronian with potential mineralization in 1872-73 as having brought it to pass that "nearly all the discoveries and developments of mines and minerals in the Huronian areas that have been indicated by the Survey have been made," and he proceeded to lash out at his critics:

> ... whether [these facts] prove the truth or otherwise of the reiterated and apparently somewhat popular statements that of recent years the Survey has paid no attention to and takes no interest in the development of the mineral resources of the country, may perhaps be left to the decision of the public and to the testimony of the sixteen volumes of reports, maps and other documents that have been published by the Survey since 1870.[28]

A more certain effect of the committee was the steps Selwyn took to organize a Section of Mines around the two mining engineers Coste and Ingall whom he had hired in 1883, no doubt with this in mind. These men were formally appointed as mining geologists on July 1, 1884, with H. P. H. Brumell as their assistant. Coste worked in eastern Ontario and Ingall in the mining areas north of Lake Superior, about which he presented a lengthy report in 1888 on the history, present status, and future prospects of the silver mining industry of the region. Here Macfarlane's Silver Islet mine at last received the notice he had craved. Brumell's work was mainly statistical, but he also prepared a useful survey of the natural gas and petroleum industries. The 1887 report also inaugurated the first successful "Statistical Report on the Production, Value, Exports and Imports of Minerals in Canada," while under Ingall's leadership after 1889 a collection of drilling logs was begun that ultimately developed into the Borings Division.

The effectiveness of the mining studies undertaken by the Section of Mines depended in large measure on the work of the regular field parties assigned to work out the basic geology in places where mining or mining development was proceeding. So geologists who were experts in particular regions were encouraged to prepare reports on potential mineral resources or mining properties in the areas they studied. The versatile G. M. Dawson prepared one in 1888 on "The Mineral Wealth of British Columbia with an Annotated List of Localities of Minerals of Economic Value," while in the following year Ells prepared a similar report on the mineral resources of Quebec as a byproduct of his surveys from Gaspé to Montreal. Such reports were designed to offer useful assistance to mining operators, capitalists, and investment counsellors.

Another experiment in the same direction was the publication, in the *Annual Report,* Volume I, of a short article by Coste on the mining laws of the Dominion and of Ontario and Quebec, with suggestions for their improvement. (How characteristic of the brash, young Coste—the choice of subject, critical tone, and cocksure assertion of solutions.) Coste attributed the slowness in opening productive mines to the laws that allowed speculators to hold large tracts of mineral lands without developing or evaluating them. He recommended the granting of long-term leases over clearly-defined areas to bona fide developers willing and able to exploit the resources, and subject to cancellation if the work requirements were not carried out.

What effect, if any, did the inquiry of 1884 have on the director's concept of the Survey's role or his approach to its work? The committee was the first investigation of the operations of the Survey by an outside agency in more than a generation. As such, perhaps, it offered the first inkling of a changing relationship between the Survey and the government resulting from the move to Ottawa three years previously. In the changes in the reports and the establishment of the Section of Mines may be seen Selwyn's efforts at meeting the criticisms, and perhaps, his realization of the need to conciliate the politicians and the public. To a degree the Survey began paying closer attention to the more settled (and politically powerful) parts of the country in its field work. By the order of priorities in the national mapping program Selwyn indicated his readiness to undertake more intensive work in areas of geological importance. However, he considered himself the sole and final judge of the Survey's operation and refused to be intimidated by pressures from regions like Nova Scotia or by delegations of mining men and politicians. He held the ideals and objectives of the Survey to be above the exigencies of politics; and, to do it justice, the government of the day left him pretty free to follow out his own plans.

Though Selwyn could no longer afford to undertake lengthy explorations over prairie and mountain, each year he visited one part or other of Canada to inspect the work and decide on future programs. These annual journeyings dramatically illustrated the improvement in communications across the continent following completion of the C.P.R. In 1882 and 1883 he examined long stretches of the north shores of Lakes Huron and Superior from a boat, but in 1884 he followed the C.P.R. construction west of Lake Superior to examine the Laurentian and Huronian exposures in the railway cuttings, and to inspect the prairies beyond. In 1887 he rode the new railway all the way to the Pacific coast, and visited a string of projects extending from Gaspé to Vancouver Island, "where some personal observation was required, either for information on points of economic importance, or in connection with the working out of the geological structure of the region."[29] These included a reported occurrence of petroleum in Gaspé, the relations of the graptolites in the rocks between Gaspé and Levis, the new copper mines at Sudbury, silver and iron occurrences near Port Arthur, coal at Lethbridge and Banff, silver-lead mines at Illecillewaet, and hot springs at Banff and Harrison Lake.

In other years he examined the north shore of the Gulf of St. Lawrence as a passenger on the government ship *Napoleon III* to, and through, the Strait of Belle Isle, noting evidence of Huronian among the prevailing Laurentian rocks. The goldfields of Nova Scotia engaged his attention in the summer of 1890 in an endeavour "to ascertain what would be the best plan of operation in commencing a detailed examination and survey" of the gold-bearing rocks.[30] A well-boring operation at Deloraine, Manitoba, that the Survey was subsidizing, and the coal beds of the Souris-Estevan district were other objects of his attentions in the early nineties.

VII-12 ALFRED RICHARD CECIL SELWYN
Director, 1869-95

Much of Selwyn's time was taken up with exhibitions and tours of scientists and businessmen in Canada. Meetings of learned societies had to be properly received, instructed, and entertained by the director of the Geological Survey of Canada. He had to oversee and support the work of his staff in scientific and learned societies in Canada and abroad. The director also had to cope with the heavy administrative demands of his organization, and with the public, increasing numbers of whom visited the museum each year, sent in questions that required answering, or brought rocks to be examined.

Oversight of the several office-based activities was an increasingly heavy burden, too, as Selwyn's letterbooks attest. The museum, the chemistry and mineralogical branch, the paleontological branch, the natural history section, the topographical and mapping section, and the "housekeeping" sections–clerical, accounting, and library–were all expanding physically and growing in complexity and importance. The library, under Dr. John Thorburn, its head since April 12, 1882, was recatalogued, and began the acquisition of books and periodicals that has made it one of the great scientific libraries of the continent; its growth echoed the proliferation of scientific research and publication throughout the western world in the last quarter of the nineteenth century. The rapid expansion of the physical holdings of the museum reflected the widened interests and the broader area from which it was drawing materials; this was also true of the paleontological division. The chemistry division's work also increased mainly because the public was making increasing use of its facilities, added to which the division was assuming important new functions. In 1881, Adams was sent to Germany to learn the new microscope techniques, and following his return to Canada he spent half his time in the laboratory studying the structures of igneous, metamorphic, and sedimentary rocks to gain a better understanding of the processes of rock formation. After Adams left, his successor, W. F. Ferrier, was appointed specifically as petrologist, and the section in 1893 was renamed "Chemistry and Mineralogy."

How was the Survey faring in the esteem of the scientific world? Was its reputation declining, as some of Selwyn's critics suggested? The answer would be 'no' judging by its role in exhibitions and congresses in the learned societies of the day, or in publications of its officers. The Survey under Selwyn participated in four major world exhibitions that kept its name and work well to the fore: at Philadelphia in 1876, Paris in 1878, the Indian and

Colonial Exhibition in London in 1886 (where, as in 1878, Selwyn was absent from Canada for practically the entire year, during which Dawson directed the Survey), and the World's Columbian Exhibition at Chicago in 1893. These brought the work of the Survey, as well as the accomplishments of the Canadian mining industry, before the influential, moneyed, social and political groups that patronized such gatherings, and earned a great deal of publicity in periodicals and newspapers. This, as Selwyn said, was a very important contribution towards promotion of the mining industry. In addition, many special studies by officers on a wide range of subjects were published as papers in learned journals, or read before important foreign scientific congresses. The Survey and its staff members also were hosts to an increasing number of conferences—a recognition of their leading positions in those particular sciences. The more widely-travelled and better-known members—Dawson, Macoun, Ami, and especially Bell—prepared many popular accounts of their journeys for the press or as public lectures. From the time the Survey arrived in Ottawa its members were prominent in the Ottawa Field Naturalists Club, and through local field trips and other activities, they introduced thousands of amateurs to the joys of natural history.

A tangible symbol of the prominent place that the leading officers occupied in Canadian science was the Royal Society of Canada, founded in 1882 under the sponsorship of the Marquess of Lorne. Selwyn, Dawson, Bell, Whiteaves, Bailey, Matthew, Laflamme, Macoun, and Hoffmann all were founding members of Section IV, the geological and biological sciences; and Selwyn and Whiteaves were the first president and secretary of the section. More than half the papers in the first *Transactions* of the society, published in 1884, were written by officers of the Survey. The calendar of publications by the members of the Royal Society of Canada, prepared in 1894, lists several hundred books, articles, and addresses produced by the officers, exclusive of their contributions to the annual reports.

The drastic revision of the program envisaged by the 1884 House of Commons committee would seem to call for new legislation, but none was forthcoming until 1890, when a new Survey act (53 Vict., Chap. 11), largely reflecting Selwyn's ideas, was introduced in Parliament by the Minister of the Interior, Edgar Dewdney. The new act provided for the creation of a new department in the federal civil service to be called the Geological Survey, led by a deputy head and director, under the parliamentary supervision of the Minister of the Interior but otherwise separate from that department. Henceforth, the director's report no longer appeared as part of the report of the Department of the Interior as it had since 1880.

Reflecting its changing function was the addition of a duty above and beyond those assigned by the previous act of 1877:

> To collect and to publish, as soon as may be after the close of the calendar year, full statistics of the mineral production and of the mining and metallurgical industry of Canada; to study the facts relating to water supply, both for irrigation and for domestic purposes, and to collect and preserve all available records of artesian or other wells, and of mines and mining works in Canada.[31]

The duties respecting mining statistics and mining records were *ex-post facto,* in the sense that they were already being carried out. What was surprising was the addition of the water resources function, which was certainly not in accord with the earlier criticisms of the Survey for not paying sufficient attention to geology and mining. Besides, it seemed a more appropriate responsibility for the Department of the Interior or the Department of Agriculture. That it was assigned to the Geological Survey no doubt reflected the work the Survey was already doing in drilling for artesian water, petroleum, and coal on the prairies.

The new act also raised the qualifications of the scientific staff in line with Selwyn's long-expressed view that geology was a profession and its practitioners must have training and experience comparable to those in medicine

VII-13 JOHN THORBURN

VII-14 THE ROYAL SOCIETY OF CANADA, 1896 MEETING
Photographed in front of Ottawa Normal School. Survey officers present include A. R. C. Selwyn, president (*front row, fourth from left*), R. W. Ells (*to the right of Selwyn*), J. F. Whiteaves (*two places right of Ells*). In the second row are F. D. Adams (*behind Selwyn*), B. J. Harrington (*slightly right of Ells*), and H. M. Ami (*in front of centre panel, right side of arch*).

or law. Now that Canada was beginning to produce qualified geologists in sufficient numbers, Selwyn could specify that new appointees to the scientific staff must be graduates of a recognized university, college, or national school of mines and serve an apprenticeship of at least two years before receiving a permanent appointment. An even more idealistic condition of employment implemented by the act was a prohibition against officers becoming involved in private mining ventures that were in any way related to their duties. Vennor had been forced to resign, and there had been a minor scandal over Coste's supposed connections with the Postmaster General, J. G. Haggart, a member of the board of the company that employed him after he left the Survey to embark on a career that made him the "grand old man" of the petroleum and natural gas industry of Ontario. Under this code of conduct, officers were forbidden henceforth to buy Crown lands, report their findings to anyone except their immediate superior, make investigations or prepare reports for private individuals, or have any pecuniary interest in any mineral activities in the Dominion. These conditions protected the reputation of the individual officers and made the Survey's reputation one of the brightest in the history of Canadian administration. This ordinance was all the more courageous in that officers' salaries at the time were, and remained for many years, at a very low level, for this action barred them from participating in a lucrative source of outside income open to geologists in other occupations, particularly in the universities.

The years passed, the Survey carried on despite financial difficulties, and the bitterness engendered by the Commons committee of inquiry in 1884 receded. However, the opinions and animosities aired at that time continued to fester and add to the general unhappiness of the men housed in the building hard by Ottawa's Lower Town Market. Old familiar figures, like Michael O'Farrell, Logan's handyman and housekeeper in the early days, and the veteran, T. C. Weston, retired. The year 1893 saw the death of a promising young geologist, W. H. C. Smith, and the next year that of Scott Barlow and of A. S. Cochrane. Cochrane had been seriously ill since 1891 and had required frequent leaves of absence. Selwyn was able to grant these with full pay, for which he received from Cochrane, his former antagonist, a warm letter of thanks for his many kindnesses. In the same period, the newcomers who made their appearance included the surficial geologist Robert Chalmers; the "surveyors and explorers" James M. Macoun, James McEvoy, James White, D. I. V. Eaton, W. J. Wilson, and N. J. Giroux; a number of draftsmen, notably C. O. Senécal, the future geographer and chief draftsman; H. P. H. Brumell and L. L. Brophy, assistants in the Section of Mines; and in the laboratory, W. F. Ferrier, the petrologist, C. W. Willimott, the rock and mineral collector, and R. A. A. Johnston, the future Survey mineralogist.

VII-15 T. C. WESTON

A noteworthy attempt by some of the younger staff to surmount the animosities of the past was the Logan Club, an institution that still survives. Its formation in 1887 was mainly the work of A. C. Lawson and E. D. Ingall, to judge from the statement in the report of the twelfth and final meeting of the first year's program, in 1888: "After passing votes of thanks to Mr. Lawson for the important part he took in starting the idea of the Club & to the Secy Treas for his Services it was decided after a little discussion to leave the calling together of the Club next fall to the Secy, Mr. Ingall."[32] Meetings were usually held in a local hotel or sometimes simply in the Long Room at the Survey, and to prevent them from proceeding at too great length, an early resolution ordered that the chairman close the meetings at 10:00 p.m. Subjects discussed in the first year included evidences of the history and origin of the Archean rocks (Lawson), "Are Archaean rocks metamorphic?" (Coste), volcanic rocks in stratigraphy (Dawson), prehistoric man (McInnes), Cretaceous fossils (Whiteaves), the Quebec Group (Ells), natural gas and petroleum (Bell), post-glacial lake basins (Tyrrell), and "The First Dry Land" (Selwyn)—in themselves a fair indication of the educational and scientific values of the club. A similar program of a dozen meetings was held in 1889-90, but then the club seems to have fallen apart—perhaps because of the departure of Lawson—or at any rate to have greatly declined. It was reconstituted as a study group from time to time, but more particularly for social purposes, mainly in connection with anniversaries and testimonials, or to arrange and manage meetings like that of the Geological Society of America in Ottawa in December 1905 (for which, incidentally, $28.00 had to be collected from the staff members when the expenses of $490.49 exceeded income). The rather chequered course of the Logan Club in these early years may be seen as indicative of the factionalism and continuing low morale in the Survey.

Meanwhile, what of Selwyn, who attained the age of 70 on July 24, 1894, and completed a quarter century as director on November 24 of the same year? As early as 1888 there had been a flurry of rumours that Selwyn would be replaced by Dawson or Bell. That there was some substance may be seen from Dawson's lengthy memorandum on the circumstances of his succession:

> As long ago as 1888, Dr. Selwyn's superannuation & my appointment as Director of the Geological Survey were in prospect. Hon. Thomas White, then Minister of Interior informed me that he had decided upon this & had obtained Sir John Macdonald's consent to it, shortly before his death. I had been in charge of the Geol. Survey for nearly a year, as Acting Director during Dr. Selwyn's absence about the Colonial & Indian Exhibition in 1886 . . . Although Mr. White intimated the approaching change to me, he apparently hesitated in taking the final step, & a few days afterwards fell ill, & on April 21, 1888, died. Thus the matter ended for the time.[33]

During the autumn of 1894, while the field work was still in full swing and the men were scattered from Cape Breton to British Columbia, Selwyn went on a three months' leave of absence in England. On September 6, a month before departing, he wrote asking Dawson to return to Ottawa as soon as possible in view of his intention to apply for leave. Dawson did not receive the letter at Sicamous, B.C., till September 21, and since he had already arranged to meet his assistant McEvoy first, he did not reach Ottawa until October 6, a day or two after Selwyn's departure. He surmised at once that Selwyn might want to retire, and on arriving in Ottawa, he noted "Find it is well understood that this 'absence' is preliminary to superannuation, and generally understood that I am to be Dr. Selwyn's successor."[34]

Rumours were circulating in the Ottawa press that Selwyn was being retired as of the beginning of 1895, though he vehemently denied this when a *Citizen* reporter addressed the question directly to him. The same newspaper, however, saw significance in the fact that Selwyn had put his home up for sale. While Dawson, as acting director, took charge of affairs, Bell returned from the field and secured an interview on October 27 with the Prime Minister, Sir John Thompson, to present his own suit for the directorship. Dawson, more casually,

VII-16 COMPOSITE PHOTOGRAPH OF THE STAFF, 1888

1 Sir W. E. Logan
2 T. S. Hunt
3 Elkanah Billings
4 Alexander Murray
5 James Richardson
6 A. R. C. Selwyn
7 G. M. Dawson
8 R. G. McConnell
9 J. B. Tyrrell
10 John Macoun
11 T. C. Weston
12 John Thorburn
13 Scott Barlow
14 Hugh Fletcher
15 Robert Bell
16 J. F. Whiteaves
17 A. P. Low
18 F. D. Adams
19 R. W. Ells
20 Samuel Herring
21 John Marshall
22 H. P. H. Brumell
23 L. M. Lambe
24 C. W. Willimott
25 W. R. McEwan
26 James White
27 R. L. Broadbent
28 Eugène Coste
29 D. B. Dowling
30 E. G. Kenrick
31 William McInnes
32 L. N. Richard
33 H. M. Ami
34 James McEvoy
35 E. R. Faribault
36 J. M. Macoun
37 J. A. Robert
38 W. H. C. Smith
39 A. E. Barlow
40 Robert Chalmers
41 John McMillan
42 R. A. A. Johnston
43 A. S. Cochrane
44 Amos Bowman
45 N. J. Giroux
46 Michael O'Farrell
47 E. D. Ingall
48 J. C. K. Laflamme
49 L. W. Bailey
50 A. C. Lawson

THREE MINISTERS OF THE INTERIOR

VII-17 THOMAS WHITE 1885-88

VII-18 EDGAR DEWDNEY 1888-92

VII-19 THOMAS MAYNE DALY 1892-96

submitted a short note on his own claims to the position to his minister, T. M. Daly, on November 9. Mr. Daly saw that the necessary papers for Selwyn's superannuation and Dawson's appointment as his successor were prepared in good time for Selwyn to be notified before he sailed back from England.

Sir John Thompson's sudden death in England on December 12, and the formation of a new cabinet under Sir Mackenzie Bowell on December 21 followed by an intense political crisis within the government, delayed the passage of the necessary orders in council. Bell, who claimed Thompson had supported his candidature, wrote a lengthy memorandum on January 5, 1895, to J. G. Haggart, now Minister of Railways and Canals, appealing for his support and pointing to his seniority, university education, work on the Royal Commission on the Mineral Resources of Ontario, his 127 reports and publications and 39 public lectures, and above all, his vast experience with the geology of Canada: "I have worked over all parts of the Dominion from Baffinland to the United States boundary and from the Atlantic Ocean to the Rocky Mountains and I am the original and only living authority on a great part of what is known of the Geology of this, the greater part of the Dominion."[35]

On January 7, 1895, an order in council was issued recommending the immediate retirement of Alfred Richard Cecil Selwyn with a half-pay pension of $2,000 per annum, as well as the promotion of Dawson to deputy head of the department and director of the Survey at a salary of $3,200 per annum. The news produced an indignant letter from Bell to the Prime Minister against the "terrible injustice," the "most undeserved slur" of passing him over after a "lifetime spent in faithful services to the country" in favour of a man eighteen years his junior in the service.[36]

But, with this exception, Dawson's appointment was received with pleasure and rejoicing. Letters and telegrams of congratulations poured in on Dawson. The General Mining Association of Quebec, which was meeting at the time, greeted the news with a standing ovation, and the newspapers printed tributes and eulogies. In sharp contrast, the retired head was almost completely ignored. Selwyn, who had still received no official notification, in the meantime landed in New York and proceeded to Ottawa by way of Montreal to appear at the Survey on the morning of January 10. He informed reporters he had heard nothing about his supposed superannuation, and so far as he knew, he was still director. As Dawson commented,

> The situation was a little embarrassing both for us & for his son Percy who was acting as my secretary, but I must say that he took it very well, although he had apparently forgotten completely his request for superannuation . . . & thought himself rather hardly treated, particularly in the matter of want of due notice. . . His attitude to me personally was & has always been most friendly. This he expressed very fully at the Annual dinner of the Logan Club which was held on Jan. 17th & was made the occasion of a complimentary dinner to him.[37]

The circumstances of Selwyn's removal from the position he had held for 25 years still make painful reading. Despite his disclaimers, he may very well have given the ministers what they took to be a definite statement of his intention to retire, and then failed to realize, or forgot, what he had done. Most likely the whole affair was simply an oversight, an example of the bureaucratic chaos that befell the federal government in the last year of Conservative rule following the sudden, tragic death of Sir John Thompson. Whatever the reason, the unceremonious way Selwyn was dumped from office, even if he was past 70, was shockingly bad manners on the part of the government and an unwarranted stigma upon the man's reputation.

VII-20 GROUP AT "EOZOON LOCALITY, COTE ST. PIERRE," about 1894
At the left are R. W. Ells, A. R. C. Selwyn, and F. D. Adams; seated on the side of the rock, J. B. Tyrrell; standing in front of rock, Hugh Fletcher. Grouped at top of the rock are E. D. Ingall, R. G. McConnell, R. W. Brock, and William McInnes.

For all his difficulties with his staff (in contrast to his private character, which was marked by "sociable, amiable and chivalrous qualities"[38]), Selwyn made a great contribution to the institution that was the centre of his single-minded devotion. He had a high regard for the personal well-being of his men, but refused to be content with poor or shoddy work. A lover of neat, orderly thought and expression, he forced his men to strive for the same high standards. While sparing with praise, he was a good judge of first-class work and of ability in his employees. He was outspoken, and argued his convictions vehemently, but he was said to bear no grudges and respected most those men with whom he had the most heated arguments.

He made a surprisingly large number of good staff appointments and did not hesitate to dispense with the services of those who did not measure up to his standards. In most cases he made sound decisions in matching his men to the assignments they were best suited to carry out, even though he had been much criticized on this point. He wrestled with governments for adequate funds and status for the Survey and its employees, he extended the work to all parts of the far-flung Dominion and adjusted its program to meet the changing needs of the times. The work under Selwyn became a blend of reconnaissance explorations and of detailed work in more important districts, of studies of general geology and examinations of specific mine fields, of reporting on geology, natural history and ethnology, of topographical and geological mapping, of collecting statistics, and securing specimens for the museum.

Though his critics persisted in attacking Selwyn as being indifferent to the needs of the mining industry, their complaints were unfounded. He did not possess, as was so often charged, a "hereditary and constitutional tendency to throw cold water on every mining enterprize,"[39] but believed the true public interest lay in encouraging solid, healthy development. As he wrote as early as 1871, the object of the investigations ought to be "the acquisition of precise and reliable information and correct knowledge of the geological structure and thereby of the actual and probable mineral resources of the country."[40] He conceived of the Geological Survey as primarily a scientific organization that could serve the mining industry and the country best by following his own order of priorities, and he took care to maintain its high reputation in Canada and the international scientific community.

One need only glance at the achievements under Selwyn's direction to recognize that he provided good, effective leadership for the Survey. In terms of exploration his directorship featured some of its greatest, most famous reconnaissances. The work, that had been carried shakily across the continent in the seventies, was extended boldly over the breadth of Canada and from North Saskatchewan River to the Arctic Circle. Excellent economic reports by Ingall, Brumell, Ells, or Dawson added a new dimension to the work, as did the work of Adams and Ferrier in the laboratory, and that of the museum staff. The period also saw the launching of the new age of detailed specialized scientific work with brilliant achievements in unravelling some of the problems of the Precambrian and Cordillera, attempting to sort out the Quebec Group, and in surficial and glacial geology. Under Selwyn the work was carried to new scientific heights. Such accomplishments, particularly in view of the hard times, limited funds, unsympathetic public, and ambitious subordinates, indicate Selwyn had a firm guiding hand, good judgment, and a clear head as regards objectives and priorities. By 1895, however, the time had come for younger, more dynamic leadership that would carry the Survey forward into the new age of material prosperity and rapid resource development that was to be hailed as "Canada's Century."

VIII-1,-2 WINTERING A SURVEY VESSEL AT THE MOUTH OF GREAT WHALE RIVER, 1898-99
This vessel, which was built for the Survey, was used by Low and Young for their investigations along the east coast of Hudson Bay.

8

Rolling Back the Frontiers of Canada

A MOST IMPORTANT CONTRIBUTION OF THE SURVEY during its long and honourable history, especially in the last half of the nineteenth century, was its part in rolling back the map of Canada. As G. M. Dawson in his report for 1896 pointed out, the Survey "is the only organization under the Dominion Government occupied with anything of the character of a general mapping of the country as a whole."[1] His successor, Robert Bell, went on to say that its field parties were among the foremost explorers of the unsettled and undeveloped parts of Canada to the point that its operations "have been the means of mapping out the principal part of what is known of the topography of Canada."[2] In his general report on the work for 1902, Bell reviewed the topographical work since Confederation in these terms:

> Previous to the confederation of the first four provinces . . . the operations of the Geological Survey were confined to those southern portions which constitute the provinces now called Ontario and Quebec. But since confederation, in addition to the maritime provinces and British Columbia, the attention of the department has been directed to surveying topographically and geologically the vast newly acquired territories . . . , including those portions of them which have been added to Ontario and Quebec. These great regions were entirely unsurveyed and but partially explored, only the main geographical features being roughly indicated on the sketch-maps . . . . The field-men of the Geological Survey have been the first surveyors of the natural or geographical features of . . . nearly one-half of the continent.[3]

Topographical map-making was incidental to the Survey's main function, which was to examine and report on the geology of Canada. As Dawson explained in a paper published in *The Geographical Journal*,

> Nearly all the geological work of the Geological Survey of Canada implies a certain amount of concurrent geographical work; for even in the better-known parts of the country the existing maps are seldom sufficiently accurate or detailed enough to serve without addition as a basis for the geological features. When, however, reconnaissance surveys are carried into new districts, the geographical part of the work frequently becomes as important as the geological.[4]

The need to make topographical surveys along with geological examinations was accepted as an unavoidable evil by Selwyn, and by his successor Dawson, who complained that the task consumed a great deal of time and effort, and was a misuse of the geologists' special skills. Both directors called on the provinces to make greater efforts to meet their own mapping responsibilities, after which the geologists could devote themselves to adding the geology to their base maps, as was done in other countries. But the provinces, having little need for surveys apart from the requirements of the land-granting departments, preferred to map their territories at a leisurely pace, while the Survey, as the pioneer preceding the lumberman, the miner, and the settler almost always operated far ahead of the land surveyor. Directors had to accept the added burden and make the best of the situation by employing men, usually with a civil engineering background, to work expressly as topographers in conjunction with the geologists. Such was the arrangement under which Tyrrell and Dowling explored much of the country north of Saskatchewan River, or McEvoy assisted Dawson in the mountains of British Columbia, or Keele accompanied McConnell in the explorations of the Yukon Territory. These cases, incidentally, demonstrate how men initially hired as topographers could gradually develop into important reconnaissance geologists in their own right.

Robert Bell was the complete antithesis of his predecessors on the question of the incompatability between geological and topographical work. Rather, he contended, "owing to the fact that the topographical features everywhere depend upon the geological structure, the geologist becomes the best topographer,"[5] and since "much of our field-work is now being done in unsurveyed and even unexplored regions, the most useful geologist is he who is also a good surveyor."[6] Every geologist ought to do his own topographical surveying as an aid to his geological work, Bell said, and reasoning from his own experience, he vigorously opposed providing topographers for geological field parties. One had only to look at the large number of maps that the Survey had produced which "have always been recognized as the most accurate obtainable in regard to topography"[7] to see how well the two activities went together. Economy and efficiency were further advantages, for as long as the Geological Survey carried on the two activities simultaneously, "the country has had the benefit of both at a remarkably small cost."[8] Combining the two activities also made for better co-operation in the office among geologists, assistants, and draftsmen, contributing to the *esprit de corps* of the entire department and to the professional pride of its staff.

Bell, in fact, occasionally even put the need to complete topographical mapping ahead of the geological work. In his first report, after discussing the virtues of carrying on topographical and geological work simultaneously, he went on to say: "The Survey has thus been a most useful institution for the general progress of the country, and as no better method could be employed, the same topographical system must continue if we are to have the whole country mapped out at all, and as the geological work adds little or nothing to the cost, we might as well have it done at the same time."[9] By 1905, he was fully committed to the principle of completing the exploration of Canada in connection with the geological mapping of the country as a whole: ". . . one of the principal duties of the Geological Survey is to produce as complete a geological map of the Dominion as possible, and as large areas still require to be explored for this purpose, a certain amount of energy must be given to this branch of our duties."[10] This goal led him in 1904 and 1905 to order a series of difficult, unpleasant surveys of the southern and southwestern Hudson and James Bays, which had little or no geological significance, so as to complete "the last link of the topographical survey by this department of the entire coast of our great inland sea."[11] Indeed, as Owen O'Sullivan, one of the field officers, reported,

> From a geological point of view there is nothing very interesting to be seen along that part of Hudson Bay coast which we traversed. Nothing but mud flats and boulders looking seaward, and marshes, dunes, ponds and muskeg, bordered by stunted evergreen woods, chiefly small spruce, looking landward.[12]

In defence of Bell's attitude it should be remembered that under the acts of 1877 and 1890 the Survey was responsible for more than geological investigations, notably in connection with the museum. The many-sided reconnaissances that investigated natural history, economic resources, and ethnology as well as the geology and topography were entirely within the terms of reference. Even Selwyn had recognized this in his instructions to field parties to bring back as much information as possible on the geography, soils, vegetation, and fauna, inhabitants and communications of the areas examined, so long as it did not detract from the geological aspects of their work. Geologists, according to their particular interests and opportunities, returned with much information and material not strictly of a geological character, especially when traversing a previously unexplored region. Thus in the course of Low's crossing of the Ungava peninsula from Richmond Gulf to Ungava Bay in 1896, "A large collection of plants was made, useful as an index to the climate and also in extending the range of many species. Collections of bird-skins, birds' eggs, small mammals, shells and insects were also made and are at present in the museum, together with a small collection of Eskimo carved ivory."[13] The surveyors were encouraged in these efforts by associates outside their organization, for example, Dr. J. Fletcher of the Entomological Branch, who in 1901 urged the members of the Survey "to make an effort to bring home with them at least a few specimens every year," noting the exact dates and localities where the insects were collected.[14] Officers even arranged sometimes to test the potential of the regions in which they travelled. In August 1896, Bell examined the growth records of certain types of seed he had sent to the trader in charge of Waswanipi post in northern Quebec on behalf of the Central Experimental Farm; and on the basis of these he reported that "almost every kind of garden crop grown in an average district of Canada" flourished there, and that agriculture was feasible in that remote outpost of northern Abitibi.[15]

The breadth of the Survey's interests is indicated also from the way Bell corresponded with a wide circle of fur traders and missionaries on the subject of Indian folklore. That he was a very good collector of such material is revealed by his instruction in May 1905 to O'Sullivan prior to his proceeding to examine the James Bay coastline:

> Please talk to all likely people including your own Indians, about the matter of legends or stories, and after they get through telling you these stories, please write them down in the back of your field book but do not attempt to make notes while they are talking, otherwise they will soon stop. You will know how to humour them about such matters. It is rather difficult to start a good story-teller but once going he will be very apt to warm up to the subject and tell you quite a number of stories before he stops. You will have to

humour them a good deal, say with a plug of tobacco to get them to start. While in camp during a rainy day, or on a cool evening around the fire is a favorable opportunity.[16]

For this sort of broad-based reconnaissance a special kind of officer was required—not a narrow specialist but a man of as broad a range of interests and scientific talents as possible. Even Dawson made a rather similar point in a widely-noted paper of 1890, "On Some of the Larger Unexplored Regions of Canada," to the effect that

> In order, however, to properly ascertain and make known the natural resources of the great tracts lying beyond the borders of civilization, such explorations and surveys as are undertaken must be of a truly scientific character. The explorer or surveyor must possess some knowledge of geology and botany, as well as such scientific training as may enable him to make intelligent and accurate observations of any natural features or phenomena with which he may come in contact. He must not consider that his duty consists merely in the perfunctory measuring of lines and the delineation of rivers, lakes and mountains. An explorer or surveyor properly equipped for his work need never return empty handed. Should he be obliged to report that some particular district possesses no economic value whatever, besides that of serving as a receiver of rain and a reservoir to feed certain river-systems, his notes should contain scientific observations on geology, botany, climatology and similar subjects which may alone be sufficient to justify the expenditure incurred.[17]

Bell, no doubt thinking about his own experiences, enumerated some of these diverse tasks after describing the geological aspect of the field party's operation:

> While working in new territory we also take advantage of the opportunity to obtain heights of banks or cliffs, hills and mountains and comparative levels of waters, grades and depths of streams and lakes, records of the temperature of the air and water and of other meteorological observations as indications of climate, notes as to the kinds and characters of the forest trees and on the flora generally; also as to the fauna, the collection of zoological and botanical specimens, making notes on the nature of the surface of the country, whether hilly or level, rocky, swampy or covered with soil, the character of the land, and on various other matters. We also enquire from the natives as to the topography, etc., of regions beyond our own explorations. Photographs are taken to illustrate the geology, scenery, the character of streams, etc.[18]

In outlining the Survey's field operations, Bell seemed to place the emphasis upon industry, versatility, and resourcefulness, rather than upon special training in, or knowledge of, geology. He stressed above all, the leadership qualities required of the party chief, rather than his academic or professional attainments in geological science. Geology to begin with, was only a part of a broad range of activities that included the whole spectrum of natural history and surveying. The kind of observations Bell asked of his geologists—"to trace out the geographical position of the various rock formations, some of which carry economic minerals, while others do not"[19]—appeared to be of a sort that could be made by any intelligent man after a little practical field training. He seemed to put the emphasis on gathering data as an end in itself and placed little importance on the party chief being trained to make on-the-spot generalizations

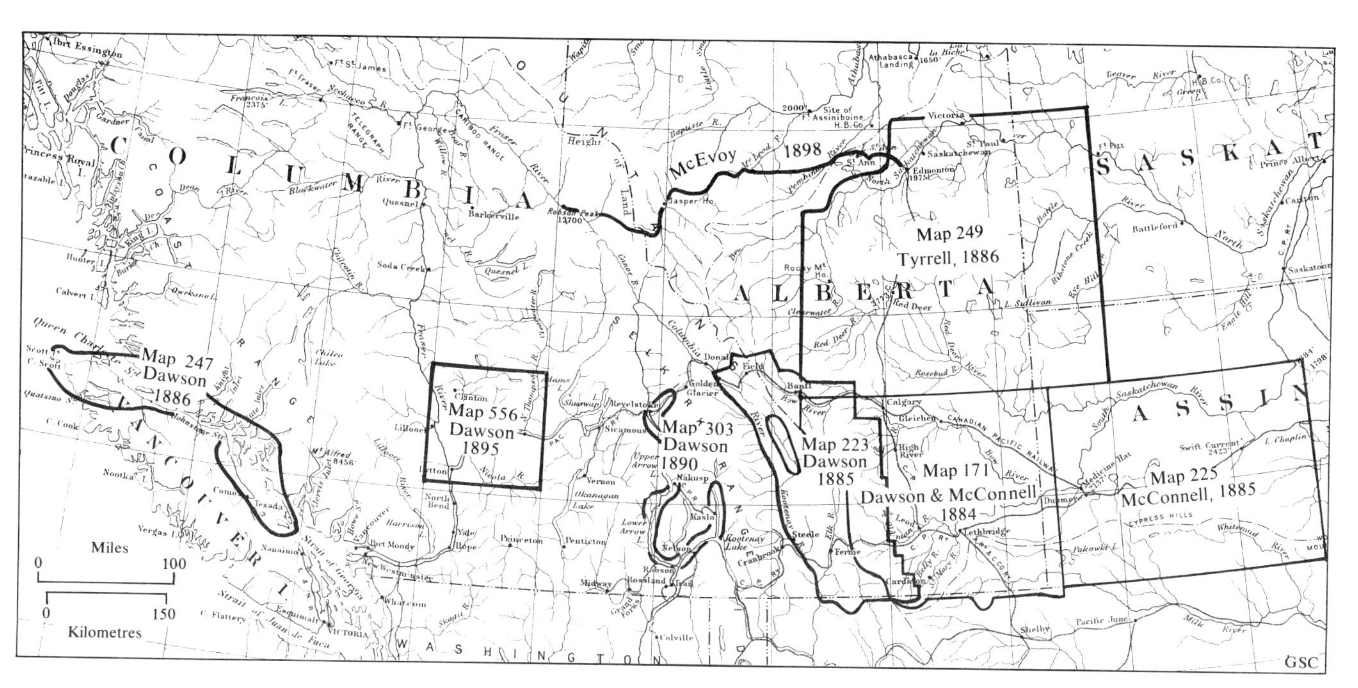

VIII-3 SOME MAJOR EXPLORATIONS IN WESTERN CANADA
Base map—Department of the Interior, *Map of the Dominion of Canada*, 1901.

or an intelligent selection of key data likely to hasten the ultimate solution of the geological problems.

By putting less emphasis on specialized training in geology as an essential condition for appointment as a party chief, Bell was able to find chiefs of field parties more easily than his predecessors had, and he was able to double the number of parties he sent to the field. Inside the Survey he found his party chiefs among the topographers and draftsmen as well as the geologists, and he hired former officers of the Survey and university professors in unprecedented numbers to direct parties under contract. All too often men who were specialists, who had developed a special understanding of a certain type of formation or of a given district, were transferred to direct the all-purpose reconnaissance surveys that Bell considered the times demanded. William McInnes was the outstanding example; after devoting some ten years to the complicated problem of the Precambrian succession in the Rainy Lake district, in 1902 he was transferred to the country west of Lake Nipigon to assist in mapping that area, following which he spent three seasons exploring the district between Attawapiskat and Winisk Rivers.

In reconnaissance surveys so much ground had to be covered that it was impossible for the geologist to gain more than a general impression of the geological features of the area in which he was travelling. The aim was to indicate the general geological features of the region traversed and point out what parts warranted further, more detailed, examination at a later date because of their interesting geological formations or possibly valuable mineral deposits. Nevertheless, the reconnaissance surveys could make valuable contributions to geological science, especially of aspects like glaciation that were readily ascertainable through cursory examinations. The evidences of glaciation extend over great distances so the farther the geologist travelled, the better he was able to perceive and come to understand the large-scale movements. Furthermore, the reconnaissance surveys continually approached closer to the seats of the mighty continental ice sheets in Keewatin, Labrador, Hudson Bay, and the islands of the eastern Arctic.

The geologists, on the basis of these reconnaissances, were able also to point out deposits of economic minerals, or formations (like the Huronian) that from past experience were considered favourable indications of the occurrence of such deposits. Thus in 1881 and 1882 Bell in making preliminary reconnaissances over the waterways between Lake Superior and Lake of the Woods paid special attention to the boundary between Laurentian and Huronian formations, on the basis of which he could point out the locale of a future mining camp:

> A very careful track-survey was next made of Red Lake itself, as its shores proved to be of great geological interest. The whole lake (which is of considerable size) lies within a wide belt of Huronian rocks, among which several of the rarer varieties are well developed, and they were found to contain some interesting minerals.[20]

They reported on coal in the Cretaceous of the prairies and British Columbia, possible precious and base metal showings in the Cordillera in the Yukon and Omineca, or in the Precambrian from Great Bear Lake to Athapapuskow Lake and Red Lake, from Chibougamau and Hamilton River. The discovery of the great iron ore deposits in Labrador was the result of a reconnaissance survey in 1893-94 by A. P. Low; the reports of another large iron ore deposit on the Yukon-Northwest Territories boundary and of a potential oil and gas field on the Peel Plateau were presented by two reconnaissance surveyors in 1904—and the list could be multiplied many times over.

Thus year after year members of the Geological Survey fanned out from Ottawa to explore new districts of Canada, and often they produced the first accurate maps of the regions they studied. They traced the rivers that flowed from the height of land north of the Great Lakes and the St. Lawrence to James and Hudson Bays, examined the mountain ranges from the Rockies along the International Boundary to the Coast Range behind the Alaska Panhandle, and made circuits of the islands of Canada's Pacific coast and traverses of their interiors. They filled in the map with the tangle of interlaced rivers and lakes, forests, rocks and muskegs that stretch between Saskatchewan River and the Arctic tundra, and sketched the interminable mudflats of the southern and southwestern coasts of James and Hudson Bays. They examined the Yukon, Peel, and Mackenzie basins and figured prominently in all the Canadian government exploring expeditions sent to Hudson Bay and the Arctic islands between 1884 and 1903. Working northward, they traced the country along the rocky coasts of Hudson Bay, penetrated the remotest sections of the Ungava peninsula and the heart of the Keewatin District, and from the Mackenzie system they began working eastward into the tundra region of the central Arctic. By 1906, having made a major contribution to the delineation of the coasts, rivers, and geographical features of most of Canada's mainland, they were in a position to extend their reconnaissances along the Arctic coasts and onto the adjacent islands. To indicate this immense achievement in a tangible way, Bell published a map in 1902 that showed the geographical work of the Survey in its first 60 years (1842-1902), in the form of coloured lines that represented the routes followed in exploring and surveying the more remote regions.

A large number of those tracks were made by Bell himself, for his own work surpassed that of the several other very able explorers who worked for the Survey in this period in terms of the area covered, the districts mapped for the first time, and the reporting of local conditions and prospects of a wide part of Canada. In a 50-year period, Bell ranged as far afield as the prairies of Saskatchewan, the oil sands of the Athabasca, and north to Great Slave Lake and Baffin Island. But mainly his work was concentrated in what is now northern Ontario and Quebec, in an arc extending from the Saguenay to Lake Winnipeg and from these points north to the shores of Hudson and James Bays. It was no idle boast when he wrote in 1904 that almost single-handedly he had done the major part of the preliminary reconnaissance work for the third transcontinental railway that was being built during these years:

> When the construction of the Grand Trunk Pacific railway was first proposed in 1902, the region to be traversed was found to be already fairly well known all the way from Quebec to Winnipeg, as to elevations, topography, soil, timber, climate, fauna, flora, etc., as well as in regard to its mineral resources, through the work of the survey which the writer had been carrying on in nearly all parts of that region during the previous thirty-five years. The results of all this work, which had been fully reported and illustrated by maps, enabled our public men to judge of the feasibility of the undertaking, and much time was thus saved in arranging for the construction of the railway.[21]

His range was equally broad—including not only geology and mineral resources, but soil, surface conditions, transportation facilities, forests, waterpowers, crops, fish, wildlife, vegetation, climate, and as has been seen, anthropology and ethnology. His outstanding contribution to exploring and mapping "an immense area of Canada previously unknown" was fittingly recognized by the award to Bell in 1906, at the close of his active career, of the Patron's medal of the Royal Geographical Society.[22]

* * *

DURING THE 1870'S THE EXPLORATORY INTEREST of the Survey was focused on the Parkland and Peace River districts of the Central Plain and on central British Columbia—the districts being considered for the route of Canada's first transcontinental railway system, the Canadian Pacific Railway. But almost at the same time as the Survey moved to Ottawa, a new management took control of the CPR project and decided to build the railway along a southerly route through little-known country that had therefore to be examined immediately from a geological point of view. The Survey at once set to work, with G. M. Dawson spearheading the drive in the foothills and Cordillera region, assisted first by McConnell, then by Tyrrell, and then by James White, all three of whom received their field initiations on these projects. As early as 1884 the Survey, based on the work of three seasons, was able to produce a "Report on the Country in the Vicinity of the Bow and Belly Rivers, North-West Territory," covering a 26,960-square-mile area "selected for this purpose because of the known and reported value of its coal deposits, and its relation to the adopted transcontinental railway route."[23]

The main goal of the work on the Plains was the examination of widely-scattered coal and lignite occurrences found at several different horizons in the Cretaceous system. Coal seams (some already beginning to be mined, as at Coal Banks, now Lethbridge) were traced for distances up to 175 miles, the stratigraphic position of the principal coal measures was determined, and the 1884 map gave the approximate distribution of the coal measures over wide areas covered with drift. Dawson also produced a map showing the distribution of the various species of trees and grasses, a useful tool for planning the agricultural uses of this part of the country. These explorations also investigated the folded and faulted Cretaceous rocks along the eastern margin of the mountain ranges from the 49th parallel to North Saskatchewan River, and traced the contact between the coal-bearing rocks of the Plains and the uplifted, highly-folded limestones of the Rockies. Observing outliers or isolated basins of Cretaceous rocks far to the west of the limestones of the front ranges of the Rockies, Dawson and McConnell predicted that coal might also be present in such isolated basins—a prediction that was confirmed as early as 1883 and often afterwards.

Farther east on the Plains, McConnell examined the border region in the vicinity of the Cypress Hills and the Missouri Coteau, hills that apparently had never been glaciated and were still capped by coal-bearing rocks of Tertiary age. Tyrrell worked in the same general neighbourhood in 1884 and 1885 investigating the region dominated by the Red Deer and Battle Rivers systems, observing extensive coal deposits and making important discoveries in the badlands of the Milk, Red Deer, and other rivers, of the fossil remains of dinosaurs and other reptiles. The veteran T. C. Weston went out to the Plains in 1883 to collect specimens.

VIII-4 DRIFT BLUFFS IN BELLY (now OLDMAN RIVER) VALLEY AT "COAL BANKS CROSSING" (now LETHBRIDGE, ALBERTA)
This reproduction of G. M. Dawson's photograph of June 27, 1883, indicates how well the technical difficulties in printing photographs had been mastered by 1885. Printed by the Artotype Process by G. E. Desbarats and Company, Montreal.

In these years, Dawson assaulted the complex of high ranges that forms the southeastern boundary of the Cordillera—zigzagging his way through this difficult, almost impassable region, starting with Palliser's old, large-scale map but finishing with his own, made virtually from scratch. His report, published in 1886, discussed the uplifting, tilting, overturning, and fracturing of strata, the effects of weathering, erosion, and glaciation, and explained the various mountain ranges, passes and valleys, lakes, and rivers he encountered. In 1886 McConnell made a geological section across the Rockies about 100 miles north of where Dawson had left off, proving that almost a complete succession of Paleozoic rocks is present. Amos Bowman, a Canadian-born mining engineer and journalist in California whom Selwyn had hired as Dawson's assistant in 1876, also worked on the geography and geology of the country across which the CPR was building its line to the Pacific coast, mapping along Fraser, Chilliwack, Tulameen, and Similkameen Rivers and securing data for a revised edition of the map of the Shuswap and Okanagan Lake districts. In a few short years the country tributary to the C.P.R. had been examined sufficiently to meet present needs.

In the meantime, Dawson was able to develop a fairly adequate picture of the geological history of the region, which work and study over the next 20 years in the main confirmed. Briefly, this view was that the oldest part of the region was an axis of crystalline, complex, plutonic Archean rocks occupying an area centred on the Selkirk Mountains, which was "the key to the structure in the entire region of the Cordillera."[24] On one side of this axis was the Laramide Geosyncline, afterwards the birthplace of the Rocky Mountains, and on the other a West Geosyncline, the future site of the Coast and Islands ranges. The region east of the axis was occupied by shal-

VIII-5 FORT CALGARY, N.W.T., IN 1881
G. M. Dawson, who reached Fort Calgary from the west following his reconnaissances in the Rocky Mountains, wrote that "The situation of Calgary is remarkably beautiful. The plateaus here retire to some distance from the river, which is bordered by wide flats thickly covered with bunch-grass and well adapted to agriculture. The river is fringed with trees, and from the higher points in the neighborhood the Rocky Mountains are still visible."

low seas in which marine sediments were laid down, while on the west and north the sedimentary strata were often mingled with large bodies of material arising from intense and prolonged volcanic activity and intrusion of igneous masses. At the close of the Cretaceous, great orogenic movements resulted in an eastward thrust of the axis which crumpled the strata farther east and uplifted the Rocky Mountain chain; the elevation of the West Geosyncline also was completed by the early Pleistocene.

Following their work in the Rockies, Dawson and McConnell, assisted by McEvoy, embarked on one of the most remarkable reconnaissance explorations in the entire history of the Survey, that carried its work across the Arctic Circle for the first time. Most of southern and central British Columbia had been examined in a general way by 1887, but the geology of the northern half of the province and the Yukon country farther north was still unknown, apart from sketchy descriptions by fur traders and by scientists who ventured through the region. The northern part of the Plains region—the district drained by Slave, Liard, and Mackenzie Rivers—was somewhat better known, but much the same held true there also. By the mid-eighties both districts were becoming accessible and were receiving the attention of hundreds of gold prospectors, mostly Americans, who flocked into the western part of the region, to the Cassiar diggings in the upper Liard basin and the upper Yukon River. Their presence brought legal and administrative problems, notably in connection with the still-unmarked boundary between Canada and Alaska. Hence it seemed important economically, scientifically, and nationally, to have these northwestern districts explored and mapped by agents of the Canadian government. Associated in the operation was the Dominion Lands surveyor, William Ogilvie, who determined the location of the 141st meridian boundary with Alaska, and mapped the routes in and

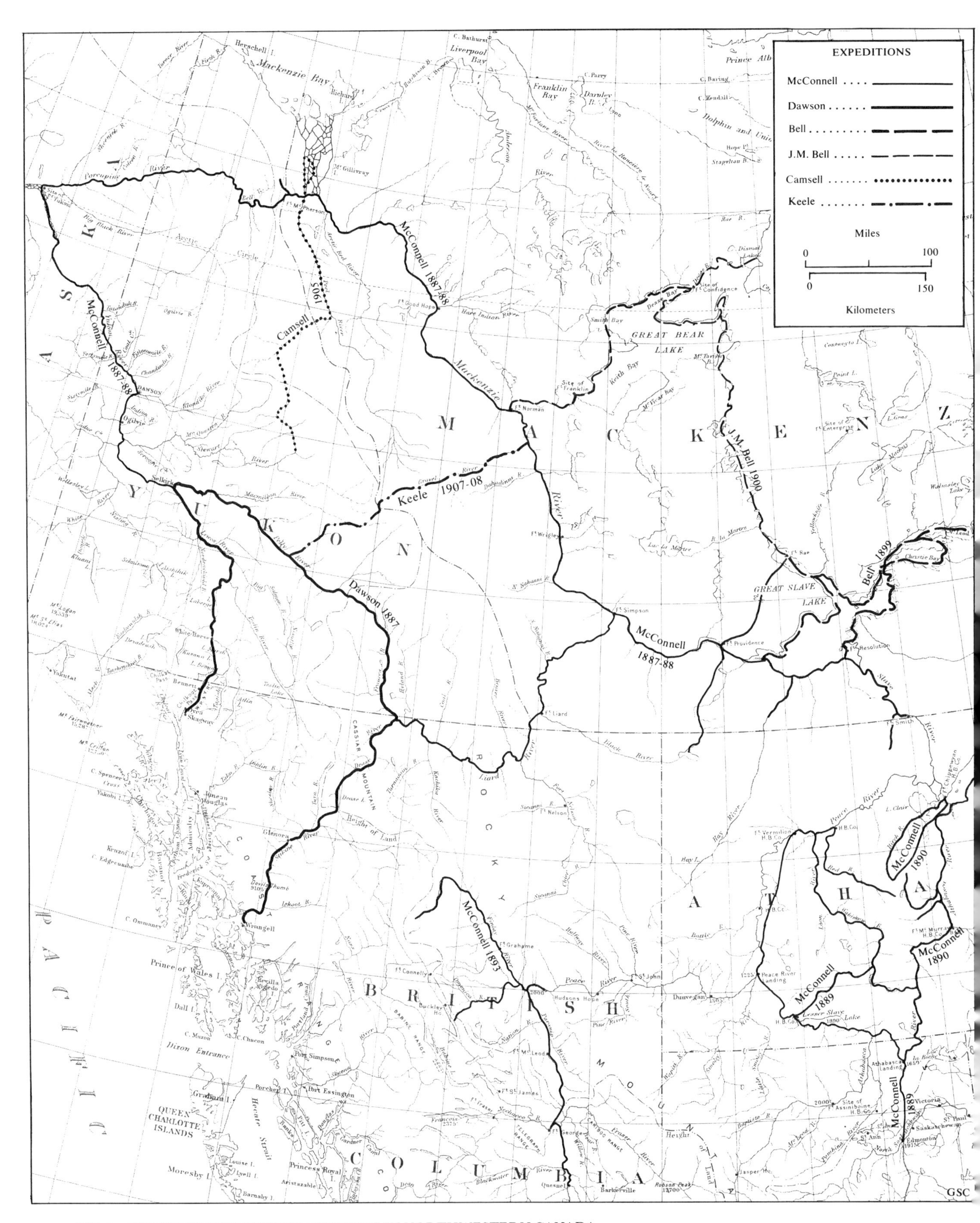

VIII-6 NOTEWORTHY RECONNAISSANCES IN NORTHWESTERN CANADA
Base map—Department of the Interior, *Map of the Dominion of Canada,* 1901. The geographical knowledge of the Peel River system shown in this map is somewhat defective.

out of the country from Chilkoot Pass down the Yukon to the Alaska boundary, up Porcupine River, over the height of land to Mackenzie River, and then all the way up the Mackenzie waterway to Lake Athabasca, where his survey connected with the existing system of Dominion Lands surveys at Fort Chipewyan.

Dawson, McConnell, and McEvoy entered the country farther south by way of Stikine River to Telegraph Creek and overland to Dease Lake and the Cassiar placer gold mining camp. From Dease Lake they descended to Lower Post, at the confluence of Dease and Liard Rivers, where they separated. McConnell proceeded down the Liard to the Mackenzie, while Dawson and McEvoy continued up the Liard to Frances Lake. From there Dawson and McEvoy made a difficult crossing to Yukon River, for they were compelled to limit their supplies to what they could carry, and proceed on foot to the Pelly branch of the Yukon. They arrived at the Pelly "after a very toilsome journey" but with supplies to spare, paid off their coast Indians, set up their canvas canoe and descended to Yukon River. At the site of Fort Selkirk they encountered Ogilvie on his way into the country. The leak-prone canvas canoe was replaced with a wooden boat, and, reversing Ogilvie's route, Dawson and McEvoy ascended the Yukon (then known as Lewes River in this southernmost section) to the Chilkoot Pass and thence to the Pacific Ocean.

They had completed a circuit enclosing almost 60,000 square miles of country, with very limited time or opportunities for geological observations. However, Dawson's report showed his usual remarkable powers of observation and deduction. He noted that the granites were intruded between the Triassic and the Cretaceous, and recognized evidences of very recent volcanic activity, including volcanic ash lying in places on soil, and lava flows on unconsolidated gravel. Indeed, when McEvoy investigated the Portland Canal-Nass River area a few years later, he discovered evidences of very recent (within the preceding 200 years, he calculated) volcanic activity that confirmed the local Indian legends. Dawson in 1887 observed that the vanished ice sheets had flowed north, not south as might be expected, and he discussed the relationship of the glaciation to the newly-discovered deposits of placer gold. Indeed, he set the stage for the recognition of that strange anomaly, the unglaciated region of the far northwest, that accounts for the survival of the rich gold placer deposits of the Klondike. A beginning had been made at the exploration of the Yukon district, which was to attain immense national importance within a decade.

McConnell faced a more arduous journey when he separated from Dawson and McEvoy at Lower Post to descend Liard River on June 23, 1887, for this was one of the most difficult and treacherous rivers in America, which even the Hudson's Bay Company had given up as a fur trade supply route. Having completed that part of his journey through the hundred miles of canyons, raging torrents, and gruelling portages, McConnell paid off his Indians and drifted the rest of the way, alone, to Fort

VIII-7 TAIYA (DYEA) INLET, LOOKING UP FROM HEALEY'S POST, 1887

At the conclusion of his summer's exploration of the Yukon district, G. M. Dawson left the country by way of this future gateway to the Klondike.

Liard. McConnell, too, found his canvas boat unsatisfactory; he had to concoct a compound from sperm-oil candles, gun oil, bacon grease, and spruce gum to keep it in service. From Fort Liard he completed his journey down the Liard to Fort Simpson in a birchbark canoe. In the autumn he devoted a month to reconnaissances of Slave River, Salt River, Hay River, and part of the western end of Great Slave Lake, and in the winter made hasty traverses to Lake Bistcho, Fort Rae, and around Fort Providence. Early in the spring he went back by dog-team to Fort Simpson to await the breakup of Mackenzie River, proceeding down-river May 28, 1888, then up Peel River. He met Ogilvie just after entering the Peel, and the two parties declared a day's holiday to exchange news and reports. From Fort McPherson McConnell portaged across to Porcupine River, down it to Fort Yukon, up Yukon and Lewes Rivers, and out over Chilkoot Pass to the Pacific. This was Ogilvie's route in reverse, except that in his 4,000-mile journey from tidewater to tidewater McConnell had had to contend with the more difficult Stikine, Dease, and Liard Rivers.

In his *Report on an Exploration in the Yukon and Mackenzie Basins, N.W.T.* McConnell admitted his examination was hasty and his conclusions very tentative; indeed, he did not attempt to indicate rock units on the maps but merely printed descriptions of his geological observations here and there along his route. Nevertheless, he outlined the general features of the region quite satisfactorily on the basis of his knowledge of similar rock formations farther south. Proceeding eastward along the Liard, for example, McConnell observed in sequence nearly parallel limestone mountain ranges, the foothills belt of long, nearly parallel ridges of tilted sandstone beds; the high plateau of flat-lying Cretaceous rocks through which the Liard had cut the deep gorge that had occasioned so much difficulty; the flat plain mantled with glacial drift embracing the basins of Mackenzie, lower Liard and Hay Rivers, and the western part of Great Slave Lake; and finally, east of the Mackenzie, Precambrian granites and gneisses which, as at Slave River rapids, extend west as far as the waterway. One of the most interesting geographical features he observed was the apparent displacement of the Rocky Mountains north of Liard River, where the range "is suddenly jogged eastward for a couple of degrees and is then continued northward along nearly the same bearing as before."[25] This northern extension is now known as Richardson Mountains. He noted the widespread distribution of Devonian rocks east of the mountains; and on the Pacific, or western, slope, the apparently unglaciated nature of the Porcupine and Yukon districts. The many seepages and showings of oil-saturated sands and shales also made him very optimistic on the region's future as a major oil-producing

VIII-8 VIEW ON LOON (now WABISKAW) RIVER, ALBERTA, 1889
On his exploration of the flat marshy country encircled by Peace and Athabasca Rivers, R. G. McConnell followed the river courses, notably that of the meandering 350-mile-long Loon River.

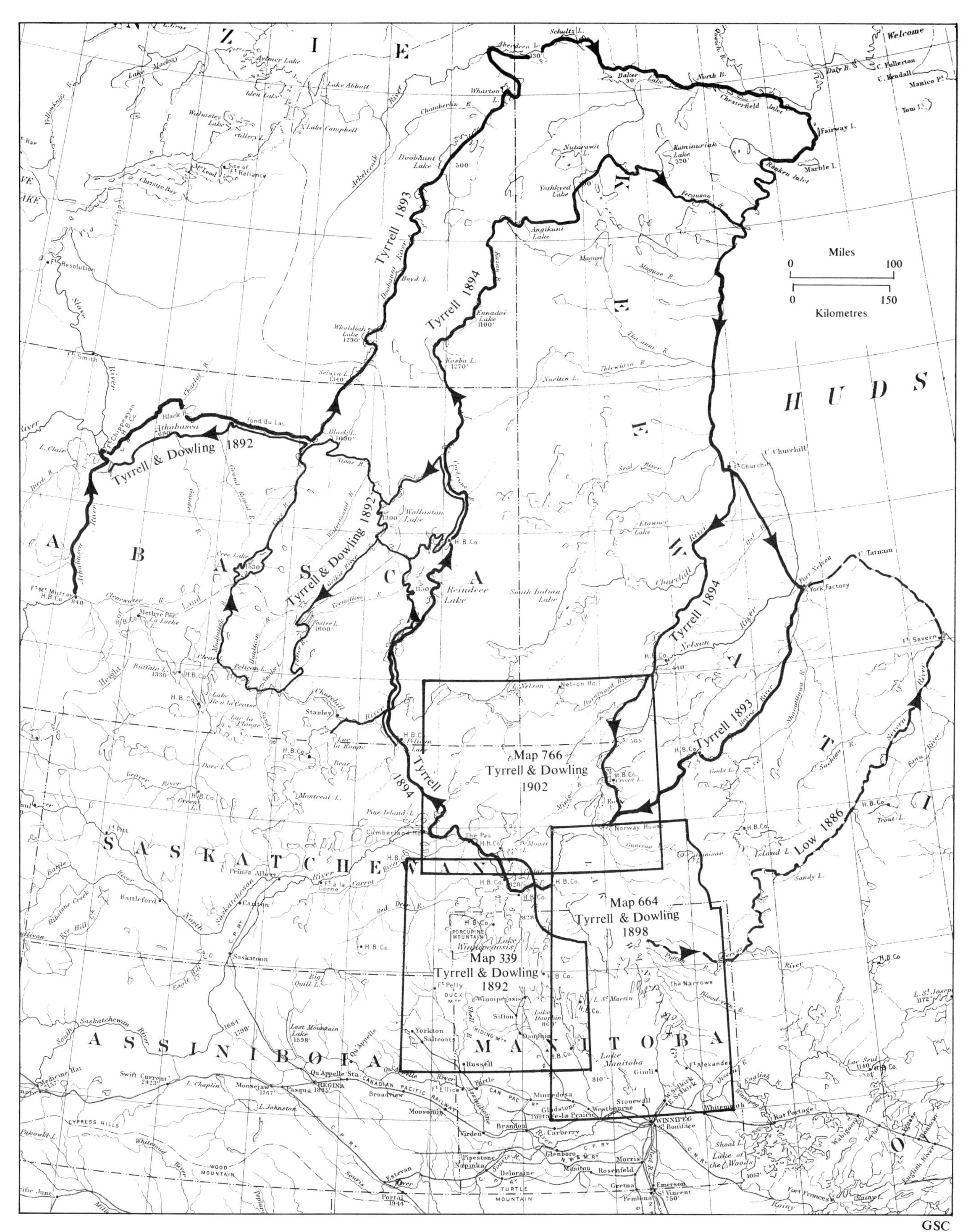

VIII-9 EXPLORATIONS IN THE TERRITORY WEST OF HUDSON BAY, 1886-1902
Base map—Department of the Interior, *Map of the Dominion of Canada*, 1901.

centre:

> The possible oil country along the Mackenzie valley is thus seen to be almost co-extensive with that of the valley itself. Its remoteness from the present centres of population, and its situation north of the still unworked Athabasca and Peace River oil field will probably delay its development for some years to come, but this is only a question of time. The oil fields of Pennsylvania and at Baku already show signs of exhaustion, and as they decline the oil field of northern Canada will have a corresponding rise in value.[26]

Following this remarkable journey McConnell in 1889-90 investigated the country between the Peace and Athabasca districts, that Bell had explored in 1882, to report on the petroleum and salt deposits of the Athabasca district. McConnell's experience was extremely frustrating, for the country was covered with forests, muskegs, soil, and shallow lakes, with few outcrops of underlying strata. In the Peace River district where the rivers had cut deep valleys through several formations, he was able to note occurrences of lignite, petroleum and natural gas, salt, gypsum, placer and tar sand. He returned again to the remote northwest in 1893 to the mountainous Omineca district of northeastern British Columbia containing the headwaters of Peace River. Entering by way of the Fraser, he descended the Parsnip branch of the Peace to its junction with the southward-flowing Finlay, spent three weeks examining the former placer gold mining camps, of its western tributary, the Omineca, then ascended the Finlay to its source 150 miles farther up the Rocky Mountain Trench. The region was hemmed in by the Rockies on the east, and by "a confused medley of nameless ranges" on the west.[27] The rivers—swift, wild torrents in places—were interrupted by frequent rapids and portages like those around Black and Little Canyons of the Omineca, or Deserter's and Long Canyons of the Finlay. He reported on the Cambrian and Cambro-Silurian limestones of the western slope of the Rockies and the Tertiary and Triassic beds lining the valley of the Finlay, mentioned fine gold in the gravel bars along the Finlay, coarse gold and silver-amalgam (mercury) in those of the Omineca and its tributaries, as well as veins of argentiferous galena, that were too expensive to mine under existing conditions. He felt that the exposures of volcanic schists and other eruptive rocks along the Omineca held out the best hope for mineable metallic mineral deposits. In these four great expeditions extending from Edmonton through the Mackenzie and Yukon basins, McConnell opened the door for future generations, to the discovery and development of the great metallic mineral and coal, oil, and gas deposits in the northwestern quarter of the Dominion.

Contemporaneous with these investigations in the far northwest was a series of surveys along the northeastern margin of the Plains region and the western part of the Precambrian Shield beyond. As early as 1879 Bell made a track survey of the country west of Hudson Bay between Norway House and Cross Lake, while his assistant A. S. Cochrane mapped the tangled network of rivers and lakes between Lake Winnipeg and Lake Athabasca. Later, in 1887 and 1892 Tyrrell and Dowling continued this work, first examining the basins of the Manitoba lakes and their bounding plateaus, then exploring still other parts of the waterways between Lake Winnipeg and Lake Athabasca. Besides the extensive gypsum deposit they reported in 1887 in the interlake country of Manitoba, Tyrrell made important contributions to our knowledge of the glacial history of the region, then being intensively studied by Warren G. Upham, an American geologist who produced an outstanding report on glacial Lake Agassiz in the Red River District of the United States and Canada.

The 1892 expedition along the waterways of northern Manitoba and Saskatchewan was followed by two celebrated journeys by the much-travelled Tyrrell into and through the Barren Lands of the District of Keewatin, whose trackless wastes had been described only by Samuel Hearne, 125 years before. Accompanying him on the first of these, in 1893, as his topographical assistant was his brother, James W., who had assisted on several Geological Survey parties, and gone on the Hudson Strait expedition of 1885-86 where he became fluent in Eskimo. Proceeding via Edmonton (after a visit to the Columbian Exhibition in Chicago, a 'must' for all Survey employees in 1893) the brothers traversed the length of Lake Athabasca to Black Lake, then journeyed north to Selwyn Lake and to Dubawnt River, fortunately finding a caribou herd that replenished their meagre store of food. When they reached Dubawnt Lake on August 7, they left the forest behind them, which made it difficult to secure fuel and keep dry and warm as they proceeded across the Barrens. The travellers' discomfiture did not apparently extend to the biting insects, which were as voracious as ever. Two days after leaving Dubawnt Lake, they encountered their first Eskimos. Despite baleful warnings from the Lake Athabasca Indians, and the discovery that James' command of Eskimo did not encompass the dialect of this area, they approached these aborigines, who were frightened out of their wits. With a few plugs of tobacco and some trinkets the explorers quickly gained their confidence and hospitality; thereafter they passed from one Eskimo camp to another, until by September 2, they were at Baker Lake and on September 12 on the shores of Hudson Bay.

Strangely enough, the Barren Lands turned out to be the easiest part of the journey. Winter was fast approaching, and travel became enormously difficult. Their canoes were of little use against the drifting ice and the bitter

VIII-10 ROMAN CATHOLIC MISSION AT DU BROCHET POST, REINDEER LAKE, 1892
This post, photographed by D. B. Dowling, was used by Tyrrell as the base for his expedition of 1894 across the Barren Lands of Keewatin.

blowing winds of Hudson Bay. The shorelines of the coast shifted long distances between high and low tides; visibility was often reduced to zero; and storms were frequent. As they inched their way along the 300 miles of coast to Churchill, their food began to run out, and finally, in desperation, they cached their equipment and specimens, and pushed on with all speed to avoid starving to death. Eventually, when some of the party were laid up with dysentery and frostbite, food and dog teams arrived from Churchill to carry them all to safety. The party rested and recuperated a month in Churchill, then they left for the interior on November 6 by dogteam to Norway House, and completed the journey to Selkirk on snowshoes. From there, just as fears began to be felt for their safety, J. B. Tyrrell dispatched a telegram which Selwyn received on January 2, 1894: "Complete success: crossed barren grounds; explored Chesterfield Inlet and west shore of Hudson Bay."[28] The expedition resulted in J. W. Tyrrell's famous travel book, *Across the Sub-Arctics of Canada.*

J. B. Tyrrell returned to the same district the following year, this time accompanied by Robert Munro-Ferguson, a young aide-de-camp of the Governor General, the Earl of Aberdeen. The plan was to follow a route parallel to that of 1893, but along a smaller arc that began at Brochet Post on Reindeer Lake and reached Hudson Bay somewhat south of Chesterfield Inlet. From Brochet Post they surveyed 141 miles of Cochrane River before reaching Kasba Lake and Kazan River. Their Indian guides left them at Ennadai Lake, but again helpful Eskimos provided guides for their further progress. By August 26 it was evident, however, that Kazan River did not lead to Hudson Bay, but north to Chesterfield Inlet. So, to avoid a repetition of the previous year's difficulties, they portaged from below Yathkyed Lake east to Ferguson Lake and down Ferguson River to Hudson Bay. They found themselves still more than 260 miles north of Churchill, and winter was fast approaching. They finally reached Churchill on October 1 after experiences similar to those of 1893 and remained there until November 28. Travelling by way of Split Lake, they reached Norway House on December 24, 1894, and Selkirk on January 7, 1895.

During 1895 and 1896, in addition to his summer explorations, J. B. Tyrrell prepared the reports of his three years' explorations of the area between Lake Athabasca and Hudson Bay. This included a review of previous explorations and travel of the district that may well have started him on his future avocation as a student of history and editor of the writings of those great pioneer explorers of the areas he knew so well—Thompson, Turnor, and Hearne. Tyrrell reported that most of the region he had traversed was underlain by the granites and gneisses of the Precambrian. He commented on the 'Huronian' rocks on the west coast of Hudson Bay, but failed to recognize that these and similar rocks observed along his inland routes were part of a more or less continuous belt extending all the way to Lake Athabasca. He guessed

VIII-11 TRACKING BOAT UP FAIRFORD RIVER (now MANITOBA), 1897
J. B. Tyrrell's party was returning to Lake Manitoba through deep mossy swamp country from an investigation of the former shorelines of Glacial Lake Agassiz.

that the flat-lying, reddish sandstone he noted at Dubawnt Lake might be continuous with the similar rocks south of Lake Athabasca and cover a vast area to the west and north.

His principal geological contributions, however, were his conclusions respecting the Pleistocene glaciation of central Canada. He considered the area the home of the Keewatin ice sheet, which, with the Cordilleran and Labradorean ice sheets, dominated the recent geological history of Canada. He suggested that the glacier had an original centre north of Dubawnt Lake from which it may have extended as far as the Great Plains; that its centre later shifted southeastward, from which it spread out in all directions, but particularly towards the Rocky Mountains and the northern prairie states; and then, in a final stage it became reduced in size and perhaps broke into several smaller glaciers with centres nearer the seacoast in the Chesterfield Inlet region. By directly examining one of the main foci of the continental ice sheets and recognizing that glaciation was not a single phenomenon Tyrrell was able to share in disposing of the older theories once and for all.

This did not occur immediately, however, for the earlier concept that the phenomena we associate with glaciation were mainly the result of floating sea ice still carried considerable strength in Canada, especially since G. M. Dawson, in everything else a model of geological discernment, lent his powerful support to the older view. In his descriptions of the glacial period in the Plains region, in particular, Dawson seemed trapped by his earlier theorizing which followed the floating ice interpretation (1875) as well as by his father's lifelong rigid adherence to that point of view. As a result of field evidence, Dawson had come gradually to accept a great ice-sheet approach for the Cordilleran region, which was his own area of work. In 1890 he described how a vast Cordilleran ice sheet covering the valleys to great depth and overriding plateaus and some mountains, had flowed, hemmed in by the Rockies, southeastward into the United States, northwestward into the Yukon to about the 63rd parallel, and in places westward to the Pacific Ocean.[29]

However, even then he had postulated that most of the evidence on the Western Plains was produced from drifting ice; and Tyrrell's findings, together with those of Low on the Labradorean glacier, did not alter Dawson's views. While he accepted their observations, Dawson contended in 1897 that the Labradorean and Keewatin glaciers were phases of a larger phenomenon which he called the Laurentide glacier. Dawson still regarded the Plains region as having been invaded first by tongues of the Cordilleran ice sheet and then by an incursion of the Pacific Ocean when the Cordilleran region was depressed about 4,000 feet, both of which resulted in debris from that region being deposited on the westernmost parts of the Western Plains. This was followed by a rise in the Cordilleran region and a tilting of the Plains region, which was depressed to a point where incoming seawater could carry glacial ice laden with Laurentian and Huronian rock debris and deposit it over most of the Western Plains. More than one such invasion occurred, resulting in at least two distinct boulder-clays separated by interglacial deposits. The Missouri Coteau, Dawson now felt, "may be regarded either as part of a continental moraine or as the marginal accumulation of an ice-laden sea."[30] But his final word remained that, in his view, changes in land levels and concomitant invasions by ice-carrying seas were the chief mechanism responsible for glacial materials in the Plains region:

> It is with hesitation that we are prepared to admit great succeeding changes in level of large areas of the continent, but the alternative explanations, attributing as they must the most extraordinary effects to glacier-ice, seem to present at least equal difficulties.[31]

Not until the new century, when Dawson was no longer on the scene, did the Survey begin to study the soils and other phenomena associated with the Pleistocene at all extensively, and that principally in parts of the country where the records are more abundant than in the Shield or Cordillera. Then, at last, the Survey was able to arrive at the proper interpretation of the phenomena and to assess correctly the key roles of the great continental ice sheets in shaping the physiography of much of Canada.

Tyrrell, following his return from Keewatin, went back to exploring the general vicinity of Lake Winnipeg in association with Dowling, the work including Gunisao River district northeast of Lake Winnipeg, the country between Lake Winnipeg and Lake Manitoba, and, especially, reconnaissance surveys of the country west of Nelson River and north of Saskatchewan River. This was the first geological exploration of what has since become the important Flin Flon-Sherridon district. Tyrrell in 1896 and Dowling in 1899 worked around Grass, Burntwood, Goose, and Kississing Rivers and Lakes Athapapuskow, Reed, and Sisipuk, a region of very disturbed rocks, which Dowling reported as offering good prospects of worthwhile mineral discoveries:

> The several large areas of Huronian rocks which are here partly outlined will at some future time be thoroughly prospected, and, as has been the case in nearly all such areas, ores of the useful and precious metals are likely to be found. As it is at present a very hasty visit has shown that many quartz veins and intrusive dyke cut these rocks, and indications of the precious metals are not wanting.[32]

Dowling was in fact anticipating the discoveries, barely 15 years later, that led to the rise of a prosperous and continuing base metal mining industry.

* * *

THE SURVEY DID NOT RETURN to the Klondike until the gold rush had begun, when Dawson sent McConnell and Tyrrell in 1898 to examine the Klondike placer gold deposits and other parts of the Yukon Territory with a view to indicating where new mineral deposits might be sought. McConnell was assigned the southeast district around Teslin Lake, and Tyrrell the southwest, in the White River district. In ensuing years McConnell returned every summer to the Yukon to report on conditions in the gold diggings and to conduct other reconnaissances in the hope of discovering new mining districts, his usual associate after Tyrrell's resignation being the topographer Joseph Keele. Singly or together, they explored Stewart and Macmillan Rivers, their tributaries and headwaters in the Selwyn and Ogilvie Mountain ranges, and the Kluane district in the southwest. On these expeditions they were mapping for the first time most of the country that they explored.

The Klondike discoveries also inspired the Survey to undertake more thorough examinations of the Mackenzie country east of the mountains. Reports, and specimens of lead, copper and gold discovered in the region, made it desirable to learn more about the district "which has remained practically unknown geologically and to a great degree geographically."[33] Robert Bell, assisted by J. Mackintosh Bell, was sent in 1899 to examine and map Great Slave Lake, after which J. M. Bell was to winter in the area and investigate Great Bear Lake and the country between the two great lakes during the following summer. Robert Bell examined the south shore of Great Slave Lake from Fort Resolution eastward and back along the north coast, while J. M. Bell concentrated on the North Arm of the lake. Measuring distances by floating boat-log, taking bearings by compass, making almost daily observations for latitude and frequent climbs to high vantage points to sketch shorelines and get some idea of the relief, they secured enough data to "enable us to construct a fairly good map of the whole of Great Slave Lake."[34] Robert Bell reported that the country southwest of the lake exposed flat-lying, nearly unaltered Devonian strata; the east end, Laurentian gneisses and granites; and the country between, including the North Arm and the chain of lakes north to Great Bear

Lake, Huronian schists with good possibilities of mineral wealth. He examined the galena and zincblende occurrence in the Devonian limestone some 30 miles southwest of Fort Resolution—the future Pine Point mining camp—but concluded that the silver content was not very high, while development of the lead-zinc mineral was not economical for the present.

J. Mackintosh Bell's exploration of Great Bear Lake in 1900 was one of the most exciting episodes in Survey history. After wintering at Fort Resolution, Bell proceeded to Fort Norman where his exploration was to begin. There he met and hired as an assistant Charles Camsell, the college-educated son of Julian S. Camsell, chief factor of the Hudson's Bay Company's Mackenzie River district. They examined and mapped Great Bear River, then after waiting till July 4 for open water, traced the major bays of the lake—Keith Arm, Smith Arm, then Dease Bay—from which they made a side trip by way of Dease River across to Coppermine River, about 60 miles distant.

While returning to the lake one of their men, Charles Bunn, disappeared; and after searching as long as they dared, they resumed their travels to avoid being caught by the winter. Near Dismal Lake they encountered a group of Eskimos who immediately turned and fled. The explorers located their camp, however, and helped themselves to some wood and meat (for they were on the verge of starvation), and then resumed their trip back to the lake. "We were evidently the first white men they had seen, as not a single article of white man's manufacture was found in their camp." [35] Camsell was told years later that the Eskimos had followed them for nearly two days until they were satisfied the white men had definitely gone from the vicinity. They had concluded the intruders were friendly because before leaving the camp they had left behind a tin plate and two steel needles, articles beyond price to the Stone Age people, as payment for what they had taken.

Resuming their investigation along the east coast of Great Bear Lake, they reached the craggy, island-studded coast of McTavish Arm where

> The high rocky walls were stained and weathered to beautiful shades of purple, red and brown, and gave, with the reflection of the precipitous cliffs in the clear northern waters, a singularly rich effect. I climbed several high hills along the eastern shore of Eda Travers bay and was able to get a good view of the country to the eastward. As far as the eye could reach stretched hill after hill, lake after lake, and forest after forest.[36]

Bell wrote that "In the greenstones, east of MacTavish Bay, occur numerous interrupted stringers of calc-spar, containing chalcopyrite and the steep rocky shores which here present themselves to the lake are often stained with cobalt-bloom and copper-green."[37] Thirty years later, this phrase about cobalt-bloom attracted Gilbert LaBine to the region expressly to find the uranium deposit that this foretold—and the radium mine at Cameron Bay was the result.

At the southern end of McTavish Arm, Bell reached Camsell River and decided to return to Great Slave Lake by that route. This proved extremely difficult; winter was approaching, and much time was lost locating the main course of the sluggish river as it wound through the series of uncharted lakes. When the men were over halfway to Fort Rae, they encountered Indian portage trails that indicated the correct route. Finally they fell in with a party of Dogrib Indians, whom they accompanied to Fort Rae, crossed Great Slave Lake to Fort Resolution, and travelled up Slave River to reach Fort Chipewyan on Lake Athabasca before the onset of winter, arriving in Edmonton on December 7, 1900.

The missing Charles Bunn, after hope had been abandoned, finally turned up in Edmonton with an incredible tale. He had sprained his ankle and been abandoned during a heavy snowstorm by his fellow half-breed boatman, then after eight days' wandering came upon on a camp of Indians who eventually took him to Fort Norman when they went there to trade. He brought forward a claim against the Survey for his regular pay of $1.50 per day and provisions until his arrival in Edmonton on March 31, 1901, but in the end received only half-pay for the period since he had become separated from the party.

Charles Camsell was appointed to the staff of the Survey in 1904 as a result of his work with the expedition and on the strong recommendation of J. M. Bell; and so began his long, important career in the public service of Canada. Following a summer in northern Ontario exploring Severn River, Camsell was sent in 1905 on an important geological reconnaissance of the Peel River basin in the extreme northwest, travelling by way of Stewart River and a northern tributary, over the height of land to Wind and Peel Rivers, to the head of Mackenzie River delta. This included a track survey of the 275 miles to the mouth of Wind River and a micrometer survey for 335 miles down Peel River to its mouth. He returned by way of Porcupine and Yukon Rivers to Dawson, along the route followed by Ogilvie and McConnell nearly 20 years before. He commented on the glaciation of the valleys of the Ogilvie Mountains and of the plateau country through which the Peel had cut its way. He observed an interesting Tertiary basin around the junction of Wind, Bonnet Plume, and Peel Rivers that contained deposits of lignite coal, reported good pros-

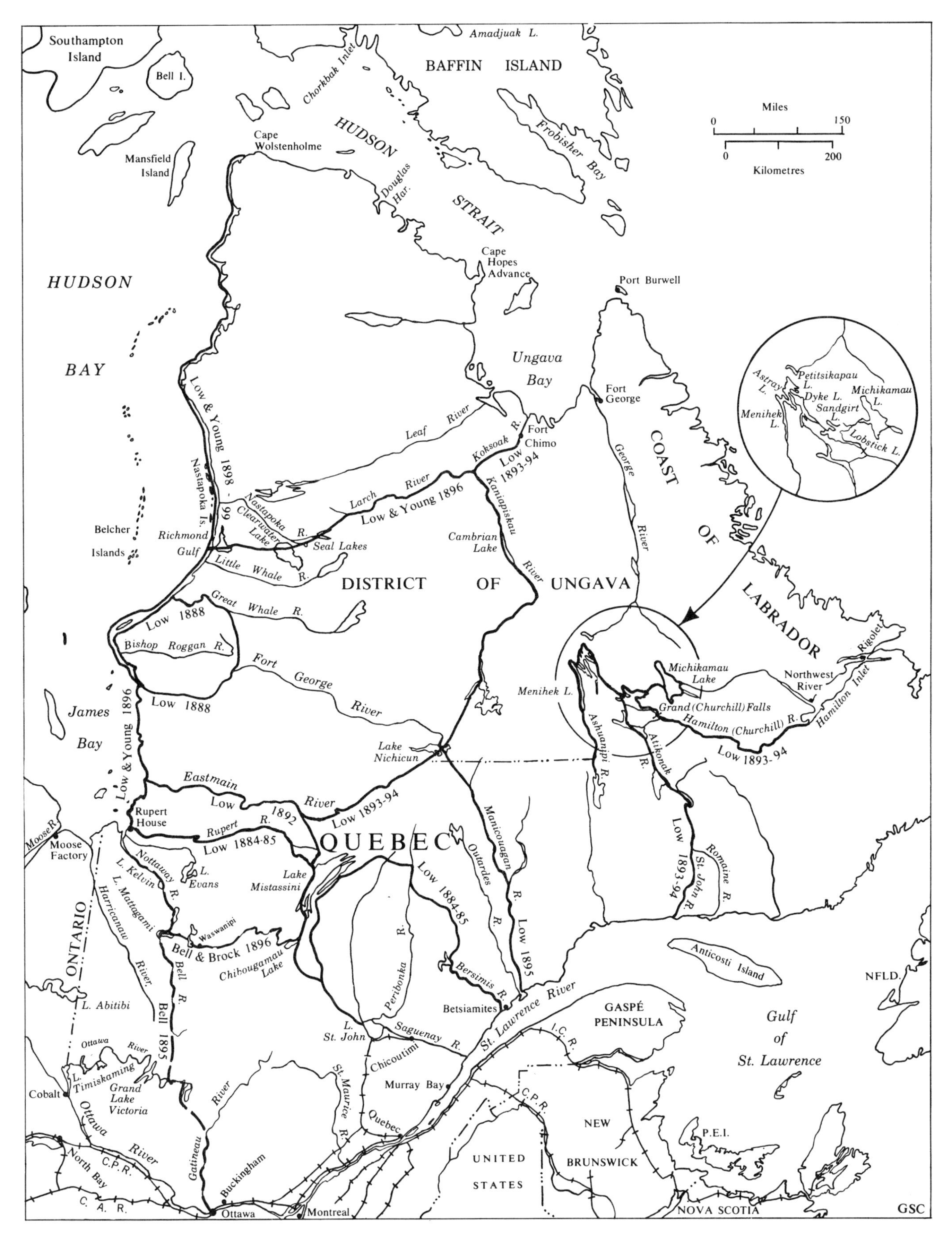

VIII-12 EXPLORATIONS IN NORTHERN QUEBEC AND LABRADOR, 1884-99

pects for petroleum in the Peel Plateau, and, like Keele, reported a large iron deposit somewhere in the country where he had crossed the height of land, once again anticipating the discoveries of present-day mining men.

The Precambrian Shield to the east of Hudson Bay, also, was the locale of several outstanding explorations by officers of the Geological Survey of Canada, especially by Albert P. Low. When Low began his work there, the district was hardly better known than the remotest parts of Africa or South America; and Dawson rated it, in 1890, the single largest unexplored area of Canada, an estimated 289,000 square miles. In a 20-year period after 1884 Low explored and mapped the major waterways and most of the coasts, outlined the principal rock formations, and made a comprehensive inventory of its resources. His first notable expedition, in 1884-85, was one of the most difficult of his entire career, though its purpose was nothing more than to explore from Lake Mistassini west to James Bay. Since the province was interested, Quebec contributed $1,500 and the Geological Survey an initial $3,000 towards the cost of the expedition. Quebec appointed an elderly land surveyor, John Bignell, P.L.S., "to take charge of one of the exploring parties under the general control of Dr. Selwyn, Director of the Geological Survey of Canada," while Low was appointed as "geologist and second officer in charge."[38]

Trouble developed almost at once. Bignell wasted most of the summer of 1884 making his way north to Lake Mistassini, while Low, eager to get on with geological work, went ahead examining the route before returning to join the Bignell party on the upper Peribonka River for the overland journey by toboggan to Lake Mistassini. Again Bignell was too leisurely for the impatient Low and showed no disposition to proceed with the survey of the lake while it was frozen, though this was the best season to complete the work and would leave the expedition in a position to start the westward explorations to James Bay in the spring. Finally, exasperated beyond measure, Low left to appeal to higher authority and, accompanied by two of the guides, made his way back to Lake St. John, Quebec, and Ottawa, to pour out his complaints to Selwyn.

The director immediately sided with his young subordinate. Complaining over the high cost and meagre results to date, Selwyn arranged with the provincial authorities that Bignell be dismissed and the expedition turned over to Low. The Quebec government agreed to transfer the expedition to the exclusive control of the Survey and to recall Bignell. On March 23, 1885, Low, assisted by James M. Macoun, left for the lake and on April 29 presented Bignell with his letter of recall. Bignell delayed until navigation was open, and took with him the bulk of the supplies, the best canoes, and the best men. But at last, after an outlay of more than $10,000 by the Survey (one-eighth of its annual budget in 1884-85) to meet the claims of Bignell and his men over and above the $1,500 the province had contributed for the purpose—and after almost a year's delay—the Survey was able to proceed with the desired work.

The examination itself was an anticlimax. Low and Macoun quickly surveyed Lake Mistassini and proved that it was not nearly so large as the exaggerated reports that compared it to one of the Great Lakes. This done, they proceeded to Rupert River, and thirteen days after leaving Lake Mistassini they were at Rupert House. They reported the country as predominantly Laurentian gneisses and associated rocks, except for southwest of Lake Mistassini, where Huronian rocks were encountered and where Richardson in 1871-72 had reported copper pyrites near Lake Chibougamau, another future mining camp. Low followed this exploration of 1885 by others between 1887 and 1892 of the east coast of James Bay and Hudson Bay to Richmond Gulf plus the islands of James Bay, Rupert and Eastmain Rivers from near Lake Mistassini to James Bay, and the lower courses of Fort George, Bishop Roggan, Great Whale, and Clearwater Rivers.

In 1893, the same year that Tyrrell made his momentous exploration of the Barren Lands of Keewatin, Low received the ambitious assignment to cross the Labrador peninsula from south to north and from east to west in two seasons, with an intervening winter in the country to prepare the work of the second season. Leaving the Lake St. John district on June 21, 1893, the party (including D. I. V. Eaton as topographical assistant) proceeded from Lake Mistassini to Eastmain River, this time ascending and surveying the river to its source, crossed to Lake Nichicun and portaged to Kaniapiskau and Koksoak Rivers which were followed down to the Hudson's Bay Company post of Fort Chimo on Ungava Bay, making a standard track survey of the route as they went. There were the usual troubles en route with guides and labourers, incessant rain, and—a contradiction—almost interminable haze from forest fires, for the whole peninsula seemed ablaze. However, offsetting these trials, was delight at observing, as they proceeded down Koksoak River, iron-bearing sediments large and rich enough to "supply the world," as Low commented.

At Fort Chimo the party took passage on the Hudson's Bay Company steamer *Eric* for Rigolet, and canoed up Hamilton Inlet to Northwest River, their winter headquarters. From January 1894 they were occupied transporting their supplies 250 miles into the interior, to the Grand Falls (now Churchill Falls) of Hamilton (now

VIII-13 VIEW NEAR CLEARWATER LAKE, DISTRICT OF UNGAVA
A. P. Low, who took this photograph in July 1896 on his traverse between Hudson and Ungava Bays, described the rock-strewn scene as "broken by rounded hills of granite and gneiss . . . bare and rocky with very little soil on or about them, the valleys being chiefly filled with boulders."

Churchill) River, their intended base for the season. There were only occasional breaks from the gruelling routine of hauling outfits, canoes, and provisions in sleds over the frozen rivers and lakes. Even Good Friday and Easter Sunday were devoted to transport work, with only the evenings for relaxation. The arrival of letters from Ottawa was a tantalizing treat, with an announcement of the Logan Club dinner celebrating the semi-centennial of the Survey. The Queen's Birthday was marked on May 24 with a dinner of plum duff and pâté de foie gras garnished with marmalade and washed down with "good old Canadian Rye slightly discoloured," "the gay & festive mosquito" joining in the festivities.[39] When the rivers were free of ice, the party began examining the country centred around Churchill Falls, notably Sandgirt, Lobstick, Michikamau, Dyke, and Menihek Lakes and the Ashuanipi branch of Churchill River. With so much to do, they devoted the season to this region instead of working westward across the peninsula. In the autumn they left by the southern, or Atikonak, branch of Churchill River, to Romaine and St. John Rivers, and so to the Gulf of St. Lawrence. When the party reached Ottawa that autumn, 5,660 miles had been travelled, "made up as follows:— In canoe, 2960 miles; on vessel, 1000 miles; with dog-teams, 500 miles; and on foot, 1000 miles."[40]

The results of Low's season in Labrador (as well as the next one, in 1895, along the upper Manicouagan, Outardes, and Fort George Rivers) appeared in 1896 as a 387-page "Report on Explorations in the Labrador Peninsula along the East Main, Koksoak, Hamilton, Manicouagan and Portions of Other Rivers in 1892-93-94-95," a work one admirer hailed as "une *somme*, une *Bible*, le plus important texte jamais paru sur le Québec-Labrador."[41]

The praise was not unmerited. Besides the authoritative report on the geology, the work was a compendium of information about the history, native peoples and their problems, missions, fish, game and forest resources, climate, and physical geography of the huge peninsula, the centre of which Low described as a great level plateau, from which closely intertwined streams descended in all directions to the sea. As well as exploring 170,000 square miles of country, Low secured all the information he could about the remainder, including a good map of the region around Lake Nichicun which two Hudson's Bay clerks had compiled in 1842, besides which he visited Betsiamites to draw on the reminiscences and check the maps of the remarkable missionary-explorer, Father L. Babel, O.M.I.

The report offered a summary and review of the different geological formations, divided into Laurentian, Huronian, and Cambrian. It gave a good account of the Laurentian gneisses which covered the bulk of the vast region (90 per cent, Low estimated), and he commented on the Huronian-type rocks in the country along lower Eastmain River and southwest of Lake Mistassini, predicting for one of these areas (that included the Lake "Chibougamoo" district), good possibilities for paying mineral deposits. Mainly, however, he reserved his enthusiasm for the great thicknesses of widely-distributed, layered 'Cambrian' sedimentary and volcanic rocks that he encountered along Koksoak and Churchill Rivers and which, he anticipated, extended along Ungava Bay to Cape Hopes Advance. Along Koksoak River below Cambrian Lake, he observed ten miles of rocks containing up to 31 and 33 per cent metallic iron, and commented that "the amount of ore in sight must be reckoned by hundreds of millions of tons."[42] Though equally large showings were not encountered along the Churchill River system, samples with up to 30 and 40 per cent iron were secured from Dyke Lake, Lake Petitsikapau, Astray Lake, and the outlet to Menihek Lakes, and he regarded the host rocks as largely the same. Low had discovered the huge belt of rocks known as the Labrador Trough and recognized its geological and economic implications.

His report also presented conclusions as to the glaciation of Labrador, based on a vast amount of evidence. He concluded that Labrador at one time had been almost completely covered by vast, thick glaciers; he saw signs of their presence in the moraines, eskers, erratics, and drift everywhere, and above all, in the striae fanning out mainly in a southwesterly and southeasterly direction from the centre. He also observed, from the marine deposits, how the country had risen as the glaciers receded, the amount of uplift being greatest in the south and west. Considering the nature of his journeys, Low's was an outstanding feat of geological assessment and interpretation, as well as a momentous exploration of one of the largest, little-known parts of Canada.

While Low was completing his investigations of the bleak interior of the peninsula during 1895 and 1896, Bell was making an important exploratory survey farther south, through a little-known but very important part of western Quebec, in the course of which Bell, "the father of place names in Canada," discovered the considerable river which has since borne his name. His purpose was to trace the Huronian rocks northeast from Lake Timiskaming by traversing the country north of Grand Lake Victoria, approximately 100 miles east of Lake Timiskaming. Cochrane, who had crossed the first 70 miles of the route between Grand Lake Victoria and James Bay in 1887, had encountered Huronian rocks and learned of a new river route to James Bay. Now Bell's party made an easy crossing from Grand Lake Victoria over a short sandy tract to Bell River, which was mapped from its source to its mouth in Lake Matagami, from which Bell surveyed the large, little-known Nottaway River to its mouth in Rupert Bay. The next summer, assisted by R. W. Brock and J. M. Bell, he returned to map various tributary rivers of the Bell and Nottaway like the Waswanipi, from which Brock ascended the Chibougamau to Lake Mistassini. Bell reported Huronian rocks along practically the entire distance from the height of land north of Grand Lake Victoria to Lake Matagami, as well as in much of the country from Bell River to Lake Mistassini, forming part of a 700-mile belt of Huronian country—"the great Belt of the system"[43]—extending from Lake Superior northeast to Lake Mistassini which today accounts for a large part of Canada's mineral wealth. He also observed two smaller bands of Huronian rocks farther north on Nottaway River at Lake Kelvin, and around Lake Evans.

Following his great report on the Labrador peninsula, Low commenced four years of reconnaissance surveys, mainly along the margins of the region he had done so much to map, ably assisted by his nephew, G. A. Young, a future chief geologist of the Survey. In 1896 they crossed the northern part of the peninsula, reaching the country from the C.P.R. by way of Missinaibi River to Moose Factory, then travelling 500 miles along the coast to Richmond Gulf, pausing to study a couple of Huronian-type showings en route. Their journey to Fort Chimo, by way of Clearwater Lake, Seal Lake, Larch River, and Koksoak River, was the first surveyed traverse of this part of the peninsula. From Fort Chimo Low again returned on the *Eric* to Rigolet, whence a schooner carried them direct to Quebec. He noted the extension north from Kaniapiskau River of the iron-bearing formation he had discovered in 1893. In the following year, 1897, he was a member of the Hudson Strait Expedition of the Canadian government in the chartered ship *Diana* which deposited him at Douglas Harbour on the north coast of the peninsula, beyond the western end of Ungava Bay, from which he explored the coast of the bay east to Fort George, then returned to Fort Chimo to await the *Diana*'s call on its way back to Quebec.

In 1898 and 1899 Low and Young conducted a two-year project of exploring and mapping the entire east coast of Hudson and James Bays, including the offshore islands. They began at Cape Wolstenholme and worked their way south to Great Whale River, where they wintered. In the winter Low explored inland northeastward to Seal Lake, which he had reached in 1896, while Young made a micrometer survey of the coast north to the mouth of Nastapoka River. They made a short exploration up Great and Little Whale Rivers in the spring, then resumed their examination of the coast, finishing the survey at Rupert House. Interestingly, as for generations past, the rocks and natural history specimens they collected still had to go by ship to London and be returned from thence to Canada. Low's report of the geology and geography of the country discussed the Laurentian and 'Cambrian' formations along the coast and islands, and referred particularly to a vast magnetite and hematite deposit on the Nastapoka Islands. Perhaps he was over-optimistic. In 1901 he left the Survey to join an American-owned Dominion Development Company and staked iron claims there, but the project came to nothing. While so employed, however, Low completed his report of the 1898-99 expedition and prepared a further report on his detailed study of the Nastapoka Islands.

* * *

THE COASTS, ISLANDS, AND WATERS OF HUDSON BAY and the lands beyond that Canada had acquired in 1880 represented another part of the Dominion in which exploration and reconnaissance surveys were bound to remain the order of the day for a long time. Once more, the Survey was in the van, and figured strongly in gathering and disseminating what knowledge Canadians were slowly gaining of the northernmost part of their heritage. As early as 1880 Selwyn authorized Bell to conclude his examination of the Nelson River district by taking passage at York Factory for England to gain in passing some knowledge of the geology of the coasts of Hudson Bay and Hudson Strait. He was to report on navigation conditions in view of the current interest in using the Hudson Bay route as the natural highway between western Canada and Europe. All Bell could do on this occasion was observe the coasts they passed and interview his fellow passengers. He had better opportunities in 1884 and 1885, however, to study the geology of Hudson Bay in connection with the Department of Marine and Fisheries' investigation of the navigability of Hudson Strait and Hudson Bay, under the command of Lieutenant A. R. Gordon. He joined the expedition as geologist, photographer, taxidermist, and medical officer, putting to use the M.D. degree he had obtained from McGill in 1878 while the Survey was still quartered in Montreal. Since the main work of the expedition was to station observers at various points to observe weather and ice conditions, during the one-or-two-day stops at each station Bell was able only to inspect the local terrain and confer with the observers and local Eskimos. Furthermore, this expedition was operating in regions outside his previous experience—the Labrador, Ungava, southern Baffin Island coasts, north of Churchill, and around Southampton and Mansfield Islands. During the 1885 tour, Bell obtained rock samples brought in by Eskimos. He also corresponded with Dr. Franz Boas, the anthropologist studying the Eskimos and exploring southern Baffin Island, about the geology of the interior of that island.

At this time (1886) Bell envisaged Hudson Bay as occupying a great depression in the centre of the vast Laurentian system, the bottom of which was probably lined with flat-lying Paleozoic strata like those displayed along parts of its coasts and islands. He expressed little hope for useful mineral discoveries among the red and grey Laurentian gneisses that predominated along the Labrador coast, both sides of Hudson Strait, and much of the coast of Hudson Bay. He was more optimistic about those parts of the Hudson Bay basin that contained Huronian rocks, such as were reported south of Ungava Bay, along the Labrador coast, along the east coast of James and Hudson Bays, and particularly between Chesterfield Inlet and Eskimo Point west of Hudson Bay. He visited the northwest coast in a few places and received "a carefully labelled collection of rock-specimens"[44] that included some from Rankin Inlet with interesting showings of metallic minerals. Here again, Bell was anticipating the later findings of Tyrrell and Low, and the attentions of modern-day mining interests. His report of 1884 also expressed a theory that Hudson Bay itself was a centre of glaciation, "a sort of glacial reservoir, receiving streams of ice from the east, north and north-west and giving forth the accumulated result as broad glaciers, mainly towards the south and southwest."[45]

Eleven years passed before the next Canadian government Arctic expedition of 1897, during which time Bell carried on his work in northern Ontario and Quebec and Low his extensive traverses of the Labrador peninsula. This time, both Bell and Low were appointed in conjunction with the expedition in the steamer *Diana* under the command of Captain William Wakeham, Bell's commission being to explore the northern, or Baffin Island, side while Low operated on the southern, or Labrador, side of Hudson Strait. Bell was landed at Ashe Inlet, at the eastern end of his survey, and had to work as far west as he could before returning there to meet the ship on September 10. In that time he made a track survey of 250 miles of the indented, rocky, island-studded coast west to Chorkbak Inlet, and travelled inland 50 miles to within sight of Amadjuak Lake. His survey produced a

VIII-14 EXPLORATIONS IN NORTHERN AND ARCTIC WATERS, 1897-1904
Base map—the map of Canada in A. C. Bradley, *Canada in the Twentieth Century* (London: Constable, 1905).

vastly improved map of this region, as well as a report on potential harbours and various phases of natural history. As to the geology, he encountered Laurentian gneisses everywhere except on the journey inland where he reached the edge of a horizontal-lying, fossiliferous Silurian trough. His study of the glaciation confirmed his earlier concept of a vast glacial river flowing out from an elevated Hudson Bay towards the Atlantic via the future Hudson Strait.

In 1903, when the Dominion government dispatched its first expedition specifically to assert Canada's authority in the Arctic islands and waters, Bell was acting director and the government turned to A. P. Low, who had recently rejoined the Survey after a short period in business, to command the expedition in the chartered steamer *Neptune*. This was to be much more than a scientific expedition; it was to patrol and establish permanent police posts for law enforcement and customs collection. Hence the vessel carried Major J. D. Moodie, a staff sergeant, and four constables, who were to operate a Northwest Mounted Police post at Fullerton Harbour on the west coast of Hudson Bay, north of Chesterfield Inlet, mainly to supervise whaling operations by Americans in northern Hudson Bay and Foxe Basin. In addition to Low, his assistant G. F. Caldwell, and the topographer and surveyor C. F. King, the expedition included Professor A. Halkett, who made the zoological collections and prepared the specimens for the museum, and Dr. L. F. Borden, who was the expedition's botanist, physical anthropologist, and physician.

VIII-15 ESKIMO FROM CHESTERFIELD INLET, PHOTOGRAPHED BY A. P. LOW

The expedition first sailed to Cumberland Gulf, on the east coast of Baffin Island, then south into Hudson Bay, to take up its winter quarters at Fullerton on the northwest coast. In the summer of 1904 the *Neptune* returned to Baffin Bay and made its way north to Cape Herschel, Ellesmere Island, "where a document taking formal possession in the name of King Edward VII., for the Dominion, was read, and the Canadian flag was raised and saluted. A copy of the document was placed in a large cairn built of rock on the end of the cape."[46] Then the ship skirted the coast, and turned into Lancaster Sound as far as Beechey Island, the wintering base of the Franklin Expedition, where another flag-raising ceremony took place. The return trip along the south side of the sound included a stop at Port Leopold, Somerset Island, for yet another proclamation of Canadian sovereignty, then continued along the Baffin coast to Pond Inlet, the main base of the Scottish whalers. Finally, the *Neptune* crossed Hudson Bay to Fullerton Harbour, returned to the Labrador coast, and home to Halifax.

On this expedition, which took the work of the Geological Survey into the Arctic islands proper, the geographical record amounted to surveys of 2,041 miles of country, including 1,175 miles of "Log and compass surveys of coast line, checked by astronomical observations, previously unsurveyed, or roughly sketched in by sailing vessels."[47] In addition, during the winter of 1903–04, Caldwell and Low traced 610 miles of Hudson Bay coastline from Chesterfield Inlet north to Wager Bay, part of the west coast of Southampton Island, and completed the mapping of Ungava Bay between George River and Port Burwell. Geological examinations were made on these land journeys, as well as at every ship's halt, and large collections were made of rocks and fossils, northern birds, eggs and nests, skins and skeletons of animals, fishes and marine invertebrates, and plants and insects. The customs, health, and physical measurements of the Eskimos were studied, and a census was taken. Observations were recorded of the weather, ice conditions and movements, tides and currents, and the habits and distribution of the whales, seals, walrus, caribou, and muskox of the region.

Low had a good background from previous geological work in the Arctic by British and other expeditions, plus his own previous and present work, with which to produce a fairly comprehensive account of the geology of northeastern North America. He saw its history in terms of a succession of rises and subsidences, during which a series of formations representing the periods from Cam-

VIII-16 A. P. LOW SALUTING THE FLAG AT CAPE HERSCHEL, ELLESMERE ISLAND, AUGUST 11, 1904

brian to Triassic were added to the west and northwest of an original nucleus of igneous and sedimentary rocks of Archean age occupying a narrow band along the eastern rim of the Arctic islands. The uplifting of the land and the changing amount of glacial ice cover—phenomena that, according to his observations, were apparently unrelated—governed the latest phase of the geological history.

Besides the usual report on his activities published in the Survey's annual report for 1904, Low prepared a more detailed and elaborate account, *The Cruise of the Neptune, 1903-04,* which appeared in 1906. As the first comprehensive Canadian report on the Arctic islands Canada had acquired in 1880, the book offered a different approach from that favoured in the widely-read writings of foreign Polar explorers. His descriptions of the people, both natives and whites, and of the conditions in the region, were realistic and frank. Thus his visit to Peary's former base at Cape Sabine evoked words of praise for Peary's courage and perseverance, but went on to add that "The pluck and daring of such men are to be admired, but the waste of energy, life and money in a useless and probably unsuccessful attempt to reach the pole can only be deplored, as no additional scientific knowledge is likely to be gained by this achievement."[48] He displayed interest, but no sense of alarm, over the contemporary expeditions of Sverdrup and Amundsen in the Arctic Archipelago. His strong nationalism was expressed in his enthusiasm over the possibilities of utilizing the natural resources of the region and developing the navigation of Hudson Bay as a future trade route with Europe. As spokesman for an official government expedition, he expressed the nation's concern with conserving the region's wildlife for the original inhabitants, and protecting the Eskimos from the inroads of diseases and changed ways of life introduced by the white man. Low was a sympathetic but realistic observer of their primitive way of life. He said of their morals, "who is to say what is right in this respect among a people situated as they are,"[49] and, regarding their relations with whites, that "As a people, they are very hospitable and kind; but like other savages would probably soon tire of continuous efforts to support helpless whites cast upon them, especially when the guests assume a superiority over their hosts."[50]

Low's expedition in the *Neptune* presents a dramatic illustration of how much and how far the Survey's exploring activities had advanced since 1881. Then it had been concerned with exploring the territory near the International Boundary in conjunction with the C.P.R.'s plans to construct its line across the southern prairies. Dawson, McConnell, Tyrrell, and others had explored and charted Souris, Milk, and Bow Rivers, and the passes beyond. In less than 25 years, reconnaissance activities had been extended northward from the Alberta foothills all the way to Great Bear Lake and the Mackenzie delta, from Rainy River to Foxe Basin, from Lake St. John to Ellesmere Island. Wherever the geologists had gone, they had carefully mapped the physical features of the districts traversed, and had contributed immeasurably to the map of Canada's territory. The expanding operations of the Survey yielded broader and fuller insights into the variety and scale of Canada's geography and natural history, as well as into the sweep of her geological evolution from the continent-building movements to the immense changes wrought in recent times by great ice sheets and the regular erosive forces of water and air. The reconnaissances provided concrete information on people, resources, and conditions of many corners of the country, and produced inventories of pockets of arable land, forests, fish and game, waterpowers, climatic conditions, and transportation facilities and routes for a large part of Canada, in addition to more specific reports on the presence, possible occurrence, or likely absence, of mineral wealth. The Geological Survey and its officers did more than any other group or agency during the quarter-century after 1881 to make Canadian conditions and opportunities known to the nation and the world.

Low's appointment to command the *Neptune* expedition was a recognition also of the Survey's role through the years in helping extend Canada's sovereignty to the full territorial limits of the Dominion. Operating usually in districts still dominated by Indians and Eskimos, fur traders and missionaries, the reconnaissance geologists prepared reports that directed the future lumberman, miner, and farmer who came to develop the natural wealth and extend the blessings (or defects) of western civilization, technology, and society to an undeveloped land. Almost everywhere they went in the north, field officers found themselves in the position of being the very first advance agents of Dominion government authority. In 1887 they led the way to the occupation by Canada of its distant Yukon frontier. From reconnaissance explorers like Dawson and Bell came frequent warnings to the government of the need to assert Canada sovereignty and authority in the interests of the nation in general, and the native inhabitants in particular. Now, in 1903-04, they were again in the lead, proclaiming the Dominion's authority over the last frontier, the islands of Canada's Arctic.

IX-1 EXAMINING A LIMESTONE CLIFF, RED ROCK, NIPIGON BAY, LAKE SUPERIOR, 1894

9

Starting the Search for Answers

> An erroneous impression prevails among many persons who have never had occasion to inform themselves, as to the nature of the work performed by this department. They imagine that geologists devote themselves largely to 'Theoretical and purely scientific' geology, instead of giving their attention, as they do, entirely to practical work, looking to the development of our various mineral resources.[1]

WHY DID DR. ROBERT BELL feel it necessary to make this statement in the opening paragraphs of his *Summary Report for 1902*? Was the Survey indeed spending part of its effort on what some people could regard as "theoretical and purely scientific geology," and neglecting other, more proper, aspects of its work? And, if so, on what were such criticisms based?

Certainly it was as true in 1902 as it was some 30 years later when it was said that "The main object of the Geological Survey, however, is to determine, as far as possible, *how, where,* and *why* mineral deposits occur."[2] What could be more practical than that? But to achieve such an objective certain basic information is required that can only be secured by the kind of field and laboratory studies that are often regarded as "theoretical and purely scientific." Bell contended, as did Logan, Selwyn, and Dawson, that in the long run these were the very studies that proved to be of the greatest help in finding mineral deposits, aiding the mining industry, and enhancing the national prosperity. Thus, by means of many studies, both in the field and the laboratory, it became possible to outline the one one-hundredth of the area of the Eastern Townships that is underlain by serpentine rocks in which alone asbestos may be found—certainly a most practical result produced by a great deal of seemingly "theoretical and scientific study."

This example indicates that systematic field mapping in itself was perhaps the greatest single scientific accomplishment of the Survey in the period after 1881, when Selwyn embarked on a standard program of areal mapping as a basis for the systematic study of Canadian geology. Previously, the work had tended to fall in either of two categories—broad-based, rapid explorations of the major features of little-known territory; or special studies of mining districts like the Nova Scotia coalfields, or of geological problems like the Quebec Group. By the eighties, however, the reconnaissance phase had been largely completed for those parts of the country that were readily accessible and capable of being developed in the immediate future. The Maritime provinces, the south shore of the St. Lawrence and the Eastern Townships, southern Ontario, the southern margin of the Canadian Shield in Quebec and Ontario, and the southern part of British Columbia were some districts that called for more thorough study in their own right and as type areas.

Dawson, in his first report as director, expressed this changing focus of interest:

> The operations of the Geological Survey in the field, constitute the basis of the entire work of the department. These naturally divide themselves under two principal heads: (1) Reconnaissance surveys and explorations, covering in a general way large tracts of country, and (2) the systematic mapping and description in detail of less extensive areas. The first inevitably precedes the second class of work, and for many years it must, in the nature of things, remain the only method possible of dealing with the vast regions of Canada which lie beyond the boundaries of connected settlement. While the exploration of new districts, in which geographical information is obtained concurrently with data on the general geology and mineral resources, may attract popular attention to a greater degree, the methodical delineation of the geological fea-

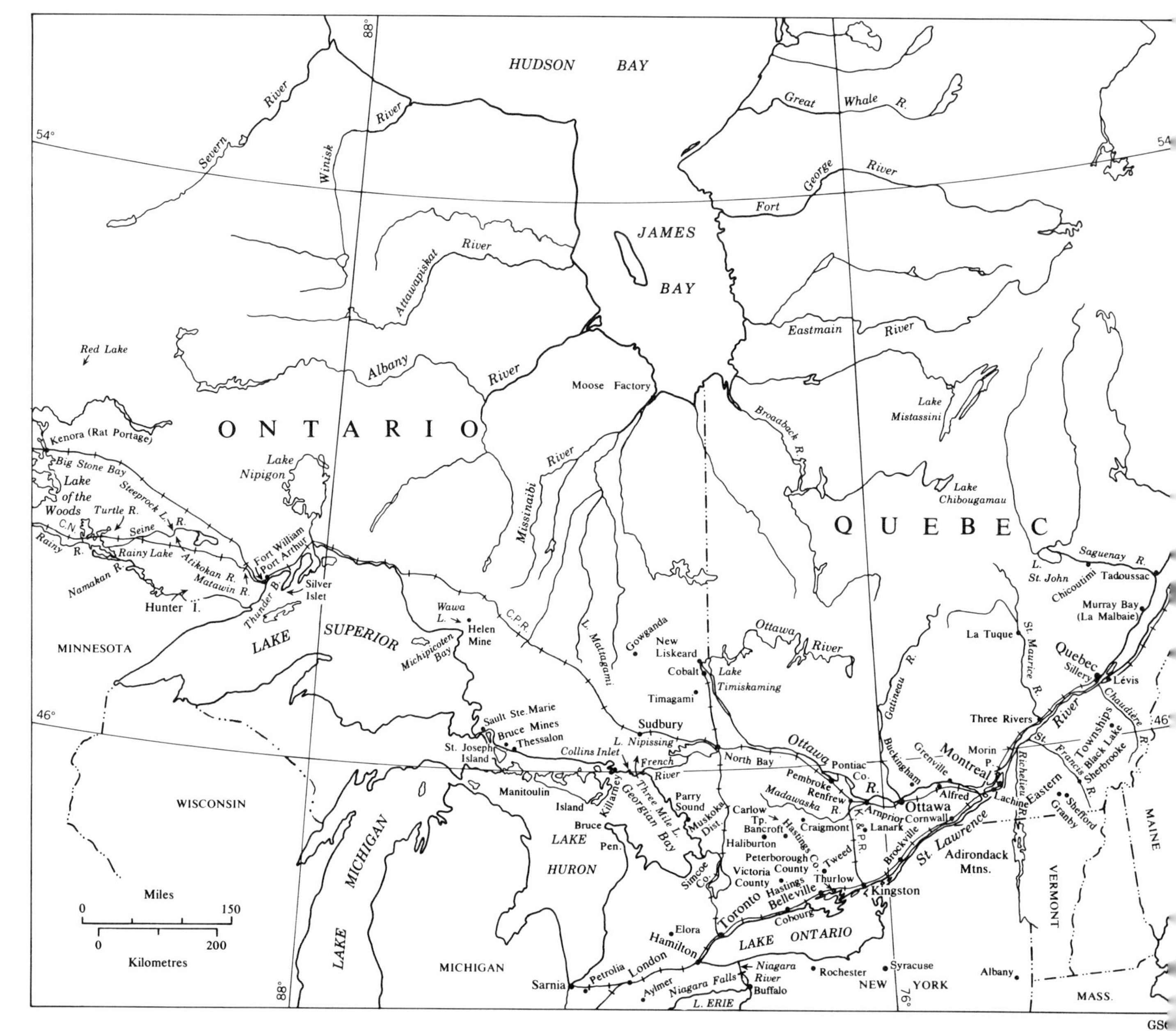

IX-2 CENTRAL CANADA, 1880-1907

> tures of the older parts of Canada must be regarded as at least equally important and as requiring no inferior ability or diligence on the part of those engaged in it.[3]

Dawson undoubtedly believed that the time had come for the Survey to shift its emphasis to the systematic geological mapping of areas shown by reconnaissance surveys as being of the greatest immediate interest—as he himself had done in British Columbia.

Hence, though the work of exploration continued, it was far surpassed, in terms of staff operations, by a new type of field activity that followed a different sort of method—the systematic mapping of tracts of land known as map-areas. To conduct such a survey required that the geological party crisscross the area with many traverses, mostly on foot, and in more than one direction, to visualize the nature, relationships, and distributions of the rock units. The results of such study were then presented in a report and a map. The amount of data secured depended on the complexity of the geology and the time allotted for the field work. A geologist might spend a single day compassing ten square miles or examining half an acre. The latter, naturally, would apply only to an attempt to "break the code" of a very complicated structure, and the findings of such study would be impossible to show on anything but a very large-scale map. The scale on which the map was published, and how much geological information it showed, depended on the purpose for which the data were being presented. But equally, the scale of the final map also governed the amount of detail the field officer needed to secure.

It was Selwyn's feeling, echoing Logan's, that the program that best suited the conditions of the time and place was a system of 4-mile mapping. Such a scale resulted in a map that was capable of showing the surface distribution of the major rocks and, by means of sections, something of what happens to them below the surface. Departures from this 4-mile standard are necessary in practice, because of varying geological conditions, or other reasons. In parts of the Precambrian the sweep of the major rock units may be so great that 4-mile map-areas may be too small to indicate more than a single unit. At the other extreme, around the Cape Breton coalfields the 4-mile scale would not yield a map sufficiently detailed to be of use, so 1 inch = 1 mile maps became standard for such areas. Such maps, however, meant that the geologists had smaller areas to cover, and more closely spaced traverses were required to collect more data per unit of area. In exceptional cases an even smaller scale was needed, for example to show folds in the gold-bearing series of Nova Scotia that were only a few hundred feet, or less than 100 feet across; these necessitated maps or figures to scales of 1 inch = 1,000 feet, or even 1 inch = 100 feet. For detail work the scales are determined as circumstances require. Other scales have also been used at various times—2-mile, 6-mile, 8-mile, or 12-mile. These scales are not simple multiples or reductions of some standard map. Each scale dictates a different level of field work and of intensity of effort.

Selwyn had to consider not only the nature of the geology and the needs of users of his maps; he had to decide how much time and effort to assign to the work of mapping the geology of Canada. Given the pressure to push ahead with the work, and the limited manpower available to him, he was inclined to think in terms of a program that gave an adequate but not too concentrated coverage. Therefore the Survey's task became to cover Canada systematically on the 1 inch = 4 mile scale, using smaller scales only when circumstances permitted no alternative.

Thus the work of systematic areal mapping advanced as fast as means permitted, in what became the longest-sustained of all Survey programs. By the early eighties parts of New Brunswick and most of Nova Scotia were being mapped systematically (much of Nova Scotia on the 1 inch = 1 mile scale), and by the end of the eighties the initial mapping of New Brunswick on the 4-mile scale was completed. In 1885 Selwyn announced that "A scheme for the geological mapping of the peninsular portion of Ontario in sheets of uniform size, like those employed in Maritime Provinces, has been laid down, and some progress has been made in the compilation of surveys to form a basis for the geological representation."[4] Hence from the mid-eighties onward a large part of the field work in Ontario (and in Quebec as well) was diverted to 4-mile mapping, apart from complex formations that were mapped on the 1-mile scale.

In the decade after 1882, the Abbé J. C. K. Laflamme did systematic mapping in the Lake St. John country and from Tadoussac to Three Rivers, while F. D. Adams, A. P. Low, and N. J. Giroux mapped other areas north of St. Lawrence and Ottawa Rivers. In the meantime, R. W. Ells, who completed his work in New Brunswick in 1882, moved on into Gaspé and then along the

IX-3 MSGR. J. C. K. LAFLAMME

south shore of the St. Lawrence to the Eastern Townships. After he had completed this, he crossed the river to map the eastern part of Ontario and along the Ottawa valley. The Frontenac Axis, which had been worked on by Vennor and Coste, was studied and mapped after 1891 by F. D. Adams and A. E. Barlow. The work by then had also been extended into the Precambrian of northern and western Ontario, beginning with A. C. Lawson, W. H. C. Smith, and William McInnes—who among them mapped the region between the lakehead and Lake of the Woods on the 4-mile program (1-mile in at least one program). Beginning in 1887, Robert Bell led off the work in the Sudbury district, assisted by A. E. Barlow, and afterwards by R. W. Brock and W. G. Miller, a future director of the Survey and a chief geologist of Ontario, respectively. This group was responsible for a Sudbury sheet in 1891 (later remapped and reissued in 1904), as well as other map-areas from Lake Timiskaming nearly to Sault Ste. Marie. Later Bell also worked, unwillingly, on the 4-mile map of the Michipicoten district.

In western Canada progress was much slower, though geological maps were published to deal with areas studied in particular investigations, the scales varying from 8-mile and 16-mile down to 1 inch = 300 feet. The regular 4-mile program was not initiated until 1890, when Dawson decided to study and map the Kamloops map-area. In spite of other preoccupations, thanks to the assistance of J. McEvoy, he was able to complete the map and publish the report in 1895 as Part B of Volume VII (1894) of the *Annual Report*. Dawson's report was a model for scope and organization, and established a standard widely followed in systematic areal studies. It provided the first modern version of a table of formations, the first modern map-legend, and a correlation table showing the author's ideas as to the age relationships of different rock groups—another device still widely used. Later McConnell, Brock, Young, and Camsell also worked on the map-areas of mineral-rich southern British Columbia, assisted by topographers J. McEvoy, W. W. Leach, and W. H. Boyd (the future head of the Topographical Division and one of the first men to use the new method of phototopography in mountain country). In addition, working along the border for distances of from 5 to 10 miles on either side was R. A. Daly, the geologist with the International Boundary Commission in the years from 1900 to 1905.

Once systematic areal mapping, even on the 4-mile scale, got well under way, the results became quickly apparent in the improved understanding of sedimentary successions and igneous intrusions; of the extent of metamorphism of various units; also what types of mineral deposits might be sought, and what were the most likely rocks and structures in which they might occur. As a result, the period from 1881 to 1907 was as successful, lively, and progressive on the side of specialized study and the mastery of geological problems as it was noteworthy for geographical exploration.

In the Maritimes there was the work in New Brunswick of Professor L. W. Bailey, who worked for the Survey every summer except when the straitened finances made it necessary to dispense with his services. He worked mainly in and around the Carboniferous basin of south and central New Brunswick, but also in the northwest to the juncture of the province with Maine and Quebec. In the early nineties he did some work in southwestern Nova Scotia as well. In that province systematic, detailed mapping occupied the entire careers of two field officers—Hugh Fletcher between 1875 and 1909, E. R. Faribault, whose career from 1882 to 1932 is the longest of any field geologist in the Survey's history. Most of Fletcher's work was concerned with the coal measures of Cape Breton Island and the mainland. Faribault's work was on the intricate folds in the gold-bearing series of the much-altered rocks along the southeast coast of the province. He soon realized that the gold orebodies occurred

IX-4 HUGH FLETCHER

like saddles along the crests of small anticlines, apparently as a result of openings developing as the strata folded, through which circulating solutions deposited quartz and gold particles. Through painstaking mapping of certain beds, details of the structure became clear, and the position of the orebodies within these structures was evident. Thereafter Faribault spent each field season mapping the complicated structures area by area and in great detail. While other geologists spent difficult months in swampy, insect-infested terrain, Faribault usually went to the field in the company of his family and bicycled to the site of his work from their summer cottage. Still, his monumental though largely routine task led to the discovery of innumerable, mostly small, orebodies. The series of maps prepared by these two geologists was the most ambitious project the Survey had attempted to that time, and it continued until Faribault's retirement in 1933. Between them Fletcher and Faribault published seventy 1-mile map-sheets, besides many plans and sections on scales as large as 1 inch = 500 feet.

Ells' mapping projects from New Brunswick into Gaspé and beyond, enabled him to refine, revise, and extend the work of Logan, Murray, and Richardson. Ells followed the trend of the rocks southwest into the Eastern Townships, carefully measuring sections and mapping as he went. Inevitably he became involved in the controversy that was still raging over the stratigraphy and age of the Quebec Group. He soon discovered that the folding there was far tighter and more complex than had previously been suspected, and that the same beds were repeated more than once in a section that had earlier been regarded as a single sequence of superimposed strata. Beds Logan had regarded as being near the top of his section turned out to be the same as those near the bottom. The order of succession in such metamorphosed rocks, in the absence of fossiliferous beds, could not be determined until the structures were understood. Ells' contribution was to work these out systematically and painstakingly, and as his understanding of the structure

IX-5 E. R. FARIBAULT

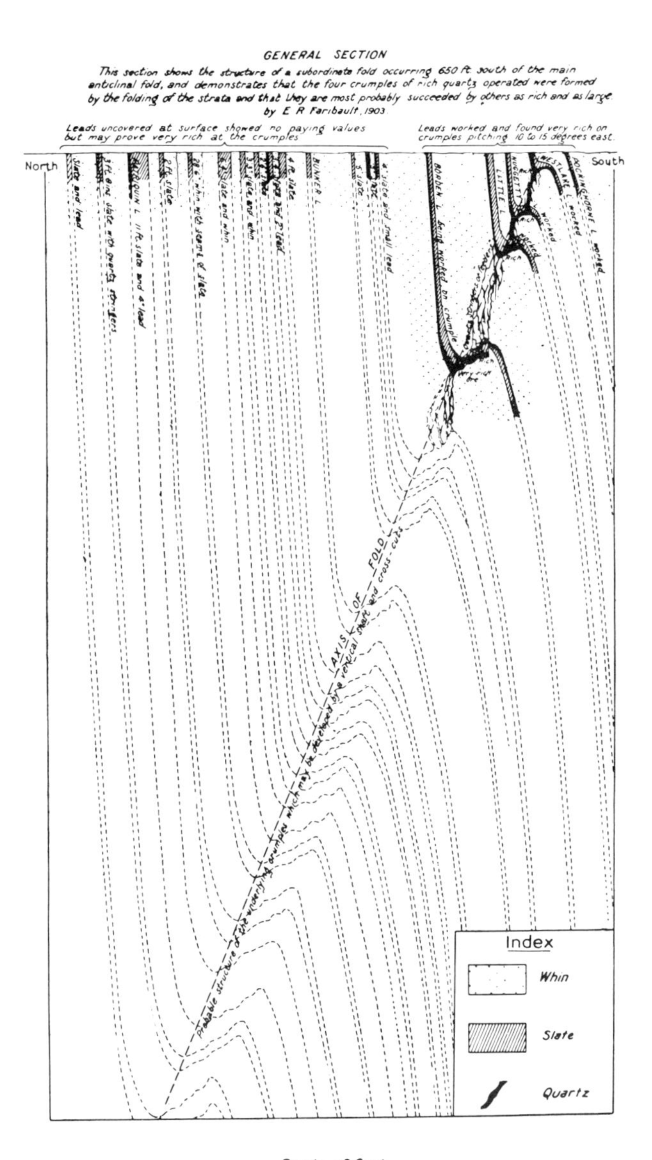

IX-6 FARIBAULT'S THEORY OF GOLD-BEARING STRUCTURES
This transverse section of West Lake mine, Mount Uniacke, N.S., was published in Memoir 20 (1912), *Gold Fields of Nova Scotia,* by Faribault and Malcolm.

grew so did he begin to make sense of the stratigraphic succession, though some of his identifications have had to be revised. His work was successfully applied across the border, repaying to some degree the assistance the Survey had received from that quarter in an earlier time. As a result of Ells' efforts, Selwyn and the future director of the United States Geological Survey, Charles D. Walcott, in 1889 were able to make a joint examination of some sections near Quebec and along the Quebec-Vermont border "with a view to uniformity of mapping by the respective surveys, and which, it is hoped, will now be secured."[5]

In the Maritimes and the settled parts of Quebec and Ontario, the areal mapping program was determined by political and economic considerations; in the more remote sections, where there was the whole region to choose from, the selection of areas for mapping was more likely to be done out of scientific considerations, though economic factors were certainly not ignored. This is well illustrated by the way areal mapping was begun in the Cordillera. Dawson had been studying the geology of the region for years, and his general ideas were already well formed. After his return from the Yukon in 1888 he was on the lookout for an area that was both accessible to transportation by way of the newly built Canadian Pacific Railway and of some current mining interest, and one in which the basic problems of the Cordillera were best displayed. His choice of the Kamloops map-area in 1890 followed two years of widely spaced traverses over a much larger part of southeastern British Columbia, plus special examinations of mining camps. His work on the Kamloops map-area, in turn, clarified and strengthened his understanding of the regional geology by compelling him to name and define rock units and piece together a sequence of geological events that extended over the entire system.

Inevitably, the work of systematic areal mapping compelled geologists to make decisions respecting the chronological sequence and typical and diagnostic characteristics of rocks they examined, and to correlate them with rocks in other parts of the country and the world. Given the many variables and ambiguities, and also the rapid development of new ideas and interpretations, it is not surprising that geologists often disagreed over the rocks they or their fellows mapped. The instance in the years 1899-1902 that achieved most notoriety concerned the geological age of the Union and Riversdale formations in Nova Scotia, which Fletcher on the basis of their stratigraphic relations mapped as Devonian, while Ami discovered fossils that led him to identify the formations as Carboniferous. The disagreement resulted in a great deal of unpleasantness, for Ami insisted on making the case public, whereupon the Survey was attacked by the *Canadian Mining Review* for suppressing Fletcher's impor-

IX-7 R. W. ELLS

IX-8 C. D. WALCOTT

IX-9 E. R. FARIBAULT AND FAMILY ON AN OUTING, PORTER'S LAKE, N.S., 1891

tant maps and geological information for "petty, almost childish" reasons.[6] Bell finally ordered the ten map-sheets to be published on the grounds that the public was entitled to have the use of the maps, and that Fletcher's work was the result of "his unremitting study and labour to the correct elucidation and representation of the geology of the region in question, having no object but the truth to maintain."[7] In the end, however, Ami's conclusions were upheld. The affair is a fine example of the reluctance of the geologists of the old school to pay much attention to evidence produced by their juniors, particularly evidence from another branch of the science. The great Logan, it will be recalled, had known better than to reject indisputable paleontological evidence, and his discovery of the tectonic feature known as Logan's Line had followed.

While formations have to be described, classified, and dated relative to one another in order to map them, why is it so important whether a particular formation is assigned to one period or another of the geological calendar, or given one nomenclature or another? To the geologist the intelligent grouping, naming, and defining of rock units is a fundamental tool, just as are definitions and terminology in other fields of study. Accurate, acute observations are of little use unless they can be precisely communicated to others, and only if they are defined and described in unmistakable terms can a geologist compare his findings with those of others. Thus it was of fundamental importance that a geologist in Nova Scotia made an intelligent grouping of certain rocks, determined their place in the time-scale, and published a clear description of their characteristic features and relationships. This information could then be applied by a geologist elsewhere in the region, or farther afield in North America or in Europe; and by the same token, what was learned about similar rock units in other parts of the world could be applied to the unit in question. Only by making clear, accurate definitions—and a continuous process of refinement to make the definitions even more exact—can the reports and maps of geologists find their fullest use. All these aspects—classification, correlation, nomenclature, refinement, and re-evaluation—are evident again and again in the study and research of Precambrian geology that was going on in this period.

* * *

IN THE UNION AND RIVERSDALE FORMATIONS there is fossil evidence to help with the questions of ages and relations. But in the Precambrian such evidence is lacking, and no means existed, until very recently, to help with age determination problems. Moreover, the great extent of igneous activity and metamorphism often make it most difficult to apply the usual principles of stratigraphic relations with full effectiveness. Small wonder that early geologists were content to classify Precambrian rocks in two great groups: granites and gneisses as Laurentian, recognizable sediments and greenstones (altered volcanic rocks) as Huronian. Little progress had been made in working out more precise definitions of either. With all his care and genius, Logan himself had only been able to subdivide the Laurentian into an oldest group, the Fundamental Gneiss; a younger group, the Grenville; and finally, the anorthosites, which Logan thought overlay the Grenville. Another series, the Hastings Series, belonged somewhere between the Fundamental Gneiss and the anorthosites.

Then to the aid of the geologist, especially the geologist working in the Precambrian, came a powerful new tool. Used first at the Survey in 1874 by Harrington, this tool enabled him to peer into the heart of these enigmatic rocks and see their internal structures, and thereby perhaps come to more accurate conclusions as to their true histories and relations. European scientists had found that when a chip of rock is ground down to a translucent wafer and viewed under a polarizing microscope it could be identified by its optical properties, and much information could be gained as to its nature and origin. Such studies were of immense aid in understanding the successions and age relations of Laurentian, Huronian, and Grenville—discoveries that had far-reaching economic as well as scientific implications. Scientists, armed with this new tool, were able in the two decades after 1881 to solve problems that had frustrated the men of Logan's generation, and to arrive at a much clearer understanding of the complicated history of the Precambrian rocks that make up so large a part of Canadian geology.

Systematic study of the Precambrian, the greatest of all Canadian geological mysteries, was inaugurated in 1882 in the region between Lake Superior and Lake of the Woods. The area had been attracting the attention of miners since the 1840's; the Silver Islet mine in Thunder Bay and the great iron mining industry in adjacent parts of the United States were evidences of its mineral wealth. Moreover, the area was being opened by road and railway by the Dominion government as part of the national task of linking the west with the older eastern provinces. Selwyn decided, therefore, to have the region studied:

> Owing to the discovery of the precious metals at the Lake of the Woods, it seemed desirable to have an area of the so-called Huronian rocks carefully worked out as a type of this system, as it occurs in the west, Dr. Bell was therefore instructed to commence a more minute examination . . . including an actual survey of the shores and islands of the Lake of the Woods and of Shoal Lake, for the purpose of showing accurately the arrangement and distribution of the sub-divisions in that region, of the system referred to.[8]

(Note how Selwyn sets out a scientific as well as economic purpose for this investigation by his reference to working out a type area of the "so-called Huronian rocks.")

Having placed his two assistants—A. C. Lawson and J. W. Tyrrell—at Big Stone Bay, the former as geologist, the latter as topographer, Bell proceeded to the more congenial work (for him) of track surveying the route to Red Lake. Coste also helped with the work at Lake of the Woods by examining the various gold occurrences. When Bell returned from Red Lake he checked the work done by Lawson and Tyrrell and directed them to work as separate parties, sent Coste to examine mineral occurrences at Rainy Lake, Rainy River, and part of Lake of the Woods, then departed on another reconnaissance, this time of Winnipeg River. At the close of the season's work the four parties re-convened at Rat Portage (now Kenora, Ontario).

IX-10 A. C. LAWSON AS A YOUNG MAN, 1881

IX-11 CONTACT OF LIGHT-COLOURED LAURENTIAN GNEISS WITH SCHISTS OF "COUTCHICHING SERIES," RAINY LAKE, ONTARIO
Lawson's photograph of 1886, like others in the *Annual Report for 1887-88* (Volume 3), was reproduced by a halftone technique by G. E. Desbarats and Son, Engravers and Printers, Montreal.

In the following year, 1883, Lawson, who had now graduated from the University of Toronto with the gold medal in natural science, was placed in charge of the work, assisted by J. W. Tyrrell and W. F. Ferrier, the Survey's future petrologist. Almost five months were spent on a detailed survey of the northern half of Lake of the Woods, particularly the Huronian belt of rocks which "by the great extent of their exposure and ease of access, [exhibit] the leading lithological and structural features of this metalliferous series, as well as its relations—here clearly shown—to the underlying Laurentian rocks, thus affording us a key to the elucidation of similar belts in less accessible regions."[9] The distribution of Huronian rocks was more accurately defined and an attempt was made to relate the occurrence of gold to the geology: ". . . it seems to be generally true that the largest and richest lodes are in proximity to intrusive masses of igneous rocks."[10] In 1884 the work was continued by Lawson while A. E. Barlow and W. H. C. Smith conducted an independent survey of the district between Lake of the Woods and Rainy Lake. Lawson also transferred his attention to Rainy Lake and its eastern feeders, Namakan, Seine, and Turtle Rivers, where he traced out a 20-mile-wide belt of supposed Huronian rocks. The results of the three seasons' work were embodied in the standard topographical and geological map of the complete Lake of the Woods map-area and in Lawson's report published in 1885, "Report on the Geology of the Lake of the Woods Region, with Special Reference to the Keewatin (Huronian?) Belt of the Archaean Rocks," which marked Lawson's emergence as one of the great original minds of Precambrian geology. The report directly attacked the current division of all Precambrian rocks into Laurentian and Huronian which Logan had inaugurated in 1857, and which subsequent explorer-geologists had translated into a licence to map light coloured, granitic rocks as Laurentian, and everything else as Huronian. Lawson argued that even Logan had been inconsistent in lumping together with his type Huronian rocks on Lake Timiskaming the Dore River greenslate rocks north of Lake Huron which were of distinctly different nature and age. Lawson contended that the schists at Lake of the Woods differed from Logan's type Huronian of Lake Timiskaming, and he proposed a new name, Keewatin, for the distinctive series at Lake of the Woods composed mainly of metamorphosed volcanic rocks with some interbedded sediments, and, Lawson believed, a conglomerate at the base.

During 1886 Lawson and Smith continued to map the country and traced the Keewatin along the canoe routes between Lake of the Woods and Rainy Lake. As they proceeded, Lawson found exposures of "an immense development of mica-schist with some fine-grained, evenly laminated, micaceous gneiss" more than two miles thick.[11] The following year they continued systematic mapping of the Rainy Lake and Rainy River districts and began a reconnaissance of the Thunder Bay region farther east and of the nearby iron ranges of Minnesota. In 1887, while Lawson filled in details for his report on the Rainy Lake map-area, Smith continued mapping the topography of Hunter's Island and Seine River map-area farther east.

When Lawson's "Report on the Geology of the Rainy Lake Region" appeared it was quite as revolutionary as his previous report on Lake of the Woods. He proposed the name Coutchiching for the new series he believed to underlie the Keewatin. In his masterly "Summary of Archaean Geology" he gave his interpretation of the formations and relationships of the various rocks of the Rainy Lake region: that the Laurentian gneisses are plutonic rocks that crystallized slowly from a viscous magma; that under differential pressures in subsequent ages they became deformed and plastic enough to assimilate parts of the overlying Coutchiching and Keewatin Series; that the Coutchiching Series of mica schists was formed from sedimentary strata deposited in a long, quiescent episode with little or no volcanic activity, and subsequently metamorphosed (the lower strata in particular) by upward percolation of heated aqueous solutions from the underlying magma; that the Keewatin Series is mostly of volcanic origin, though it also includes some sedimentary rocks and other material so metamorphosed that its origin is uncertain. The advent of the Keewatin was marked by widespread and intense volcanic activity, marking a profound change from the previous conditions of sedimentation. The Keewatin consists of a lower part of basic lava flows, and an upper part of acidic flows, indicating a change in the character of the volcanic activity, although the two parts are not clearly separable. Both, however, were earlier than the Huronian.

By 1890 Lawson began to question further the position of Laurentian. In 1887 he asserted that in his region the Laurentian, which consisted entirely of Logan's "Fundamental Gneiss" (there being no Grenville there), was composed entirely of intrusive material, and that there was no sign of basement rocks earlier than the Coutchiching and overlying Keewatin, although the Coutchiching gave evidence of having been deposited on a granite. In 1890 he proposed that the granite, as well as some of the Coutchiching and Keewatin, had disappeared, "absorbed by liquefaction in a sub-crustal magma which later crystallized . . . as the Laurentian."[12] Accordingly, the Laurentian in reality was younger as a solid formation than was the Coutchiching-Keewatin. In effect Lawson was defining Laurentian simply as the oldest remaining granitic rocks of deep-seated origin.

His work was not accepted without some resentment. His biographer reported that Lawson's ideas were considered rank heresy and that "his report of 1887 was blue-pencilled out of all resemblance to what he intended to say. It was only after a hard fight, his first of many geological contests, that he salvaged the essential part for publication."[13] The same source, however, credited the International Geological Congress, which Lawson addressed in 1888 at its London meeting, with giving him much encouragement. There his views were accepted as a point of departure for further studies of Precambrian geology, and Keewatin and Coutchiching entered the vocabulary of Survey publications. Indeed, soon, like the earlier term Huronian, the terms began to lose their original meaning through improper application far from the type area. Tyrrell, for example, applied Keewatin to rocks he observed at Rankin Inlet, 1,000 miles away, and later, in 1895, to schists and quartzites he encountered along the east coast of Lake Winnipeg. However, Lawson's ideas, especially his method of attack on the problems, had opened the way to a better understanding of Precambrian geology. He had thrown new light on the confused picture that was contemporary Precambrian geology in Canada, even though in some cases he was trapped into oversimplified generalizations. He had demonstrated that the basic principles of stratigraphy could be applied even in Archean rocks, and, combined with petrography, could go far to determine the relative ages and origins of the rock units.

Lawson was soon lost to the Survey. He had been attending sessions at Johns Hopkins University and in 1888 was awarded the Ph.D. degree. His appearance before the congress in London was followed by two more papers in 1889 to the American Association for the Advancement of Science, meeting in Toronto. Early in 1890 he resigned and moved to Vancouver to become a consulting geologist. But academic life beckoned, and later that same year he joined the faculty of University of California as assistant professor of mineralogy and geology, and served nearly 40 years until becoming professor emeritus in 1928. He continued his researches almost to the end, publishing his last article in 1950, two years before his death at a very advanced age. Throughout his long life he maintained a connection with Canada and with Precambrian geology, in addition to his many other interests. He returned several times to work in the area west of Lake Superior after 1910, contributing an article to the summary report in 1911, a memoir on the geology of Steeprock Lake (No. 28, 1912), a second memoir on *The Archaean Geology of Rainy Lake Re-studied* (No. 40, 1913), and a paper, "A Standard Scale for the Precambrian Rocks of North America," which he presented at the Twelfth International Geological Congress when it met in Toronto in 1913.

Lawson's work in the region eastward from Lake of the Woods was continued from 1890 onward by William McInnes and W. H. C. Smith, and after Smith's untimely death, by McInnes alone. McInnes' report on the Seine River and Lake Shebandowan map-sheet (1897) was his first major publication. Mostly McInnes extended the mapping of Lawson's rock units: "In the main, the conditions found by Dr. A. C. Lawson in the Rainy River region have been found to hold good for this region, and his admirable description of the relations between the series there will apply generally here also."[14] However, Lawson's series was not so clearly demarked in McInnes' area as it was farther west. McInnes, as he moved eastward, found two other formations. One was the Animikie rocks of the Port Arthur district which McInnes ranked provisionally as lowest Cambrian; these rocks were already well known in adjoining parts of the United States. The other was a completely new group between the Keewatin and Animikie that McInnes named the Steeprock Series, which comprised nine units with a total thickness of 5,000 feet.

Contemporaneous with these studies of the Precambrian in western Ontario a similar enquiry into Laurentian was going forward in Quebec and eastern Ontario by another brilliant scientist, Frank Dawson Adams (1859-1942). Adams had studied chemistry and mineralogy under Harrington at McGill, then joined the Survey in 1879 as lithologist and assistant chemist, succeeding Harrington. He received six months' leave in 1881 to study petrography at Heidelberg under the great authority, Professor Rosenbusch, and with some added experience quickly became the Survey's leading specialist in the study of crystalline rocks. Though Adams resigned from the Survey at the end of 1889 to become Logan Professor of Geology at McGill, he continued to work for the Survey in the summer field seasons for some years, and he lent his skills and knowledge unstintingly to the Survey and to Canadian geology for the remainder of his long life. In 1885 Adams was assigned to the district north of Montreal to ascertain "the true character and relations of the great masses of anorthosite which occur in it and which have been supposed by Sir William Logan to constitute an upper member of the Laurentian system."[15]

Logan's first Precambrian work had been in much the same area, and this had become his type area for the Laurentian. As Adams' initial task was to study the anorthosites there, his was not so much a problem of correlating between different districts (as was Lawson's) as one of studying the nature and origin of the rocks Logan had described. For several years he studied the anorthosite body north of Grenville, Quebec, that Vennor had examined in 1880. By 1888 Adams concluded that the anorthosites "have proved to be either eruptive masses cutting through the gneisses, or masses interstratified with the latter, but probably still of eruptive origin,"[16] and therefore they were not a regular part of the Laurentian. In 1891 Adams further investigated the Morin anorthosite and other rocks in the Grenville area, as did Ells, though Adams' report was not published until 1896 and Ells' not until 1900.

Adams' "Report on the Geology of a Portion of the Laurentian Area lying to the North of the Island of Montreal" (1896) offered his revision of Logan's earlier findings based on his own field work from 1885 to 1891, which in its effects was comparable with Lawson's great reports of 1885 and 1887. In this report, Adams dealt with the three major rock units of Logan's Laurentian. His exhaustive and noteworthy discussion of the anorthosites, which contain orebodies of titaniferous magnetite, demonstrated clearly their intrusive nature and that they were probably some of the youngest Precambrian rocks in the area.

The other two units were a much greater challenge, and herein lay Adams' greatest contribution. The Grenville consists of an immensely complicated assemblage of granites and foliated gneisses (i.e., banded or layered as the result of metamorphic action), with here and there lenses and bands of limestone. After careful petrographic studies of thin sections and chemical analyses, Adams concluded that the Grenville consists of sedimentary and volcanic rocks intensely metamorphosed and intruded by younger granites. He was able to show that at the height of thermo-metamorphism some components of the series became plastic and actually flowed, while others remained brittle and broke into widely separated fragments. This novel idea explained many Precambrian structures and is generally accepted today.

In similar fashion Adams examined the Fundamental Gneiss, which Logan had regarded as the oldest rocks in the region and part of the original crust of the earth, as implied in the name he gave them. Adams discovered under the microscope that they were mostly crushed fragments of still older rocks, recrystallized to their present form by subsequent metamorphic processes. This discovery destroyed the idea that these rocks must be ancient basement, and indicated, instead, that they probably were derived from relatively younger rocks, possibly the Grenville, whose original character had been almost completely obliterated. Adams' conclusion, expressed in 1896, was that the Fundamental Gneiss "forms part of the downward extension of the original crust . . . perhaps many times remelted and certainly in many places penetrated by enormous intrusions of later date," into which "when in a softened condition, there have sunk portions of an overlying series, the Grenville series."[17]

IX-12 F. D. ADAMS

After 1891 Adams moved on to study the Precambrian of eastern Ontario. In 1892 and 1895 he worked in the Haliburton map-area, and in 1893 in a 3,500-square-mile area in Victoria, Peterborough, and Hastings counties. He reported the widespread presence there of the limestone-bearing Grenville Series with which metallic mineral deposits seemed to be associated, some Laurentian gneisses, and a small area of the Hastings Series, which resembled the Grenville but included many rocks of eruptive origin. From his study in Haliburton he made a good start towards solving the problems of the Grenville rocks in Ontario. Since the Grenville in Ontario is generally less altered than its counterpart in Quebec, Adams was in a better position to ascertain the original nature of many of its component parts. Indeed, he was able to conclude that the lightly metamorphosed rocks of the Hastings Series were simply less altered parts of the Grenville.

Adams and Lawson had made the first great advances in the study of the Precambrian of Canada, Lawson mainly through working out the stratigraphy and field relations of different rock types, and Adams principally by using the new tools and concepts of petrography, the study of thin sections of the rocks under the microscope. The work of both men was powerfully assisted by the studies being carried out in the same years by Adams' assistant and apt pupil, A. E. Barlow, particularly his outstanding work on the area east of the Sudbury map-area, which he studied and mapped between 1892 and 1895. Barlow's field work included the area between Lake Nipissing and the northern end of Lake Timiskaming, comprising the Nipissing and Timiskaming map-areas. Most of the area was Laurentian gneiss, apart from the Huronian rocks that extended east from Sudbury to underlie a large part of the Timiskaming map-area; there were also smaller areas of Paleozoic limestones and sandstones on the islands of Lake Nipissing and the northern part of Lake Timiskaming.

Barlow added his support to Lawson's view of the Laurentian in 1892 by reporting that a similar relationship existed along the eastern border of the "original Huronian" area in the vicinity of Lake Timiskaming. He expressed the same conclusion—that Laurentian was in a molten or plastic condition at a period subsequent to the hardening of the Huronian sediments and that the crystalline Laurentian rock was therefore younger than Huronian strata.[18]

This idea that Laurentian was merely an intrusive rock of Precambrian age, as advocated by Lawson, Barlow, and Adams, was upsetting and not altogether accepted by the older generation of geologists. In the summary account of the geology of Canada which G. M. Dawson prepared in 1897 for the British Association for the Advancement of Science meeting in Toronto, Dawson remarked that

> The rocks composing the Fundamental Gneiss often assume the physical relations of eruptives, of later date to the Huronian, along the lines of contact; but in how far this may really be the fact, and in how far this appearance may be attributed to a certain amount of re-fusion of previously existing basal rocks, has not been determined.[19]

Many of the younger geologists, in the Survey and elsewhere, however, were taking up the ideas promulgated earlier by Hunt, and were agreed that at least some of the Laurentian was the product of rocks buried so deeply that they had been fused and returned to the upper crust by cutting upward through rocks that may have originally overlain them. The University of Toronto geologist A. P. Coleman stated in 1898 that while the Laurentian appeared in many places to invade the Huronian, in others, especially in the United States south of Lake Superior, the Huronian did indeed rest on a "Basal Complex." Perhaps, he suggested, where the Huronian formed very thick strata, as in western Ontario, the Laurentian was "buried deeply enough for hydrothermal fusion" in which "the old basement and various amounts of the overlying strata were fused, thus producing the Laurentian granites" which ". . . crept upward where the load of overlying rock was smallest, the heaviest Huronian beds meanwhile settling slowly into synclines between the rising batholiths."[20] On the other hand, where the Huronian was thinner, the floor on which the strata rested may not have been fused but retained its original form.

The state of the controversy over the nature of the Laurentian by the end of the century was expressed by R. W. Ells in a report of 1899 on his work in the counties adjoining Ottawa River:

> . . . within the last twenty years there has been a gradual change in opinion . . . as the result of much careful and detailed work, both in the field and the laboratory, so that it is now very considerably established that much of what has been regarded as altered sediments and so described in the earlier reports must now be accepted as altered igneous rock. Under this head must now be placed the greater bulk of the gneissic rocks which form so large a portion of the Laurentian system as well as much of the pyroxenic and feldspathic rocks . . . so often associated with the crystalline limestones.[21]

The Huronian attracted much interest in the period, especially from the standpoints of working out its areal distribution and temporal position in assemblages and of determining its characteristics. This, indeed, had been one of the reasons for starting the work in the Lake of the Woods district in 1882 which gave rise to Lawson's important studies. Because of a long series of wrong identifications, the name "Huronian" had come to include volcanic rocks, gneisses, and schists whose sedimentary origin was beyond dispute, so that the group embraced a very wide range of rocks of vastly different ages, origins, and petrological characteristics. Since some of these rocks contain metallic mineral deposits, the whole complex was credited with having great economic importance and was sought after by geologists who were unaware of the diversity of rocks embraced in the formation. Any step that could separate the components of the complex so that each could be identified and studied, would have immense practical as well as scientific interest.

The first task, that of exploring and discovering Huronian occurrences, was the work mainly of reconnaissance geologists like Tyrrell and Dowling, Low, and especially Bell, who returned for most of the seasons between 1887 and 1901 to work in the Precambrian districts of northeastern Ontario, particularly around Sudbury and through the regions south and west to Georgian Bay, its coast and islands. Bell's work on the Sudbury map-area was "to ascertain more precisely the northern extension and distribution of the great mineral bearing belt of Huronian rocks,"[22] and this he did. But the geology of the district is enormously baffling and his mapping was confined to the time-honoured units and did little to pry out the secrets of Huronian geology. The geology of the regions south and west of Sudbury in the French River and Manitoulin map-areas was relatively straightforward, consisting of clearly marked Laurentian, Huronian, and Paleozoic (Silurian) formations. The most remarkable feature of the French River area was a belt of red granite, which Bell named the Killarney belt, between one and three miles wide, extending from Collins Inlet to Three Mile Lake and bordered by Huronian quartzites or schists on the north and by Laurentian gneisses on the south. On the evidence of the deformation its intrusion produced, it was younger than the Huronian quartzites, and formed "part of the line of the great break . . . which, farther to the north-eastward, forms the division between the Laurentian and Huronian systems."[23] In 1893 the work was extended to the North Shore map-area, west of the Sudbury map-area, which Bell reported as containing Huronian rocks surrounded by masses of granites like those of the Sudbury district. Bell was still uncertain whether to classify the great granite area as Laurentian or Huronian, and he was becoming aware of the necessity to divide the Huronian into two divisions, though he was not clear as to the limits of either group.

Concurrent with the delimiting of occurrences of Huronian went more detailed work on the nature of the Huronian formations themselves, beginning with Lawson's of the years 1882-84 in the Lake of the Woods district that had turned into his great investigation of the Archean. He had, however, begun the questioning of the identification of the Huronian by his predecessors, with his claims that Logan himself had been inconsistent in applying the name both to his type area by Lake Timiskaming and also to the Dore River greenslates north of Lake Huron—which last rocks were comparable with the "so-called Huronian" Lawson was investigating at Lake of the Woods and to which he gave the name "Keewatin Series." He also disputed Professor R. D. Irving's (of the

United States Geological Survey) identification of the Animikie Series rocks in the Thunder Bay district as being "identical lithologically and stratigraphically, with Logan's typical Huronian;"[24] indeed, McInnes in 1897 ranked these provisionally as lowest Cambrian.

The main studies of the Huronian by the Survey in this period were those of A. E. Barlow, who worked with Bell from 1887 in northeastern Ontario, then later on his own in the region between Lake Nipissing and the northern end of Lake Timiskaming, comprising the Nipissing and Timiskaming map-areas. He collected 700 samples in 1892 alone, and he spent three months of the winter of 1892-93 in Montreal, under Adams' direction, studying them under the microscope. In 1893 he was accompanied to the field by Adams to determine the geological boundary between the Grenville and Huronian, in the Timiskaming district, which was "examined in further detail and can now be mapped with greater precision."[25] In 1895 he was accompanied by W. F. Ferrier (whom Barlow succeeded as Survey petrologist on his resignation in 1898), and "great care was taken in the delimitation of the diabases, gabbros and other eruptives, which rocks had been found to contain the nickeliferous pyrrhotite and chalcopyrite in the Sudbury district to the south-west."[26] Eventually, in 1899 there appeared Barlow's excellent 302-page "Report on the Geology and Natural Resources of the Area included by the Nipissing and Temiscaming Map-Sheets comprising Portions of the District of Nipissing, Ontario, and of the County of Pontiac, Quebec." The district from Lake Timagami to Lake Timiskaming and east and north of that lake, was an interesting mixture of granites, diabases, gabbros, Huronian conglomerates, greywackes, slates, quartzites, and Paleozoic sediments. Barlow retained the term Huronian as appropriate for the rocks of this region which was an extension of Logan's type area. His studies showed that the lower members of the Huronian were of volcanic origin, and he concluded that "the earlier part of the Huronian period in this district was evidently a time of intense and long continued volcanic activity,"[27] followed by a period of sedimentation accounting for the quartzites and conglomerates. He regarded the intrusive diabases and gabbros also as Huronian because of their apparent relationship to the Huronian volcanics, although he noted similar basic intrusions in the Laurentian gneisses and granites, and younger dykes cutting both Laurentian and Huronian. His explanations reflected the latest thinking in all these fields though, like Adams, his emphasis was on the petrology, rather than stratigraphy.

IX-13 A. E. BARLOW
Barlow died at the height of his career when the *Empress of Ireland* sank in the Gulf of St. Lawrence in the summer of 1914.

Barlow continued his investigations of the Huronian in his studies of the Sudbury mining camp and of the Timagami district in 1904. Eventually, thanks to his work and that of other geologists, notably Professor A. P. Coleman, Canadian geologists came to agree on the division of the Huronian into lower and upper formations. In Lower Huronian should be included mainly the limestones, basic and acidic schists, principally of igneous origin, with the iron-formation at its top. The Upper Huronian, separated from the Lower Huronian by an interval of pronounced erosion, would consist of sediments and volcanics and have as its base a conglomerate holding water-worn fragments of the Lower Huronian. The merit of this decision was that both the iron-formation and the conglomerate were widespread throughout the Huronian areas, making the two groups easy to trace. Coleman suggested that the Upper Huronian corresponded with Logan's original Huronian, and he excluded from it the Animikie, which was younger still. He considered that the Lower Huronian extended down to include at least part of Lawson's Keewatin. The American geologists gradually accepted this demarcation between the two divisions of the Huronian, though they continued to disagree as to the extreme limits of both lower and upper divisions, and in particular to urge that

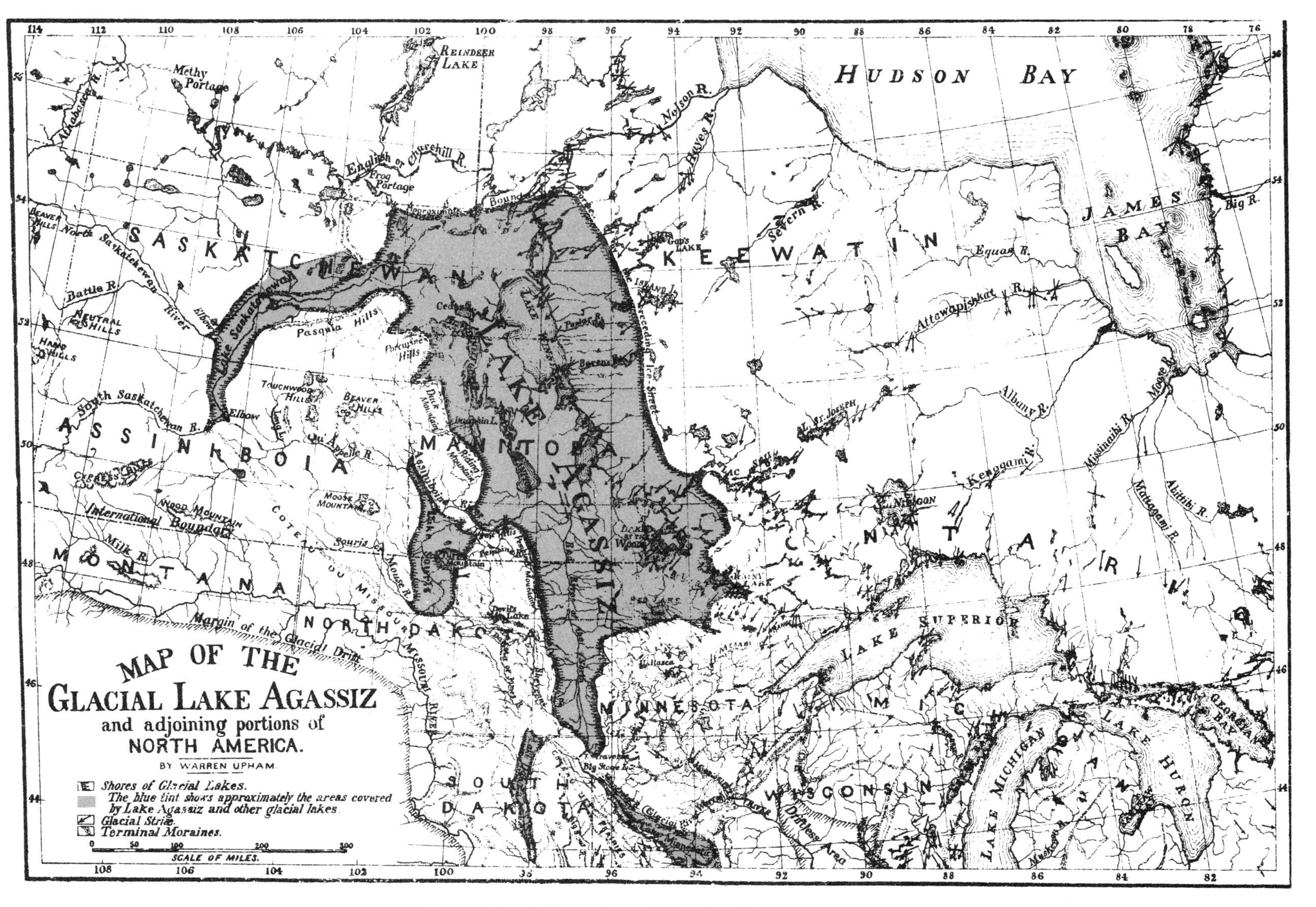

IX-14 FROM *ANNUAL REPORT*, VOLUME 4, 1888–89

the upper division should include younger formations like the Animikie, which were important constituents of the geology of the region south and southwest of Lake Superior.

Precambrian nomenclature was not a problem for Canadian geologists alone. It was an American problem too, since the northern part of the United States included important areas of Precambrian rocks in the Adirondacks of New York and in the states south and west of Lake Superior. It was also a worldwide problem, the solving of which was one of the functions of the International Geological Congress, founded in 1876. Selwyn was anxious for the Survey to play a role in both spheres. He prepared and implemented for Canada in 1881 a standard legend and colour scheme for geological maps, but his proposal that this be adopted internationally was not followed. The long-standing collaboration between Canadian and American geologists begun by Logan was also maintained by Selwyn, to be demonstrated, for example, in the joint survey of 1889 along the southern Quebec-United States border in the Appalachian region.

In 1892 Selwyn responded favourably to a suggestion from Professor C. R. Van Hise of the University of Wisconsin, who was in Ottawa addressing the Survey's Logan Club, that "it would be of great advantage to geology, if a common scheme of nomenclature and mapping could be adopted by the Canadian and United States Geological Surveys. To this end it will be necessary that geological series extending into both countries be mapped by parties of the two surveys working in unison."[28] A conference was held in the summer of 1893, during which some of the type occurrences on each side of the border were visited. But not until 1903, when C. D. Walcott, the director of the United States Geological Survey, appointed Van Hise and C. W. Hayes to meet with Bell and Adams, did the committee come into being, along with subcommittees for the Adirondack and Lake Superior regions.

The Lake Superior subcommittee met in August 1904, and began six weeks of visits to key locations on both sides of the lake, including the Gogebic, Mesabi, and Vermillion districts in Michigan, and Rainy Lake, Lake of the Woods, Port Arthur, and the north shore of Lake Huron in Ontario. In the Rainy Lake district they studied the Coutchiching and Keewatin, and at Thunder Bay the Keweenawan and Animikie. They noted the major discontinuity within the Huronian in Logan's type area and the major unconformity at its base (both had been reported by Logan). They also erected a new term, Thessalon, for a series of greenstones and schists that had been intruded by a granite that was in turn unconformably overlain by Lower Huronian. The Thessalon was thus much older than the Huronian and could not be a part of it.

In 1905, after due deliberation, the committee published its tentative findings for the Lake Superior Precambrian succession. At the top was the Keweenawan, embracing the Nipigon and some American formations; below and separated from it by a major unconformity was the Huronian. Indeed the view was expressed that the Keweenawan might even be lower Paleozoic. The Huronian was divided into three parts by two unconformities. The Upper Huronian included the Animikie (which satisfied the Americans), while the Lower and Middle Huronian accorded with the Canadian views for the type area and also accommodated many of the formations on the American side of the border. The upper part of the Coutchiching was also considered to be Huronian. Below the Huronian, and separated from it by another major unconformity, was the Keewatin of Lake of the Woods, and included with it were some of the American formations and the lower part of the Coutchiching. Finally the committee, in deference to Logan, retained the name Laurentian, but applied it to all pre-Huronian granites and gneisses regardless of whether they were older or younger than the Keewatin schists.

Later in 1905 the International Committee on Geological Nomenclature met in Ottawa to decide on a program for examining the Precambrian in eastern Ontario and New York. The group consisted of Adams, Professor Coleman, and, for a time, Barlow, representing Canada; and Van Hise with Professors H. P. Cushing and J. F. Kemp representing the United States. In July-August 1906, various parts of the Adirondack Mountains were visited, and then the party crossed to Canada to examine the Precambrian of eastern Ontario. From Tweed to as far as Bancroft along Hastings Road, they saw "one of the finest sections of Pre-Cambrian rocks to be found anywhere in America."[29] As before, the committee prepared its report, then after circulating it to the members, had it published in *Journal of Geology* in April 1907.

That report concerned itself with the two groups in Logan's Lower Laurentian, his Upper Laurentian being set aside as consisting essentially of anorthosite intrusions. The committee ruled that since the Fundamental Gneiss was the major formation in terms of size, it should be termed Laurentian, which corresponded to the decision reached by the subcommittee for the Lake Superior region. The upper sedimentary and volcanic series, which had been invaded in part at least by the igneous rocks of the Laurentian, was given the name Grenville Series, also in accordance with Logan's nomenclature. The term Hastings Series was abandoned as being Grenville in a less altered form, a conclusion Adams had reached from his work in Haliburton in 1895-96. Wisely it was stated that

> The committee consider that it is inadvisable in the present state of their knowledge to attempt any correlation of the Grenville series with the Huronian or Keewatin, so extensively developed in the region of the Great lakes. The Grenville series has not as yet been found in contact with either of these, and until this has been done and the relations of the several series have been carefully studied, their relative stratigraphical position must remain a mere matter of conjecture.[30]

It is noteworthy that most of the major groupings and subdivisions of units were in accordance with the conclusions already reached by Canadian geologists and were given the names being used by the Canadian Survey. This indicated that the Americans on the committee accepted the Canadians' leadership in this field.

* * *

THE CONCERN TO STUDY AND EXPAND scientific knowledge, so well displayed in the work on the Precambrian, and to a lesser degree on the Appalachian and Cordilleran regions, was little in evidence in respect of regions like the Prairies or southern Ontario. Despite Selwyn's espousal of a deep-well drilling program at Deloraine, Manitoba, or Dawson's in connection with the Athabasca oil sand formation in northern Alberta, both directors seem to have concentrated the Survey's efforts on regions characterized by igneous and metamorphosed rocks that were favourable to the deposition of metallic minerals, and to have neglected the little-disturbed sedimentary formations. The trouble unquestionably lay in the financial constraints on the Survey, and the consequent inability of the directors to employ enough specialists in paleontology and surficial geology. In his own field work Dawson had certainly not shown such narrowness of interest.

Nevertheless, while great advances were being made in paleontological studies in other countries, the Survey remained relatively static in this field. The activity of the small staff seemed confined to helping field officers with their stratigraphic problems, identifying fossils, working out the taxonomy of new species, and attending to the fossil collection and the museum. Apart from Ami's rather limited work around Quebec City and in Nova Scotia, little was done by the Survey to study the paleontology and stratigraphy of the Paleozoic and younger rocks that underlie so great a part of western Canada.

The Survey, it is true, did pay a little more attention to surficial geology, to working out the sequence of glacial movements, their effects on the physiography and on the changing levels of the land, and the like—subjects that were by then receiving a good deal of study in the American midwest and elsewhere. Apart from one field officer (of some twenty) who concentrated on the Pleistocene, it relied on the subject being investigated and reported on in the course of general field operations. Such reports provided descriptions of various phases (such as glacial striae), though the treatment varied with the geologist's ability and interest in the subject, or its relative importance for the area he was examining. Bell, Dawson, Low, Tyrrell, and Dowling paid considerable attention to glacial geology during their broad reconnaissances, evidences of which were usually readily apparent to the trained observer. Tyrrell in particular, since he worked on the northeastern fringe of the Prairie region, devoted a good deal of effort to this subject, as his field work of 1887 to 1890, especially his report for 1888, indicates. From the extensive knowledge he afterwards acquired of conditions all the way from the Churchill basin to Baker Lake and Chesterfield Inlet, he was able to make an important contribution to the general glacial history of the midcontinental region, as were Low and Dawson as a result of their work in Ungava and British Columbia.

But despite these efforts, the need was for specialists in the latest methods and interpretations, who could make exhaustive, authoritative studies of particular areas or phenomena. What was needed may be seen from the study of glacial Lake Agassiz in the Red River region that the American geologist Warren Upham conducted between 1879 and 1887, a project initiated under the noted authorities N. H. Winchell and T. C. Chamberlin. Upham investigated the part of his subject-area that lay in Canada, and the Survey, in return for paying part of his expenses for that season, received from him a *Report on the Exploration of Glacial Lake Agassiz in Manitoba.* This was a study of the effects of the lake (really a succession of lakes), covering a total area of 110,000 square miles, formed of meltwaters from wasting ice sheets that had extended far into the United States, and that endured until the ice sheets retreated sufficiently for the present-day drainage pattern north to Hudson Bay to develop. Upham's detailed report on the evidence and effects of this important episode in the Pleistocene was submitted in 1889 and can be credited with giving the Survey's geologists a better understanding and fuller appreciation of this side of the work. Tyrrell, for one, undoubtedly learned a great deal from Upham's work and their discussions from 1887 onward. Another, later example of the Survey's use of an outside specialist was the employment of ex-Canadian J. W. Spencer, who since 1888 had been studying the effects of the Pleistocene on the Great Lakes region, to prepare a report on *The Falls of Niagara* (1907).

The achievements of the Survey's own specialist should not be minimized, however. Robert Chalmers was a sturdy, stocky, tenacious, largely self-trained New Brunswick schoolmaster, naturalist, and writer on scientific subjects in Maritimes newspapers, who had drifted into working with the Survey as a summer assistant in 1882 and 1883 before joining the permanent staff in 1884. He was a gifted amateur of wide experience in a field that was reaching specialist status, and besides, he was only one man coping with an important subject as vast as all Canada. To 1895 he worked in the Maritimes; not until the late nineties did he do some work in central Canada, first in the Eastern Townships, then in southern Ontario. Only once, in 1906, did he make an excursion into western Canada—from Winnipeg across the prairies, then by rail to the Pacific coast at Burrard Inlet.

His work, given these limitations, was of a high order for the time. In the Maritimes he studied New Brunswick, then adjacent parts of the other provinces, publishing several reports, notably that of 1895, "Report on the Surface Geology of Eastern New Brunswick, North Western Nova Scotia and a Portion of Prince Edward Island," which was a thorough, competent work summing up a dozen years of studies. It offered conclusions respecting the sizes and movements of the glaciers and marine ice, the rises and falls of the land and resulting invasions of the sea, the postglacial erosion, the origin of the remarkable tides of the Bay of Fundy, and the like. He also made a useful study of the placer gold deposits of the St. Francis and other rivers of southern Quebec, while after 1901 under Bell's administration he undertook studies of the Pleistocene in the St. Lawrence valley and the Ontario peninsula.

However, in the fields both of stratigraphic paleontology and of surficial geology, the Survey did not seem to be keeping pace with the latest scientific advances, nor to be giving enough attention to these important subjects. This was all the more apparent after 1900. The public was growing increasingly interested in the problems of its environment, and feeling a need for expert information and guidance. The efforts to revive the faltering petroleum and natural gas industry in southern Ontario and to develop one in western Canada indicated a growing need, also, for more stratigraphic and paleontological field studies, undertaken in a more aggressive, enterprising manner. The overcoming of both these deficiencies, however, had to wait until the Survey secured the means and the staff to undertake these studies in more detail—and that, in turn, had to await a more propitious time in the years after 1907.

Paralleling the careful work on the Precambrian Shield aimed at reaching a proper understanding of the relations of the various component parts of that vast geological province, went similar efforts to understand the equally complex, economically vital process of ore deposition. The former would enable the prospector to concentrate his efforts on those formations known to be favourable to the occurrence of valuable minerals; the latter would help him locate the ore-bearing components of a metalliferous complex and indicate how they could be best exploited. It was widely believed that elements like gold, copper, and lead were mostly carried to their present positions in hot, dilute aqueous solutions, and that magmas, bodies of molten rock within the earth's crust, supplied the heat and probably much of the water and its contained chemical constituents. As these magmas worked their way towards the surface they cooled and crystallized into igneous rock, expelling most of their contained water as they did so. When this hot water invaded the surrounding rocks, at suitable places and in suitable rocks it released its mineral content. The Survey's task, exemplified by Brock in particular, was to indicate areas of intrusive rocks that might have released mineralizing solutions, to determine which of the surrounding rocks were the most favourable hosts for those solutions; also to recognize and trace the fractures and other structures where mineable concentrations of ore might have been deposited.

Although many mineral deposits seemed related to hydrothermal solutions, there were others that did not appear to be explainable in that way. The petrologists Adams and Barlow were grappling tentatively with indications that certain types of ore deposits were themselves of igneous origin. Thus Barlow, in March 1891, in reporting on the nickel and copper deposits at Sudbury, described the ores as having a "common genesis" with the enclosing diabase rock, and that "the ores and associated diabase were therefore in all probability simultaneously introduced in a molten condition, the particles of pyritous material aggregating themselves together in obedience to the law of mutual attraction."[31] Soon afterwards Professor J. H. L. Vogt of the University of Christiania (now Oslo) announced an important breakthrough—the principle of the magmatic origin of certain ores. Adams immediately saw its implications, and his eager mind seized on the theory. On January 12, 1894,

IX-15 ERRATIC BOULDER NEAR WATERTON LAKES, ALBERTA
G. M. Dawson, who photographed this huge erratic in 1881, assigned it to the Huronian and conjectured that ice sheets had transported it from the Precambrian Shield, over 1,000 miles to the northeast. Recent study of many erratic boulders scattered about western Alberta places their origination somewhere in the Cordilleran Region to the west.

he gave a paper on the subject before the annual meeting of the General Mining Association of Quebec, which was published in the *Canadian Mining Review* for that month. The theory was that in a magma that contained unusual amounts of iron, nickel, copper or other elements, minerals of which those elements were the principal constituents might remain in the parent magma as it cooled, and be sufficiently concentrated in parts of the final rock to be profitably mined. Such massive titaniferous iron ores were present in the anorthosites of Quebec, which, for metallurgical reasons, could not be exploited at that time; while a prime example of the second type was the important nickel-copper sulphide deposits at Sudbury. Adams pointed out that

> These deposits not only occur in connection with the igneous rock but actually appear to form part of it, the ore occurring distributed through the rock and the heavy bodies of ore, merging gradually into it in many places, so that it is impossible to tell where the rock begins and the ore body ends. The ore body in fact seems to be merely a portion of the igneous mass in which the ore, which is one constituent of the normal rock, is concentrated sufficiently to form workable deposits.[32]

It was especially fitting that the opportunity to re-examine the Sudbury deposit in the light of this new theory should fall to Barlow in view of his earlier work there as Bell's assistant, his wide experience in dealing with Precambrian rocks, and his special training in petrography. In 1901, he was sent back to Sudbury to examine in greater detail the geological features he and Bell had studied between 1887 and 1890: "One of Dr. Barlow's principal objects was to ascertain whether or not the relative richness in nickel of the pyrrhotite ores was in any way related to the composition or other characters of the different eruptive country-rocks in which they occur."[33] He was relieved from his laboratory duties at the Survey by a financial agreement with the Mond Nickel Company, one of the major owners of mining properties in the district. His assignment to revise the Survey's map of 1891 continued into the summer of 1902, after which he prepared a meticulous report that presented a comprehensive study of the district, including its history, geography, and statistics of development, to provide the public with a complete account of this important mining area in one handy work.

His report, which within a dozen years was acclaimed as having "now become a classic in the literature of ore deposits,"[34] began by indicating the dramatic advances that had occurred in the science in general, and the vastly greater amount of information available about the Sudbury deposits in particular, in the ten years since Bell and Barlow's first investigation. Thus the map of 1891 had portrayed the intrusive mass in which the ore was situated in the form of a lopsided horseshoe with a gap along the southern margin; prospecting in later years had disclosed the southern, or major, belt of the nickel-bearing eruptive to be situated there. Moreover, Bell's explanation of the deposit had been along traditionalist lines—that the metallic mineral had been intruded into mixed fragments of quartz diorite and greywacke that often has the appearance of a conglomerate, that the orebodies occurred in lenses, often in the diorite, and their concentration was possibly connected with the nearby diabasic dykes.[35] Furthermore, Bell's map of 1891 had failed to make any distinction between the nickel-bearing norite and rocks that resembled the norite but were altogether barren of valuable minerals. In fact, "the presence of the norite as a distinct geological unit was not suspected until long after this first work was completed, the associated greenstones being considered as portions of the norite, which had been metamorphosed by the intrusion of the younger granite masses."[36] As recently as 1902, indeed, the latest geological map of the Ontario Bureau of Mines had failed to rectify either of these inaccuracies.

Barlow had come around to a somewhat modified position as regards the role of magmatic differentiation in the formation of the Sudbury nickel-copper deposit. As he confessed, the great attractiveness of the hypothesis when it was first enunciated, plus the need to secure its acceptance, had meant that "too much emphasis was perhaps given to the idea . . . as in itself, giving an adequate explanation of all the phenomena witnessed,"[37] His own position, as a result of further investigations, was that "the manner of formation of the ore bodies [was] much more complex than was at first supposed,"[38] and that

> More recent and detailed examination of the various ore bodies, has shown that while the first hypothesis of a segregation of these [metallic] sulphides, directly from the magma, is in the main, the true explanation of their present position, other agencies, which are usually grouped together under the name of secondary action, have contributed rather largely, to bring about their unusual dimensions.[39]

He had been led to this view by the inability of the simple magmatic differentiation theory to account for an "abnormally large amount of original or primary quartz"[40] uniformly distributed throughout the norite or gabbro. This led him to conclude that the sequence of events must have been along these lines:

> There can be no doubt, however, that much of the sulphide material was introduced simultaneously, as an integral portion of the same magma, along with the other minerals of which the norite or hypersthene-gabbro is composed. There can, moreover, be little doubt of the abundant presence of heated solutions and vapours, which were capable of dissolving out, and under certain conditions, of redepositing these sulphides. Such agencies certainly began their work before the whole magma had cooled, bearing their heavy burdens of sulphide material, most of which was obtained from the magma in the immediate vicinity, to occupy the various cavities and fissures as fast as these were formed. The whole of this action was practically completed before the intrusion of the later dykes of the olivine-diabase which are now regarded by the writer, as the end product of the vulcanism to which the norite masses owe their intrusion. In certain of the deposits, the various hydrochemical agencies accompanying dynamic action have been more active than in others, . . . but in others, as for instance, the Creighton mine, magmatic differentiation has been the main and almost sole principle, determining and favouring the development of this the largest and richest sulphide nickel mine in the world.[41]

The use of the theory of magmatic differentiation in accounting for the diversity of igneous rock types in a given area became increasingly frequent in Survey reports. Adams, Dresser, and Young used it to explain the unusual rock types encountered in the cores of the conspicuous Monteregian Hills, of which Mount Royal is the best known. D. D. Cairnes, on his first independent exploration in 1906 on behalf of the Survey in the Tagish Lake district along the Yukon-British Columbia border, reflected the impact of magmatic differentiation on contemporary thinking: "Newer than the granites is a somewhat complex series of porphyrites, porphyries, diorites, gabbros, & c., which apparently represent rocks from the same magma, but which differ considerably in character on account of segregation, cooling under different conditions, &c."[42] By the time Daly had concluded his study of the International Boundary, he was able to list numerous examples of magmatic differ-

IX-16 J. A. DRESSER
Born in 1866 in the Eastern Townships, Dresser could remember seeing "an old man with a great white beard and a straw hat, tied on with red ribbon" hammering on a limestone ledge, and hearing his teacher call out, "Look children look! There is Sir William Logan."

entiation and to develop the subject in a general and theoretical way by devoting to it two chapters of his Memoir No. 38, *Geology of the North American Cordillera at the Forty-ninth Parallel* (1912). Indeed, Daly added a new element to the theory by pointing out that the most acidic layers near the top of the intrusive body were sometimes due in part to the absorption by the magma of silica from the intruded quartzites, indicating the ability of magmas to digest large amounts of foreign material as they moved into place, with resulting changes in their own composition.

Such examples indicated how the important new insight into the nature of igneous activity was quickly put to use as a scientific tool to help solve many hitherto insoluble problems, some theoretical, some practical. Barlow's recognition, for example, that many of the nickel deposits at Sudbury were magmatic ores and his working out the implications were of immediate and tremendous importance to the mining industry, and a striking illustration of the way in which supposedly "theoretical and purely scientific" geology goes hand in hand with important practical results.

X-1 GEORGE MERCER DAWSON
Director, 1895-1901

10
A Matter of Direction

WITHIN A BRIEF SPAN of 12 eventful years after Selwyn's retirement, the Survey experienced the leadership of two very contrasting personalities. The brilliant, diminutive George M. Dawson, who became director on January 7, 1895, at the comparatively early age of 45, was barely six years at the helm when, after but a single day's illness, he died on March 2, 1901. Robert Bell was immediately appointed acting director by simple ministerial letter, without promotion or increase of salary, on what was obviously an interim basis. But he carried on in this uncertain, unsatisfactory state for over five years, until on April 1, 1906, with two days' notice, his interim appointment was terminated, and A. P. Low gained the directorship. Low's period of command was brief, for in a year's time the Survey was merged into a newly-created Department of Mines and he became its deputy minister. Hence, for practical purposes, the last years of the Geological Survey Department can be regarded as coinciding with the Dawson and Bell regimes, the short periods that each was given to manage the Survey's affairs.

Two such very different men, possessing such widely contrasted personalities, present an interesting commentary on the role of leadership in an institution like the Survey. How far did the personality and training of each affect the style and performance of the Survey? Their contrasting characters make this a good opportunity to examine how far the personality of the man at the helm could affect the operations of the department. In this sense, the story of the Survey in these years may contribute something to the perennial historical debate over the rôles of the individual and of the environment in determining the course of events.

The central quality of Dawson's directorship was conservatism, the avoidance of new departures. He had been close to Selwyn and presumably had played some part in the changes introduced since 1884; as director, his goal was to improve the operations, not change the system. Besides, the extremely difficult financial situation that extended, unhappily, for the whole of his directorship made it impossible to introduce changes, even had Dawson so desired. These difficulties were at their worst in the final days of the Conservative governments of the nineties, but the situation was no better under the Liberal government that came to power in 1896. Times continued bad, enjoining economy on the new government, and the Survey suffered. Then when conditions began to improve, the minister responsible for the Survey, Clifford Sifton, for reasons of his own, would not increase the appropriation, either in proportion to the government's improved means or in line with the expansion of the mining industry and consequent growth in demands for the services of the Survey. Hence Dawson was forced to carry on operations as best he could in a time of expanding commitments, shepherding his limited resources and aiming for quality in performance.

This goal was not incompatible with Dawson's own tastes, for versatility, tireless effort, striving after excellence, were part of his nature. Everyone who met him was impressed by his acute intelligence, fine logical faculty, great capacity for research, industriousness, and uncommon ability to master details and reach sound conclusions. The aptitude for hard work revealed in his field work was matched by the effort he put into his other tasks. He was an omnivorous reader, and his reports always were thoroughly researched. In the field he secured every scrap of useful information, not merely about the geology, but on the natural history, native peoples, economic conditions, and history of the districts. In his first field operation, while on the International Boundary survey of 1873-75, he published a paper on the locust invasion of 1874 in Manitoba, based on his observation and study of that phenomenon. When the Klondike gold rush occurred, he prepared a useful paper on the early history of the Yukon, a byproduct of his extensive reading, interviewing, and correspondence in connection with his exploration of the region in 1887. The same high standards characterized these minor works as they did his major writings. His ethnological reports, for example, based on relatively short visits to the Salishan, Kwakiutl, and Haida, made him a recognized authority, "one of that handful of original observers whose work affords the foundation for scientific knowledge of the North American natives."[1] His work in every field to which he turned his hand commanded respect and was studied with profit.

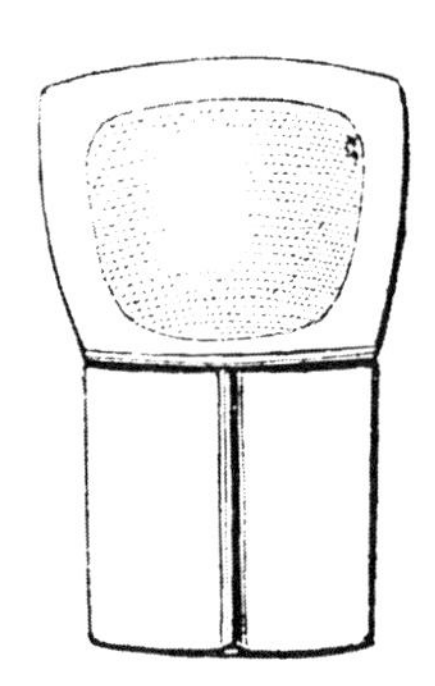

X-2 A NORTH PACIFIC INDIAN "COPPER"
From Dawson's report on the Haida of Queen Charlotte Islands, an important contribution on a remarkable people undergoing acute cultural change.

Dawson was also careful, moderate, and wise in expression, of a naturally courteous and conciliatory disposition that was reflected in his management of the Survey's relations with politicians, civil servants and public, as well as his diplomatic missions on Canada's behalf. He was a thorough Canadian nationalist in the imperialist sense of those times, enthusiastic about Canada's prospects and solicitous for her interests. Though he was quiet and reserved in person, he was also extremely genial, cheerful, and entertaining among friends. At the Rideau Club he was a most welcome dining companion: "In conversation he was witty and humorous to a degree, while at the club, or at a public dinner, his sallies were wont to keep the table in a roar. A quick wise smile lit up his countenance when indulging in good-natured badinage, of which he was very fond; and when talking upon serious subjects his eye flashed with the intelligence which made his most trivial discourse luminous."[2] Another commentator observed that "The secret of Dr. Dawson's widespread popularity, no doubt, lay in his downright unselfishness and in his sunny and sympathetic nature."[3] A man of high standards, Dawson did not accept mediocrity gladly. It went against his grain to court public acclaim or turn his reports into public manifestoes. While he did not regard the work of the Survey as a "mystery" reserved for an elite, he felt it could not, or should not, aim at catering to the lowest levels of public interest. He refused to stoop to the geological illiterate in the reports, arguing there was a limit how far it paid to "popularize" them. He resisted the drive to have the publications given away automatically to all who requested them, because he felt applicants were unlikely to value or make good use of something they had not paid for. In these and similar ways he was perhaps a little out of step with the new times, much as such attitudes were valid in the late nineteenth century.

For six years the hard-working, brilliant Dawson remained immersed in the many details of directing the Survey or advising the government and the public from his wide experience and wider knowledge. He found it necessary to be in Ottawa during the ever-lengthening sessions of Parliament preparing publications, answering inquiries, escorting distinguished visitors through the museum, and coping with the many cares of his position. "My own time, during the year," went his report for 1895, "has of necessity been employed chiefly in connection with the office and in the rearrangement of some parts of the executive work, as well as in supervising the printing and publication."[4] In 1896, he pointed out that the special session of the new Parliament, and the "difficulties arising from the want, during some part of the season, of any appropriation to cover the work in progress in the field,"[5] had kept him in Ottawa all summer. A different sort of involvement—"Preparations for the representation of the mineral products of Canada at the forthcoming exhibition in Paris"[6]—accounted for much of his time in 1899. And so it went. Indeed, it was becoming increasingly apparent that administering the Survey was a full-time operation that left little time for inspections of field activities, and none whatever for field work. Only occasionally could he go over his own maps and samples, or think about the geology of southeastern British Columbia, on which he had been working before he was summoned to Ottawa.

Dealings with ministers occupied much of his time, both with the old and new governments. He also had to visit other ministers besides his own on Survey matters, or in response to requests for advice on such questions as mining regulations for the Yukon Territory or the export of natural gas from southwestern Ontario. Sometimes such audiences were occasioned by trifles: "Called on Hon. Mr. Scott about 25 cents worth of maps which a deadbeat in N.S. [is] trying by writing to Ministers to force us to send him free!"[7] There were also many calls to and

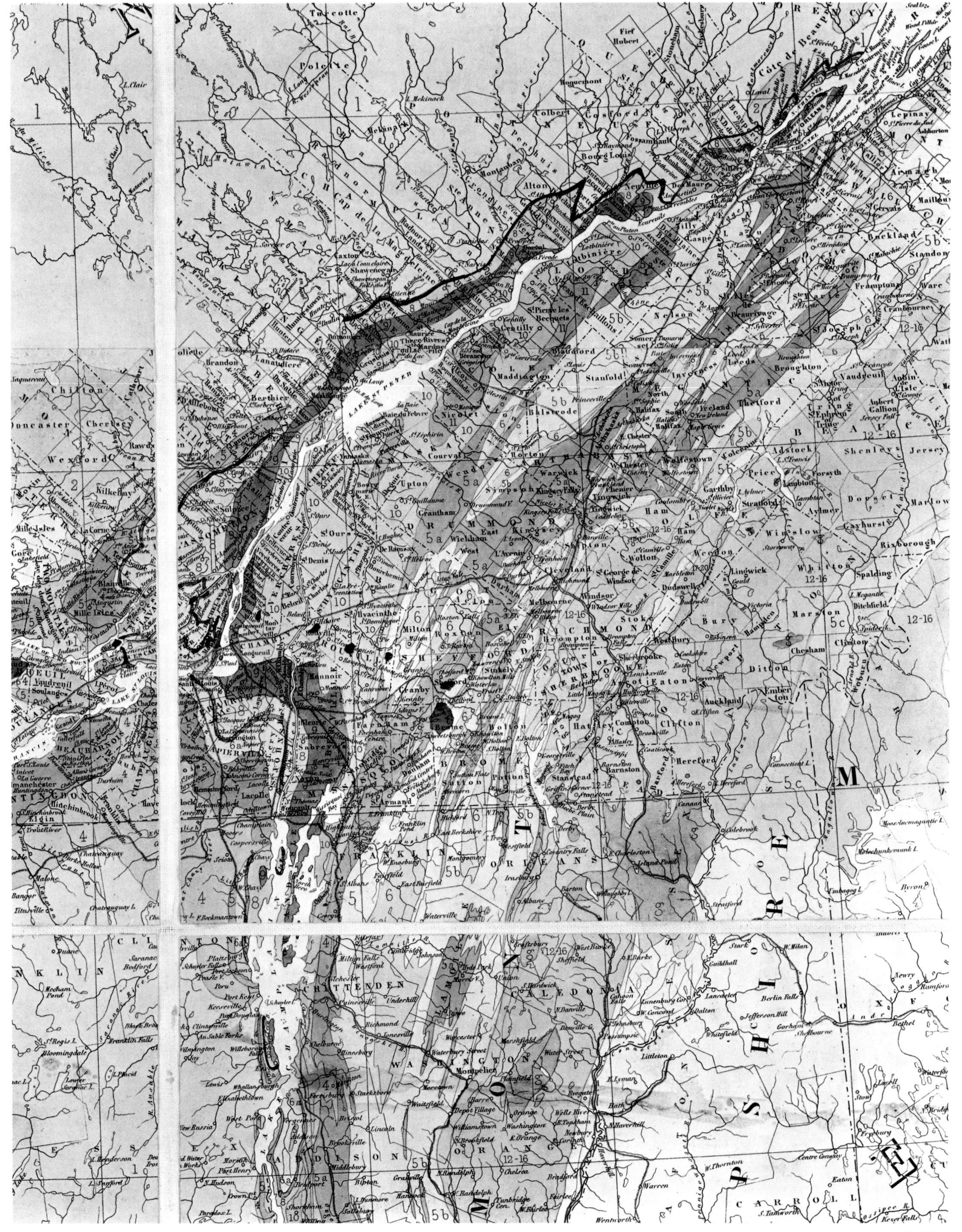

Tuscotte
Polette
Fief Hubert
Stoneham
Roquemont
Colbert
Gosford
Mekinack
Côte de Beaupré
Laval
Montauban
Bourg Louis
Alton
Neuville
Caxton
Shawenegan
Lotbinière
Gentilly
Nelson
Leeds
Broughton
Thetford
Stanfold
Somerset
Maddington
Bulstrode
Nicolet
Doncaster
Wexford
Kilkenny
Chilton
Berthier
Brandon
Lanaudière
Upton
Grantham
Wendover
Simpson
Kingsey Falls
Tingwick
Warwick
Horton
Adstock
Shenley
Jersey
Forsyth
Lambton
Stratford
Aylmer
Dorset
Marlow
Winslow
Weedon
Lingwick
Whitton
Spalding
Rixborough
Ditchfield
Marston
Chesham
Clinton
Woburn
Ditton
Bury
Hampden
Westbury
Dudswell
Windsor
Melbourne
Durham
Cleveland
Shipton
Wotton
Wickham
St Ours
St Denis
Milton
Roxton
Shefford
Granby
Farnham
Brome
Bolton
Potton
Stanstead
Barnston
Hereford
Barford
Compton
Clifton
Hatley
Auckland
Eaton
Sherbrooke
Richmond
Stoke
Lacolle
Hemmingford
Beauharnois
Vaudreuil
Soulanges
Huntingdon
Hinchinbrook
Elgin
Chateauguay
Malone
Franklin
Orleans
Caledonia
Chittenden
Washington
Montpelier
Orange
Addison
Middlebury
Bristol
Lincoln
Carroll
Shelburne
Gorham
Lancaster
Littleton
Bath
St Armand
Champlain
Plattsburg
Keeseville
Ferrisburg
Vergennes
Bridport
Shoreham
Randolph
Chelsea
Wentworth
Warren
Conway
Eaton
Berlin Falls
Stratford
Colebrook
Canaan
Derby
Barton
Irasburg
Albany
Waterville
Fairfield
St Albans
Lake St Peter
5a
5b
5c
6
8
10
12-16
M

MAP 1.

The crumpled, folded, metamorphosed, mineral-rich rocks of the Eastern Townships have been studied intensively since the time Logan's original work, and improved knowledge of the geology of the region has resulted. Illustrated here are a few successive maps of t oldest core rocks of the main anticlinal axis, extending from Sutton Mountain northeast towards Gaspé.

In his magnificent, rare "Geological Map of Canada, 1866" – *see Map 1* (foregoing page) – Logan assigned these rocks to the Lower Siluri "Quebec Group," shown in purple (5a – Levis and 5B – Lauzon) and yellow (6 – Sillery).

Later geologists, beginning with A. R. C. Selwyn, have given an earlier date to these rocks. In 1877-78 Selwyn classified them "(Huronian?)". The 1 inch = 45 miles "Map of the Dominion of Canada . . . from Surveys Made by the Geological Corps, 1842 to 188 (Map No. 411, published in 1884)–*see Map 2 below* – listed the formation as "Pre-Cambrian" (marked "AB"), flanked by Cambrian rocks (C So did R. W. Ells, who mapped the entire district between New Brunswick and Montreal on the 1 inch = 4 miles scale. In his map of t Eastern Townships, Montreal Sheet (Map No. 571 of 1895, *see Map 4 opposite*), Ells identified the rocks in brown (AB) "Huronian?-Pre-Cambrian," flanked by Cambrian (C), Cambro-Silurian (D), or Silurian (E), all coloured in varying shades of grey a yellow.

More recently, improved understanding of the complicated history of the district in Paleozoic time and of the 1200-million-year interval th preceded the Paleozoic has made it possible to work out the succession and distribution of the region's rocks with greater confidence. On t forthcoming Gatineau River Sheet by W.H. Poole—of which part of the manuscript is reproduced as *Map 3* below—the Sutton axis roc are assigned to the Hadrynian (the era immediately preceding and conformable with the Cambrian) and to the Cambrian.

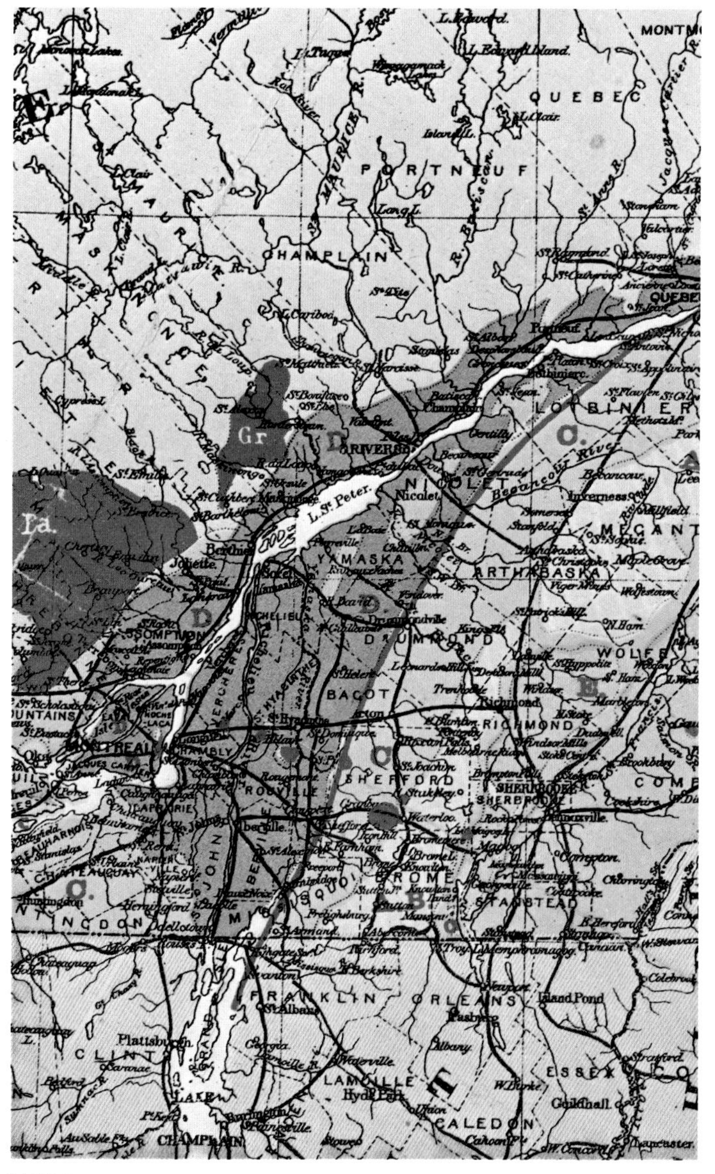

MAP 2.

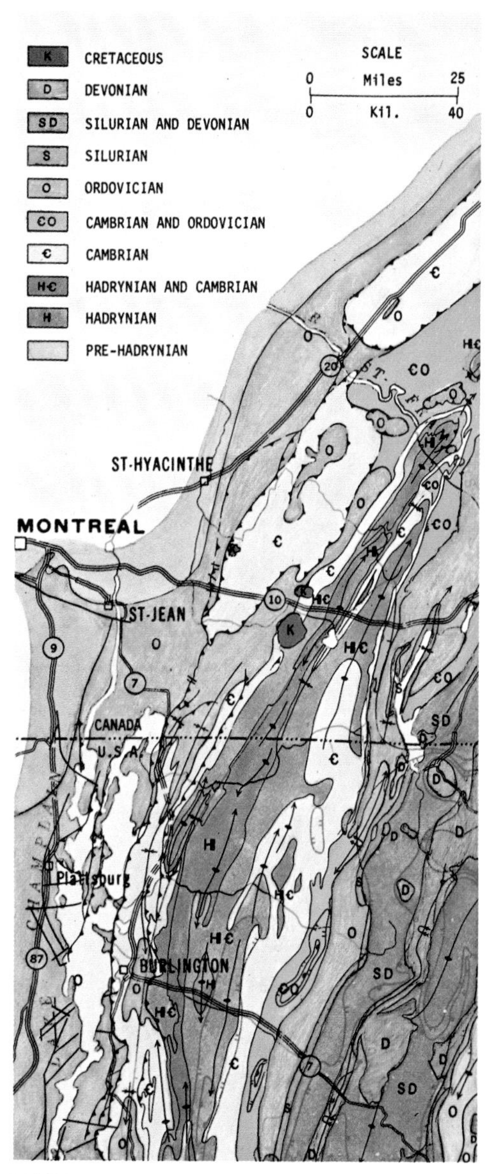

MAP 3.

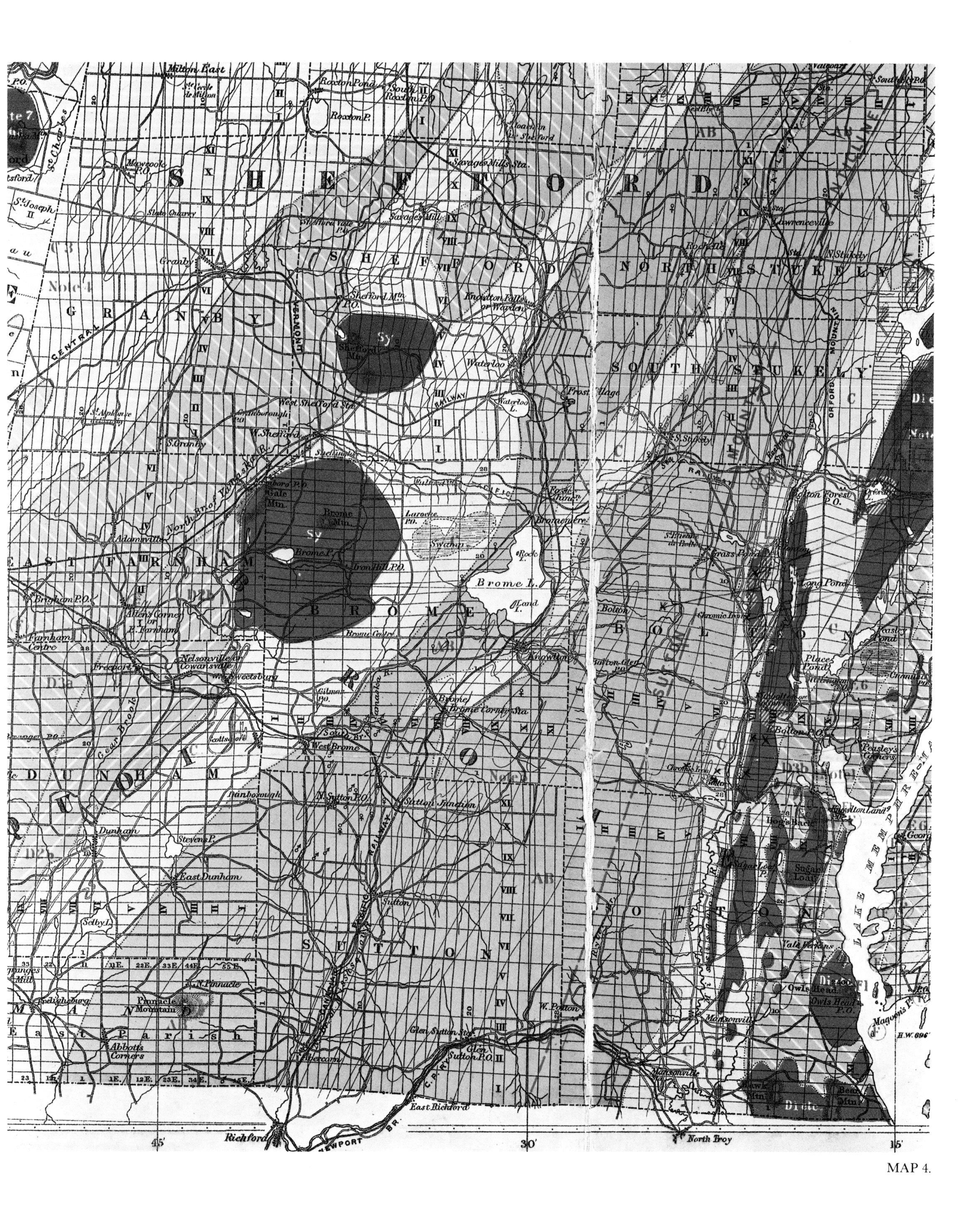

SHEFFORD
GRANBY
NORTH STUKELY
SOUTH STUKELY
EAST FARNHAM
BROME
BOLTON
DUNHAM
SUTTON
POTTON
Brome L.
Waterloo
Knowlton
Richford
North Troy
LAKE MEMPHREMAGOG

MAP 4.

A GUIDE BOOK MAP, 1913

The map reproduced below is one of the many impressive, simplified maps that were prepared by master draftsman A. Dickison and his colleagues for publication in the guide books of the XIIth International Geological Congress in 1913. Taken together with the Haliburton map (*Map 6, see following page 264*), it demonstrates the variety and the high standards of scientific and cartographic excellence of the Survey's maps in the second decade of the present century.

This sketch of the spectacular Rocky Mountains near Banff, Alberta, is taken from Guide Book No. 8, *Transcontinental Excursion C1, Toronto to Victoria, and return, Via Canadian Pacific and Canadian Northern Railways, Part II.*

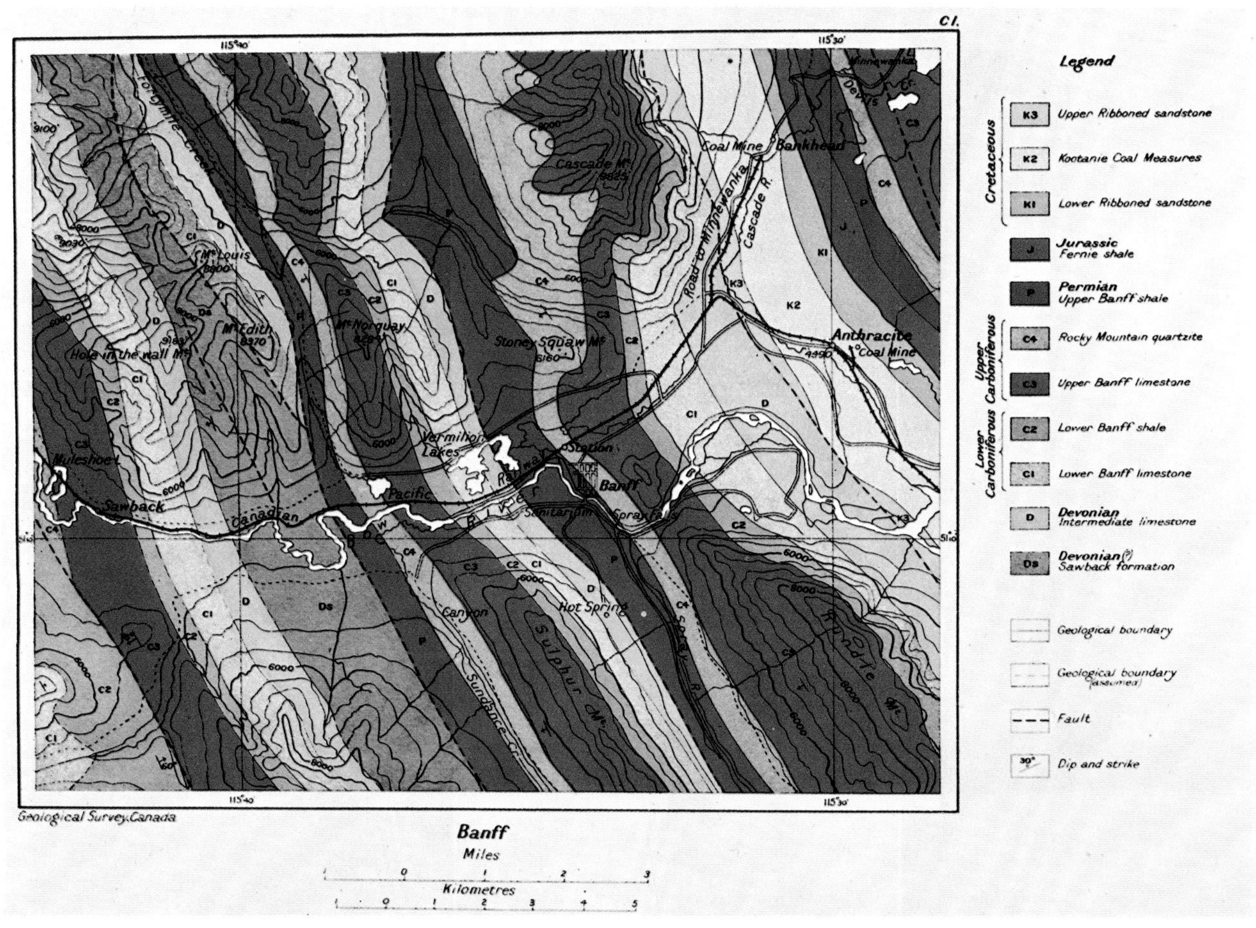

MAP 5.

X-3 THE MARITIME PROVINCES AND EASTERN QUEBEC, 1870-1930

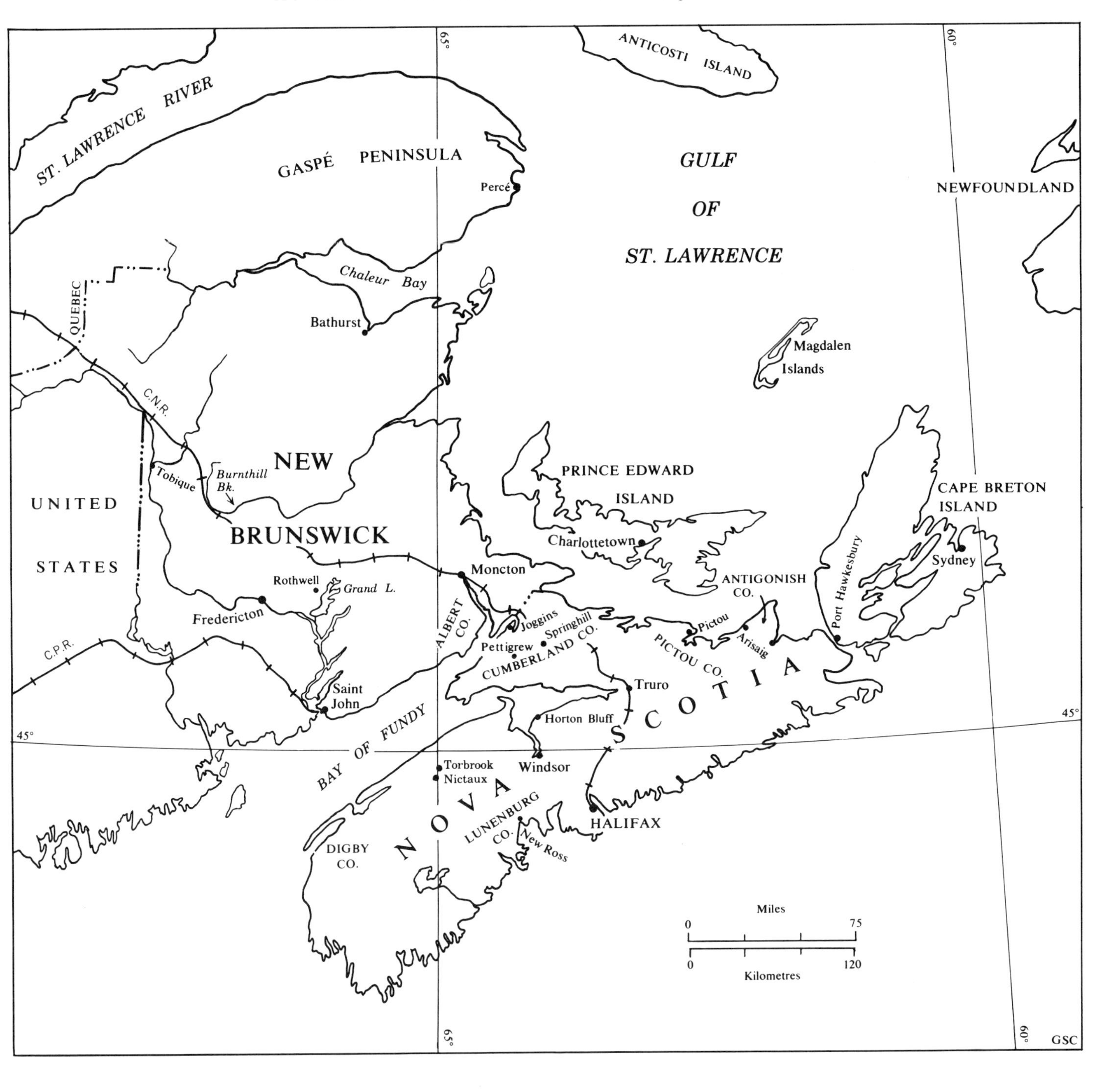

from ministers and other members of Parliament seeking advice on mining matters, for "both the late and present cabinets seem much interested in mining!"[8]

Dawson also had frequent dealings with his fellow officers in other branches of government—on exploration and mapping with the Surveyor General, Edouard Deville; with the Queen's Printer, S. E. Dawson (no relation); with the Dominion Architect over the Survey building and plans for a new national museum. With these and others he discussed such varied subjects as the Yukon Territory and British Columbia, the Bering Sea seal fishery question, territorial boundaries, the protection of the caribou and wood bison, and Indian policy. One subject of moment with which he became involved in 1895 was the possible entry of Newfoundland into Confederation. While negotiations were in progress, James P. Howley, director of the Newfoundland Geological Survey, wrote inquiring about its—and his—position under the new union. Dawson accepted the proposal with enthusiasm, and advised his minister that the Survey should naturally be extended to Newfoundland; that its role should be written into the union agreement, as with British Columbia; and that since the island had an established, reputable survey, Howley should be invited to carry on as a member of the Canadian Survey. But though Dawson was willing, the governments were not. After a disappointing negotiation characterized by Canadian niggardliness, Newfoundland and its survey resumed their separate ways, and Canada did not gain its tenth province for another 54 years.

Early in his directorship, Dawson faced the first major change of government since the Survey moved to Ottawa. The election of June 23, 1896, brought one casualty. Dismissed at the end of April 1897 "for political partisanship" was L. L. Brophy, the convivial assistant in the Section of Mines who had worked actively for the former government in the election. "To do him justice he does not 'squeal,' as many other active partisans do when their time comes. He says he is surprised he was allowed to stay so long," wrote the Montreal *Witness*.[9] It was rumoured that Dawson's own position was none too secure, because of Sir William Dawson's influence with the Conservative party and his role in the Manitoba School question. But to all appearances the director was not affected, and with the exception of Brophy the Survey experienced only minor interferences from the change in government. Dawson received a note on one occasion from the minister complaining that officers were employing men and securing supplies from people who were active opponents of the government. He went up to the House of Commons, found Sifton, and "Gave him list of requirements for parties in the way of supplies, men, etc. in different parts of Dominion. As far as possible asked his instructions in order to avoid political complaints. All this sort of thing is quite new in our service and very disgusting."[10] Filling the appointments to field parties brought an annual tug-of-war between university professors trying to find valuable experience for their geology students, and politicians trying to place protégés on the government payroll, with the latter usually gaining the day. Dawson commiserated with Professor A. P. Coleman on their common problem:

> I quite realize, however, the advisability of trying to get the best men and graduates when possible for such summer employment, and it has not been my fault that more of them have not been taken, for in that way one could keep track of likely men to fill vacancies or undertake special work. Unfortunately, as you are no doubt well aware, other considerations frequently come in, and first year students and other young fellows with very little education of any kind are often those most favored by fortune.[11]

Dawson also experienced gentle, not at all unreasonable, requests from Prime Minister Laurier to try to find another French Canadian to succeed to the place of N. J. Giroux, who died in 1896. The lone candidate did not meet the requirements of the Survey act, however, and in the end Sifton accepted Dawson's recommendation and appointed R. W. Brock from among the temporary employees. Later, in 1899, Laurier sent around another candidate with a medical education but no training in either geology or drafting. When Dawson suggested he would qualify only for a temporary, low-paid position, the application was withdrawn. Laurier was not the least insistent in either case. As Dawson recorded in his diary, "Laurier is reasonable, but seems to think that French Canadians have been barred from this service. Could not tell him that insofar as this the case it depended on the fact that they were not educated in science, or with few exceptions."[12]

More serious was the failure of the new government to treat the Survey at all generously as regards the annual appropriation, though admittedly in this respect it was only copying the previous administration. Dawson took over in a time of acute difficulty, and scarcely had he begun wiping out the accumulated deficits of past years when the new government seemed to frustrate all hope of substantial increase. In no year in which Dawson was in charge did the annual expenditure of the Survey attain even the level of 1894-95 ($117,456), the year of his appointment. Salaries and wages remained stationary, as did the numbers of staff, comprised of 31 or 32 permanent and 15 to 20 temporary employees. The expenditure on field operations never quite regained the level of 1894-95 ($68,692) either, being as little as $45,227 in 1895-96 and only $61,941 in his last year. It is remarkable, therefore, that the Survey persisted in spending over $7,000 a year for five years on borings in northern

Alberta. The high costs of some of the reconnaissance expeditions, notably Tyrrell's of 1893 which was over $6,500, also produced outcries against such enterprises and encouraged work closer to settled areas.

X-4 J. B. TYRRELL IN MILITIA DRESS
Tyrrell was a captain in the Governor General's Foot Guards, Ottawa.

What with outstanding liabilities from that year, 1893-94, plus a reduction in the operations grant in 1894-95 from $60,000 to $53,000, Dawson actually had only $33,910 to spend on this side of the work before he too would be going into debt. In his own words,

> With this small sum a very limited amount of field work was accomplished in 1894-95, very little printing could be undertaken, and with all possible economy a deficit is inevitable at the close of the year. It will in fact only be possible to pay the salaries of temporary employees to the end of the year by holding over unpaid accounts due the Printing Bureau. . . . A large mass of reports and maps in MSS have accumulated in the meantime, for which enquiries are constantly being made by the public.[13]

Just when he had succeeded in reducing the deficit to something like $8,268, the news came that for the coming year, 1895-96, his appropriation would be cut to $45,000. He had to interview Prime Minister Mackenzie Bowell to have the Athabasca boring continued. A little later the project of sending Low back to Labrador was before the cabinet, from which the minister, T. M. Daly, "phoned down that he had better go to Labrador if expense can be covered by our vote. Replied that it could, but would mean taking about $1,000 from Printing."[14] And the same hand-to-mouth situation continued. The next year Dawson had to give his personal cheque for $175 to pay for a 72-pound meteorite brought in from Thurlow, Ontario, since "the Survey has no available funds at present."[15]

The constant shortage of funds had very adverse effects on staff morale, and Dawson's era saw several serious resignations from the Survey. Adding to the critical situation was the fact that many important, attractive opportunities were arising in Canada for men of their ability. The growth of the universities, and especially the expanding mining industries, opened many new positions to men whose training and expertise were of immediate value. It is surprising, indeed, that the Survey did not lose all its trained personnel, for the salaries, already low, continued to fall further and further behind the rapidly rising pay scales in industry. The raises for the technical staff were annual for the most part, amounting to $50, but these were not assured; no increase had been given for 1892-93. This was not rectified until 1894-95 when a double increase of $100 was given, which still did not make up for the $50 lost in each of the previous two years. When the officers were again overlooked in 1896-97, nineteen of them signed a memorandum to Dawson expressing their discontent: "We submit that the continued uncertainty about our salaries is discouraging, and we ask that some steps be taken to improve our position in this regard."[16] Most of the men received no increments either that year or the next, and only in 1899-1900 and 1900-01 were increments of $100 and $200 given to restore some semblance of fairness. The assistant directors, Bell, Hoffmann, and Whiteaves, remained at a standstill since they were already at their $2,400 maximum; as did Dawson, who had been given a raise to $3,200 when he became director, still well below Selwyn's $4,000.

Though Dawson was not to blame for the unhappy situation, he was sometimes criticized for failing to help his staff secure their proper due. Mrs. J. B. Tyrrell in her autobiography remarks that Dawson "for all his brilliance as a geologist, seemed to lack understanding of, or interest in, the members of his staff and their welfare. He was rigid in his economies and would not make any recommendations for promotion or for increases of salary, even when they were overdue by reason of long and good service."[17] No doubt Edith Tyrrell saw matters from her husband's point of view, not from that of the much-harried director, trying to make ends meet. Dawson's diary often reflects his concern: "Cole, who has been for some time working under Ingall, informs me that he is about to resign, having got a much better paid appointment in B.C. Constant difficulty of this kind in endeavouring to run office with professional men at starvation wages—most detrimental to good work."[18]

As a result, Dawson's directorship saw a considerable turnover of staff. N. J. Giroux died in 1896, and L. L. Brophy was dismissed in 1897. The resignations for those years were more serious: H. Brumell to join a company interested in graphite mines in Buckingham, Quebec; H. Y. Russel to take up work in western Canada; and D. I. V. Eaton, the topographer who accompanied Low to Labrador in 1893-94, to join the Royal Canadian Regiment. In 1898, besides Cole, W. F. Ferrier, the Survey's petrographer, resigned to enter a mining company at Rossland, B.C., at a reported salary of $6,000 per annum, nearly four times his current $1,850 salary.

Receiving much publicity was the resignation at the end of January 1899 of the celebrated explorer-geologist J. B. Tyrrell, occasioned by dissatisfaction over his position and prospects, as well as his low salary. The situation began almost a year previously, in February 1898, when McConnell was offered the position of provincial mineralogist of British Columbia at a salary of $4,000, more than double his current $1,850. Dawson, anxious to retain him, induced him to stay by offering a token increase of $200. On April 21, McConnell finally made his decision to remain. Even that modest increase caused some discontent among such men as R. W. Ells and H. Fletcher, whose own salaries were now no better than McConnell's who was some years their junior. Tyrrell, who was McConnell's contemporary and had been receiving the same salary, was most put out, as Dawson recorded (May 7, 1898): "Tyrrell seems considerably miffed about McConnell's getting increase which makes his salary greater than Tyrrell's. He so expresses himself. Tried to explain the special cause and the way it looks to outsiders."[19]

To add to Tyrrell's dissatisfaction, McConnell's decision meant he resumed his assignment to direct the Survey's belated entry into the exciting Klondike country in 1898, which Tyrrell had hoped for himself. Now Tyrrell would have to go as a sort of second-in-command and supernumerary to McConnell, something that did not suit the ambitious, publicity-seeking Tyrrell in the least. He went to the Yukon in a dour mood, and by the autumn had decided to become a mining consultant, mine manager, and investment counsellor in the Yukon. Dawson was annoyed at Tyrrell's letting the resignation get into the newspapers before there was any opportunity for reconciliation. Tyrrell had a rather disappointing ten years in the Klondike, but later, in the 1920's, he became interested in the new mining boom in northern Ontario, took a leading share in the management of the Kirkland Lake Gold Mining Company, and became a millionaire, able to indulge his interests in early fur trade explorers and in other professional and educational causes.

Occurring also in 1899 was the departure of the chief draftsman, James White, who assumed the position of geographer in the Department of the Interior. White, who had been appointed a member of the Geographic Board on behalf of the Survey, was offered the new appointment at an increased salary, somewhat to Dawson's annoyance at having his staff lured away by another government department. Dawson could only delay the move until White had completed his current work; his place in charge of the mapping division was taken by C. O. Senécal.

Another serious, and presumably unexpected departure, was that of A. P. Low, announced in a letter of February 17, 1901:

> I herewith beg to tender my resignation as geologist from the technical staff of the Geological Survey Department. It is with great reluctance that I take this step which will sever the very cordial relations which I have always had with yourself, Sir, and with the other members of your staff and I will always look back with pleasure upon the many happy if unpecunious years spent with my scientific confreres of the Geological Survey Department.
>
> My present reason for leaving the Department is due to an offer made me by the Dominion Development Company to undertake the location of their mineral claims on the Hudson Bay. The salary and other inducements offered are greatly in excess of my present salary in this department, and as I seriously believe in a brilliant future in the mining possibilities of that region, I have considered it my duty to my family, to take advantage of the present chance, and to try to earn a competency for them, which would be impossible in my present position; for as you well know the present salaries paid to the technical staff of the Department allow only a bare living to the members, and do not compare at all with what any member of the staff might expect if he were to engage in any outside mining business.[20]

Low, who had spent 1900 at the exhibit in Paris, applied for three months' leave, his resignation to take effect at the end of June 1901. He promised to complete the report of his previous explorations of 1898-99 along the east coast of Hudson Bay, and turned in the report to Bell, who had succeeded Dawson in the meantime. Also coming in 1901 shortly after Bell assumed the acting directorship, were the resignations of J. C. Gwillim, who had succeeded Tyrrell in 1899, to take effect May 31, 1901, and J. McEvoy, on August 31, 1901. McEvoy's position with the Crowsnest Pass Coal Company carried a salary of $5,000. Needless to say, these departures received a great deal of notice in the press, all of it siding with the Survey and critical of the shortsightedness of the government—and all of it to no avail.

X-5 JAMES WHITE
Dominion Geographer, 1899-1909
Secretary, Commission of Conservation, 1909-20

X-6 C. O. SENÉCAL
Chief Draftsman, 1899-1931

Against these departures there were relatively few additions to staff but several promotions from the temporary to the permanent list, including D. B. Dowling, J. F. E. Johnston, and C. O. Senécal in 1896, Brock in 1897, and J. M. Macoun in 1898. Newcomers added to the temporary staff were John McLeish, who succeeded Brophy as statistical assistant in the Section of Mines in 1897, Theo C. Denis, who replaced Cole as assistant to Ingall in the same section (1898), and four topographers and draftsmen: Joseph Keele and W. W. Leach (1898), and J. Mackintosh Bell, and W. H. Boyd (1900).

In view of the bad financial situation and the loss of qualified staff, Dawson, much against his will, was obliged to confine his activities to geological work and to neglect other Survey concerns. Ethnological work, in which he had considerable personal interest, was one casualty. As he told the newly-elected M.P. from Edmonton, Frank Oliver, in September 1896:

> I quite agree on the advisability of the Government undertaking some Ethnological research, particularly in the west, and have consistently advocated this as well as worked to some little extent at it myself. Mr. Wilson favours the idea of connecting such work with this Department, as being the principal Scientific Dept. of the Gov't. There is some reason in this, but the money voted for geological work and for maintenance of Museum, etc., could not be expended in ethnological research. I fear a special vote would be necessary and doubt whether the Gov't will regard this favourably.[21]

He found it impossible to make any grants towards ethnological work or to purchase ethnological and anthropological collections, however greatly he feared these collections would be lost to the aggressive, well-financed museums of the United States and Europe.

Even more vexing, because unavoidable, was the continuing expenditure of the Survey's limited resources on topographical surveying, which Dawson called "the principal difficulty in connection with the geological work."[22] Something he could control, as far as the limited funds allowed, was the choice of work. He strongly favoured systematic mapping in settled areas and the districts contiguous to them, especially those shown by earlier reconnaissances as likely to be of economic importance. Reconnaissance surveys, he felt, were of little real help to the mining industry and were not an effective use of the Survey's limited resources. J. B. Tyrrell's exploration of Keewatin had just been completed amid complaints that its geological results were slight, its economic value negligible, and the cost enormous. His second expedition, of 1894, was made possible only by the support of the Governor General, who was said to have contributed a large part of the cost. In a private letter to J. W. Tyrrell, who had accompanied his brother on the first expedition of 1893, Dawson unburdened himself on the subject in no uncertain terms:

> In regard to your remark about the Survey receiving a large share of credit, you may possibly be aware that the only "credit" heretofore resulting from the considerable popularization of these northern trips has taken the form of grumbling that the Government should send expeditions to such regions when various districts nearer home still remain to be examined![23]

Yet the Survey, notwithstanding, found itself sponsoring two field expeditions on either side of Hudson Strait and Hudson Bay, and of Great Slave and Great Bear Lakes and the country between them. Undoubtedly in undertaking these, Dawson was responding to nationalist pressures of the day.

In his disposition of the field parties, Dawson deployed his limited staff in as many different parts of Canada as he possibly could, fairly evenly across the country, nearly always in or near regions where mining was in progress or was a distinct future possibility. He saw the Survey as advancing the understanding of the geology of Canada, but he felt the work might as well be carried on in accessible districts which presented fewer logistical difficulties and less expense than remoter sites, and which were likelier to be immediately helpful to the development of Canadian mining. Though he selected his field projects with such practical considerations in mind, he never compromised the scientific character of the program. Like Logan and Selwyn, he firmly believed progress could only be based on the discovery of truth, but unlike Selwyn, Dawson was careful in his reports to show the connection between so-called scientific geology and the progress of the mining industry. Besides, he genuinely aimed at selecting field projects that were at once of practical and of scientific merit, for that was one way he could make his limited forces go further.

Another way was in being careful in his choice of men, and in assigning the right men for particular tasks. Bell, who was exceptionally gifted at exploration, was allowed to continue in that line of endeavour through most of the period, being sent to Hudson Bay, Great Slave Lake, and Bell and Nottaway Rivers, though not to the Klondike where he dearly wanted to go. The Klondike was reserved, instead, for McConnell, who had done good work on Cordilleran geology and was Dawson's successor in that field. True, Bell spent two seasons in the Michipicoten district of Ontario, because of which he complained bitterly that he was being persecuted. But Bell's assignment was partly the result of his lateness in going to the field which made it necessary to place him in an accessible district, and partly his tardiness in completing the 4-mile map of what was developing into an important mining district. Dawson kept W. McInnes on his work on the Precambrian west of Lake Superior, and Fletcher and Faribault in Nova Scotia, since all three were making satisfactory progress in mastering the geology of their districts. His care in fitting men to the best possible positions was shown, too, in his use of McEvoy, Keele, and Leach, who were recruited as map-makers, worked as topographers with field parties, then finally undertook some geological work of a not-too-difficult kind in which careful measuring and mapping was the main need, or in areas where the basic stratigraphic problems had been solved.

He also employed part-time men, mainly from the universities, both as an economy measure and to assist with certain assignments where the men had previously worked in the field or were specialists in the particular geological problem. He continued to make use of Professors Adams and Bailey, and in 1900 hired Professor J. A. Dresser to make a petrological study of Shefford Mountain in the Eastern Townships. In 1899 he persuaded Professor A. Osann, a noted petrologist from Mulhausen, Germany, to come to Canada for the summer to make a special petrographic study of the Grenville gneisses and to train Canadians in the latest techniques. Comparable was the employment of the noted American vertebrate paleontologist Henry F. Osborn to make preliminary studies of the great dinosaurs from the Mesozoic of Alberta, and to train Lambe to become a specialist in that field.

Dawson also encouraged consultations among geologists working on a particular area or field. Osann's work was of considerable help to R. W. Ells, E. D. Ingall, and A. E. Barlow in their particular tasks in the Precambrian of the Ottawa region. The director's visits to parties in the field (for example to Nova Scotia and New Brunswick in 1897) was for the purpose of bringing his advice to bear on problems encountered by the men at their work. He also encouraged visits by specialists in petrology or paleontology, to help solve some of the field problems. As he wrote, "An extension of consultary relations of the geologists working in adjacent fields, is also to be desired and it is hoped that this may be more fully attained in the future."[24]

Of course, sending one geologist to examine the work of another sometimes led to disputes when they failed to agree, and occasionally they fell to quarrelling about who was right. What should the director do in the event of such differences of opinion? Dawson accorded with his predecessors that it was incumbent on him to be the final arbiter, since he ultimately was responsible for everything published under the Survey's name. All work, Dawson felt, had to meet a double test, of having "a permanent value to those actually interested in employing it upon the ground, while at the same time commanding the respect and confidence of the scientific world."[25] This applied especially to the Survey's maps and reports, which had to be accepted as authoritative by the scientific and professional world as well as by the public. Dawson stood behind everything that went into the reports, taking care they were lucid, well-organized, thorough, and above all, as accurate as he could make them. Hence he mastered the geology of Canada as completely as he could, and spent as much time as possible on inspections in the field to see for himself:

> It was found possible also to make some excursions in the field, in connection with the general control of the work there in progress and for the purpose of correlating the surveys and observations of different members of the staff. This forms a very necessary part of the scheme of the Survey, which implies a uniformity of plan in the definition and mapping of the formations of all parts of Canada. It is recognized that such supervision should not be confined merely to the published matter of the several reports, but should include an actual knowledge of the main facts, of such a character as to enable the Director to assist the individual workers in reaching a concurrent rendering of their results for presentation to the public.[26]

In sum, Dawson's direction was characterized by an orderly, disciplined approach; by balance, moderation, and common sense that rejected extremes; and by uniformly high standards of achievement without ostentation. He preferred doing a little very well, to doing more that attained only to the level of mediocrity. Schooled by life in the virtue of facing up to one's burdens, he did not struggle openly against the financial restraints under which the Survey suffered, and the limitations these placed on the work. He shrank from playing politics, empire-building, or striking out for popular acclaim. Perhaps he stood too much on his dignity, and did not capitalize enough on his great reputation and popularity to force better treatment from a rather hard-nosed minister who respected little besides power in those with whom he had to deal.

For Dawson had amassed an immense store of credit in the world of science and in the popular imagination, on which he might well have drawn—though he was quite incapable of this. Because he did everything so well, honours and recognition came easily to him. They included degrees from Princeton as well as Queen's, McGill, and Toronto, the Bigsby gold medal of the Geological Society of London (1891), election as a fellow of the Royal Society, and in 1897 the Patron's medal of the Royal Geographical Society. He was elected president of the Royal Society of Canada in 1893, and president of the Geological Society of America in 1900, to which he read a paper on "The Geological Record of the Rocky

X-7 GROUP OF GSC OFFICERS AT WEST BAY, PARRSBORO, N.S.
Left to right—G. F. Matthew, Hugh Fletcher, W. F. Ferrier, A. P. Low, and G. A. Young. Probably about 1898, the year Ferrier left the Survey and Young was a youth of 20.

Mountains Region in Canada," that represented the final summation of his work in Cordilleran geology, his main field of study.

That paper, in fact, was his farewell to scholarly work, for he died almost as soon as it was published. On Friday, March 1, 1901, he went home ill with a mild attack of bronchitis, which suddenly became acute the following afternoon, and he died early that evening, Saturday, March 2. His death was attributed mainly to his physique which prevented his lungs from expanding properly, and made him liable to very serious effects in the event of pulmonary or bronchial trouble. Added to this was the sedentary life he had been forced to lead since becoming director.

The many tributes at the news of his sudden passing indicated the high respect in which he was held everywhere, and the great loss his death meant: "It is indeed now sad to think that all he had done was after all but 'an earnest' of all he might have done had he been spared to the great world of science in which he had already won so honoured a name."[27] Another aspect of his contributions was summed up by the *Geographical Journal:* "Dr. Dawson's whole life was a remarkable instance of entire devotion to duty, to the promotion of his chosen science, and to the welfare of the great Dominion whose resources he did so much to discover and render available."[28] And he was not mourned by the educated world alone. In British Columbia, where he was particularly esteemed, he was already something of a legend for his arduous labours in the field, so it was natural for Joseph Ladue in 1896 to name his townsite Dawson City, though Dawson had never been within 150 miles of the place. Dawson was something like the patron saint of the prospector and developer, whose sentiments were expressed in this ode to "Dr. George" by Captain Clive Phillips-Wolley, published in the *British Columbia Mining Record* for April 1901:

> Hope she has fooled us often, but we follow her Spring call yet,
> And we'd risk our lives on his say so and steer the course he set,
> Down the Dease and the lonely Liard, from Yukon to Stikine;
> There's always a point to swear by, where the little doctor's been,
> Who made no show of his learning. But, Lord! what he didn't know
> Hadn't the worth of country rock, the substance of summer snow.
> I guess had he chosen, may be, he'd have quit the noise and fuss
> Of cities and high palavers to throw in his lot with us.
> He'd crept so close to Nature, he could hear what the Big Things say,
> Our Arctic Nights, and our Northern Lights, our winds and pines at play.
> He loved his work and his workmates, and all as he took for wage
> Was the name his brave feet traced him on Northland's newest page—
> That, and the hearts of the hardfists, though I reckon for work well done,
> He who set the stars for guide lights, will keep him the place he won,
> Will lead him safe through the Passes and over the Last Divide,
> To the Camp of Honest Workers, of men who never lied.
> And tell him the boys he worked for, say, judging as best they can,
> That in lands which try manhood hardest, he was tested and proved A Man.[29]

* * *

ROBERT BELL, who became acting director in 1901, was a complete contrast in almost every respect. Nearly ten years older than Dawson, he was sixty in 1901; his career in the Survey began as a boy of fifteen in 1857, to become full-time in 1869, in Logan's last year as director. He had explored more of Canada than any other man of his generation, and his knowledge of the country—its geography, natural history, travel conditions, and inhabitants—was unrivaled. He was serious, hard working, and persistent, always gave his best to the Survey and was thoroughly devoted to its interests. He was exceedingly ambitious, too, and felt that Logan's mantle belonged of right more to himself than to either Selwyn or Dawson. In Bell's view, both men had dragged the Survey along undesirable paths, far from Logan's practical objectives of mapping the national domain and assisting the mining industry in every possible way. He had described Selwyn as "ignorant, incompetent, and unscrupulous,"[30] and he thoroughly hated Dawson, his lifelong rival. Dawson's succession seemed to ensure the long continuance of Bell's frustrations, but by sheer accident, in 1901 Bell practically achieved his ambition. All that was needed now to make his appointment a proper one as director was for him to commend himself to the government through his good management of the Survey.

While Bell's opinions of his qualifications and past services did not seem to have been greatly daunted by his failures to achieve the recognition he craved, his long history of disappointments had given him a strong persecution complex which his uncertain position was bound to aggravate. His setbacks, he had convinced himself, were the results of other people's actions and never of his own shortcomings. The failure to obtain a professorship at the University of Toronto in the seventies arose from prejudice against a native-born Canadian and betrayal by persons he trusted. The disappointments in his Survey career were the result of Dawson's machinations and the management of Selwyn by the elder Dawson in his son's interest. He was even convinced that Dawson's wishes for Bell to undertake a study of the economic minerals of northern Ontario or to finish his mapping along Lake Superior, reflected a desire to cast him in the shade, and he complained to Sifton that Dawson was deliberately pushing him off again into "the most obscure country possible to do the least important work of anyone on the staff."[31]

Now that he had at last become acting director, he was convinced that the government's continual failure to appoint him director arose from libels being spread against him—that he was out of sympathy with the Liberal government, that he could not manage the department efficiently or effectively, or that he was unable to get along with the staff. To counter these sinister forces and promote his rightful claims, Bell sometimes was driven to unscrupulous, vindictive courses of action, dabbling in politics for his own ends. He carried his complaints against his superiors to the press or to the minister, and after he became acting director, to the press and the M.P.'s against the minister. Often he was his own worst enemy, too outspoken, too unreasonable in his arguments, too unrealistic with his demands, and sometimes so crude in his tactics as to alienate would-be supporters. He was pathetically eager for recognition, to the point where he abandoned any reserve or show of modesty.

Bell's basic insecurity, aggravated by his difficult situation as acting director, intensified his search for personal honours and for public acclaim. To demonstrate his talents beyond a shadow of doubt, he worked hard to win a good public image for the Survey. Where Dawson seemed content for its good works to advertise themselves, Bell's annual summary report was prefaced with a justificatory piece designed to convey the Survey's great accomplishments forcibly to the reader and demonstrate its entire devotion to the public well-being. To prove his superiority over his predecessor, he set out at once to establish new records of performance. "During the past summer," he wrote of the 1901 season, "31 parties were out, as against 13 employed the previous season, and the amount of work accomplished is believed to have far exceeded that of any previous year in the history of the department."[32] To hammer home the message, he himself took to the field then and in most subsequent years, adding this burden to the other onerous tasks of managing the Survey, but demonstrating that he "tried harder."

Bell also tried hard to win the loyalty and devotion of his staff by securing higher wages and salaries for all, as well as more generous expense allowances to relieve some of the hardships of field work and permit greater attendance at conferences and meetings, while he accorded more recognition to the non-geological part of the staff. Thus he rallied most of the employees to his side and the series of resignations was halted, though some of the younger, ambitious specialists, perhaps, did not feel they received sufficient consideration from the acting director. Bell also carried his desire to build up a body of support to the point sometimes of backing his friends on scientific questions on personal grounds, irrespective of the merits of the cases themselves. The results were cries of favouritism and unfair treatment, and stories of the Survey racked by jealousy and its discipline undermined, that militated against Bell's hopes of being appointed director.

Given the necessity of enhancing his own reputation, together with his lack of sympathy with the work of his predecessors, it is not to be wondered that he set out to redirect the program in line with his own ideas and what he took to be the desires of the government and people of Canada. The time was opportune. The nation was entering on an age of rapid expansion and industrial growth, climaxed by the building of two transcontinental railways and a number of other important lines that opened up vast tracts of land and their resources. The government, too, had abandoned its earlier frugal ways and had become carried away by the torrent of free-spending optimism it itself had unleashed. The Survey began receiving a share of the new largesse, though nowhere in proportion to the growth in mineral production in the country. Nevertheless, where the expenditure under Dawson averaged only $111,000 per annum, under Bell it averaged $145,000 per annum and reached as high as $170,941 and $170,484 in his last two years.

Much of this improvement resulted directly from Bell's requests for extra money for particular purposes. Almost immediately, he wrote Sifton for a special increase in the mapping appropriation since the field work had got far ahead of mapping and publication of results. Some $6,000 was earmarked for the purpose in 1901, and extra draftsmen were hired to catch up with the work. The following year the extra allotment was raised to $9,000 "for the express purpose of engraving a considerable number of Geological Survey maps which had fallen in arrear during the last few years,"[33] besides which $4,798 was spent for compiling and plotting. Still larger sums were allotted to these purposes in later years. Again, Bell turned the demand for extra work in British Columbia into a special grant of $19,000 in 1903-04 and succeeding years, which enabled the work in the Cordillera to be expanded, while work already being done there could also be credited to this special account, leaving more for the rest of the country too. He also succeeded in 1906 in obtaining a special grant of $3,000 "to provide for the cost of an estimation of the present value of the gold gravels of the Klondike,"[34] under which McConnell was able to make an interesting study of the gold placers of the low and high levels in the area.

Bell also obtained special grants for scientific equipment and facilities, for special library purchases, and to secure large collections of anthropological material for the museum, including one in 1904 of D. H. Price of Aylmer, Ontario, that Dawson had been unable to purchase. In this way a valuable collection, consisting of some 9,000 specimens, was retained in Canada for the national museum that was now building. In all these ways, by continually pressing the government for grants for small, specific objectives, Bell succeeded in raising the Survey's appropriations piecemeal, until he could show a very considerable improvement over the previous situation. While some of the grants carried obligations that detracted from the overall program, the net effect was that the Survey could expand its operations in a great many directions.

He also successfully fought off a move in 1902 by James White, now the geographer in the Department of the Interior, to consolidate map-making in a geographical branch under his leadership. The Survey was up in arms at the suggestion, as was Bell, who felt it completely negated the work and character of his department. He regarded the geologist, his professional assistant, and the draftsman, as parts of an integrated team, the first two of whom worked in the field on both the topography and the geology, then became a trio after returning to the office: "The presence of a number of geologists and topographers under the same roof is mutually beneficial and promotes an *esprit de corps.* The men take a pride and an interest in their work, and they regard the professional credit which it may bring them, if well done, as of more importance than the pecuniary remuneration they receive."[35] White's proposal was not implemented, and during Bell's administration, thanks to the efforts of the drafting staff, the backlog of maps was somewhat reduced.

Also added to the technical staff and the Survey's

X-8 ROBERT BELL
Acting Director, 1901-06

budget—not with Bell's approval, however—was a metallurgist. Donald Locke was hired expressly at Sifton's wish in the autumn of 1902 in connection with an investigation of the Sudbury nickel-copper mining camp that A. E. Barlow was making, with the financial backing of the Sudbury mine owner Ludwig Mond and the approval of the minister. Sifton had an interview with Dr. Hoffmann, the veteran chemist and metallurgist, to arrange for Locke's use of the assay room, and later intervened personally and returned Locke to the office when Hoffmann tried to put a stop to his using the assay room. Despite Hoffmann's long years of service and his rank of assistant director, Sifton had him ignominiously suspended and fined four days' pay. Later, when Locke resigned in 1903, M. F. Connor was appointed in his place during Bell's absence and without his knowledge. Bell, after conferring with Hoffmann, tried to dismiss Connor, explaining to Sifton in a letter of March 11, 1904, that it was the practice to send the small amount of metallurgical work out to "established plants" at McGill and Queen's. In reply he received a ferocious, menacing letter from the minister:

> It was your duty as acting Deputy to ascertain whether I desired the continuance of his services before taking action which would have resulted in dispensing with his services had my attention not been called to it. . . .
>
> I may add that I find in your letter evidence of the lack of a satisfactory spirit and lack of a desire to loyally carry out and make the best of instructions, which, unfortunately, is having an injurious effect upon the work of the Geological Survey.[36]

These strong actions on Sifton's part would seem to indicate, besides his bad temper, a deliberate policy of using the Survey to hold in reserve a metallurgical specialist as part of his plans for reorganizing the federal government's role in the field of mining.

Another area in which Bell made a special effort was the publications. There was an obvious need for a full-time editor to relieve the directors and their secretary, Percy Selwyn, of a task that occupied much of their time. Bell at last secured such a man, Frank Nicolas, who apparently dropped in, without any pre-arrangement, on December 5, 1904, went to work, and after a week's trial, was hired. Bell envisaged a very responsible role for Nicolas, not merely copy editing, but checking the accuracy of the facts and the desirability of the conclusions expressed. Low added the editing of the maps, and requisitioning of supplies and services to Nicolas' other responsibilities. The large *Annual Report* volume, containing the director's summary report and the Section of Mines' annual report on mineral production, as well as a varying number of individual reports, continued to be the main effort. The summary report, issued as a separate as early in the new year as possible, included reports of the work of the various divisions, but consisted mainly of ex-

X-9 PART OF THE ORNITHOLOGICAL COLLECTION, THIRD FLOOR, SURVEY BUILDING

tracts from the reports of field surveys of the past year. In Dawson's day these were usually presented as extended quotations from the officers' submissions, apparently selected and edited by Dawson. Bell gave his officers more freedom to prepare the whole of their reports in their own words and under their own names, with the result that they varied considerably in length, approach, and treatment. He also gave them greater leeway to publish what they wished, with the result that in at least one case (J. M. Macoun's report on the agricultural prospects of Peace River district, 1903) Bell was put in a very embarrassing position for not having censored or suppressed the work as being injurious to the government's immigration program.

A source of great and growing complaint, as 20 years before, was the lengthening delays in the appearance of the annual report volumes and the Section of Mines reports. There were only three annual report volumes in the Bell years, for 1901, 1902 and 1903, and 1904, and these were published in 1905, 1906, and 1907, respectively, the delays arising from the tardiness of some of the contributors. To take the last of these, Volume XVI for the year 1904, the summary report for that year was dated 1905, four of the field reports bore 1906 imprints, while the Section of Mines' report, dated February 1905, treated developments for the year 1903, and some of its latest statistics were for 1901. By the time that final volume for 1904 appeared in 1907, its successor, the summary report for 1905 had been out almost a year, and that for 1906 was about to appear, if indeed it had not already done so. Obviously the need was not only for a full-time editor, but for a drastic revision of the publications program.

Besides broadening the structure of the department by adding new staff and placing mapping, editorial, and other operations on a more satisfactory basis, Bell also concerned himself with the critical problem of accommodations. Here again, he was the beneficiary of the good times and the new-found generous mood of the government. Dawson had given a great deal of thought and effort to the problem, but with no result. Beyond question, the present headquarters were most unsatisfactory, for the building was grossly overcrowded, and the basement and some of the upstairs rooms were liable to flooding. The possibility of its collapsing from age and from the heavy weight of materials stored in it led the Department of Public Works in 1897 to put up sixty posts to prop up the floors and give added support, so that any visitor "who happens to go into that section where these props are most numerous can hardly fail to have the sensation of being in a standing forest of dead timber."[37] Because of a fear of fires spreading from adjoining properties a fire hydrant was installed in the courtyard, and pipes fitted with hose attachments were run to every room in the building. To relieve the overcrowding, at least for the present, Dawson had succeeded in having the government rent the adjoining building to the north, though delays in remodelling the extension confined its use to storage space alone. Dawson realized this Baskerville annex was only a stopgap; he put his hopes for a long-term solution on the construction of a new building on another site to meet the Survey's space requirements and serve as a proper home for Canada's national museum, which the Survey's museum was rapidly becoming both in fact and in the public mind. He was in frequent communication with the Dominion Architect and the Minister of Public Works about plans for a new museum, his preference for the site being the north end of Major's Hill Park, in the vicinity of Nepean Point, overlooking the Rideau Canal and Ottawa River.

Nothing had been done beyond a few preliminary sketches when Bell took charge. But the death of Queen Victoria on January 22, 1901, was followed by a government decision to build a new museum as a memorial to the late Queen. On May 20, a sum of $50,000 was put on the supplementary estimates "to authorize commencement of the construction of the Victoria Memorial Museum."[38] Though annoying delays continued, the work gradually was put under way. A new site was chosen in November 1903, the construction contract was let in November 1904, and completion was promised for December 1907. Accordingly, the Survey bent every effort to round out the collections in anticipation of moving to more spacious quarters, while Bell began negotiations with the minister respecting the administration of the new national museum, particularly the matter of the curatorial and security staffs' positions in relation to the Geological Survey department.

In the meantime, however, something had to be done about the present quarters, notably the overcrowding. A committee of staff was appointed to reallocate the space made available by the occupation of the annex for offices. It reported that eight offices could be provided in the remaining two floors (the third floor had been leased for some years and was already being occupied), and suggested that the seven men in the "long room immediately at the rear of the library" could be transferred there, with the eighth office reserved for a photographic dark room, at Senécal's suggestion: "In fitting the Annex for office accommodation I would suggest that a room on the level of the drafting office be reserved for photographic purposes (blue prints, enlargements etc.) including a small dark-closet for developing. Photography is now indispensable in a drafting office."[39]

X-10 MRS. JANE ALEXANDER
GSC Librarian, 1908-12

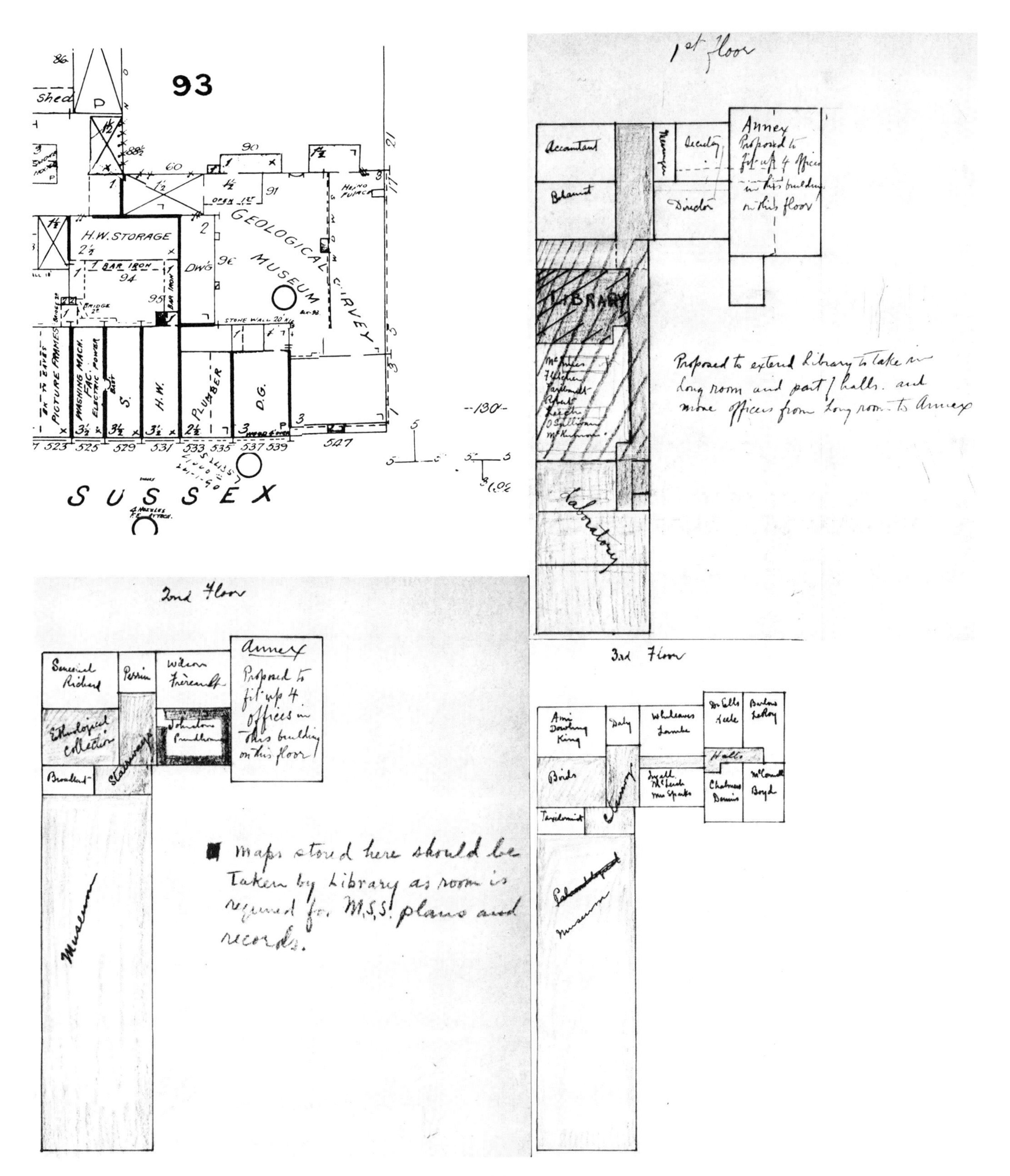

X-11 GROWING PAINS AT SUSSEX STREET IN THE BELL YEARS

Top left—Part of the insurance plan of the City of Ottawa, updated to May 1895, showing the ground plan of the area around the Survey building, which was 160 feet long and had a frontage of 60 feet at the time. Expansion entailed occupying the building "D.G.," with a 35-foot frontage and 50-foot depth.

Top right and bottom—Sketches of the proposed rearrangements. The details of the individuals' offices, collections and facilities assist one to visualize the physical situation of the Survey in the early years of the present century.

It was the disgraceful situation in the library, above all, that had precipitated the committee's report. As Mrs. J. Alexander, the assistant librarian, described the situation, "... for years, books have just had to be put down wherever a vacant space presented itself, and constant moving of the books in the endeavour to crowd a few more into shelves already too full, has made any proper arrangement quite impossible ... the lower shelves are completely blocked by books piled up on the ledges, and the assistant librarian is not strong enough for the constant lifting about of heavy books which this entails, to say nothing of the amount of time wasted."[40] It no longer was possible to find any book with certainty through the card catalogue, and everyone had to seek out his requirements for himself. No wonder that re-cataloguing the holdings was recommended, for "The absolute lack of system in the library at the present time makes the members of the staff dependent on the already over-burdened assistant librarian on whose memory alone they must rely if a work not already located on the shelves by their memories is to be found."[41] The committee also recommended trebling the library space by having it absorb the long room and the corridor alongside both rooms, an arrangement that would cut off the laboratory from the rest of the building and make it necessary to enter it from the courtyard. The books could then be removed from the floors and shelved in proper order, with room remaining to display periodicals and new books, for some study tables, for the map collection, and for expansion until the move to the new building.

This was done, and by 1905 Bell was proposing a further expansion by renting the building beyond the Baskerville annex. He also took steps against the fire hazard by hiring three firemen in 1904 to furnish round-the-clock protection, in addition to the night watchman. That another sort of protection was also needed was indicated by Ami's report of frequent thefts from the museum cases and his suggestion that a member of the Dominion Police be secured as a guard.

Altogether, the end-product of Bell's five years was a Survey that was quite transformed in its physical and administrative structures, occupying an enlarged, improved, better-equipped premises and actively preparing for the great move to the new quarters rising just beyond the south limit of the city of Ottawa. There were one-third more persons on the payroll now, among them draftsmen, library assistants (the staff had risen to four), a dozen clerks, stenographers and "typewriters," tradesmen and labourers, watchman and firemen, as well as an editor, a French translator, an accountant, and the director's secretary. On the technical side, too, the Survey was expanding and evolving into a collection of more distinct sections and units, as the staff became more highly specialized and embarked on new fields of activity. There was the expanded work of chemical analyses, petrological examinations, collecting and preparing minerals for displays and kits, and the newly-begun metallurgical work. The museum activities were also beginning to take on a more distinct form, with their own *raison d'être* apart from serving as an adjunct to the geological work. The emergence of a vertebrate paleontologist, the expansion of the natural history side with the appointment of J. M. Macoun (formerly an assistant to geological parties) to conduct his own studies as assistant naturalist, and the greatly accelerated process of acquisitions of anthropological material, all seemed to point to a museum section with a life of its own. The mapping section, basking under Bell's special favour with an enlarged, more specialized staff, was moving into the comparatively new field of photographic work and developing a special relationship *vis-à-vis* the field geologists. Even the paleontologists felt a similar stirring; it was in response to such a tendency that Bell made the suggestion, pooh-poohed by Sifton, to accord J. F. Whiteaves the title of chief paleontologist.

* * *

ALL THESE SEEMED RATHER OMINOUS developments to the new breed of university-trained geologists who appeared largely overlooked in the expansion of the organization in so many new directions. The number of geologists among the technical officers of the civil list actually declined, from fifteen in 1900–01 to fourteen in 1906–07. Besides, they and Bell seemed completely at odds in their concepts of field work, especially as regards systematic work and the investigation of fundamental geological problems. Not that there was any disagreement about the importance of field work—in fact, it was almost an obsession with Bell to send out as many parties as possible, even though the data collected in a single season far exceeded the staff's ability to process them during the coming winter. To multiply field parties in 1901 Bell sent out every available geologist and topographer with his own party, even summoning Brock home from Germany—where he had gone to study petrology, with

Dawson's approval—for the purpose. In subsequent years half a dozen members of the drafting staff also were given parties. Above all, Bell turned to outside help, hiring four to six such party chiefs each summer, among them men who had worked for Dawson and Selwyn, such as Professors Adams of McGill, the Abbé Laflamme of Laval, Bailey of New Brunswick, and Dresser of St. Francis College—plus W. A. Parks and T. L. Walker of the University of Toronto, Ernest Haycock of Acadia, and H. S. Poole, Nova Scotia mining inspector and professor at Dalhousie. Arthur Webster, who had left the Survey almost 30 years before, was hired for one summer's work; as were G. F. Matthew, Bailey's long-time associate; J. W. Spencer, a former assistant of Bell's who had subsequently embarked on a university teaching career in the United States; and A. F. Hunter, an amateur historian and naturalist of the Georgian Bay district. These part-time workers, particularly the geology professors, were men of high scientific attainments, who could be employed at relatively cheap rates since the universities paid their regular salaries. At the same time, the summer work gave the professors extra income, enabled them to practise their scientific skills, and provided training for their students.

By now, indeed, positions on the Survey were much sought-after for summer employment for geology students and others. Some 300 applications per annum were received by Bell, but few of the places on the thirty parties went to the best qualified young men, who might be encouraged to join the Survey after graduation. In 1904 Professor T. L. Walker was moved to publish a letter on "The Geological Survey as an Educational Institution" in which he complained that "for ten years no advanced student in mineralogy, geology or mining, from the University of Toronto, has been able to secure a place on any of these expeditions," and concluded with:

> . . . the Geological Survey of Canada is, and has been from the beginning, an educational institution. In its generous distribution of its reports and of geological specimens it is following the right path; but it has almost ceased to co-operate with the Canadian universities, their best students finding it next to impossible to get field training on the summer parties of the Survey. Some even apply to the United States Geological Survey for permission to join their field parties. Advanced students want field training, and would in some cases be willing to go as volunteers without pay. For Toronto students at least the door has been shut for ten years.[42]

Bell could only sympathize with this complaint and hope that patronage would vanish and leave the Survey free to fill the places on professional grounds.

How was it possible for Bell to find chiefs to head thirty or more parties, when Selwyn and Dawson had never been able to secure more than half that number? Partly it was the result of the increased number of trained geologists on the staffs of the universities, partly the Survey's reputation, and partly Bell's wide connections. But mainly it was because Bell expected a type of work from the parties that did not require men of very high qualifications. To all appearances he saw the collecting of the data as an end in itself, to be reported and recorded on the map. The geologist should devote himself to mapping the topography of the district, since a main purpose was to complete the map of Canada. Mapping was also the basic element in studying the geology of a district. In his letter to Brock in Germany, Bell went so far as to say:

> . . . during the remainder of the present season you will have to do your own topography at the same time as the geology. In fact this is the proper method for every member of the staff to adopt. It seems to me almost impossible to separate the work and do the geology without the topography. Except in the case of yourself and Mr. Low I do not remember any attempt at separating the two branches of the work ever being made before, since the Survey began, and hereafter we shall have to return to the old methods which were obviously the best.[43]

Quite apart from the gratuitous, rather offensive, attack on Brock and Low, this letter indicated a viewpoint completely at variance with that of earlier directors. Unquestionably Bell was harking back to his own many years as a reconnaissance geologist and explorer, carrying on all-purpose surveys that embraced not merely the rock formations but all other aspects of the region's natural and human history and economic problems. He was an excellent surveyor and an all-round naturalist, with little formal training in geology, certainly far less than the young men emerging from their university courses or than Dawson had achieved through constant study. He lacked sufficient understanding of, and sympathy for, the role of expanding the frontiers of geological science, on which the Survey's professional reputation largely rested, on which the morale of its specialist staff depended, and from which the country in the long run might derive far greater dividends than from mapping remote parts of Canada.

Equally, besides his emphasis on mapping, Bell was concerned in his field work plans to meet the country's needs for the demarcation of all natural resources that were capable of early development—a view that carried him into investigating soils, forests, waterpowers, fish and game, as well as minerals. These were properly parts of the Survey's mandate under the acts of both 1877 and 1890, and there were added reasons for devoting some attention to these aspects because of the needs to collect material for the Victoria Memorial Museum and to assist the government to open up the new frontiers north of the CPR, and beyond the prairies. But to trained geologists, such enterprises naturally seemed a detraction from the true purpose of the Survey and a negation of it.

Furthermore, many of Bell's geological assignments involved mere reporting and outlining concentrations of possibly mineable minerals. Small wonder that the Survey began to seethe with an undercurrent of discontent against the acting director.

Nor did Bell act as a final arbiter and authority on geological questions as his three predecessors had done. He was not at all trained for such a role, nor was he especially interested in studying current geological literature and giving constant, thorough attention to revising and editing the reports. On scientific questions he accepted the authority of persons he respected and trusted, such as Ells, Fletcher, or Adams. Without the unifying force of a single expert authority, the Survey appeared to lack an underpinning of common standards and bases of action. As each geologist became his own authority and carried on in his own way, the work tended to become diffused in emphasis, approach, and results.

The diffused appearance of the Survey's effort in the field was further intensified by the character of the assignments, which tended to be made in the light of a current interest or in response to pressures from the mining industry or the government, giving the work an *ad hoc* unplanned quality. The work skipped about, as Bell attempted to appease various regions and interests. The undertaking of an examination of Prince Edward Island was a good illustration of how political, rather than scientific, considerations sometimes affected the assignment of work. McInnes was removed from a productive study of the problems of the Precambrian to make general reconnaissances in geologically-barren country north of Lake Winnipeg, or west of Hudson and James Bays. Two geologists were transferred from the Maritimes across the continent to British Columbia to study coals that had little in common with those they were familiar with in eastern Canada. Naturally, the results were most uneven. One might single out such work as Brock's in the Kootenay, Chalmers' in Eastern Townships and Ontario, or the early work of G. A. Young, Charles Camsell, and D. D. Cairns, and especially Barlow's at Sudbury, as most noteworthy scientific achievements of the staff; but there was much wasted effort as well. All in all, Bell's management of field operations was less than distinguished as regards its geological accomplishments, though respectable enough from the standpoints of geography and resource assessment and evaluation. In this the Survey was reflecting only too accurately Bell's strengths and weaknesses, and his concepts of the work.

Partly because of the geologists' disappointment at Bell's order of priorities, and partly because of Bell's sometimes tactless and abrasive personality, he quickly found himself at odds with a number of the men, among them the most ambitious, prominent, politically-influential staff members. These were encouraged in insubordination and intrigue by Bell's ambiguous, uncertain position. When he became acting director the entire staff signed a letter wishing him well and avowing their support for all his endeavours, but within a year five of the number, in a letter to the minister on a different matter, criticized him obliquely by urging the appointment of a head "who is not dominated by the tradition of the past, and who will regard the members of his staff as responsible professional men."[44] Among the five signers were three of the four geologists with whom Bell had his least satisfactory dealings—Barlow, Brock, and Daly—whose motives could have been personal ambition as much as resentment of their treatment of Bell's hands.

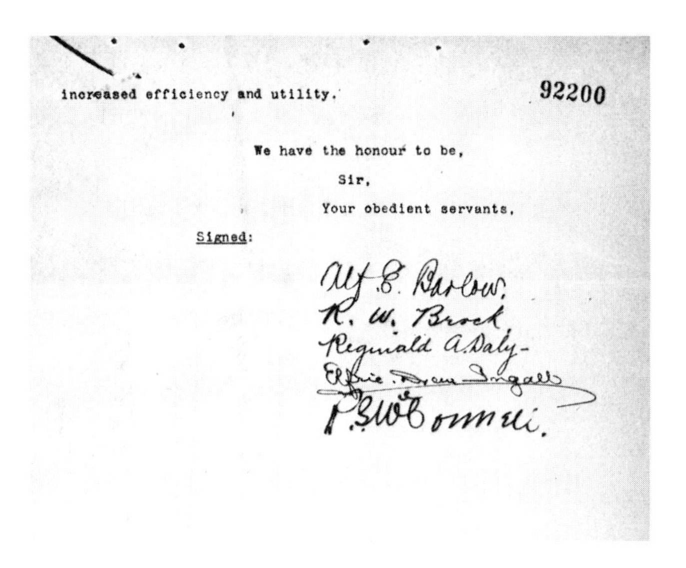
increased efficiency and utility. 92200

We have the honour to be,

Sir,

Your obedient servants,

Signed:

Alf. E. Barlow.
R. W. Brock
Reginald A. Daly
Elfric Drew Ingall
P. S. W. Donnell.

X-12 SIGNATURE PAGE OF THE OFFICERS' LETTER TO SIFTON

This four-page letter of March 1, 1902, vigorously criticized Bell's program as lacking in scientific orientation and content.

In the case of Barlow, the difficulty may have arisen from Barlow's direct link with the minister, who took a great interest in his work in Sudbury and elsewhere; this interest had the effects of encouraging Barlow's hopes and exciting Bell's fears. By 1904 Bell was complaining about Barlow's wilful disobedience of instructions and not doing his work properly. With the Canadian-born, American university professor R. A. Daly, the clash undoubtedly arose from Daly's ambiguous position as geologist of the International Boundary Commission surveying the 49th parallel across British Columbia. Daly drew his pay from the Survey and had an office there. But his actual superior was the Chief Astronomer, W. F. King.

X-13 R. A. DALY
Born and educated in Canada, Daly spent a long academic career (1898-1942) in the United States, apart from his years on the International Boundary survey, 1901-07.

Bell demanded that Daly should report on his activities, particularly explaining his absences, while Daly complained that "For nearly four years I have had to lose much time and energy laying the needs of the indoor work before Dr. Bell, only in most instances, to meet with his refusal to assume these legitimate charges on geological work."[45] In the end, Bell received an order in council removing Dr. Daly from his control, to which he added the hope that this would also mean Daly's removal from the building. Daly's connection with the Survey ended in 1906, together with his five-year term of appointment.

The strained relations between Bell and Brock probably began with Bell's peremptory summons home in the middle of the 1901 field season, and the accompanying lecture—if it did not, indeed, date from Brock's years as Bell's summer assistant. The antagonism increased when Brock after his return from the field took up an appointment at Queen's University with the consent of Sifton but without informing Bell or securing his permission. Brock began spending seven months of the year at Kingston and five months in the field with the Survey. The arrangement lasted four years in all, without any formal leave of absence. While Bell fumed, the Kingston School of Mining compounded the iniquity by subletting Brock's services for three months to enable him to arbitrate a mining company merger at Rossland. Again, in 1905, when Brock was proposed for the economic study of the Rossland district, Bell complained that it was not a proper task for the Survey since it was for the benefit of private mining interests. He urged that in fairness to the others "among whom dissatisfaction has long been felt at his anomalous position"[46] Brock should resign from the Survey. Brock probably took the position at Queen's University mainly to be out of Bell's way, for immediately Bell ceased to be acting director, Brock returned to full-time work at the Survey.

Bell's difficulties with H. M. Ami were devoid of any element of rivalry since Ami had no designs on the directorship, but were based on temperamental personal factors. Almost everything about Ami must have rubbed Bell, a serious, humourless, duty-driven Presbyterian, the wrong way—Ami's vivacity, his *bon vivant* ways (Mrs. Ami was a wealthy woman), his physical delicacy and frequent illnesses, his curious interests, his devotion to Dawson. A good paleontologist much given to controversies, Ami first ran afoul of Bell over his efforts to carry a feud with Fletcher on the age of the Riversdale and Union formations in Nova Scotia to the geological societies and journals after Bell had pronounced on the side of Fletcher. When Ami defiantly went to Rochester to give his accusatory paper before the Geological Society of America, Bell analyzed the time book and discovered to his horror that Ami in the previous four years had been absent on 287 regular working days. He thereupon addressed the following instructions to Ami:

> Please take notice that from this date forward you are required to initial the time book and to record legibly the hour of your arrival in the morning, of going to and returning from your midday meal and of leaving the office in the evening, that you are to remain in the office during office hours, that you are in every other way to conform to the rules and regulations of the Civil Service and of this Department in the same manner as other members of the staff.[47]

Later, while checking Ami's expense claims for 1901 Bell noticed that Ami had been at Murray Bay at the height of the season, and in Quebec and Montreal during the festivities in honour of the Duke and Duchess of Cornwall, and he could not be convinced that all three were pure coincidences. He continued to challenge Ami's absences and even doubted his medical certificates. At the same time, Ami complained, Bell abetted by Whiteaves virtually stopped him from doing any paleontological work whatever by putting him to work on the Pleistocene of the St. Lawrence and Ottawa valleys and holding back his publications.

X-14 H. M. AMI

This seeming tempest in a teapot may have had serious repercussions, for Ami, a man of some considerable political influence, carried his complaints to Sifton and to Laurier. He counterattacked by threatening to sue Bell for defamation, and pleaded with Laurier to remove his oppressor: "My dear Sir Wilfrid, For God's sake, if for no other reason, do give us a new Director for this Department."[48]

These difficulties with a few of his staff should not be taken to mean that Bell could not get along with the members of the department, though this was the excuse used by those in authority to justify their refusal to appoint him director. He seems to have got along as well as could be expected with most of the employees, and certainly there were no further resignations to compare with those under Dawson, even though the disparity between their pay and what they could earn outside still was very considerable. O. E. LeRoy, who was appointed in 1902, did take on an important post with the Chinese imperial government in 1903 but he rejoined the Survey three years later; Leach also left for a time but soon returned, as did Low, who had resigned in February 1901. In just two years, his project having fallen through, Low applied to rejoin the Survey and Bell made arrangements for him to recommence work immediately, the assignment being the preparation and management of the important *Neptune* expedition to the Arctic. In 1904 Bell appointed Charles Camsell and G. A. Young to the staff, Camsell after valuable service with J. Mackintosh Bell on Great Bear Lake in 1900 and a summer reconnaissance west and south from Fort Smith in 1902; and Young after serving with his relative Low in Labrador and Ungava between 1896 and 1899 and doing work on his own for the Survey in the following summers while he carried his education through to the Ph.D. from Yale (1904). The 1905 appointments were equally notable, for they included W. H. Collins, D. D. Cairnes, and W. A. Johnston, all of whom had served earlier apprenticeships as field assistants. Still in this training stage but soon destined to join the Survey were M. E. Wilson (who headed his own party in 1907), G. S. Malloch, and R. Harvie. Indeed the pattern was being established whereby the summer appointments were being used to train geological and mining university students with a view to recruiting the best of them for permanent careers with the Survey.

Through all this, Bell's own position continued in the same unsatisfactory state as when he had succeeded Dawson in 1901. Year after year Bell used every possible means to persuade or compel the government to make him director, while the ministers refused to be moved. In 1902 he applied to Sifton for an increase in his salary from $2,400 to $3,200, the amount Dawson had received and the amount carried in the appropriation for the director's salary. But his request was rejected. For his part, Bell refused to apply for any smaller increase, feeling that would compromise his claim. In 1903 Bell circulated a letter to a number of M.P.'s urging his appointment as director, and a member of the Opposition raised the matter in the Commons: "Is there any reason why the appointment should only be of a temporary character?" To this Sifton replied, "All I care to say at the moment is that I have not felt like recommending his permanent appointment up to the present time."[49] By then it was obvious that a major consideration was Sifton's plan to reorganize the Survey as part of a wider program of revising the Dominion's role in mining. Pending the implementation of that plan, Bell was being retained as acting director to permit his easier removal at the appropriate time.

Sifton's resignation from the cabinet on February 28, 1905, brought a renewal of the question of the directorship, the *Canadian Mining Review* expressing the hope that the government would appoint F. D. Adams "to re-organize the Survey," as a person who was acceptable both to the staff and the mining industry, and a man of great talents who would "make no unworthy successor, to even the late Dr. G. M. Dawson."[50] Bell, too, began making every effort to secure the appointment, writing to Laurier and visiting the ministers to enlist their support when the new Minister of the Interior, Frank Oliver, brought the matter before the cabinet, as he was supposed to do, on August 4, 1905. When, once more, no action ensued, Bell prepared to take the matter to the House of Commons in the early months of 1906, confident, he wrote his wife, that right would win out, for "I have now entered my sixth year of managing the Survey & have made no mistakes."[51] Little did he know, though, that the minister had already decided against him. As Oliver wrote Laurier on March 13, 1906:

> The present Acting Director, Dr. Bell, has been very much dissatisfied with his present position, and has pressed strongly for appointment as Director.
>
> In this connection I regret to say that while I have every confidence in Dr. Bell as a scientist of the highest class, and therefore one whose services are valuable and desirable in that capacity, my judgment is that he does not possess that administrative ability which is necessary in order that good results may be secured from the work of subordinates. I regret very much having to reach this conclusion, but it has been reached after careful and full consideration of this matter in all its bearings.
>
> . . . I have no wish to in any way discredit Dr. Bell, or to put him at any disadvantage, on the contrary I am most anxious that he should not be discredited or put at any disadvantage either financially or otherwise.[52]

When Robert Borden, the Leader of the Opposition, raised the matter of Bell's promotion in the House of Commons a week later, Laurier replied for the government along the lines of Oliver's letter. Bell, he said, was a man of unquestionably superior merits as a scientist, but his relations with his staff were not so cordial as they might be, and perhaps Bell did not possess the faculty of managing men, in which his predecessor had been so eminently successful. When Dawson died, the government was faced with a serious dilemma; there were good reasons for and against appointing Bell. Consequently, the government had made the acting director appointment. But this was about to change: "The minister has come to a conclusion, I think, after conference with me—but which, before it is made official, I think should not be given to the House. But in a few days the appointment should be made."[53] The Opposition continued to argue the point, J. E. Armstrong, M.P. for Lambton East, urging the government not to take the word of one or two men about Bell's shortcomings but canvass the situation very thoroughly: "I believe if he investigates he will find that a very large majority of the officials who are engaged under Dr. Bell are sympathetic with him, and are his tried friends."[54]

The government's decision reached Bell in the form of a letter of March 29, 1906, from the Minister of the Interior, informing him he had been passed over, but cushioning the blow:

> I have to inform you that an Order-in-Council has been passed giving you the title and position of Chief Geologist of the Geological Survey Department with salary of $3,000.00 a year.
>
> The Annual Report or Reports which have been prepared by you *or under your direction* will be issued under your name.
>
> The Government desires in relieving you from *the purely administrative duties* which you have discharged during the past five years, in addition to your valuable scientific work, to give you a better opportunity to follow up the scientific researches in which you have such a deep interest, and from which the country may expect to see so great benefit.[55]

It was in this fashion that Bell learned that the government had decided to replace him in two days' time by A. P. Low, and that as a sop he was being offered the title of chief geologist and the $3,000 salary that had been carried in the year's budget for the director. Low's salary as director was soon raised to $3,500.

Thus Bell's long ordeal ended in yet another crushing blow to the high expectations that always pervaded his thoughts. Now, at the age of 65, he would revert to a subordinate place in the Survey, under a man twenty years his junior, a man who, in Bell's opinion, was wanting in talent or achievement and owed his place entirely to political influence. The stage was set for a train of further humiliations that led to his retirement on November 30, 1908, amid a storm of denunciations, and a subsequent round of appeals for redress that did not end until his death in 1917.

Why was Bell cast aside so abruptly after five eventful years as acting director, just when he was approaching retirement age? Clearly the government kept him in a state of suspension while it matured its plans and found an agent who would be able to implement them. When that time came, the government decided to rid itself of Bell, and chose to put its decision on his administrative incapacity rather than his scientific competence. As much as anything, though, it was precisely doubts over Bell's scientific attainments and concern that had kindled most of the unrest within the Survey, and thus had given the government its grounds for Bell's removal. Nevertheless, in his five years as acting director Bell had moved the Survey forward in a number of necessary and irreversible directions, and his successors, for some time at least, much as they denounced him, had to continue along the lines Bell had indicated.

XI-1 A SURVEY PARTY FORDING THE NORTH SASKATCHEWAN RIVER IN 1903

11
The Survey and Canada's Century

As THE TWENTIETH CENTURY DAWNED on Canada, an era of prosperity transformed the country, as though by magic, from a disheartened, stagnating land of five million population in 1895 to a proud, confident, healthy country of eight million inhabitants, able by 1916 to prove its mettle in the tests of world war and international diplomacy. Encouraged by the vigorous immigration program of the Dominion Government, about two million newcomers moved into the country to occupy her farmlands, swell her industrial cities, and people the new settlements being planted in the wilderness by the railways, the forest and the mining industries.

Constantly improving technologies removed many of the obstacles hampering resource development and established many new uses for the products of Canada's mines. Nickel is a good case in point; the industry had to await improved mining methods, the solving of very difficult metallurgical problems, and the appearance of an almost unlimited market in the armaments industry before it attained its full potential. Similar factors account for the growth of such other mineral industries as copper and zinc, mica and asbestos. New processes and markets, high metal prices, and the influx of capital caused existing mines to be updated, and brought new mines into production to add their output to meet foreign demand. Because of the population growth, the rapid industrialization, and the high rate of new construction in Canada, the domestic market was not to be ignored either.

Above all, the remarkable increase in railway construction, highlighted by the building of two new transcontinental lines, contributed enormously to the progress of the mining industry. The resources of many new districts across Canada became accessible to development for the first time, and the mining industry, which is peculiarly dependent on cheap transport to move large tonnages of low-value freight from mines to smelters, refineries and markets, was put on a vastly improved footing. From Cape Breton to the Klondike, mining suddenly swung into high gear. In just six years the output of the industry rose from \$20.5 million in 1895 (itself a not inconsiderable increase from \$10.2 million in 1886, the first year for which reasonably complete statistics were secured) to a peak of \$65.8 million in 1901 before reaching \$90.0 million in 1909. True, the figures are swollen by the large gold winnings from the Klondike from 1897 onwards, which attained a peak of \$27.9 million in 1901. But this was not the whole story. Of greater significance than Klondike gold was the increasing production from other mining areas—the Kootenays, the Eastern Townships, and especially the Canadian Shield. This region was emerging as an important source of metallic minerals with the opening of Sudbury mining camp, the unexpected discovery of silver at Cobalt, and the development of a string of smaller mining camps along Lake Superior and west to the Manitoba border. Governments strove to develop in Canada a coal and steel complex, and the economic advance of the country was reflected in the increased production of coal from Nova Scotia and New Brunswick, as well as by the opening of coal mines in Saskatchewan, Alberta, and British Columbia. But even though the annual coal production exceeded ten million tons and in dollar value stood far above any other mineral, Canada still imported half her requirements.

The rise of a confident, powerful mining industry

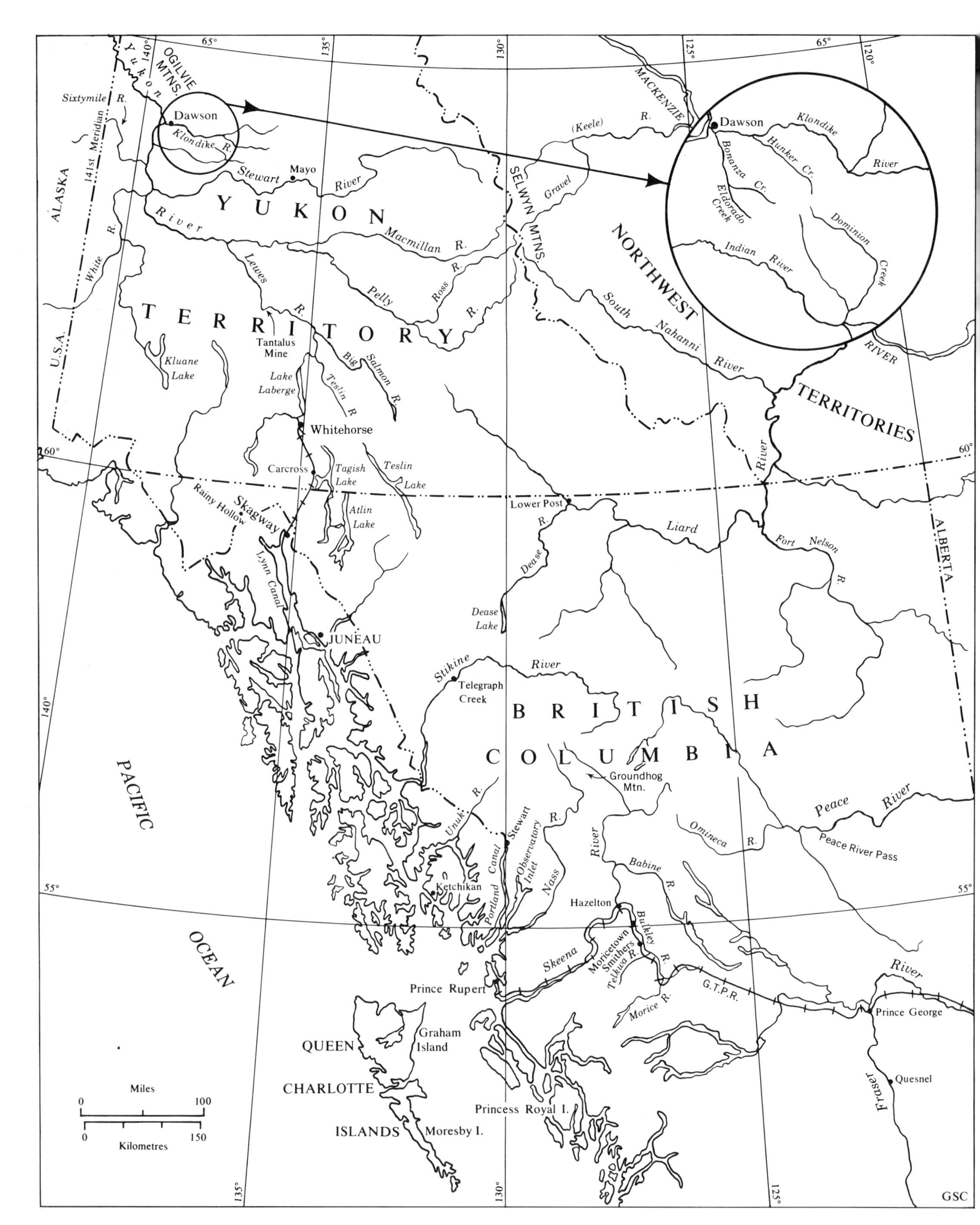

XI-2 NORTHERN BRITISH COLUMBIA AND YUKON TERRITORY, 1895–1920

brought demands for new kinds of governmental assistance and leadership. The Survey did what it could with the limited staff available, in part through the Section of Mines, under Ingall, but mainly through its traditional program of field work and mapping. Its limited manpower was set at work in areas, or on formations and structures, that were considered most relevant to the current needs. Though Dawson in 1900 had only seventeen parties, they were deployed with reference to areas and subjects of current mining interest and activity, primarily to solve the geological problems of these fields. Bell, too, proclaimed his adherence to the goal of serving the mining industry along the line of tracing formations considered likely to carry economic minerals, mapping the geology of as much of the country as rapidly as possible, and getting the information before the public and the industry.

The Klondike gold rush that followed the discovery of rich placers in the Yukon in August 1896 was a fitting prelude to the years of buoyant optimism and material growth that characterized the Canada of the early twentieth century. The gold rush suddenly turned the eyes of the world upon Canada as a source of valuable mineral wealth, a country with a great future; it gave Canadians a sense of destiny as a "nation of the north" that could turn its back on the International Boundary and set to work developing vast hinterlands all the way to the Polar Sea. The Survey had an early acquaintance with the Yukon district, and with the sending of McConnell and Tyrrell to the Yukon in 1898 it inaugurated a continuing connection with the territory which still persists. In the late summer, having made their way into the country by separate routes, the two men met in Dawson to begin a hasty joint investigation of the gold-bearing creeks of the Klondike. Their report, not altogether accurately, described the productive part of the district as covering approximately 1,000 square miles, from which the gold was secured from four main creeks—Eldorado, Bonanza, Hunker, and Dominion—with an aggregate length of about 30 miles of productive gravels.

Quoting J. E. Spurr of the U.S. Geological Survey, they attributed the gold as deriving from thick veins or narrow stringers in a quartz schist, probably of Cambrian age, situated somewhere in the immediate vicinity, from which it was transported by the streams and small local glaciers that filled parts of the valleys.

McConnell resumed his examination of the Klondike gold placers in 1899, assisted by J. F. E. Johnston. By now he had time to investigate the district much more thoroughly, and could report more fully on its geology and on the placer deposits of the various creeks. His geological review listed four Paleozoic stratified series, with eruptive granites from the Tertiary. He collected a large number of specimens from gold-bearing quartz veins, in hopes these might provide some clue as to where the placers originated. For he considered it most improbable that the gold-bearing veins were completely eroded, and held out the hope that "productive veins or zones of country-rock will eventually be discovered."[1] He was also optimistic that the district had many productive years before it, since very little work was being done outside the Klondike goldfields or in the Yukon Territory as a whole. Alas, he was wrong, at least as far as gold was concerned. While the Yukon does have a great mining future, it is mostly in base metals, silver ores, and fossil fuels.

XI-3 HYDRAULIC OPERATIONS ON AURIFEROUS "WHITE CHANNEL" GRAVELS AT AMERICAN HILL IN THE KLONDIKE, YUKON TERRITORY

A little later McConnell returned to the Klondike for other thorough investigations of the problem of the placers, and produced two major reports published in 1905 and 1907, the first on the geology, topography, and low-level gold-bearing gravels, and the second on the high-level gravels, the so-called White Channel gravels. In the first, the geological statement was now much more precise, the series and formations being carefully described, and efforts were made at correlation on the basis of fragmentary fossil evidence. He noted that the gold must have been derived from quartz veins cutting older schists which had been eroded and concentrated by the swift-flowing streams; he suggested that the original gold content of the quartz was probably no more than a few cents to the ton, and hence the veins might not be productive even if discovered. However, he estimated an ultimate yield of some $200 million for the whole field and that there were about 500 million cubic yards of mineable gravel on Bonanza, Hunker, and other creeks, provided the camp moved into hydraulic mining on a large scale. Later, in 1906, after hydraulic operations had begun, he returned to spend a season measuring all the important bodies of bench and creek gravels and estimating the eventual amount of recoverable gold.

Though most of the time was occupied in measuring, sampling, and collecting mining records, the surveys were something more. It was now realized that the phenomenon of the Klondike placers resulted from the fact that the district, unlike virtually all of Canada, had not been glaciated, enabling the rich placers accumulated over Cenozoic time to survive to modern times. The great bulk of the rich placer was deposited in the late Tertiary, in the beds of much wider, more gradual streams, and the lower parts were covered by a thick gravel bed when the Klondike River assumed its present position. The country was afterwards uplifted and the upper parts of the former valleys were cut through by the present swift, narrow creeks, which washed away some of the gold from the earlier beds, while more was left behind in the former channels high on the benches, or embedded in the hills formed by the overlying Klondike River gravel bed. Hence the places where the gravel was rich in the creeks were related to similar patterns of richness in the upper-level gravels. Thus it was possible to point out where still-undiscovered pockets of White Channel gravels might be found. By careful measurement, McConnell calculated that at least $60 million-worth of auriferous gravel remained unmined on the Klondike creeks (against $119 million already mined), the gold content of which varied from 34¢ to 15¢ per cubic yard. His two reports were remarkable predictions of the future trend of mining development in the area, and of the gold that remained to be secured from the rich placer creeks.

Besides the reconnaissances of 1898, McConnell with a series of assistants—notably Keele in and after 1901—made a number of explorations in the territory, including the Kluane district, Salmon, White, Sixty Mile, Stewart, and Macmillan Rivers. They studied the geology, assessed the mineral possibilities, and assisted the operations of the many scattered prospectors, miners, and developers. Among mining prospects frequently studied were the coal mines along the Yukon and the large copper belt near Whitehorse, which became an important producer for a time, and which McConnell described in a later report. This belt is now being worked on a large scale by open-pit methods. D. D. Cairnes made the Tantalus coalbeds and Tagish Lake copper deposits his major objects of study in 1906 and 1907, commencing with these investigations his work as the Survey's Yukon specialist. An interesting reminder that the Survey had an even wider role in assisting Canadian development than its geological work and service to mining was a visit to the Yukon in 1902 by John Macoun, who spent most of July and August examining the vegetation as indicative of the climate and the agricultural possibilities.

Paralleling the rush to the Klondike was a smaller one to Atlin Lake in the extreme north of British Columbia, where a not-inconsiderable goldfield with points of resemblance to the Klondike was discovered in 1898. With the Whitehorse copper belt farther north, it was another early example of the presence of mineral wealth in the northern part of the Cordillera. J. C. Gwillim made a preliminary examination and survey of the region in 1899, and he and W. H. Boyd mapped the goldfield in 1900. Gwillim's report, completed in May 1901, was the first to detail the nature, structures, and succession of any part of the northern Cordillera. His observations respecting the occurrence and position of the placer gold and its derivation from mineralized quartz veins were similar to those of McConnell.

Yet another study of the geology of the Coast Range was undertaken in 1903 of a district some 200 miles south of Atlin, near the headwaters of Unuk River, where prospectors were displaying some interest in placer gold and ore deposits. In view of the recent discoveries along the eastern margin of the Coast Ranges at the head of Portland Canal, Unuk and Stikine Rivers, and near Carcross, the Survey took the occasion to employ the American geologist F. E. Wright, who was investigating the Juneau mining district of Alaska, to conduct a similar exploration of this region immediately across the border in Canada. In his excellent report, drawing on his knowledge of the Ketchikan and Juneau districts and the past record of prospecting and mining in the region, Wright discussed where to look for minerals in the light of the

XI-4 VIEW OF DAWSON CITY, YUKON TERRITORY, IN 1907

geological structure of the Coast Range and its history:

> From these considerations it is inferred that the sedimentary belt to the east of the Coast Range granite in the Unuk river section merits investigation and may reward careful prospecting for orebodies. The difficulties of transportation which have been encountered heretofore will be materially decreased by the completion of the wagon road to Sulphide creek. Prospectors will then be able to devote a large part of their energy to the search for and development of metalliferous veins in the region.[2]

More important even than the Klondike was the series of mine fields emerging as large producers of metallic minerals along the southern part of the Canadian Cordillera, in the Kootenay country where Dawson had just finished his last field work. Almost coinciding with his departure, the mining activity in the region made a remarkable surge which McConnell noticed as soon as he took over the work in 1895:

> The facts . . . show the wonderful expansion of mining enterprises taking place in the West Kootanie district and the extent and richness of the deposits carrying silver and gold there. One of the most notable points brought out, is the occurrence, lately ascertained, of ores of exceptional value in parts of the granitic area, which has heretofore been almost disregarded by the miners. . . . Ten years ago this district was almost an untrodden wilderness, but it is difficult now, with the means at the disposal of the Geological Survey, to keep pace with the march of discovery.[3]

Easing the Survey's work in the district—and perhaps subtly altering its direction—was the close relationship, amounting almost to a collaboration, that existed between the mining fraternity and the representatives of the Survey who came to work in the Kootenays. By a sort of exchange, the geologist viewed the results of prospecting and mining operations and was thus helped to a better understanding of the geological structure, while the miner gained the benefit of the geologist's advice. An officer who mastered the geology of a given locality was in a position to advise prospectors and managers on the best possibilities—what kind of rocks to examine for possible ore deposits, where more ore-bearing veins might be encountered, in what directions the mine could be extended, and the like. Brock's reports indicate that mine operators asked for—and received—advice on practical points of mine development with a liberality that is altogether remarkable, for Brock held that this was probably the best way in which the Survey could be of assistance.

In 1895 McConnell had been assisted by H. Y. Russel, in 1896 by W. W. Leach with McEvoy in charge of the topographical work, and in 1897 the trio became a quartet when R. W. Brock joined the group as assistant geologist. In the three seasons the work shifted from Slocan, to Rossland and Trail in 1896, then east to the Salmo country. For the season of 1898, Brock took over the

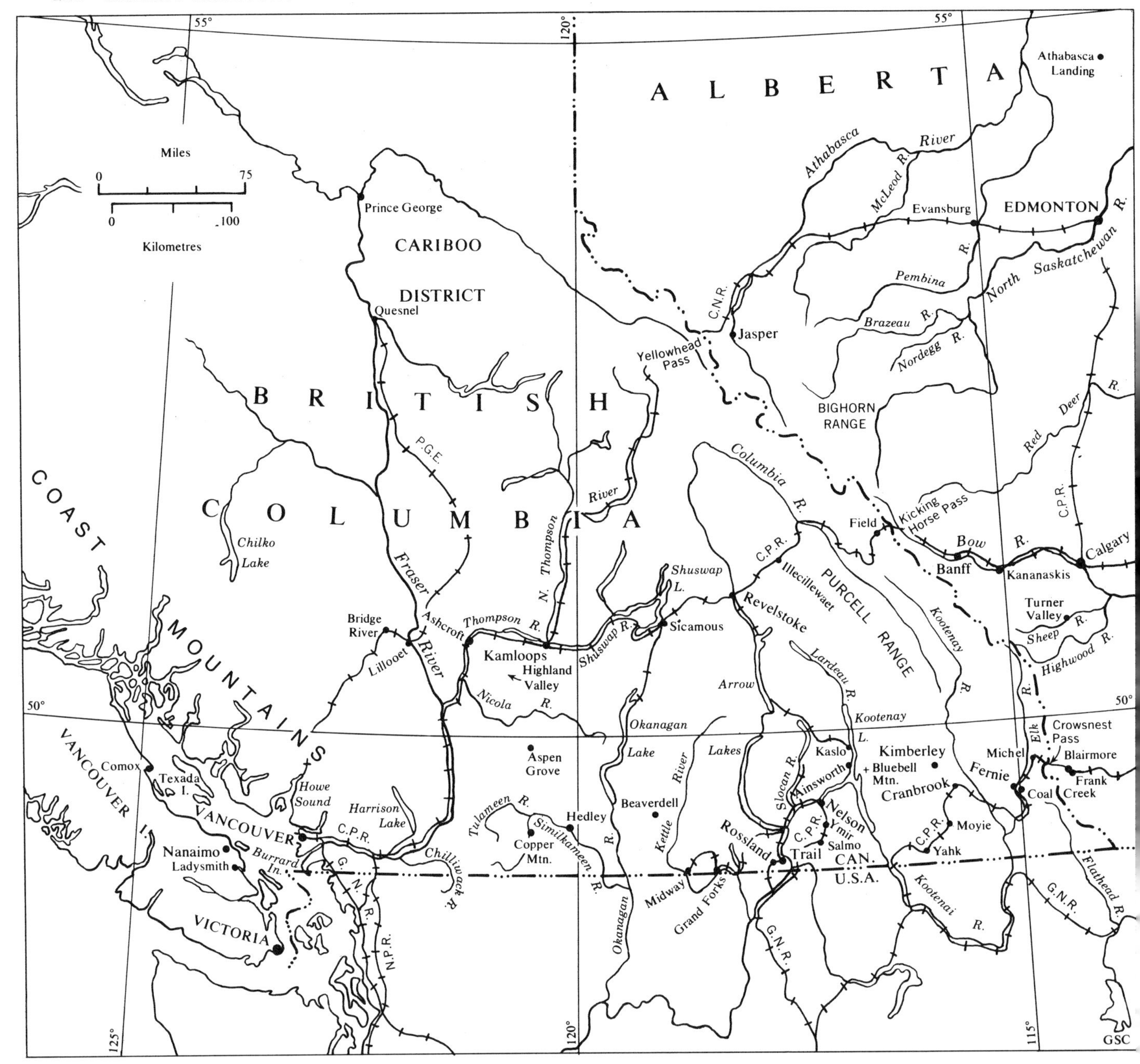

XI-5 SOUTHERN BRITISH COLUMBIA AND WESTERN ALBERTA, 1870-1920

West Kootenay sheet, assisted by Leach as topographer. They worked over the country between Arrow Lakes and Columbia River, on either side of Lower Arrow Lake between Slocan River and Christina Lake; and finally, west from Columbia River to Kettle River. In addition, McEvoy began work on the western part of the East Kootenay sheet in 1899 from a centre at Cranbrook. These were triumphant years. Helped by higher silver and lead prices, the Slocan district went forward rapidly in 1898, thanks to an infusion of English capital. Large mines and good prospects were being developed in East Kootenay at Moyie and Yahk, but especially around Cranbrook and Kimberley, the latter with the North Star and Sullivan mines and their massive orebodies. The opening of a big smelter at Grand Forks led a number of large developed properties to begin production.

XI-6 W. W. LEACH

In 1901 and 1902 Brock carried the work westward into the Okanagan map-area, beginning with the Boundary Creek district, an area where mines possessed huge, low-grade, copper-gold orebodies, mostly operated under the same ownership as the smelters to reduce the smelting expense, and with labour-saving machinery like steam shovels and giant crushers. Smelters were the main bottleneck; but with three modern ones operating in 1902, over 500,000 tons of ore was mined.

In 1903, 1904, and again in 1906, Brock and W. H. Boyd worked in the area north of the West Kootenay map-area, the Lardeau district, particularly along Lardeau River and Kootenay Lake to Kaslo. In 1905 Brock was requested to undertake a thorough examination of the mining camp of Rossland which had fallen on uncertain times since the phenomenal boom years of 1896-99. The Rossland Board of Trade asked for a "structural geological survey" of the local mines, and Brock, Boyd and Young were assigned to the task of helping Rossland regain its former pattern of growth. While Boyd mapped the district in detail and Young studied the general geology, Brock concentrated on the "underground and economic geology of the camp," particularly "the outlying workings, in the hope of discovering differences of conditions that would throw some light on the causes of mineralization, and thus afford a clue that might be of service in prospecting for orebodies in the district."[4] In 1906 the same trio returned, mapping was completed, and again the rocks were studied in localities a little removed from the zone of intense mineralization in an effort to solve problems without this added complication.

In his report after the first season, Brock reviewed the history of the camp, the ores themselves, mining and treatment, and the economics of the operation—costs of mining, of treatment, profits. Given the circumstances, the most valuable part of the report was probably that section labelled "Future Prospects" because advice and exhortation were exactly what the camp needed most. Brock urged the operators to be bold and take courage—the mineralization was produced by ascending hydrothermal solutions, so ore must continue downward to considerable depth and miners should not be deterred by adverse indications but keep searching for it; companies should undertake more lateral development; a great deal of promising ground above the 900-foot level had still not been sufficiently prospected. More effort should be made to find new veins; prospectors should follow their extensions because they might well carry other ore shoots farther on. He concluded with what could best be described as a pep-talk: a camp that had produced $34 million in metal in a decade certainly was not finished; a great deal of ore remained to be found and developed. Past mining methods had been inefficient and profits had not been what they should, but now that many of the developmental costs were met it would be possible to operate with greater economy and efficiency. With good management and skill, the camp could win through to better days.

Meanwhile, farther west, the Survey responded to the reports of mineral discoveries by sending R. A. A. Johnston in 1904 to examine the Aspen Grove copper claims south of Kamloops. In 1906 C. Camsell started work on the Tulameen 4-mile map-area in which were the small, rich copper veins of Law's camp up Tulameen River, the placer deposits farther downstream which had once been one of the world's major sources of platinum, and the massive deposits at Copper Mountain on the upper Similkameen River. Farther down the Similkameen were the gold deposits of Hedley, discovered in 1896 and from 1903 to fairly recent times, an important gold producing area. Before Camsell's work, except for Dawson's reconnaissance survey of 1877 "when there was not the slightest suspicion of such valuable ore occurring," nothing was known of the geology of the region.[5]

Farther west again, the coast north of Vancouver, where important developments were taking place, was investigated in 1906 and 1907 by O. E. LeRoy and J. A. Bancroft. LeRoy confined his attention mainly to two principal areas, Howe Sound, where the Britannia Copper Company was developing an enormous low-grade copper lode; and Texada Island, where two mines were shipping copper, gold, and silver ores to smelters at Tacoma and Ladysmith, Vancouver Island. Bancroft visited and reported on some 35 prospects along the coast north from Powell River.

* * *

IN ONTARIO, ANOTHER BOOMING MINING PROVINCE, the Survey was inhibited from playing the same role of closely co-operating with the mining industry by the attitude of the provincial Bureau of Mines, established in March 1891, "to collect and publish information and statistics on the mineral resources and the mining industries of the province."[6] From the beginning, the bureau had set out to play a large part in the development of the mining industry of the province, including geological, mineralogical, and metallurgical investigations. As the industry grew and the province began deriving large revenues from it, the research work of the bureau was greatly increased. Geologists and mining engineers of the provincial universities were regularly employed to examine rocks in general and the structures of existing mine fields. With men of the calibre of Professors A. P. Coleman and W. A. Parks of Toronto, W. G. Miller of Queen's, and A. B. Willmott of McMaster reporting in the single year 1899 on the copper region of the upper lakes, the Nipissing-Algoma boundary, corundum and other minerals, the Michipicoten Iron Range, copper in the Parry Sound district, and the Lower Seine gold mine, Ontario seemed adequately served at least as regards investigations of its mining districts.

With two organizations, the Survey and the Ontario Bureau of Mines, engaged in studying the interesting and important geology of the province, a degree of uneasiness was bound to develop. The Survey did not want to be drawn into an apparent competition and tended to avoid areas of active mining operations where the bureau seemed directly interested. A decade elapsed after 1891 before it returned to Sudbury, and then it came principally at the invitation of a private mine owner and with the minister's approval. For the most part, it was the Ontario bureau that advised the miners of the Lake of the Woods, Rainy Lake or Fort William districts, and to which the operators at Sudbury normally turned for assistance.

Something of a crisis did arise over work in the Cobalt area in 1904, shortly after the spectacular silver strike there. Bell sent Professor W. A. Parks, who frequently worked for either organization, to study the geology of a district from Lake Timiskaming northward, including "the geological conditions of occurrence and general extent of the deposits of ores of silver, nickel and cobalt discovered along the right of way of the Timiskaming and Northern Ontario Railway."[7] After approximately a fortnight, Parks learned that the provincial geologist, W. G. Miller, had been sent "to carry on the same investigations which I had been instructed to undertake;"[8] and Parks was then directed "to examine the country northward to the height-of-land, paying particular attention to the extent of the silver-bearing series, but not neglecting the features usually dealt with in a general geological report."[9] Miller published his report on his work at Cobalt in the December 1903 issue of the *Canadian Mining Review,* then as a pamphlet, and finally, in the annual report of the Bureau of Mines for 1904. Parks, besides discussing his work north and east of Cobalt also published fairly lengthy descriptions of some of the properties at Cobalt, dealing with the nature of their orebodies, work done, some analyses of ore samples, and the economics of the industry—all of which seemed rather an infringement on Miller's field. Bell also intervened in 1906 by publishing a report of his own on the Cobalt mining district, based on visits there in November 1905 and April 1906, which report seemed intended to promote the boom by stressing the richness of the ore and the great wealth being earned by the mines. His report referred to some of the major geological features of the district and the possible extent of the silver-bearing formation, how deep the deposits might be expected to continue, and the like. This gratuitous and apparently pointless, intervention at Cobalt was certainly not appreciated by the Bureau of Mines. It was quickly followed by an agreement as to the respective fields of activity, which Low, Bell's successor, published early in 1907 in the *Summary Report for 1906* under a heading "Relations between the Geological Survey and the Ontario Department of Mines." This recited that Low had reached a "mutual understanding" with T. W. Gibson, the deputy minister,

> . . . concerning the operations between the federal and provincial departments as to the scope and relations of each in order that they may work in harmony and avoid duplication of surveys. As the control of mines and mineral industries is vested in the prov-

inces it is advisable that the Geological Survey should acquiesce as far as possible in the wishes expressed by provincial authorities as to the mining investigations they consider advisable.

During this conversation with Mr. Gibson, afterwards confirmed in writing, it became evident that the provincial department, while eager to reserve for itself all investigations into the economic mineral resources of Ontario, was willing to supply the Geological Survey with complete mineral statistics of the provinces at the earliest possible date and in a form suitable for publication in our Mines Section report, provided that the Survey would discontinue the collection of mineral data in Ontario. . . .

It is proposed for the future to confine the work of the Survey in Ontario to the compiling and publishing of the systematic series of geological map sheets of the more settled parts of the province and to reconnaissance surveys in the northern portions.[10]

Hence while more and more parties were deployed in British Columbia, relatively few were active in Ontario, and these were mainly sent on reconnaissances in the undeveloped areas, or in districts like southern Ontario, the Ottawa valley or the Frontenac Axis where there were few major mining operations. Under Dawson, for example, the main activities in northern Ontario were McInnes' west and north of Lake Superior, and Bell's mainly in the Michipicoten district. Though both were inspired by a wish to assist the development of mining, neither afforded much direct assistance to particular mines compared with what Brock was doing in the same years in the Kootenay. Both concentrated mainly on topographical and geological mapping, outlining the boundaries of the Precambrian formations and evaluating their qualities and characteristics, including economic minerals.

McInnes' report on Seine River to Lake Shebandowan district (1897), as has been shown, was concerned mainly with the stratigraphy and relations of the Precambrian formations. However, he also described two ranges of high-grade, extensive iron ore deposits at Matawin and Atikokan, indicated where the cleanest and best deposits might be found, and provided analyses showing that the ore approximated 65 per cent metallic iron and was free of titanium. He observed at Steeprock Lake that "The beds from which the blocks of rich float were derived seem to be largely covered by the waters of the lake, and were not discovered,"[11] thereby anticipating the remarkable Steep Rock iron development 40 years in the future. In 1900 he traced the iron-bearing formation in the region southwest of Port Arthur, using formations in adjacent Minnesota to get an idea of the succession and the position of the iron beds near the base of the Animikie formation. After 1900, though he worked in the undeveloped country north of the CPR, he made several further visits to Atikokan at the request of the Canadian Northern Railway which was developing the property.

Bell also returned to very familiar ground in 1898 and 1900 when he was sent to the Michipicoten area because

XI-7 HEDLEY, BRITISH COLUMBIA, IN 1908—A MAJOR NEW GOLD MINING CENTRE

of a gold discovery at Wawa Lake that had led the Ontario government to proclaim the area a gold mining district. His report for 1900 described the great changes in the area with the opening in the previous year of the rich Helen iron mine, eight miles northeast of the mouth of Michipicoten River. Bell could not resist working out an estimate of the ore tonnage.

An exception to the general avoidance of work in the productive mine fields of northern Ontario was the work Barlow was able to do at Sudbury, for which he was commissioned by Dr. Ludwig Mond, the British chemist who developed a radical new extraction process for nickel, secured mining lands at Sudbury, and entered the field with the Mond Nickel Company. Mond was prepared to pay Barlow a large fee to do a thorough study of the Sudbury field as a member of his organization. Faced with the prospect of losing Barlow, Bell agreed to give him a leave of absence on condition that Barlow remained available to do petrographic work for the Survey, and that the results of his labours for Mond should become available to the Survey. In this fashion in the seasons of 1901 and 1902 Barlow was able to undertake his momentous study and develop a theory of the mineralization that indicated the structure of the ore deposits and greatly eased the problems of locating and mining the productive ore. The Survey received a first-class geological report on the structure and petrology of the region and a comprehensive discussion of the operations of all the companies, their mining, concentrating and refining processes, production and markets. Barlow's report was so useful that within three years a second edition was needed, the original one of 1904 having run out of print.

After completing the work at Sudbury, Barlow, assisted by LeRoy, moved on in 1903 to resume mapping the Timagami district, where the Mines Branch of the Department of the Interior had located iron ore with the aid of a Swedish-devised magnetometer. He was followed by Young and W. J. Wilson in 1904 and 1905, who investigated the country southeast and west of Lake Timagami, respectively. In those years Barlow prepared a report on the corundum deposits at Craigmont, Ontario, in 1904, while in 1905 he was a member, with Keele, of the Dominion government's zinc commission that investigated the problems of developing that industry in southern British Columbia. Next, in 1906 he continued the detailed work of tracing the Cobalt formations begun by Parks, but on the Quebec side of Lake Timiskaming—a work carried on in subsequent years by M. E. Wilson.

The Survey also continued to investigate the mineral resources of the Frontenac Axis without any apparent protest by the Bureau of Mines. Since the early eighties a series of investigations had been made by Coste, Adams, Barlow, and others mapping the geology of the 4-mile map-sheets, some primarily from scientific, others from 'practical' motives, yet all serving the one objective. Thus work on the Haliburton map-area satisfied Adams' great interest in the petrological characteristics and relations of the Precambrian rocks; but in addition, he reported on occurrences and prospects of iron, gold and lead, and discovered a remarkable deposit of the unusual rock nepheline syenite, used principally as a decorative stone, and an important ingredient in the glass-making industry today. A little later Ferrier found corundum, a valuable mineral used for emery wheels and other abrasive materials, in a rock sample from the same general area in Hastings County and located the deposit in Carlow township in 1896. Before long corundum was discovered in an extensive band of country extending east from Carlow along the general line of Madawaska River almost to the Ottawa, and the Ontario Bureau of Mines ordered the area closed to sales of Crown land. The Survey's reports discussed the problems of establishing the industry, advised prospectors on how to recognize corundum, and tried to develop foreign markets for the mineral.

Another investigation of the Frontenac Axis' mineral resources in the same period concerned the feasibility of establishing an iron and steel industry at Kingston, based on ores from many small scattered deposits in the surrounding district. Following the completion of the Kingston and Pembroke Railway and other lines, the Kingston Board of Trade requested the Survey to make an evaluation of the "quantity and quality of the deposits of ore likely to become tributary to Kingston smelters, as might justify the necessary investment of capital there."[12] Dawson had Ingall investigate over forty separate magnetite and hematite deposits, many of which had shipped ore in the past though none was operating then. He concluded there was enough iron in many irregular deposits to supply a 100-ton smelting operation, though care would be needed to avoid too high levels of sulphur and phosphorus. He updated this study in 1899.

In Quebec the main centre of Survey work in relation to the mining industry was in the Eastern Townships, where Chalmers, the Survey's specialist on the Pleistocene, was sent in 1895 to try to revive the decaying placer industry on the Chaudière, St. Francis, and other rivers. He concluded that the gold was present on two levels—a postglacial level which carried a little gold; and preglacial gravels that contained the principal gold placers, which were buried beneath a heavy layer of boulder clay deposited during the glacial era. After the clay was deposited, the rivers began to dig down to their former levels; where they did so, an enriched gold-bearing gravel, combining the product of the recent age with that of the original Tertiary gravels, resulted. The thick overburden made the placers of the Eastern Townships very difficult to discover and mine. Furthermore, the gold did not occur in a continuous paystreak, but in pockets that had to be discovered one at a time. Chalmers suggested prospectors should drill from the surface and dig down to the gravels only where pay values were indicated.

In 1902 Bell employed Professor J. A. Dresser to trace the copper-bearing belts of the Eastern Townships by petrological and stratigraphic methods. In his first season Dresser found three principal belts of basic eruptives that carried copper values, in the second he further limited these copper occurrences to certain ancient volcanics, and in 1904 he discovered indications of gold as well as of copper in the volcanic formation. His final report (1906) classified the numerous copper occurrences in three main types, showing the characteristics and geographical locations of each. He described and accounted for the various Precambrian volcanics (copper-bearing), younger gabbro diorites and diabase and serpentines, still younger granites (later Devonian), and later again, dykes. The Monteregian Hills were part of a totally unrelated series, arising from an entirely different magmatic source. In 1907, he began a similar investigation of the serpentines "with a special reference to their economic products—asbestus, chromic iron, copper, talc, antimony, &c.,"[13] in which he examined around eighty mines, prospects and occurrences, his final report appearing only in 1914, in the new memoir series.

Thus the Survey had investigated all three major economic minerals of the Eastern Townships region in the light of the latest scientific knowledge of the time. The contrast indicates the different relationship that prevailed with the provincial mining agency. The provincial inspector of mines, J. Obalski, had no staff and did little beyond visiting mines, reporting on analyses of ore samples, and trying to promote markets for Quebec minerals. His report for 1904 was almost the first to describe a piece of field activity—and that was a description of his visit to Lake Chibougamau with Peter Mackenzie to see the asbestos Mackenzie had discovered the previous summer while prospecting for iron. Not until Theo C. Denis left the Survey to become superintendent of mines and the mineral production of Quebec climbed to $9.0 million (in 1911), did the division begin undertaking field operations of its own.

The Chibougamau discovery was also investigated by Low in 1905 as a result of "a largely signed petition . . . from prominent citizens of Quebec addressed to the Rt. Hon. Sir Wilfrid Laurier, asking that a geologist be sent from the Geological Survey to examine the newly discovered mining region of Chibougamau lake."[14] Low's report consisted of a general description of the area, its resources and the access to it, and a detailed description of the economic geology, particularly gold, veins of which he had seen; and iron and asbestos, occurring in at least two serpentine bands each over five miles long and upwards of a mile wide. He concluded that while the asbestos was comparable with that found at Black Lake, profitable mining was out of the question until a railway reached the district. Far from resenting this investigation of a mineral resource or even the discouraging advice, Obalski quoted some ten pages from Low's report in his own report for 1905.

Nor did the Maritime provinces feel any tenderness about the Survey's role; in fact, there the shoe was on the other foot. More than in any other region, in the Maritimes the Survey encountered a sustained demand for its work, and more local interest (not always of an enlightened kind) was shown in its activities there than elsewhere. The mining industry of the Maritimes owed its continued prosperity, and in some cases its existence, to the untiring efforts of the Survey, which was also looked on to inaugurate new mining industries that would enhance the regional economy.

XI-8 A MICMAC INDIAN CAMP, NOVA SCOTIA, 1891
Photographed by E. R. Faribault.

A sizable proportion of the work of the staff was earmarked for that region, where three to seven parties worked each season, besides shorter visits by specialists made to help with particular problems. Prince Edward Island was ignored, apart from brief forays by Ells and Chalmers in the eighties and nineties, until 1901 when Bell arranged for a local man to survey the island. As he explained, the inhabitants felt neglected by an institution their taxes helped support, there might be a chance of discovering coal, gold or petroleum, and there was botanical and natural history work that needed to be done. Finally, the proposed investigator, Lawrence W. Watson, was willing to work from his own home for only $350.00. With expenses, the work cost $367.52, for which Watson hunted for fossils, found no economic minerals aside from peat bogs and beds of "mussel mud," collected several hundred plants, and commented on the need to control the spread of the sand dunes by planned forestation. However, his suggestion that "coal areas may underlie the Post-Carboniferous rocks of certain districts at a depth not too great for mining"[15] put another face on the matter; Ells had to be sent to make a careful stratigraphic study in 1902. He concluded it was unlikely that a coal-bearing formation could be encountered closer than 2,500 feet from the surface, the outside limit of successful mining operations; or that it would carry more than thin, unworkable coal seams. A boring might be worthwhile to determine the succession, but it would cost money and would have to be carefully supervised. The government was determined to go ahead, however, and a number of borings for coal and oil were undertaken in 1908-09.

From Prince Edward Island, Ells proceeded to Gaspé to investigate "the geological structure of the basin in which the oil occurs,"[16] at the request of companies that had been searching for oil since 1889. Ells reported a Devonian horizon was present, but the strata were steeply inclined and very fractured and faulted, hence "the outlook can scarcely be regarded as favourable."[17] This report apparently put an end to the drilling for the time; at any rate the district was inactive when Ells visited it again in 1909 in connection with his investigation of the oil-shales of eastern Canada.

Collaboration among the survey, the mining industry, and the provincial mines branch reached its highest peak in Nova Scotia in the work of Fletcher and Faribault. As time passed, each man gained greater experience in the intricate geology of his area, and his usefulness to the mining industry increased until he was accepted as indispensable to the progress of the coal and gold mining industries.

This applied particularly to Faribault, who mapped the gold-bearing saddle-like reefs on a 1 inch = 200 feet scale and was able to delineate many such formations both large and small. This information, much of which was communicated verbally to the owners of properties, was of immediate practical value. He also took an interest in the mining side, studying deep-mining methods, the application of electric power, and new extraction processes, with a view to having them adopted in Nova Scotia. Good machinery, methods, and practices observed in one place were held up for other mines to copy. The provincial government consulted him on questions affecting the industry, while his reports on deep-mining methods caused the government to offer subsidies to encourage the sinking of deep shafts, and to request Faribault to assess the applications for assistance.

Fletcher was similarly regarded with great affection and respect. In 1903, when he gave a paper on the limits of the workable coals of the Cumberland field, one contributor to the discussion said that "about all Nova Scotia got from the confederation, was the work of the geological survey" and that if the federal government could not pay its personnel what they were worth, "I am sure Nova Scotia could afford to over bid the Dominion for Mr. Fletcher's services in this province, so that we would have the results of his work firsthand."[18]

Fletcher concentrated on the stratigraphy of the major coalfields of the province in Cape Breton, Pictou, and Cumberland, continuing the study that began in 1872 when he went to Cape Breton as assistant to Charles Robb and ended with his death in 1909 from pneumonia contracted while in the field. Through careful mapping and study of the strata he became the leading authority on the stratigraphy of the Paleozoic formations of Nova Scotia, able to advise companies where they might extend the workings or drill to reach lost seams. He proved that a certain coal seam in the Sydney district, which had been estimated to be over 1,400 feet below the surface, was only about 450 feet below ground at that point. Near Springhill, he succeeded in tracing the lowest seams worked at the mines more than 2½ miles beyond the point they had been mapped previously. This kind of work was assisted by means of borings, with a hand drill, that ranged from a few feet to as much as 146 feet—a method that was held up as economical and efficient in tracing the strata below ground. In 1904 his advice led to the sinking of a borehole at Pettigrew to a depth of 2,340 feet through unproductive cover to a seam of coal ten feet thick. Bell singled this out in his summary report as demonstrating the "value of exact geological work." He added further that this discovery was worth incomparably more than the cost to date of all Fletcher's geological work in Nova Scotia, and yet was "only one among many practical proofs of the

XI-9 MEMORIAL PLAQUE TO HUGH FLETCHER
Unveiled at Nova Scotia Technical College, Halifax, on May 21, 1926.

great value of his investigations, which are now represented on a considerable number of published maps showing his topographical and geological surveys of a large portion of the province."[19]

Fletcher's maps, like Faribault's, were eagerly awaited as guides to mining operations; Fletcher frequently revised them during his regular visits to add to his records of exact positions of all boreholes and the strata they cut. He gave expert witness at lawsuits and trials, and was consulted by the provincial government on matters relating to mining policy. The summary report for 1909 referred to the painstaking care and accuracy of all his work and stated he was "greatly esteemed both on account of his geological knowledge and charm of personality."[20] One who knew him remarked that "Personally, the late Hugh Fletcher had much charm of manner; he was stalwart and active; courteous and kind; and modest to a degree that one would scarcely expect from a man of so much ability. Among men of science he ranks as a keen observer of nature and an excellent stratigraphical geologist."[21]

New Brunswick, even more than Nova Scotia, looked to the Survey for guidance to reverse the long record of mining disappointments and failures. Indeed, the province seemed content to leave everything for the Survey and other governmental agencies. A major obstacle to mining development, L. W. Bailey was convinced, was simply the lack of enterprise of the mining community—its unwillingness to prospect for new deposits and orebodies while present mines were being exhausted, and its failure to keep abreast of new technological advances that might make the exploitation of known occurrences economic. This was the central refrain of the report he prepared in 1899, at Dawson's request, on the mineral resources of New Brunswick. More active prospecting was the crying need—prospecting to follow up and utilize the findings of the Survey on the geological formations of the province. His report, therefore, outlined the geological formations of the province from "Laurentian" upward, gave their locations, and evaluated the potential of each as a possible source of economic minerals. Then Bailey proceeded to discuss each mineral—location, mining history, past results, and future prospects.

Given the crucial importance of a local mineral fuel industry, Bailey advocated a systematic series of borings to settle whether the coal formations of Nova Scotia extend to New Brunswick, and also utilization studies of the bituminous shale deposits. Bell took up the suggestions by having Bailey and Professor H. S. Poole of Dalhousie College study the coal problem and Ells the bituminous shales. Bailey and Poole concluded that the flat-lying, narrow coal seam at Grand Lake in the Carboniferous basin of New Brunswick was the equivalent of the lower seam being mined in Cumberland County, Nova Scotia, and that there was no equivalent of the upper seam. There appeared little chance of that upper seam appearing east or north of the lower seam, nor any place where the lower seam might be thicker, though this should be checked by drilling. In the meantime, efforts should be made to develop the known deposit at Grand Lake by strip mining or by parallel trenches. Ells investigated the extensive Devonian bituminous shale beds of Albert County, and to evaluate its economic potential, compared it with a similar operation in Scotland. He recognized the problem was economic—of extracting the hydrocarbons at prices that made them competitive with liquid petroleum, and its derivatives—a task to which his son and assistant in 1902, S. C. Ells, was to devote his working career with the Mines Branch in the oil sands of northern Alberta.

* * *

THESE SAME OIL SANDS after having been explored by Bell (1882) and McConnell (1889) entered their first development phase in 1894. Following Bell's first traverse of the region, a government reserve had been proclaimed over the area to prevent the premature disposal of a valuable natural resource. But now, with a railway built to Edmonton, inquiries were beginning to be received about lands there, and the Dominion government needed guidance as to what policy to follow. The central question was whether the vast Cretaceous sands beds indicated a proportionately large petroleum field in the underlying Devonian. Nothing would so stimulate settlement of the district as news of a proven oilfield. Too, the Survey needed to know more about the trends of the strata at depth. These considerations induced Selwyn to order an experimental boring, with his minister's approval, at Athabasca Landing, a most convenient site from the logistical standpoint, where McConnell anticipated the oil sands would be encountered at a 1,200–1,500-foot depth and the Devonian a couple of hundred feet farther on.

By October 1894, the driller, W. A. Fraser of Petrolia, had reached a depth of 1,011 feet, and all was going well except that some formations proved thicker than expected. In 1895 and 1896 Fraser carried the well to a depth of 1,700 feet, where, because of drilling difficulties, it had to be abandoned, while still not down to the desired strata. For 1897 Dawson proposed to drill two wells, one nearer the bituminous sands at Pelican River, where the overburden would be much less, the other near Victoria on the North Saskatchewan, where productive beds should be reached at about 2,000 feet. The Pelican well at 750 feet penetrated oil sand, then at 820 feet started a large gas flow that made it impossible to continue. Leaving the well open over the winter for the gas to clear, Fraser returned in 1898 but had proceeded only a few feet more when it became necessary to abandon the well entirely. The Victoria well was equally disappointing; three seasons of drilling carried it to 1,840 feet, far short of its objective, when the drill became blocked and this too had to be abandoned. For an outlay of $38,131 (more than one year's total field expenditures at that time), the Survey had done little more than gain some stratigraphic information, demonstrate how difficult it would be to discover oil in the oil sands district, and indicate that northern Alberta was still some way from becoming the locale of a major oilfield. A gasfield of unknown dimensions had been discovered, and much had been learned about coping with some of the obstacles faced during drilling operations. The Dominion government thought it had its answer too; the area was opened to prospectors for petroleum under general regulations in 1901, which permitted a discoverer of petroleum in paying quantities to buy 640 acres of land surrounding his discovery at a flat rate of $1 per acre.

This disappointing experience was quickly forgotten, as the Survey found new tasks in western Canada in connection with the region's coal resources. Everywhere fuel was suddenly becoming an acute problem. Factories, railways, paper mills, smelters required increasing quantities of cheap fuel, and regions that wanted to progress had to be able to furnish this to attract new industry and capital investment. To these years, then, belonged the Survey's efforts to discover and assist the development of coalfields, oil and gas, bituminous sands and shales, and peat bogs.

Above all, the Survey gave its attention to opening up the coalfields of the Rockies. In the 15 years after 1896 no fewer than five new railway lines were directed towards these, either as their terminus or as an important stop on a through line to the Pacific coast. These lines required large coal tonnages for their own operations, and besides they hoped to haul millions of tons of coal as paying freight.

A commission from the Dominion government drew the Survey into the Crowsnest Pass coalfields in 1900. Both Dominion and British Columbia governments had sought a railway between the Kootenay district and the rest of Canada to replace the present lines of communications with Spokane, Washington, and Butte, Montana. The province offered a 250,000-acre coal lands subsidy for a railway into the Kootenays and chartered a railway company to build the line. The CPR took over the charter and the land subsidy, then made an agreement with the Dominion government for a large cash grant to help with the building of the line. In return, the railway reduced its freight rates (the famous Crowsnest Pass Agreement), and also handed over 50,000 acres of the land grant to the Dominion. The Survey was designated to select this acreage.

There was considerable pressure on Dawson to make the selection in 1899, but he refused to be hurried, and only in the summer of 1900 did a party, consisting of McEvoy and Keele and A. O. Wheeler of the Dominion Topographical Survey, proceed to make the survey. By this time, 1,000 tons of coal was being mined daily at Coal Creek, while other mines were being opened at Michel and Morrissey Creek. McEvoy estimated there were some 230 square miles of coal measures (about 150,000 acres), that provided at least 100 feet of workable coal, mostly in two seams, each 46 feet thick. He calculated that the basin contained some 22,000 million tons of coal, submitted his memorandum with the suggested selection of lands in February 1901, then resigned and took a position with the Crowsnest Pass Coal Company.

Faced with this emergency, Bell withdrew Leach and Denis from their regular work to finish the survey and complete the selection. They found a promising northward extension of the field between Elk and Fording Rivers. Leach returned to the Crowsnest in 1902 to examine the newly-opened coalfields on the eastern slope of the pass at Blairmore, and at Frank, where a company which had gone into production the previous year was already producing 500 tons per day. The field held very good prospects—plenty of good quality coal, a railway, and a location where the coal could be mined by working up an incline into the mountain.

Unfortunately, in just two years, tragedy was to strike in the early dawn of April 29, 1903, when an enormous mass of rock broke away from Turtle Mountain and overwhelmed a corner of the town of Frank and the outside workings of the mine to claim around 70 lives. McConnell and Brock were immediately ordered to investigate the catastrophe and report on its causes. They explained the slide as caused by the breaking away of a tilted mass of limestone from the underlying shale and sandstone along a zone of weakness between the two. The triggering action might have been the effect of a large earthquake in the Aleutians in 1901, in a small degree from the mine workings, but principally from alternate freezing and thawing along the fracture zone between the two rock masses. There had been a series of warm days and cold nights, and on the fateful night the temperature had fallen to zero.

Leach traced the Cretaceous coal formation some distance to the north in 1901, while Poole examined the coalfield and mine in the Kicking Horse Pass district along the C.P.R. main line. In 1903 Bell assigned Dowling to undertake a thorough investigation of the geology of the coal-bearing formations of the Rocky Mountains in Alberta. Dowling, who had worked in a very wide variety of employments, had already investigated the Souris and Turtle Mountain coal beds. Now, in 1903 he was embarking on the major work of his career, in which he became the leading authority on the geology of the Plains area and of coal resources in Canada, and "laid the foundation for the study of stratigraphy and structural geology in southern Alberta."[22]

XI-10 GOLD WASHING WITH A "GRIZZLY," 1898
This small hand operation on the North Saskatchewan River above Edmonton was photographed by G. M. Dawson.

Dowling began by tracing the coal beds from south of Bow River, in the Sheep Creek and Highwood River districts, northwest to the Cascade coal basin, where Gwillim was prospecting for the CPR, and on to Panther River. In 1904 he followed the strata of the Cascade coal basin south to Kananaskis, then north beyond Panther River to the Costigan seam and the headwaters of Red Deer River. In 1905 he worked mainly south of the CPR main line, examining the structure of the Elk River coal basin north from Fernie. He found that Elk River coal beds did not appear to extend beyond the Rockies to the Kananaskis valley, which was the physiographic extension of the Elk River valley. In the same season D. D. Cairnes, on his first work with the Survey, observed good lignite seams in the Foothills in the Kootenay formation, hitherto believed to exist only within the mountains.

In 1906 and 1907 Dowling, assisted by G. S. Malloch, explored the northward continuation of the Cascade and Costigan coal basins and searched for other coal seams "in the hope that workable coal seams of the better class of coal might be found nearer the proposed route of the railways through the Yellowhead pass."[23] He outlined the coal areas of the Cascades, Palliser, and Costigan basins and discovered a new basin of good coal-bearing rocks exposed in the Bighorn Range, the future Nordegg-Brazeau district. In 1907, while Malloch completed mapping the coal basins south of the Bighorn, Dowling began with the Brazeau, traced out a number of valuable seams along Bighorn and Brazeau Rivers and predicted that "the development of mines in this district will both extend settlement to another large fertile area west of the Saskatchewan and open up a large lumbering area."[24] He left by descending Rocky River to the Athabasca, observing further exposures of coal at Prairie Creek, which he felt would be useful for a cement industry near the mountains on the transcontinental railway being built through the Yellowhead Pass. The series of coal mines and the cement plant were all in operation by 1914.

The concern over Canada's fuel problem extended also to British Columbia, where in 1904 and 1905 Ells and R. A. A. Johnston examined the coal outcrops of the Nicola basin south of Kamloops. In 1905, Ells' main work was to examine the coal deposits and other mineral resources of Graham Island, the northernmost of the Queen Charlotte Islands that Richardson had studied over 30 years earlier. The principal coal deposits were found in Cretaceous strata and were mainly of bituminous grade that sometimes passes into anthracite. However, Ells indicated that the beds were narrow, uneven, and very broken by faults, and that poor transportation was an added difficulty.

Accompanying Ells to Nicola in 1905 was Poole, who spent the main part of that season investigating the Nanaimo-Comox coalfield of Vancouver Island, principally to get at the probable dimensions of the mineable coal, the closing of a major mine having started rumours of an impending shortage. Poole reported the usual difficulties: the heavy vegetation and thick overburden, extreme variability of the seams (unlike Cape Breton where, he said, "the beds carry a fairly uniform thickness for miles"[25]), and reluctance to supply information about past borings. Nevertheless, he ventured an estimate:

> Still, if it be desired that a conjecture be hazarded of the quantity of coal exceeding a thickness of two feet, and within a vertical depth of 4,000 feet, an estimate of 600 million tons, though based on most incomplete data, would seem conservative and yet at the same time sufficiently large to allay apprehensions of any immediate shortage in the output.[26]

In the north of the province, Leach was occupied in 1906 and 1907 examining the Telkwa-Bulkley River country south of Hazelton on the upper Skeena River, along which the Grand Trunk Pacific was building towards its terminus, Prince Rupert. He also devoted considerable effort searching for economic minerals, notably coal in sediments of Cretaceous age or younger, but also silver and copper ores associated with the older volcanic rocks. In the Telkwa valley, subsequent volcanic activity and the associated folding and faulting had converted the coal into a semi-anthracite, but made for seams that were irregular and uncertain. In 1907 Leach examined this and other properties, but concluded that

> It is now fairly certain that no great coal field exists in the Bulkley Valley district from Hazelton to the headwaters of the Morice, but many comparatively small, isolated areas are known in which coal varies from a lignitic to a semi-anthracite. In some of these areas the strata are greatly disturbed, much faulting and folding being in evidence.[27]

But the development of coal mines in Nova Scotia or Alberta still could not meet the most urgent fuel needs of the populous, industrialized, coal-poor central provinces. Hence the possibilities of developing the many peat bogs of the fuel-deficient parts of the country were not forgotten, and such bogs were described in many accounts and listed among the economic minerals. Chalmers was the expert, though several others, Ells in particular, were also authorities. A comprehensive review prepared in 1904 by Chalmers estimated the bogs to aggregate at least 37,000 square miles and listed 127 specific bogs in accessible districts as possible candidates for development. The subject, which Hunt had treated in the early days, has recurred again and again throughout the histories of both the Survey and the Mines Branch, especially in times of fuel shortages.

XI-11 THE AFTERMATH OF THE FRANK SLIDE, 1903
A 1.03-square-mile-area of the valley and the town was buried to a depth of from 3 to 150 feet.

The work of Chalmers on the Pleistocene, ranging over present and past landforms, was almost all of immediate practical value, and as the public came to appreciate the importance of expert advice on environmental questions, Chalmers and others began to be consulted on such matters. The contractor building the great transcontinental railway bridge across the St. Lawrence at Quebec asked Ami, who was frequently working on the strata at Sillery, to check the formations beneath and adjacent to the piers and abutments. Ells, like others before him, was called upon to explain and report on landslides. Civic administrations, increasingly in need of larger, safer, more secure water supplies, turned to geologists for advice on where to find better sources of potable water. Fletcher advised the people of Truro, Chalmers the water commissioners of London, while F. D. Adams helped secure a survey of the underground water resources in the neighbourhood of the city of Montreal. Geologists were also asked about satisfactory sources of construction materials—sands, gravels, and clays—and to evaluate the qualities of soils for agricultural purposes.

From the Eastern Townships, Chalmers moved on to southern Ontario, a region largely neglected apart from periodic studies of the oil and gas and salt industries. There he began a study of the wells and borings and also of the surface geology. Like the true specialist, he welcomed and was stimulated by the challenge of working in the new setting. He studied the rises and falls of the land and the changing shapes of the Great Lakes as explaining features of the present terrain. Tracing the raised beaches along the shores of the lakes and the rivers flowing into them, he concluded there had been at least three major changes in the levels of the land, reflecting two distinct glacial periods separated by a long interglacial interval of moderate temperatures.

During his second season in 1902, A. F. Hunter also began a detailed investigation of the raised beaches, selecting his home district, Simcoe County. Later in 1904, Hunter examined the beaches along the Blue Mountain Escarpment, and in 1907 he traced shore lines from Georgian Bay to Ottawa River along the south margin of what has been called the Nipissing Gateway. Both Chalmers and Hunter tried to work out the implications of the tilting and later rectification of parts of the landmass to the present angle of inclination.

Some of these implications were indicated in J. W. Spencer's study of Niagara Falls, which demonstrated the connections of geology with hydro-electric power resources, and with the preservation of scenic values for tourists and future generations. Spencer had pursued the subject for many years; he studied how the various rises, falls, and tiltings of the land altered the drainage patterns of the Great Lakes, and had defined and traced the Algonquin beach from Georgian Bay around the lakes. All this culminated with his study of the phenomenon of Niagara Falls—how it was formed, its rate of progression, the factors shaping its development—which led to a wide variety of related sciences:

> The recession of the Falls through the different strata is the ordinary limit of research required of the geologist. But the changes, in the volume and currents of the river, in the height of the Falls, and in the effects of the buried valleys, determined by causes acting far from the great cataract, opened a new field of investigation, as did the application of more precise methods of research than were formerly followed, so that the Falls of Niagara have given rise to a chapter in science, belonging entirely to themselves, which had not hitherto been understood, and which could not have been interpreted by any outside standard.[28]

Further studies followed, including in 1904 the first survey of the water flow on the Canadian side of the falls, whose findings completely altered the accepted views of the physics of Niagara River. Bell hired Spencer to supply a detailed account of the interesting phenomenon and important natural resource, the value of which was only then being grasped. Though Bell was fiercely criticized for the arrangement with Spencer, Low continued and extended the study. The 500-page report, *The Falls of Niagara*, the most comprehensive and important contribution to the subject to that time, appeared in 1907.

Its implications for future public policy were obvious; the probable effects of the diversion of water for power purposes, and the accurate mapping of the boundary which here divides a natural resource of immense material value, are only two of the most important aspects.

The report discussed the hydro-electric power potentials of the falls and the whole of Niagara River; it enlightened Canadians about the extensive water diversions by American power companies; and it provided a wealth of accurate information for future popular accounts and tourist literature. On the side of geological history, Spencer reported on the evolution of the Great Lakes basin, the history of the Niagara gorge, and the two buried channels of former rivers—one of them a former channel of the Niagara, the vestige of which is the Whirlpool below the falls.

Minerals, soils, water-power, and more besides, were involved in the efforts by the Survey to assist the railway-building projects of the boom years of 'Canada's Century.' As in the past, its men were called to make reconnaissance surveys of the sum total of natural resources along the proposed railway routes as a help in planning the location of the roads, to advise on possible sources of traffic, and to assist the government generally with its plans to settle and develop the territories opened by the railways. Bell's own contribution to this work in the region between Quebec and Winnipeg was enormous.

As early as 1898 Dawson had sent McEvoy to explore the route from Edmonton west through Yellowhead Pass. McEvoy's report stressed the advantages of the Yellowhead Pass route for railway construction, particularly that it was free of snowslides and was an easy link between the important Edmonton and Fraser River districts. He also spoke favourably of its agricultural possibilities, as far as they were indicated by Mr. Swift's farm and ranch near Jasper House. In somewhat similar fashion Bell sent James M. Macoun in 1903 to investigate the agricultural potentialities of the Peace River region, which by then was attracting much attention as the possible route of the Grand Trunk Pacific Railway and was beginning to draw a few farming pioneers. Macoun's report was very severe. While he conceded that agricultural conditions appeared satisfactory in the flat lands along the river valley and at Fort Vermilion in the north, he emphasized these were limited areas and that the greater Peace River district was a different matter. Because of the severe climate, notably the frequent early frosts, "notwithstanding the luxuriant growth that is to be seen almost everywhere, the upper Peace River country, to which so many eyes are now turned, will never be a country in which wheat can be grown successfully. That this grain will mature occasionally there is no doubt, but that it will ever become the staple product of any considerable area I do not believe."[29]

For such an opinion to come from the son and pupil of John Macoun, who, in the seventies, had been enthusiastic about the possibilities of the region and whose long-questioned predictions respecting the southern prairies were being so abundantly confirmed, seemed ironic. Small wonder it was so bitterly attacked in the press and in Parliament, and earned Bell a reprimand for having let it escape into print without suitable "revision."

After the routes of the two new transcontinentals and the Hudson Bay railway had been located, the Survey was called to make specific explorations of those parts that were through relatively unknown territory. In 1906, six field explorations, and in 1907 three more, were assigned expressly to such work. Moreover, many other field assignments, like Dowling's or Leach's in the western Canadian coal basins, were inspired by the same general purpose. The National Transcontinental-Grand Trunk Pacific project, which was the main beneficiary of these efforts, received the following explorations: R. A. A. Johnston, through central New Brunswick (1906); J. A. Dresser, eastern Quebec to the St. Lawrence (1908); O. O'Sullivan, LaTuque west (1907); W. J. Wilson, east from Bell River (1906 and 1907); W. H. Collins, Lake Nipigon west (1906 and 1907); R. Chalmers and John Macoun, across the prairies (1906); and R. G. McConnell, Prince Rupert to Hazelton (1912). In addition, McInnes and O'Sullivan explored the line of the Hudson Bay Railway from Saskatchewan River to Fort Churchill in 1906, and McInnes worked in similar country (the lower Saskatchewan River and Carrot River districts) in 1907. These were strictly reconnaissances, dealing with the general geology and economic minerals of the regions traversed, also timber, arable land, fish and game, hydro-electric power and other resources of the districts. The reports that discussed rock formations in the main were those of men like W. H. Collins, who were working in country of exposed Precambrian rocks. McInnes and W. J. Wilson tried, but their districts simply did not offer much scope for bedrock geology.

Chalmers and Macoun were not concerned over this lack. Chalmers, who was making his acquaintance with the surficial geology of the western plains and the lower mainland of British Columbia, discussed clays and brickworks, and qualities of the soils in different parts of the prairies, the relief, erosive work of the rivers, glaciation, drainage features, groundwater supply; and in British Columbia, where he went as far as Burrard Inlet, marine deposits and shorelines. For Macoun, who had traveled by wagon across the prairies to Edmonton along his route of the seventies, it was a triumphal

XI-12 NEW BREAKING ON WESTERN CANADA'S PRAIRIES
A ploughed field and sod house near Eagle Creek, Saskatchewan, photographed in 1908 by D. B. Dowling.

return:

> The conclusions regarding the fertility of the soil which I published in 1872, 1879 and 1880 have been practically illustrated by the results obtained by actual experiment. At this time it is conceded by all observers that the growth of grain throughout the whole of what was formerly called the 'Fertile Belt' is no longer an experiment, but an actual fact and can be relied on for all time.[30]

He traveled through what appeared "an almost continuous wheat field," then farther west through districts that were still not settled, but soon would be. At length he reached Edmonton, from whence the elderly botanist took the train back to Ottawa.

These were only a few of the many and varied ways in which the field operations assisted the mining and general economic development of Canada. In all his summary reports, but particularly those of 1904 and 1905, Bell strove vigorously to impress on his readers the strenuous efforts the Survey was making to assist the mining industry, and to hasten the attainment by Canada of the golden destiny promised by its resources and institutions. His 1903 report claimed for the Survey a major share of the credit for the remarkable growth of the mineral industry during the previous 15 years:

> It is, therefore, marvellous that such great services could have been rendered the country at such a small cost, by the above-mentioned liberal publication, by striking displays of our economic minerals at so many International Exhibitions and in the Museum at headquarters, all simultaneously with the vigorous prosecution of the examinations of mining districts and of general geological and topographical surveying over half a continent, for the most part lying in a state of nature. The comparatively rapid progress which has been made, in spite of artificial hindrances, in the development of our mineral resources, now yielding upwards of $60,000,000 a year, is due to the above efforts more than to any other cause.[31]

In 1904, after outlining the field work done during the year, he stressed that "it was nearly all of thoroughly practical character, intended to promote the discovery and development of the mineral wealth of the Dominion," and went on to launch into an extended passage that went from Logan's advising the laying out of the Woods Location on Lake Superior on which the later Silver Islet mine was found, to Fletcher's recent advice on the placing of a boring for coal at Pettigrew that "may prove to be worth many millions of dollars."[32]

Ironically, despite these invaluable efforts, the Survey was still being subjected to a most dangerous attack on this same score of its allegedly inadequate performance in relation to the needs of the Canadian mining industry. This helps explain Bell's frantic efforts to justify the role of the Survey. For never before in its history had such fierce pressures been put on it by the mining industry, the public, and the government, to undertake work of immediate economic utility. Indeed, Bell's successor, Low, was even more forthright in proclaiming his faith in the "practical" role of the Survey, as his report for 1906 made abundantly clear:

> By the advice of the Minister of the Interior the field work during the past season was mainly confined to economic subjects, the investigations being carried on in (a) mining districts developed, (b) mining districts under development, and (c) districts along proposed routes of new railways. No parties were sent into the far north or into regions difficult of access. . . . As will be seen, the field parties may be divided into those performing exploratory work and those devoting their time to economic geology. Although the work of the former division frequently includes that of the latter, a tentative classification of the parties under these divisions shows seven engaged on economic geology, six on economic work of an exploratory character, six on exploratory work of a more or less economic nature, and five in special work relating to the mineral and natural resources of the country.[33]

XII-1 CLIFFORD SIFTON
Minister of the Interior, 1896-1905

12

Into the Department of Mines

THE CONTROVERSIES OVER THE SURVEY'S FUNCTIONS as among the geographical, scientific, and economic aspects of its work, and the scramble for the directorship of the years after 1901 were the outward symptoms of a difficult struggle for survival that the institution faced as it entered its seventh decade of service. At a minimum, it faced a move to alter its existing pattern of operations, its traditions, and its power to determine its own role; at a maximum—the complete dismantling of its organization, the reassignment of its staff, and the abolition of the historic institution. The Survey had faced challenges in the past, but what was new and dangerous about the present threat was that it came, not so much from the public as from its political superior, Clifford Sifton, the Minister of the Interior in the Laurier government, who was firmly convinced that the Survey's principal functions should be transferred to a new institution of his making. While he moved towards that goal, he kept the Survey in a deprived, uneasy state that had unhappy effects on its personnel and deplorable results upon its work. These strengthened the criticisms of the Survey's performance and wore down the resistance of its staff against accepting the revised role demanded of it. The period finally ended in 1907 with the Survey losing its independent status as a department of government and becoming a branch in a newly-established Department of Mines.

By the end of the nineteenth century, the Canadian mining industry was entering a new phase in its history—as a major producer of wealth rapidly approaching the $100-million level, and comparable with forestry or agriculture as a source of export earnings. The important copper mines of the Eastern Townships were supplemented by the beginnings of copper production from Sudbury, and the rise of new base metal mines in southern British Columbia. All of these found markets in the United States in varying states of manufacture—as sorted ore from the Kootenays, as copper-nickel matte from Sudbury, and as smelted metal from the Eastern Townships. The national interest lay in having as much manufacturing as possible done in Canada. Other mining industries, notably graphite, phosphate and iron ore, faced serious economic problems because of the discovery and development of rich deposits of these minerals in the United States, while operators needed advice on more efficient mining and processing techniques and on markets. Prospective mines, notably at Sudbury and in southern British Columbia, were hampered by difficult metallurgical problems associated with the separation of nickel and other metals at Sudbury, or of lead, zinc, and silver in British Columbia. These were some of the areas where the need for governmental assistance was felt.

By the end of the century, also, the industry was becoming a significant political force not only because of its greater size and economic importance, but because it had become much better organized to exert pressure on governments. Paralleling the inauguration of mining bureaus in the provinces, the General Mining Association of Quebec was established in 1891, the Mining Society of Nova Scotia in 1892, the Ontario Mining Institute in 1894, and the Federated Canadian Mining Institute in January 1896, to be followed by the Canadian Mining Institute in 1898. At their annual meetings, besides hearing papers on important or interesting aspects of company operations and technological and scientific developments the members found a forum in which to express their collective views or to mobilize the industry to press governments for favourable concessions.

XII-2 ROAST YARD OF THE CANADIAN COPPER COMPANY, COPPER CLIFF, ONTARIO, ABOUT 1900
In the early years the copper-nickel ores of Sudbury were concentrated by simple roasting in the open air on piles of local timber. The dense fumes of sulphurous gases that resulted denuded large tracts of land of their vegetation.

The work of these associations was supplemented by the mining journals, notably the *Canadian Mining Review*, edited since 1886 by B. T. A. Bell, a Scottish-trained engineer who came out to Canada as a young man, worked briefly on the CPR, then quickly drifted into newspaper work. He soon became the foremost spokesman of the industry, wielding such a vigorous pen that he was said to be involved on an average in one libel suit per year. He helped organize the association in Quebec and became its secretary, he founded the Ontario Institute, he had a strong influence with the Nova Scotia association, and he became the first secretary of the Canadian Mining Institute in 1898. By then he had also taken over full ownership and control of the *Canadian Mining Review*, making himself the unquestioned voice of the entire industry. He completed the merger of the various interests early in the new century by securing the removal of the Canadian Mining Institute offices and library from Montreal to Ottawa, handy to the federal government. Until his death on March 1, 1904, at the early age of 42 from the effects of an accident, he was a most potent force in the industry and in presenting its views before the Dominion government—or as some critics alleged, impressing the views of the Minister of the Interior on the mining industry, in exchange for favours received.

Of course, mining was essentially a field of provincial jurisdiction, as most provinces were the owners of the mineral resources and regulators of mining operations within their boundaries. Yet because its needs were so great, and because of its size and importance, the mining industry looked particularly to the Dominion government for help, and began to claim its participation almost as a right. The federal government was urged to play a larger role to aid the mining industries by its trade and transportation policies and through the Geological Survey. It was called on to collate and co-ordinate provincial activities, help improve mining and processing on the technical side, and create a cabinet ministry responsible for the needs of the mineral industries. After all, the mining men argued, fisheries was far less significant to the nation as a whole but had its own minister, while agriculture, also an area of provincial jurisdiction, was represented in the cabinet. They began pressing for a larger federal role in the mining and metallurgical sides of the industries, either within or outside the framework of the Geological Survey.

For the Survey, this all had a familiar ring—Logan had been faced with similar demands, and in the fifties and sixties, when there were far fewer pressures on it, the Survey had done a good deal to advise operators on mining, metallurgical, and technological advances in Europe and America. But the enlargement of its area of operations and the necessity of coming to grips with important geological problems left the Survey, with its limited resources, little scope for dealing with the difficulties of particular mining industries, regions, or individual mines. The House of Commons committee of 1884 had commented on this seeming neglect:

> The attention of the Survey to the mineral and economic resources of the country—its gold, copper, iron, phosphate, lime, gypsum, manganese, &c.—appears to be much less than it formerly was, even although the importance of the subject, and the means of acquiring and publishing information in reference to it, have largely increased.[1]

It had attributed the alleged neglect to misdirected effort, and urged that the Survey's program be changed: "In the opinion of the Committee, the primary object of the Survey should be, to obtain and disseminate, as speedily and extensively as possible, practical information as to the economic mineral resources of this country, and scientific investigations should be treated as of only secondary importance, except where necessary in procuring practical results."[2] Its chief recommendation was for the Survey to establish a division devoted exclusively to the interests of the mining industry, headed by a mining engineer with the rank of assistant director, to report on mining and metallurgical developments and prepare statistical and other records of the progress of the industry. But the recommendations of the committee remained little better than a pious wish. Selwyn and Dawson complied by appointing mining engineers and establishing a Section of Mines, though they did not make its head an assistant director. The field officers paid as much attention as they could, and in the course of their surveys reported extensively on mining operations and deposits of economic minerals. From time to time an officer prepared a special study of a particular mining region—Lake Superior, British Columbia, Quebec, or New Brunswick.

The appeal of the mining industry for federal government assistance was reciprocated by the dynamic new government elected in 1896. But what could the hard-pressed Survey do about accepting added responsibilities? Dawson in his first annual report addressed to Sifton, his new superior, stated that "The fundamental work of the Geological Survey is that of providing geological maps and reports of the several parts of the country, such as to be of value to the explorer, the miner and others," then went on to say that the Survey was finding it impossible even to keep up with this responsibility because of the increased interest that was "simultaneously shown concerning all parts of the vast area of Canada." Far from assuming new responsibilities, Dawson called on the provinces to help out at least by undertaking their own topographical mapping "as the provincial revenues are those actually benefited by the sale of mining lands and royalties on output,"[3] to enable the Survey to proceed more quickly with its primary task. When in September 1896 a project of establishing a "Dominion Mining Bureau" in Montreal was mooted, Dawson presented the Prime Minister with a memorandum filled with such questions as: What would it do that was not already being done by the Survey or the provinces? Would the government be responsible for its operations? Would not its location in Montreal raise similar demands by other centres?—and the like. Ingall, who attended the meeting, reported the project had little support, while Dawson recorded in his diary, with evident satisfaction, that "Laurier apparently dismissed the subject."[4]

That the government remained restless over its inability to accommodate the desires of the mining industry was indicated from a later entry in Dawson's diary of April 7, 1899, "Suggestion made of the appointment of Mining expert to Dept. of Interior. I do not think this would be a good move in interests of Survey which the Minister has not learnt to use in that way, but did not think it advisable to push objections strongly as the project may fall through." What Sifton was proposing was drastically to alter the role of the Superintendent of Mines in his department, currently occupied by the famous western pioneer William Pearce, and limited strictly to administering mining land concessions, collecting fees and royalties, and enforcing the mining regulations in the lands under Dominion control. Under a new "mining expert" the office would become responsible for compiling mining statistics, examining technical processes, preparing reports on particular industries, and publicizing the opportunities in Canadian mining. This was precisely what the Section of Mines already was attempting to do; which accounted for Dawson's lukewarmness, and his comment about the minister's perhaps deliberate failure to make use of the Survey's facilities.

No doubt Dawson knew something about the candidate for the proposed appointment, for Sifton was thinking as much of the man as of the position. While a student at Victoria University (at the time located in Cobourg, Ontario), Sifton had been powerfully impressed by his professor of physics, Dr. Eugene Haanel, and a friendly relationship was formed that survived the passage of years. Haanel, a native of Breslau, Germany, had come to Canada in 1857 as a youth of 16, and embarked on a varied teaching career that included teaching modern languages at two small colleges at Michigan, becoming professor of natural science at a third Michigan college, then the same at Victoria University from about 1873 until 1888, when he transferred to Syracuse University in New York. He was said to be a highly effective teacher, and his departure for Syracuse was necessitated by the entry of Victoria University into the federated University of Toronto, at which time it had to abandon its program in the sciences. He was an original member of the Royal Society of Canada, Section III (Mathematical, Physical and Chemical Sciences) but had resigned or been dropped from membership by 1889, perhaps because of his change of employment or allegiance.

Sifton's elevation to the Dominion cabinet in November 1896 was greeted by Haanel in a letter that referred pridefully to how his faith in his pupils at Victoria was being confirmed: "The older I grow the more I cling to the early friendships formed with the noble young men of Victoria especially those of the V. P. Association, and I always feel that I share in their success and I certainly rejoice in the great work they are accomplishing."[5] It is difficult to know how far Sifton was impelled to revise the role of the Dominion government in the mining industry by motives of personal friendship, and how far by a genuine desire for reform. Probably both were equally real; the kind of reforms he envisaged were entirely in keeping with his pragmatic, energetic, materialistic temperament, and his particular brand of nationalism which looked worshipfully on American methods and practices for appropriate models in dealing with the problems of Canadian development. On the other hand, his friendship for Haanel was reflected in his determination that the new role should be implemented in a separate, independent command for Haanel. And it was to Haanel that Sifton turned for advice as to what that role should be.

That Dawson failed to dissuade his minister from his purpose is evidenced by the fact that from February 23, 1900, onward, when Haanel received favourable assurances as to his prospects from Sifton, he "made (quietly) arrangements to make the change from here to Ottawa."[6] The ability of the Survey to meet the threat was vastly worsened by Dawson's sudden death on March 2, 1901. Dawson's political acumen, powerful connections, and fame were such as to have made it most difficult, even for a minister with Sifton's drive, to take over or supplant the Geological Survey, or transform it into the kind of mining bureau that Sifton and Haanel wanted. For one thing the mining industry was solidly behind Dawson and the Survey, as indicated by the kind of resolutions the Canadian Mining Institute had passed at its annual meetings. The *Canadian Mining Review* was also generally appreciative and laudatory of Survey efforts. It greeted the appearance of the *Annual Report for 1897* (in 1900) with a favourable review, to the effect that the 1,000 page volume "again bears testimony, if any were

XII-3 EUGENE HAANEL

needed, to the great value and utility of this important, and sometimes overlooked, branch of the public service."[7] Later that year, however, with praiseworthy impartiality, it reprimanded both the Bureau of Mines of Ontario and the Survey for neglecting work in eastern Ontario, for being more interested in exploring remote hinterlands than helping industry at hand, and for failing in their reports to enlighten the public sufficiently on the mineral resources of Canada.

Dawson died on the eve of the annual meeting of the institute which opened its meeting by passing a resolution of condolence, and closed it by urging the minister to appoint as his successor the distinguished F. D. Adams. On the evening of March 7, 1901, it dispatched the following telegram to Sifton: "Canadian Mining Institute at a large and representative meeting tonight unanimously recommends the appointment of Dr. Frank D. Adams, an old member of the Survey staff and Logan professor geology at McGill, as successor to the late Dr. G. M. Dawson."[8] With such backing, his strong interest in petrography and mineralogy, his well-deserved reputation for scholarship, and his long association with the Survey, Adams would undoubtedly have made an excellent choice to meet the challenges of the time—of winning and keeping the respect of the officers and of the scientific community, while guiding the Survey to make its work more directly applicable to current mining needs. Similar advice was forthcoming from A. P. Low, who wrote Sifton on March 5 that "The only man in Canada approaching Dr. Dawson's celebrity is Dr. F. D. Adams of McGill University. From many conversations with foreign scientists last year in Paris I know personally that he is well known and is looked upon in Europe and the United States as the foremost scientific man in Canada today."[9] However, offsetting this, was the letter of another correspondent advising Sifton to disregard the institute's recommendation, as having been rigged by a small faction of the membership.

Sifton had little need of this advice; his informant was advised that the resolution was gratuitous interference, and if the institute continued to allow itself to be used, its legitimate influence would be lost. His own mind was made up. Already he had heard from his private source. Immediately the news of Dawson's death reached Syracuse, Haanel had hurried to write Sifton "whether the sudden removal of Dr. Dawson will not offer an excellent opportunity for reorganizing the Survey on lines more suitable to the needs of your Department and would not the inauguration of the Dept. of Mines come in quite naturally with the institution of a new regime in the Survey Office?"[10] For the present, then, Sifton met the immediate problem of the management of the Survey by notifying Dr. Robert Bell by letter to take over as acting director "till further notice."[11] No doubt he hoped eventually to make some provision for Dr. Haanel's talents inside the government service but this would obviously need time; meanwhile Bell could serve as a *locum-tenens*.

In this unsatisfactory, inadequate fashion Bell finally achieved his heart's desire. Without the authority of a director, without the backing of his political superior, without even an increase of salary, he was condemned to spend five years struggling to make his place permanent by proving himself indispensable to the Survey, the mining industry and the nation, and trying to placate an implacable superior. And all the time, because his position was so obviously precarious, his authority over the staff and his power to direct the Survey's affairs were being undermined. Ambitious officers were encouraged to attempt by every possible means—through political influence, especially—to win the coveted prize of the directorship for themselves. As Bell confessed to Adams on February 2, 1906,

> Owing to the fact that the Directorship of the survey has been hung up for five years in the sight of all the office seekers in the Dominion, and the number of aspirants has gone on increasing and with this increase a corresponding increased misrepresentation of myself in order to create a vacancy for one of these aspirants to fill, fully half of my time is taken up in counteracting and fighting off the hungry wolves who are jumping for the prize.[12]

One is tempted to feel that any wrongs Bell may have done Selwyn and Dawson were requited with interest during his own five years as acting director.

At first, however, all seemed to go well. Now that he had come into a position of power, Bell began hearing from long-absent friends like A. W. Webster, who had left in 1880, about the prospect of rejoining the Survey. From far-off Rhodesia was heard another voice from the past, that of Wallace Broad, with a reminder of how he too had suffered in Selwyn's day from the "sinister Dawson influence" and had been punished for his "opposition to the clique."[13] J. Galbraith, the principal of the School of Practical Science, Toronto, sent off a letter to Sifton urging he appoint Bell to the vacant directorship, but to no avail. Most important of all, was a manifesto of support for Bell, signed by everyone at the Survey premises, including even the clerks, typists, and caretaker. It assured Bell that he might hope for internal harmony within the organization, but even more, it was a demonstration of solidarity in the face of insidious rumours of ministerial designs against the Survey:

> We, the undersigned, Members of the Staff of the Geological Survey of Canada, wish to express our satisfaction at your appointment as head of this Department, and desire to assure you of our hearty and unanimous co-operation in the work of the various branches in connection therewith.
>
> The fact of your always having taken an interest in every branch of the work carried out by this Department will be an assurance that, under your care and guidance, our labours will be sympathetically and efficiently directed, and the future of the Department in all its branches will be assured.

J. F. WHITEAVES	G. CHR. HOFFMANN
R. W. ELLS	JOHN MACOUN
HUGH FLETCHER	WILLIAM McINNES
R. CHALMERS	W. J. WILSON
JNO. MARSHALL	W. H. BOYD
LAWRENCE M. LAMBE	R. L. BROADBENT
THEO. DENIS	A. T. McKINNON
W. SPARKS	HENRI LEFEBER
S. HERRING	J. F. E. JOHNSTON
LOUIS N. RICHARD	OLIER E. PRUD'HOMME
R. G. McCONNELL	J. N. O. McLEISH
P. H. SELWYN	C. OMER SENÉCAL
F. G. WAIT	J. M. MACOUN
CHAS. W. WILLIMOTT	THOS. BURKE
JNO. THORBURN	W. W. LEACH
B. URQUHART	J. ALEXANDER
D. B. DOWLING	E. F. GOODMAN
M. H. BARRY	ALF. E. BARLOW
HENRY M. AMI	ELFRIC DREW INGALL
JAS. McEVOY	A. P. LOW
E. R. FARIBAULT	ROBERT A. A. JOHNSTON[14]

That a real threat existed was indicated on May 15, 1901, when Sifton, replying to questions in the House on the Survey's publications program, went on to observe:

> The remarks of the hon. gentleman are pretty much the same as those I made the late director of the Survey some time ago. . . . While the Geological Survey has done very good work, the time has come when some little reorganization of the department is required. I have given some attention to the subject, and I hope to be able to suggest changes which will improve that branch of the service in the direction indicated.[15]

A few days later, on May 20, Haanel suggested a division of labour between the Geological Survey and the proposed Department of Mines:

> If the Geological Survey confines its attention to Natural History, Palaeontology, Stratigraphy, Microscope Lithology and Topography while all strictly practical matters of importance to the development of the country and its resources in the way of mines, building materials, economic minerals, assaying and analyses and the arrangement and exhibit of specimens in the museum, be handed over to the Dept. of Mines both the purely scientific and the practical demands of the offices will be met.
>
> The more I think of your suggestion: to affiliate the schools of mines of the Dominion with the Dept. of Mines, by giving its graduates on passage of a special Govt. Examination an opportunity to serve on the staff of the Govt. office, the more I am impressed with the exceeding wisdom of such a procedure.[16]

The project of a Dominion School of Mines was an old chestnut that had been discussed as far back as 1871 but had died a natural death from the disinclination of the provinces to support the project even to the extent of discussing the subject. In Haanel's view, then, his organization would handle all matters of direct concern to the mining industry and gain all the opportunities for dealing with the public. The Survey, for example, should advise the Mines Branch of occurrences of potentially economic minerals, which would send its officers to examine the sites and then report the discoveries to the public. The Survey, in effect, would be subordinated to Haanel's organization (the future Mines Branch) and see its work taken over and presented to the public under the guise of being the work of his organization. Haanel later had the effrontery to propose exactly this to McInnes, whose support he tried to enlist.

To make such radical changes in the face of the powerful and influential Geological Survey at the behest of an unknown, 60-year old professor not even living in the country, and with an exclusively academic background was too much even for a dynamic minister like Sifton. For the present, the establishment of a Department of Mines was out of the question, and so, apparently, was the injection of Dr. Haanel into the Survey. Accordingly, on June 5, 1901, a position was created for him in the Department of the Interior, Sifton's main department, as superintendent of mines, with a salary of $3,000 per annum (higher than Bell's $2,400, incidentally), and Haanel shook the dust of the United States from his feet. To make room for him, Pearce was transferred to a position of chief inspector of surveys. The report of the Department of the Interior for 1901-02 explained the change in the following terms, one phrase of which, respecting the scope of his duties, was a clear invasion of work the Section of Mines had long been doing:

> . . . it was felt that provision should be made for the appointment of a special technical officer whose scientific knowledge and practical experience in mining matters would fit him to take charge of this particular branch, such officer to advise the department upon the requirements in connection with this service and prepare reliable information for publication. Professor Haanel, who was latterly employed as Professor of Physics at the University of Syracuse, in the state of New York, had previously, for some fifteen years, held the chair of science at the Victoria University, Cobourg, Ont. He is [i.e., had been] a member of the Royal Society of Canada, an expert mineralogist, and otherwise specially qualified by scientific knowledge and attainments to take charge of the important position to which he has now been appointed. He has already rendered very valuable services in connection with the establishment of the new Dominion assay office at Vancouver, and as he will be specially charged with the compilation of accurate information and official statistics with regard to mines and mining industries generally throughout the Dominion, there is no doubt that he will thus be in a position to supply a long felt want in this respect.[17]

That work, of organizing and establishing the Dominion Assay Office in Vancouver, was Haanel's first task, and he set about it in a manner that belied his age. Business interests in British Columbia and the Yukon Territory hoped that the assay office would arrest the export of gold bullion to Seattle and San Francisco, which was credited with being largely responsible for deflecting Yukon business to those centres. The government wanted the office opened quickly to show its support of the business and mining interests concerned with the Klondike, while Haanel, for obvious reasons, was determined to demonstrate his technical and managerial abilities.

In less than five months Haanel assembled the necessary equipment for weighing, assaying, and refining the gold, hired a staff of five, leased a building on Vancouver's Hastings Street and arranged for the necessary improvements, drew up the instructions, scale of fees, and system of formal reports, and opened the office on July 26, only 11 days behind the original objective. In the following month he visited the Yukon, reported on mining methods, plans and machinery in use, described the mineral occurrences, ran tests and assays, and prepared general statements on the situation in the Klondike and its future prospects. He also prepared an elaborate 40-page report which included sketches of equipment in use in the Klondike, a large floor plan (in colour) of his assay office, and seven photographic plates depicting the various rooms and their apparatus.

In addition, while he was in New York collecting the equipment for the assay office, he visited various museums "with a view of preparing plans for the Victoria Memorial Museum," and in due course handed a set of "sketches of floor plans for the museum" to the chief architect of the Department of Public Works.[18] As drawn to scale by A. A. Linnell of the Department of the Interior "Under the direction of Dr. Haanel, Supt. of Mines," they show a very conservative exterior reminis-

XII-4 THE BALANCE ROOM, DOMINION ASSAY OFFICE, VANCOUVER

cent of the East Block of the Parliament Buildings, with relatively small windows, and mansard roof and dormers, surrounded and surmounted by ornate ironwork railings, and crests. An interesting feature of the interior was that the director of the Geological Survey and the superintendent of mines would each have offices on opposite sides of the entrance hall, reflecting the "separate but equal status" Haanel foresaw for himself and the agency he headed. Haanel's plans would not seem to have had any significant effect on the design that was eventually adopted for the museum in 1903.

Meanwhile the Canadian Mining Institute, through its council, was being drawn into line with Sifton's plans when at its meeting on November 14, 1901, it decided to appoint Sifton as its patron in view of his great interest in the mineral industries, his generosity in granting it $1,000 of public money for the years 1900 and 1901 (subsequently raised to $3,000 for each of the next two years), and because of "the likelihood that there would at no distant date be organized under the administration of his Department an efficiently equipped Department of Mines for the Dominion."[19]

It was in response to hints of this sort, and as a counter-pressure from the Geological Survey that on March 1, 1902, Barlow directed his letter to Sifton, signed also by Brock, Ingall, Daly, and McConnell:

> We are given to believe that, in order to form a nucleus for the new Department of Mines, it is planned to remove from the staff of the Geological Survey those of its members who are most familiar with the mining regions of the Dominion. We respectfully submit that this will mean serious injury to the efficiency of the Survey instead of the substantial aid and encouragement which it so sorely needs. Is there no other way to avoid that result?
>
> Again, if there be not cordial relations between the Director of Mines and the Director of the Survey, there would inevitably come up questions of function and of jurisdiction which would demand a large amount of special legislation and administration.
>
> So intimately connected must the work of the two departments be that separate, they could not exist indefinitely, for this would occasion a dissipation of energy and a waste of funds.[20]

The letter proceeded to their own proposal—that the Department of the Geological Survey should be reorganized so as to include both scientific and economic aspects of geology under the title of "Department of Mines and Geology," in which the geologists, under a new director, would be freed from the necessity of having to spend so much of their field time on topographic work.

* * *

IN THE LIGHT OF THE DELICATE, uncertain situation, the next annual meeting of the Canadian Mining Institute, held in Montreal on March 4, 5, and 6, 1902, became a test of strength between the supporters of the Survey and those of Haanel. Each group strove to demonstrate that its opinion was that of the majority, and to

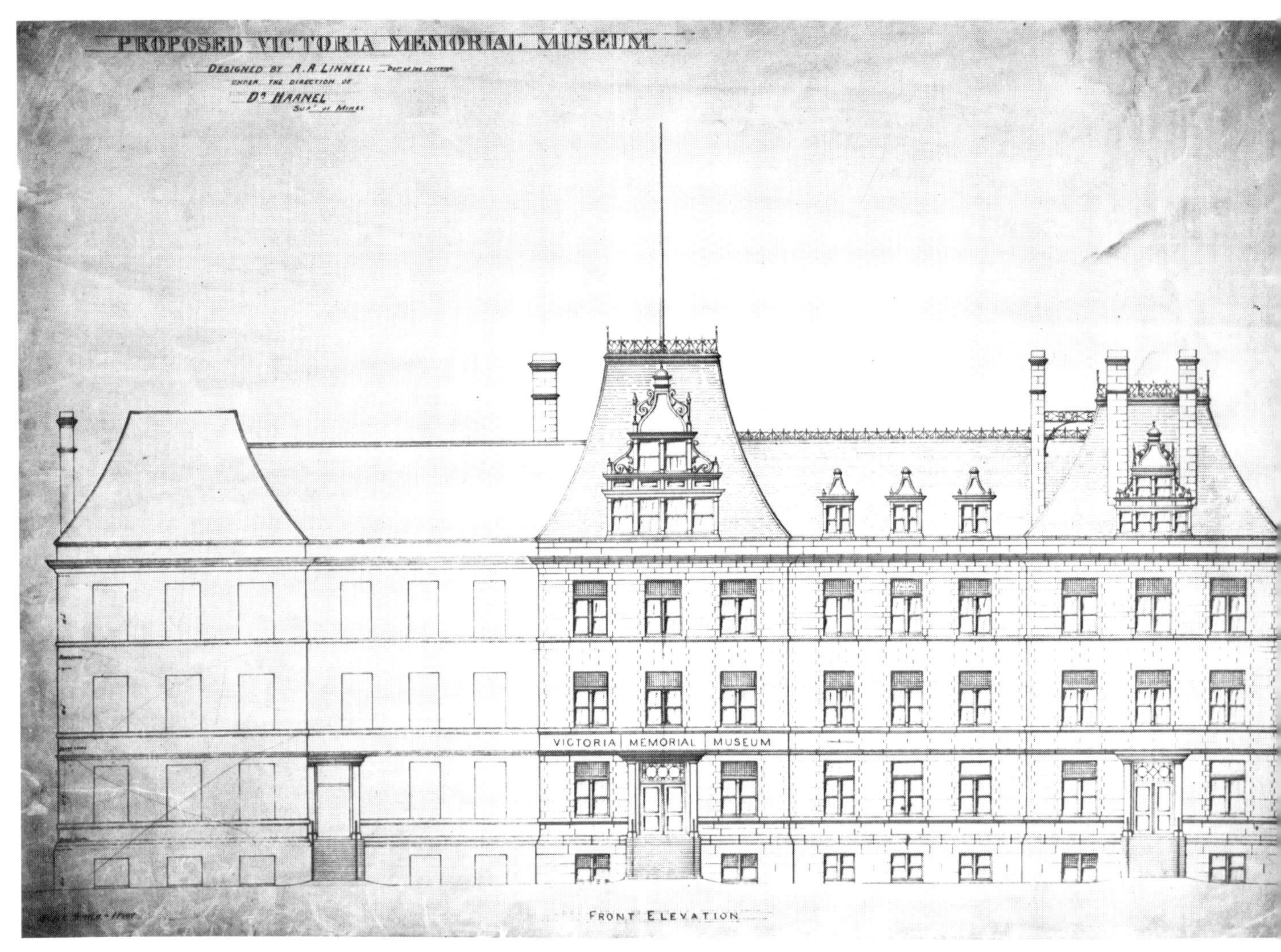

XII-5 HAANEL'S DESIGN FOR THE VICTORIA MEMORIAL MUSEUM

present Sifton with a resolution indicative of that preference. The discussion was led off by a paper by J. E. Hardman, a Montreal mining engineer and past president of the Institute, on "The National Importance of Mining" which began with the premise that the Dominion government had a role in support of the mining industry which the Survey was not adequately fulfilling. He criticized the Survey for not being so fully oriented towards aiding the mining industry as was that of the United States; the work he said, was not "devoted to such branches as are of immediate use to the greatest number of the country's citizens."[21] The field work should not be so heavily concentrated on undeveloped districts but should mainly investigate areas of interest to the industry, for example by providing detailed studies of camps like Rossland or Sudbury that would help solve geological problems and encourage investment and stimulate production. It should prepare technical reports on iron and nickel dealing with every aspect of those industries from geology to marketing. A properly constituted bureau of mines should collect and supply accurate statistics relating to the production of metals and minerals throughout the country. Publications should be a separate branch under a separate chief, and be concerned with publishing bulletins on known mineral localities, on the growth of each industry, and on each major metal or mineral, prepared by experts and written in a manner that would interest the mining public and general public.

The lively discussion that followed on the theme of "Government Aid to Mining" revolved around the resolution that was to be forwarded to the minister, and particularly what place the Geological Survey should hold in the Department of Mines that nearly everyone now expected to see established. Some spoke of a Dominion Department of Mines including "a full staff of geologists;" this viewpoint was developed most fully by McEvoy and Adams, who urged that the Survey should be put in a better position to do the many kinds of work required of a national agency—that topographers and mining geologists should be added to the staff, as well as specialists from outside to perform particular pieces of work requiring their skills, that university facilities should be utilized for ore-dressing and metallurgical laboratory work, and that a proper statistical department should be established. Adams concluded with:

> I believe, then, that the Government has in the Geological Survey the nucleus of a department which might render the most important and valuable assistance to the mineral industry of the whole Dominion. The Survey, however, needs to be *reorganized and extended,* and converted into a "Department of Mines and Geological Survey." (Applause.)[22]

The opponents felt the Department of Mines should be separate and distinct from the Survey and take over the statistical and mining responsibilities. B. T. A. Bell, for example, spoke of the necessity of establishing a Department of Mines "entirely separate and distinct, if need be, from the present Geological Survey."[23] When a resolution was introduced by W. Blakemore that called on the Dominion government to increase its aid to the mining industry and establish a strong and practical Department of Mines, Robert Bell objected that the word 'practical' was a slur on the Survey by implying that its work had not been of a practical character, when what could be more practical than such work as that of McConnell in the Yukon, Gwillim in Atlin, or Brock in the Boundary district? His plea "If you can't do us any good don't do us any harm,"[24] coupled with Dr. Haanel's reminder that he considered a separate department absolutely necessary and that his report to Sifton "will encourage him on this subject and greatly strengthen his hands,"[25] caused the friends of the Survey, notably Coste and Adams, to introduce an amendment that the proposed department should include the present Geological Survey and all other necessary branches. B. T. A. Bell tried desperately to have the amendment ruled out of order, but the meeting insisted on its going ahead. Eventually Blakemore had to withdraw his resolution and second Coste's amendment, which then carried by a large majority:

> Resolved that the Canadian Mining Institute in Annual Session assembled desires to direct the attention of the Federal Government to the magnitude and importance of our mining industry which during recent years has developed so rapidly and respectfully urges an increase of Government aid wherever possible and the establishment of a strong and practical department of mines or of a department which shall be devoted to the interests of the mining and metallurgical industries and which shall include the geological Survey and all other necessary branches.[26]

The effort to use the institute had failed; notice had been served that the Survey had a strong following among the leaders of the mining industry, who would not let it be sidetracked in any governmental reorganization, however much the minister might desire this.

For the next year matters remained at a standstill, while each side worked to strengthen its position. The *Canadian Mining Review* gave Haanel all the publicity and encouragement he could have desired. Its April 1902 number, for example, carried a full page photograph over the caption "Dr. Haanel will, in every likelihood, be the Director of the new Department of Mines and Geology, at present being organised by the Hon. Clifford Sifton, M.P., Minister of the Interior," while the accompanying article added that "He strikes us as a shrewd, broad-minded officer, fully alive to the importance and necessities of our rapidly expanding mining industries. He possesses an excellent technical training for his work, and, we should judge, will make an excellent administrator of the new Department of Mines."[27]

Within the Survey, meanwhile, the external threat caused a rallying around Bell to protect the *status quo,* as witness Fletcher's plea to Bell not to be provoked into resigning as acting director: "Stick on for the sake of all of us who do not wish to have Dr. H. recommended for the position of successor to Logan, Selwyn and Dawson by B.T.A. Bell and his confederates."[28]

The long-expected move came during the Parliamentary session of 1903, when Sifton carried his plans a step further. On July 17, during the debate on the Survey's estimates, he announced that the vote was being reduced by $10,000 from $60,000 "in pursuance of an arrangement that I will explain later on for the purpose of doing the work more especially connected with mining. I intend to introduce a resolution on the subject."[29] Questioned about Bell's and the Survey's positions under any new arrangement, he went on to explain his own attitude towards the Survey and to hint at his intended course of action:

> I anticipate being able to make some changes which I have had in mind for the last two or three years in connection with the survey and possibly the permanent appointment of Dr. Bell might be facilitated by these changes. What I propose to suggest is that that which may more properly be called economic work, or work directly connected with mining, shall be separate from the other work and that change I propose to carry out by means of legislation this session . . . .
>
> I may say to the committee something that I will say at greater length when I make the definite proposition I intend to make, that while I have no doubt of the great value of the work that has been done by the Geological Survey, the result of my observation and supervision of the work during the last seven years has led me to the conclusion that sufficient attention has not been given to making the results of the work of practical value and easily available to the public . . . . My purpose in organizing a mining branch is primarily to bring about the result that we will have in succinct and in somewhat popular form, all the information of a practical and economic character made readily available, and the duty of the branch would be to put that in shape . . . . That is the principal fault I find with the survey as it is organized at the present time. Of course it is very difficult to make changes; every-

> body who is connected with an institution is inclined to think that that institution is managed on just about the proper line, but after six or seven years experience I am clear in my opinion that a change is not only desirable but necessary if the country is to get value for the work it is doing.[30]

Since Sifton was still in England when his reform program was presented in the House of Commons on October 12 the Prime Minister, Sir Wilfrid Laurier, spoke to the item in the Department of Interior's estimate, "salaries and expenses of Mines Branch,"[31] being the $10,000 sum by which the Survey's appropriation had been reduced. Reading from the memorandum Sifton had left for him, Laurier explained that "The general and immediate object of the work of this branch [is] to be the collection and publication of data regarding the economic minerals of the country, and of the processes and activities connected with their utilization." Its duties would be statistical in part, involving studies of production, consumption, exports and imports, economic data on costs, freight, and markets, the information to be published annually or more frequently as practicable. It would also prepare and publish bulletins and monographs on particular minerals,—make assays and analyses of ores, test building materials, describe mining methods and treatment processes in a form suited to the needs of the public commercially interested in the matter. Much of the information would be secured by collating material scattered through published sources, plus visits to secure information on current developments. The branch would require a staff of clerks and mining geologists to collect data, prepare the reports, and furnish reliable assays and analyses of the economic minerals. "In other words this is to establish a mining bureau on the pattern of a similar bureau in Washington."[32]

XII-6 SIR WILFRID LAURIER IN 1906
Prime Minister of Canada, 1896-1911

That Laurier was not completely familiar with the situation emerged when he went on to say that "I think it is to be an annex of the Geological Department."[33] But he rectified this by admitting that Dr. Haanel would head the proposed branch, whereupon the Opposition charged that the appropriation was legislating a new branch of government into existence and demanded to know why the government was proceeding in this extraordinary fashion. In reply, Laurier actually claimed that this was the government's preference since the branch was an experiment. Questioned why it was necessary at all since most of the activities were already being performed, he was somewhat at a loss, but he saw the proposal to prepare a volume on "The Mineral Resources of Canada" as a new function. Laurier was reminded that this new agency was a very different undertaking than Haanel's past work as superintendent of mines, and Haanel's fitness was questioned on the score of his age and purely academic background. The question of his nationality was even raised, the speaker claiming he understood Haanel was a naturalized American. (Haanel took the oath of allegiance on August 8, 1901, at the same time as he was sworn into office.)

Now that a Mines Branch had been established in the Department of the Interior, competition between the two agencies began in earnest. The Mines Branch tried to give an impression of vigour and powerful influence with government, and made every effort to ingratiate itself with the mining industry. The Survey, through its field work, publications, and the work of its Section of Mines, strove to demonstrate it could do everything the Mines Branch had been established to do, and do it better. It set about emphasizing aspects of its work that had been insufficiently stressed in the past, by the way of proving that no satisfactory reorganization of the work could take place without its participation. Bell in his annual reports hammered away at the theme of the many great services performed for the mining industry over the years, and the complete immersion of the Survey in activities that were of the greatest possible benefit to the national economy. Since both agencies were working in behalf of the same economic interest, and since their spheres of activity overlapped a great deal, there was a considerable risk of duplication of effort as each organization sought to assert its priority to the widest possible sphere of operations.

The government does not seem to have had any clear idea of drawing the lines between its two agencies, and perhaps the minister was not averse to letting competition proceed. He seems, for example, to have encouraged the Survey to move into the field of metallurgy, an area in which the Mines Branch was strongly involved, first by foisting Locke, then Connor, on it. Perhaps this was because of Barlow's presence in the Survey. Sifton took an interest in and encouraged Barlow's great work at Sudbury, the report of which was a model of the kind of statement the industry wanted the government to provide, and he secured the metallurgist to aid Barlow in his work. When Barlow investigated the iron beds at Timagami in 1903, Sifton arranged for him to be assisted by Erik Nystrom of the Mines Branch, an expert in the use of the magnetometer—to the great annoyance of Bell, who complained that Barlow had contravened instructions in employing Nystrom, and in not doing "certain needful and important practical geology" Bell had ordered, but instead, had "kept his assistant and party working at what were virtually preliminary mining operations, mostly confined to a single small area of private property."[34] Again, in 1905, when the Mines Branch was made responsible for an investigation into the zinc industry of British Columbia by a special commission, Barlow and Keele were "detached from the Geological Survey Department to investigate certain undeveloped deposits and prospects."[35] Since Sifton also designated Barlow to accompany Haanel to the convention of the American Mining Congress in Portland, Oregon, in August 1904, there is ground to speculate that Sifton perhaps had plans for Barlow in the reorganization he was still contemplating. Barlow's staying with the Survey until after the new arrangements had been completed in 1907, then retiring almost immediately, are perhaps evidence of such hope or expectation from Barlow's side too.

An earlier example of reluctant co-operation between the Survey and its rival occurred on April 30, 1903, when McConnell and Brock were delegated to assist in investigating the causes of the rockslide at Frank, in the

XII-7 ELECTROLYTIC LEAD REFINERY AT TRAIL, B.C., IN 1907
The complex ores of the huge Sullivan deposit were mined only for their lead content at first, the lead being smelted and refined at this plant, while the zinc was discarded with the slag for lack of a suitable recovery process. The "Commission Appointed to Investigate the Zinc Resources of British Columbia and the Conditions Affecting their Exploitation" reported on the problem in 1906, but a satisfactory process did not become available for another decade.

XII-8 THE PRAIRIE PROVINCES TO 1920

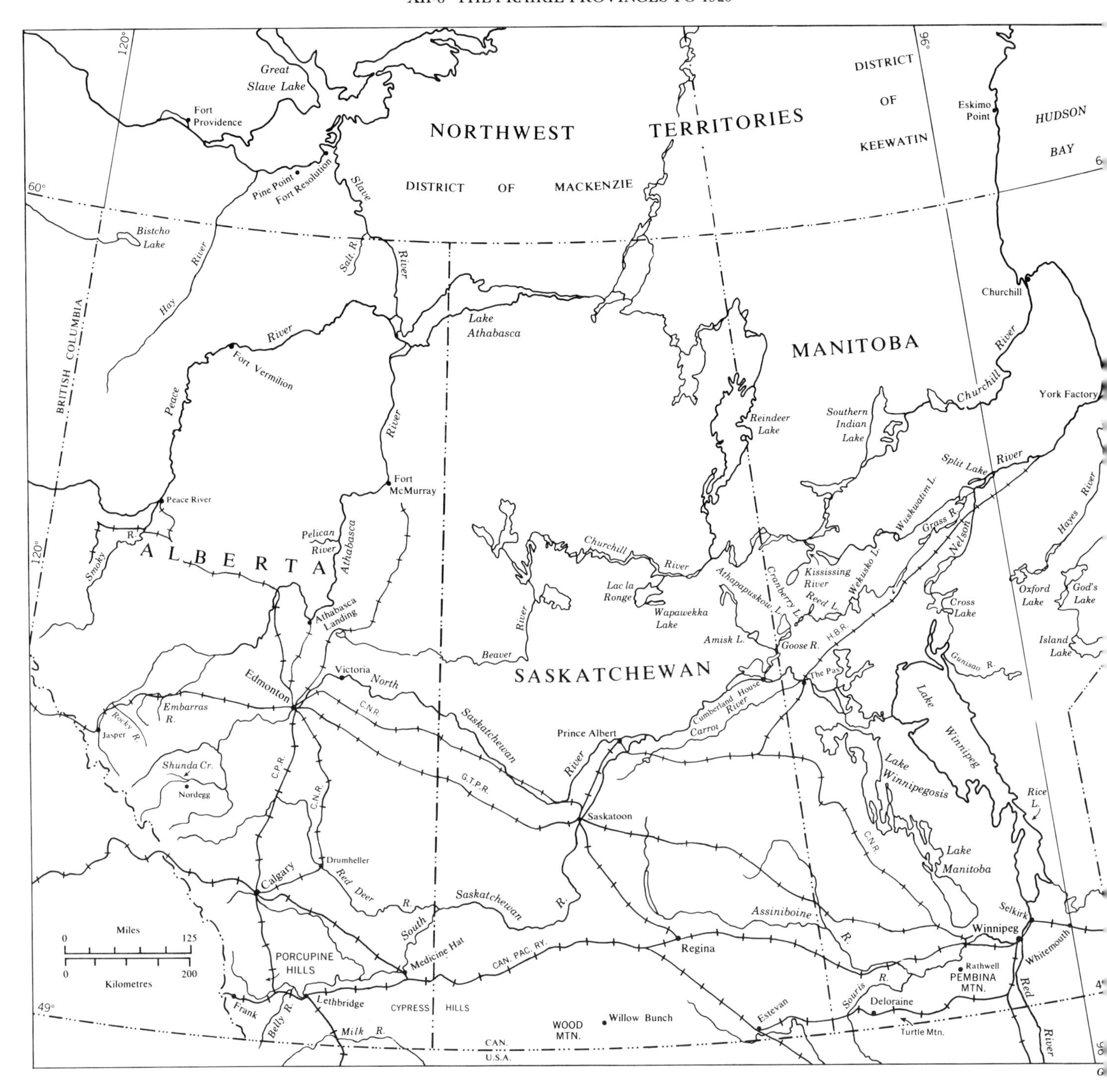

Crownest district of Alberta. Since this was Haanel's responsibility as superintendent of mines, the geologists received their instructions from him. These—characteristically printed in full in Haanel's report—had an authoritarian ring that could not have made any friends for Haanel in the Survey:

> Messrs. McConnell and Brock.—
> You are to proceed at once to the scene of the disaster and investigate thoroughly the nature and causes of the catastrophe. If a landslide, determine as carefully as possible the condition of the strata before the avalanche, and the causes which led to the breaking away of the mountain mass, whether due to imperfect or careless mining operations, the explosion of fire damp, exceptionally bad ground, or similar causes, or whether the results were due to the destructive effects of atmospheric weathering, or other causes.
>
> Determine with especial care the present physical condition of the mountain, with a view of determining the likelihood of a recurrence of similar destructive phenomena.
>
> Report as soon as possible, consistent with thorough examination, the results of your investigations.
>
> Eugene Haanel
> Superintendent of Mines[36]

The geologists did this work before proceeding to their regular assignments, Brock to the Kootenays and McConnell to the Klondike, and handed in their report to Haanel to be published in his report. What they may have thought is not recorded, but in May 1904 Haanel advised Sifton against using Survey officers in conjunction with the work of the Mines Branch, since, he argued, this would antagonize the Survey and perhaps drive it into open opposition to the government's plans for reconstruction.

With these few exceptions, each organization went its separate way. The Survey continued to aid the mining industry as best it could with the means available. Most directly challenged was the Section of Mines, which existed for many of the same purposes as the Mines Branch. Almost all the energies of Ingall's small group went to preparing the annual reports on the mining industry in Canada, securing statistics on operations, production, employment, exports, and statements of conditions in particular industries that had to be brought up to date each year. These reviews were sometimes expanded and enlarged as a result of specific studies; the report for 1902 offered a fairly extensive presentation by Denis on the coalfields of Canada that gave complete listings of mines, production, operating methods, markets, analyses of different coals, a bibliography of publications, and maps of the coal-bearing regions. Denis prepared similar reports on salt production in eastern Canada, on zinc, and on petroleum and natural gas fields. A rather futile geological examination of the Bruce Mines district was undertaken after 1901; intended to aid the possible re-opening of the historic mines, the study failed to make any significant contribution to geological knowledge or advance the mining industry, and was a considerable drain on the limited energies of the group.

The work of the Section of Mines, of course, was only a minute fraction of the Survey's aids to mining. As the work of Fletcher, Faribault, Chalmers, Ells, Brock, and McConnell indicated, a large and growing proportion of the Survey's undertakings during these years under Bell's leadership, was of immediate and direct benefit to industry. In fuel problems alone, for example, there were the very extensive studies of Fletcher in Nova Scotia, Bailey and Poole in New Brunswick, Chalmers and Ells of peat resources in eastern and central Canada, Dowling along the eastern margin of the Rockies, and Poole, Ells, and Leach in the coalfields of the Cordillera. And even the much-criticized explorations and reconnaissance surveys were very helpful in gaining for the Survey a wide measure of popular support that must have stood it in good stead in its struggle for survival and recognition.

The reports, in fact, were the show-window by which the work and value of the Survey were judged, so there was much discussion about their style, quality, time of appearance, and distribution. In 1901 Sifton himself criticized the reports as too prolix and too technical in their language; later, another M.P. complained that they contained "mountains of information" but that it was like searching for a needle in a haystack to find what one was looking for.[37] The tardiness of publication was another cause for complaint, particularly the annual report of the Section of Mines, which at best did not appear till eighteen months after the year ended.

Some effort was made to meet all these objections. Dawson had tried to employ as few technical terms as possible, while under Bell, who had little inclination to pursue "scientific" geology and a great desire to please the public, the reports took on a more descriptive and narrative form, so that problem solved itself in part. The ability to use past reports was improved by indexing the volumes of annual reports; and, beginning in 1905, with the appearance of the summary report not long after the close of the year. The very short reports for 1906 and 1907 appeared just a few days after the New Year. To make the accumulated research of the past more readily available the *Summary Report for 1903* printed a lengthy list of Survey publications and maps on economic minerals, dating back to 1855, classified by minerals and localities. Moreover, to demonstrate that the staff was able to prepare geological, technological, and economic reports on Canadian minerals without the interposition of the Mines Branch, the Survey in 1903 began issuing a series of mineral resources bulletins that reached a total of fifteen by 1907 when the series was discontinued. Ells was responsible for five of these, on apatite, copper ores of eastern Canada, graphite, mica, and asbestos.

The creation of the Mines Branch in October 1903 was a considerable blow to the hopes and expectations of Haanel. The elderly professor who had so impressed the young Sifton had failed to commend himself to the Survey even in its sad, director-less state, and Sifton, his principal, had been forced to settle for the more limited objective of a branch in his own major department. Any thought of solving the impasse from the other side, in a manner more in keeping with the desires of the Survey personnel, was barred by Haanel's stubborn determination. Offered the position in November 1903 of curator of the new national museum that at last was being realized with the beginning of work on the Victoria Memorial Museum, he rejected the offer on the grounds that it would look as though he was being removed for having failed in his task. The museum did not get a curator for nearly 20 years nor a truly independent status for 40 years (with what incalculable effects on its future progress!), while Haanel persisted in the Mines Branch as its director until December 1920, when he finally accepted retirement at the age of 79.

In the meantime, Haanel strove to give shape and purpose to the Dominion government's second agency for helping the Canadian mining industry. The assay office in Vancouver, installed with such *éclat* in 1901, went into an abrupt decline, but he quickly found fertile new causes in the investigation (mainly in Europe) of metallurgical and technological processes that might be applicable in Canada. He went to Europe at the head of a commission to examine electric reduction techniques and equipment, and followed up his study by tests of several processes at the iron works at Sault Ste. Marie. Studies were undertaken to develop uses for various mineral resources in different parts of Canada, and the branch's own mineral reports were published as well as yet one more study of iron ores (Ingall had made two within the previous decade). The experiments at outlining iron deposits by means of magnetic surveys with which Barlow co-operated at Timagami, were also carried out in the Lanark and Renfrew districts of Ontario, in New Brunswick, and in Nova Scotia. Though Haanel built up a small staff of mining engineers, most of the work was done by experts hired for specific tasks. For example, a mining engineer of Montreal, Fritz Cirkel, drew up the branch's reports on mica, asbestos, and graphite—which duplicated the Survey's bulletins on the same subjects. Haanel also participated in the zinc commission in British Columbia which made a notable attack on the metallurgical problems of separating zinc, lead, copper, and silver in the complex ores of the Kootenay district.

That Haanel was persisting in his original objective was evident from frequent reminders in his reports of the importance of the work. Thus, in the three pages of his report for 1904-05 devoted to the American Mining Congress at Portland, Oregon, in August 1904, Haanel cited in full its resolution that the United States government establish a Department of Mines and Mining, explaining disingenuously that "I insert the above statement of the proceedings of the congress as being of some interest in showing the trend of thought upon this subject in the United States at the present time."[38]

Not everyone took the Mines Branch so seriously as Haanel did in the lavishly illustrated reports of his branch's activities. In June 1906, after Sifton had resigned from the ministry, the Opposition took the Mines Branch to task by pointing out that almost two-thirds of its $18,000 expenditure in the previous year had been spent on printing and lithographing, and not more than 200 days of actual practical work in the field had been performed. It was asserted that there was no excuse for a separate Mines Branch at all, which had been set up only to accommodate "a political pet of the former minister,"[39] and whose work could be done quite satisfactorily by the Geological Survey Department. Sir G. E. Foster complained that its reports were based entirely on information laboriously gathered and prepared by the Geological Survey, which could just as well have got out the pamphlets. The government sprang to the defence of the scientific work of the Mines Branch, called its printing program sound public relations, and saw a need for it so as to do the things the Survey could not, or would not, do. In the following year it similarly rejected a further charge that work of the Mines Branch on the electric smelting of iron ores at Sault Ste. Marie had never been carried beyond the experimental stage, and was without practical effect whatsoever.

* * *

SIFTON'S DEPARTURE FROM THE CABINET on February 28, 1905, did not significantly affect the government's attitude on the question of aids of mining, or on the necessity of carrying the reorganization beyond the position reached in 1903. His eventual successor, on April 8, 1905, Frank Oliver, the pioneer newspaper editor from Edmonton, had no special reason to be partial towards the Survey. The year before, in 1904, he had led an angry attack on J. M. Macoun in the agricultural committee because of his unfavourable report on the Peace River district, which Oliver regarded as unwarranted, unjustified, and a dangerous blow to the development of western Canada. Still earlier, Oliver had blamed the Survey, under Dawson, for foolishly choosing to drill for petroleum in northern Alberta at places where there was no evidence of oil, instead of at a location where it was known to be present:

> Mr. Oliver. I assert most positively that the late director of the Geological Survey, Mr. Dawson, directed that the boring should be carried on at two certain places, at which neither Dr. Dawson nor any one else ever suggested that there was any evidence of oil.
>
> Mr. Blain. That is rather a reflection of the late Dr. Dawson.
>
> Mr. Oliver. Dr. Dawson had reasons of his own. He had the borings made upon a geological theory which in that particular case did not happen to be correct . . . . I have very little faith in our Geological Survey Department. This is only one instance from which I have formed that opinion.[40]

That it was not the function of the Survey to drill for oil and that Dawson was looking for something else, never seems to have entered Oliver's head. Dawson's failure to find oil was simply one more evidence to this western pioneer that learning and commonsense were very different commodities, and that the members of the Geological Survey, however great their scientific attainments and qualifications, were not the people to deal effectively with practical problems. He said as much in response to the Conservative attacks upon the Mines Branch:

> The mines branch is a complement or development of the geological branch. The work of the geological branch is to discover what minerals there are in the earth and the work of the mines branch is to find out the most economical and effective means of utilizing them. It may be said that in both cases the work might be left to the Geological Survey but you do not always find great scientific ability and great practical ability combined. Frequently men of the highest scientific attainments have least practical ability. Therefore there is nothing improper in giving the geological branch its full measure of scientific work and putting into the hands of the mining branch the work of demonstrating the practical value of the discoveries made.[41]

Oliver's connection with the mining industry was not to be of long duration, however. As early as August 1905, it was being rumoured that the industry would at last be given its own Minister of Mines—Senator William Templeman of British Columbia, the Minister of Inland Revenue—and once more the structure of the department became a subject of debate. The *Canadian Mining Review* boomed the Mines Branch by reporting favourably on its work, and reflecting in unfavourable terms on the Survey, the extreme being reached in an editorial of August 1905, entitled "A National Department of Mines."[42] This alleged that the Survey's reports were late, of little practical value, and very inferior to those of the United States Geological Survey, that under Bell it was riddled with patronage and favouritism, with much of its time being wasted on topographical surveys. As for the Section of Mines, its reports were so tardy that they were only pieces of ancient history. Consequently the new department should be based on the Mines Branch, not the Geological Survey, though under another head than Haanel:

> A well organized Department of Mines would, then, include a staff consisting of mining geologists, field geologists, topographers, chemists, and the ordinary editing and clerical staff. It should have as its head a director, young, energetic, tactful, a good organizer, a capable administrator, and last, but not least, a man of high scientific standing and attainments. And we have every reason to believe that Government has already taken cognizance of these requirements.[43]

In its December 1905 number the *Review* took up the same refrain, lecturing both organizations for having published booklets on asbestos and mica, and the government for failing to put a stop to the unnecessary rivalry:

> But the absolute futility of this sort of thing must be apparent, and it is high time that it ended. Instead of antagonism, there should be harmony and co-operation, and surely an arrangement is possible by which this more satisfactory state of affairs might be brought about, for there is ample scope and work for both institutions. But the sooner the present chaotic condition of affairs in connection with the administration of the Geological Survey and other matters is settled the better for all concerned.[44]

A more moderate opinion, to which the government was bound to pay heed, was expressed in March 1906 at the annual meeting of the Canadian Mining Institute in a paper by H. Mortimer Lamb "On the Advisability of the Establishment of a Federal Department of Mines."[45] After outlining the histories, functions, and achievements of the two organizations he observed that the establishment of the separate Mines Branch was an indication the Survey "has not in recent years, on its strictly economic side, kept pace with the growing requirements of the mining industry."[46] He went on to propose as the ideal solution to the problems of the government and of the industry a merger of the two agencies in a single department under a minister of its own:

> Our aim here is merely to point out that the mining industries of Canada might at the present time be greatly assisted if the work

> of the Geological Survey and the Mines Branch of the Department of the Interior was taken up seriously by the Government, correlated, systematized, extended, and made to conform to modern requirements. The duplication which now exists would thus, in the interest of economy, be avoided, and the whole work would be put on a proper businesslike basis.
>
> If this were done, it is certain that the mining interests of the country would be well served and that the action of the Government would receive the hearty endorsation of every one interested in mining, and that furthermore, as the value of the work became increasingly evident, the Government would feel justified in providing additional means for its prosecution, so that a larger staff of properly paid and thoroughly efficient men, aufait with the modern methods and results of science applied as to the study of these economic problems, could be permanently employed by the Government in the development of the mineral resources of our country.
>
> . . . if the happy result above indicated could be insured by the appointment of a Minister of Mines who would have direct supervision of this work, the expansion of the Geological Survey into a Department of Mines and Geological Survey, would receive the support of the whole mining community.[47]

Lamb's reasonable, persuasive address coincided with long-overdue action by the government to solve the problem of the vacant directorship of the Survey, the prize for which Bell had never given up striving. The government had turned to a younger man—not Brock, not Barlow, but A. P. Low. Bell's notification by ministerial letter followed, that effective April 1, 1906, Low would become director and Bell would resume his rank of assistant director with the title of chief geologist. Now, at last, the situation inside the Survey had been resolved, and the Survey was in a position to take its rightful place in the negotiations leading to the reorganization of the Dominion's role in mining and to make itself a significant element in the new structure. During the following year Low played a prominent part in working out the details of the arrangements leading to the establishment of the Department of Mines on a federal principle, with the Geological Survey and Mines Branch as distinct units retaining their separate identities, traditions, and functions.

The government's plan finally emerged on March 25, 1907, after a gestation period of six years, when a bill to create a Department of Mines was introduced in the House of Commons for first reading by Mr. Templeman, the Minister of Inland Revenue and presumed minister-designate of the new department, who had left the Senate and been elected an M.P. in the previous year. On April 19, when the bill came up for second reading and discussion in committee of the House, Mr. Templeman described the department as consisting of two branches, a geological branch, a mines branch, and probably an eventual administrative branch. Because of the strong wishes of the members of the Survey to retain their historic name, the two parts would be called the Geological Survey Branch and the Mines Branch.

Much of the discussion ranged around the matter of the personnel to be appointed to the leading positions in the new departments, since they could vitally affect its orientation. A fortnight earlier newspaper reports had circulated about the country to the effect that the Commissioner of the Yukon, W. W. B. McInnes, had been offered the position of deputy minister of the new department. Now Templeman announced that Low had been offered the deputy ministership, and that another officer might be appointed as director of the Survey, or Low could retain both positions if he desired. No appointment had as yet been made to the Mines Branch, but he presumed it would be headed by Dr. Haanel, who would therefore be subordinate to Low, or whoever became deputy minister of the department. The question of Bell's eligibility for the directorship was raised once more, and Templeman expressed a willingness to consider his qualifications, though Bell's relations with Low, he thought, would make for difficulties. There were some complaints over the choice of Low. His very recent appointment as director of the Survey and his absence at such an important time (perhaps news of his physical breakdown had come out) were argued as reasons for

XII-9 WILLIAM TEMPLEMAN
Minister of Mines, 1907-11

postponing the legislation. But the minister would brook no further delays, and the bill went forward to the final reading and royal assent on April 27, 1907, as "An Act to create a Department of Mines" (6-7 Edward VII, 1907, Chap. 29). Templeman was appointed Minister of Mines on May 3, 1907, but also retained his Inland Revenue portfolio.

The Mines Act of 1907 defined the functions of both branches, mainly continuing each in its previous field of operation. The Mines Branch was to be responsible for collecting and publishing full statistics of the mineral industry in Canada, also data regarding economic minerals and their utilization, the former bone of contention between the branches. It would also make detailed investigations of mining camps and areas containing economic minerals to determine modes of occurrence and character of the orebodies, and carry on chemical, mechanical, and metallurgical investigations on behalf of the mining industry. The Geological Survey remained responsible for practically all its former functions, including examining and surveying the geological structure and mineralogy of the country, studying water supply for irrigation and domestic purposes, mapping and reporting on the forest resources of Canada, and carrying on ethnological and paleontological investigations. Its power "to make such chemical and other researches as will best tend to ensure the carrying into effect the objects and purposes of this Act"[48] left a possibility of duplicating the work of the Mines Branch, as did the definition of the areas of their field operations. The Survey's association with the museum was reaffirmed in the clause that "The Department shall maintain a Museum of Geology and Natural History for the purpose of affording a complete and exact knowledge of the geology, mineralogy and mining resources of Canada," plus the provision that it should "collect, classify and arrange for exhibition in the Victoria Memorial Museum such specimens as are necessary to afford a complete and exact knowledge of the geology, mineralogy, palaeontology, enthnology, and fauna and flora of Canada."[49]

The provisions respecting the Survey staff employment were extended for the department—that is, technical officers were subject to the provisions of the Civil Service Act, and had to be science graduates of Canadian or foreign universities, of the Mining School of London or Paris or other recognized school of equal standing, or of the Royal Military College. A former difficulty over the part-time employment of professional people was reflected in a provision empowering the government on the concurrence of the minister "without reference to any examination, or to the age of the person . . . temporarily [to] employ person at such remuneration as is deemed expedient."[50] The present officers of the Geological Survey could be assigned "to the branch in which it is deemed desirable that their services shall be utilized," with their usual rate of pay and tenure. Thereafter those employed in one branch might be assigned "to any other section in the same branch," but might not be directed from one branch of the department to the other.[51]

The provisions respecting the conduct of members of the Survey were extended to the new department: no employee was permitted to purchase land except for the purpose of personal residence, locate bounty land warrants or land scrip as agent for anyone else, make investigations or reports relating to the value of the property of individuals, "hold any pecuniary interest, in any mine, mineral lands, mining works or timber limits in Canada," nor disclose, except to his superior "any discovery made by him or by any other officer of the department, or any other information in his possession in relation to matters under the control of the department or to Dominion or provincial lands, until such discovery or information has been reported to the Minister, and his permission for such disclosure has been obtained."[52] Even at the height of that "get-rich-quick" age of easy public morality, the same stringent restrictions that had so successfully kept the Survey and its staff free of scandals, were reaffirmed and extended over the entire department.

The establishment of the Department of Mines was the culmination of a six-year process begun with Haanel's injection into the Department of the Interior as Superintendent of Mines in 1901, though that step reflected a state of mind on the part of the minister, Sifton, that probably went back to his accession to office in 1896. The drive took on greater substance two years later, with the creation of the Mines Branch. For the next four years the competition was intensified as Haanel strove to establish a viable agency with a satisfactory range of activities and functions that would ensure its survival. On the other side, perhaps deliberately, the Survey was kept short of funds, badgered into undertaking a variety of unrelated activities, attacked by an organized lobby, subjected to interference from outside, and its acting director openly humiliated. Its morale was reduced to a point where its staff became resigned to being led into an amalgamation in the new department of government.

The long delay in carrying the enterprise to a completion, that outlasted Sifton's tenure of office, was occasioned principally by the Survey. The strong sense of pride in its identity that prevailed among the staff and those who had left it to go into industry made them unwilling to contemplate amalgamation, or any curtailment of their traditional freedom to be all things in the field of geology (and other sciences besides) and to serve the mining industry in any way it could. Its mistrust of the planned reorganization was intensified by the minister's insistence on placing Haanel at the heart of his proposed system. One can well imagine the different reaction that might have greeted Sifton's proposal had it been presented by a member of the Survey or a former officer like Adams. The establishment of the Department of Mines, in which Haanel's place was confined to his own Mines Branch, was a concession to the Survey officers' feelings, as was the elevation of Low, the director of the Geological Survey, to the headship of the new department. The Survey had received guarantees that its interests would be protected under the new arrangement, and Low was able to lead it, without incident, into the new structure.

The Department of Mines was probably a fair distance from Sifton's original intention. He, like many of the critics, might have desired nothing more than to have the existing Geological Survey Department, renamed perhaps, oriented more completely towards the 'practical' role of directly assisting the mining industry as the United States Geological Survey did; with Haanel at its head or holding some senior responsible position in the new organization. Instead, what emerged was not an amalgamation but a loose federation of the two agencies, each with its program and staff carrying on as before, the deputy minister serving as a kind of co-ordinating agent or referee. Haanel was left to carry on in the way he had started, developing his staff, perfecting his organization, and moving further into the realm of practical metallurgy and technological aid. The Survey, too, was left much as before, with the same staff, building, divisions, ancillary services such as the library and museum, and its traditional modes of operation in the field, laboratory, and office.

Yet the amalgamation of the Geological Survey into the Department of Mines meant a significant change in its future. It was now relieved of the Section of Mines and of a function it had never been able to carry out to anyone's satisfaction, chiefly because of the persistent lack of funds and personnel. It was divested of the embarrassing responsibility that more than any other had brought it under attack. Instead, it now could concentrate more fully on its fundamental task of mapping and studying the geology of the vast country. It still was handicapped by the continuing lack of topographical surveys and by its responsibility for collecting anthropological and ethnological material, which was now a high priority duty in view of the establishment of the Victoria Memorial Museum. But the industry could now take its metallurgical, technical, and economic problems to the Mines Branch, while the Survey was strengthened in its original role of "causing a Geological Survey of Canada to be made."

The changed administrative position of the Survey was a major departure from the 60-year tradition of largely autonomous operation, when the Survey had first been a sort of Crown corporation, then a separate department of state. Now, inside the Department of Mines under a minister specifically designated to this purpose, the Survey was bound to be faced with a different kind of political control and be under more direct political oversight than previously. The department would be headed by a deputy minister who would be the ultimate day-to-day authority over the operations of the Survey, a man who had to keep in mind the needs and interests of both the Survey and Mines branches.

XII-10 ALBERT PETER LOW
Director, 1906–07

As a branch in a department of government alongside another co-equal branch, the Survey was in a very different situation than when it had its own budget and was free to choose which of many responsibilities it should stress in its operations. It would have to compete with the other branch for a share of the departmental appropriations, and for the ear of the deputy minister and minister for decisions as to which program to pursue, and with what intensity. It would have to face sharing a vaguely-defined field of operations, of having to defer to the Mines Branch in certain areas, or repel its intrusion into others. Thus the Survey was under constraints after 1907 that had not been felt in the earlier years, though Bell had begun to face something of the kind from the competition of Haanel and the interventions of Sifton.

One cloud remained to darken the horizon, to hinder the advance of the Survey into the bright future of being able to concentrate on its primary role of studying the geology of Canada. This was Low's tragic illness, which had already begun to manifest itself at the time the Department of Mines was organized, and soon became serious enough to reduce his ability to occupy his important position almost to zero. Once again, the Survey faced a vacuum of leadership, until the problem was resolved with Brock's appointment as director and Bell's retirement. But Low's continuing incapacity hampered the Survey from gaining the full advantage of having a man of their own interest as the first deputy minister in the formative years of the new department. Brock's position as acting deputy minister could not give him the same force to deal with either the minister or the Mines Branch as Low might have wielded had he retained full possession of his faculties.

Thus the Geological Survey of Canada, with the passage of the act establishing the Department of Mines entered a new phase in its history, under an act that sketched out its inner structure and administrative position and its relationships with other branches of government for the next 29 years. The new act established an admininstrative unit which has survived in essence to the present, though the Geological Survey has produced further offshoots, and it itself has been transferred bodily from department to department as Mines and Resources, Mines and Technical Surveys, and Energy, Mines and Resources succeeded each other. Through all these changes, however, the Mines Branch and the Geological Survey have continued along parallel courses, side by side under the same administrative roof, and practically next door to one another in their present-day physical locations.

PART THREE

Through Uncertain Times 1907–1950

XIII-1 REGINALD WALTER BROCK
Director, 1907-14

13
The Survey Reorganized

THE REORGANIZATION OF THE SURVEY to meet the needs of the twentieth century was predicated upon an outstanding man being available to take it in hand and transform its work and personnel, a man who commanded the respect of the government, the staff, the mining industry, and the scientific community. Such a man the Laurier government felt they had found in Albert Peter Low, the renowned explorer of Labrador and commander of the Arctic expedition of 1903-04, then in his vigorous prime at the age of 45. Unfortunately, despite his considerable ability, and the high regard in which he was generally held, his effectiveness was almost at once arrested by a very serious illness that was diagnosed as cerebral meningitis and publicly attributed to overwork, but was whispered as being the result of a social disease. The editor of the Survey offered the following report:

> In April, 1906, Mr. Low was appointed Director of the Survey, and immediately started on a stretch of hard work that came near to being his undoing, both physically and mentally. Late hours at the office and an attack of "grip" were the cause of the evil, and in January last [1907] Mr. Low was reported dying. But the constitution of a giant and the pluck that had pulled him through so many narrow escapes came to his rescue and after a month's anxious waiting the Survey heaved one big sigh of relief when the daily bulletin read "out of danger."
>
> The illness has left its mark. The remarkably youthful appearance that characterized the Director is no longer seen .... But, fortunately, this is all. The clearness of thought is still there, the ability to decide quickly what line to take, and the determination to take it and stick to it are still there; above all, he still retains the friendship and respect of the staff—who, as one of them remarked the other day—would "work their fingers to the bone to do old Low a turn."[1]

Whiteaves assumed the management of the Survey for approximately a month during this first onset of Low's illness. Low made enough of a recovery for the government to proceed with the legislation creating the new Department of Mines in April 1907, though even then his absence was cited in the House of Commons as reason for delaying the legislation. Nevertheless, the bill went through, and Low, in addition to directing the Survey, was appointed deputy minister of the new department. Whether he had much of a hand in the departmental administration is problematical, for by May 28, 1907, Whiteaves took up the acting deputy ministership and retained it until November 1, 1908. On that day the position was handed over to R. W. Brock, who combined the acting deputy ministership with his directorship of the Survey until Low was finally retired at the end of 1913. Throughout this five-year period Low does not seem to have been in any condition to perform his duties, though he appeared at the Survey from time to time. Tragically, he lived on and on, the mind of a child inhabiting his once-powerful frame, to 1942, when he died at the age of 81.

Under the circumstances, the actual direction of the Survey and the department from 1907 to 1914 fell into the capable hands of Brock, who had returned to full-time employment in the winter of 1906. Early in December 1907 the government appointed Brock acting director of the Survey in Low's place, and Brock prepared and signed the *Summary Report for 1907*. In the public mind this appointment was taken as being to the directorship, and Brock was referred to as "director" in an editorial in *Toronto Saturday Night* for December 12, 1907, and as "the new director" in an editorial in the *Canadian Mining Journal* of December 15, 1907. A month before, that journal (which had absorbed the *Canadian Mining Review* with the March 15, 1907, issue) had published a powerful editorial on the subject, "Needed—a Director:"

> Reorganization must be left to a young and vigorous man. His appointment must not depend upon politics. His hands must be free to smite and to build .... The man chosen must be not only a geologist—he must be a fearless, just and capable manager. There is restlessness and dissatisfaction among the younger members of the Survey staff. The new Director must deserve and win the respect of these men, and must also have the earnest co-operation of the older men.[2]

It greeted with joy Brock's eventual appointment as director, and commented that his past work had "subordinated the academic to the economic," and that he was neither a "savant" nor a "specialist," but "a virile young Canadian, who knows Canada's needs at first hand."[3] The brief biographical notice emphasized his practical work, observing that "Geology as a branch of academic speculation is one thing. Mining geology is a quite different thing. Of the former Canada has had enough. Of the latter all mining men fully appreciate the necessity." It went on to stress Brock's high reputation in mining circles, and to call him one of "the most brilliant and substantial of North American practical, economic geologists."[4]

Reginald Walter Brock was only 34 when he formally attained the directorship on December 1, 1908, after a year's apprenticeship. But men's careers were launched at an early age in those days, and Brock had already had time enough for at least two useful careers equipping him for the new responsibilities. He had devoted most of 17 years to geological study, beginning with field work in the summers of 1891 to 1895, when he accompanied Willet G. Miller on Robert Bell's parties mapping the geology of the Sudbury district and north of Lake Huron, plus a sixth season (1896) under Bell's direction, exploring on his own between Abitibi and Lake Mistassini. He started his work in British Columbia in 1897 assisting McConnell, then was appointed a technical officer on the staff by Dawson and replaced McConnell in the Kootenays. He familiarized himself with the structures of the region and its mining camps, acquired fame as a specialist on ore deposits, and became so highly regarded as an expert in the economic aspects of mining that important decisions and large investments were being made on the basis of his advice and he was invited to arbitrate important business matters. He gained the respect and co-operation of mine owners, of local business groups, and of the officers of the provincial government. Eventually he was favourably known throughout the Canadian mining industry as one officer of the Survey whose work was completely directed to fulfilling their needs.

By the time he was 30 Brock had also made a reputation as a teacher and administrator. He began his university studies at Toronto, where he attended two years, then stayed out of university for a year and a half during which time he continued his work as a field assistant, clerked, and was a newspaperman for a time. In January 1894 he entered the school of mining recently opened in association with Queen's University under W. G. Miller, and graduated in June 1895 with an M.A. and the medal in metallurgy. He studied in Heidelberg in 1895-96 under Rosenbusch, and again in the January term of 1901 until his recall to Canada by Bell. He lectured in mineralogy at Queen's in 1896-97 before joining the Survey, and returned to Kingston from 1902 to 1906 as professor of geology and petrology in succession to Miller, who had just joined the Ontario Bureau of Mines as Provincial Geologist. Though Brock was too high-strung, nervous, and abstracted to be a good formal lecturer, he was an excellent teacher because of his scholarship, judgment, intelligence, forthrightness, and enthusiasm. He aroused intense devotion among his students and launched many of them on distinguished careers in geology and mining in Canada and the United States. Under him, Queen's became a leading centre for geological studies, and Brock established invaluable connections with academic and scientific affairs.

Reference might also be made to Brock's athletic prowess at university, which made him a formidable field companion even in his later years; his brief newspaper experience; or his many important contacts in the mining industry and government. All of these contributed to his remarkable career as director. Another sort of asset, some political strength, arose from his marriage in 1900 with Mildred, youngest daughter of B. M. Britton, Liberal M.P. for Kingston (1896-1902), and granddaughter of Luther H. Holton, the Montreal businessman and Liberal politician of pre-Confederation days who had been one of Logan's firmest supporters during the difficult sixties. Brock's own political attitudes accorded strongly with those of the Laurier government, especially his optimistic belief in Canada's progression to greatness through the development of its vast natural wealth, and his nationalistic feeling that Canadians were capable of matching the achievements of the most advanced nations of the world. He fought resolutely to keep politics out of the Survey's affairs, but his ties and sympathies with the government of the day only strengthened his relations with the Laurier administration, and may have helped overcome the unfavourable impression of the Survey that some ministers might have held previously.

Brock was able to bring the Survey into the modern scientific age, and assemble a group of brilliant young specialists in the many fields of activity with which it was concerned. Though he was a man of wide-ranging interests himself, he was keenly aware of the need for greater specialization, and he built up a staff that included men

The map below is part of the 1 inch = 4 miles map of the Haliburton Sheet, Eastern Ontario, that embodied the results of many years of careful work in field and laboratory on the Precambrian rocks of this classical district by F. D. Adams and A. E. Barlow. The patterns of magmatic intrusion, differentiation, and metamorphism are clearly indicated by the colours and the swirling lines of foliation. They make this map – Map No. 708, drafted by J. Keele and published in Memoir No. 6, *Geology of the Haliburton and Bancroft Areas, Ontario* (1910) – a model of its kind and an important landmark for future studies of the varieties, structures, and history of Precambrian rocks.

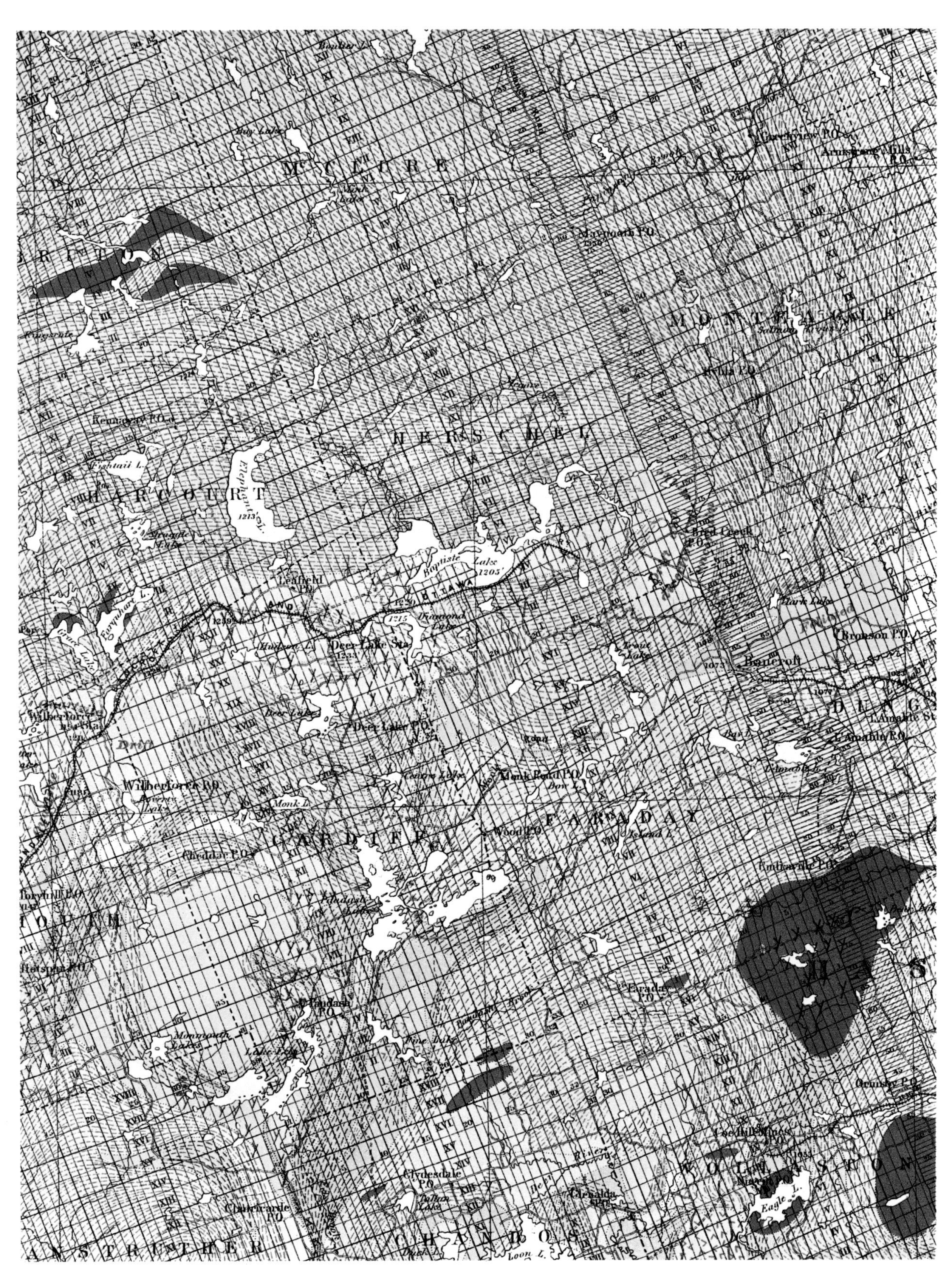

MAP 6.

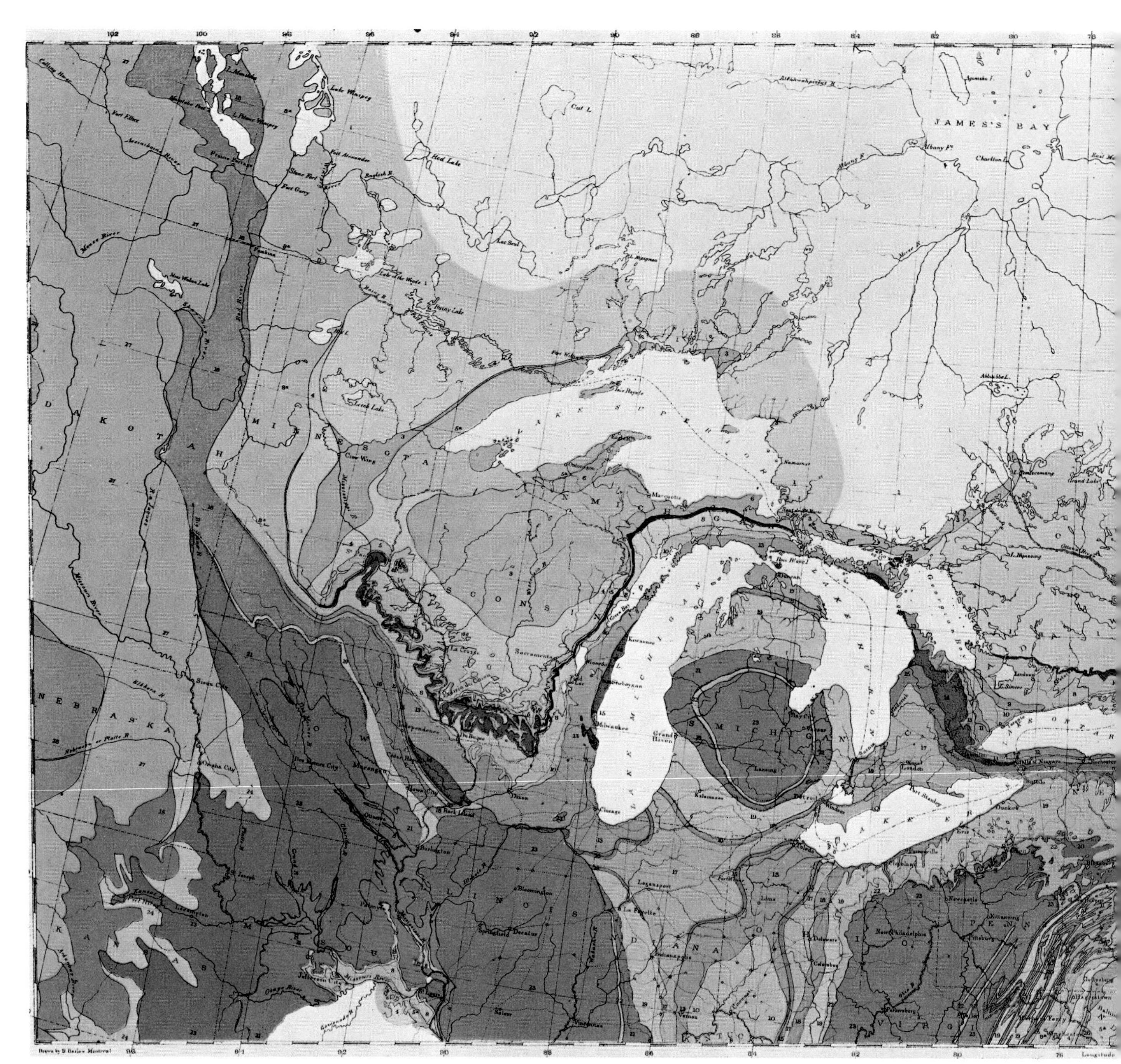

JAMES'S BAY
Albany Fr
Charlton I.
Cat L.
Red Lake
Lake Winnipeg
Fort Garry
Pembina
Lake of the Woods
Rainy Lake
Red Lake
Crow Wing
LAKE SUPERIOR
Marquette
La Crosse
Sacramento
Dubuque
Independence
Sioux City
Omaha City
Des Moines City
Marengo
Iowa City
Rock Island
Dixon
Chicago
Milwaukee
Kewaunee
Green Bay
Grand Haven
Kalamazoo
Lansing
Detroit
Bloomington
Springfield
Decatur
La Fayette
Logansport
Indianapolis
Lima
Delaware
Columbus
Cleveland
New Philadelphia
Newcastle
Kittanning
Pittsburg
Parkersburg
Erie
Buffalo
London
Port Stanley
Goderich
Toronto
Lake Simcoe
Lake Nipissing
Grand Lake
D A K O T A
M I N N E S O T A
W I S C O N S I N
M I C H I G A N
I L L I N O I S
I O W A
M I S S O U R I
K A N S A S
N E B R A S K A
I N D I A N A
O H I O
L A K E M I C H I G A N
L A K E H U R O N
L A K E E R I E
L A K E O N T A R I O
Drawn by B. Barlow Montreal
Longitude
102
100
98
96
94
92
90
88
86
84
82
80
78

MAP 7.

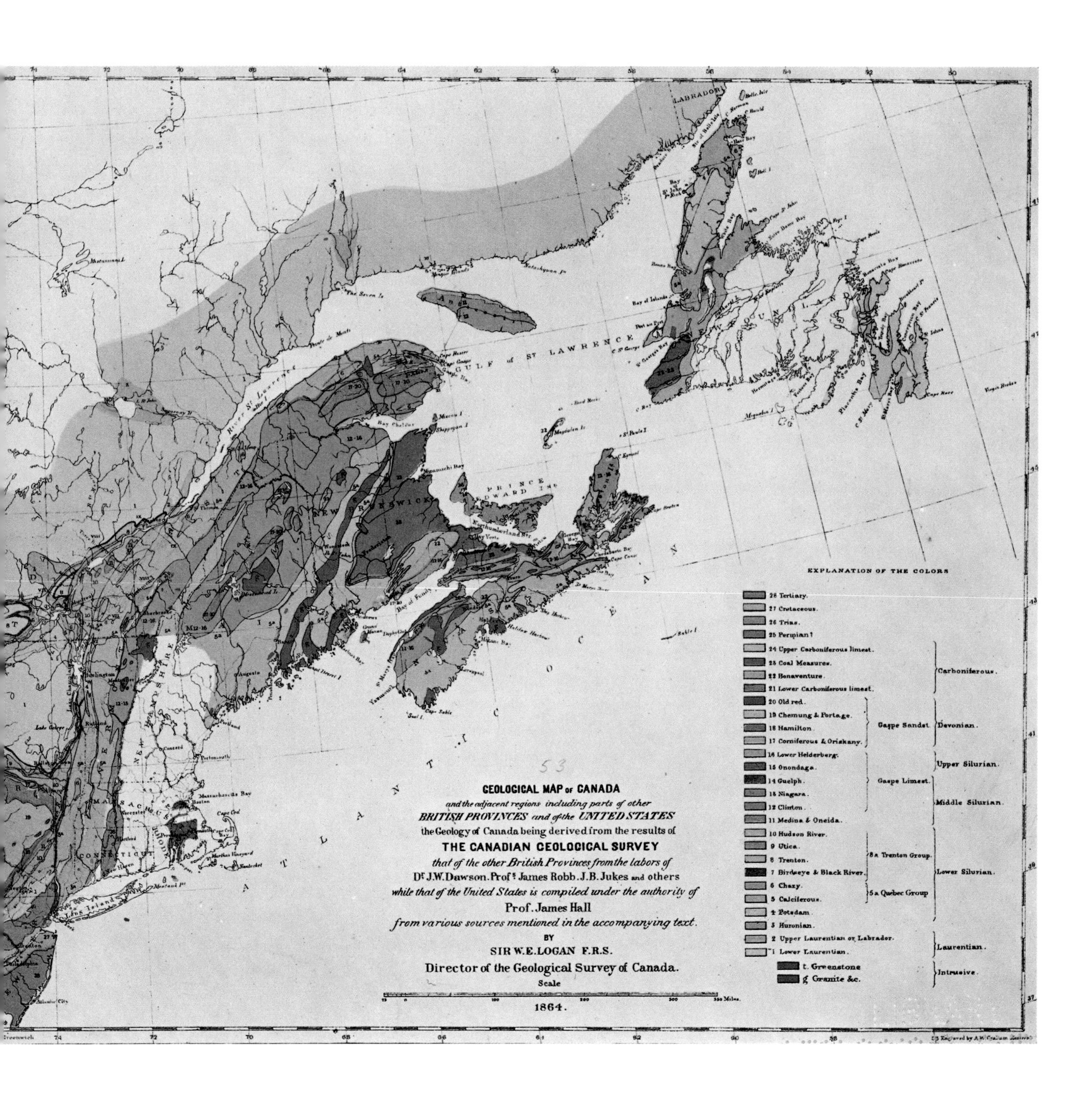
GEOLOGICAL MAP OF CANADA
and the adjacent regions including parts of other
BRITISH PROVINCES and of the UNITED STATES
the Geology of Canada being derived from the results of
THE CANADIAN GEOLOGICAL SURVEY
that of the other British Provinces from the labors of
Dr J.W. Dawson. Profs James Robb. J.B. Jukes and others
while that of the United States is compiled under the authority of
Prof. James Hall
from various sources mentioned in the accompanying text.
BY
SIR W.E. LOGAN F.R.S.
Director of the Geological Survey of Canada.
Scale
1864.
EXPLANATION OF THE COLORS
28 Tertiary.
27 Cretaceous.
26 Trias.
25 Permian?
24 Upper Carboniferous limest.
23 Coal Measures.
22 Bonaventure.
21 Lower Carboniferous limest.
Carboniferous.
20 Old red.
19 Chemung & Portage.
18 Hamilton.
17 Corniferous & Oriskany.
Gaspe Sandst.
Devonian.
16 Lower Helderberg.
15 Onondaga.
Upper Silurian.
14 Guelph.
13 Niagara.
12 Clinton.
Gaspe Limest.
11 Medina & Oneida.
Middle Silurian.
10 Hudson River.
9 Utica.
8 Trenton.
7 Birdseye & Black River.
8a Trenton Group.
6 Chazy.
5 Calciferous.
5a Quebec Group
4 Potsdam.
Lower Silurian.
3 Huronian.
2 Upper Laurentian or Labrador.
1 Lower Laurentian.
Laurentian.
t. Greenstone
g Granite &c.
Intrusive.
LABRADOR
GULF of ST LAWRENCE
NEW BRUNSWICK
PRINCE EDWARD ISLE
NEWFOUNDLAND
ATLANTIC OCEAN
Bay of Fundy
River St Lawrence
CONNECTICUT
Engraved by A.W. Graham

MAPPING THE HURONIAN

The Bruce Mines district on the north shore of Lake Huron, Ontario, was the locale from which the geological name "Huronian" was derived, though whether this, or the district along Lake Timiskaming farther east, was Logan's type area has created difficulties for future investigators. On the basis of field surveys made chiefly by Alexander Murray, Logan contributed a "Map Shewing the distribution of the Huronian Rocks between Rivers Batchehwahnung and Mississagui" (*see Map 8 opposite*) to the Atlas volume accompanying the *Geology of Canada*, 1863. It showed the main geological features of the region: the major fault, now called Murray Fault, which passes south of Ottertail Lake, and the major anticlinal fold passing through Bruce Mines. Later geologists have been concerned with subdividing and refining Murray's units, mapping their distribution more carefully, and correlating them with Precambrian rocks elsewhere.

One such geologist, W. H. Collins, carefully traced the rocks between Lake Timiskaming and Bruce Mines and related them to known metal-bearing units in the Cobalt and Sudbury mining districts. This field work is reflected in the increased detail and accuracy of his map of the Bruce Mines Sheet (scale 1 inch = 2 miles), Map No. 1969 published with his Memoir No. 143, *North Shore of Lake Huron* (1925). In this map (*see Map 9*) the Huronian is presented as a succession of eleven formations ranging upward through the Bruce Series (B to B6) and the Cobalt Series (Co1 to Co4).

The same area has been studied in recent years by M. J. Frarey, part of whose manuscript map is shown as *Map 10*. Though the scale is the same as on the Collins map, this latest map is based on closer traverses, covers formerly inaccessible sections, and incorporates results derived from aeromagnetic surveys and from increased geological knowledge generally. Thanks to these, it has been possible to correct some identifications and to refine the mapping of some formations.

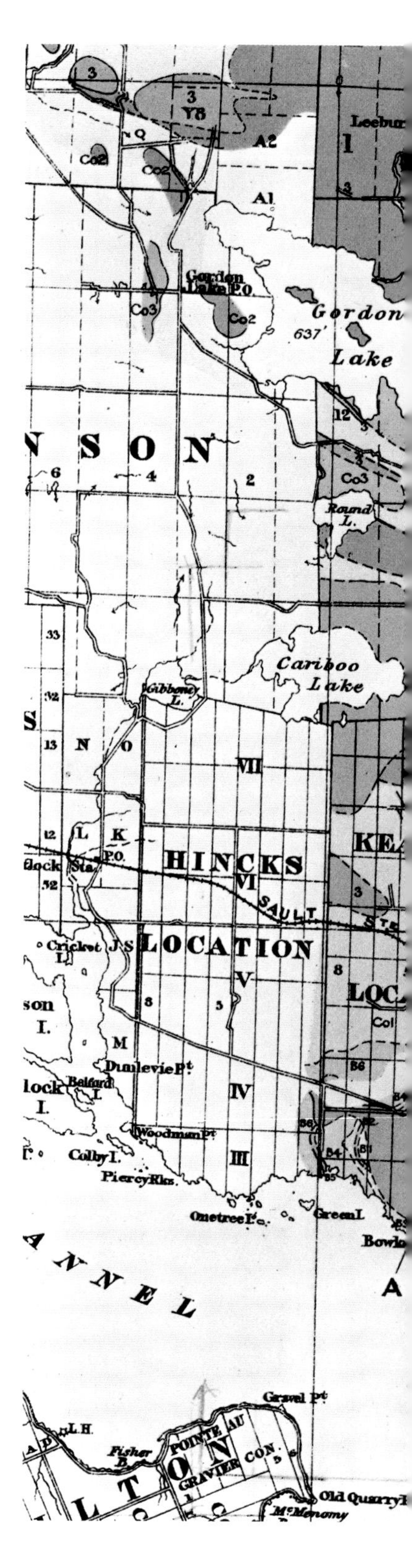

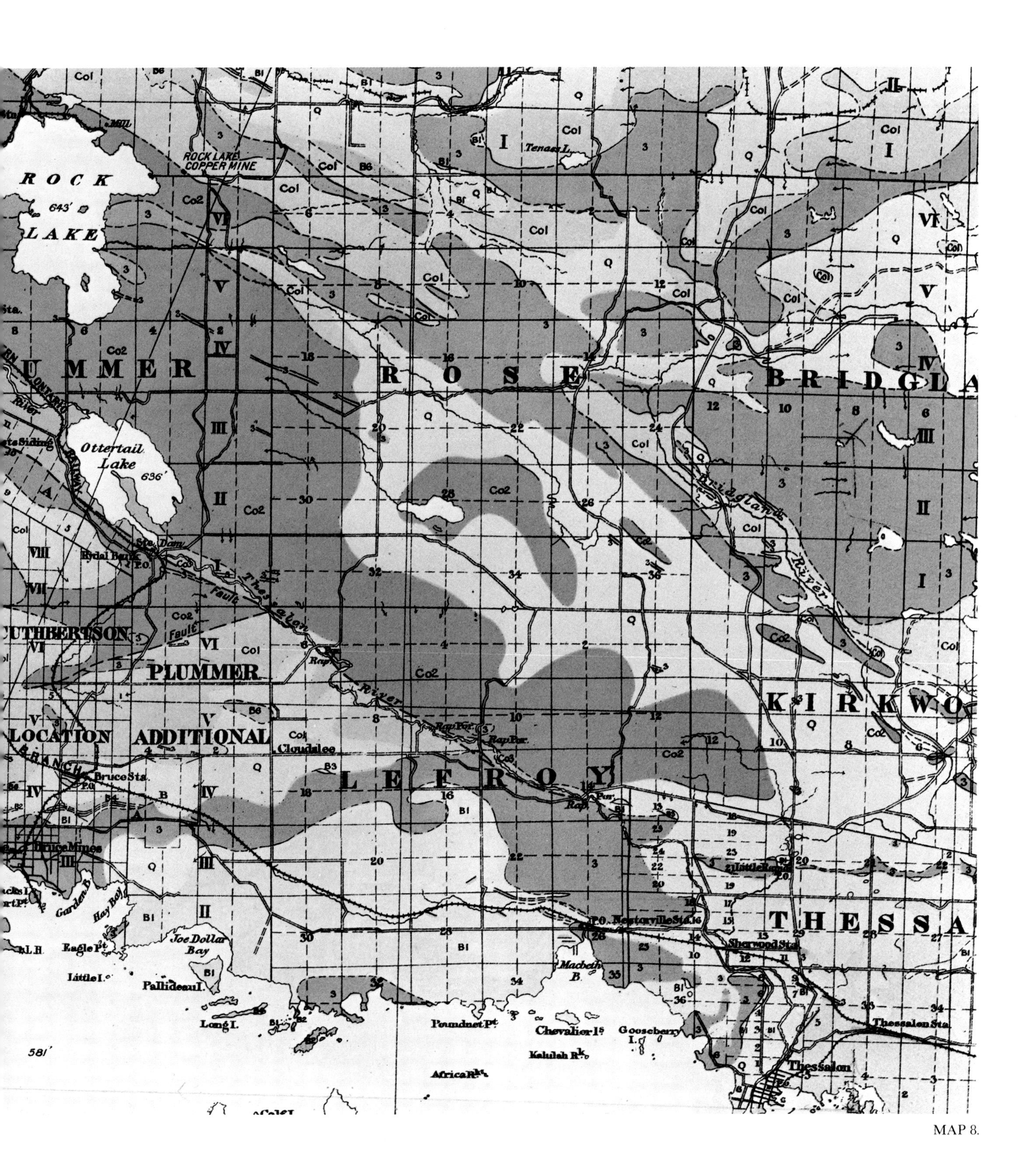
ROCK LAKE
ROCK LAKE COPPER MINE
ROSE
BRIDGLA
Ottertail Lake
CUTHBERTSON
PLUMMER
LOCATION ADDITIONAL
Cloudslee
LEFROY
KIRKWO
THESSA
Bruce Mines
Thessalon
Thessalon Sta.
Sherwood Sta.
Nestorville Sta.
Joe Dollar Bay
Pallideau I.
Long I.
Poundnet Pt.
Chevalier I.
Gooseberry I.
Africa Rk.
Kahlah Rk.
Macbeth B.
Garden B.
Hay Bay
Eagle Pt.
Little I.
Tenass L.

MAP 8.

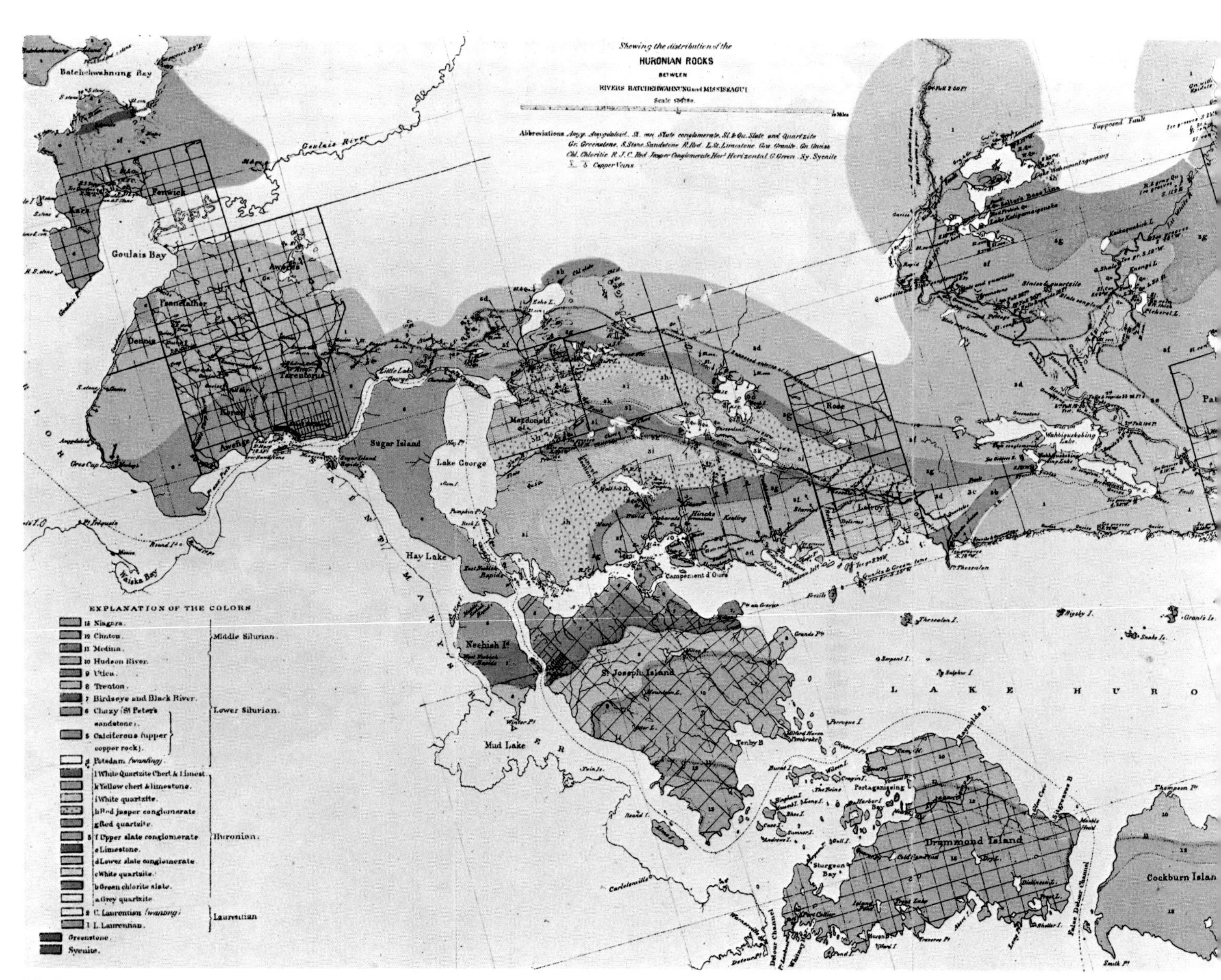
Showing the distribution of the
HURONIAN ROCKS
BETWEEN
RIVERS BATCHEHWAHNUNG and MISSISSAGUI
Abbreviations Amyg. Amygdaloid. Sl. & Qu. Slate and Quartzite
Gr. Greenstone. S.Stone. Sandstone R.Red. L.St. Limestone. Gn. Gneiss
Chl. Chloritic R.J.C. Red Jasper Conglomerate Hor. Horizontal G. Green. Sy. Syenite
Copper Veins
Batchehwahnung Bay
Goulais River
Goulais Bay
Fenwick
Awenge
Pennefather
Dennis
Little Lake George
Sugar Island
Lake George
Macdonald
Rose
Hincks
Laird
Kirkwood
Supposed Fault
Wahbiquekobing Lake
Campement d'Ours
Hay Lake
Waiska Bay
Neebish Id
St Joseph Island
Mud Lake
Tenby B
Portaganissing
Sturgeon Bay
Drummond Island
Cockburn Island
L A K E H U R O
EXPLANATION OF THE COLORS
13 Niagara.
12 Clinton.
11 Medina.
Middle Silurian.
10 Hudson River.
9 Utica.
8 Trenton.
7 Birdseye and Black River.
6 Chazy (St Peter's sandstone).
5 Calciferous (upper copper rock).
4 Potsdam (wanting).
Lower Silurian.
l White Quartzite Chert & Limest.
k Yellow chert & limestone.
i White quartzite.
h Red jasper conglomerate
g Red quartzite.
3 f Upper slate conglomerate
e Limestone.
d Lower slate conglomerate
c White quartzite.
b Green chlorite slate.
a Grey quartzite
Huronian.
2 U. Laurentian (wanting)
1 L. Laurentian.
Laurentian
Greenstone.
Syenite.

MAP 9.

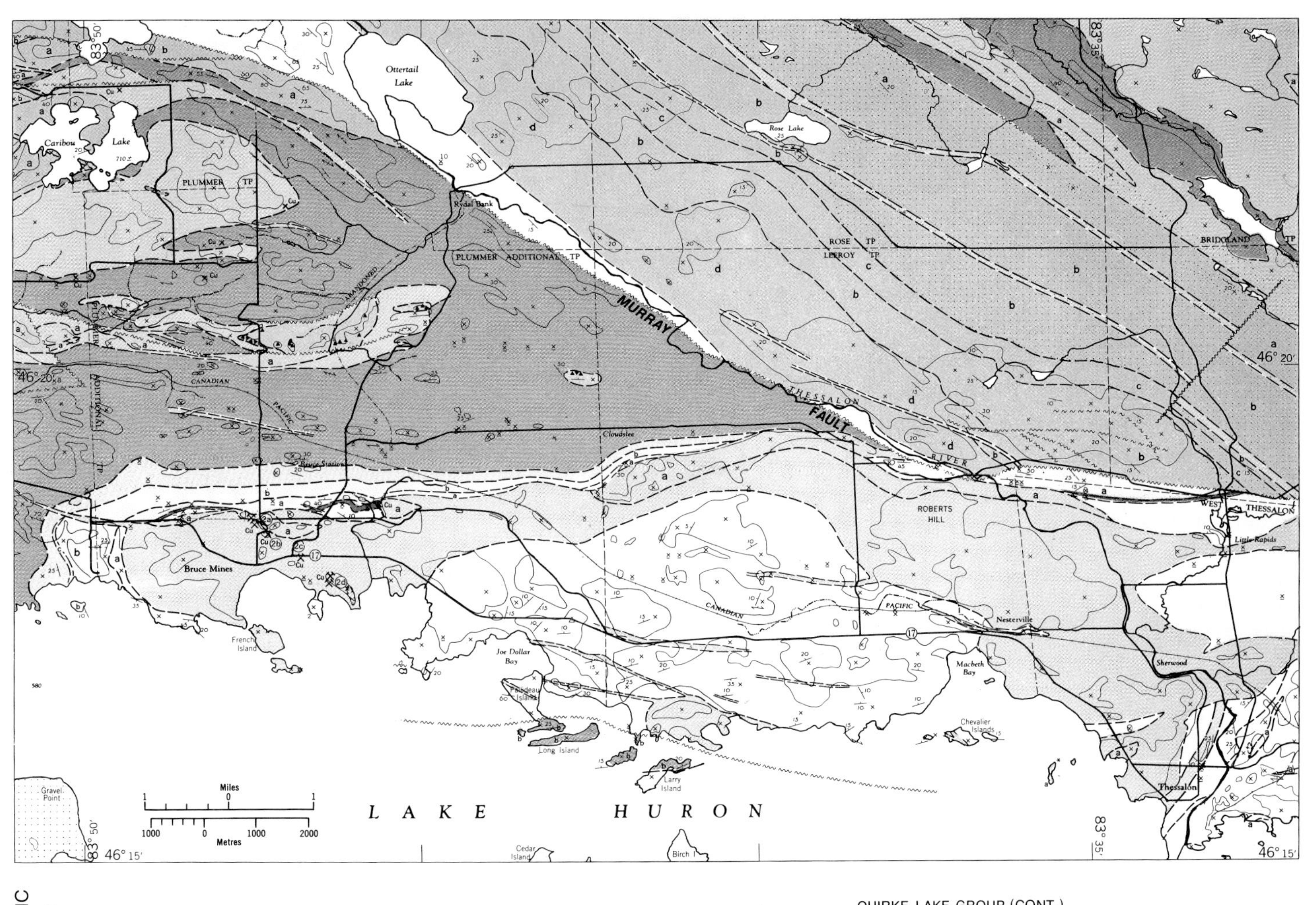

Ottertail Lake
Caribou
Lake
Rose Lake
PLUMMER TP
Rydal Bank
PLUMMER ADDITIONAL TP
ROSE TP
LEFROY TP
BRIDGLAND TP
MURRAY
THESSALON
FAULT
RIVER
CANADIAN
PACIFIC
Cloudslee
Bruce Station
Bruce Mines
ROBERTS
HILL
WEST THESSALON
Little Rapids
Nesterville
Joe Dollar Bay
Macbeth Bay
Sherwood
Thessalon
French Island
Long Island
Larry Island
Chevalier Islands
Gravel Point
Cedar Island
Birch I.
LAKE HURON
Miles
Metres
83° 50'
83° 35'
46° 20'
46° 15'

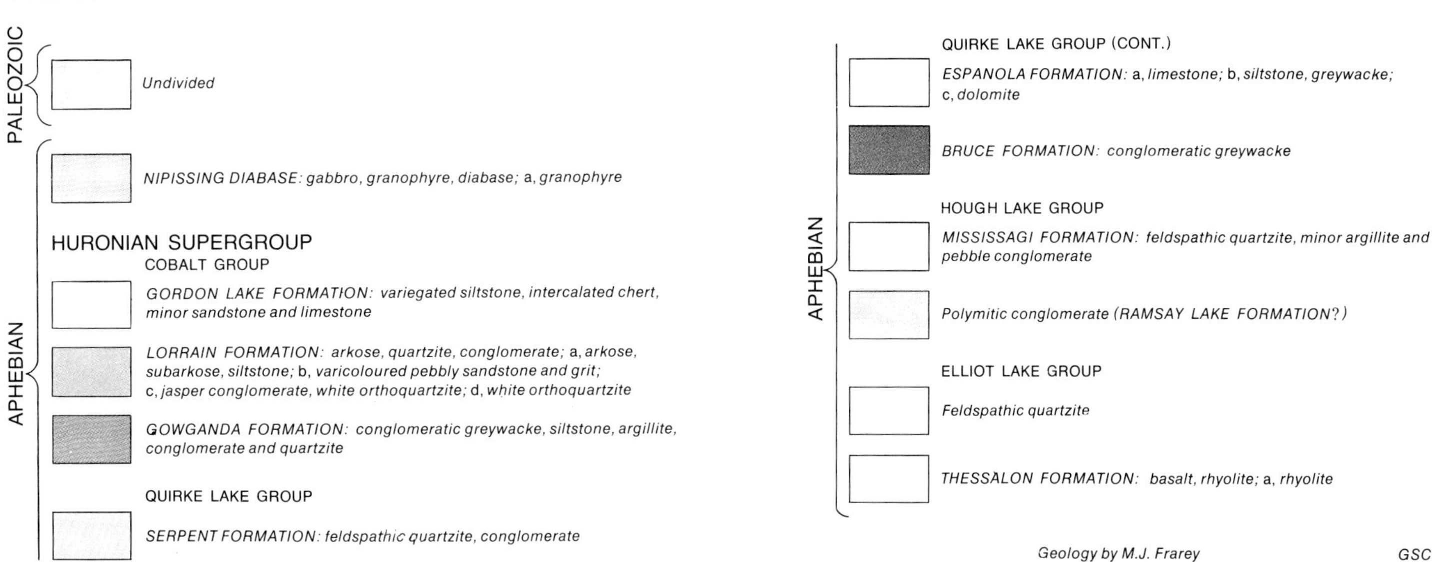

PALEOZOIC
Undivided
APHEBIAN
NIPISSING DIABASE: gabbro, granophyre, diabase; a, granophyre
HURONIAN SUPERGROUP
COBALT GROUP
GORDON LAKE FORMATION: variegated siltstone, intercalated chert, minor sandstone and limestone
LORRAIN FORMATION: arkose, quartzite, conglomerate; a, arkose, subarkose, siltstone; b, varicoloured pebbly sandstone and grit; c, jasper conglomerate, white orthoquartzite; d, white orthoquartzite
GOWGANDA FORMATION: conglomeratic greywacke, siltstone, argillite, conglomerate and quartzite
QUIRKE LAKE GROUP
SERPENT FORMATION: feldspathic quartzite, conglomerate
QUIRKE LAKE GROUP (CONT.)
ESPANOLA FORMATION: a, limestone; b, siltstone, greywacke; c, dolomite
BRUCE FORMATION: conglomeratic greywacke
HOUGH LAKE GROUP
MISSISSAGI FORMATION: feldspathic quartzite, minor argillite and pebble conglomerate
Polymitic conglomerate (RAMSAY LAKE FORMATION?)
ELLIOT LAKE GROUP
Feldspathic quartzite
THESSALON FORMATION: basalt, rhyolite; a, rhyolite
Geology by M.J. Frarey
GSC

MAP 10.

SPECULATIVE POSITIONS DURING RETREAT OF LAST ICE-SHEET

Map by V. K. Prest, 1967. Numbers represent thousands of years before the present.

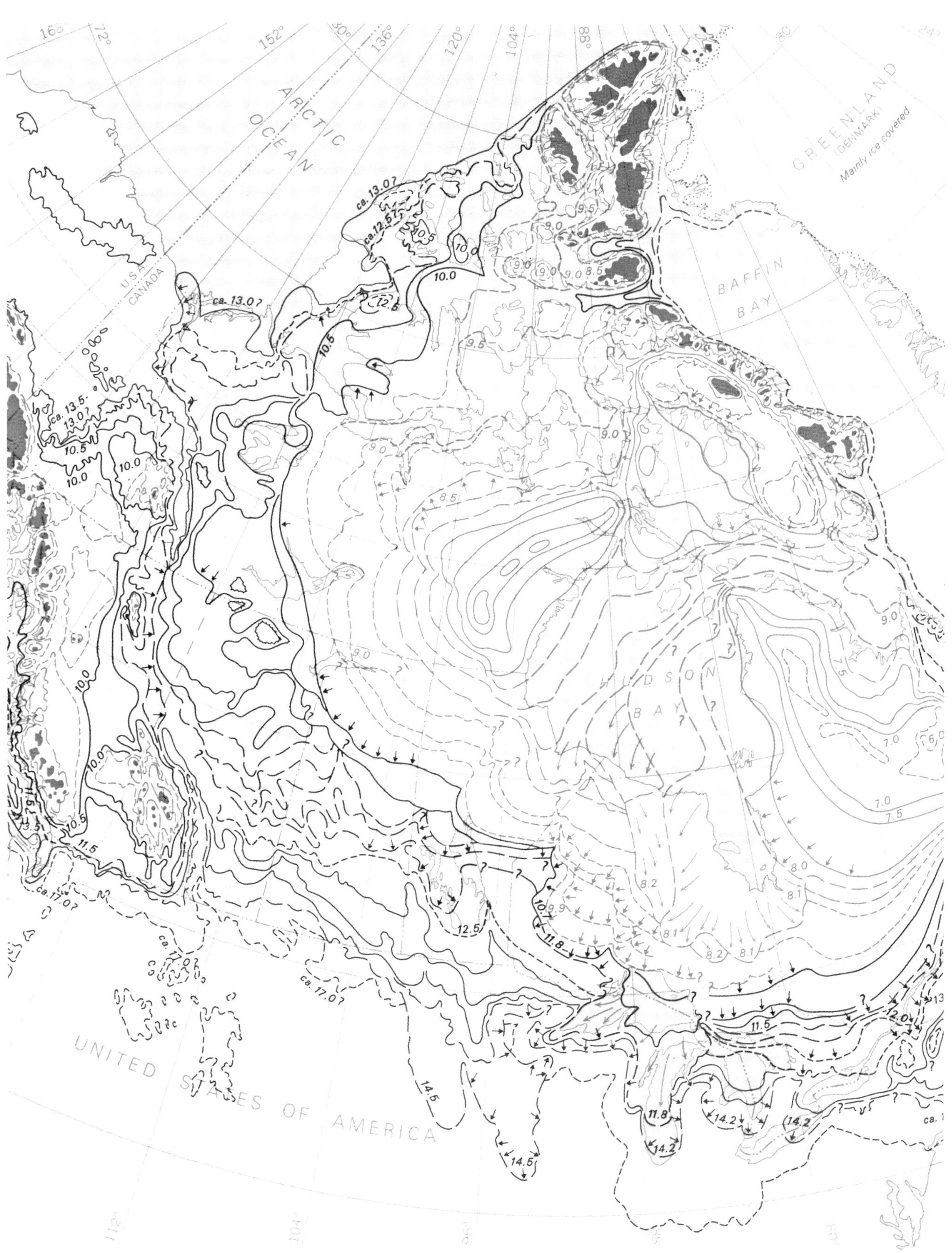

MAP 11.

trained in every major field of geology who could be set at work on any problem that might arise. He was a man of grand ideas, rather untidy at paperwork, and impatient with details. He set high goals for himself and for the Survey, but was very practical in his manner of achieving them. Strong and forceful where necessary, he had a highly developed sense of public relations, and valuable skill at achieving his objectives through personal contacts. His enthusiasm, integrity, and considerateness made each man strive to do his best, and built up morale to a high pitch.

As acting director, director, and acting deputy minister, Brock worked indefatigably in the Survey's interests. There were the usual tasks of managing the operations of the Survey and the department, onerous enough in themselves, but made doubly so by the need to reform and reorganize, to regain a favourable reputation with industry, government and the public, and to hire large numbers of new staff. These he passed off as being merely the routine work of the Survey. He undertook numerous extra tasks besides, such as drafting a model mining law for the Dominion government, planning the Survey's move to a new building, and arranging the meetings of the XIIth International Geological Congress held in Canada in 1913. He was constantly visiting and inspecting work in the field, from borings at Prince Edward Island to gravels of the Klondike, and most frequently in the mining camps of northern Ontario and British Columbia. He published detailed reports on these visits, his interpretations of the geology and mineralization, along with hints to prospectors and descriptions of mining conditions and opportunities, written in his usual encouraging vein, as an incentive to greater investment and development.

Two of his longest trips in 1908 and 1909 were to the Kootenays and Yukon, in the company of William Templeman, the Minister of Mines and a British Columbia M.P. There the minister saw at first hand the degree to which the industry supported the labours of the Survey. In 1910 Brock made a similar trip along the Hudson Bay route in the company of the Governor General, Earl Grey, which gained the Governor General's sympathies for the needs and aspirations of the Survey. He made many visits to the United States, to recruit staff and attend meetings of professional organizations. He also visited the United States Geological Survey and various museums to study their arrangements and organizations and to establish useful contacts for the future. By taking his minister along on one such trip, in May 1910, he won Templeman's support for his plans of extending the role of the Survey in relation to the Victoria Memorial Museum, which was then nearing completion. In Canada, he regularly attended the council and annual meetings of the mining institutes and the provincial mining organizations. The International Geological Congress involved Brock in the full complement of meetings and planning sessions; frequent visits to Toronto, Montreal, and other cities; preliminary inspections of routes to be traversed by the Congress; and participation in three of the longest excursions.

Great as were the Survey's and Brock's preoccupations with the XIIth International Geological Congress, an even more momentous concern of the Brock years and

XIII-2 YORK BOATS RETURNING FROM OXFORD HOUSE (MANITOBA) IN 1910
An episode of the Governor General's tour to Hudson Bay by the Nelson River route, photographed by R. W. Brock.

one that affected the organization and operations of the Survey for many years to come, was the move of the Survey to larger quarters in the Victoria Memorial Museum building. This entailed far more than the upheaval caused by sorting, cataloguing, packing, unpacking, and displaying hundreds of thousands of items of various sizes, shapes, weights, and degrees of fragility—rocks, minerals and fossils, plants, insects, animals, birds and ethnological materials, as well as books, pamphlets, plans, maps, papers, and records. It occasioned a drive to build up the collections into more complete representations of the infinite variety of the Canadian scene, made curatorial functions important to almost every section of the Survey, and led to a noteworthy expansion of the staff and organization to serve hitherto-neglected natural history and human history sides of its responsibility. Brock hired additional specialists in the several important fields of museum activity, and he established two new divisions—Biological and Anthropological, each comprised of three sections—for the purpose.

This represented a momentous change, for the Survey had just surrendered responsibilities for assisting the mining and metallurgical sides of the industry in order to specialize in its fundamental role of mapping and studying the geology of Canada. Now it acquired the responsibilities and distractions of managing the nation's greatest museum in its crucial formative period, plus the direction of a large part of the Dominion government's efforts in biological, natural history, and anthropological investigations in all parts of Canada. For the Survey this meant a diversion of money and effort away from the geological work, while the National Museum suffered from being dependent on a branch of government with largely different interests and values. However, on the plus side, the museum was provided from the beginning with the most complete collection in Canada of specimens in many fields, the result of the Survey's early and widely-diffused collecting activities. The Survey was also a reputable organization that understood the meaning of scholarly work and high scientific standards; and in its director, Brock, the museum specialists could find a constant, sympathetic friend who saw that their work was started on the proper scientific footing.

That the building on Sussex Street was hopelessly inadequate and a fire hazard was long known. As early as 1894 the Survey's collections numbered over 120,000 items, most of which could never be displayed, for the precious results of its 50 years of collecting had to be presented in two large rooms, each about 108 by 35 feet, on the second and third floors of the building plus a smaller room on each floor. Brock therefore looked to the new museum building to overcome all the Survey's space requirements for some time to come:

> The 'Survey' has long since outgrown its present quarters. The congestion is such that not only is it impossible to display or store further Museum material, but it is also impossible to properly carry on the office work in connexion with maps, reports and the various materials brought in for investigation . . . . When in the commodious quarters which the new building will provide, a marked increase in the usefulness and efficiency of the 'Survey' will be possible.[5]

However, the emphasis behind the new building was not the Survey's need for more space in which to operate, but the need for more display space and safer storage facilities for its museum. From the earliest ministerial writing on the subject, government thinking had envisaged the building housing other agencies of government as well.[6] When the question was before Parliament in 1901, these were listed as the Supreme Court, Exchequer Court, National Art Gallery, and the Fisheries Exhibit. For good measure, the Ottawa city council added the Royal Society of Canada and the Archives Branch of the Department of Agriculture to that list of candidates. A further complication—and a cause for more delays—was the choice of site. Dawson and other early advocates of the new museum had favoured a site on Major's Hill Park, overlooking the Ottawa River and facing the Houses of Parliament across the Rideau Canal. Other suggestions for allegedly more convenient locations were soon brought forward by interested parties, however. By the middle of 1902 the government apparently had made up its mind; at all events, Haanel informed Brock on July 2, 1902, that "the Government, I understand, are now in favour of purchasing the Stewart Property for the location of the Museum, although nothing definite has been done in the matter."[7] Though the government continued to give evasive answers to questions in Parliament down to October 1903, Haanel's information proved correct. By an Order in Council of November 27, 1903, the nine-acre subdivision of Stewarton was purchased for $75,000 from the trustees for the Stewart estate, the architect's plans were prepared, and the building was given for tender in the autumn of 1904. The successful bid, the lowest of five, by George Goodwin of Ottawa for $950,000, was very considerably under the government's own estimate of $1,250,000.

By 1906 the construction was under way, and the Survey was storing up materials for the anticipated move. In proper patriotic spirit the government insisted that Canadian sandstone be used throughout, but in almost every other particular, negligence was the rule. The plans were so defective that a large room (70 by 70 feet, and 22 feet high) built at the base of the tower was approached by two long, winding narrow staircases leading to a pair of 30-inch-wide doors. The original hot water heating system was quite inadequate, and the charwomen had to use cold water because someone had forgotten to provide

XIII-3 THE VICTORIA MEMORIAL MUSEUM UNDER CONSTRUCTION, SEPTEMBER 1907
The circular structure in the background was afterwards occupied by the auditorium and the library, each two stories high.

XIII-4 THE COMPLETED VICTORIA MEMORIAL MUSEUM
After the imposing central tower was pulled down as a hazard in 1915, the building was left with its present rather squat appearance.

hot water for cleaning purposes; only in 1912 was additional heating and plumbing equipment installed to overcome these difficulties.

But these were only minor defects. In determining to proceed on the Stewart property, the government had not consulted the Survey and had actually chosen to build on land that was geologically unfit. As the unfortunate contractor learned to his sorrow, the site was underlain by about 140 feet of unconsolidated blue clay into which the building quickly began to settle. The heavier outer walls sank below the floors, which were held up by the inside walls. Cracks appeared in the pavement of the ground story and in the upper stories, and in the basement the tiled walls developed dangerous cracks and sheer planes. The weight of the tower caused the walls supporting it to come away from the floors and opened leaks in the roof that made it necessary to keep buckets under drip spots. Despite all efforts at patching, the crack widened and the tower began to develop a list; eventually in 1915 the tower had to be pulled down before it collapsed of its own accord. Already in 1912 the *Contract Record* had concluded that "Patchwork on such an important Government building, which cost the country millions, is discreditable, and it is to be hoped that an expert report will be obtained by the present government."[8]

Meanwhile, the Survey had been making its own plans to occupy the Victoria Memorial Museum. As early as 1904 Bell had sought to carry the point (against Sifton) that the curators and staffs (including administrative personnel) should be under the control of the museum, which in turn was under the control of the Survey. Under Brock this principle was conceded, and the Survey gained a very large added area of responsibility. The act of 1907 that established the Department of Mines specifically empowered the Survey for the first time to work in the ethnological field by adding to its prescribed functions "to carry on ethnological and palaeontological investigations," and also affirmed its connection with the new museum: "to collect, classify and arrange for exhibition in the Victoria Memorial Museum such specimens as are necessary to afford a complete and exact knowledge of the geology, mineralogy, palaeontology, ethnology, and fauna and flora of Canada." However, the act also spoke of the Survey maintaining "a Museum of Geology and Natural History for the purpose of affording a complete and exact knowledge of the geology, mineralogy and mining resources of Canada"—which appeared to imply a specialized collection distinct from the Victoria Memorial Museum.[9]

Relying on these powers, Brock continued the same integral relationship between Survey and museum that had existed since the 1850's. After visiting American museums, he decided "to equip the New Victoria National Museum on the lines now being followed by the American Museum of Natural History."[10] No doubt he also established good relations with the administrators of those museums that enabled him to induce a remarkably talented group of scientists and technicians to join the Survey after 1910.

As the building neared completion in the winter of 1910-11 (two years behind schedule), some discussion arose as to the Survey's right to occupy it. By then the east wing had been definitely conceded to the National Gallery of the Arts, but the Survey intended to occupy the remainder of the space. On December 24, 1910, a correspondent, H.J.W., defended the Survey's claim in the *Ottawa Evening Journal* by pointing out that the Survey had collected all the museum's material and had managed it with conspicuous success in the past. A critic, writing on January 7, 1911, however, heaped scorn on the Survey's involvement with anthropological collecting and researches, expressed doubts that the Survey could make effective use of more than 10 per cent of the space, and wondered why Canada's Geological Survey had to operate a comprehensive museum at all.

Consequently the Survey, feeling a certain sense of urgency lest others might impinge on the space, hastened to occupy the building. First were the geologists and topographers direct from the field in the autumn of 1910, who spread out their tents and other equipment to dry on the vast floors of the rotunda and empty exhibition halls, and commandeered choice locations for their temporary offices. Brock, too, hurried to bring the building under proper occupation. By the end of 1910, he reported that "the greater part of the offices and collections had been placed in the new building, where the valuable material will no longer be in danger of destruction from fire,"[11] and outlined his future plans for the building:

> The Museum will include the illustrative material acquired by the various divisions of the Survey, namely, mineralogy and geology, biology, and anthropology. It will, therefore, be a complete natural history museum. In the old Museum, mineralogy and geology were dominant, the other divisions being only sparingly represented. It is the intention to increase the economic material in the mineralogy and geology division so that Canadian ores and their products, their mode of occurrence, etc., may be thoroughly represented, and to greatly strengthen the biological and anthropological divisions . . . . For the present it is the intention to restrict the Museum to Canadian material . . . in order to make it, first of all, the great Canadian Museum, whose collections in Canadian material will surpass all others. When this has been accomplished in all divisions it may be advisable to enlarge its scope, and make it a world museum.[12]

The halls were quickly filled with displays and exhibits,

particularly from the large number of vertebrate fossils being assiduously collected from the badlands of Alberta. As soon as the augmented staff of preparators could get the material ready and satisfactory display cases could be secured, exhibits were mounted. The Hall of Fossil Vertebrates opened on January 20, 1913, and proved to be most popular, being crowded every Sunday afternoon from the start. Anthropological materials were being collected so rapidly that as early as 1912 the Anthropology Division concluded it had insufficient space for adequately representing the five main aboriginal cultures. It limited its hall to only three of these (Eastern Woodlands, Eskimo, and West Coast) and appealed for a second exhibition hall to display the Plains and Plateau-Mackenzie culture areas, "material from which it is expected will be coming in in increasing quantities."[13] In the following year Brock reported on the lack of space for the anthropological displays, and on delays in obtaining badly needed storage and display cases: "It is most discouraging to zealous officials to be unable to perform their necessary duties on account of the lack of the essential tools and facilities."[14]

* * *

THUS THE SURVEY acquired a new domicile that was to be its home for almost half a century, until 1959. The concomitant need to expand its work into the new museum fields gave added impetus to the reorganization of the staff that had already been started on the geological side, facilitated by the noticeable increases in the annual grants. The now-solicitous Liberal government wished to support the new department it had created and the men—Low and Brock—it had placed in charge. It was also responding to the public's desire to promote the growth of the mining industry, and to the heightened political power of that industry. Low had made the point very forcibly in 1906—that with a sevenfold increase in mineral production between 1886 and 1905, there had come only a 50 per cent increase in federal government spending on behalf of the industry. The argument grew even more cogent as production doubled in the next decade, from $68.8 million in 1905 to $137.1 million in 1915. Responding to the wild enthusiasm of these years, the Laurier government and the Borden government that followed, steadily increased the amounts allotted each year to the Department of Mines and the Geological Survey, as indicated in the table.[15]
Two main items—salaries and explorations—accounted for the overwhelming bulk of the expenditure, though their combined percentage declined from 90 per cent of the total (as in Bell's time) to between 67 and 81 per cent in the Brock years when publications, library purchases, and museum acquisitions became more significant spending items.

In less than a decade, the money made available to the Survey (Column 3) was more than trebled, and the actual outlay (Column 2) increased by 287 per cent. By 1910, in fact, Brock could report that "the Survey is being liberally provided with funds for its work,"[16] but that the inability to obtain enough qualified people was the chief limiting factor. From 1909 large amounts reverted to the Treasury at the end of each year, the levels being as high as 30 per cent of the original appropriation in 1910-11 and 1911-12. It says much for Brock's integrity (especially in such lax times) that he did not find ways to spend all the money that was placed at his disposal. Instead, he continued to hunt for first-rate men and expanded services only when they increased the effectiveness of the Survey, preferring to let funds lapse rather than use them to less than full advantage.

Year	Mines Department Expenditure	Geological Survey Expenditure	Granted to the Survey	Unexpended and Lapsed
	$	$	$	$
1905-06 (year to June 30)	–	168,376	180,323	11,946
1906-07 (9 mos. to March 31)	–	145,672	155,856	10,184
1907-08 (year to March 31)	268,698	219,011	227,575	8,564
1908-09	366,313	254,989	281,877	26,887
1909-10	381,706	263,196	349,956	86,759
1910-11	401,863	271,294	391,889	120,594
1911-12	459,398	288,432	424,226	135,794
1912-13	598,327	356,188	440,377	84,189
1913-14	733,530	440,760	522,230	81,469
1914-15	759,862	482,283	555,237	74,954

With the means at his disposal, his organizing flair, and his strong, determined character, Brock set about reorganizing the activities and structure of the Survey in accordance with current needs and opportunities. The retirements from the staff and emergence of new fields of operations enabled him to revivify and reorganize in keeping with the advances in science and technology.

The reorganization was mainly along the line of grouping staff in units according to their specialties. Earlier there had been minimal specialization and each man was responsible for every kind of work in his study area with some small aid from the paleontologist, chemist, and mineralogist. With a staff now comprised largely of specialists, some sort of organization by fields was needed if their work was to be properly supervised and related to the over-all program. The plan also steered members of the staff into fields for which they were best suited, and relieved the director from day-to-day details to concentrate on the lines of broad policy and on representing the Survey before the government, the industry, and the public.

The *Summary Report for 1906*, besides the inevitable reports of field work, presented reports on chemistry and mineralogy, the Section of Mines, paleontology and zoology, natural history, publications, drafting and engraving, library, and the accountant's and secretary's reports. These were not formally recognized as divisions, however, though three of the units—chemistry and mineralogy, paleontology and zoology, and natural history—were headed by the assistant directors Hoffmann, Whiteaves, and Macoun. The loss of certain functions to the Mines Branch made some reorganization necessary. The transfer of chemistry left behind a Section of Mineralogy that was raised to the Mineralogy Division in 1909, while the departure of the statistical part of the Section of Mines still left a Water and Borings Section. Following the death of Whiteaves in 1909 the unnatural combination of paleontology and zoology was terminated. Zoology was merged with natural history to form a Natural History Division in 1910 (renamed Biological Division in 1912), and a Palaeontological Division was set up in 1910 to deal with the principal unit of Whiteaves' empire. Similarly, the mapping and engraving section was raised to a Draughting Division in 1909 (renamed Mapping and Engraving Division in 1910 and Draughting and Illustrating Division by 1913); a Photographic Division was inaugurated in 1910 to take over all photographic work. Established in the same year to take over a function that was emerging to new importance was the Anthropological Division. A Topographical Division was created in 1908 to implement the principal reform in field practice envisaged by Brock—the separation of topographical from geological work, with each placed in the hands of specialists. However, not until 1911 was the reporting of topographical work grouped into a single report of the division chief. On the other hand, and contrary to the trend towards differentiation, the paleontology and mineralogy divisions were re-incorporated into the Geological Division—perhaps the result of the increasing amount of field work carried on by their staffs, and the greater dependence of the geological field parties on their findings. From 1912 onward, the reports of these two former divisions, and those of the Water and Borings Section, were presented among the reports of the Geological Division, following those of the geological field parties.

Such, in brief, were the main outlines of the new organization developed under Brock. The *Summary Report for 1913* gives the complete chart, which also reflects the intake that year of 21 new staff members, representing an increase from 56 in 1909 to 97 by 1913—an increase that was one more reason for establishing a number of units for purposes of control:[17]

Director

Administration:
General: Secretary; 5 stenographers.
Distribution: Chief; publication clerk; distribution clerk.
Stationery: 1 clerk.
Cabinet-maker: 1.
Messengers: 1 mail clerk; 4 messengers.

Geological Division:
1 geologist in charge of field work; 1 geologist in charge of office work; 11 geologists; 10 junior geologists; 4 palaeontologists; 2 preparators; 1 clerk; 1 stenographer; 1 mineralogist; 1 assistant; 1 collector; 1 stenographer.

Topographical Division:
Chief topographer; 1 triangulator; 7 junior topographers; 1 modeller; 1 custodian of instruments.

Biological Division:
Botany: 1 botanist; 1 assistant botanist; 1 stenographer. Zoology: 2 zoologists; 3 preparators and collectors; 1 stenographer.

Anthropological Division:
1 chief; 1 junior ethnologist; 1 preparator and collector; 1 stenographer; 1 archaeologist; 1 preparator and collector; 1 stenographer.

Draughting and Illustrating Division:
1 chief draughtsman and geographer; 11 draughtsmen; 1 clerk.

Photographic Division:
1 chief photographer; 1 assistant.

Library:
1 assistant librarian; 1 cataloguer; 1 stenographer.

The setting up of the divisions was intended to facilitate the actual work done and put more authority into the hands of responsible division chiefs. Brock also instituted a system of committees to deal with important matters that concerned staff members from more than one division. In 1908 a geological committee was set up "to carefully consider all geological reports, and the geological colouring of maps before printing,"[18] and a map committee—"an editing committee for maps . . . to standardize the maps, settle upon scales, and to critically examine all maps before they were allowed to go to the engraver."[19] In 1909 a library committee was set up, and in 1912, "for the purpose of considering matters connected with the Museum," a museum committee.[20]

Finally, to improve the management of the large and vital Geological Division, which was greatly enlarged in the scope and extent of its work and staffed by many young geologists, Brock placed two fairly experienced geologists in charge of the work—O. E. LeRoy over the field work, and G. A. Young over the office work, including the production of maps and reports. The men were relieved from systematic field mapping themselves, "to assist in general supervision, and to act in consultative and advisory capacity toward the staff."[21] This was also opportune because of the heavy amount of field work involved in the preparations for the International Geological Congress; Brock received much-needed assistance with the oversight of the geological work. In fact, so numerous and diverse were the field operations by this time that it was quite beyond the abilities of one man, even a Logan or a Dawson, to manage them, let alone find time to direct the other affairs of the Survey and the Department of Mines. In 1912, for example, there were 44 reported field operations, ranging from the western boundary of the Yukon Territory to Lunenburg County, Nova Scotia, and representing every geological region of Canada and the whole range of geological periods. They included broad reconnaissances, areal surveys, detail mapping, and concentrated studies of specific geological problems; among them, too, were paleontological investigations (predominantly collecting expeditions to secure fossils), mineral specimens and borehole record collecting operations, investigations of economic minerals, and the reports from the laboratories.

XIII-5 G. A. YOUNG

The two specialties merged in the Geological Division after 1912, paleontology and mineralogy, were each too wide by now to be encompassed by a single man. Brock therefore built up his paleontological staff with men specially fitted to work in places where the need was greatest. Besides temporary assistance from such noted American authorities as H. W. Shimer, A. F. Foerste, C. R. Stauffer, and C. D. Walcott, and the addition to the permanent staff of E. M. Kindle, P. E. Raymond and L. D. Burling (each responsible for a period of geological time), he assigned M. Y. Williams to do field work on problems of Paleozoic succession, hired a preparator, E. J. Whittaker, and appointed Alice E. Wilson to the invertebrate paleontology section as clerk to catalogue, label, mark and arrange specimens, and prepare guide books in connection with the reference collection and the displays.

L. M. Lambe, who had recently completed his training as a vertebrate paleontologist, was retained as the specialist in that field. Besides their value in fine dating of strata owing to the rapidity with which vertebrates evolved, vertebrate fossils dramatized the marvellous versatility of evolution, and they were excellent museum display material. So while Lambe handled problems of identifications and successions raised by specimens submitted by field geologists, assistants were procured for him who were collector-preparators. Brock hired Charles M. and George F. Sternberg and had fitted out for them a large laboratory in the basement of the new museum, equipped with all necessary labour-saving equipment, such as an overhead trolley system and hoisting block capable of moving two-ton weights; a blast furnace for forging mounts; electric drills and other tools to work in metal, wood, and rock; and delicate dental tools and brushes for cleaning. With this equipment the museum,

XIII-6 L. M. LAMBE

from 1912 onward, rapidly built up an extensive collection representative of "the wonderful diversity of the dinosaurian fauna of the Belly River formation" in Alberta.[22]

Still another field developing during these years was paleobotany, the study of fossil plants, which was particularly important for succession problems in the non-marine sedimentary basins of Nova Scotia and New Brunswick. This became the main work of W. J. Wilson, who had joined the Survey in 1901 and had been placed on general surveying work by Bell. After 1908 he was allowed to concentrate on paleontological material from the Maritimes, and in and after 1911 his report bore the heading "Palaeobotany." About this time he began to receive assistance in the field from Walter A. Bell, then a graduate student of Queen's and Yale. Whereas the paleontological staff under Whiteaves had numbered only three and for a time in 1910 was reduced to Lambe alone, by 1914 it numbered nine full-time scientific employees, as well as several summer assistants.

Following the transfer to the Mines Branch of the chemists and the standard analytical work the Mineralogical Section had to be reorganized and re-equipped with chemical and other apparatus. Besides identifying specimens from the field or submitted by the public, men were occasionally sent to the field to investigate mineral occurrences. Also rocks were gathered regularly to prepare mineral samples for schools and other institutions (the standard collection in 1908 included 150 different rock specimens) and other purposes. Some of the material was collected by the field geologists, a little was supplied by private individuals, but most was secured by the collector, R. L. Broadbent, and his successor, A. T. MacKinnon. The chief, R. A. A. Johnston, often working without assistance, conducted most of the tests, studied and reported on the properties of new, rare, or important minerals, and published catalogues and handbooks for reference purposes. After he received E. Poitevin as an assistant in 1913, it became possible to proceed more effectively to reorganize the collections and displays in the new museum, to undertake more field work (particularly Poitevin), and to make specialized studies of subjects of interest, such as the collection of meteorites.

The Geological Division included another section that was destined to have a long history—the Water and Borings Branch, also termed in the reports "Water and Borings Records" (1910), and "Borehole Records" (1911-13). This vagueness of nomenclature should serve as a reminder that the functions of the various units were far from rigid; also that it was not until the 1930's that specific meanings were given to the words "section," "division," or "branch" as administrative categories in the federal civil service. It had long been recognized that deep borings afforded much valuable information, and the Section of Mines had collected all the drilling logs and core samples it could. When the Section of Mines was transferred, its chief, E. D. Ingall, stayed with the Survey. Brock established the section to put Ingall's ability and experience to good use, and to continue this important aspect of geological study. Ingall instructed the operators to keep careful logs and furnish copies for the Survey. To encourage them to send in drill-samples he supplied small cloth bags, franked for free postal delivery, and promised to treat their information as secret or confidential if they so desired. A search was begun for oil well records that were in the hands of government bodies, companies, and individuals; and the material and data previously collected by Brumell, Chalmers, and others were restudied. Ingall regularly went to sites of borings in the Ottawa district to observe and advise operators, and in 1913-14 he detailed S. E. Slipper to do similar work in connection with the search for oil then underway in southwestern Alberta. The work progressed slowly; Ingall complained at first about the indifference and suspicion of the well drillers. But gradually a satis-

XIII-7 C. M. STERNBERG
Four Sternbergs came from Kansas to work for the Survey in 1912 and 1913 as vertebrate fossil collectors and preparators. Charles M. alone remained after 1918, continuing with the Museum until his retirement in 1951.

XIII-8 R. A. A. JOHNSTON

factory representation of records began to arrive, reflecting the hope Ingall had expressed in 1910 "of an extension of usefulness . . . as the operators are reached, and a wider appreciation established amongst them of the need and value of the work."[23] Records were being received from a wide variety of operations by 1914, and Ingall was able to publish a comprehensive, fairly detailed picture of drilling activities across Canada.

Associated with the Borings Branch was the important field of water resources, included in legislation from 1890 onward as a Survey interest. The Survey had engaged in such work from the 1880's in connection with Selwyn's drilling program in southern Manitoba, and the surficial work of Chalmers and others was very often concerned with groundwater and subterranean drainage problems. Most of the accurate information Canada possessed of the water and water-power resources of the undeveloped parts of the country came from the reports by Survey's reconnaissance explorations. Brock, in his report for 1908, referred to the need for more action in this field:

> The investigation of the water resources of the Dominion is becoming more and more pressing. In Central Canada information regarding the distribution and amount of water-power is urgently needed. In the North West provinces the problem of a water supply is one of the most serious in connexion with the settlement of the country. The Geological Survey, in accordance with the Mines Act, has been entrusted with the investigation of water supply and water resources, and provision must be made for systematic work in this field.[24]

Brock, however, did not feel impelled to increase the efforts beyond what was normally being done through surficial studies and collecting borings records. No doubt he did not want to duplicate too closely the work of the Irrigation and Dominion Water Power Branches of the Department of the Interior, or the agencies of the various provinces, or especially the important investigations of the Commission of Conservation after 1910 under its committee on waters and water-powers. This last-named agency alone published five volumes on the subject, beginning in 1911 with *Water-Powers of Canada* compiled by L. J. Denis. With so many groups in the field, there seemed little need for the Survey to do more in this direction.

The creation of the Topographical Division reflected Brock's most significant innovation in the system of field surveys. In contradistinction with Bell, Low and Brock returned to the idea of setting geologists to work intensively on specific areas of limited size, but of major economic or scientific interest. The new emphasis on solving geological problems rather than merely compiling and recording data meant that the geologist's time should be concentrated on the geologically-significant and intricate features of his area, with his assistants, perhaps, doing the simpler parts of the map-area. Further, since the supply of qualified geologists was limited, it was imperative that each should be given every facility to accomplish as much as he could in each field season on the

XIII-9 E. D. INGALL

XIII-10 R. H. CHAPMAN
The United States major who trained the Survey's topographers is standing behind a plane-table with telescopic and open-sight alidades.

subject for which he was specially trained. The most obvious aid, as was long realized, was to start work with a good base map, the way geologists in other countries did; for this purpose the Survey had been employing topographers as early as Logan's time. Moreover, topographical surveying had reached a stage where mapping was best done by highly-trained personnel. Hitherto topographers had worked as members of a geological party; from now on they were to head topographic parties of their own. To develop a full-fledged organization for the purpose, Brock in 1908 set up a Topographical Division:

> A topographer was appointed to have general supervision of the topographical work in the field, and the compilation of this work in the office. The rapid opening up of the country has increased the need for accurate topographical maps, and the development of this part of the 'Survey's' work has reached a stage which makes it desirable to give it distinct recognition. This is secured by the creation of a topographical division under the supervision of a topographer. It is hoped that greater uniformity and accuracy in methods will follow the improvement in the official status and organization of this branch of the 'Survey's' work.[25]

Brock chose his long-time associate, Walter H. Boyd, to fill this position. If the division was to be a success, a good staff had to be recruited, and the topographers had to be trained in the most modern techniques. In 1909, through the good offices of the United States Geological Survey, the services of Major R. H. Chapman were acquired for this purpose. A veteran of the exploration and mapping of southwestern United States, Chapman introduced American mapping techniques and methods of field work, along with a measure of its military discipline that included flying the Survey's red and white flag in camp, and decking the men out in khaki jackets, trousers, and Stetson hats as dress wear. The flags, belt buckles, and brass buttons were supplied, but the men bought their own uniforms. These innovations, intended to enhance discipline and develop an *esprit de corps*, never spread beyond the Topographical Division. Even there they quickly vanished after Chapman returned to the United States at the end of his third season in 1911. Geologists continue to argue the merits of at least flying a Survey flag to make their whereabouts known to prospectors and others who might be looking for them. A more permanent legacy was the men who received at least one season's training in Chapman's methods—K. G. Chipman, F. S. Falconer, A. G. Haultain, S. C. McLean, and B. R. MacKay. In time McLean specialized still further in control work, establishing base reference points by triangulation.

Since the division did not produce enough maps to keep pace with the requirements of the Geological Division, Boyd was continually adding to his staff. Like the geological staff, they were trained as assistants for a period of years, and then given permanent employment if they had proved themselves. In 1910 Boyd directed a large-scale operation, which involved most of the future staff, at Turtle Mountain, the site of the Frank slide, and in the Crowsnest Pass district generally. Other parties he sent on assignments as far apart as Moncton and Vancouver Island, and north all the way to White River in the Yukon, and along the Arctic coastline of Canada. Whereas the division in 1911 numbered only five, including Boyd, by 1914 it had grown to a dozen, including three topographers, six junior topographers, the triangulator, and an editor. With these increases in staff, the division was able to produce more maps and provide greater assistance to geologists, though the field geologists were never entirely relieved of preparing base maps.

The actual production of the topographical or geological maps was the task of the Geographical and Draughting Division (its name by 1914), which was headed by C. O. Senécal, who had been the geographer and chief draftsman since 1899 when James White transferred to the Department of the Interior. The division had been considerably expanded under Bell, and Brock had no intention of enlarging it further. The staff, which averaged thirteen during these years, included a clerk, and from nine to thirteen draftsmen. Although Senécal complained of the burden of the work and called for additional staff, the only changes were replacements for men leaving the division. What the staff was capable of doing was displayed during the emergency year of 1912 when, besides its regular work, it made up the 140 maps for the Guide Books for the International Geological Congress, even though some of the men were incapacitated by ty-

XIII-11 CHARLES CAMSELL AND A. J. C. NETTELL AT THEIR CAMP, TALTSON RIVER, N.W.T., 1914
An added point of interest is the rarely-photographed Geological Survey flag of that period.

phoid fever. The maps for the Guide Books were assigned to a small special unit of the staff, headed by A. Dickison; the effectiveness of their work may be seen in the finished product.

A further burden that began to weigh heavily on the division in and after 1912 was preparing pictorial drawings:

> During the year, in addition to the large number of drawings, sketch maps, diagrams, sections, etc., which were made by the staff of this division, to illustrate geological memoirs, a large amount of artistic or free-hand drawing for the Anthropological Division, also devolved upon the draughtsmen. This kind of work will henceforth necessitate the exclusive attention of one of the draughtsmen, who is specially trained in that line.[26]

By 1913 this was being done, "the services of one special draughtsman being now reserved for artistic drawing"[27] to meet the needs of the Anthropological Division. Another addition was an engraver, in response to Brock's complaint in 1912 that the division's finished maps were piling up "on account of the inadequacy of the present arrangement for reproduction. This should not be. If we were given engravers so that the engraving could be done in the same office as the map-making, much of this trouble would be avoided."[28] Brock fared better than Bell, who had tried in vain to obtain an engraver; Senécal's staff in 1913 included an "office copper engraver," who worked on a few of the maps.[29] By the 1920's the Printing Bureau had accepted the argument and farmed out four of its engravers on permanent loan to the Survey.

Brock's innovative temperament was reflected best in his efforts to exploit the newest photographic techniques. His action in setting up a Photographic Division, even though the staff amounted to only two technicians, indicated the great importance he attached to this branch of the work. For some years Senécal had been urging the establishment of a photographic darkroom in connection with his section; he was successful, to judge from his report in 1907:

> In closing this statement, I may be allowed to call your attention to the fact that a large quantity of preparatory work for the compilation of maps is now advantageously made by photography. Reductions, enlargements, copies, etc., are done very expeditiously by this method, but the quantity of work increases from year to year, and now that well equipped photographic rooms have been placed at the disposal of the draughtsmen the employment of a professional photographer under my supervision is earnestly recommended.[30]

While Brock proceeded with refitting and improving the photographic laboratory, clearly he did not want to have this work put completely under the drafting section, inasmuch as "Photography is now being used with advantage in almost every branch of the work."[31] The Survey's experience with photography went back to the early sixties when field men often took photographs of places and formations encountered in the course of their explorations. In the early years, however, the cumbersomeness of the equipment and the inability to publish photographs without first making engraved copies restricted its use. The amount of photographic work required then was relatively small, and for years was done by keeping H. M. Topley, the photographer for the Department of the Interior, on a $300 annual retainer. But equipment became more convenient, and the uses for photography doubled and redoubled. Besides being used to illustrate geography and geology and by the draftsmen, it was applied by topographers in new mapping techniques in which hundreds of overlapping photographs were taken from key reference points. Fossils, instead of being drawn laboriously by hand, were now photographed under strictly controlled conditions, and their taxonomic descriptions were augmented in this way. So even though it had been the draftsmen's needs that led to setting up the photographic laboratory, Brock decided on a new division to consolidate the many varied activities of developing plates and negatives, photographing specimens, making blueprints, reducing and enlarging prints for the mapping division, preparing lantern slides, and processing the masses of photographs now being brought back from the field by the topographers. By 1914 the two technicians, ensconced in the new laboratory at the Victoria Memorial Museum, as their year's work produced 12,207 prints, developed 4,220 films and plates, and made 1,734 photostats and 236 lantern slides.

Brock also took in hand the very important matter of publishing reports, which had occasioned so many complaints in the past, by improving the organization that handled this work and instituting a number of new publications. He did not go so far, however, as to set up a formal publications division. Distribution of the publications had formerly been a function of the library, indicating their importance as an exchange medium. But the staff list on Brock's organization chart of 1909 showed two publications clerks in the Administrative and General Division. In 1912 the geological compiler, Wyatt Malcolm, was transferred to this section and given charge of all the publications. In addition, the scientific editing required by the International Geological Congress was done by other officers, Camsell and McInnes in particular.

More noteworthy were the reforms in the publications themselves. The major problem in the past had been the long delays between the actual field work and the results being reported to the public. The annual reports had been two and three years behind the work described, so the public, with reason, complained that their value was largely negated by the delays. The first major reform to meet this problem was instituted in 1906 under Low, when the summary report, containing the director's review of work and presenting preliminary material from the reports of each field party and office section, was published early in 1907. This slender volume of less than 200 pages had been hurried through the press without maps in an effort to make it available as soon as possible. The summary report for 1907, prepared under Brock's supervision, actually appeared by January 15, 1908. It had only 125 pages and lacked two reports, but all the same was acclaimed on the score of its early appearance. In time, however, the summary reports began to repeat some of the tendencies of their predecessors. They began to present maps and plans, and to grow in length and detail, successively to 220 pages (1908), 307, 314, 412, 544 (1912), and 417 pages in 1913. As a result of the accompanying delays, the last-named report was not ready until early 1915.

The summary reports did not leave any room for the final reports of the officers after they had completed their study and analysis of the results. For a few years, particularly between 1906 and 1910, the Survey published such final reports as separates, as part of a loosely-defined series that had been used since 1847 for miscellaneous publications and which had been almost monopolized by the paleontological writings of Billings, Whiteaves, and Lambe. In 1909 Brock announced a new series intended to convey the final, complete, and considered work of the scientific staff—the Memoirs. The last separate report was published in 1911, and the first seven memoirs appeared in 1910, inaugurating a series that still exists and numbers over 350 volumes to date. The serial numbers did not necessarily denote the order of publication, for sometimes a memoir was delayed for revisions or maps, or in press. The list of memoirs in press on May 5, 1915, for example, included numbers 34, 46, 62, and 63 as well as the current numbers running from 68 to 76.

XIII-12 THE BADLANDS OF SOUTHERN ALBERTA, COLLECTING GROUND FOR DINOSAUR REMAINS
Sections of the fossil skeletons are wrapped with strips of burlap dipped in fluid plaster, then packed in boxes for shipment to Ottawa.

The memoirs were the final reports of studies that might have taken years in the field and office to prepare; some were used as doctoral dissertations. A number of the first memoirs have become classics, among them Adams and Barlow's Haliburton and Bancroft area (no. 6, 1910), Daly's North American Cordilleran report, in three volumes (no. 38, 1913), Lawson's restudy of the Archaean geology of the Rainy Lake district (no. 40, 1914), Camsell's on Hedley mining district (no 2, 1910), and Faribault and Malcolm's on the goldfields of Nova Scotia (no. 20, 1914). Not all memoirs dealt with geological subjects; the first 76 included 63 geological, ten anthropological, two biological, and one topographical subject.

Even the memoirs did not cover the entire range of scientific output of the staff. Often in the course of an investigation a scientist gained valuable insights into important problems of wider applicability than to his particular research. To deal with these in a memoir meant digressing from the subject of study and the possible entombment of a contribution to general theory that deserved to reach a wider audience. Such articles could be published in scientific journals, but there they often experienced long delays, severe editing for the purposes of the particular publication, and were credited to the individual authors rather than to the Survey.

Accordingly, a new publication was instituted in 1913—the Museum Bulletin—to carry articles that were too general for inclusion in the memoirs or summary reports, and that were too valuable to risk being lost in them. Museum Bulletins were issued irregularly and contained one or more separate papers; the first bulletin, published in 1913, included a dozen on geology, three on biology, and one on anthropology. Most of the 14 bulletins that were published by 1915 dealt with geological subjects, but with a sizeable admixture of biological, anthropological, and ethnological topics. The bulletins publicized important work of the Survey; they were not restricted by concerns of space or expense of preparing illustrations from presenting the work as it deserved; and the diversity in subjects showed the wide range of scholarly activity of the staff.

Even these three series did not exhaust the diversity of works published. There was also the first edition of *Geology and Economy Minerals of Canada* (1909) which was prepared by G. A. Young and was the forerunner of the Economic Geology Series; the first prospector's handbook *Notes on Radium-bearing Minerals* (1914); the first museum guide book (1914); and the publications for the International Geological Congress.

To sum up, in the five years from 1910 to 1914, inclusive, the Survey published four director's summary reports, six separate reports, 52 memoirs, seven museum bulletins, eight guide books in eleven parts, one prospector's handbook, and one museum guide book, besides editing and seeing to the publication of the three-volume *Coal Resources of the World* and the *Compte Rendu* of the International Geological Congress. A most impressive quantity of work in such a short period. As usual, however, the achievement found critics, though, in general, it was the publications of the Mines Branch that roused most comment over their allegedly limited value to professional groups. Whereas one writer referred gratefully to the high quality of Survey publications and the "well-advised generosity" of the director's distribution program,[32] another, writing in March, 1915, and reflecting the newborn austerity induced by the Great War, expressed this opinion of the publications of the Department of Mines:

> The numerous changes in the past ten years in the classification of publications have resulted in the production of a bewildering series of reports, memoirs, bulletins, handbooks, etc. An over-indulgence in classification of the publications has resulted disastrously for the reader. The earlier system of annual volumes was much more satisfactory.[33]

Memories are short, indeed, and it is impossible to please everybody. What a different criticism that correspondent would have expressed just a decade previously, in 1905, had he had occasion then to comment on the Survey's publications!

* * *

THE BUILDING OF THE VICTORIA MEMORIAL MUSEUM encouraged the Survey to extend operations in two peripheral areas that were traditional adjuncts to its regular work—the fields of natural history and anthropology. The explorations of Dawson, Bell, Low, Tyrrell, and others had resulted in considerable knowledge of the flora, fauna, and original inhabitants of the sparsely settled parts of Canada, as well as a major collection of anthropological, archeological, and ethnological material that was greatly augmented by purchases of important private holdings. Similarly, these explorations (reinforced by the work of John Macoun and his assistants) had

XIII-13 EDWARD SAPIR

XIII-14 HARLAN I. SMITH

yielded many specimens of plants, insects, mammals, birds, and fish. The museum was credited with having the best collections in the biological and human categories in Canada, and by the nineties was already being referred to as "the National Museum." For the Survey to increase its activities and organization in these directions was simply a matter of continuing a long-standing tradition, though it was also a practical necessity if the Survey was to maintain its control over the new museum and the building in which it was housed.

Hence, as the building approached completion, Brock laid new stress on the importance of the Survey's taking action in the field of anthropology:

> In the new Museum one of the most popular sections is likely to be the ethnological exhibit. Very little investigation has been made in Canada of the native races, and what has been done has mostly been under the auspices of foreign institutions. The opportunities for such studies are fast disappearing. Under advancing settlement and rapid development of the country, the native is disappearing, or coming under the influence of the white man's civilization. The older people who are familiar with the folk lore or traditions of the tribe are dying off, and the rising generation under the changed conditions is acquiring a totally different education.
>
> If the information concerning the native races is ever to be secured and preserved, action must be taken very soon, or it will be too late. It is a duty we owe to the Canada of the future to see that such material is saved.
>
> The work on the Eskimo of the Arctic, undertaken this year in conjunction with the American Museum of Natural History, is hoped to prove only the beginning of a serious effort to acquire the valuable ethnological data that is yet to be had.[34]

The last reference was to some small financial assistance Brock had extended to the 1908-12 expedition of the future-celebrated Vilhjalmur Stefansson and his colleague, R. M. Anderson. Though attempts have been made to credit Stefansson with awakening the Survey's interest in scientific study of native civilizations, this concern had been in evidence as far back as Selwyn's time; however, Brock was correct in stigmatizing it as "spasmodic and entirely secondary."[35] That there was a sharp upsurge in public interest was indicated by the formal listing of anthropological work for the first time among the Survey's duties under the act of 1907. In 1909 the Royal Society of Canada, the British Association for the Advancement of Science (at its Winnipeg meeting, August 1909), and the Canadian branch of the Archaeological Institute of America, combined to petition the Dominion government to sponsor such activity as a national concern. The resolution of the British Association, for example, recited

> That it is essential to scientific knowledge of the early history of Canada that full and accurate records should be obtained of the physical character, geographical distribution and migrations, languages, social and political institutions, native arts, industries, and economic systems, of the aboriginal peoples of the country. . . .
>
> That it is therefore of urgent importance to initiate, without delay, systematic observations and records of native physical types, languages, beliefs and customs; to provide for the preservation of a complete collection of examples of native arts and industries in one central institution. . . .
>
> That the organization necessary to secure these objects and to render the results of these enquiries accessible to students and to the public, is such as might easily be provided in connection with the National Museum at Ottawa, which already includes many fine examples of aboriginal art and manufactures, and might easily be made a centre for the scientific study of the physical types, languages, beliefs, and customs of the aboriginal peoples.[36]

XIII-15 MARIUS BARBEAU

Brock, in his report for 1909, referred to the support of these institutions as proof that the time "seems opportune to begin serious and systematic work in Canadian ethnology and archaeology,"[37] and promised that when the collections were moved to the new building, "a scientific, trained ethnologist" would be appointed "to take charge of the collections in the Ethnological Hall and to direct work in connexion with ethnological and archaeological investigations."[38] The appointment of Dr. Edward Sapir to take charge of the newly-established Anthropological Division followed in 1910, his work being outlined as:

> The plans of the Anthropological Division include field work among the native tribes of Canada for the purpose of collecting extensive and reliable information on their ethnology and linguistics, archaeological field work, the publication of results obtained in these investigations, and the exhibition in the Museum of specimens illustrative of Indian and Eskimo life, habit and thought.[39]

In 1911 the division was divided into an archaeology section under Harlan I. Smith, and an ethnology (later ethnology and linguistics) section, under Sapir, who was assisted by C. Marius Barbeau. In 1912 a third section, physical anthropology, began to emerge, though it was 1914 before F. H. S. Knowles was appointed to the permanent staff in charge of this work. Two more assistants were added in 1913, W. J. Wintemberg to the archaeology section, and F. W. Waugh to the ethnology section, besides which the division was rounded out by two stenographers, giving a staff by 1914 of eight.

The six field officers spent their summers in the field, working feverishly to collect information and artifacts from what were felt to be fast-dying civilizations. They were supplemented by other scholars, whose work was underwritten as part of the program for speeding the work forward. Since Canada did not yet have enough scholars of its own to study the native cultures in this time of emergency, Brock did the next best thing—he saw to it that the civilizations were competently recorded and that the National Museum received its fair share of materials representing the native arts and cultures. Since ethnological research was essentially a one-man operation, in which an individual lived in and became as much as he could a part of the native community, it was desirable to sponsor as many different field studies as possible. A relatively large number of sponsored studies were carried on simultaneously with those by the permanent staff, under the supervision and oversight of the heads of the several sections and in conformity with their over-all programs.

While field work was the most urgent side of the work of the division, Sapir and Smith were conscious of the need to publish their findings and prepare effective museum displays. Smith, in his first report (1911), for example, divided the work into two main groups: "activities for diffusing archaeological knowledge by such means as museum exhibits, guide books, and lectures, and those for increasing such knowledge, as by exploration, original research, and systematization."[40] Publications included, besides the contributions to the annual reports, several memoirs, and articles in the museum bulletins and in scientific journals. The collections themselves required a great deal of work—preparing specimens, cataloguing, procuring display cases, and arranging the exhibits. Some idea of the complexities of these tasks is given by Sapir in 1912:

> Thus far the museum work of the scientific staff of the Division of Anthropology has been seriously hampered by the lack of a regular preparator or technical assistant, as the purely scientific and office work of the staff makes it difficult for them to do full justice to the proper care of museum material. The necessity for an anthropological preparator, whose duty it would be to treat (clean, fumigate, and poison), sort out, number, catalogue, store, and keep in constant good care the ever increasing anthropological collections of the museum, is imperative. Provision might also well be made for a skilful mechanic for the division, who could be employed to repair or reconstruct material in poor or fragmentary condition, prepare models and groups illustrating various phases in the life of the natives, and do such other technical work as might be required.[41]

To meet these needs, the staff of the division was increased, and the preparation of collections and displays proceeded so well that by 1914 the exhibits were in good order.

Before 1909 natural history had been under the dual control of John Macoun, naturalist, and J. F. Whiteaves, who combined the oversight of the zoological collections of the Survey with his paleontological work and by virtue of his seniority was responsible for the museum. Macoun, the self-taught naturalist, was primarily a botanist, though he also published other works in the wider field of natural history, notably a pioneer catalogue of Canadian birds. His botanical and zoological collections were supplemented by his assistants W. Spreadborough and his son, James M. Macoun, who had previously doubled as topographical assistant with Low and other geologists. Exploring geologists also brought back many specimens of the flora and fauna of every region of Canada, which Macoun and others studied, classified, and catalogued. Thanks to the long years of studying and collecting, the Survey's knowledge of the biological assets of the country was unrivaled and its collected material outstanding. In 1913, for example, the herbarium was described as the largest and most complete in Canada, was often consulted as a reference collection, and was used as a repository for types of newly-discovered species.

After the death of Whiteaves, Macoun assumed control over the whole natural history field, and in 1910 he became head of a newly-established Natural History Division, two years later renamed the Biological Division. Here again, as with the anthropologists, the challenge of preparing new displays for the national museum occupied a large part of the energies of the staff. There was even a similar emphasis on the need for haste:

> The Canadian country is changing rapidly from an unsettled state to that of civilization and cultivation. This is having a most profound effect upon our flora and faunal life, and vast changes are being brought about in our biotal conditions. The old order is passing away, in many places has already passed, without leaving a record of its being behind. If the next generation is not to charge us with being indifferent to their interests we must improve every opportunity of making record of present conditions. The time for this work is now, for every day means some loss on the pages of our records that can never be filled.[42]

Efforts were made to complete collections and fill gaps in order to make the collection truly national, and more suited to excite the curiosity of visitors. In 1908 Macoun was putting the collections in order, working up a flora of the Ottawa district for local interest, and beginning a long-term collection of the marine and terrestrial fauna and flora of southern Vancouver Island. In 1910 he started a similar collection in Nova Scotia, and J. M. Macoun undertook a collecting expedition of Arctic species from the northwestern coast of Hudson Bay, between Churchill and Wager Bay. As with anthropology, some collecting and studying was done by part-time specialists, who supplemented the efforts of the regular staff—by collectors such as Spreadborough, or a trained zoologist like R. M. Anderson, Stefansson's colleague in the Arctic coast work of the period 1908-12, or the Americans J. Alden Mason and Francis Harper, who made biological investigations of parts of the Mackenzie

XIII-16 W. J. WINTEMBERG

XIII-17 P. A. TAVERNER

basin. Other staff members were the preparators, the veteran taxidermist Samuel Herring, and C. H. Young, his assistant since 1907.

The division displayed the same expansion of staff and differentiation of function characteristic of other divisions. In 1911 an assistant naturalist, P. A. Taverner, was hired, who reported that year on the zoological section, mainly on vertebrata. He later became the Survey's ornithologist. C. H. Young was assigned to the preparation of invertebrate material. In 1913 R. M. Anderson was added to the staff as a mammalogist, and C. L. Patch, the future herpetologist, as yet another taxidermist. A colourist joined the preparatory staff in 1914, and it was proposed to establish a tanning plant for the section as being advantageous from the standpoints of economy, care, and control.

The botanical section did not grow to the same extent as the zoological; it consisted of the two Macouns. The senior Macoun practically retired in 1913 when he began living year-round on Vancouver Island from where he sent back reports and collections to Ottawa, and the report for the division in 1914 was made by James M. Ma-

XIII-18 THE SURVEY'S NATURALISTS IN THE FIELD
The aged John Macoun is flanked by the specimen preparator, C. H. Young (*left*), and William Spreadborough, the assistant collector (*right*).

coun in place of his father. Entomology constituted virtually a third section, under the appointed curator, the Dominion Entomologist, Dr. C. G. Hewitt. In 1913 the appointment of a full-time entomologist to the staff of the Survey was requested, and in the next year a plan was broached to consolidate the Department of Agriculture's collection with the Survey's at the Victoria Memorial Museum.

In sum, R. W. Brock as director built up the reputation of the Survey from a faction-ridden, rather moribund institution into a vigorous, active, alert government agency staffed by outstanding people, which reaffirmed its highest scholarly and scientific traditions and could compare with the best in the world in many of its widely-ranging fields. As director from 1908 to 1914, Brock demonstrated a rare talent for organization that marked every phase of his career—studying and teaching at the department of geology at Queen's, recruiting and training a universities unit in the Great War, establishing the faculty of applied science at the University of British Columbia, and directing geological survey projects in Palestine and Hong Kong. Though Brock led the Survey for less than seven of its 130 years, he completely overhauled its structure and rearranged and redirected its activities along lines it was to follow for more than a generation. Not a brilliant scientist in the tradition of Logan or Dawson, R. W. Brock is nevertheless justly regarded as second only to Logan in the way he placed his particular imprint on the Survey.

XIV-1 INTERNATIONAL GEOLOGICAL CONGRESS GROUP IN FRONT OF CENTRE BLOCK, HOUSES OF PARLIAMENT, OTTAWA, AUGUST 1, 1913

14

The Survey on Display

THE CHANGE OF 1907 that made the Survey a branch in the Department of Mines afforded an opportunity to remake its public image and revise its program. The new leaders, A. P. Low and R. W. Brock, could appeal for a fresh mandate to meet the challenges and opportunities of the twentieth century. Canada was enjoying unprecedented prosperity. New railways were opening whole kingdoms to settlement and industrial development, and the mining industry was soaring to new heights. Now that the people of Canada had embraced the philosophy of national expansion, they were more conscious of the Survey's past contributions to Canadian growth and more ready to assist it adequately to conduct its future work.

The age of the great explorations was largely past, though the prestige and good will from that role were a valuable legacy as the Survey prepared to carry on with types of work more suited to its changed situation. The mandates to investigate the geology of Canada and maintain its national museum made the Survey the foremost scientific government agency, embracing a wide range of disciplines—geology and mining, topographical surveying, anthropology and ethnology, zoology and botany. These responsibilities Brock took very seriously. Though he showed a lively awareness of the need to conciliate the mining industry, he also was concerned with raising the scientific attainments and technical competence of his staff to the highest possible level. He felt keenly the stirring intellectual climate of his age and saw the enormous advances being made in geology and other sciences. He recognized the supreme importance of securing the most highly qualified men in every field if the Survey was to make significant and original contributions. Under Brock the Survey underwent a bright renaissance and reaffirmation of its best scholarly, scientific traditions.

But a first-rate staff could only be recruited, as Low and Brock had good reason to know, if the Survey was able to offer adequate salaries. Low's plea for an increased appropriation in 1906 centred on this need:

> Geologists are made, not born, and several years must be spent in the making. At present there are few trained men outside the service who are capable of undertaking the work performed by the staff of this department. Owing to the small salaries paid, in comparison with the pay of private individuals and corporations, those who are so trained refuse to accept government employment . . . .
>
> In the past the practice in the department of appointing young men at exceedingly low salaries and of advancing them by the ordinary fifty dollars annually has worked disastrously. It became the custom for students to enter the service for the excellent training afforded them and after a few years to leave and take positions where the salary was much greater than that paid by the government.[1]

Brock echoed this complaint two years later, adding that it was false economy not to secure anything but the best staff possible. He therefore specified education to the Ph.D. level as the standard of qualification—a quite unheard-of requirement in the Canadian public service of 60 years ago, and one that the Survey has retained to the present day:

> For the proper prosecution of the work, only picked, well trained, expensively educated men can be successfully employed. They should properly have seven years of college training (graduate and post graduate), and in addition, wide experience in the field . . . . The stamp of man required by this 'Survey' for responsible work is equal to that required for a chair in a university.[2]

Otherwise, Brock said, the Survey would not only fail to attract the type of men it needed but would be raided by other employers and deprived of its best men, leaving only "the culls" to carry on. When an experienced officer resigned in 1911 to take a better-paying position, Brock renewed his complaints, pointing out that besides the loss of the man's skills, the Survey had also lost the knowledge and experience he had gained on particular districts and problems.

XIV-2 M. E. WILSON

Fortunately, as the finances improved it became possible to raise the salaries to a level where, even though they did not match those offered elsewhere, they were high enough that good men could be held by such incentives as national patriotism, a sense of public service, devotion to the ideals of scientific truth, and by the *esprit de corps* among the staff. The public image of the Survey was so improved that a career there became the life's ambition of many of the ablest young geological and engineering graduates of the time. To meet the minimum material requirement, a salary "sufficient to enable the official to live without financial worry, and to properly educate his children,"[3] Brock was able to apply a new salary schedule instituted in 1908 across the Dominion Civil Service that equated the work in the several parts of the organization. There were three grades, I, II, and III—each with an A and B division — with floors high enough to attract desirable recruits and increments large enough to retain their services. Grade III was essentially for clerical and labouring staff, with floors of $800 (IIIB) and $900 (IIIA), and division IIB ($1,000) for skilled tradesmen such as the draftsmen. University graduates, beginning geologists for example, started at $1,600 (IIA), moved upward into IB ($2,100) as they demonstrated special skills and gained experience, and eventually, achieved the grade of IA ($2,800), which was reserved at first for assistant directors. In 1910 division IA was opened to senior geologists and to heads of important divisions, so that their salaries could advance beyond the $2,800 level. Young men thus could look forward to increases through promotions (up to $500 between IB and IA alone) and through regular increments, which were of $100 per annum in grade I and $50 in grades II and III. This system proved capable of attracting and retaining the services of a competent group of men in the face of serious competition, and no further troubles arose in this regard until the unsettled years after the Great War.

University training was only part of the making of a good geologist. In 1904, when the universities added their complaints to those of the Survey against the system of political influence that had almost monopolized appointments to the field parties, a system of summer training was instituted from which the Survey expected to secure its staff needs. In 1906, Low reported,

> . . . efforts are being made to recruit the field staff by an agreement with the several mining schools of Canada, whereby places will be given on the summer field parties to a number of the best qualified students with a view to partly training them for the work of the Geological Survey during their college vacations and ultimately giving them permanent positions upon the staff if they are found to be adapted to the work . . . . In this manner it is expected that the staff of the Geological Survey will be recruited from the best and most efficient graduates of our mining schools; the process may be slow, but it is the best and only one by which a really efficient staff can be secured.[4]

Students in the upper years of honours geology and mining courses were offered places on field parties where they could finance their studies while gaining practical experience. The Survey obtained an excellent opportunity to size up the men, train them to its requirements, and influence them into joining the service after graduation. By 1911 Brock was pronouncing the system a complete success:

> . . . the Survey has been enabled to try out each year about sixty prospective technical men especially selected by their respective professors. The more promising of these are encouraged to proceed with their training, the geologists to take doctorate degrees in geology in post graduate universities, the topographers to train in the Survey. During the past year a number of such men have finished these courses and received appointments to the staff. Next year a further number of such specially qualified men will be available.[5]

This method of training became the accepted practice, and every important field party had one or more geological assistants who later joined the service.

As Brock mentioned, some of the men were encouraged to proceed to graduate work in the United States in the winters, after they had received practical training in the field to the point of taking charge of parties. At the universities they could also use the results of their summers' work for their Ph.D. theses. Morley E. Wilson, who was already on the staff, for example, received leave with pay for three winters to study at Wisconsin, Chicago, and Yale. The report of his field work of 1910-11, written up during the winter of 1911-12, was accepted by Yale as his doctoral dissertation, which the Survey published in 1914 as Memoir 39, *Kewagama Lake Map-area, Quebec.* Other young men financed their education through graduate school by working for the Survey during summer vacations.

Now that the Survey had a more attractive salary and recruiting system, and larger appropriations for staff, Brock embarked in earnest on a vigorous hiring program, to fill the large number of vacancies created by the many retirements and deaths during the years after 1906, and to expand the organization. These years were a sort of watershed when control was passing from an older generation to a younger one, which was to dominate the affairs of the Survey for the next 30 years. Many of these changes came in the first two years. Five staff members were transferred to the Mines Branch and another six, mostly clerical and administrative personnel, to the staff of the new department. Resignations in 1907 and 1908 included those of the veterans Robert Bell, G. C. Hoffmann, and A. E. Barlow, C. W. Willimott the mineral collector, John Thorburn the librarian, and F. Nicolas the editor (who later returned to the Survey, in the 1920's). H. M. Ami, plagued by ill health, resigned in 1910, after seeing the controversy with Fletcher resolved in his favour. J. A. Dresser resigned in 1911 and P. E. Raymond, paleontologist, in 1912. Deaths included those of Robert Chalmers, Hugh Fletcher, and J. F. Whiteaves in 1909; R. W. Ells and the mineralogist R. L. Broadbent in 1911; and Mrs. J. Alexander the acting librarian and J. D. Trueman in 1912. Trueman, a junior geologist about to be appointed to the permanent staff, was drowned while working in the Atikokan district. W. W. Leach died in 1913, and G. S. Malloch perished tragically early in 1914 while with the Canadian Arctic Expedition. All told, about half of the approximately 70 full-time staff of 1906 were replaced by 1914.

The most important of the departures was that of Robert Bell, which spanned two controversy-laden years. It began when Bell was replaced as director by Low and was appointed as chief geologist under the vague directive of March 29, 1906, which advised him that the reports prepared by him or under his direction would be issued under his name. With the onset of Low's illness early in 1907 and then Low's appointment as deputy minister of the Department of Mines, Bell was emboldened to revive his claims to the directorship, even going so far as to denounce Low to Sifton, his own erstwhile adversary:

> On the other hand, with the exception of political pull (from having married the daughter of the principal boss in Liberal politics here) Mr. Low has no known qualifications above hundreds of others, either in education, special knowledge, administrative ability, temper, disposition, or anything else. He has just got over the first stages of a serious brain disease and it is still uncertain that he will recover his mental faculties. There is no reason but politics why he should have all, or his choice of all that is going. The Survey will be entirely ruined very soon, if this be allowed.[6]

A more direct appeal in July 1907 to the Minister of Mines to be appointed director on the grounds that Low's continuing illness was creating a serious vacuum brought down on Bell a proper snub:

> I beg to acknowledge receipt of yours of the 14th November applying to be appointed to the position of Director of the Geological Survey, owing to the illness of Mr. Low, to whom you refer as the "Acting Director".
>
> Mr. Low is the Director of the Survey. I am not advised that his illness is permanent, and therefore the question of considering a successor is not at present under consideration.[7]

The government solved the matter by appointing Brock acting director early in December, and the trouble with Bell soon came to a head in the form of a missing letterbook, and a bill for a conference Bell had attended in Cobalt in the summer of 1907. Percy Selwyn, the director's secretary, claimed Low had expressly forbidden Bell to attend that conference as "it was considered that Dr. Bell's presence at such gatherings was more detrimental than beneficial to the best interests of the Geological Survey."[8] Bell claimed there had been no senior person at the office to authorize him to attend the conference, and that in any event he was empowered to make such decisions on his own, both of which claims Selwyn disputed. Bell carried the matter of his expense claim all the way to the Prime Minister, who referred it back to Brock, who thereupon prepared a memorandum giving Selwyn's side of the affair.

In the meantime, Bell's superannuation was being taken in hand. As early as October 15, 1906, a recommendation retiring him as of November 1, 1906, when he would be 65, was prepared but not acted on. Another of April 1908 proposed to retire him as of May 31, 1908; this was replaced by a more equitable one of May 20, 1908, recommending that his superannuation take effect

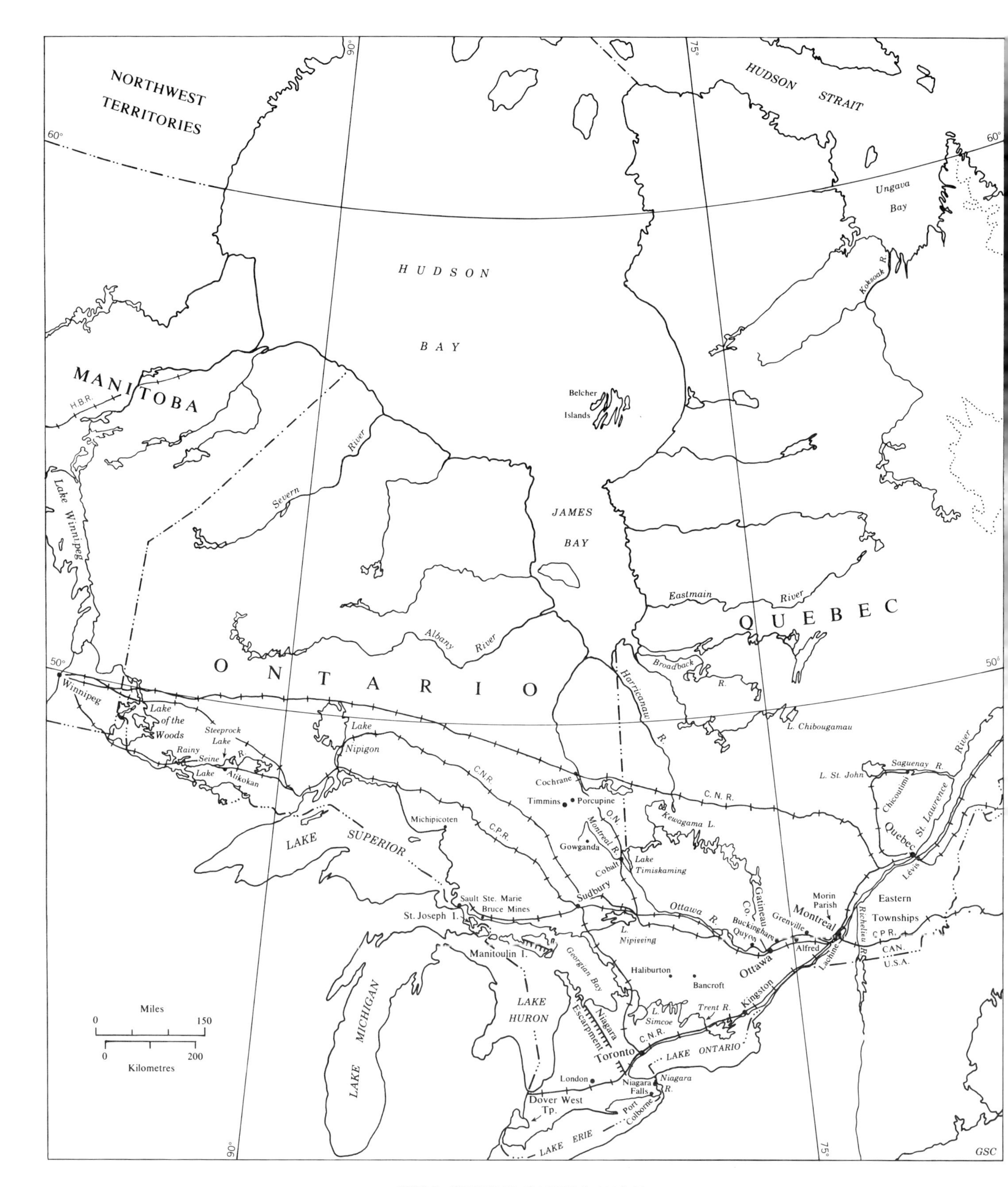

XIV-3 CENTRAL CANADA, 1907-20

November 30, 1908, with terminal leave from June 1, 1908. Only after Bell had formally departed was Brock elevated from acting director to director of the Geological Survey, effective December 1, 1908.

After his retirement Bell continued to bombard the Prime Minister with claims for arrears of salary for his five years as acting director (1901-06) and with requests to be appointed to a suitable position in connection with the conservation activities of the Laurier government. His last publication was an obituary of T. C. Weston, 1832-1910, which appeared in the *Bulletin* of the Geological Society of America for 1911. Robert Bell died in relative obscurity after a prolonged illness on June 17, 1917, at Rathwell, Manitoba.

Brock, in addition to making good the staff losses of the period 1907-14, also increased and strengthened the staff in line with Canada's needs and the Survey's many responsibilities. As he reported in 1909:

> This staff is too weak, numerically, to begin to cope with even the most pressing work in a country that is so extensive as Canada and that is so rapidly being opened up. To overcome this in some degree, outside assistance is engaged for geological, topographical, and ethnological field work . . . .
>
> But the official staff must be strengthened to meet the growing needs, especially in those divisions that are relatively weakest, such as the topographical and palaeontological. Since Dr. Whiteaves' death all the palaeontological work has devolved upon Mr. Lambe. The geological division, which may be taken as representing the effective 'fighting strength' of the Survey, must, of course, be added to; at present it is scarcely larger than that assigned by the United States Geological Survey to work in Alaska alone. Still it is relatively over-large for the topographical division, which should be strong enough to keep it supplied with base maps. Until the topographical division has been brought up to such a strength, thoroughly satisfactory work can not be done. To do this is not merely a matter of funds; in fact, the greatest difficulty is in securing the right type of qualified men. A number of young men are now in training for both the topographical and geological divisions. When the Survey is installed in the new Museum building, additional museum assistants will be necessary. . . .[9]

Brock proceeded with such great care that it seemed almost as if each new employee, down to the typists, had been hand-picked for his or her position by the director. About 85 newcomers joined the Survey between 1906 and 1914, comprising about 70 per cent of the enlarged staff of 120. Geologists promoted to the regular staff after apprenticeships as summer assistants included Robert Harvie, Leopold Reinecke, and S. J. Schofield in 1911, and Harold C. Cooke, C. W. Drysdale, A. O. Hayes, J. J. O'Neill, Bruce Rose, and M. Y. Williams in 1913. Other geologists, still in the 'pipeline,' who were leading their own parties by 1914 and were to join the permanent staff soon after, were F. J. Alcock, W. A. Bell, J. D. MacKenzie, S. E. Slipper, J. S. Stewart, and T. L. Tanton. Topographers recruited in similar fashion after a period of field training included K. G. Chipman and W. E. Lawson in 1910; A. C. T. Sheppard, 1911; and J. R. Cox, F. S. Falconer, E. E. Freeland, D. A. Nichols, A. G. Haultain, B. R. MacKay, and S. C. MacLean in 1913. The quality of the drafting staff was strengthened by bringing a number of men over from Scotland as well as by hiring Canadians who had first demonstrated their proficiency in selection competitions.

The Palaeontological Division was strengthened by hiring specialists in the major geological periods and in particular sorts of fossil materials, plus technicians to collect, prepare, and display specimens; the scientific staff expanded from one man (Lambe) in 1910 to nine by 1914. Eugene Poitevin, a graduate of the École Polytechnique, Montreal, joined the staff in 1913 as assistant to the mineralogist R. A. A. Johnston. The needs of the Museum were met by hiring other specialists in the natural history and in the human history field, where a truly remarkable group of men was recruited, whose work reflected great credit on the Survey and the National Museum. Some of the new technical staff also deserve mention as indicating the growing diversification and specialization of the Survey organization, in particular the geological compiler, Wyatt Malcolm; the custodian of instruments, G. D. Barrowman; a number of photographers; and the librarians, Miss M. Calhoun who succeeded Mrs. Alexander as acting librarian in 1912, and Mrs. F. B. Forsey, a future librarian, who joined the staff in 1913.

Yet, despite energetic recruiting, Brock still did not have adequate staff to carry on the immense tasks of the Survey. Consequently, outside scientists were invited to work on a part-time basis for one or more seasons. A sizable proportion of the field work, especially after 1910, was conducted by such part-time personnel—professors at Canadian and American universities, members of state surveys and of the United States Geological Survey, and other specialists in geology, anthropology, and natural history.

That the assistance of such persons was so readily obtained demonstrates the close ties Brock had formed with the most progressive elements in the academic world. American geologists, who were foremost authorities on ore deposits, continental uplift, glaciation or paleontology, had often turned to Canada for places in which to investigate their subjects. As for the biological and anthropological fields, Canada was almost a virgin field for studying and collecting. Such experts could be set to work on important subjects at small cost and with excellent chances of their making significant contributions. Just as important, the staff with whom these outside specialists were associated could learn from them the latest concepts and methods. The *Canadian Mining Journal*, reflecting the needs of that time as well as the reciprocity agitation of 1911, paid graceful tribute to this interchange in an editorial, observing that "It is gratifying to note that the most prominent geologists on the continent are glad to be associated with our Geological Survey. This fact is in itself a constructive compliment to this branch of the Federal Department of Mines .... The establishment of reciprocity in brains between Canada and the United States is quite as important as tariff-tinkering."[10]

* * *

BEHIND THE HIRING of such highly-qualified personnel was Brock's desire to change the style of operations from that followed by earlier directors, especially Bell. Now the Survey was concentrating on areas that were accessible for economic development and, for the most part, were past the preliminary investigation stage. Further work, therefore, meant progressing to detailed mapping and special studies of the principal features and problems of the area, kinds of work that could be carried out efficiently only by the trained specialists Brock was now recruiting. Besides, the natural and social sciences had advanced to the point where the many-faceted reconnaissance survey was no longer adequate. Investigators were expected to interpret as well as describe the phenomena they observed, and to carry their studies to levels and standards acceptable to specialists all over the world. Each aspect, therefore, needed to be studied by specialists who could produce significant reports in their various fields. The age of the general observer was past; the day of the giants of geological exploration—men like Dawson, Bell, Tyrrell, McConnell, and Low—was substantially ended.

However, highly-trained specialists, to be effective, had to be permitted to concentrate on their particular fields of study, and not have their time taken up by work for which they had no special aptitude or skill. Whereas Bell had sent out as many parties as he could, mostly to undertake multi-purpose regional studies chief among which was topographical mapping, Brock's parties were sent out to do a variety of specialized jobs. In 1913, for example, of 64 parties in the field, 11 were assigned to topographical work, 15 to natural history and anthropology, 4 to economic geology projects, and 34 to other geological work, including paleontology, petrology, and other specialties. Whereas the parties under Bell had mapped the geology as only a part of their field work, Brock's parties assigned to geological work went to their fields of study equipped with the latest tools and with expert knowledge best suited to understand the geological or other features and problems they would most likely encounter.

A marked difference, therefore, was the very noticeable withdrawal of the Survey from the broad, reconnaissance-style operations that were so much a feature of the Selwyn and Bell eras. Brock was disinclined to favour multi-disciplinary, unspecialized work, or work in districts that were remote from development. When such an enterprise was undertaken, for instance, the Canadian Arctic Expedition of 1913, Brock employed a team of specialists and insisted that the scientific work be of specialist quality and in a restricted, specifically-defined area. For the rest, any reconnaissance-type surveys that persisted into the Brock regime were mainly holdovers from the Bell years—most notably, the difficult and remarkable exploration of 1907-08 by Joseph Keele across the Cordillera at the 62nd parallel, from the Yukon to the Mackenzie by way of Pelly, Root, and Gravel (now Keele) Rivers.

Another such operation was the veteran William McInnes' exploration of the basin of Churchill River in the little-known area between Lac la Ronge and Southern Indian Lake between 1906 and 1910. This was his last field work; thereafter he was assigned to editorial and administrative work in connection with the Congress. As a result, his final report, *The Basins of Nelson and Churchill Rivers* (Memoir 30) did not appear until 1914. It was curiously old-fashioned because of the large amount of space devoted to such non-geological subjects as geography, soils, forests, rivers and water-powers, fish and game; and because the map resembled the familiar track survey, the geology being limited to the watercourses. McInnes, however, displayed his sound geological sense in the way he correlated the main Precambrian formations with those of Ontario and recommended prospectors to turn their attentions to areas marked Keewatin or Huronian, because of the parallels with proven mining districts. He showed the Keewatin as extending north from beneath the Paleozoics in several tongue-like strips that embrace Wapawekka, Amisk, Athapapuskow, Cranberry, Reed, Wekusko, Cross, Oxford, and God's Lakes, and upper Severn River—most of which have since become mining centres.

XIV-4 VIEW OF GRAVEL (NOW KEELE) RIVER, N.W.T.
Photographed by J. Keele on his difficult exploratory traverse from Yukon to Mackenzie Rivers.

In directing the geological field work Brock never abandoned the long-standing objective of making the Survey a leading scientific force in the attack on the geological problems of the day. Like Logan, he saw no contradiction between the twin goals of professional excellence and helping industry in material ways. He considered it axiomatic that "intelligent and efficient mining must have as a foundation a sound knowledge of geological conditions."[11] Like Dawson, he expected to reconcile the Survey's scientific purposes with the necessity for doing work of immediate or prospective economic value: "It does not necessarily follow that the scientific side would be neglected; for the economic problems furnish as fruitful scientific results as purely academic studies, and some problems of apparently only academic interest have a most important economic value."[12]

Low and Brock were both genuinely interested in the economic aspects of geology, and their own field work had been strongly inclined towards studies of mineral occurrences and of mining development. They spent as much time as possible visiting mining camps and attending mining conventions. They made friends with the mining men, became better informed about their points of view and problems, and they also reminded them of the services the Survey was performing on their behalf.

Low was a hard-headed realist, even something of a pessimist; Brock, in contrast, was an inveterate optimist completely in tune with a time when government and people accepted Canada's dazzling future as an article of faith and did all they could to realize that destiny. He tended to stress the glowing prospects more than the problems to be guarded against. His optimistic, exuberant outlook is well portrayed in this review of the highlights of a government-sponsored tour of the Canadian mining districts by a group of European mining engineers:

> The main impressions made upon the visitors were the extent of our territory; the scale and magnitude of the operations in the districts at present accessible; the rapidity with which a district is developed as evidenced by Cobalt, and the industries along the Crowsnest Pass railway, which were practically non-existent a decade ago; the modern plants and methods and the calibre of the men at the head of the successful properties.
>
> The magnitude of the operations of the Dominion Coal Company, the richness of Cobalt, the extent of the nickel-copper industry of Sudbury, the immense coal resources of the Crowsnest pass, the Trail smelter with its ingenious lead refinery, the first of its kind in the world, and the labour-saving appliances of the Granby mine and smelter were among the most striking and interesting features noted by the visitors.[13]

Even when he had to report failure he discerned a silver lining:

> While it is disappointing that no available mineral resources should have been discovered during this work on Prince Edward Island, it should be borne in mind that rarely in a limited area are rich agricultural and mineral resources associated, and of the two the island possesses in its agricultural land a more lasting and valuable asset.[14]

Both Low and Brock stressed again and again that the effort was concentrated on districts where mining activity was under way, and that the Survey was meeting all requests from Boards of Trade and other interested groups it could. Brock's reports abound with such phrases as: "work undertaken by the Survey is essentially practical and economic in character,"[15] "in general only such work should be undertaken as gives promise of affording direct economic results"[16] "This year, as last, almost all of the work undertaken by the Survey has been along strictly economic lines,"[17] or "The guiding principle in the selection was to choose those districts in which the work would be likely to prove of most immediate or of greatest value."[18]

The economic implications of field work also were emphasized in the capsule lists of the work of each party, especially such immediately realizable benefits as coal basins discovered and outlined in Alberta and the Yukon, promising silver-lead deposits in British Columbia, or reports of a tungsten occurrence in Nova Scotia. Even investigations of clay and shale deposits were valuable, Brock reminded his readers, because "the clay industry of a well developed country forms one of its principal industries, both as regards the value of the output and the number of men employed."[19] Short, enthusiastic comments were made about promising developments like oil prospects of Alberta, natural gas near Moncton, or the new mining camp of Porcupine, as well as exhortations to prospectors and developers to persevere. Even failure, on occasion, was adduced as an argument for continuing:

> For the encouragement of prospectors [in the Yukon] it may be noted that, up to a certain point, the greater the number of veins that prove barren or almost so, the greater are the chances that some occur that are rich, for the reason that the fewer sources there are for the gold, the richer these sources must be.[20]

Brock was careful, however, to emphasize that the Survey did not propose to do more for industry than to help it with expert and disinterested advice.

XIV-5 STURGEON-WEIR RIVER, SASKATCHEWAN
Photographed by W. McInnes in 1910 in the course of his prolonged multi-resource survey of northern parts of Manitoba and Saskatchewan.

The bulk of the economically-oriented areal mapping activity was carried out in western Canada, the region most actively developed in the decade before the Great War. Mining was a major concern for the Cordilleran region, and the mining community and provincial government were glad for the Survey's help. A series of able young geologists, assisted by separate topographical parties, was assigned to map the geology of the southern part of the province and advise local mining interests—Charles Camsell, in the country between the Fraser and Lake Okanagan; Leopold Reinecke, a South African currently completing a doctorate at Yale University, in the Beaverdell district near Midway; O. E. LeRoy, in the Slocan and Nelson districts until administrative duties took him off field work; C. W. Drysdale, on the details around Rossland, Franklin and Ymir mining camps; and S. J. Schofield, in the Cranbrook district.

Camsell, after completing his report on the Hedley district, proceeded to the Tulameen district, once a famous source for platinum. To this he added a further distinction that injected a momentary touch of excitement—a report of diamonds in a sample of chromite, "the first recorded discovery of diamonds in Canada either in the solid rock or in placer."[21] Hopes fell quickly when it was learned that the largest was no bigger than the head of a pin and difficult to extract from the enclosing rock:

> Charles Camsell as all of you know,
> Out West to find diamonds did go;
> All he got were so small,
> They were no use at all,
> So he's waiting now till they grow.[22]

So ran the ditty about Camsell composed for the Survey's annual dinner of 1912. Indeed, we know now that it was all a mineralogist's mistake, and there were never any diamonds at all!

Another centre of activity, mostly in support of actual or prospective mining operations, was the island-strewn west coast of British Columbia, a region that was easy to develop because it was accessible by ship. The American C. H. Clapp, at that time an instructor at Massachusetts Institute of Technology, made a competent, thorough study of the bedrock geology of southern Vancouver Island, of the coalfields at Nanaimo and Comox, and of the surficial geology. J. A. Bancroft and R. P. D. Graham completed the reconnaissance of the coast to Prince Rupert. McConnell, in the seasons from 1908 to 1913, investigated Texada Island, the vicinities of Portland Canal and Observatory Inlet, a gold mine on Princess Royal Island, and finally, Rainy Hollow, in the northwest triangle of British Columbia west of Lynn Canal. The coal resources of Graham Island, one of the Queen Charlotte Islands, were examined briefly by Clapp, then in 1913 and 1914 received a thorough study by J. D. MacKenzie. In view of the construction of the Grand Trunk Pacific across the middle of the province from Yellowhead Pass to Prince Rupert, W. W. Leach, G. S. Malloch, and McConnell also examined the country adjacent to the railway, as well as other prospects farther afield, for example, the remote Groundhog Mountain coal deposits, 100 miles north of Hazelton. North again, D. D. Cairnes worked in several locations in the Yukon Territory for the decade after 1907 when he replaced McConnell as the Survey's specialist in that region.

On the Prairies, where D. B. Dowling after 1904 made himself the leading authority on the regional geology and mineral occurrences, the work was to map formations that contained workable deposits of coal, petroleum, natural gas, gypsum, and salt, and especially, to assist the expansion of the coal mining and petroleum industries. Greater local markets and prospective export markets for coal inspired an active search for good quality coal measures. Dowling defined three separate coal formations of Cretaceous age in Alberta. The lowest, in the Kootenay formation, brought to the surface by the uplift of the Rockies, was characterized by thick, readily-workable seams of high-grade bituminous coal approaching anthracite in some areas. Next were the Middle Cretaceous, Belly River formation coals, mainly in the foothills, not subjected to so great pressure as the Kootenay but still of good bituminous quality. Finally, came the younger Edmonton-Laramie formation coals that underlay most of southern Alberta and Saskatchewan—low-grade bituminous and lignite coals, easily and cheaply worked, sometimes by strip mining, and acceptable for local domestic uses, though often too friable to stand transporting any great distances.

Dowling, assisted by Malloch, worked his way northwestward along the eastern margin of the Rockies from the Bow River basin to beyond the Athabasca headwaters, tracing and measuring sections of the Kootenay coal measures, the objectives of several railways currently under construction. By 1911 he traced two coal basins in the approaches to Yellowhead Pass, right on the routes of the two new transcontinental railways. The Grand Trunk Pacific by then was also building a branch to coal beds in the headwaters of Embarras, Pembina, and McLeod Rivers that Dowling had outlined in 1909, while the Canadian Northern in 1913 began building a branch to another coalfield Dowling had discovered on Shunda Creek. Similarly, Dowling, Leach, MacKenzie, and J. S. Stewart traced the coal formations from the Crowsnest Pass southeast in the Porcupine Hills of Alberta, and due south to the Flathead valley in the extreme southeastern corner of British Columbia. The Edmonton-Laramie coals received less attention, though Dowling examined favourably-situated deposits at Pembina River crossing (Evansburg) west of Edmonton, and around Edmonton, while B. Rose studied those of the Willow Bunch and Wood Mountain districts of southern Saskatchewan. In 1909 Dowling issued a comprehensive study, *The Coal Fields of Manitoba, Saskatchewan, Alberta and Eastern British Columbia,*[23] which included the first estimates of coal reserves and grades. Coal resources were featured by the International Geological Congress in

XIV-6 LEOPOLD REINECKE

XIV-7 D. B. DOWLING

1913; Dowling prepared the estimates for Canada, and demonstrated his authoritative knowledge of the geology of the Plains region from Winnipeg to the crest of the Rockies by his contribution to the guide books prepared for the Congress.

Not long after the Congress a new era began for western Canada with the discovery of oil at Sheep Creek, the future Turner Valley oil field, southwest of Calgary. There had been previous natural gas discoveries, and Brock, in his report for 1909, had referred to the probability of both oil and gas being found in quantity in the Cretaceous rocks, particularly in the Dakota sandstone, a formation just above the Kootenay. The 1913 find was made by a well drilled at the crest of an anticline which hit a small pocket of oil at a depth of 1,550 feet in the Belly River formation (Middle Cretaceous). A year later, in November 1914, the Dingman well tapped a good supply of oil and gas at 2,700 feet, in what Dowling believed was Dakota sandstone. Naturally, these discoveries intensified Survey activities. During 1914, S. E. Slipper and J. S. Stewart worked out the stratigraphy of the foothills south and west of Calgary; B. Rose carried on a reconnaissance in the district farther north that Dowling had worked over not so long before; and A. McLean studied the succession at Pembina Mountain, where Cretaceous formations were also present. Dowling directed and interpreted the work in Alberta, attempting to indicate the most promising places to drill on the basis of his studies of the regional structure. The inauguration of the oil industry demonstrated the need to secure as many well drilling records as possible, and to gain a more accurate knowledge of the characteristic fossils as an aid in identifying the various formations encountered in the course of drilling. Both of these activities, as a result, received a strong boost.

The Maritimes was another region where the Survey's efforts were bent, in the main, towards helping build up new mining industries that might alleviate some of the severe regional economic difficulties. This is readily apparent from a program undertaken in 1908-09 to determine whether coal measures, or other worthwhile resources, occurred at depth on Prince Edward Island. The local government had appealed for help, "pointing out that, although under the terms of confederation geological investigations should be carried on in Prince Edward Island, little had been or could be done, and suggested that, in lieu of surface studies, geological sections by means of bore-holes be obtained."[24] Five holes were started, but because of seepages and cave-ins only two got through the Permo-Carboniferous boundary to depths of 1,910 and 1,670 feet. Neither had come anywhere near the presumed coal-bearing formation, which would not have been mineable economically at such depth. Nor was there any sign of other mineable minerals; all the Survey received for its considerable outlay, besides silencing the criticism that it was failing in its duty to the island, were the drill logs and the samples taken every ten feet.

Other prospective developments were carefully watched and inspected. When natural gas was discovered in the Moncton district, G. A. Young and W. J. Wright made a report and a topographical survey was ordered. R. W. Ells continued to study the oil shales and visited Europe to investigate processes that could be introduced in New Brunswick. Young and Wright also examined other discoveries: iron ore near Bathurst and placer gold at Tobique, N.B.; a prospective gold mining district in the Cape Breton Highlands; and, arising from Faribault's continuing work of mapping the gold reefs of Nova Scotia, his reported finds of cassiterite (a tin metal) in granite at New Ross, and of scheelite, a tungsten mineral grown newly important in the manufacture of electric light filaments.

The Survey's economic work was less in evidence in the populous and prosperous St. Lawrence region, although the current exploration for oil and natural gas in southern Ontario was aided by its paleontological studies. The Pleistocene investigations also improved the knowledge of conditions of deposition of materials such as sand, gravel, clay, peat, and friable soils. Cities, anxious to locate satisfactory water supplies and to obtain advice on which subsoils were dangerous to build upon, sought out the Survey officers. So, too, did the Ontario Highway Commission, when it wanted road-building material capable of standing up to the growing automobile traffic. Reinecke, assigned to this study in 1913, concluded it might be economical to import crushed trap rock by boat from Bruce Mines to Lake Ontario. The higher transportation cost would be more than offset by the superior wearing qualities of the stone as compared with the gravels located along the western and northern shores of Lake Ontario. The Eastern Townships district, too, received further study of the origins and locations of copper, gold, and asbestos at the hands of J. A. Dresser and his successors, and by A. Mailhiot, of granites that could be used for a Canadian polished stone industry.

This district, as well as other regions of Canada, became the locale for a wide-ranging investigation of clay and shale deposits by the great specialist, Professor Heinrich Ries of Cornell University, who was assisted by J. Keele. They examined clay and shale beds across the country that were being exploited at the time or that held prospects of being developed, took samples back to Ottawa and recommended the best uses for the materials examined, or advised operators on the kinds of material they should seek. Ries and Keele worked for a season in the Maritimes in 1909, and on the Prairies in 1910 and 1911. Then Ries continued in the Prairies and British Columbia while Keele applied his knowledge in Quebec and southern Ontario. Since their work consisted chiefly of laboratory tests and analyses, it is not surprising that it was eventually turned over to the Mines Branch, along

XIV-8 HEINRICH RIES

XIV-9 JOSEPH KEELE

with Keele, who joined that branch in 1916 as chief of the Ceramics Division.

By the end of 1909, as a result of the efforts of Low and Brock, the Survey had been so completely rehabilitated in the esteem of the mining industry that the *Canadian Mining Journal* came round full-circle to side with the Survey against a threat of Parliamentary interference:

> At present the staff of the Survey is composed of men who rank with the best in any corresponding body anywhere. To a remarkable degree their reports are economic. The whole point of view of the Survey is coloured by the needs of the mineral industries. The officers of the Survey are members of the Canadian Mining Institute. They are constantly in touch with mining problems.
>
> It is also unqualifiedly true that the Survey enjoys the implicit confidence of the mining fraternity. This is a precious asset, an asset that must be guarded with utmost care. An irresponsible utterance in the House of Commons, baseless, irreparable criticism of a disappointed grafter, can do damage. . . .
>
> The Survey is doing its duty. Parliament has not done, is not doing, its duty in respect of the Survey.[25]

* * *

NOW THAT THE CONFIDENCE of the mining industry had been regained, Brock could turn more openly and completely to the difficult task of restoring the Survey's reputation in the international scientific community, as his report for 1910 hinted:

> The geological and topographical work undertaken by the Geological Survey during the past season, has, as usual, been economic in its bearing, most of it directly so; but a little has been on the broader problems of Canadian geology whose solution is required for the interpretation of the facts gleaned in the detailed examination of the mining districts.[26]

Among the scientific problems to which the Survey addressed itself during these years was a better understanding of the evolution and present structure of the Cordillera and Precambrian. The Cordillera still had been little more than touched on, though Dawson had made a good beginning by outlining the basic geology. Discoveries were leading to constant revisions of the succession and new interpretations of recurring phenomena. Furthermore, the work of correlating the geology of the region with that of the continent was scarcely begun. However, the situation improved as the succession became better understood and fossil beds were discovered.

With the Precambrian the position was different, because of the efforts since the 1840's to work out a general scheme for the southern part of the Shield. But even though an international correlation committee had recently set up a standard succession and nomenclature, much remained unresolved, including even the nature and relative ages of the two basic subdivisions, Laurentian and Huronian.

Other urgent problems concerned the nomenclature and succession of the Paleozoic, which was so important an element of the geology of the Maritimes and the St. Lawrence Lowlands. Initially the New York subdivisions had been adopted in Canada, but the Americans, after further study since the 1830's, had modified original definitions and established some entirely new formations. Moreover, further work was revealing that the units in Canada were not exactly like their supposed equivalents in the United States. Thus W. A. Johnston found (1908) that two formations on either side of the Frontenac Axis formerly regarded as identical actually contained quite different fossils; and when he referred the question to workers in the United States, he learned that they had encountered the problem earlier and had largely revised their original succession. Information obtained by oil-well drilling was raising new problems while solving others, and more refined paleontological work was upsetting some of the ideas of the earlier stratigraphers. A pressing need existed, therefore, for a revision of the earlier work in the St. Lawrence Lowlands in the light of the newly-won understanding of some structural and paleontological problems.

Higher up the geological time-scale was another subject of growing interest and study, the evolution of life forms in the Mesozoic and Cenozoic, as evidenced by the vertebrate fossils being unearthed in the Red Deer valley of Alberta in particular. The Pleistocene also was demanding additional study. Notwithstanding the good pioneer work of Logan, Bell, and Chalmers, much new information that had been discovered in the United States had to be assimilated into Canadian work on such subjects as the effects of glaciation on the Great Lakes basin.

A whole new category of scientific work demanding a great deal of attention was related to geological processes rather than periods, and called for the study of various phenomena: the characteristics and results of igneous intrusions, how different kinds of rocks are influenced by the conditions under which they are formed or altered, what controls affect the deposition of metallic minerals, how a complex mineral like asbestos is formed, and how the residues of prehistoric seas are altered into useful mineral. All these, while mainly general scientific problems, had an economic aspect as well, since solving them would indicate where to prospect for specific minerals, and how to go about exploiting occurrences once they had been found.

Long-range scientific objectives had been intertwined with short-term economic ones during the previously-described field work as well. The latest knowledge and theoretical findings were applied to explain phenomena encountered in the field, and new theoretical concepts were tested against field evidence, sometimes leading to the revision of former findings. The geological work in British Columbia presents a good example, for it added to the slowly developing knowledge of Cordilleran geology. To R. A. Daly's earlier section (1902-05) across the region at the International Boundary were added systematic studies of many parts of the province south of the main line of the Canadian Pacific Railway, culminating in a section along the line from Banff to Vancouver in 1911-12 by J. A. Allan, R. A. Daly, B. Rose, C. W. Drysdale, C. Camsell, and N. L. Bowen. Camsell and A. M. Bateman made a further survey west from Ashcroft to the eastern margin of the Coast Range at Chilko Lake. Farther north, most of another section had also been made along the Grand Trunk Pacific Railway, and along the coast, the inlets and many islands had also been studied. One major revision was of the geology along the southern border. While Daly was revising Dawson's part of the Shuswap Sheet along the Canadian Pacific Railway in 1912, not always for the better, S. J. Schofield was discovering that Daly had made errors in mapping the Purcells and that Daly, moreover, was mistaken in assigning the Purcell Series to the Cambrian instead of the Precambrian. Such revisions of former work are inevitable, though they do not often come so speedily as in this instance. Daly, a great thinker and teacher, was impatient with humdrum tasks like mapping boundaries and sorting out local successions, so the details of his mapping were often at fault. Still, his contributions to geological science were great, and he always examined evidence that contradicted his own findings impartially and without embarrassment, and even welcomed it as proof of the progress of scientific truth.

In eastern Canada, J. A. Dresser's re-study of the Appalachians in Quebec produced an advanced interpretation of the evolution of that region. He saw the geological history as shaped by three great tectonic movements: an early (Ordovician) intrusion of a basic magma out of which were derived chromic iron and an ultrabasic fraction that altered to serpentine and then in part to asbestos; a Middle Silurian intrusion of diabase that carried copper and gold minerals; and, in Permo-Carboniferous time, the Appalachian deformation that twisted and metamorphosed the earlier formations.

These results were derived from regular geological mapping, but surficial work of the Survey was a subject of special study. The work, which also had strong economic connotations, was inspired by a desire to update theory and practice in the light of major advances, especially in the United States. Brock turned to that country in 1908 for two leading experts, F. B. Taylor, who had worked on the evolution of the Great Lakes since 1893, and Professor J. W. Goldthwait of Dartmouth College, besides which the noted F. Leverett later made a study of the raised beaches of St. Joseph's Island, near Sault Ste Marie. Taylor and Goldthwait, assisted by W. A. Johnston of the Survey, traced the old beaches of Lake Algonquin and the course of the great river that had drained the Great Lakes through Lake Simcoe and the Trent River system. Afterwards Taylor made an extensive study of the deposits left in southern Ontario by the retreating glaciers. Goldthwait turned his attentions after 1910 to the raised beaches along the Ottawa, St. Lawrence, and Richelieu valleys that record the incursion of the marine Champlain Sea, which extended past the city of Ottawa and far into upstate New York. Later he studied the surficial and glacial history of the Maritimes. Indeed, he became so enamoured of that region that he prepared a bulletin to open the eyes of Nova Scotians and visitors to the scenic wonders of their province.

The role of Canadians in surficial geology included the continuing work of A. F. Hunter in tracing the raised beaches in the Georgian Bay district; a study of the district around Montreal by J. Stansfield of the Survey; and the work of W. A. Johnston, who gradually emerged as the Survey's specialist in the Pleistocene. The decade Johnston spent in detailed mapping of the Simcoe-Peterborough district allowed him to meet many glacial geologists and acquainted him with interesting problems in the field. Afterwards he applied his knowledge across Canada, beginning in 1913 with a study of the lacustrine sediments of glacial Lake Agassiz in the Rainy Lake district, and the surrounding glacial drift.

The need for revision and updating was even more acute in the field of paleontology, as the great increase in deep-well boring and the bitter disputes over the successions in the Maritimes indicated. With the approach of the International Geological Congress, it was imperative that the confusion should be resolved once and for all, and the succession worked out with impartiality and accuracy.

Besides hiring permanent staff from the United States, Brock also recruited some American specialists for sea-

XIV-10 BROW OF THE NIAGARA CUESTA AT HAMILTON, ONTARIO

sonal work; among them C. R. Stauffer, who re-examined the Devonian in southwestern Ontario for three seasons. P. E. Raymond in 1911-12 subdivided the Ordovician of the Ottawa valley, by splitting the historic Chazy, Utica, and Trenton of Logan into eight new formations. In 1912 M. Y. Williams, one of Brock's bright young men who had just finished similar work in Nova Scotia, began a re-study of the successions along the Niagara Escarpment. He was joined for a time by Professors C. Schuchert, W. A. Parks, and T. L. Walker, and Dr. A. F. Foerste, and together they worked out the Ordovician and Silurian of the eastern part of Manitoulin Island. Williams and Kindle in 1913 began a three-season study of the Silurian and other formations to the east of the Escarpment, that remarkable cuesta formed by the resistance to erosion of the massive Devonian limestone bed that there reaches the surface. Elsewhere, the American H. W. Shimer in 1910 worked out a provisional succession at Banff of seven sedimentary formations from Cambrian to Permian, four of which were in the important Cretaceous. In 1910 and 1911, Dr. C. D. Walcott, then secretary of the Smithsonian Institution in Washington, investigated "one of the most remarkable deposits of [Cambrian] fossils ever discovered" at Field, B.C., in the Kicking Horse district.[27] In 1914 the young Survey paleontologist F. H. McLearn began work on the succession in the Blairmore map-area to establish the conditions under which the various Cretaceous formations were laid down. Kindle also worked one season in the interlake district of Manitoba on the Silurian and Devonian rocks, as did Raymond on the fossil beds at Levis.

In the Maritimes the problem was to test and revise the successions of many previously-mapped formations in the light of more recent findings. The trouble was of long duration. The earliest work, notably of Sir J. W. Dawson in the 1850's, had mainly relied on paleontological evidence in assigning the various formations to their respective positions on the geological time-scale. Then, after Confederation, Fletcher and others did a magnificent job mapping the complicated structures, but worked out the successions mainly from stratigraphic and structural evidence. The upshot was that formations formerly given to the Lower Carboniferous were re-assigned to the Devonian, to the accompaniment of noisy, angry complaints from paleontologists. When Ells in 1908 adopted Fletcher's determinations for his own work in southern New Brunswick, the area of the controversy spread.

In the meantime, much important work had been done on the flora and fauna of the Carboniferous and other systems in the coal districts of the eastern United States and elsewhere, resulting in new information that was relevant to the Maritime region. From the time Brock became director, a new determination to settle the question was in evidence. Ami was assigned in 1908 "to make an independent palaeontological investigation" of some of the rocks of the coal formations in Antigonish County;[28] in the same year, W. J. Wilson, the Survey's paleobotanist, began collecting and studying fossil plants from the Bay of Fundy region and southern New Brunswick "with the object of definitely fixing the geological horizons of the formations found there."[29] Soon he, too, reported that rocks Ells classed as Devonian were actually Lower and Middle Carboniferous, as Dawson had determined. Similarly, Lawrence Lambe also began a close study of fossil fishes from the Albert Shales and other formations to determine the geological age of the shales and "certain supposedly correlative beds in Nova Scotia."[30] A single fish bone, obtained at Horton Bluff and sent to New York for examination, conclusively identified the formation as Lower Carboniferous.

When M. Y. Williams began his work in the Arisaig district in 1909, he found what he termed "probably the best selection of Silurian rocks in eastern America;"[31] in 1910 he checked Fletcher's report and map of 1886 of the district and found that Fletcher had stood up quite well: "the work . . . as far as it concerns stratigraphy and the larger questions of geological structures, has served to confirm the views of Fletcher. "[32] However, he could not accept the correlation nor the formational names, and instead, adopted Ami's classification and nomenclature. Also joining in the work was the brilliant young W. A. Bell, who in the years 1911-14 re-examined from a paleobotanical point of view the famous Joggins succession and Carboniferous zone and the Windsor and Horton series, other centres of the controversy. Beginning in 1913, A. O. Hayes made a study of the extremely complicated geology of the Saint John district, where even eminent paleontologists such as G. F. Matthew and C. D. Walcott were in disagreement. After three seasons Hayes had made some progress identifying 15 different rock units, but was only at the stage of selecting key locations for further detailed study of the structural geology.

An unusual type of investigation for the Survey, by which it re-entered briefly Hunt's long-neglected domain of geochemistry, was the pioneer work by Professor R. C. Wallace in the interlake region of Manitoba. Wallace, who was to have a long, distinguished career in the academic and scientific life of Canada, made an important, effective contribution by his study of the origins and relations of the local deposits of anhydrite, gypsum, dolomite, and salt. This study became the basis for a paper at the International Geological Congress.

XIV-11 A. C. LAWSON

The main scientific concern of the period, however, was with two aspects of Precambrian geology—the order or succession of the main formations, and the character and states of the various rock types. The great work in the second area was that of F. D. Adams of McGill, and A. E. Barlow, both former Survey members, who continued to work seasonally on the complicated assemblage of granites and metamorphosed sediments of the Frontenac Axis. The Haliburton-Bancroft district was an ideal place to study Logan's Grenville, for the district was comparatively accessible, and displayed "progressive metamorphism" to a degree that made it almost a laboratory of Precambrian petrology. Their years of study were apparent in the classic *Geology of the Haliburton and Bancroft Areas* (Memoir 6), published in 1910. It described the many geological processes on the basis of extremely careful mapping of the structural folds, faults, and foliation of the rocks, and on analyses of the wide range of rocks and minerals, coupled with observations on their origins and derivations.

A. C. Lawson had made an important contribution in working out the Precambrian succession in the 1880's, and set in train much study and discussion by other scientists that had culminated in the reports of the international committees. Lawson disagreed strongly with these reports, however. The committee, although it had followed his views on the Laurentian and the Keewatin, had rejected his conclusion that the Coutchiching Series was a single unit below the Keewatin and had assigned part of it to the Lower Huronian, well above the Keewatin. Lawson therefore accepted Brock's offer to re-examine the Archean formations west of Lake Superior. In 1911, assisted by J. D. Trueman, R. C. Wallace, and H. C. Cooke, he carefully re-examined the relations of Coutchiching, Keewatin, and Laurentian (Lawson's order) in the Rainy Lake district, then made a reconnaissance east along Seine River to Steeprock Lake. There he traced a younger series above the Laurentian, the Steeprock Series, in which he discovered fossils, "so far as I am aware, the oldest well defined organic remains now known to science."[33] Above the Steeprock was another, younger sedimentary formation that he named Seine Series, and above that, still another series of Laurentian-like igneous rocks.

In his report, *The Archaean Geology of Rainy Lake Re-studied* (Memoir 40, 1913), Lawson reaffirmed his view that Coutchiching was the oldest sedimentary formation, and correlated the Steeprock and Seine Series with Huronian in other districts. He pointed out that the Laurentian as previously defined constituted two formations from widely separated times. He proposed retaining the name Laurentian for the earliest of these, the gneissic batholiths that broke through the Coutchiching and Keewatin, and renaming the younger series overlying the Seine Series, Algoman. He also proposed a major break in the Precambrian following the Algoman episode, since it was succeeded by a "protracted interval of profound erosion," the 'Eparchaean Interval,' which he regarded as the true break in early geological history."[34] Once again, Lawson had made a valuable contribution in revising the Precambrian succession, elements of which continue in the present-day interpretation.

The two strands of Precambrian studies were taken up in these years and continued for the next generation by two young Precambrian geologists, W. H. Collins and M. E. Wilson. Both men came from similar rural Ontario backgrounds and proceeded from small-town collegiates to the University of Toronto, where they were distinguished students. After they had worked as summer assistants, Bell appointed them to the regular staff, Collins in 1905 and Wilson in 1907. Both completed their Ph.D.'s in the United States with the support of Brock, Collins in 1911 and Wilson in 1913. Wilson's work showed the influence of Barlow (whose assistant he had been) in its careful mapping and close study of rock types and conditions of their formation. Two early examples are his explanation of the banding or layering in Laurentian gneisses (1912), and his explanation of the Abitibi (Keewatin) volcanics, based on the study of the pillow lava form in which they sometimes occur (1913). Almost all his field work before 1913 was in western Quebec, east and northeast of Lake Timiskaming, where he encountered a belt of older greenstones and sediments, which afterwards were proved to be highly mineralized and gave rise to the Noranda mining camp. He cautiously gave these a non-committal name of Abitibi Group as he did not feel a term like Keewatin should be extended so far from its original locale without much better proof of the equivalence. After 1913 Wilson worked mainly in the Buckingham district near Ottawa, where he could fully exercise his talent for careful, patient, detailed work mapping the complexities of the 430-square mile area, with its many mineral showings.

Collins began his life's work, an extended study of the Huronian in the district between Lake Timiskaming and Georgian Bay to trace the succession between the two formations Logan had identified as Huronian, in 1908. To 1913 the work was centred in the district between Cobalt and Sudbury, but in 1914, assisted by T. T. Quirke and W. E. Cockfield, he extended the work of correlation beyond Sudbury towards Sault Ste Marie. By 1912 he was beginning to reach conclusions similar to Lawson's. At Gowganda, he reported, a Keewatin basement was intruded by Laurentian batholiths, overlain by Huronian sediments, then finally cut by post-Huronian diabase sills. These conclusions agreed with those for the sequence at Cobalt of Dr. W. G. Miller, chief geologist of the Ontario Bureau of Mines, but differed from those of Professor A. P. Coleman for Sudbury, in which Keewatin was followed immediately by Huronian. Collins presented these conclusions in his reports, in his Ph.D. thesis which became Memoir 33, *Geology of Gowganda Mining Division* (1913), and in his paper to the International Geological Congress meeting in Toronto in 1913.

* * *

THE WORK OF LAWSON AND COLLINS, and that of other geologists across Canada, in the years after 1910 was inspired by a determination to set Canada's geological house in order before the world's leading authorities arrived to inspect the country as members of the International Geological Congress. The guide books give ample evidence of the great care that had gone into reviewing questions of succession and nomenclature all across the country in the light of the latest scientific evidence. The Congress was a major force impelling Brock, and giving him the excuse to raise and update the scientific standard of geological work in Canada.

As early as 1897 Dawson had been asked by Thomas Macfarlane and Monseigneur Laflamme to support their going to the seventh Congress in St. Petersburg to campaign for a future meeting in Canada. Laflamme was sent to Russia by the Canadian government though Dawson was less than enthusiastic at the suggestion, feeling such a meeting would disrupt field operations too much and that the delegates would take the first opportunity to cross over into the United States, as members of previous conferences brought to Canada at government expense had done.

Perhaps Dawson's pessimism was justified at the time, but after 1900 the economic climate changed rapidly and the attentions of the world became fixed upon Canada as a land with a glowing future. Robert Bell went to the Vienna meeting in 1903 with the government's blessing and a commitment of a grant towards the expenses of a meeting of the Congress in Canada, only to be outbid by Mexico, whose agent, Aguilera, made lavish promises amounting to as much as $300,000 and secured the tenth Congress for Mexico. Bell had to console himself by accompanying Laflamme in 1904 to Stuttgart to the 14th International Congress of Americanists, where together they won the right to host their next Congress at Quebec in September, 1906, under Bell's presidency. After Prime Minister Laurier had agreed to a $4,000 grant in aid of the meeting, a delicate matter of protocol almost upset the arrangements. Invitations normally were sent from government to government to gain official backing and support for some delegations, but the Canadian government still did not enjoy the diplomatic status to approach foreign governments on this basis. Bell was informed that "the Cabinet do not see their way to authorize any invitation being sent on behalf of the Government of Canada to other Governments to take part in this Congress. They feel that to do so would be to establish a precedent which might be very embarrassing in the future."[35] However, the meeting took place as arranged.

In fact, while Bell was presiding over the Americanists at Quebec, Low, the new director, proceeded to Mexico City to renew his country's bid for the International Geological Congress. En route he made a long inspection trip through British Columbia that gave him "some idea of the great extent and value of the mineral resources of the southern part of that province"[36] and viewed the aftermath of the San Francisco earthquake. He arrived late at the meetings "owing to wash-outs and accidents on the Mexican Central railway,"[37] but had a useful time meeting with other geologists, and doing public relations work. Since it was felt the Congress could hardly meet twice in a row in North America, Sweden was chosen for the locale of the 11th Congress in 1910, but Canada's claims for the 1913 meeting were well to the fore. At Stockholm, where Canada was represented by W. G. Miller, A. P. Coleman, and F. D. Adams, the Congress agreed to hold the next meeting in Canada.

There had been similar conferences held in Canada, notably meetings of the British Association for the Advancement of Science in Montreal in 1884, Toronto in 1897, and Winnipeg in 1909, but nothing quite so important for the geological community had ever happened before. Brock, acting for the Canadian government, convoked a meeting in Toronto of leading Canadian geologists on December 2, 1910, to plan the arrangements. Adams was elected president and Brock, secretary-treasurer (to be assisted by a full-time hired secretary or manager). An executive committee was named that also included LeRoy and McInnes for the Survey; W. G. Miller and T. C. Denis, representing the provinces of Ontario and Quebec; Professors Coleman and Parks of the University of Toronto where the sessions were to be held; and J. B. Tyrrell and G. G. S. Lindsey, representing the mining industry. A. E. Barlow was afterwards added to the committee, meaning that seven of the eleven members were current or former officers of the Survey. The meeting decided that the major subject should be the coal resources of the world rather than the topic "the fracture system of the earth's crust" that had been handed on from the previous conference.

At the second meeting, on February 18, 1911, the committee began planning the heaviest of all the Survey's burdens in connection with the Congress—the excursions. It was decided to arrange these in three groups: one before the sessions at Toronto, involving travels about the Maritimes and Quebec; a second one of short visits to places within a radius of 100 miles of Toronto, to be held around the time of the sessions; and finally, excursions to northern Ontario, western Canada, British Columbia and along the Pacific coast to the Klondike once the meetings had finished. Subcommittees were set up to arrange the details of the trips and to select local guides and authors for the accompanying guide books. Brock immediately began to direct the field work for 1911 towards the districts and places to be covered by the tours, and thereafter the program became increasingly devoted to providing satisfactory accounts of the geology of the parts of Canada that would be visited.

The committee also selected the subjects for discussion, approached experts to submit papers on those topics, and invited other contributions. All the scientific subjects, like the main official subject, the coal resources of the world, were of major importance to Canadian geology: the differentiation of igneous magmas, the influence of depth upon metalliferous deposits, the origin and extent of Precambrian sedimentaries, the subdivision and correlation of the Precambrian, the interglacial periods, and Paleozoic seas.

A problem peculiar to the time and place was the need to arrange the program in conformity with the provisions of the Lord's Day Act and in line with the strong convictions of the Lord's Day Alliance. This had to be taken seriously, for any person could bring proceedings for alleged violations of the provisions of the act. No meetings or tours could be scheduled for the Sundays, though "no objection can be made to walks on Sundays in the Country or elsewhere to points which may or may not be of interest geologically."[38] The committee was at pains to assure the secretary of the Alliance, Mr. Rochester, with whom Adams was in correspondence that "the Committee of the Congress would be sorry indeed to intrude its work upon the Lord's Day," that they would be careful to live up to the spirit of the act in every way and give their guests every opportunity "to attend Divine Service at various churches."[39]

Associated with organizing the excursions was the exacting task of preparing the guide books. As Brock indicated in 1912,

> The approaching meeting of the International Geological Congress in Canada has thrown a vast amount of extra work upon the Geological Survey, particularly upon the geological and draughting divisions. The excursions in connexion with the Congress extend from Cape Breton island to Vancouver and Dawson, and cover practically all the main routes of transportation, including the new transcontinental railways . . . . Some idea of the magnitude of this task may be gathered from the fact that this entails the preparation of about 140 special maps, in addition to those already available as ordinary Survey publications. Such convenient and well illustrated booklets on the geology and resources of the main railways and waterways have long been needed, and will prove of great value to all travellers in Canada who desire to secure an intimate knowledge of the country. The material for these guides is being prepared for each district by the geologists who have done the field work in the district under review. Mr. Charles Camsell is acting as general editor.[40]

Practically all the geological field assignments that year instructed the men to prepare some part or other of the guide books. Contributors were asked to set out the general geology in their introductions, then go on to fuller descriptions of the part of the district covered en route, with most attention being given to places and formations actually to be visited and studied. The work of men such as Taylor and Goldthwait on the Pleistocene shorelines and of Raymond on the Ordovician of Montreal and Ottawa was prominently featured in the completed books. The deployment of strings of field parties, such as the five-party group along the route of the Canadian Pacific main line through British Columbia, or the similar deployment of Leach, Schofield, LeRoy, and Camsell along the Crowsnest Pass-Kettle Valley route, were for the purposes of the Congress, as the guide books attest.

A great deal indeed was expected of the authors of the major regional reviews, which sometimes were of considerable length—F. B. Taylor's 70 pages on Niagara Falls and gorge, D. D. Cairnes' 70 pages on the Skagway-Dawson district and route, R. A. Daly's 55-page introduction to the geology of the Cordilleras, and similar extended studies by Dowling, Faribault, and Young. These were more than summaries; many contained much new material, for the authors took a fresh look at the country in the light of the latest knowledge and concepts. Adding to the writers' care was their awareness that their work would be scrutinized by the foremost experts in the world, who would be seeing for themselves the evidence on which their comments were based. The guide books were landmarks in the progress of Canadian geology in their own right and are still occasionally consulted.

The Survey was equally responsible for the chef-d'oeuvre of the visit of the Congress to Canada, the three-volume *Coal Resources of the World*, which was put under the general editorship of McInnes, assisted by Dowling and Leach, who also contributed the Canadian submission. The completed work ran to 1,200 printed pages and an atlas of 70 maps, but the presentation was very uneven. Japan, for instance, presented very full detailed reports on the home islands, Korea and Manchuria, whereas the United States supplied only figures and tables. Efforts had been made to impose uniform standards for classifying quality of coal and estimating quantities in the various deposits and beds, but these were not always successful.

From the Canadian standpoint, the study afforded comparisons between Canada's immense resources and those of other countries, and indicated where Canada might hope to find markets for her coals. The estimates advertised the then little known fact of the extent of Canada's coal resources, which were estimated as being of the order of 1,234,269 million tons (80% of it in Alberta), comprising 16.7 per cent of the world's total coal in "Actual, Probable and Possible Reserves." This was 60 per cent greater than the total given for the whole of Europe, about equal to the total for the continent of Asia, and was exceeded in the world only by the United States, whose totals were three times as great as Canada's. However, two thirds of the Canadian total was not in the "Actual," but only in the "Probable" and "Possible" categories. Moreover, while the total ascribed to Canada included more than 30 per cent of the world's sub-bituminous and lignite grades, it included only 7 per cent of the bituminous grade and less than one half of 1 per cent of the world's anthracite coals—something that did not augur too well for Canada's ability to make as great use of the gigantic quantities as appeared at first glance. Objections were raised later, also, that the estimates were highly unrealistic in what they listed as mineable coal. Coal "in seams containing not less than 1 foot of merchantable coal occurring not more than 4,000 feet below the surface," and coal "in seams containing not less than 2 feet of merchantable coal occurring . . . not more than 6,000 feet below the surface" was ranked as mineable and included in the totals.[41] Thirty years later a revised study by B. R. MacKay was required to reduce these estimates to a much more realistic, though still enormous, figure. The 1913 survey also made no allowance for economic factors and for technological changes in relation to competing fuels and industrial materials, even though these have mainly determined the actual course of coal utilization in Canada.

The Survey was also responsible for editing the transactions of the meeting. The *Compte-Rendu de la XIIe session, Canada, 1913,* a massive 1,000-page volume prepared under the editorship of W. H. Collins, contained the full record of the Congress: organization, planning, statements and business proceedings, and especially the reports of the scientific papers and discussions. Business included discussions and reports of subcommittees on perennial general scientific concerns: the reviews of problems in paleontology, petrology, glaciology, fossil man, a stratigraphic lexicon, and a standard geological map of the world. There were suggestions, too, of the need for a new institute to study volcanoes and for an international committee to co-ordinate Precambrian work around the world. There was also the usual awarding of prizes, discussions of constitutional amendments, and preliminary planning of the XIIIth Congress in Belgium, which was destined not to be held until 1922.

The *Compte-Rendu* gave the registered attendance as 467 members from 23 countries, principally the United States (155), Canada (144), Great Britain (38), Germany (33), France (19), and Russia (17). It also described the activities in Ottawa and Montreal that preceded the Toronto sessions. More than 200 delegates were addressed in Ottawa by Prime Minister Borden, then a memorial to Logan was unveiled in front of the Survey headquarters, the Victoria Memorial Museum building. It was a plaque affixed to a Laurentian boulder, with a design by Henri Hébert giving Logan's head in profile, together with the inscription:

Sir William E. Logan, Kt., LLD, FRS
1798-1875
The Father of Canadian Geology
Founder & First Director of the
Geological Survey of Canada.
1842-1869
Erected by the Twelfth International
Geological Congress
Canada
MCMXIII

This boulder and plaque now stand on the court in front of the present Survey building in Ottawa. A few days earlier, during the excursion to the Maritime provinces, a similar plaque had been affixed to the cliff face at Percé, Gaspé Peninsula. In Montreal, the delegates attended a special convocation at McGill, where five distinguished geologists from as many countries were awarded honorary degrees. They also visited Caughnawaga Indian Reserve for an entertainment of Indian sports and tribal customs, then shot the Lachine Rapids in a steamer. Sunday brought a walking tour of Mount Royal, followed by the trip to Toronto.

There were 27 excursions, totalling 25,000 miles in all, and ranging from less than a day (like those from Montreal to the Morin anorthosite, or from Toronto to Niagara Falls), to the great continent-spanning tour C-1 between Toronto and Victoria that lasted 33 days and covered 6,000 miles, going and returning by different routes on the Canadian Pacific and Canadian Northern railways. A similar excursion, C-2, that paid special attention to mining districts, also crossed the continent, travelling through the mountains by the Crowsnest route of the Canadian Pacific and returning by Grand Trunk Pacific from Yellowhead Pass to Winnipeg, with a side trip to Sudbury, Cobalt, and Timmins before returning to Toronto. Some of the participants on C-1 took a further excursion, C-8, from Vancouver to Dawson, with a side trip from Prince Rupert inland to Moricetown (C-9). On their return journey from Vancouver they took the Crowsnest Pass route and swept through the northern Ontario mining area. These excursions had a strongly international flavour; the one to the Yukon, for instance, included 52 passengers from 13 different countries.

XIV-12 THE LOGAN BOULDER

Along each route the guide book information was supplemented by a series of excursion leaders and local guides, that relied heavily upon Survey personnel. Some 51 of the 97 listed leaders and guides were full-time or part-time employees of the Survey who had been working in the districts. Many of them emerged from their current field work by prearrangement to conduct the excursion through their district. Former officers, like Barlow, Tyrrell, Gwillim, McEvoy, and even the aged L.W. Bailey gave similar assistance. Adams, besides presiding at the Congress and participating in the tours about Montreal and Haliburton, took charge of excursion C-1, to the immense satisfaction of the travellers, one of whom, Pierre Termier, director of the Geological Survey of France, recorded this appreciation:

> Le Professeur Fr.-D. Adams avait assumé la charge écrasante de diriger cette caravane de 120 géologues sur un parcours de plus de 8000 km. Grâce à la prudence, au dévouement, à la science de cet excellent chef, le programme s'est accompli à souhait. Dans le détail des courses, nous étions guidés par des savants du plus haut mérite . . . . Pour raconter tout ce que nous avons vu, il faudrait plusieurs volumes.[42]

The general verdict was that the excursions were "splendidly organized."[43] The transcontinental excursion, C-1, for instance, travelled in a special train of eleven new railway carriages, including five sleeping cars and two baggage coaches, "one fitted with shelves to hold the specimen boxes supplied each member"[44]—for all of which the passengers were charged only regular half-fare. During their stops the travellers were the guests of provinces, cities, local corporations, and wealthy businessmen. They were given every opportunity to view geological sites and mining operations, to ask questions of the managers and scientists, and to collect rock samples for themselves or their sponsoring agencies. Nor was the visit all 'business': there were civic receptions, mountain climbs, Indian displays, opportunities for botanizing, and simple sightseeing, as at Banff or Muskoka. There were the normal tourist adventures, like the lady who took a fancy to a bear cub at a northern Ontario stop, bought him from his Indian owner, and shipped him back to Toronto in a crate labelled "a bear called Congress."

The main work during the week in Toronto was the reading and discussion of scientific papers, mainly on the pre-arranged themes, except for the coal resources, which received little discussion, presumably because the monograph had exhausted the subject. There were enough other submissions to form the nucleus of three additional sessions: one on economic and chemical subjects, another on tectonics, and the third a miscellany without any unifying theme. The 69 papers were presented by contributors from 20 different countries, with the United States accounting for 21, followed by Canada and Britain with 7 presentations apiece. The most heated debate was initiated by Lawson's proposing his Rainy Lake-Steeprock succession as a universally applicable "standard scale."[45] The claimed universality was questioned by most of the other scholars present as not agreeing with what they knew about the areas they had studied. The North American specialists, particularly Coleman and the American professor J. A. Leith, questioned the details of Lawson's succession that went counter to the international correlation committee ruling, such as Lawson's reaffirmation of Coutchiching as the lowest Precambrian formation. Some foreign geologists, however, agreed that Lawson's great unconformity, the Eparchaean Interval, accorded with their own observations. One Irish geologist, agreeing with Lawson's discussion of the confusion over the Laurentian, suggested that Logan's Laurentian was an obstacle to the proper understanding of the Precambrian and should be scrapped. Collins played a relatively neutral role, sticking closely to describing his findings at Cobalt and Sudbury. An English geologist, B. Hobson, commented that:

> The impression made upon the writer by the discussion on American pre-Cambrian classification was that hardly two authorities agree, and that the grouping under the name of Huronian of three or more systems, separated by great unconformities, only leads to confusion.[46]

Clearly the discussion of the Precambrian by such a broad group of specialists was of immense benefit to all those present, and to the progress of the work around the world. The Finnish geologist J. J. Sederholm carried "Some Proposals concerning the Nomenclature of the Pre-Cambrian, etc." to the business meeting of the Congress in the form of a resolution which was adopted with only minor amendment:

> The International Geological Congress expresses the hope that the governments of those countries which possess contiguous areas of Pre-Cambrian rocks will promote the comparative study of such areas by forming international committees that will include representatives of the Geological Surveys of all the countries concerned, for the purpose of correlating the Pre-Cambrian formations in the different countries.[47]

Here again, in an important area of science, Canada and the United States were already pointing the way to the world.

With the return to Toronto on October 5, 1913, of the last excursion the International Geological Congress meeting in Canada came to a close. It had been a pleasant, profitable, but most onerous duty, for on the Survey had fallen the major work of making the Congress the resounding success it was. Brock's *Summary Report* handed out the credits for a job very well done:

XIV-13 AN IGC PARTY ON TOUR
A visit to the International Coal mine at Coleman, Alberta.

From the Director, who was General Secretary of the Congress, down, almost every member of the staff performed important duties writing guide books, editing publications, planning and supervising the details of excursions and acting as guides and leaders, attending to the manifold duties connected with the meetings, etc. Credit cannot be given to each individually for his share in the success of the undertaking, but special attention might perhaps be made of Mr. O. E. LeRoy, in connexion with the excursions; of Mr. G. A. Young, in connexion with the Maritime Provinces guide books and excursions and the meeting in Toronto; of Mr. C. Camsell, in editing the guide books; of Mr. A. Dickison, in preparing the special maps; of Mr. W. McInnes and his associates, Mr. D. B. Dowling and Mr. W. W. Leach, in editing the Coal Resources of the World, and of Mr. W. H. Collins, in editing the papers for the Transactions. The Coal Resources of the World consists of three quarto volumes and a large atlas; the Transactions will occupy about 1,200 pages. The guide books consist of thirteen well-illustrated volumes, of which two were issued by the Ontario Department of Mines, and eleven by the Geological Survey. For these the Draughting Division of the Survey, under Mr. C.O. Senecal and Mr. Dickison, prepared about one hundred and fifty special maps. The Department of Public Printing, Ottawa, deserve the highest credit not only for the excellent printing but also for filling, in the short time at their disposal, a rush order of this magnitude.[48]

The meetings did much to bring Canadian geology and the Survey into the mainstream of world science. Eminent scientists from other lands were given the opportunity to see Canada for themselves, to meet the Survey's officers and take the measure of their interests and abilities, and to form many strong, lasting friendships. Foreign guests took back samples of Canada's rocks and minerals to study and to add to their countries' museum collections, and sometimes (like the Japanese delegates) arranged for the Canadian museum to receive comparable samples of their countries' rocks and minerals. The visitors helped Canadians to understand their own geology better by describing similar phenomena from their own lands and by giving them the benefit of their experience. In return, they carried away ideas that they could apply at home, so everyone was the winner.

In spite of these many benefits, the Survey's greatest reward lay in its own exertions. The Congress was both an occasion and a justification to concentrate more of the effort on the theoretical and scientific side of the work, and not to become completely engrossed in supplying the needs of the mining industry. In preparing for the meetings, the Survey had been forced to catch up on such neglected basic work as making a section through the Cordillera, revising the geology of the Paleozoic and Pleistocene, and commissioning Lawson's special study of the Precambrian. Its performance in 1913 demonstrated that the Geological Survey of Canada was a well-integrated, progressive institution, led and manned by capable scientists and administrators. The Survey had proved convincingly that it could hold its own with the best in the world.

* * *

IN 1911 THE PATRONS OF THE NEW SYSTEM and of Brock—the Liberal government of Sir Wilfrid Laurier—were defeated at the polls and replaced by the Conservatives under Sir Robert Borden, the one-time defenders of Bell and the critics of Sifton's scheme for 'reforming' the Survey. But the Conservatives had also criticized the niggardly treatment of the Survey in those years, and had alleged interference with its efficient workings from political motives, so they could hardly reverse their stand of supporting the work when properly conducted. And by 1911 there could be no question that the Survey was beyond reproach in this respect. It had found a new role within the Department of Mines that functioned smoothly, and it was managed by an efficient, young director who believed in the merit principle.

The change of government from Laurier to Borden, and from Templeman to a succession of Conservative Ministers of Mines—W.B. Nantel (1911-12), R. Rogers (1912), W. J. Roche (1912-13), and L. Coderre (1913-15)—disrupted the close rapport that had grown up between Brock and Templeman. But to outward appearance it did not seem to have affected Brock's management of the Survey or the department. The annual votes of money continued to increase beyond his ability to spend them, and he seems to have been left to conduct departmental affairs unhampered.

Meanwhile, the burdens of the dual offices of director and acting deputy minister were beginning to tell on Brock. He was kept busy, year after year, hiring, improving, setting up logical units, perfecting the organization, attending to dealings with government and public, and making decisions that would improve the functioning of the department. The duties of acting deputy minister must have been a heavy burden for Brock, even though the two branches of the department, Geological Survey and Mines Branch, operated as separate entities to the point that there was no annual report of the Department of Mines, and the Mines Branch was almost an independent command under the aged E. Haanel. The management of the Survey, too, was more onerous for lack of a deputy minister with whom the director might share the difficult problems that filled the years after 1908. Illnesses occasionally impaired Brock's efficiency; for example, an attack of diphtheria while he was visiting Prince Edward Island in June 1909, and typhoid fever in the autumn of 1911 that cut short a trip along the Pacific coast north from Prince Rupert.

The summary reports mirror Brock's growing preoccupations. From 1908 to 1911 he published lengthy accounts of his own travels and his interpretations and assessments of geological and mining developments across Canada. But after 1911 there was no director's report per se beyond a bare list of his movements, and the expansive policy statements figured less prominently. Occasionally, too, an unusual querulous note appeared in his reports; in his complaints over the lack of help with the engraving of the maps, the failure to fit out the photography laboratories as quickly as he would have liked, the shortages of cases and other essential museum supplies, or the delays in the publication of reports and maps.

Above all, Brock felt a strong personal grievance over his own position. He had accepted the position of acting deputy minister, he wrote Borden, "on the understanding that when it was certain that Dr. Low would not be able to perform the duties of office, I was to be appointed Deputy Minister."[49] But although Low clearly was unable to assume his responsibilities, he was retained in his position for years; and for years Brock continued performing the duties without salary. Finally, in December 1913, after the International Geological Congress was out of the way, Brock submitted a formal request for the appointment to Borden and Coderre. The eight-page memorandum reviewing his achievements as director and his claims to the deputy ministership included the threat that unless he received the appointment mass resignations from the Survey would follow:

> I am obliged to bring to your attention the condition of the Department of Mines. It has now drifted so long without an active Deputy Head that the situation has become intolerable. We are about to lose some of our best men unless action is taken and a man is appointed whom it is known will maintain the standards and ideals by which the Survey has been raised, in the last few years, to its present state of efficiency.
>
> The break-up would have taken place before this had the Survey not been bound up in the International Geological Congress. Now that this is over, we cannot, under present conditions, hold these men. A few will leave almost immediately, others as outside opportunities occur . . . .
>
> Personally I cannot afford to continue in office while the Institution disintegrates, and consequently unless the above action is speedily taken, shall be reluctantly forced to place in your hands my resignation, as well as that of Mr. LeRoy.[50]

That Brock felt it necessary to resort to such a threat, when he had so strong a case in his favour, clearly indicates how far his alienation had proceeded. In response either to the threat or to the justice of his cause, the government, by two orders in council of December 31, 1913, and January 6, 1914, retired Low on pension for 31 years' service and appointed Brock to the position he had so long exercised in fact, deputy minister of the Department of Mines.[51] The *Bulletin* of the Canadian Mining Institute, after a regretful glance at the necessity for Low's retirement, congratulated Brock on his promotion and called for a reorganization of the department in the interests of the mining industry, in particular to bring "closer co-ordination of work between the two branches of geology and mines."[52] A. E. Barlow, in his presidential address of March 1914 to the Institute referred to Brock's promotion in these words:

> It is, however, satisfactory to record that at last a Deputy Minister of Mines for the Dominion has been appointed to succeed Dr. A. P. Low, whose health has not permitted him, for some time past, to give due attention to the exacting duties of that office. Dr. Low's successor is Mr. R. W. Brock, for over six years Director of the Geological Survey. Mr. Brock is a member of the Council of the Institute, and has always taken a keen and active interest in the work of the society. He has, moreover, always kept in close touch with the conditions and requirements of the mining industry of Canada. In accepting the office of deputy minister of mines he has assumed heavy responsibilities. The successful administration of the Mines Department means much to the mining industry. Mr. Brock, I need scarcely say, enters on his new duties with the good wishes of the members of this Institute.[53]

Brock's promotion, however, was not to be of long duration. Early in August 1914, completely unexpectedly, he resigned from both the deputy ministership and the directorship of the Survey to become the first dean of the Faculty of Applied Science at the newly-organized University of British Columbia. Brock was quite worn out by the heavy work of the two positions in Ottawa and his restless spirit was looking for a new outlet for his creative talents. He wanted a change from administrative work to his other interests—academic life and organizing scientific work. On August 4, 1914, Canada had been drawn into the fast-spreading Great War, with results that were apparent to a man with as many international contacts as Brock possessed. In the words of M. Y. Williams, one of Brock's colleagues at the university, in an obituary following Brock's untimely death in an airplane crash in 1935:

> Dawson's memory, his [Brock's] love for, and belief in, Canada's western province, and his interest in education were all luring him away from . . . a path . . . often made weary by political interference and bureaucratic stagnation. As few others saw it, Brock also sensed the coming world struggle, and he felt it would be easier to free himself from his new duties than from his national post of responsibility—for he never doubted but that duty would involve him in overseas service. And so it was that he resigned as Director of the Geological Survey and Deputy Minister of Mines, the highest Canadian position attainable for a geologist, to become Dean of the Faculty of Applied Science in a University in the building.[54]

No doubt Brock also detected a growing conservatism in the attitudes of the government that was to be re-enforced by the war that Canada had entered. Certainly he had been the man best equipped to meet the Survey's needs in 1908, at the very crest of the boom period. In his six years as director, Brock made over the Survey largely in conformity with his own ideas, into an organization that had no more immediate need for his innovative skills. In an age of stability, other, more patient, talents were needed. The time had come for consolidation, for welding the system Brock had created into a functioning

XIV-14 R. G. McCONNELL

whole. To fill this need, the government turned, not to 'Pete' LeRoy, whom Brock had recommended, but to R. G. McConnell, the senior officer, with service going back to the spring of 1879.

There was a measure of irony in McConnell's appointment, for in 1901, when Bell had been made acting director, it was McConnell who had written to Brock complaining that "Length of service is the only thing that counts now à la British army before the [Boer] war—."[55] But different times call for different men, and in the years that lay ahead, McConnell's steadiness and self-effacing devotion served the department, the Survey, and the country well. He entered upon his new position of deputy minister in August 1914[56] with the good wishes of his fellows and of the mining industry, which saw in him "one of the 'stalwarts' of the old brigade of the Geological Survey," termed his promotion "a fitting reward for long and honourable service,"[57] and announced that "Among his colleagues and associates, Mr. McConnell is held in high esteem and respect and his promotion will be as popular with them as it unquestionably is with the Canadian Mining fraternity."[58]

XV-1 WILLIAM McINNES
Director, 1914–20

15

The Survey Goes to War

WHEN THE GREAT WAR BURST on an incredulous world on August 4, 1914, adventure-starved Canadians thronged to the colours to get to the front before it was all over. Government planning was concerned with processing as many troops as possible, and shipping them overseas as quickly as it could. Little thought was given to organizing for a long war, let alone turning Canada into an important arsenal of allied victory. Canada was in something of an economic slump at the beginning, until the warring nations settled down to a long war of attrition enormously wasteful of material and men. By 1915 the economy was in full swing, and increasing quantities of nickel, copper, lead, zinc, and iron were being mined for mushrooming war industries. Special emphasis was placed on securing supplies of key minerals—antimony, molybdenum, manganese, chromite, and cobalt for specialty steels, magnesite for refractory brick, and platinum and mercury. Despite shortages of fuel, transport, and especially labour, mineral production increased until by 1918 its value stood at $211.3 million, as compared with $128.8 million in 1914.

A great technical advance ushered in a remarkable expansion in base metal production—the application of oil flotation methods to separate the metallic components of sulphide ores, particularly the complex lead-zinc-copper ores so prevalent in British Columbia and other parts of Canada. Other experimentation was under way in the electrolytic refining of copper and zinc, and in manufacturing steel electrically. An aroused public demanded that smelting and refining processes should be carried out in Canada in the interest of national self-sufficiency, and made the International Nickel Company its principal target. Reports that nickel was reaching Germany via the still neutral United States added fuel to the agitation, until the company finally capitulated and consented to set up a nickel refinery at Port Colborne.

The fuel shortage that developed after 1915 was an especially grave problem. Demands for fuel rose because of the accelerated tempo of industry and the large volume of freight moving by rail and ship, but coal production actually declined after 1913 because of shortages of miners and engineers. The deficiency was most acute in industrial central Canada; the needs could have been met from mines in Nova Scotia and Alberta but for the transportation difficulties. So Canada's dependence on American coal supplies mounted until imports reached a peak of 21.6 million tons in 1919, far exceeding the domestic production for that year. Interruptions of supplies from the United States, uneasiness over the adverse trade balance, and a strengthened sentiment of economic nationalism inspired a long campaign in the early twenties to develop suitable power supplies in central Canada, or to make it economical to ship coal from the mines of the Maritimes or the Prairies.

Both branches of the Department of Mines contributed to the war effort to the best of their ability. The emphasis was on securing maximum mineral production from existing mines through improved technology and scientific research into mining and metallurgical problems. Consequently, the laboratories of the Mines Branch were busy helping develop the flotation and electrolytic refining processes, processes for briquetting lignite coals and manufacturing peat for fuel, and practical concentration processes for minerals like molybdenite. As the mining industry progressed into more advanced levels of refining and secondary manufacturing, the likelihood grew that the Mines Branch would continue to play an increasing role in relation to the mining industry and the national economy.

In contrast, the Survey's activity of mapping geological formations and solving geological problems seemed rather irrelevant in a country waging all-out war. Hence attention shifted to helping discover and develop mineral resources needed by the war effort, especially in areas that were accessible to transport facilities and early development. Almost every Summary Report began by emphasizing the economic and military significance of the work; that for 1914 announced that "Special attention was devoted to regions which promised to be of interest and detailed investigations of several producing areas was made."[1] The report for 1916 was longer, but no more explicit:

> Although the operations of the Geological Survey have been somewhat curtailed owing to conditions arising out of the war, it was realized that the work was of so great importance to the welfare of the country that its activities must not be needlessly hampered. The explorations and investigations were directed, even more than in the past, along lines promising to lead to economic results and special work was done in the investigation of materials required for the prosecution of the war and in materials required in the industries, the supply of which had been cut off by the war.[2]

Though work was mainly directed to these practical ends, the traditional objective of mapping the geology of Canada was not completely forgotten. The Survey carried on as best it could during the war, despite serious shortages of staff and the added upheaval of suddenly having to vacate the Victoria Memorial Museum and disperse to offices and storage buildings scattered throughout Ottawa.

During the night of Thursday, February 3-4, 1916, the Centre Block of the Parliament Buildings was destroyed by fire, and new quarters were needed at once for the Senate and Commons which were then in session. The Museum building was selected as the most suitable location from which to carry on the business of the nation. While the Department of Public Works threw up partitions for the purposes of the new occupants, all pitched in to pack and move their offices and collections out of the building. So quickly was this effected that the House of Commons met on the afternoon of February 4th in the auditorium, while the Senate was able to meet on the following Tuesday, February 8, in the cleared-out west exhibition hall—appropriately enough, that formerly assigned to the fossils. Later more space was requisitioned; the mineralogical and anthropological halls were appropriated for the use of Parliament, as well as many of the offices. The Parliament continued to occupy the Museum until early in 1920, when it could return to a rebuilt Centre Block.

Meanwhile, most of the Survey staff was assigned offices in old government buildings on the north side of Wellington Street west of Bank Street; the mineralogists were stationed at 227 Sparks Street; while the Biological Division found new quarters on Nepean Street. Only the paleontological, drafting, and photographic units, and the library stayed in the Museum throughout the period.

The several parts of the Survey were separated from one another, and some from their displays and stored materials as well. The Anthropological Division, all of whose exhibits were in storage, temporarily discontinued its field work and concentrated on research and publication. The Mineralogical Division fitted out a laboratory for analyses and research, and set up an exhibit of minerals important for the war effort. The Biological Division, which retained its space in the Museum until January 1918 and moved in more leisurely style, was able to take the herbarium as well as the study collections, and continued its scientific work. In fact, it fell heir to the collection of the Fisheries Exhibit when the building in which that was housed was torn down. A source of serious difficulty was the separation of the offices from the library; to offset it a small temporary branch library was set up at 221 Wellington Street.

The return of Parliament to its rebuilt quarters did not completely reunite the Survey. The mineralogical laboratory remained in its Sparks Street location, and still later at the old Survey building at Sussex and George Streets, which locations were convenient for receiving inquiries and rock samples from the general public. Part of the Distribution Division also remained downtown at 347 Wellington Street. In rearranging the space at the Museum, the National Gallery was given one of the Survey's exhibition halls for its expanded holdings—that of vertebrate paleontology, leaving the Survey more than ever short of display space. The *Canadian Mining Journal* was scandalized by this disrespectful treatment of the work of Canada's geologists. A suggestion that the skeleton of the great horned dinosaur be boxed in and left among the paintings if it was too fragile to be moved caused the journal to comment:

> Possibly some itinerant scientist will come across the "boxed up" dinosaur and express himself as Samuel Butler did about the pantless Discobalus in Montreal. If the right of the Palaeontological Branch to continue its exhibits in the Victoria Museum is to be taken away, then it were a pity to have disturbed the dinosaur from its age-long rest in Alberta, and in justice, the bust of Logan and the boulder on which it is placed should be removed from the entrance to the Museum. Perhaps, if Ottawa does not want the dinosaur, the Province of Alberta, or the United States, might welcome it. That is where most of the geologists have gone anyway.[3]

During the hasty evacuation many paleontological specimens became scattered, were mislaid, or lost their labels, and it was many years before the bulk of the collection was returned to full use.

More serious than those physical inconveniences were the serious losses of staff to the armed services. The offical policy regarded Canada as playing a twofold role—to provide as many troops as possible for the war in Europe, and to supply the armies and civilian populations of the allied countries. But during the Great War little or no effort was made to give any special or deferred status to the technical and scientific staff. The men were left to follow their own consciences—whether to enlist as proof of their loyalty to king and country, and of their manhood, or carry on with their regular work of helping discover and develop resources vital to the war effort. Enlisting in the army was the socially acceptable course, encouraged by every propaganda instrument at the disposal of state and society. In Ottawa, as in English Canada in general, a man needed great strength of will to resist these pressures and continue in his regular work, no matter how important or necessary it was.

Consequently, from the spring of 1915 onward, the men of the Survey began enlisting in the army, and the department merely tried to make do without their services. For the permanent staff it made up the difference between army pay and the salaries they would have earned had they remained at their work—including their regular increments. Among this group were the geologists O. E. LeRoy and S. J. Schofield, the topographers A. G. Haultain, W. E. Lawson, A. C. T. Sheppard, S. C. McLean, J. R. Cox, and E. E. Freeland, the anthropologist D. Jenness, the draftsman S. C. Alexander, and the relief map maker, L. N. Richard. Some junior men, employed full time but not yet on the permanent staff—the geologist W. J. Wright and the topographers R. Bartlett and R. C. McDonald—were also given extra pay.

Another group lost to the Survey, that the Survey did not subsidize, were the assistants who enlisted—the geologists W. A. Bell, J. D. MacKenzie, George Hanson, and George S. Hume, and the topographer W. H. Miller—who would have joined the permanent ranks much earlier than they did. Bell, for example, had been doing excellent work since 1911 and was ready to join the full-time staff as soon as he completed his Ph.D. at Yale University. By enlisting in January 1916 in the Canadian Field Artillery he was unavailable for work for three seasons and his educational program was also set back; it was December 1920 before he was appointed to the permanent staff.

XV-2 A SITTING OF THE HOUSE OF COMMONS IN THE VICTORIA MEMORIAL MUSEUM LECTURE HALL, 1918
Sir Robert Borden, the Prime Minister, is seated to the Speaker's right, behind the second desk from his front, and Sir Wilfrid Laurier, Leader of the Opposition, directly across from Borden, in the third seat to the Speaker's left.

Sheppard and Richard were released after training but most of the remainder saw action on the Western Front. Major W. E. Lawson died of wounds in France on August 29, 1918, and J. D. MacKenzie was so severely wounded that though he returned to work he died on December 16, 1922, as a result of his wartime injuries. O. E. LeRoy, answered Brock's call to arms by joining the 72nd Seaforth Highlanders in Vancouver and became a captain in the 196th (Western Universities) Battalion which Major R. W. Brock was organizing. To get to the front he took a position of lieutenant in the 46th Battalion, regained the captaincy for his part in the attack on Lens in June 1917, but was severely wounded on October 27 leading his men at Passchendaele and died the following day. A likely candidate for the directorship had he lived, LeRoy was greatly mourned, and after the war a LeRoy Memorial Fellowship in Geology at McGill and a scholarship at the University of British Columbia were established in his memory. The operations were crippled further by the loss of two more important field officers; D. D. Cairnes died in Ottawa on June 14, 1917, after a short illness, and C. W. Drysdale was drowned while at his work in southeastern British Columbia on July 10, 1917.

Further reducing the effective work force was the frequent assignment of men to such wartime agencies as the War Purchasing Commission, the Munitions Resources Commission, the War Trade Board, or to the Department of Militia and Defence. Field work was seriously crippled also by the drying up of the supply of part-time assistants. With declining enrolments in geology and mining courses in Canadian universities, field parties began to be manned by American university students. Many parties were led by such university professors as J. A. Allan, R. C. Wallace, and A. P. Coleman, by former officers like J. A. Dresser, and other outside specialists.

The shortages of personnel and the need to meet the demands of a country geared to war production militated against Brock's successors attempting any drastic changes in the structure or program, if indeed they had any great desire to do so. R. G. McConnell, the new deputy minister of the Department of Mines, was a self-effacing, conscientious administrator who left the Survey and the Mines Branch to continue as two completely separate organizations and made no effort to integrate them. He considered himself first and foremost as a geologist, and even managed to do some field work in the years 1915, 1916, and 1918. Probably he also retained a strong voice in the management of the Survey where no director was formally appointed till 1919.

It was true that William McInnes was appointed to the charge of the Survey with the title "Directing Geologist" on April 1, 1915, but he received no raise in pay, and except for the announcement of his promotion, McInnes' name did not appear anywhere in the *Summary Report for 1915,* which carried only a letter of transmittal from McConnell to the minister. That for 1916 showed McInnes' signature at the end, while those for 1917 and 1918 were headed, a little less diffidently, in bold face, "Report of the Directing Geologist, by William McInnes." His last report, for 1919, finally bore the heading: "Report of the Director, by William McInnes," indicating that in April 1919 McInnes at last had assumed the rank. During the entire period of his headship, McInnes received no allowances for travel or for any field work. Presumably he restricted his activity to Ottawa, principally to editorial duties. Under McConnell and McInnes between 1914 and 1920 the leadership of the Survey was masked, in complete contrast with the open, direct, personal sort of leadership that characterized it since its founding by Logan, and above all, under the dynamic, assertive, firebrand Brock.

The chief organizational change was an attempt to break down the enormous Geological Division into which Brock had lumped the paleontological, mineralogical, and borings sections. The Borings Records Sec-

XV-3 MRS. F. B. FORSEY
Survey Librarian, 1918-41

XV-4 O. E. LeROY

XV-5 J. D. MacKENZIE

tion received the style of Borings Division in 1916; in 1917 the Mineralogical Section became a division; then in 1919 vertebrate paleontology, invertebrate paleontology, and paleobotany were combined as the Division of Palaeontology. By 1919, therefore, the divisions were: Geological, Palaeontology, Mineralogy, Borings, Topographical, Biological, Anthropological, Photographic, Distribution, and Geographical and Draughting, plus the library. Furthermore, an attempt was made to establish a regional structure for the Geological Division in place of the functional one that Brock had instituted, under which LeRoy and Young divided oversight of the field and office operations between them. In 1916 the following arrangement was announced:

> In order to ensure closer supervision of geological parties in the field and to keep more closely in touch with the needs of the different parts of the country, the Dominion has been divided into districts, which have been placed for purposes of geological investigation, under the supervision of the following geologists:
> *E. R. Faribault,* geologist in charge of Nova Scotia division.
> *G. A. Young,* geologist in charge of Eastern Quebec and New Brunswick division.
> *W. H. Collins,* geologist in charge of Pre-cambrian (Ontario and Quebec) division.
> *D. B. Dowling,* geologist in charge of Great Plains division.
> *C. Camsell,* geologist in charge of Northern Exploration division.
> *O. E. LeRoy,* geologist in charge of British Columbia division.
> *D. D. Cairnes,* geologist in charge of Northern British Columbia and Yukon division.[4]

To what extent this system was implemented is hard to tell. It did not appear to provide for the western part of the Precambrian, which received a fair amount of attention in the period. Besides, events quickly blurred some of the divisions. With the enlistment of LeRoy in 1916 and the death of Cairnes in 1917, Camsell added the two Cordilleran divisions to his own Northern Exploration division. On the other hand, Dowling seems to have functioned as a proper supervisor and co-ordinator of the work in the Plains region.

Coinciding with this reform a new publications policy was introduced in 1917. In place of the summary report that presented the work of all the divisions, that for 1917 appeared in six parts—a director's or general report, Part A, and five regional reports, Parts B to F, for the Cordillera, Interior Plains, Precambrian of Manitoba and Saskatchewan, Ontario and Quebec, and Maritimes (the geographical boundaries vary slightly in certain years). The system, introduced "for the sake of greater economy in printing,"[5] continued to 1933 with only minor revisions. A few special studies were incorporated into this format as an occasional Part G, for example, "The Platinum Situation in Canada, 1918" by J. J. O'Neill (1918), and two studies of Paleozoic and Mesozoic formations along the route of the future Ontario Northland Railway from Cochrane to James Bay, by M. Y. Williams and J. Keele (1919). From 1920 onward, the general review of Survey activities began appearing in the *Annual Report of the Department of Mines,* first launched in the fiscal year 1920-21; Part A of the Survey's summary report therefore included the geological reports on the Cordilleran region with the other four regions being renumbered from B to E to correspond.

These reports, of course, were only summaries of the field work. Separate studies presenting the matured work of the staff members continued to appear during the six years 1915 to 1920 inclusive, for a total of 66 memoirs and 21 museum bulletins. About one third of the memoirs were studies of economic minerals and in mineralogy, anthropology, vertebrate paleontology, and biology. Half of the 87 publications were issued in 1915 and represented work of the previous period. From 1916 onward no more than ten memoirs were published in any one year, and only three appeared in 1920. The depletion of the staff and the general tone of austerity during the war years is nowhere better reflected than in this meagre publication record of the years 1916 to 1920.

An important innovation that was destined to become a permanent legacy of the McConnell-McInnes regime was the opening of district offices to enable the Survey to conduct its field work more efficiently and be immediately helpful to local mining interests. The founding of the British Columbia office was justified in the summary report as being designed "to keep more closely in touch with prospecting and mining development throughout the province and in Yukon Territory, to work more closely in co-operation with the Provincial Department of Mines, and to act as a local distribution office for reports, maps and other geological information."[6] Privately, McConnell explained the proposal in greater detail in a memorandum to the Minister of Mines:

> The advantages to British Columbia mining men and prospectors of such a branch office are obvious. We are constantly receiving communications here relative to problems met with in prospecting or in the development of mineral deposits. A resident geologist would be in a position to deal quickly with such matters, and difficulties of structure, stratigraphy or ore occurrence could be discussed with him without the long delay involved in communicating with the central office.
>
> The establishment of this Branch would conduce also to closer co-operation between the Dominion and Provincial Mines Departments.
>
> The branch office would be supplied with a complete set of geological and mining maps and reports, and mining men consulting these would have the services of a qualified man to interpret, if necessary, the data contained in them.
>
> The proposed plan would enable the Department to keep more closely in touch with the progress of development and with new discoveries, and field work could be planned and carried out with greater advantage.
>
> Mining booms, some of a harmful nature, occur periodically. One of the duties of the resident geologist would be to inquire into these without delay and issue an official statement respecting them.
>
> It is as much the duty of the Department to discourage the investment of capital in unwise mining ventures as to encourage the investment when the prospects seem hopeful.[7]

No doubt McConnell was preaching to the converted, for the minister of the time was Martin Burrell, M.P. for Yale, British Columbia.

Actually, a branch had been opened earlier in Alberta, where S. E. Slipper had remained over the winter of 1913-14 to keep in touch with the current oil boom southwest of Calgary, advise drillers, and collect stratigraphic data and drill samples. The office was in Calgary but was moved to Edmonton when drilling proceeded into northern Alberta. After Slipper resigned at the end of 1917 the office was unmanned until the spring of 1919 when J. S. Stewart was put in charge. In the summer of 1919 Dowling relieved Stewart for three months to enable him to accompany a party that was prospecting for petroleum on the British Columbia government's account in the northeastern corner of that province, in the foothills south of Peace River.

The office was greatly appreciated by Edmonton residents; the Board of Trade in a resolution of December 8, 1919, expressed its "appreciation of the valuable work being performed by the new office of the Geological Survey and Mines Branch."[8] But when Stewart resigned from the Survey at the end of February 1920, to the great annoyance of local business and mining interests the office was not reopened. Eventually the permanent office for the Interior Plains was located in Calgary rather than Edmonton.

The British Columbia office had greater staying power. With no metropolitan rivalries beclouding the issue, it could be safely assigned to Vancouver. McConnell's memorandum was dated April 24, 1918, and as early as May 8, Charles Camsell, the designated appointee, was applying for a living allowance or a raise in pay to offset the higher living costs of Vancouver, and the more advanced social position he would have to maintain there. The office, which Camsell arrived in Vancouver to open on May 27, was in the Pacific Building, at the corner of Howe and Hastings Streets. It was equipped with a good library, including full sets of the publications of the Survey and Mines Branch, the British Columbia Mines Department, and United States and state geological survey publications bearing on problems encountered in British Columbia, plus subscriptions to the major periodicals in mining and geology. Considerable use was made of the office by prospectors and mining engineers, and Camsell noted that the office made it easier to co-ordinate the Survey's work with that of the provincial Department of Mines. The number of samples to be tested was on the increase, and for lack of facilities to do the work on the spot these often had to be sent to Ottawa for analysis. Camsell appealed for chemical apparatus and reagents to make simple analyses, while the industry called for a small ore-testing plant. The office was an immediate success in terms of the volume of useful work it was able to do on behalf of the British Columbia mining industry.

* * *

IN THE FIELD it took some time to effect the changeover from the normal activities program to one of all-out support to the war effort. The war began in the middle of one field season when all the parties were fully committed, while the program for 1915 was drawn up in the winter of 1914-15 before there was any real conception of how deeply Canada was involved. By 1916, however, the Survey was reacting to the seriousness of the struggle. Some of its men were enlisting, with more to follow in 1917 when the 1916 season's work was finished. Though the field parties were still deployed across the country, the actual assignments were those best calculated to help the war effort: investigating and reporting on problems of mine development, and discovering deposits of economically or strategically important minerals. The searches were pursued in every region, though the odds against new discoveries being developed in time to aid the war effort tended to increase the farther one went from Ottawa. Molybdenum occurrences in British Columbia, for example, were examined, but the mineral was mined on a large scale only at Quyon near Ottawa in 1916, the Mines Branch lending its facilities for the purpose. Tungsten was reported from the Yukon and northern British Columbia, and in the Lake of the Woods country, but the deposit that was most thoroughly investigated was at Burnthill Creek in central New Brunswick. Chromite had been mined extensively in the Eastern Townships until the Canadian product was unable to compete with cheaper foreign supplies. Now Harvie, aided by Poitevin, joined in efforts to reopen old mines and discover new deposits, and by 1918 chromite production valued at almost one million dollars was recorded.

Platinum and mercury were important metals for the war effort, supplies of which were difficult to obtain from traditional sources. Since platinum had been mined in the Tulameen placers, Camsell and Poitevin made a further search with the aid of two government-owned drills. In 1918 O'Neill conducted a country-wide study and pinpointed 31 locations, mostly in British Columbia. His conclusion was that a considerable quantity of platinum actually was being mined in Canada but most of it was exported mixed with other metals since it did not pay to separate the platinum. Thus the copper-nickel matte produced at Sudbury was a main source, but Canada received no credit for platinum production since the matte was refined in Britain or the United States. Mercury was also known to occur in British Columbia; Camsell visited a number of locations on Kamloops Lake and Victor Dolmage another on the west coast of Vancouver Island, but nothing came of it. Occurrences of magnesite—formerly imported from Hungary and Greece—were investigated in British Columbia, but the one source developed was that reported by M. E. Wilson in the Grenville district of Quebec. A plant put in operation in October 1917 produced 39,365 tons, valued at over one million dollars in 1918. Infusorial earth sources were also reported by Faribault (1914) and Drysdale (1916) from opposite ends of the country.

Much of the 'practical' work in the war years merely continued earlier activities. Investigations that combined surficial studies with marketing analyses were made of the resources of workable clays and shales, exploitable gravels, cultivable soils, and potable subsurface waters. J. Keele, who had been examining usable clays and shales since 1909, made an unsuccessful search for kaolin clays in the north of Gatineau County in 1916, and in 1919 studied white Cretaceous clays in the James Bay Lowland. Reinecke had been loaned to the Ontario Highway Commission in 1913 to study possible sources of sand and gravel required for highway construction in the Toronto district. After a second year in the Toronto area in 1914, he spent two seasons in eastern Ontario and the Montreal district. When he transferred to Saskatchewan in 1917 two assistants continued the work in the Montreal district under his supervision while Wright did similar work in the Moncton district in 1915. Since Canada was rushing headlong into the automotive age and the provinces were embarking on modern highway building, expert advice on such questions was a very important service to the public.

Two specialists in surficial geology, W. A. Johnston and J. Stansfield, carried out a series of studies that constituted the tentative beginnings of a soil survey of Canada. Johnston investigated the soils and other surficial materials in the Lake Simcoe, Rainy Lake, and Ottawa districts in 1914 and 1915, then embarked on three seasons of classifying land in Manitoba for the benefit of intending settlers—in the Whitemouth district, north of The Pas, and south and west of Lake Winnipegosis. Similarly, after two seasons investigating the London, Ontario, district, Stansfield transferred his work to southeastern Saskatchewan in 1917 and 1918. These studies were matched by a noteworthy investigation by Dowling of underground water resources. In 1915, while mapping the Cretaceous of the Milk River district, Dowling realized that the extensive Milk River sandstone stratum had the necessary configuration and stratigraphic position to be an important source of artesian water. He mapped the limits of this area, while the Survey tested his work by having two wells drilled, both of which struck large flows of water at depths of 640 feet. His information was extremely useful and opportune to farmers, for this semi-arid district was being occupied by grain growers, encouraged by the high prices and patriotic pleas to increase wheat production. Dowling's work was a prelude to the large-scale program of mapping the underground water resources of the prairies that the Survey carried out in the twenties and thirties.

An equally important need was to discover new economic fossil fuel deposits, especially for the fuel-poor industrial districts of central Canada. Most of the work, however, was undertaken in the Plains region where prospects of discoveries seemed better than in the east, and besides, the region had faced a fuel problem of its own in its dependence on American petroleum products. Dowling, Slipper, Stewart, Rose, and McLearn constituted a team that brought together the findings of stratigraphic, paleontological, and historical (well-drilling records) studies to provide an improved understanding of the mineral resources of the region. From having traced the coal-bearing formations for so many years before the war, Dowling possessed an excellent grounding on the geology of the foothills region, which he augmented by using later reports and discoveries. Rose and Stewart mapped the Cretaceous formations paralleling the mountains south to the International Boundary. Also in 1917, J. A. Allan ran two sections along the valleys of the North Saskatchewan and Red Deer Rivers across the prairies which indicated the strata dipped evenly to the west and did not form an anticline in eastern Alberta, as had hitherto been assumed. Slipper collected the drilling records and compiled accurate profiles of several Alberta oil and gas fields. McLearn, who was beginning a distinguished, productive career with the Survey, studied Jurassic and Cretaceous fossils of the Crowsnest Pass region in 1914 and 1915, and in 1916 in an effort to throw some light on the vexing problem of the oil sands, he examined the exposures along Athabasca River from Athabasca town north to below Fort McMurray. When oil was discovered in the Peace River district, he investigated the valley and canyon of Peace River in 1917 and the lower Smoky River in 1918.

Dowling watched all these developments and analyzed and collated the results. In 1919 with Slipper and McLearn he published a memoir that summed up the results to date in the form of drilling records, along with a series of subsurface contour maps showing the locations and depths of the various oil and gas-bearing strata, as a guide to future endeavours by the industry.[9] That the oil companies found the Survey's information valuable was demonstrated in 1919 when the Imperial Oil Company placed rigs at two points that Stewart's memoir of 1919 on the foothills of southern Alberta indicated as favourable localities.

Finding new coal deposits was of lesser concern in western Canada, since the major problem of the industry was recognized as being economic, of reaching large markets outside the region. Still, a new deposit of particularly well-located, high-quality coal was always useful. Investigations of the coal-bearing formations along the margin of the Rockies continued, mainly by Rose, while north beyond the Athabasca, J. MacVicar in 1916 discovered a good quality, wide-seam deposit containing more coal than the whole of Nova Scotia's known resources. Dowling continued to study the geology and mining operations of the developed coalfields, including the mines of the Brazeau and Alberta Coal Branch districts he had helped discover, and the field recently brought into production along Red Deer River at Drumheller. Farther east, A. McLean investigated the Cretaceous of the Pembina Mountain-Wood Mountain district between 1914 and 1918, turning his attention from 1916 onward to the large lignite coal deposit near Estevan, which was the subject of extensive utilization studies by at least three government agencies.

From the standpoint of public acclaim, perhaps the most important economic work on the Paleozoic was that of M. Y. Williams in southwestern Ontario. By 1917 he was able to prepare detailed contour maps of some of the oil-bearing formations that revealed the folds under which pools of oil might be found, notably in the Lower Devonian. Moreover, he was responsible for an important new insight on oil-bearing formations. In 1917 in Dover West Township a well had reached a pool of oil in what was definitely a syncline, not an anticlinal structure. Williams concluded that the petroleum, in the absence of water to force it upward, simply flowed to the lowest point where it was trapped by an impervious bottom stratum. He was able, therefore, to recommend to drillers to look for oil and gas in the Trenton formation under certain conditions: "So long as the formation is free from water or only partly saturated, the synclines should be sought, rather than the anticlines."[10] In an important address to the organization meeting of the Natural Gas and Petroleum Association of Canada on June 18, 1919, in London, he described six horizons in southwestern Ontario in which oil and gas might be found, and advised drillers to explore several localities where no discoveries had been made thus far but which appeared geologically interesting. When a successful well resulted from his advice, it was hailed by the *Canadian Mining Journal* as tangible evidence of the good work being done by the Survey:

> To Dr. Williams' study of this field and his application of his knowledge in directing the position of wells, much of the success which has been obtained is due.
>
> Geological research always pays, and pays exactly in proportion to the effort and time expended thereon.[11]

In the Cordillera and Precambrian the emphasis was

XV-6 M. Y. WILLIAMS

also on practical work after 1916, with the reports focusing more and more on mineral occurrences and on economic problems of development. Cairnes and W. E. Cockfield devoted a large part of their attention each year to the lead-zinc-silver complex near Mayo, the future Keno Hill mine, while in booming southeastern British Columbia, S. J. Schofield, C. W. Drysdale, and M. F. Bancroft mapped in the Ymir, Rossland, Ainsworth, Bluebell, and other mining camps, and attempted to indicate the areas with the best prospects of containing ore. In 1916, when Drysdale was the only officer working in British Columbia, he examined a wide range of mines and prospects that ranged from Portland Canal to Bridge River and the Highland Valley copper deposits near Ashcroft, in addition to doing his main work on the Slocan map-area. An indication of the preparations for post-war mining activity was the assigning of B. R. MacKay to study the Cariboo goldfield, which was being made more accessible by the construction of Pacific Great Eastern Railway.

In central Canada W. H. Collins was set to work in 1917 studying mineral occurrences and deposits in the country between Lake Timiskaming and Sault Ste Marie where he had previously conducted his long-range correlation studies. He investigated Bruce Mines, where copper was once more being mined, as well as many smaller sulphide mineral deposits that might be worth mining through using oil flotation techniques. Collins reported that the only chance of developing many deposits into mines was to install a local smelter, and this would require special study as to its feasibility. In 1918 he searched for mineable iron ore and pyrite deposits and examined recently reported gold discoveries in the Michipicoten district. M. E. Wilson continued his studies along the Ottawa River, with special reference to mineral deposits, several of which, thanks to their convenient location and the impetus of wartime need, were developed as mines.

The chief region to attract attention as a new source of metallic minerals was the western part of the Precambrian, particularly northern and eastern Manitoba, where F. J. Alcock and E. L. Bruce continued the earlier work of Tyrrell and Dowling. Alcock operated on a roving commission that took him into the Precambrian north of Lake Athabasca in 1914 and 1916, where he reported iron formations that would not pay developing under existing conditions. A rumoured rich silver discovery also caused Camsell to make a quick trip to the Fond du Lac district early in 1915 to expose the inherent falsity of the promoters' claim before too many investors were victimized. Bruce worked in the Flin Flon district, north of The Pas, where a gold discovery had been reported late in 1913. In view of the district's increasing importance as a potential major producer of base metals and gold, Alcock joined Bruce in 1917 and 1918. Reported gold discoveries also accounted for examinations of the formations at Rice Lake and in the vicinity of Lake of the Woods by J.A. Dresser in 1916 and J.R. Marshall in 1917.

The concern for assisting the mining industry did not preclude some useful results of general significance being achieved, chiefly in revising successions and working out correlations. In the Cordillera, for example, Schofield, Bancroft, and Drysdale continued working on the succession problem in the area of Daly's earlier work and attempted to correlate the two sections through the region displayed along the International Boundary and the C.P.R. main line. While studying the ore deposits of the Slocan in 1916 Drysdale reached quite sophisticated conclusions that were an interesting early example of thinking along metallogenic (which he called metallographic) lines. He claimed four distinct rock terranes of different ages could be traced in the Kootenays, each characterized by different metallic ore associations: silver-lead-zinc; gold quartz and antimony; copper-gold; and molybdenite and tungsten. The presence and abundance of these ores depended on the local conditions at the time of emplacement of the granitic intrusions. He found in the south Lardeau Mountains "the key to the regional structure"[12] from which he correlated the successions in each direction, between the Slocan and Cranbrook map-areas, and set down seven points of disagreement with Daly's conclusions on the structure, succession and character of the rocks of the region. He also drew up

Kootenay Metallographic Belts.

METALLOGRAPHIC BELT.	MAIN TYPE OF DEPOSIT.[1]	ROCK TERRANE AND FORMATION.
Silver-lead-zinc	Fissure veins	Post-Cambrian terrane (Slocan, Nis-konlith, and Pend-d'Oreille series).
	Replacement (blanket) veins	Pre-Cambrian terranes (Purcell and Ainsworth series).
Gold quartz and antimony (Kaslo schist)	Fissure veins	Lower Cambrian terrane (Summit, Lower Selkirk, and Kaslo schist series)
Copper-gold	Differentiates and veins	Purcell sills of hornblende gabbro (Lower Cambrian ?)
	Replacement lodes	Rossland volcanic group (Triassic (?) terrane)
Molybdenite and tungsten	Stockworks and pegmatite veins	Post-lower Jurassic terrane (Nelson granite stocks).

XV-7
Drysdale's remarkably advanced theory of the metallogeny of the Kootenays, 1916, which he did not live to check, develop, or extend. The term "metallographic" is now archaic.

a tentative scheme of the correlations between the two earlier sections through the mountains on the basis of his third one, which caused him to hope that

> The following general notes taken on the trip across the ranges are here recorded in the hope that although fragmented they may assist in the solution of the broader regional problems bearing upon the stratigraphy and structure of the oldest rock terranes in British Columbia .... With the steady accumulation of geological field data bearing on this subject, it is hoped that in the near future the Kootenay district will yield critical evidence sufficiently convincing to narrow down the various alternatives so prevalent at present to a single tenable correlation table which will be in harmony with all the field facts and stand the test of time.[13]

Unfortunately, Drysdale never lived to realize this ambition; in the following summer on July 10, 1917, while at work in the district he and his assistant W. Gray were drowned in upper Kootenay River, ending a very promising career.

The work on the Plains also produced a better understanding of the stratigraphy and relations of rocks over the entire region, and assisted in correlating the successions over enormous distances–from Manitoba to the Rockies, and from the United States to the Peace, Liard, and Mackenzie Rivers on the north. The paleontological work of McLearn, Kindle, and Burling was especially significant in this regard. In addition, Kindle embarked on a general study of the present-day processes of sedimentary deposition to gain new insights into how past strata were deposited. In Ontario, Williams' work on the Silurian that occupied him from 1912 to 1915 ended with a better working out and mapping of the Lockport, Cataract, Richmond, and Guelph formations.

Scientific studies that had begun earlier and were completed in the war years included those of A. O. Hayes on the complex geology of the Saint John City map-area, concluded in 1915, and Collins' effort to correlate the succession across the important country between Sudbury and Sault Ste Marie. His method was to select seven intermediate localities that were particularly significant in determining the Precambrian record and examining them intensively. This work, too, was finished in 1916, after which Collins was asked to apply it to the problems of the metallic deposits of the region. M. E. Wilson continued his careful analysis of the relations and characteristics of the Precambrian formations in these years with special attention to the mineral occurrences.

The war years even saw some long-range work of the sort that was officially deplored, such as exploration of areas that were entering the earliest prospecting phase and were still far from immediate development. Reconnaissances along completed railways or railways under construction including those by MacKenzie (1915) and O'Neill (1917) along the Grand Trunk Pacific in northern British Columbia, Camsell (1917) and Reinecke (1918) along the route of the Pacific Great Eastern, T. L. Tanton (1916-17-18) along the Canadian Northern Railway north of Lake Superior, W. A. Johnston (1917) along the Hudson Bay Railway, and Williams and

XV-8 C. W. DRYSDALE

Keele (1919) along the Temiskaming and Northern Ontario route to James Bay. River surveys included those of Harricanaw River and west to the Quebec border (Tanton, 1914-15), of Broadback River (Cooke, 1914-15-16), and Churchill River from Southern Indian Lake to Hudson Bay (Alcock, 1915). Still other reconnaissances were Camsell's through the Omineca district, northern British Columbia (1915), Cooke's of the District of Nipissing west of Montreal River (1917-18), and A. P. Coleman's of the Torngat Mountains of northern Labrador (1915-16).

The most important exploratory surveys were in the Mackenzie River watershed. In 1914 Camsell traversed the country between Lake Athabasca and Great Slave Lake by canoe via Taltson River, observing the geography, geology, and natural history of this Precambrian country (in association with the naturalist Francis Harper), and in 1916 he made another visit around Peace, Slave, and Salt Rivers to examine the gypsum and salt occurrences for possible exploitable deposits of potassium salts, in which he was unsuccessful. Great Slave Lake was investigated as a result of reported oil seepages at Windy Point, on the north shore of the lake. A. E. Cameron, who examined the Paleozoic strata around the western end of the lake in 1916 and 1917, held little hope for a discovery of oil, feeling that the structure was unfavourable to the accumulation of significant amounts of petroleum. Today with the recognition that Devonian coral reefs make excellent petroleum traps, the picture has changed, and active, successful exploration is now in progress in these regions. In 1917 Kindle and E. J. Whittaker were also in the Great Slave Lake district, collecting fossils and working out the succession and structure. By 1918 the Survey thus was well on its way to extending the examination of the Interior Plains region to the Arctic Ocean.

* * *

THE SURVEY HAD ALREADY BEEN INTRODUCED to the geology of the Arctic coast in 1914 in what was the largest, most ambitious of all its exploratory programs of the period. This was the Canadian Arctic Expedition, organized early in 1913, which did not actually begin work until early in 1914, not long before the outbreak of war. This highly contentious enterprise, that was to have so many ramifications over the next generation, resulted from Brock's collaboration with the then-controversial Vilhjalmur Stefansson. Stefansson had made a remarkable exploration of the Central Arctic district between 1908 and 1912 with a single associate, the naturalist Rudolph M. Anderson. The expedition, which studied the civilization of the primitive Eskimos of Coronation Gulf and Victoria Island, was financed mainly by two American foundations, but also received a small grant from Brock—$200 plus $332 expenses. In return Stefansson furnished Brock with reports of his activities during the five years, as well as very good Eskimo artifacts. Needless to say, Brock was delighted at such returns from so trifling an investment.

Stefansson immediately began planning another expedition to explore the Beaufort Sea and the lands flanking it, taking soundings of the seabed and reporting on the country traversed and inhabitants met. This entailed a

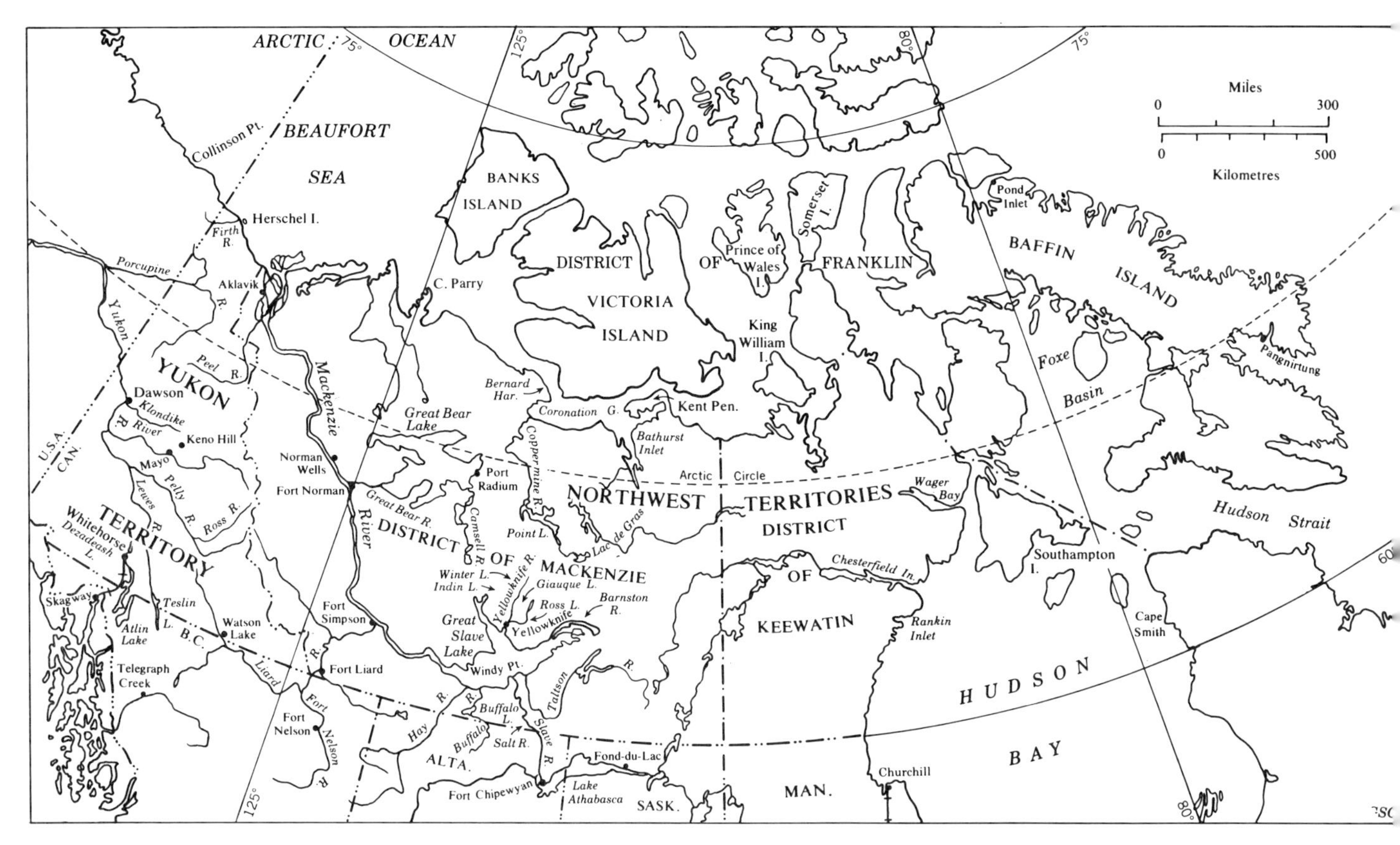

XV-9 NORTHWESTERN CANADA, 1914-40

ship, or ships, under the direct control of the expedition as well as a scientific staff. He had promises of grants of $45,000 from his sponsoring societies, even though they were somewhat miffed by the publicity associated with his earlier expedition, notably the sensationalized reports of "blond" Eskimos. However, Brock was delighted to provide more aid on the basis of Stefansson's previous work, and immediately recommended that the government contribute $25,000, hinting it might even be expedient if Canada took over the project completely. As he wrote the Minister of the Interior and the Prime Minister, "Participation to some extent is advisable in case any new lands should be discovered, a contingency that is not beyond the range of possibility. It is practically the one remaining place in the world where great geographical discovery is possible."[14]

The government decided it should become the exclusive sponsor of the expedition, and on February 22, 1913 an order in council appointed Stefansson commander and gave general oversight of the enterprise to the newly-created Department of the Naval Service, which was to co-ordinate the efforts and interests of the several departments concerned, particularly the Geological Survey. Stefansson, who was "to have full responsibility, and to have the choice of the men going on the expedition; and of the ship, provisions and outfit needed for the trip,"[15] rapidly collected a remarkably able body of specialists, to whom Brock added four present employees of the Survey—G. S. Malloch and J. J. O'Neill, geologists, and K. Chipman and J. R. Cox, topographers. He also took over three of the men hired by Stefansson who had specialties in which the Survey was interested (for its museum): R. M. Anderson, naturalist, and the two anthropologists, Henri Beuchat and Diamond Jenness.

To Brock, the expedition was a unique opportunity to introduce competent scientists into a remote part of Canada that already was beginning to attract the interest of others besides whalers, policemen, missionaries, and anthropologists. In 1911 a party of L. D. and G. M. Douglas and A. Sandberg had made a reconnaissance of the Coppermine River District and brought back encouraging reports of possible mineable copper occurrences.[16] Brock expected the expedition, now that it was wholly financed by the Canadian government, to carry out a regular scientific program, adhering to the Survey's standards of accuracy, and making a systematic, thorough investigation of the area blocked out for study. Even before the order in council was passed he was writing the deputy minister of the Naval Service, G. J. Desbarats, about the scientific work he expected to have done. Brock's concept was a far cry from Stefansson's, which was a romantic approach to exploration—traverses during which the men lived off the country and let the party's movements be guided by the migrations of natives and game, while it collected reports on the resources and inhabitants.

The plan that emerged was a compromise between Stefansson's desired freedom to explore and Brock's determination to derive the maximum scientific benefit from the expedition. There would be two parties—a northern one, led by Stefansson, to work on and around Beaufort Sea, and a southern party, under Anderson's direction and mainly comprising the personnel employed by the Survey, to work along the Central Arctic coast. Brock sent Desbarats instructions regarding "the work under the jurisdiction of the Geological Survey"[17] for inclusion in the instructions to Stefansson, which set out that the northern party should devote itself to the various sciences but be guided by Stefansson. He was far more explicit, however, respecting the work of the southern party, outlining that it would concentrate on geology, geography, anthropology, and biology and would comprise "primarily the investigation and areal mapping of the copper-bearing and associated rocks of the mainland between Cape Parry and Kent peninsula, and for approximately one hundred miles inland, and on southern and eastern Victoria Land."[18]

The work was to be of highest scientific quality, and generally would follow field reconnaissance methods of the Survey, with the parties operating independently out of base depots. Brock made specific assignments to the members of the party—O'Neill was to take charge of the geological work, and Chipman assisted by Cox, of the geographical and topographical work; Beuchat and Jen-

XV-10 THE SCIENTIFIC STAFF OF THE CANADIAN ARCTIC EXPEDITION AT NOME, ALASKA, 1913
(* = employee of the Survey; † = died during the course of the expedition). *Front row, left to right:* A. F. MacKay† (surgeon), R. A. Bartlett (ship's captain), V. Stefansson (commander), R. M. Anderson* (zoologist and deputy commander), J. Murray† (oceanographer), F. Johansen* (marine biologist). *Back row, left to right:* B. Mamen† (assistant geologist), B. M. McConnell (meteorologist), K. G. Chipman* (topographer), G. H. Wilkins (photographer), G. S. Malloch*† (geologist), H. Beuchat*† (anthropologist), J. J. O'Neill* (geologist), D. Jenness* (anthropologist), J. R. Cox* (assistant topographer), and W. L. McKinley (magnetician).

ness were to divide the anthropological work among themselves; Anderson would study mammals and birds, while Frits Johansen (the Danish marine biologist hired by Stefansson), would attend to the fields of marine biology, entomology and botany if he were with the southern party. The Survey's control of the scientific work was unequivocally affirmed in the *Summary Report for 1913,* written in 1914:

> Scientific work within the scope of the Geological Survey was placed under the jurisdiction of the Survey. The expedition was divided into two parties, the northern exploration party under Mr. V. Stefansson, the leader of the expedition, and a southern scientific party under Dr. R. M. Anderson of the Geological Survey, whose field of operations was to be in the neighbourhood of Coronation gulf, Coppermine river, and Victoria Land .... The scientific equipment for this regular work and salaries, together with expenses in joining the expedition, are borne by the Survey. Expenses in the field and intruments required for special services are being furnished by the Expedition. The officers of the Survey are working under the direction of and reporting to the Geological Survey.[19]

The transportation arrangements for the expedition confirmed Brock's view, inasmuch as the two small vessels of the expedition commanded by Anderson and Chipman, carried most of the supplies, equipment, and personnel of the southern party on the journey north from Victoria. The larger ship, *Karluk,* captained by the celebrated Newfoundland skipper, Captain R. A. Bartlett, was designated for the northern party and contained most of its personnel and equipment under Stefansson. In effect, Brock had succeeded in separating from the original program an almost completely self-contained, multi-discipline scientific exploration of the Arctic coast east of the Mackenzie district, the area in which he was interested.

The sequel is well known. The ships travelled separately into the Arctic, where the *Karluk* became caught in heavy pack ice north of Cape Barrow, Alaska, and was frozen in the drifting pack. When the ship was carried near the Alaska coast, Stefansson, Jenness, and two others went ashore to hunt caribou; in their absence the ship drifted away, where, 1,000 miles to the west, the crew and passengers finally abandoned the now-sinking vessel to make their way across the ice to Herald Island, and then to Wrangel Island. There, after Bartlett had made a heroic dash to Siberia to summon help, the men were finally rescued, on September 7, 1914. Unfortunately, only a dozen men of the original 28 marooned on the *Karluk* survived. Those lost included the Survey employees with the *Karluk,* Malloch and Beuchat, both at Wrangel Island.

The loss of the ship with all the equipment and most of the personnel of the northern party, and part of the southern party as well, threw both programs into violent conflict. Stefansson as commander tried to use some of the remaining equipment and men to recommence the work of the northern party. The members of the southern party, led by Anderson, insisted that their primary responsibility was to carry out their own assignment and not jeopardize it in a probably hopeless effort to salvage the northern program. Their attitude owed much to their personal distrust of Stefansson and their dislike of his revised plan. They honestly questioned the wisdom and value of the proposal to cross Beaufort Sea on the

XV-11 HEADQUARTERS CAMP OF THE SOUTHERN PARTY, CANADIAN ARCTIC EXPEDITION, 1914-16, AT BERNARD HARBOUR

XV-12 J. J. O'NEILL

drifting ice floes, taking soundings, and living off the resources of the sea, which they regarded as being nothing more than a useless journalistic stunt with little or no scientific merit. While this debate was going on in the north, in Ottawa officials were discussing the expedition's future and reaching much the same conclusion as the men. In letters that Desbarats addressed to Stefansson on April 30 and May 5, 1914, he advised him that

> It was felt that as the greater part of the expenditure for the expedition had been incurred on account of the southern party it was essential that this party should show results for the money so spent. It was, therefore, considered wise not to weaken or cripple the party in any way and to provide in the fullest way for transportation of their supplies and instruments.[20]

The men, however, grudgingly assisted what they regarded as Stefansson's pointless, foolhardy plan on condition he did not interfere with their work, and they helped him start his travel over the ice on March 22, 1914. But Stefansson complained that they failed to give him adequate support to achieve his program then, and later wrongfully withheld essential supplies and transport. However, Stefansson triumphed over the many obstacles; by the end of 1917, when he left the Arctic, he had successfully completed his experiment of living 96 days on the sea ice, had corrected the existing maps of the Arctic Archipelago, and had discovered new land for Canada, one part of which he named Brock Island.

From their winter quarters at Collinson Point, Alaska, the southern party began its work in February 1914, when O'Neill and Cox surveyed the coastline from the Alaska boundary east to Mackenzie River, made geological traverses up Firth River and across Herschel Island, and mapped the main branches of the Mackenzie delta. During the open-water season of 1914 their supplies and equipment were brought to a new winter quarters at Bernard Harbour, within the area they were assigned to investigate. During 1915 they made detailed surveys of the coast and rivers, carrying the work south and east along Coronation Gulf and the intricate system of deep, narrow fiords and innumerable islands of Bathurst Inlet. Altogether, by May 1916, they had mapped 600 miles of the Arctic coastline of America, plus another 100 miles between the west branch of Mackenzie River and the Alaska boundary, and portions of several rivers.

O'Neill studied the geology of the region, supplementing the coast survey with traverses up the major rivers. The formations ranged from Precambrian granite near Coronation Gulf to new land being formed daily by silt carried by rivers draining the vast Mackenzie basin. Of principal interest was the copper-bearing rocks about Bathurst Inlet, long known but first investigated in any detail by O'Neill. He found that they consisted of a succession of lava flows bearing immense amounts of copper in the aggregate, but of very low grade. His judgment, however, was optimistic: "When it is considered that copper values are found over the whole area of more than 1,000 square miles, it seems highly probable that there may be sufficient concentration in parts of the area to permit of economic mining."[21] The government, in the light of O'Neill's report and the inaccessibility of the region, proclaimed a reserve over the entire district in 1918, and closed it to staking till further notice.

The museum scientists were equally busy investigating the same large area on behalf of their particular sciences, Anderson studying animal and bird life, Johansen the vegetation, insect life, marine and freshwater fishes, and invertebrates. Jenness, on his own since Beuchat was carried away in the *Karluk*, studied the Eskimos of Cape Halkett, Alaska, in the winter of 1913-14, then accompanied the expedition east, where he spent two years among the Copper Eskimos of Coronation Gulf and southwest Victoria Island. He followed one band on its migration to Victoria Island in the summer of 1915, studying their language, habits, folklore, philosophy of life, making phonograph records of their songs, and

XV-13 R. M. ANDERSON

XV-14 DIAMOND JENNESS

gathering notes on religious performances and pastimes like string and shadow figures.

On November 17, 1914, as a result of the war, instructions went out from Ottawa ordering the southern party to discontinue its work in the autumn of 1915, but the news failed to arrive until after the close of navigation in 1915. So it was decided to carry on the work in the summer of 1916 and to leave during the short navigation season. In the meantime, the men occupied themselves preparing their field notes and completing their studies of the region. The men, their specimens, and a number of passengers for Herschel Island left Bernard Harbour on July 13, 1916, and reached Nome without any undue delay or difficulty on August 15. There they left the schooner to be sold and took steamship passage to Seattle and Victoria. Chipman, on the other hand, proceeded overland via Coppermine River, Great Bear Lake, and Mackenzie, Slave and Peace Rivers to the railway at Peace River town.

In Ottawa, they quickly threw themselves into the work of preparing reports and studying the specimens emanating from a region that had never been previously collected, some representing unrecorded genera and species. Plans were concerted to publish the results of the work of both parties in a series of volumes, half at the Survey's expense, half by the Department of the Naval Service. A five-man committee (including Anderson and J. M. Macoun of the Survey, and C. G. Hewitt, the Dominion Entomologist) was appointed to select the specialists who would prepare individual reports, and oversee publication. Eventually 73 persons contributed to the published report, which attained a total of fourteen volumes by 1946, when the project died with about 75 per cent of the work completed. Production of the anthropological reports was delayed by Jenness' enlistment in the Canadian Field Artillery, though after his return he published no fewer than five of the volumes, making him the major contributor in the series. Most of the scientists secured to write the biological and physical reports could spare little time from other important responsibilities, while Macoun and Hewitt, charged with the major responsibilities for the volumes on botany and insects, died in January and February 1920. Besides, the Department of the Naval Service was abolished in 1922, and the Department of Marine and Fisheries, which inherited its share of the responsibility, lacked the interest or drive of its predecessor. The Survey was not inclined to push the publications beyond the volume on geography and geology by O'Neill, Chipman, and Cox, published in 1924, in which it was primarily interested. The Museum had little or no means of its own to expedite publication of the anthropological and natural history volumes.

Besides, the animosity between Stefansson and the Survey personnel, notably Anderson, grew, because of what they considered distorted, even libelous accounts of the proceedings that Stefansson began broadcasting after his return, mainly in his book *The Friendly Arctic* which appeared in November 1921. When he arrived in Ottawa in late 1918 claiming as his right to supervise publication of the reports, the members of the Survey refused to cooperate with him. Stefansson was angered to learn that the scientists intended to report only on the materials and data secured in the Arctic and would thus ignore the work of the northern party almost completely. Anderson proposed that Volume 1 should be earmarked for a general introduction and narratives of both sections, written by their respective commanders. In the end, however, Stefansson preferred to present his role in the enterprise from his own point of view. Anderson, for his part, took the stand that he could not prepare a detailed narrative until all the results of the scientific work were complete, and that in the meantime the report published by the Department of the Naval Service in 1916-17 would suffice for a popular general account. So Volume 1 was never written; nor was Volume 2 on mammals and birds (assigned to Anderson and P. A. Taverner); nor were parts of other volumes.

* * *

EVEN WHILE THE FIGHTING was at its peak along the Western Front, Canadians began turning their thoughts to meeting the situation that would follow the war's ending and the soldiers' return. The nation, conditioned by the war, had developed new social and political attitudes. The war had demonstrated the power of organized action and the benefits that might come through rationalizing institutions and procedures. It had exalted the state as director of the national effort, and the public would not gladly accept a return to more passive laissez-faire attitudes of prewar days. It had introduced the concept of public planning, based on thorough research and application of scientific technology. The new outlook of the mining industry was indicated in part in D. H. McDougall's presidential address of March 8, 1920, before the Canadian Mining Institute:

> Canada is not a country where wealth is easily gained, but it is a country not yet fully known, not half-prospected where in the past sincere work has almost always reaped a satisfying reward. What we, and our children, will get out of Canada will be in exact measure to what we put into Canada in the way of brains and work. We have great national wealth, but none to waste. We have problems and limitations, but, if these are properly tackled we can lead the world in many things. Only, we must give up talking thoughtlessly of our "boundless" natural resources, and prepare, by fostering science and encouraging scientific workers, to get the best out of our country . . .[22]

The war had also demonstrated the vital importance of the mining industry in supplying the means of war, and had pointed to Canada's assuming a "proper place in the Empire as a producer of minerals."[23] the industry was determined not to let its newly-recognized importance be forgotten. Its spokesmen reminded the government that of all primary industries, mining had the greatest potential for rapid expansion and generating new wealth; it could discover new Cobalts and Flin Flons overnight that could pour hundreds of millions into the economy. New processes for treating difficult ores greatly enhanced the chances of developing successful mines. Returning veterans would need work and the opportunities that a rapidly expanding mining industry could provide. The new railways had thrown vast territories open to development, embracing parts of every province from New Brunswick to British Columbia and parts of the territories beyond. Minerals would provide welcome freight and assist struggling lines to recover their construction costs. The national debt, swelled by the war and the previous outlays on railway building, needed to be met by new tax sources such as mining could provide. Everything, in short, pointed to the desirability of pushing vigorously with the expansion of mining, based on a complete and accurate inventory of Canada's mineral resources.

As the war ended and providing for the veterans became a matter of immediate concern, thoughts began to be expressed that governments, in conjunction with the mining industry, should launch state-aided prospecting programs and undertake an accelerated geological surveying effort to point out potential mine fields. The Dominion government was urged to give the Geological Survey and the Department of Mines the means to undertake a much more ambitious geological mapping program that would proceed in advance of prospecting. Men, principally those with a background of outdoor living, should be trained at schools of mining and similar institutions, and be sent to the field as prospectors with mining engineers and state geologists directing their work. Survey officers should be empowered to take charge of the work of the teams of prospectors in their districts.

But the federal governments of the decade from 1911 to 1921 were dishearteningly indifferent to all such plans, as the record of the Mines ministry indicates. No fewer than ten men occupied the position during the period, most of them the least distinguished or able members of the cabinets. In 1917 alone, five men held the portfolio: P. E. Blondin (to January), E. L. Patenaude (January to June), A. Sevigny (June to August), A. Meighen (August to October), and M. Burrell (from October). Burrell's appointment was welcomed, even if he would have to divide his time with his other department, Secretary of State, for he had shown considerable organizing ability as Minister of Agriculture, while his representing a mining constituency seemed proof that he would take his Mines ministry seriously. But though Burrell spoke encouragingly to the Canadian Mining Institute (March 8, 1918) he pleaded preoccupation with winning the war as reason for postponing any new initiatives or reforms. A similar attitude, persisting through his administration and those of his successors Meighen (December 31, 1919-July 10, 1920) and Sir J. A. Lougheed (July 10, 1920-December 29, 1921), led the industry to renew its pleas for an effective Minister of Mines to the Prime Minister-elect, W. L. Mackenzie King.

Far from embarking on a grandiose program of expanded activity, the Survey faced a danger of having its functions threatened by some of the innumerable bureaus, boards, committees, and commissions that had mushroomed during the war as the government's response to various contingencies. Several of these studied mineral, forest, soil and water-power resources—fields assigned to the Survey under the act of 1907—and Survey personnel had worked with them on a temporary basis. But there was fear after the war that some of these might be given permanent status that could impinge on the work of the Survey. The Canadian Mining Institute's memorandum of February 8, 1918, to the Prime Minister listed some of these rivals and urged the government to consolidate its work connected with mining in the Department of Mines:

> All this dissemination of work and effort, this duplication of work, and misplacement of confidence, leads to large expense, useless extravagance and waste, and a destruction of the virtue, opportunity and effectiveness of a Department long in the field, well equipped and capable of doing the work that is being taken away from it. One can, of course, destroy the efficiency and reputation of any Department of the Government by robbing it of its duties, and having them performed by others supplied with moneys which should have been voted to the Department thus impaired. Every branch of mining, everything pertaining to minerals, all laboratory and technical work, and all research in respect to these matters, should be committed to the various divisions of the Department of Mines, each presided over by a qualified technologist.[24]

Some overlapping and duplication had always existed, which the Survey had recognized by tacitly abandoning serious work in the forest and waterpower resources fields while concentrating on geological and topographical ac-

XV-15 KIGIUNA, A COPPER ESKIMO, PHOTOGRAPHED FULL FACE AND IN PROFILE

The caption statement, "This man's eyes were dark brown suffused with a milky tinge, so that he represents one of the so-called light-eyed or blond Eskimos," challenges the sensationalized accounts of a race of Blond Eskimos in the Central Arctic, and obliquely, Stefansson on whose reports the accounts were based.

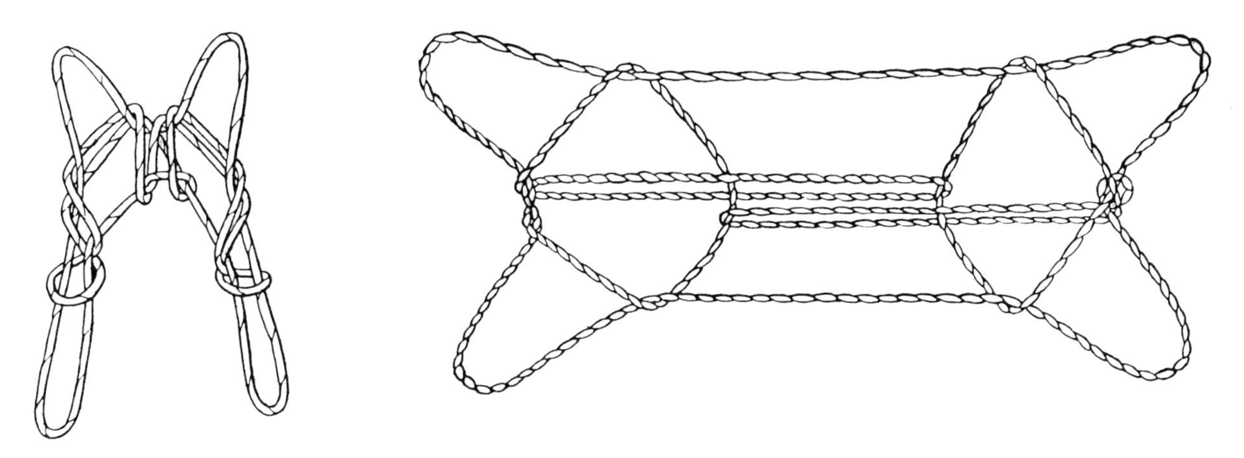

XV-16 TWO COPPER ESKIMO STRING FIGURES
Described and figured by D. Jenness. *Left:* "The Pair of Long Trousers." *Right:* "Two Men Hauling on a Sled" or "Two Men Having a Tug-of-War in a Dance House."

tivities as well as on the new sciences that had fallen to its lot by virtue of its function of managing the Victoria Memorial Museum. The act of 1907 had not even drawn a clear line between the Survey and the Mines Branch; the same memorandum referred to continuing difficulties in their relationship: "... the reason why it is believed that the division of the functions of the Department of Mines ... is erroneous, is that in practice it has led to overlapping, to divided authority, and to friction arising out of sectional jealousies."[25] It also referred to the duplication in the activities of the Natural Resources Intelligence Branch of the Department of the Interior, whose publications borrowed copiously from the work of the Survey.

In a different category was the Commission of Conservation, established in 1909 with power to investigate "all questions which may be brought to its notice relating to the conservation and better utilization of the natural resources of Canada, to make such inventories, collect and disseminate such information, conduct such investigations, inside and outside of Canada, and frame such recommendations as seem conducive to the accomplishment of that end."[26] The Commission of Conservation did some work in the mineral resources field, but it was mainly in areas of concern to the Mines Branch—such as studies of mining methods, mineral utilization and mine safety practices—rather than those of the Survey. An organization that impinged more directly was the Munitions Resources Commission, which grew out of an earlier Shell Committee and was appointed in November 1915 to conduct "an enquiry respecting the supply and sufficiency of raw materials in Canada required for the production of munitions of war, and the best means of conserving the same."[27] For three years it employed Survey and other personnel to make inventories and special studies of strategic minerals.

Both of these agencies fell by the wayside after the war. The Munitions Resources Commission reported in the spring of 1918, and did no further investigations, though in 1919 a similar Imperial Munitions Resources Bureau was formed to do comparable work on an empire-wide basis. The Commission of Conservation fell victim to pressures to reduce the welter of duplicating agencies in the name of economy, being discontinued by an adverse vote of the House of Commons in May 1921—to the regrets of many of its supporters.

A third wartime agency, destined to survive down to the present as the National Research Council, was the Honorary Advisory Council for Industrial and Scientific Research. It had been set up to conduct fundamental researches in the sciences and undertake investigations applicable to industry or to the better use of resources. An appointed body, it operated through specialized associate committees, for example, one on mining and metallurgy chaired by F. D. Adams and comprised of eight professors, three provincial mines officers, and thirteen mining men. These committees studied Canadian resources, industries and problems, and provided funds for specific projects of research. In September 1918 this council moved to transform itself, with government approval, into a permanent National Research Council to combat the lack of scientific research in Canada.

The National Research Council Bill successfully cleared the House, but, surprisingly, was defeated in the Senate, again in the name of economy. Eventually, however, in

1924 the National Research Council came into being. As early as 1921 the Honorary Advisory Council was offering scholarships for writers of theses on geological subjects, and its grants in the fifty years unquestionably have powerfully assisted the training of two generations of Canadian geologists.

One important institutional innovation of the period that did not meet with such opposition and has also survived in a flourishing state to the present was concerned with the area of statistical information. The war had demonstrated the need for accurate statistics in every field, and had made scientists conscious of the inadequacies of previous governmental efforts. The call for a uniform, impartial, national statistical organization was answered to almost everyone's satisfaction by the establishment of the Dominion Bureau of Statistics (now Statistics Canada) under Dr. R. H. Coats. When John McLeish finished his report on Canadian mineral production in 1920, a function that the Survey had initiated in the 1880's passed entirely out of the hands of the Department of Mines.

* * *

FAR MORE IMMEDIATE AND URGENT for the Survey than these potential rivals was the grave threat of disintegration from within, because of the wholesale desertion of its staff. While Canada was at war the Survey had to cope with lengthy absences of several important staff members, and found itself immensely poorer for the regretted deaths of LeRoy, Lawson, Cairnes, and Drysdale. It was almost impossible to secure full-time additions to staff, the only new appointments being E. L. Bruce in 1915, H. V. Ellsworth in 1918, and the biologist F. Johansen, who went on the payroll for two years after his return from the Arctic in 1916. Few men were proceeding to postgraduate degrees in geology or mining, and those who completed their doctoral dissertations were quickly snapped up by private companies, particularly if they had the cachet of having worked for the Survey. Three men, J. K. Knox, I. E. Stewart, and A. McLeod, who had worked as summer assistants, were lost in this way in 1917 to oil companies. Despite the hiring of outside specialists, the Survey was compelled to curtail its field operations for lack of qualified staff.

The close of the war did nothing to improve the situation; in fact, staff continued to be lost at a rapid rate through deaths, retirements, resignations, and transfers. The paleontologists Lambe and W. J. Wilson died on March 12, 1919, and August 24, 1920, respectively. James M. Macoun, the naturalist, died early in 1920 after a serious operation and a lengthy convalescence, soon to be followed by his venerable father, John Macoun, who died on July 18, 1920, at the age of 88, on Vancouver Island where he had been living for some years, still on the Survey's payroll. S. Herring, the taxidermist since 1884, died on March 12, 1919, and Marc Sauvelle, the head French translator, in March 1920. Resignations and retirements included J. A. Allan, S. E. Slipper, and J. Stansfield, geologists; G. F. Sternberg, vertebrate fossil collector and preparator; F. H. S. Knowles, physical anthropologist; and the Survey's accountant, John Marshall, who had accompanied the Survey to Ottawa in 1881 after serving since 1872 with the Logan estate. The major transfer was Keele—topographer, exploring geologist, and latterly, specialist on clay and shale—who joined the Mines Branch in 1916, but continued to work in the field with the Survey.

McConnell and McInnes found it almost impossible to fill these vacancies. It was not a problem of finance, for the annual vote rose from $555,000 in 1914-15 to $672,000 in 1920-21, and the Survey, as before, continued to return large sums at the end of each fiscal year, to an amount of $107,000 in 1920-21, the very year in which the staff problem was most acute. The difficulty was the lack of men with the requisite skills who might be recruited to work in the Survey, especially at salaries far below those offered by competing employers. In fact, there was serious danger that most of the present staff would also leave unless the salaries were drastically revised. Governmental salaries had failed to keep pace with the cost of living during the war and after. Taking the cost of living at the end of 1913 as 100.0, the index reached 158.9 at the end of 1918, 183.7 at the end of 1919, and in July 1920, at the height of the inflation, it stood at 199.9, almost exactly double the prewar level. But in the meantime salaries increased only by the usual $100 increments plus an occasional "bonus" when a man was promoted. Between 1914-15 and 1920-21 basic salaries rose in the order of only 30 per cent, and moreover, they bore an "abatement" of 5 per cent instituted in 1914 and continued into the early twenties. The men's living standards were thoroughly jeopardized by the failure of wages to keep pace with the rising living costs.

XV-17 ESKIMOS SPEARING SALMON DURING THEIR MIGRATION UP THE STREAMS, JUNE 1916

Even more were the wages out of line with their "market value," as the *Canadian Mining Journal* pointed out (February 6, 1920):

> The maximum salary of the senior members of the Geological Survey is $4,200. There are men in the survey with nearly forty years experience, and international reputations, who have reached this giddy pinnacle of affluence, and have had the pleasure of seeing many men get rich through their labours. There are men, and have been men, in the Geological Survey of Canada whose labours and deductions have added millions upon millions to our national wealth, and it has been their lot to sit by like cloistered monks vowed to penal poverty, seeking presumably,—otherwise their salaries are not explainable—to free themselves from breaches of the tenth Commandment.[28]

To increase the difficulties, many new employment opportunities opened for geologists and mining engineers in the period immediately following the war. Rich, powerful organizations, anxious to exploit the mineral resources of underdeveloped parts of the world, did their best to secure technical and scientific staff and raided government surveys without compunction. The offers began while the war was still on. Camsell, for example, received three offers by May 1918, one from a New York company that involved a five-year contract at $6,000 a year, plus a 10 per cent interest in properties taken up through his advice, plus freedom to do consulting work for about a quarter of the time. He reckoned that offer as worth about $7,500 a year overall, as against his Survey salary of $3,400. He wrote McConnell that he had asked the company to keep the offer open to the end of the year, and warned that many similar propositions were being circulated and that the Survey had better prepare by improving wages or face a real crisis when the war ended. Oil companies were in the fore, but metal mining companies had also come to recognize the value of employing qualified geologists to direct prospecting work and advise on problems of development. Other governmental agencies, provincial or state, were also expanding their services and were on the lookout for experienced, qualified men. Finally, the universities were expanding to meet postwar enrolments and needed geologists to staff their departments of geology, mineralogy, and mining.

An aggravating, unsettling element was added in 1918 when Parliament decided to undertake a reform of civil service classifications, appointments, and promotions, previously governed by the Civil Service Act of 1908. An American consulting firm, hired by the Civil Service Commission in August 1918, worked out a hasty reclassification and presented it to the government in June 1919. The classifications were immediately attacked for over-complexity (more than 1,700 individual classes of jobs were listed and rated), and for being especially unfair to technical and professional staff. Further, the suggested salaries did not take into account the high cost of living, though the plan was to allow for inflated costs by adding a bonus to every salary. Great discontent was felt everywhere in the civil service where professionally qualified staff were concerned—for instance in the Pure Food and Drug Division and the Chief Architect's office, as well as in the Survey—but the branches had no option but conform to the overall salary pattern. In the face of the storm of protests the report was withdrawn and a new one was presented that contained revisions suggested by the deputy ministers, which was then circulated about the departments. To many in the Survey, the circular of December 29, 1919, announcing the new salary schedules, was the last straw. A concerned McConnell wrote C. D. Walcott in Washington on April 9, 1920, about the dangers of the situation: "We are losing the greater part of our geological staff—thanks to the Civil Service Commission and their experts on classification, and our troubles are not over yet."[29]

The blandishments of the mining companies and the universities were felt most strongly by the junior and middle ranks of staff. A few promotions helped improve the morale of some men, but others were resentful at their own slow promotions and were easy prey to attractive offers. The recruiting agent of a single oil company, who visited Ottawa early in 1920, was reported to have persuaded six officers to join his organization to work in such places as Trinidad, Mexico, North Africa, and California. An exodus from the Survey began in 1920 that saw the resignations of W. J. Wright (March), L. Reinecke (April), L. D. Burling, B. R. MacKay, J. J. O'Neill, and B. Rose (May), E. L. Bruce (June), S. J. Schofield (December), J. S. Stewart (February 1921), A. O. Hayes (March 1921), M. Y. Williams (April 1921), as well as the topographers J. R. Cox (May 1920) and F. S. Falconer (September 1920). Most of the geologists took up work with private oil companies, particularly Whitehall Petroleum Company and Sinclair Oil Company, at minimum starting salaries of between $5,200 and $6,200, while three found positions in Canadian universities, Bruce at Queen's, Schofield and Williams at British Columbia.

The Civil Service Commission simply wrung its hands, saying it could never hope to compete with the very high salary levels offered by private companies and should not be expected to try. It pointed out that the United States Geological Survey had lost 17 per cent of its scientific staff in 1918, mainly to oil companies. Reinecke, who replied to its statement, argued that the new classification had not indicated any significant salary increase, and gave his opinion that most of the retiring staff would have continued with the Survey had they received as little as 25 per cent raises.

The departure of so many experienced officers was a serious loss which the *Ottawa Journal* stigmatized as "the worst blow ever received by the Geological Survey—the most powerful single instrument in the past development of the natural resources of Canada. Coming at the present crisis in the progress of the nation this is nothing short of a calamity."[30] It was pointed out that while most positions in the public service required only a minimum of training, and even responsible positions in the financial, statistical, or legal branches could be filled by many qualified persons, a geologist was trained "for special purposes, and for work in particular districts,"[31] so that even for one man to resign was a great loss to the country.

XV-18 A. O. HAYES

The resignation of Hayes, for example, was instanced as depriving the country of a man who had done good work revising the geology of the Saint John sheet, and highly original work on the Wabana iron ore deposit (Memoir 78, also his Ph.D. dissertation for Princeton) that represented an outstanding scientific achievement:

> His work in this connection is accounted worthy of note in every bibliography on iron-ores that has since been published. His original speculations on the part played in iron-ore deposition by low forms of animal life have received recognition in later works. The application of these speculations to the origin of sedimentary iron-ores may quite conceivably lead to the discovery of important new deposits, and may also be of great assistance in determining the extent over which known deposits may be expected to be found . . . .[32]

Since then Hayes had devoted three years studying the important Sydney coalfield, work that would have to be begun afresh by a new man. Still, Hayes had been forced to resign:

> Why? He is thirty-seven years of age, married, with ten years service in the Survey; the highest scholastic qualifications, an enviable scientific record and unique acquaintance with some of the neglected problems of the geology of the maritime provinces, and a salary of $2,400 per year: C'est pour rire.[33]

The resignation of M. Y. Williams, who had refused several offers from oil companies but was unable to resist Dean Brock's urging to become professor of paleontology at the University of British Columbia, was similarly deplored. His departure (which, fortunately, was not complete, for he continued to work summers for the Survey) deprived the Survey of an important authority on Silurian geology and a leading petroleum specialist, "who has taken a lively interest in the community life," and "was also President of the Professional Institute of the Civil service, to which organization he has given much time and energy."[34] The reference was to Williams' prominent role in establishing the institute in 1919–20, first as its secretary, and then as president, filling out the term of the first president, C. G. Hewitt, who died in February 1920.[35]

The immediate, powerful reaction to the resignations was a heartening indication of the esteem in which the Survey was regarded. The Canadian Mining Institute had been sufficiently concerned over the salary situation as early as March 1919 to address a resolution on the subject to the government. Still later, when the new classifications and salary scale were reported, its secretary quickly fired off a telegram of protest to the Prime Minister, and the council set up a committee to investigate the situation. When, early in 1920, the resignations began, the council despatched strong letters to the Minister of Mines and the Prime Minister suggesting a reclassification of the scientific staff of the department, as did the officers of several branches of the institute across the country. The *Bulletin* of the institute advised the government to make the gesture of raising salaries to a point where they would not be ludicrously out of line with those offered by private industry, and hinted that speedier promotions might help retain the younger men who otherwise "are forced out of the public service just as they are entering upon their period of maximum useful-

XV-19 THE SCIENTIFIC AND PROFESSIONAL STAFF OF THE SURVEY, FEBRUARY 1921
Front row, seated, left to right: H. V. Ellsworth (mineralogist), W. S. McCann (geologist), A. C. T. Sheppard, S. C. McLean, and K. G. Chipman (topographers). *Second row, seated, left to right:* W. A. Johnston (geologist), R. M. Anderson (zoologist), W. H. Boyd (topographer), E. M. Kindle (paleontologist), C. O. Senécal (draftsman), E. R. Faribault (geologist), P. Selwyn (secretary), W. McInnes (museum director), C. Camsell (deputy minister), W. H. Collins (director), D. B. Dowling (geologist), E. D. Ingall (mining engineer), R. A. A. Johnston (mineralogist), M. Y. Williams (paleontologist), J. A. Robert (draftsman), and W. Malcolm (geological compiler).
Third row, standing, left to right: M. E. Wilson (geologist), A. F. Clark (draftsman), E. Poitevin (mineralogist), E. E. Freeland, and A. G. Haultain (topographers), W. A. Bell (paleontologist), J. R. Marshall (geologist), F. H. McLearn (paleontologist), C. H. Freeman, R. Bartlett, and R. C. McDonald (topographers), D. Jenness (anthropologist), D. A. Nichols (topographer), A. Dickison (draftsman), L. L. Bolton (administrator), R. Harvie (geologist), C. H. Young (taxidermist).
Back row, left to right: F. Nicolas (editor), G. A. Young, T. L. Tanton, and V. Dolmage (geologists), A. P. Dowling (engineering clerk), E. J. Whittaker (paleontological collector), A. Anrep (peat specialist), M. F. Bancroft (geologist), P. A. Taverner (ornithologist), S. G. Alexander (draftsman), W. H. Miller (topographer), E. Sapir (linguist), M. F. Connor (metallurgist), and H. I. Smith (archeologist).

ness to the country and to science."[36] In the early months of 1920 the *Canadian Mining Journal* published a series of editorials attacking the government and the Civil Service Commission for failing to recognize the Survey's important contributions to Canadian development, and pointing to the utter unfairness of the present salaries.

The most impressive gesture of support came from British Columbia where the new Minister of Mines, William Sloan, introduced a resolution in the legislature calling on the Dominion government to rectify the situation:

> Be it resolved, that this Legislative Assembly of the Province of British Columbia expresses its appreciation of the great value of the work of the geological survey of Canada as assisting the mineral development of this province, and views with apprehension the depletion of the survey staff especially at this time, when, with the world entering upon a period of reconstruction, it is most important to our mining industry that all possible knowledge of British Columbia geology shall be made available.[37]

Besides expatiating on the Survey's important role in the development of his province, Sloan also raised the claim that the Dominion was contractually obliged to provide continuous and satisfactory geological services under the province's Terms of Union with Canada.

Faced with these and similar reactions across the country, the Dominion government ordered a further review of the salaries of the professional civil service. In April 1920 the Parliamentary committee examining the scientific expenditures of government was told by McConnell that the widespread resignations from both branches of the Mines Department had arisen because the staff had come to feel "that scientific men are placed at a disadvantage in the government service, and are taking the only course open, namely, getting out to private work."[38] Belatedly, efforts were made to rectify the salaries. Cost of living bonuses ranging from 5 to 40 per cent of salary according to the individual's salary level and dependants, that had already been introduced for the lowest grades, were extended to the upper grades in 1920-21. The new scale that went into effect April 8, 1920, offered good raises to most of the staff, and there were a few promotions besides. In addition, some men received retroactive pay increases back to April 1, 1920, and a few even to April 1, 1919.

The departures were made good by appointing a large number of men to the permanent staff. Some were returned veterans, but most were men who had served apprenticeships as field assistants as long ago as 1914. In the order of their official hiring the new officers were: F. H. McLearn (July 1, 1920), W. E. Cockfield, V. Dolmage, W. S. McCann, and T. L. Tanton (October 1, 1920), W. A. Bell (December 4, 1920), F. J. Alcock (April 15, 1921), G. Hanson and G. S. Hume (May 1, 1921), and J. R. Marshall (June 21, 1921). At the same time the outward movement was arrested because the persons most susceptible to such offers had departed, because some of the inequities had been rectified, and because after years of uncertainty the salaries had finally been stabilized. Besides, the collapse of the oil boom and the general economic recession that set in during the autumn of 1920 reduced offers of outside employment, while general declines in wages and in the cost of living made current salaries more attractive than previously.

When the crisis broke in the spring of 1920 McConnell, who was 63, felt very acutely the resignations and the threat to effective operation of the department that they posed. Probably this was what moved him to apply for superannuation at the end of May, 1920. His letter to the minister, Arthur Meighen, contained an urgent request for an interview: "Before acting on this, I hope you will give me an opportunity of discussing with you the present unsatisfactory conditions of the Department, and the causes that have led to it."[39] He was able to claim nine months' terminal leave on the grounds that in his forty years service he had received only two weeks sick leave and a single study leave, in February-March 1911, to visit Vesuvius and other active volcanoes in the Mediterranean. His retirement went into effect on March 7, 1921, at the expiry of a leave beginning on June 7, 1920. Four days after McConnell left, Charles Camsell was appointed acting deputy minister, and by June 18, 1920, he was in Ottawa to begin making arrangements for his succession and experiencing the first duties of his new office in the form of addressing a luncheon meeting of the American Institute of Chemical Engineers in the Chateau Laurier hotel.

Since he had taken charge of the Vancouver office in the spring of 1918 Camsell had been managing three of the seven regional divisions, though the duties were not especially onerous in view of the shortage of staff. However, he completely shocked McConnell by one request to absent himself for six or eight months during 1919 in order to explore the Lena River district of Siberia in the company of a Russian who was supposed to have the support of the British Foreign Office. Instead of going to war-torn Siberia Camsell spent a busy, productive year in charge of affairs related to his British Columbia post, assisted in the winter of 1918-19 by Dolmage. On Camsell's transfer to Ottawa, J. D. MacKenzie assumed the direction of the Vancouver office, to be followed by Dolmage after MacKenzie's death in 1922.

The new deputy minister was already celebrated for his interest in northern exploration that befitted a "son of the north" who had been born at Fort Liard when the Mackenzie waterway still echoed to the paddles and the

XV-20 L. L. BOLTON

songs of voyageurs, and who had himself played a prominent role exploring the Mackenzie basin from Lake Athabasca and Peace River to Peel River and the delta. In token of this record, as well as by reason of his position, on April 21, 1921, he was appointed a member of the Council of the Northwest Territories, reconstituted after sixteen years to administer the new land that seemed on the threshold of modern development by virtue of the recent petroleum discovery in the Fort Norman district. In 1919, in collaboration with Wyatt Malcolm, Camsell had published a popular memoir, *The Mackenzie River Basin*, and in 1920 he was echoing Dawson's lament at the unwillingness of Canadians to complete the exploration of their country. He saw the introduction of the airplane as greatly facilitating this work, since almost all the 1,000,000 square miles of unexplored territory could be reached by airplanes using the lakes and rivers as landing sites. Though the passage on the Survey's intended northern exploration program occurred in the Survey director's part of the 1921-22 report of the Department of Mines, there can be no doubt that it was Camsell's interest that shone through:

> The exploration of northern Canada is a special function of the Geological Survey .... It has, therefore, been the inalienable and important function of the Geological Survey of Canada, since its foundation, in 1842, as the scientific assistant to the prospector in his search for mineral wealth, to accompany him in his pioneer work and to be the chief Government organization engaged in the exploration of the vast unagricultural parts of Canada.
>
> Between 1915 and 1920, when the internal energy of the country was turned to production rather than to new developments, Geological Survey field operations became largely restricted to problems of mineral production in districts within operating reach of transportation. Exploration work was almost suspended. Now, with a return to normal conditions, public attention is being directed once more to the search for mineral deposits, and the Geological Survey is also resuming exploration work.[40]

One of Camsell's earliest duties was to find a director for the Survey in place of McInnes who accepted a new position created for him, first director of the Victoria Memorial Museum and editor-in-chief for the Department of Mines. He was succeeded at the Survey by W. H. Collins, whose appointment took effect on December 1, 1920. Also retiring on December 15, 1920, was the aged Dr. E. Haanel, director of the Mines Branch, who after a considerable interval was succeeded on October 1, 1921, by John McLeish, once the statistician of the Survey's Section of Mines. Entering, too, into a position of prominence was L. L. Bolton, a mining engineer with the Mines Branch who had transferred to the Survey in 1916 as an administrative officer. He was designated secretary to the Department of Mines, and on April 1, 1925, assumed the new position of assistant deputy minister of the Department of Mines, ranking ahead of both directors, and next only to Camsell himself. With a new team in command and a new Liberal government of W. L. Mackenzie King in power in Ottawa, a stable work situation, a largely new, young staff, and a renewed purpose of serving the nation in time of peace, the Survey embarked upon the interwar decades of the twenties and thirties.

XVI-1 CHARLES CAMSELL
Deputy Minister of Mines, 1920-36; Deputy Minister of Mines and Resources, 1936-49

16

The Twenties – Operating in a Decade of Prosperity

GRADUALLY CANADA RECOVERED from the postwar recession, and the middle and late twenties brought a revival of prosperity that reached feverish heights by 1928. For the mining industries the twenties were a period of great economic growth, thanks to the rising standard of living and the remarkable increase of mass-produced consumer goods, notably in the automotive and electrical appliance fields, and especially in the nearby United States. These created huge domestic and foreign markets for many of Canada's minerals—copper, lead and zinc, nickel and iron, silver and gold, and refined aluminum—and in Canada, for foreign supplies of oil, coal and iron ore.

The Canadian mining industry was in an excellent position to take advantage of these opportunities. Just before the war large tracts of country had been opened up for development by the round of railway building that penetrated and traversed new parts of the Canadian Shield and Cordillera. Cheap rail transportation made it profitable to develop many mineral deposits that formerly had been too remote or expensive to exploit. The new railways also served as an advanced base from which to launch a wave of prospecting activity still farther afield. Newly-developed transportation media—the automobile, outboard motor, and especially the airplane—had now become available to extend the range of prospecting and developmental activities. The northern parts of the provinces and even the Hudson Bay and Mackenzie basin sectors of the Northwest Territories became accessible as never before. A vast treasure-house of mineral riches had been unlocked from which Canadian and world needs could be met.

Foreign capital, mainly from the United States, flowed into the country in great amounts to exploit its wealth of minerals. Large-scale mining enterprises, such as Noranda or Flin Flon, began to take shape. These entailed multi-million dollar investments on mining, concentrating, smelting, refining, and other facilities, which meant that the properties would have to remain in operation to recover initial expenditures. Nationalist pressures within Canada accounted for the setting up of domestic refining facilities for metals like nickel and copper, and the emergence of integrated mineral-using industries. Altogether, the mining economy of Canada began to assume a more permanent, stable aspect, and the country began to lose its former character of being mainly a handy source for high-grade ores and concentrates that could be called into production or halted at the whim of foreign interests.

The mining industries had also reached a higher lèvel of technology. Astonishing breakthroughs had been achieved in the geological, mining, and metallurgical fields, and rule-of-thumb methods were no longer acceptable. In every phase of the work the dependence was now mostly on university-trained specialists, and there was a heightened appreciation of the place of science in economic development. With control passing into the hands of a generation of professionally-trained persons, the work of the Survey became appreciated by industry as never before. The reports, once criticized as too technical and esoteric for practical use, were now being used as they were meant to be, by people who understood and could benefit from them.

On the political side, however, the period was one of hesitancy and uncertainty. Federal governments had only a precarious hold on office and could not undertake dar-

XVI-2 BUILDINGS OF THE HUDSON BAY MINING AND SMELTING COMPANY
Flin Flon, Manitoba

ing or decisive initiatives. Concern over the national debt, as well as persisting regional and linguistic divisions within the country, kept the governments of W. L. Mackenzie King from espousing a role of active leadership, and restricted them to narrow, short-sighted, immediately practical objectives. Meanwhile the provinces, which were prospering mightily from the development of their natural resources, eagerly stepped into the breach, and assisted the mining industry by highway construction and scientific work. Provincial mines branches and research councils became increasingly active, while the universities began sponsoring research projects as part of their graduate programs in the earth sciences. The Survey's previously dominant position in geological work in Canada was being rapidly overcome.

For the Survey the twenties were a time of difficulties and eclipse. In place of its previously ample financial position and freedom from control it became subject to the government's caution, preoccupation with economizing, and acute sensitivity to shifting public pressures. The departmental management set out to terminate the quasi-independent status the Survey and Mines Branch had enjoyed since 1907 and to impose an overriding program that reflected the government's desire to build up the mining and manufacturing sides of the industry. While the work of the Mines Branch was stressed and it was encouraged to undertake new activities by being given the facilities for the purpose, the Survey was kept from utilizing the new opportunities that lay before it, and from fulfilling its responsibilities to an expanding mining industry. Whereas the Survey's expenditures in 1920-21 were double those of the Mines Branch, the two branches had come almost level by 1929-30, while in the thirties the Mines Branch went far ahead.

The new deputy minister, the former Survey officer Charles Camsell, quickly fell in line with this program, and over the next generation he put it into effect. Where initially he had been eager to aid the industry to expand into northern Canada with the assistance of the airplane, and had spoken of the historic role of mining as the great colonizing force in Canada, soon he threw himself into the government's efforts to build up a powerful mining industry by attracting domestic and foreign capital to exploit Canada's mineral resources. But while he followed the lead of his political superiors and strove to attract foreign investment to the mining industry, he was also to a degree an economic nationalist:

> Our mineral trade with foreign countries, however, is not as satisfactory as it should be. In raw minerals we are importing about $70,000,000 worth more than we export, though in partly manufactured minerals such as refined metals we have a favourable trade balance of about $20,000,000. It is in the fully manufactured materials of mineral origin that our condition is most unsatisfactory, the balance being about $150,000,000 worth against us annually. Heavy importations of coal, oil, and iron ore are the cause of our unfavourable position with respect to the raw materials, and in manufactured materials our condition emphasizes the necessity of establishing such industries as will take the products of our mines and convert them into finished articles, at least for our own consumption, rather than into exporting the raw materials for manufacture outside the country and later importation in the finished state.[1]

As deputy minister he was concerned to win the greatest possible support for the Department of Mines, and his reports were full of requests for additional money to enhance the contributions the department could make to the national economy. Good public relations were important, so he took care to increase the reporting of departmental activities in the press, attended conferences abroad where he made many contacts with important leaders in business and government, travelled widely around Canada visiting mining camps, talking with company officers, and dealing with officials of the provincial and federal governments and with politicians of all parties. He had an excellent personality for this kind of work, being able to get along with all sorts of men, and he became "particularly respected for the quiet, straightforward, sympathetic, and friendly dignity that he showed to all, whatever their station in life."[2]

Camsell's significance for the Survey lay in the way he emphasized and built up the activities and functions of the Mines Branch and thereby changed the relative positions of the component parts of the Department of Mines. Since the department was established in 1907 there had been comparatively little contact between the two branches, and very little oversight on the part of the successive deputy ministers. Now the department was administered by a man who desired to make the department function as a unit and to co-ordinate the work of the two branches. Almost from the outset Camsell put greater emphasis on the work of the Mines Branch and on its need for additional equipment and staff. Behind this lay his feeling that the Mines Branch was in the best position to render the greatest service, given the present need to develop efficient mining, refining, and manufacturing technology capable of competing on the world stage.

XVI-3 WILLIAM HENRY COLLINS
Director, 1920-36

Perhaps the Survey contributed to this emphasis by failing to adopt any bold or dramatic programs that would impress the politicians or the public. Camsell claimed he could not always obtain the kind of information that would lend itself to favourable publicity for the Survey, to the point that on one occasion he even took the matter before the staff in an open meeting. The trouble in part was that the character of the work did not lend itself to dramatic flourishes; but in part it was the inability or unwillingness of the director, W. H. Collins, to co-operate with Camsell to advertise the Survey. The cautious Collins was incapable of gambling that undertaking some type of activity that Camsell approved, such as a drive to promote northern work, would be reciprocated with greater grants for the regular work.

The failure to bring the Survey forward to the same degree as the Mines Branch weakened the morale and undermined the confidence of the staff, who tended to see the difficulty almost entirely in personal terms—in Camsell, who had forgotten his Survey background and had found new friends in the Mines Branch, or was only concerned with serving his own career; or in Collins, who was too stubborn to co-operate with Camsell for the good of the Survey. These may have been partly correct, but it was also understandable for a deputy minister to consider that under the conditions of the times the services of the Mines Branch needed to be greatly expanded—and expanded more than those of the Survey.

William Henry Collins, on whom fell the burden of keeping the Survey relevant to changing times and in a healthy state, was a first-rate geologist, whose important contributions to the geology of the Precambrian in northern Ontario were fittingly recognized by his being chosen president of the Geological Society of America in 1934. He saw the Survey as his life work, and his goal was to keep the Survey at the peak of excellence. He was proud of the institution and his men, glorying in their achievements. As T. T. Quirke wrote of Collins:

> It is hard to think how or when a national service could have a more devoted and enthusiastic officer. His pride in the organization of which he was a part was a very fine and sincere influence in his whole career. He gloried in the achievements and recognition of any former or contemporary officers of the Geological Survey of Canada. The "Survey" he saw as a perpetuation of all the efforts of all its members, not for past or present generations alone, but for the life of the nation. He recognized that the work of an individual, the contributions of one generation of taxpayers, form only transitory and minor parts in the structure of a national service. Without these contributions, however, the work could not go forward, and as the one man for whom he was himself responsible he gave his best all the time to advance the interests of the service

> and the country he served. Those close to him during the years of war know how deeply and truly patriotic he was. All his professional life it was his ambition to make the Geological Survey of the greatest value to his country. To him his official capacity was not a personal career; it was a public service. Even when he was elected President of the Geological Society of America, his main gratification was in the thought that it was a graceful compliment to the Geological Survey of Canada, of which he was Director.[3]

He was very kindly in his personal life, disposed to help others, and was genuinely interested in the well being of his staff, for example taking the trouble to arrange afternoon outings and picnics for the wives of Survey personnel staying in Ottawa in the summer. An effort also was made to invite every employee to the Collins home at least once a year, and there were regular bridge meetings in the winters.

A man of very high standards, he would not tolerate shoddy work or incompetence:

> As an administrator Dr. Collins lacked certain qualities that make for success, but he possessed others in a superlative degree. Though meagre in praise he was always fair, weighing each problem with judicial mind according to its merits. He made no snap judgments, but decisions reached were held with great tenacity. Recommendations for appointments and promotions were based on accomplishments and merit. No man was ever more conscientious in the administration of trust funds than he in the handling of the public purse, and few administrators have accomplished so much with so little.[4]

Efficiency and economy were his main concerns as director. He tried to operate the Survey in a businesslike fashion and get as much value for the public's money as possible. In his reports he often boasted how, with no more money than in the previous year, or with even less money, he had been able to increase the number of field parties or get other worthwhile results. He used the same argument of economy as a reason for keeping control of the Museum in his own hands as acting director. The Mineralogy Division complained at his unwillingness to buy necessary laboratory equipment, and the field staff at his reluctance to employ aircraft to assist in the work. He got a reputation for extreme parsimoniousness, not to speak of pettiness, for such actions as rationing the pencils and erasers used in the library or by the staff. In field work the same concern to secure the maximum value per dollar led him to oppose exploratory or Arctic work until compelled to do so.

A very brilliant if narrow man, Collins was always inclined to go directly to his objective without digression or embellishment. He was deliberate in his actions, taking time to think out his ideas and sentences. With the plans for field work he made up his own mind on a comprehensive program after discussions with the officers individually, then tried to implement the plan in its entirety, though sometimes he was overruled in parts by the deputy minister or by political pressures.

As director he considered it his duty to do the best—according to his judgment—with the means that were available. But expanding or redirecting the work of the Survey was something he left completely for his superiors. In his report for 1927-28, for example, he argued he had gone as far as he could with the available means; if anything more was wanted from the Survey more money would have to be provided. It did not occur to him to begin a new activity and use that as an argument to secure more money as Brock had done so successfully. Straightforward and aboveboard, he contented himself with presenting his cases in the most concise, reasonable way he could, and if such statements failed to convince he had nothing further to fall back on. Being of a shy, "retiring, somewhat uncommunicative" nature[5] and preoccupied with his own research, he gave almost no attention to publicity or public relations, and would not cultivate Camsell or the government. His relations with Camsell were extremely formal, limited often to memoranda rather than direct, personal approaches and meetings that he, as the subordinate, ought to have sought. Perhaps the difficulty in their relationship was one of temperament and style, Camsell not liking Collins' economizing and his caution about undertaking new programs or activities.

XVI-4 CHARLES STEWART, M.P.
Minister of Mines, 1921-30

XVI-5 SOME OF THE NEW FIELD OFFICERS, 1920-30
Top row, left to right: George Hanson, G. S. Hume, J. B. Mawdsley.
Bottom row, left to right: C. H. Stockwell, H. S. Bostock, T. L. Tanton.

Collins would not give himself completely to his administrative duties. He regarded his Precambrian studies as his prime contribution as well as his personal gratification, so he worked in the field every summer as long as he could, and continued with research and writing to the very end of his life. He was out of touch with Survey matters during the summers when he might have been out meeting people in the industry or visiting mines around the country. Because of his retiring nature he failed to get the views of other people very readily, and new ideas were slow to reach him. His preoccupation with research and his ill health meant that an assistant was needed to look after the Survey on a day-to-day basis. This was done with the appointment on November 1, 1924 of G. A. Young as chief geologist. But Young was a very difficult person to get along with, and as Collins did not see eye to eye with Young, the divided authority was an added source of difficulties.

All in all, Collins must be judged an unfortunate choice for the directorship, notwithstanding his great talents and many virtues, some of which inhibited his being a successful director. His failures were due primarily to his over-conscientious nature that led him to interfere unduly in the affairs of his subordinates, his lack of interest and skill in political relationships, and his narrow, unimaginative concept of the role of the Survey. He lacked the optimism and vision to carry the Survey forward to the opportunities of the years of high prosperity and economic expansion, an age of rapid technological changes exemplified by the radio, automobile, motorboat and airplane in communications, and such scientific advances as geophysics and radioactivity. Collins' failure to make effective use of these opportunities is the real measure of his shortcomings as director.

* * *

MANY STAFF CHANGES OCCURRED under Collins, but the high quality of earlier years was maintained by appointing men who had gained the Ph.D. and usually had worked for the Survey as assistants in previous summers. The difficulties that had prevailed at the start of the decade were eased by the improved salary scale, the reclassification of employees, and the granting of permanent status to thirteen men, thus making them eligible for salary increases. One continuing difficulty arising from the reclassification, however, was its putting of the draftsmen on such a low salary level that men resigned and the organization had trouble with its maps until adequate staff was recruited, first by importing trained men from Britain, then after 1927, by training many of its own personnel under the apprenticeship system.

Following the rash of resignations after the First World War when more than a dozen men left for positions with oil companies and universities, the Survey was able to turn to several men trained by its staff who were now ready to accept professional positions. The newcomers included persons of the calibre of F. J. Alcock, W. A. Bell, W. E. Cockfield, H. C. Cooke, V. Dolmage, G. Hanson, G. S. Hume, J. D. MacKenzie, F. H. McLearn, and T. L. Tanton. Through the twenties other geologists joined the staff to replace those who retired or died, and to expand the staff slightly. Among these were C. E. Cairnes (May 29, 1922), and B. R. MacKay (October 11, 1923), who was certainly no newcomer. After resigning in 1920, MacKay had spent three years in India and Burma working for an oil company, but with the expiry of his contract he returned to Canada and was invited back to the Survey. Others added were F. A. Kerr and J. F. Walker in 1924, J. B. Mawdsley in 1925, C. H. Stockwell in 1927, H. C. Gunning in 1928, G. W. H. Norman and H. S. Bostock in 1929, and in 1930—the last newcomers for some time—D. F. Kidd, A. H. Lang, and R. T. D. Wickenden.

After the salary situation became stabilized there were few departures from the staff, and those were mainly through superannuation or death. J. D. MacKenzie died late in 1922 as a result of his severe war wounds, J. Keele of throat cancer in 1923, E. J. Whittaker in 1924, and the venerable D. B. Dowling in 1925. R. A. A. Johnston retired in 1922 and was succeeded as head of the Mineralogical Division by E. Poitevin. Percy Selwyn, after serving since 1895 as secretary to five directors, retired in 1924, as did the geologist R. Harvie. E. D. Ingall, the head of the Borings Division, retired in 1928, while the end of 1929 saw the resignations of two young, valued geologists—Dolmage and Mawdsley. Dolmage, who was in charge of the Vancouver office for most of the twenties, left to go into private consulting business on October 3, 1929, practically on the eve of the Wall Street Stock Exchange crash. He faced some difficult years until the gold boom of the thirties got under way. Mawdsley, who had done most of his work on the Precambrian of Quebec, was more immediately fortunate in that he left to inaugurate the Department of Geology at the University of Saskatchewan and continued there until his death in 1964. He frequently worked for the Survey in after years and was a good friend in its dealings with the university and the Saskatchewan government.

The work of the Survey was strengthened also by the contributions of leading scientists from the Canadian universities and of promising graduate students. M. Y. Williams, who resigned in 1921, worked almost every summer thereafter as a stratigraphic paleontologist—in 1921 and 1922 in the Fort Nelson and Mackenzie River districts, then for four seasons, 1923-26, in southwestern Alberta where he helped bring in the Coutts-Sweetgrass oil field. He continued on various projects down to the 1940's, particularly at Manitoulin Island and in the Alaska Highway district. Others of his colleagues at the University of British Columbia who worked for the Survey in the twenties were R. W. Brock, W. L. Uglow, S. J. Schofield, and H. V. Warren. P. S. Warren of the University of Alberta, S. R. Kirk and J. S. DeLury of the University of Manitoba were also in this category. An American specialist often employed by the Survey was T. T. Quirke, of the University of Illinois, who worked on the Precambrian north of Lake Huron from 1919 and collaborated with Collins on a number of studies. A problem with employing academics, however, was the difficulty in securing their reports, owing to their other commitments.

Another facet of the summer program from the mid-twenties was the use of graduate students, not as assistants but on semi-independent field work under the direction of more senior men who were familiar with the areas being investigated. For example, in 1925 the staff at work in the Cordillera was augmented by C. H. Stockwell, R. H. B. Jones, H. T. James, and C. S. Evans, who worked under the direction of Cockfield, Hanson, Dolmage, and J. F. Walker respectively. In the same category were the numbers of graduate students completing research for their doctoral dissertations each year, some assisted now by National Research Council grants. The Survey either accepted their projects and assigned the men to some appropriate area under a specified geologist, or else it secured graduate students to undertake studies of areas or problems it wanted investigated.

The students and the geology departments appreciated the opportunities for training and paid research, and relations between the Survey and the universities were strengthened. It secured valuable connections with promising young geologists and a source of cheap labour, men who had strong personal incentives to do the best job they could, and who brought to the study their own talents and ingenuity plus the specialized knowledge of their academic supervisors. A good example was the report of C. O. Swanson, "The Genesis of the Texada Island Magnetite Deposits," a detailed study the findings of which could be applied to most iron deposits throughout the Cordillera and to similar deposits around the world. Swanson, a graduate student of the University of Wisconsin under C. K. Leith, assisted a party that worked in the district in 1923. Through such means the Survey at little cost was able to place a number of parties in the field each year to do important pieces of specialized research. Unfortunately, all too often, when the men had graduated and joined the Survey full time they were set to mapping and studying regular field areas, leaving their specialized knowledge largely behind them as a spare-time interest–to the long-term detriment of the Survey as well as of the individual.

In his direction of the Survey's work Collins, as a 'geologist's geologist' laid his greatest stress on the traditional aspects of field work–areal mapping–and tended to hold back those divisions that did not seem to contribute directly to this objective. This was most evident in his relations with the paleontologists, which Collins and Young both regarded as something of a frill. "Damn it! A good field man can work out his succession in the field without worrying about shells!!"[6] was Young's retort to a Cordilleran geologist's complaint about the difficulty in getting satisfactory reports on fossils to guide him in solving the complicated geology of that region. The Palaeontological Division was kept understaffed, underpaid, and the talents of the members were used for the purposes of the field mapping program.

Collins' ideal in his field work programs was comparable with that of Dawson–to promote the mining industry while advancing scientific studies. Note this explanation for the work done in 1922:

> FIELD OPERATIONS
>
> Thirty-nine parties carried on field work in various parts of Canada during the summer of 1922. In large part this work is intended to assist prospectors and explorers in the discovery and development of the mineral resources of the country by mapping the rock formations and investigating and describing the mineral-bearing possibilities of each formation. Increasing attention also is being devoted to investigating and compiling inventories of Canada's resources of coal, iron ore, petroleum, and other minerals. The topographical parties, while engaged in preparing base maps for geological use, mainly in areas containing mineral deposits, are preparing these maps on scales and according to standards which make them contributory to the systematic primary mapping of the country.
>
> Concurrently with these operations that have an immediate economic purpose in sight, an opportunity is afforded to most parties to carry on more purely scientific studies, the results of which each year help to transform the exploration of our mineral resources from a haphazard search to one guided by increasingly well understood principles. It has become recognized, for example, that the great series of granitic intrusions, known as the Coast batholith, that extends the length of British Columbia, was the source of a large share of the metalliferous deposits in the province and that this relationship should be a governing consideration in directing prospecting operations. Opportunities are also afforded to prosecute studies in mineralogy, palaeontology, and the other related sciences, and to collect mineral specimens, fossil remains, and other materials for the Victoria Memorial Museum.[7]

Like Dawson, Collins did not support reconnaissance trips into remote areas, for financial as well as scientific reasons; also because he regarded such regions as too unlikely to attract the interest of prospectors and miners. In fact, the airplane was rapidly bringing such areas within the purview of the industry; by the end of the twenties the Survey was being quite outdistanced by the new-style prospecting and Collins was making belated efforts to catch up.

Instead, Collins' policy was to concentrate on specific map-areas, especially where ore deposits were known or might be expected to occur, and where solutions to some vexing geological problems might be found. The emphasis was on helping outline districts of potential mineralization as an aid to prospecting, or on a more localized level, working out structures as an aid to mining development.

The pragmatic emphasis was carried further under Collins when he assigned certain geologists to report on the known abundance and future possibilities of certain minerals as contributions to national mineral inventories. These reports on talc, arsenic, oil and gas in western Canada, fluorspar, and lead and zinc, were published as volumes in a new Economic Geology Series. In addition, officers collaborated on two major general publications, *Prospecting in Canada*, and *Geology and Economic Minerals of Canada* for the benefit of prospectors, miners, and others. Perhaps Collins had a further motive in this sort of work—of upholding the Survey's claim to share in this activity in the face of similar efforts in the same direction by the Mines Branch. No attempt was made to work out a system of joint-authorship, but the oversight of the department, through the deputy minister and editor-in-chief, was used to keep duplication within bounds:

> Owing to the interlocking functions of the Mines Branch and the Geological Survey and the close relationship of the work of these two branches there is occasional danger of overlapping, but this is overcome by more frequent consultation between the heads and a desire on both sides for closer co-operation, which has resulted in increased efficiency without increased expenditure.[8]

The attention to working on subjects and areas that were of present or prospective importance by no means precluded its being done to highest scientific standards and seeking to advance general knowledge. Collins' own work—as well as that of Quirke—on the Precambrian north of Lake Huron, or M. E. Wilson's specialized work in the Precambrian, were fully in that tradition. Moreover, in all the work the emphasis was advancing from simple description to inference and deduction, and the published work often emphasized the theoretical geology as well as the economic implications. A report on a mining camp, for example, would dwell on the problems of the genesis of the ore minerals, and on the chemical or physical factors that concentrated them into workable deposits—information that made it possible to predict the size and richness of the orebodies under investigation, and to indicate likely places to search for others. The range of the inquiries during the twenties may be seen in the contributions to the Memoir Series, then perhaps in its heyday. There were memoirs on the structural geology of the Rocky Mountain foothills, on the origin and accumulation of gas and oil, on salt deposits, on surficial geology, and on glaciation. There was even one memoir, *Studies in Geophysical Methods* (No. 165, 1931), to record the Survey's involvement in 1928 and 1929 with these new techniques for obtaining information about underlying rock formations. The effort was limited to cooperating in companies' tests of their methods on areas containing faults and orebodies, in order "to afford to the public an impartial account of this new, promising method of prospecting."[9] The tests indicated that the methods needed to be improved greatly if they were to serve the needs of regional geologists. It would be another twenty years before the Survey would make regular use of geophysical methods and contribute to their further development.

In spite of the arrival of the airplane and its development in the twenties into a promising scientific tool, the actual conduct of field work in the decade was little advanced over Logan's eighty years earlier. Getting to and from the field area was now much easier, it was true, for trains ran where Selwyn once had travelled by Red River cart and Dawson by pack train. But once the geologist arrived at or near his assigned map-area things were little improved from the early days. Most of the

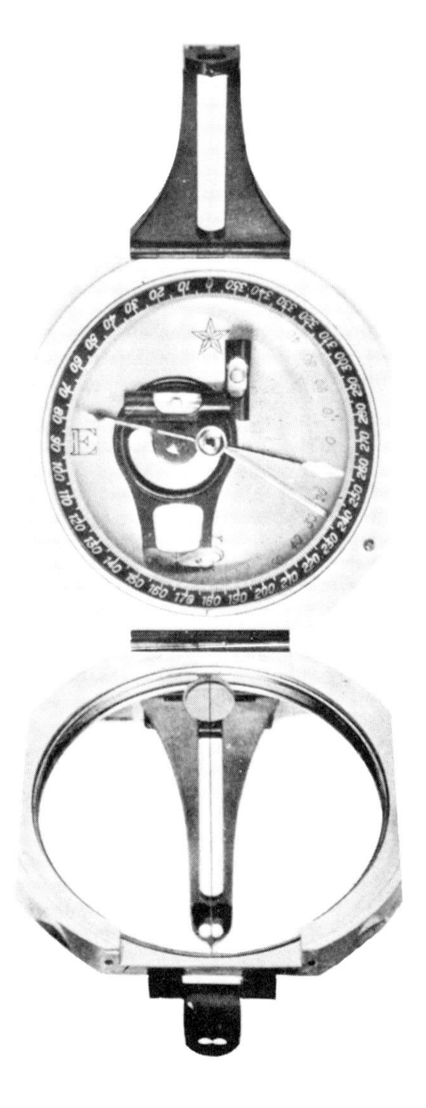

XVI-6 BRUNTON POCKET TRANSIT

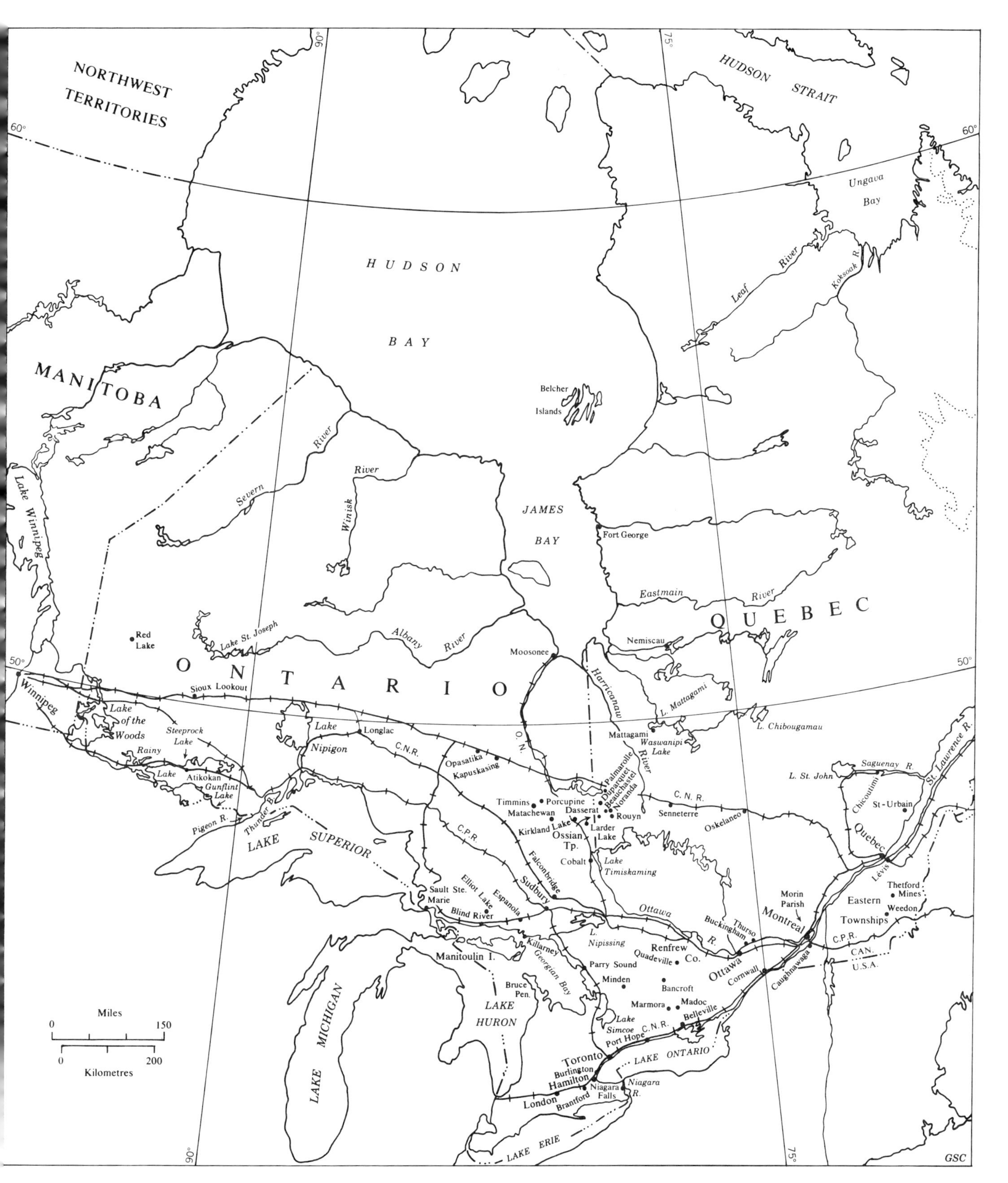

XVI-7 CENTRAL CANADA AFTER 1920

XVI-8 THE WESTERN PROVINCES AFTER 1920

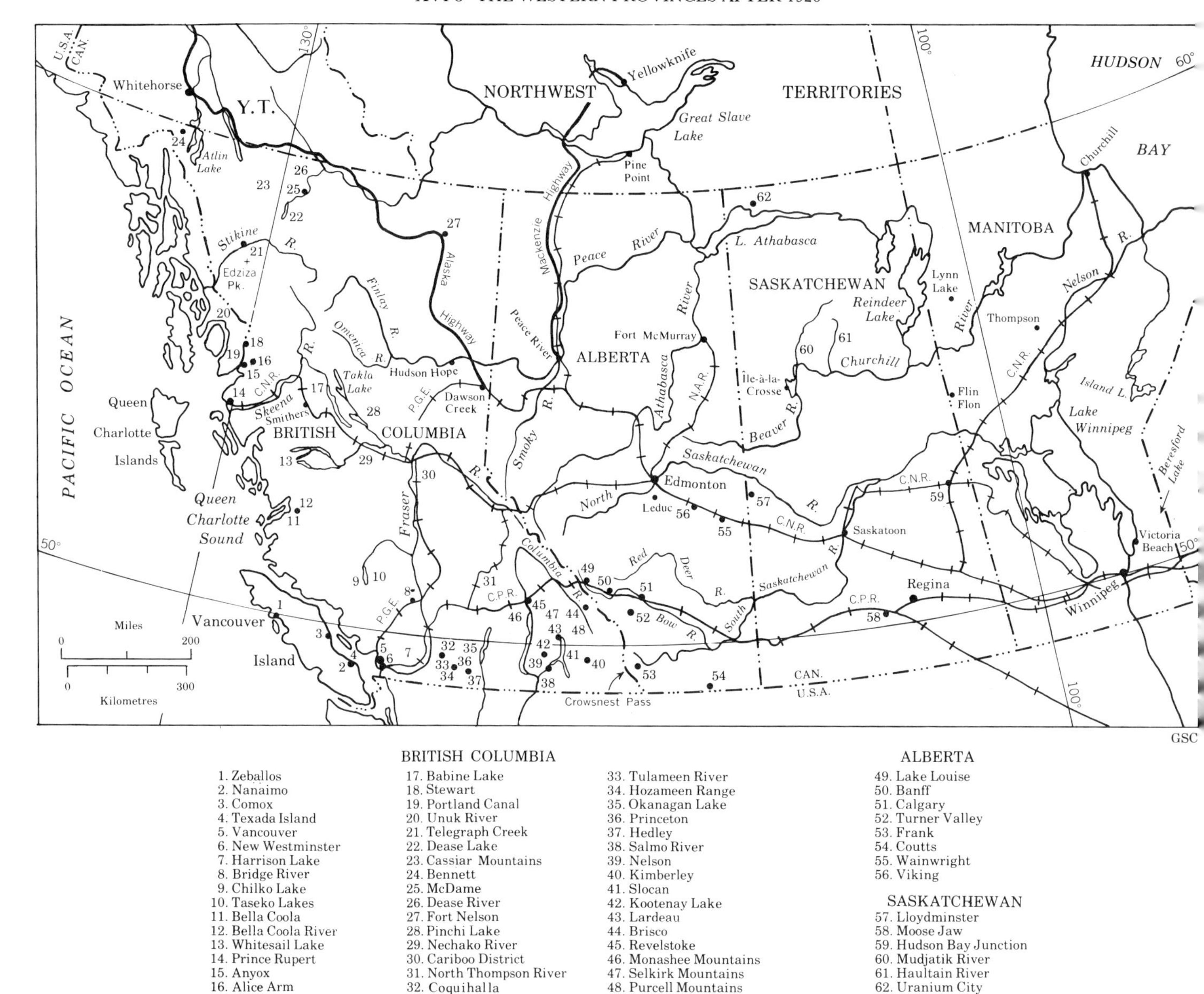

time he still travelled on foot, except where waterways were plentiful enough to get from outcrop to outcrop by canoe. Supplies were mainly carried by canoe or horses, and considerations of weight and bulk meant that the traditional, monotonous beans and bacon were still common camp fare. Backpacking remained the only method of reaching the less accessible parts of mountainous areas. Instruments had been improved, though they differed little in principle from those of the pioneers. The Brunton compass, used by every field man, was only an improved version of the prismatic compass of earlier years. Probably the greatest single improvement was the generally wider availability of good base maps, with which geologists could accomplish more and better work in a given time. Only with the appearance of the airplane and its adoption for geological mapping as well as transport, were the field methods appreciably transformed.

Despite the much greater need for concentrated, intensive work in all parts of the country, financial stringencies prevented the Survey from placing in the field as many parties as the proper conduct of the work required. With the return to normal conditions after the First

World War, the number of field parties increased to 33 by 1921 and to 39 in the following year, which remained more or less the average for the decade, though the number ranged between a high of 47 (1924) and a low of 26 (1927). The fluctuation was mainly in the number of temporary chiefs employed, for the number of permanent officers conducting geological field parties held quite consistently around the 22 mark.

Collins, a 'hard rock' geologist whose entire career revolved about the study of the Precambrian, put the main emphasis of the Survey's field work on the mapping of that enormous region. His plan of attack was to divide the Precambrian into large 60,000 square mile blocks, and to assign a geologist to one or other of these units on the basis of their importance, the plan being to map each of these on the 1 inch = 8 mile scale and to accompany the maps with reports explaining the geology and its relationships to ore deposition. The areas first selected were northwestern Quebec, northeastern Ontario (the map of which Collins finished by 1923), the Lake Superior district, and northern Manitoba. After Young had been appointed as chief geologist Collins, notwithstanding his heavy office duties, returned to his first love, the Precambrian, and resumed his many years' systematic mapping of the westward extension of the Huronian rocks from the Sudbury-Timiskaming district. His work was supplemented by that of several younger men, notably Ellis Thomson from 1920 to 1926, and H. M. Bannerman from 1926 to 1929. For the first time a thoroughly competent study was made of these important rocks, the succession was working out, subdivisions were proposed, and the distribution of these units was mapped. Many years later, when it was discovered at Blind River that the basal conglomerate of one of these units contained important uranium deposits, the exact location and extent of this bed could be traced out directly on Collins' maps.

For most of the years after 1925 Collins worked on the problem of the origin of the immense nickel deposit of Sudbury, first in the Espanola district immediately on the west of it, then from 1928 to 1935 at Sudbury itself where, assisted by Robert Thomson, W. A Jones and E. D. Kindle, he began remapping the basin on a 1 inch = 1 mile scale as an aid to prospectors, with particular attention to "the faults which are believed to cross the nickel basin and to have had an influence on ore deposition."[10] His conclusion, that the Sudbury series of sediments is of pre-Huronian age and overlies the Keewatin volcanics without evidence of unconformity or important time interval, was an advance over earlier proposals.

The problem of the Huronian, of which the Sudbury deposit is a part, was very vexing and engaged the attention of other specialists, including Pentti Eskola of Finland, who was invited to apply his experience of similar rocks to this problem in 1922. R. C. Emmons, a graduate of the University of British Columbia who later made a worldwide reputation as a professor at the University of Wisconsin, started his great career working for the Survey from 1923 to 1927 on these rocks in the Sault Ste. Marie district. Quirke, the American university professor, worked for many years for the Survey on the Huronian southeast of Sudbury along Georgian Bay, where the typical Huronian rocks disappear abruptly to be replaced largely by gneisses and granites. Between them, Quirke and Collins investigated this problem, and in 1930 published Memoir 160, *The Disappearance of the Huronian*, jocularly referred to as the Survey's only mystery story:

> The Huronian formations have, therefore, a total length from west to east of 500 miles. At the eastern end they attain their greatest thickness, comprising quartzites, conglomerates, greywackes, and limestone that aggregate 23,000 feet in thickness. Then they vanish. A definite and fairly straight line can be traced from Killarney, on Georgian Bay, northeastward to the foot of Lake Timiskaming, on one side of which are these sediments and on the other a vast expanse of granitic and gneissic rocks of quite different appearance . . . .
>
> What has become of the Huronian beyond this line?[11]

The authors could find no evidence for a fault contact of a magnitude necessary to fulfil the conditions. Then as they began an examination of the abundant inclusions throughout the gneisses the answer gradually came to them:

> The investigation . . . was undertaken principally to ascertain, if possible, what became of the Huronian formations east of the line . . . .
>
> It has been found that they do not end at this line . . . .
>
> The gneisses are, in large part at least, transformed Huronian sediments. They graduate into massive granitic material, which they resemble closely in mineral and chemical composition and from which it is not practicable to separate them. It seems, indeed, quite likely that the granite is simply part of the transformed sedimentary material which became liquid, and recrystallized, whereas there is equally good evidence that the gneisses were produced, in large part, from the sediments without departing from the solid state.[12]

The theory of granitization is now so widely accepted that this conclusion seems self-evident, but at the time it was reached, it was largely unknown or disregarded in North America. So clear a statement by two of the leading authorities on the Canadian Precambrian had a resounding effect and, in fact, set the study of the Canadian Shield at last on a firm footing.

Besides this work in the Huronian district, much was also done in other parts of the Precambrian of Ontario and Quebec to support the mining industry. Tanton worked from 1919 to 1929 around Pigeon River and Thunder Bay, a region in which current interest was primarily directed towards iron deposits. The formations were not unlike those that included the huge iron ranges in the United States, and iron ore was also known on the Canadian side too. Discovering workable deposits depended on improved understanding of the conditions of deposition. The problem was again studied in 1924 by J. E. Gill in the Gunflint area of Ontario just across the border from the huge iron deposits of Minnesota. Here again Gill attempted "to form a fair estimate of the possibilities of the occurrence of ore-bodies and to outline favourable localities."[13] Yet another investigation for the same purpose was that of the Belcher Islands in 1921 by Young. These remote islands in Hudson Bay were known to be composed of beds of which iron formation is a significant part, and there was, as always, the hope that somewhere the iron would be sufficiently concentrated to be economically workable. Young made similar studies in British Columbia in the early twenties in an effort to help determine the possibilities for an iron and steel industry along the Pacific coast, as did J. F. Wright in the Belleville district of eastern Ontario in 1920.

The Precambrian rocks of southern Ontario and southwestern Quebec contain many deposits of unusual but valuable minerals, which have been mined in a small way for years. Throughout the twenties M. E. Wilson studied the Grenville rocks which contain corundum, mica, talc, graphite, and apatite. He worked mainly south of Ottawa River continuing the work he had started around Buckingham, Quebec. Earlier in his career, around 1908, Wilson had traced the belts of sediments and volcanics which contained the Porcupine and Kirkland Lake mining camps eastward into Quebec, but he had failed then to find the eastward extension of the Huronian formations of Cobalt. Since this time, however, important copper and gold discoveries had been made around Rouyn and farther east, and it was clear that the region, that was becoming known as the Quebec Gold Belt, deserved far more thorough study. Beginning in 1922 Cooke worked along the belt east from Porcupine mine where, it was reported, "Geological investigation has shown fairly conclusively that the gold ore deposits are connected in origin with small intrusions of granitic rocks and are chiefly localized in the vicinity of these intrusives."[14] Until 1929 he continued detailed mapping round Larder Lake and other nearby areas. In 1922 Cooke introduced W. F. James to the region, and from then until he resigned in 1927. James continued to study area after area along this highly productive zone. In 1924 James was joined by Mawdsley, and after James left the Survey, Mawdsley carried on alone. Together Cooke, James, and Mawdsley made a fine contribution to the economic geology of this part of the Precambrian, published in 1931 as Memoir 166, *Geology and Ore Deposits of Rouyn-Harricanaw Region.*

In northern Saskatchewan and Manitoba the work was aimed more directly at assisting prospectors; several men studied various areas, notably the lithium-bearing pegmatite dykes that were attracting attention, and also around the Mandy and Flin Flon mines. The most extensive continuing work was by J. F. Wright, who studied several standard 1-mile areas around Beresford Lake, Manitoba, from 1922 to 1929.

For many years Collins opposed extending work into the Arctic because of the added expense and the diversion of manpower from what he regarded as more immediately valuable work in accessible districts. However, it was known that there were several large belts or areas of metamorphosed sedimentary and volcanic rocks that elsewhere sometimes contained important ore deposits. After completing his work in southeastern Manitoba, Stockwell explored the Reindeer Lake area in 1928, then in 1929 he started systematic work on one such belt running northeast from Great Slave Lake. L. J. Weeks also began work at Rankin Inlet on Hudson Bay, at the east end of another belt that Tyrrell had traced from Lake Athabasca to Hudson Bay.

The Cordillera, like the Precambrian, was an increasingly important mining centre in the age of base metals and gold. In 1918 the Survey had recognized its importance, as well as the practical difficulties of its distance from Ottawa, by establishing a regional office. For most of the twenties this was managed by Dolmage, assisted by Cairnes and Kerr. After Dolmage resigned in 1929, Cockfield began a twenty-year period as officer in charge, assisted by Walker until the latter's appointment to the British Columbia Department of Mines in 1935.

These men informed themselves each year of all the findings of the work in British Columbia by quizzing each man as he passed through Vancouver on his way back to Ottawa. The officer in charge then was able to answer inquiries about any area in which the men had worked, and to make that information available locally long before reports were published. He also became the confidant of mining men from all over the Cordillera seeking advice, or talking over problems. The position required knowledge, ability, tact, and a high ethical standard, and it is a tribute that no hint of a breach of confidence was even so much as whispered. In addition, there were each year thousands of visitors requesting information, as well as a very large number of rock and mineral specimens to identify.

In the Yukon sector, field work in this period was at first

XVI-9 CAMP 4, PREMIER GOLD MINES, STEWART, B.C.

confined to the silver-lead-zinc deposits of the Mayo district, and the copper deposits near Whitehorse. All this work was performed by Cockfield, except in 1925, when Stockwell worked at Galena Hill while Cockfield worked between Atlin Lake and Telegraph Creek in northern British Columbia in connection with the Cassiar mining boom. After Cockfield's move to Vancouver in 1929, Bostock was transferred from southern British Columbia and remained in charge of the work in Yukon until his retirement in 1965.

In the early twenties the theory was widely held that occurrences of granite and orebodies are closely related, and mineral deposits are likely to be found close to granite bodies. This attracted attention to the flanks of a huge body of granite, the so-called Coast Range batholith which was believed to extend unbroken from near Vancouver north all the way to southern Yukon. The field programs were planned with this in mind. From 1925 onward Cockfield systematically studied area after area near the Yukon end of the batholith, while farther south, in the extremely rugged and inhospitable country between Prince Rupert and the Yukon border, where only the eastern margin lay in Canada, Kerr worked from 1925 to 1929. Next he investigated the many mineral deposits along the Prince Rupert line of the CNR, a task in which he was followed after 1935 by E. D. Kindle. At Portland Canal, J. J. O'Neill's work of 1919 was taken over by Hanson, who worked around the head of Portland Canal, across high mountains and icefields to Alice Arm, then down along the east margin of the batholith to the Prince Rupert branch of the CNR. He left this region for a time (1925 and 1926) to work a little farther east around Babine Lake and Smithers. The work in the area immediately south of the CNR line was begun in 1920 as a summertime activity of Brock, and was continued from 1924 to 1928 by J. R. Marshall.

Farther south again, MacKenzie reconnoitered the eastern margin of the batholith around Taseko Lake, then was followed by Dolmage who explored a different section of this little-known region in five of the six seasons between 1924 and 1929. The one break, in 1927, was for a reconnaissance trip up Finlay River, the northern source of Peace River. On the west side of the batholith work was actually started in 1919 by Dolmage, who extended his work to Vancouver Island and the mainland coast and islands in 1920. The Vancouver Island work was resumed by Gunning between 1929 and 1933 in connection with gold discoveries there.

Meanwhile work was being done near the batholith, and east of it, along the southern border of British Columbia where Camsell had worked before becoming deputy minister. C. H. Crickmay studied the complicated stratigraphy around Harrison Lake in 1924 and 1926 while Bostock in 1926 mapped a series of one-mile areas from Hedley (where the Nickel Plate mine was the subject of his thesis) to the southern Okanagan valley. Most of the work in this region was done by Cairnes, who started in the Coquihalla area in 1920 and worked his way across the mountains to Princeton. From 1925 to 1928 he was occupied with a detailed study of the geology and many silver-lead mines in the Slocan area of the Kootenays, but in 1929 returned to the Okanagan valley. After finishing his work near the boundary, in 1934 Cairnes returned to work in the Bridge River mining camp, a famous gold producer of British Columbia. These geologists made a start on solving the complicated succession of volcanic rocks so widespread all through the region and ranging in age from late Paleozoic to late Tertiary.

The work in the Kootenays on the Precambrian and early Paleozoic was carried on by M. F. Bancroft between 1918 and 1923, then by Walker who, in turn, was joined between 1927 and 1929 by Gunning. Evans, who worked from 1925 to 1927 on a map-area near Brisco that embraced parts of the Purcell Mountain Range on the west and the Rocky Mountains on the east, provided a sound basis on which to establish later work in the western Rockies with his delineation of the complicated pattern of thrust faulting. Walker carried on the work, concentrating on the gold camps of Salmo and Bridge River until his resignation from the Survey.

Another centre of interest was the historic Cariboo district, an important gold producer since the early days, now facing exhaustion of the known deposits. In 1919 MacKay started the investigation, to be followed in 1921 by W. A. Johnston, who had just completed studies of Fraser River delta. Like all placer deposits, the investigation had two aspects: tracing the gold-bearing gravels, and locating the ore veins from which the gold had originally come. Johnston soon found that most placers were remnants of earlier deposits that had escaped removal by the Pleistocene ice sheets, and by working out the pattern of the pre-glacial drainage he deduced where further concentrations of gold might be found. Only a limited number of placers remained unmined, so the future development of the district depended on discovering lode veins. W. L. Uglow of the University of British Columbia, who had spent a successful field season on North Thompson River, joined Johnston to carry on a parallel study of the rock formations. A little later, in 1925, Johnston undertook a similar study of the placers of the Dease Lake area.

All these Cordilleran explorations were closely related to current mining interests—gold and base metal producers and potential producers; the Nanaimo and other coal deposits; occurrences of iron in the interior and along the coast of Queen Charlotte Islands; titanium, tungsten and antimony; and studies of iron, lead-zinc, talc, and other minerals.

Work in the Plains region from the International Boundary to the Arctic Circle was closely related to the growth of the oil and gas industries of the region. The industry was of special concern in view of Canada's alarmingly large petroleum imports, while its drilling records were the main source of information on the geological succession. From the discovery and development of the Turner Valley oil field, the Borings Division under Ingall had kept close contact with the resulting activities. As Ingall pointed out, "failing the collection of such illustrative sets of samples, our knowledge of the geological conditions in depth over the Great Plains and other large areas of Canada must be limited to surmise based on the surface study of very scattered and imperfect outcroppings of the various formations."[15] The Borings Division received material brought up as chips by the drilling water. The chips were stored in small bottles labelled with the number of the well and depth of the sample so that trays containing rows of these bottles gave a picture of the strata. As early as 1923 some 60,000 samples had been received. These were later to be of immense direct benefit to oil company geologists, but in the early stages what the Survey needed was information about geological principles—the kind of rocks and structures in which oil might accumulate and where these were to be found. By the twenties the Survey staff often used the office of the Dominion Lands Branch in Calgary for a convenient headquarters. With the great expansion of the oil and gas industry in the 1950's, the Calgary operations developed into a permanent office and eventually into a division, the Institute of Sedimentary and Petroleum Geology.

By 1920 a good foundation had already been laid for the understanding of the geological structure of the Plains region. Stratigraphic sections in the Rocky Mountain foothills had been examined and measured, and the broad elements of the succession of the Paleozoic and Mesozoic strata were known, with many points defined thanks to the samples secured from the wells drilled in many parts of the Plains. Unfortunately, the task of correlation is difficult and never ending. G. S. Hume was associated with this work for a generation. In 1920 he travelled far down the Mackenzie River to the Norman Wells oil field, as part of the shortlived (1920-23) investigations of the oil prospects of the Mackenzie valley by A. E. Cameron, Dowling, Whittaker, and Williams. In 1923, when a flow of oil was struck in a well drilled north of Wainwright, Hume examined this promising area. In subsequent years he continued to work at Wainwright, at Turner Valley, and along the Alberta Foothills, carefully and patiently working out the intricate structures, and time and again pointing out the most promising places to explore. Without Hume the development of Canada's western oil and gas industry would certainly have been considerably delayed. Indeed, by persisting with work all along the foothills during the thirties, Hume deserves much of the credit for keeping the industry alive through difficult times.

In addition to Hume's continuing study of oil and gas potentialities, there were further studies of the coalfields by MacKay, who investigated those of the Alberta Foothills and adjacent parts of British Columbia between 1925 and 1934. A by-product of this work was a further examination of Turtle Mountain, where the continued loosening of overhanging rock again raised concerns for the town of Frank, site of the deadly rock slide of 1903. Other studies of the Interior Plains region were made by

Williams, W. S. Dyer, P. S. Warren, McLearn, Evans, S. R. Kirk, and B. Rose. L. S. Russell and C. M. Sternberg collected and studied large reptilian fossils as well.

The desire to discover new sources of fossil fuels to relieve the nation's dependence on imported supplies spread far beyond the Interior Plains. In Ontario interest turned to the sedimentary basin along the shores of Hudson and James Bays, where McLearn examined coal-bearing Mesozoic rocks lying on the edge of the Paleozoic basin south of James Bay in 1926. Also in 1926, W. A. Bell was asked to report on the oil possibilities of some of the Carboniferous rocks of Cape Breton. In a well-reasoned discussion Bell pointed out that conditions were unfavourable for the occurrence of petroleum, much less for its accumulation into economically workable pools.

In the settled parts of central and eastern Canada the basic geology had been well established and the provincial mining agencies were active, so Collins and Young tended to concentrate the effort in other regions. However, some work continued in the St. Lawrence Lowlands and Appalachian regions to refine earlier conclusions or study special aspects; for example, Alice E. Wilson began an extended investigation in 1925 of the Paleozoic formations between Ottawa and St. Lawrence River. Alcock also worked for six seasons between 1921 and 1929 in the Gaspé Peninsula, mainly around Chaleur Bay. In the thirties, as an aid to the mining industry in its time of difficulties, the Survey resumed work in the asbestos district of Quebec under Cooke, while C. S. Evans and others attempted to extend the small but important oil fields of southwestern Ontario.

The geology of much of the Maritime Provinces, which also was reasonably well known, was an excellent subject, for detailed studies by students desiring projects for Ph.D. dissertations. It is no surprise, therefore, that the literature of the region contains the names of many geologists who worked only briefly for the Survey before assuming honoured places in geological and mining circles. For the Survey the work in Nova Scotia in the interwar period was mainly by E. R. Faribault and W. A. Bell. Faribault continued his painstaking systematic study of areas of the southeastern part of the province, gradually extending from the east coast to the Annapolis valley and the Nictaux-Torbrook iron district. In 1929 Wyatt Malcolm brought out a second memoir on *The Gold Fields of Nova Scotia,* derived almost entirely from Faribault's work. Bell carried on his careful, detailed studies of the Carboniferous formations based on the paleobotany of the region. In 1927 Norman started general mapping of the important Cape Breton coalfields. Young, W. S. McCann, Dyer, Kerr, M. E. Hurst, and others also worked for one or more seasons on aspects of Maritime geology, especially on areas believed to contain workable amounts of metallic minerals, or on mapping extensions of existing coalfields.

The Survey's management tended to neglect the important subject of Quaternary geology, Collins himself admitting that "The Geological Survey has for many years paid more or less incidental attention to this branch of geology."[16] W. A. Johnston alone was permitted to continue working at his specialty, completing a fine detailed study of the Fraser River delta region, presented in two memoirs of 1921 and 1923, that proved invaluable to the Vancouver Harbour Board. In 1921 he also visited Lake Louise in Alberta to study the debris left by the glaciers of today to see the processes by which the Pleistocene deposits of Canada were formed. A practical application of such work was the study in 1929 by H. E. Simpson of the water supply situation around Regina—a subject that was to be greatly extended in the following years. Another facet of Quaternary geology was the continuing work on the peat resources in co-operation with the Mines Branch and other agencies. The investigation was successful in that a commercial enterprise was developed at a bog near Ottawa, but it failed to secure much of a market. The Survey, as its contribution, made an inventory of bogs across Canada, the field work being carried out by A. Anrep, while a Finnish expert, V. Auer, came

XVI-10 L. J. WEEKS

out in 1926 to compare the bogs of eastern Canada with those of his homeland. Anrep continued the work until 1932, when it became a casualty of the depression.

One region where Collins and Young would not willingly commit their limited field staff was the Eastern Arctic. To every pressure from Camsell or by other branches of government, the answer was that the Survey could not afford such an enterprise unless it were given extra funds specially earmarked for the purpose of northern exploration. However, a start was made in 1925 when Weeks sailed on the C.G.S. *Arctic* on her annual cruise to the Arctic islands. Many observations were made, though each stop was too brief to allow more than a glimpse at the more obvious problems. The following year, 1926, Weeks again sailed on the *Arctic,* this time with M. H. Haycock, to winter at Pangnirtung, Baffin Island. Unfortunately, while hunting ducks, Weeks was wounded by the accidental discharge of a shotgun which severed the main artery and nerves in his right arm close to the shoulder. Notwithstanding the serious wound that left his arm totally paralyzed and in grave danger of being frozen, Weeks carried out a long winter's traverse by dogsled to make a modest beginning on the geology of Canada's Eastern Arctic. Later, between 1929 and 1932 he worked along the Hudson Bay coast south of Rankin Inlet, while in 1933 Gunning accompanied the Eastern Arctic patrol ship and spent three weeks studying the geology of the Cape Smith district on the east coast of Hudson Bay.

* * *

THE OTHER DIVISIONS OF THE SURVEY carried on their work through the twenties and thirties in keeping with the needs of the time. The Mineralogical Division seemed to grow in importance, though its professional staff was small and it was separated from the rest of the organization. The division maintained its office on Sparks Street, and after 1929, when the Mines Branch moved into its new building on Booth Street, part of the division's facilities returned to the historic Sussex Street building. Primarily a service agency, testing and reporting on materials brought in from the field or submitted by the general public, the division spent most of the time on identifications and analysis. By the late twenties the number of visitors seeking mineral determinations had greatly increased, and they were reported to be "showing greater knowledge of geology and mineralogy and a greater capacity for appreciating the results of scientific investigation of the samples submitted,"[17] a result that the chief, Poitevin, attributed to the increasing popularity of the prospecting classes and schools of mines. The vivacious Poitevin, with experience, developed into a marvellously knowledgeable authority, able unerringly to identify and give the sources of minerals from every part of Canada.

The division looked after the mineralogical display at the Museum; in 1924-25 the enormous collection of minerals dating back to Logan's time and stored in boxes since 1910 was finally sorted out by W. F. Ferrier. It operated a very considerable educational program, collecting and preparing specimens for display kits that were given away or sold to schools, universities, and prospectors. From time to time the division also prepared collections of minerals for display at various exhibitions in Canada and abroad. Finally, the staff did special research as time permitted, publishing articles on new minerals they isolated, such as camsellite, collinsite, or robertsonite, or preparing lists of Canadian mineral occurrences.

The really important work of the division in these years was Ellsworth's pioneer researches on radioactive minerals. As early as 1921 he was examining Canadian occurrences of rare earth and radioactive minerals, especially in the pegmatite dykes around Parry Sound, Ontario. A very precise analytical chemist, he began working in every spare moment on an experimental method being developed in England and elsewhere for determining the ages of radioactive minerals in Precambrian rocks by measuring their uranium-thorium-lead ratios. The principle is simple: from the time a radioactive element is fixed as part of a mineral, its breakdown proceeds at a fixed rate regardless of external conditions. In a closed chemical system it is possible for a scientist to determine how long ago the mineral was formed, provided the minute quantities of the products of the reaction can be measured, and the decay rate of the radioactive elements can be determined. After publishing some of his results in the *American Journal of Science,* Ellsworth was approached in 1925 by the director of the Geophysical Laboratories in Washington, and with the approval of Collins entered into co-operation with his American counterpart, D. M. Tenner. Together they worked to establish reliable decay factors and to identify and make adequate allowances for sources of error. Meanwhile Ellsworth also worked on a major report, *Rare-element Minerals of Canada* (Economic Geology Series, No. 11, 1932). By the end of the twenties he was a recognized world authority in the field. The discovery of radium ore at Great Bear Lake about this time put the Survey squarely on the threshold of a momentous scientific development, one that was of particular importance to Canada—the working out of the complicated history of the Precambrian by means of age determination methods.

In the long run, the really significant contribution of the Mineralogical Division during the period was Ellsworth's excellent, original work. Unfortunately, he never was given a proper opportunity to put his talents to proper use, nor did he receive the recognition or salary his achievement warranted. He was obliged to carry out the full complement of his regular duties in the division, without assistants, research facilities, or other support. It says much for his devotion to his work that he did not seek another position, or simply give up the fight and stick to his routine tasks. By not supporting him when his work stood in the forefront of world science, the Survey lost an outstanding chance to do something distinctive and original. But as in so many other cases, the Collins and Young management failed to appreciate their opportunity.

The Borings Division, under Ingall, was still active in collecting. The main difficulty was with practically-trained well drillers with little knowledge of geological formations beyond the immediate needs of their work and who were interested only in getting their work done as quickly and effortlessly as possible. The companies, which had everything to gain from an improved knowledge of the geology, were more willing to co-operate, however. Some governments also required drillers to submit all deep well boring records to the Survey. The Survey was anxious to secure uncleaned samples, since washing them might destroy important fragments of fossils or microfossils that could be useful in identifying the strata. Best of all were drill core samples that brought up large segments of the actual strata.

Collecting the samples was only a preliminary part of the task, but not until the end of the twenties was much effort made to begin analyzing them. An assistant, S. J. Fraser, was hired for this work, and after 1927, Wickenden, a specialist in microfossils, worked as a student assistant while completing his doctorate for Harvard. Such testing was essential to separate the many petrologically-identical Cretaceous shales into their various units as an aid to identifying potential oil and gas structures with greater precision.

XVI-11
H. V. ELLSWORTH

While the deep borings were the main interest from the economic and scientific points of view, there was a rising interest on the part of the public in matters like water supply, records for which could be secured from shallow wells. Hence interest was displayed in studying the Quaternary formations that might carry or hold water to meet the needs of overcrowded or semi-arid regions. By 1928 the Survey was receiving requests from departments of health, civic waterworks engineers, and railways for information on their problems. During 1929 the division made an effort to catch up by hiring three student assistants to plot the positions of the wells on maps and analyze the strata penetrated by them. In the next few years this work was to be vastly extended under the impetus of the droughts of the early thirties in the southern prairies.

The Paleontological Division carried on through the twenties under E. M. Kindle with little notice, and with a considerably smaller staff than in Brock's day. The three specialties of paleobotany, and invertebrate and vertebrate paleontology, were carried on by a largely new staff comprised of Bell, McLearn, and Whittaker (succeeded by Loris S. Russell). The fifth member, Alice E. Wilson, obtained her Ph.D. in the mid-twenties. The work involved preparing reports identifying fossils brought in from the field to establish the stratigraphic succession; and research work on paleontological problems and collections. Many new species were being discovered that required careful analysis and classification, while the searches for oil and gas were beginning to put greater emphasis on studying fossil environments (paleogeography). Fossils, properly understood, could offer

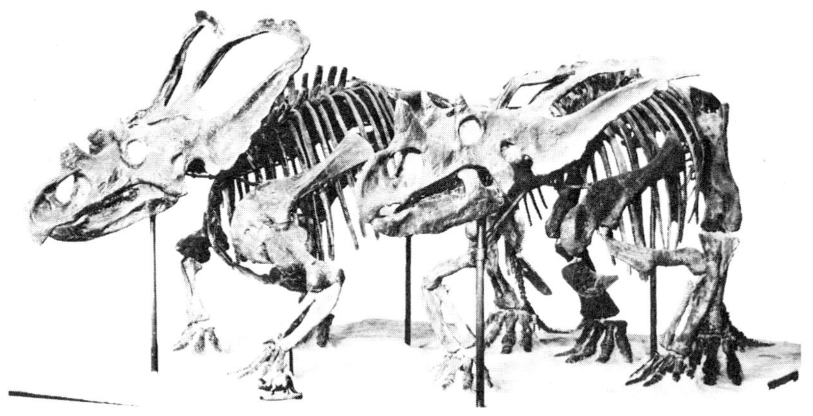

XVI-12 HORNED DINOSAUR DISPLAY, NATIONAL MUSEUM

clues as to shallow or deep water, water with high or low salinity, water with good or poor circulation, or reef structures—factors that are significant for the formation of oil or gas.

Apart from analyses of submitted materials, each member of the division studied his own specialty when he could—Bell, the flora of the Nova Scotia coalfields; Miss Wilson, the Ordovician and other periods in the Ottawa region; and McLearn, the stratigraphies and faunas of various localities in western Canada such as the Peace River canyon. For the Museum C. M. Sternberg collected Mesozoic vertebrates around Red Deer River in Alberta, with occasional visits to Nova Scotia or elsewhere to secure fossil fishes or other specimens.

Collins and Young had little sympathy for paleontological researches, and Kindle, perhaps, lacked the force to impress them with the importance of the work, or even to secure rank and pay for his staff comparable with the field geologists. In 1929 and afterwards the staff was set to regular areal mapping work. In that year Bell did detailed mapping of the Sydney coalfield, Kindle in the Gaspé region, McLearn in southern Saskatchewan, and Miss Wilson part of the Thurso and Cornwall areas. Indeed, McLearn prepared a report on the geology and mineral resources of Saskatchewan (1933) and another on the clays of southern Saskatchewan, while Russell prepared one on oil and gas possibilities in southeastern Alberta. Here again, Collins' narrow outlook failed to recognize the value of work, or people, engaged in fields other than his own. McLearn, like Ellsworth, was a first-rate scholar who should have been permitted to concentrate on his special studies where he could make significant contributions, instead of being turned into doing areal geological mapping. He even had to publish his researches largely at his own expense, in the *Canadian Field Naturalist*.

W. H. Boyd's Topographical Division had developed a smoothly-functioning routine by 1920 which was to be largely disrupted by the end of the decade through such innovations as the automobile, radio, and especially air photography, plus pressures from outside to reorganize the mapping agencies of the federal government. The well-trained staff, built up to execute topographical surveys required in connection with geological investigations, tried to keep a season ahead of the geological work, so that the geologists could work from finished maps. Since the maps were also used for other purposes, they had to follow regular standards, methods, and styles, and be based on regular areas bounded by meridians and parallels. The need for such conformity led in 1922 to the establishment of a Board of Topographical Surveys and Maps to co-ordinate the topographical work of the Survey, Department of National Defence, and the Department of the Interior. Each was represented by two members on the board, Collins and Boyd being the Survey's nominees.

The work was of two types. Geodetic, or control, work involved determining by astronomical means, key or reference points from which normal surveys could be measured. The two specialists, S. C. McLean and R. C. McDonald, travelled across Canada and based their primary locations on the existing geodetic net, thus providing subordinate or intermediate nets of locations that supplemented the main ones the Geodetic Survey was preparing. The traditional ground surveys were conducted by from seven to nine parties each year, who worked to the scales determined by Survey needs, most being on the 1 inch = 1 mile, 2 miles, 4 miles or 8 miles, with occasionally a more detailed 1 inch = 800 feet, etc., being used. Contour intervals depended mainly on the mapping scales and the relief, intervals of 25, 50 (the most common) or 100 feet being usual. By the end of the twenties the increasing use of air photographs as a basis for topographical mapping had begun to undermine this well-developed system that had served the Survey so well.

The Geographical and Draughting Division, and the Map Engraving Division, which together attended to the preparation of the maps, were also facing problems of technological change. The two divisions were headed by C. O. Senécal and A. Dickison until Senécal's retirement in 1931, when the two were merged under Dickison. Aside from such staff problems as the shortage of draftsmen, the drafting division had to cope by 1929 with the increasing availability of air photographs for purposes of base maps. The Map Engraving Division, too, was being affected by the appearance of new methods of reproduction using lithography or photography to supplement the slow, laborious, highly-skilled art of copper plate engraving.

Since publications were the Survey's main means for reaching the public, and its reports had long-term as well as immediate value, great care was taken of this aspect of the work in this period as in all others. There was an editor-in-chief for the Department of Mines (William McInnes in 1920) and under him were editors for each branch, the Survey's being F. J. Nicolas. The publications covered a great range–memoirs, bulletins, parts of the annual summary reports, museum bulletins, plus a variety of separate or individual items. McInnes oversaw the publications of both branches, plus such others as the volumes and parts of the Canadian Arctic Expedition, or the reports of the Dominion Fuel Board. After McInnes died Nicolas became editor-in-chief as well as editor for the Survey. Young, who had been the technical editor for the Survey from 1913, took over more complete control of the technical editing from his new position of chief geologist. The editorial process involved discussions between author and technical editor until the work was prepared to the satisfaction of both of them and of the director as well, then faced further revision by the editor-in-chief (chiefly literary and typographic) and by the deputy minister, the revised work being sent back for approval of the author and Survey officers. The proofs passed through a similar chain, so it was a wonder anything was ever published.

Innovations included indexes to the separate reports and summary reports, 1905-16, and 1916 to 1925, also to the paleontological publications, all three prepared by Nicolas. More important was the new Economic Geology Series, each volume of which summarized the information about a mineral or group of minerals, based on previous reports and other sources, supplemented and updated by field study. These drew together information regarding the genesis, modes of occurrence, distribution and possible location of new deposits through drawing together the information about existing deposits. The first three, issued by 1927, included reports on talc and soapstone, iron ores, and the second edition of *Geology and Economic Minerals of Canada*. Other volumes on arsenic, oil and gas, lead and zinc, rare-element minerals, fluorspar deposits, and a prospector's handbook followed. An annoying bit of censorship involved a volume of 1930, *Gold Occurrences in Canada* (Economic Geology Series, No. 10) by Cooke and W. A. Johnston. A mine manager having complained to the minister that the authors had failed to deal with his mine, Camsell passed the information along to Collins 'suggesting' that Cooke should attend to the matter, that four pages be reprinted with appropriate reference to the property, and held up distribution until the revision was made.

On December 1, 1920, the Victoria Memorial Museum–that long-time connection of the Survey–became ostensibly a separate branch of the Department of Mines with its own director, W. McInnes. As director of the Mu-

XVI-13 WILLIAM McINNES
Director, Victoria Memorial Museum, 1920-25

seum, McInnes had immediate control over the Biological, Anthropological, and Palaeontological Divisions, under R. M. Anderson, E. Sapir, and E. M. Kindle, respectively. The separation was more apparent than real, however. The Museum and Survey used the same clerical staff, the same library, and such facilities as the photographic and drafting, editorial and publications units. Paleontology and mineralogy, which had links with the Museum, were classed and treated as belonging to the Survey half of the partnership. Furthermore, the Museum had no separate existence as far as finances were concerned; its field explorations were listed in the Survey's accounts, the Survey's general expenditures included items obviously for the exclusive use of Museum personnel, such as phonographs, or equipment and specimens for the Museum. The expenses for the operation of offices and Museum were lumped together, and the Museum publications were included with those of the Survey. In brief, the Museum had no financial basis and was operated as part of the Survey's budget, under Survey control, as were the personnel of the Museum.

Moreover, even the separation of having different heads was illusory. Since McInnes was gravely ill, Collins exercised control over the Museum as well as the Survey, and after McInnes died Collins simply annexed the role of acting director to his other functions. Officially he did so as an economy measure and on a temporary basis, but in fact he continued to occupy the two positions until his retirement, a matter of ten years.

Strangely contradictory therefore was the action of the Dominion government by order in council of January 5, 1927, in proclaiming the Museum as "The National Museum of Canada." The only apparent change that followed was that a separate series of annual reports began to be issued for the Museum, which Collins prepared along with the shorter report on the Museum he submitted with his Survey director's report in the Department of Mines' annual report. In 1927 he also prepared a paper, *The National Museum of Canada,* part historical review and part outline, of the relationship—past, present and future—of the two organizations. In this, among other things, he pointed out that "An independent Museum equipped in all its present departments of natural science would probably cost between $100,000 and $150,000 a year more than it does at present."[18]

In concentrating on the superficial savings Collins was certainly ignoring the real costs of the arrangement, principally to the Museum. He was a geologist first and foremost, and even in his control of the Survey he was strongly driven to concentrate the effort on geological mapping to the detriment of such aspects as paleontology and mineralogy. How much less consideration would he pay to the needs of anthropologists or naturalists who contributed nothing to advancing the geological knowledge of Canada? When money was lacking to do the geological work satisfactorily there was no doubt that the Museum would be the place where strictest economizing would prevail.

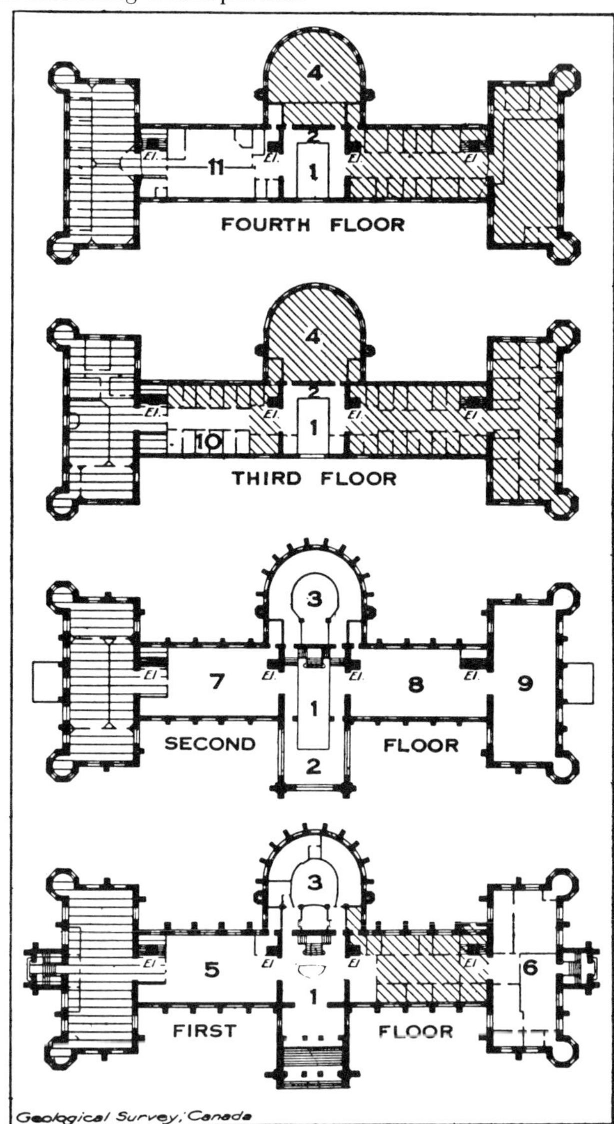

XVI-14 FLOOR PLANS OF THE VICTORIA MEMORIAL MUSEUM, 1927

The horizontally-shaded part is occupied by the National Art Gallery and the oblique shaded parts by the Geological Survey and administrative offices of the Department of Mines. 1. Entrance hall, which reaches the full height of the building. 2. Galleries around entrance hall on second, third, and fourth floors. 3. Lecture hall, with gallery on second floor. 4. Library, two stories in height. 5. Paleontological exhibits. 6. Offices and collections of biological division. 7. Biological exhibits. 8,9. Anthropological exhibits. 10. Offices of anthropological division. 11. Workrooms for preparation of biological and paleontological exhibits. Elevators are represented in solid black.

XVI-15 SOME OF THE WORK OF THE MUSEUM DIVISIONS IN THE 1920's
Top left: THE TOTEM POLES OF GITWINLKUL, UPPER SKEENA, B.C. Photographed by Marius Barbeau about 1928.
Centre left: CAMP OF A NASKAPI FAMILY. Photographed by F. W. Waugh.
Lower left: A SARCEE CHIEF INSIDE HIS TIPI. Photographed by Diamond Jenness.
Right: SOME ILLUSTRATIONS FROM P. A. TAVERNER, *BIRDS OF CANADA* (1922). No. 1, Belted Kingfisher; No. 2, Osprey. (Drawings in colour by C. E. Johnson and F. C. Hennessey.)

Still, the two groups appeared to get along well enough. They collaborated in the Museum's lecture series which was begun in 1920 in the auditorium vacated by the House of Commons. Each year a series of lectures on scientific subjects was offered for adults in the evenings of the week, and for school children on Saturday mornings, the latter being so popular as to require to be repeated twice and even three times a morning. The series was managed by a committee of two, Harlan I. Smith for the Museum and Dowling for the Survey (succeeded by M. E. Wilson in 1925). The presentations were given by members of both staffs at first, and afterwards by other governmental bodies as well.

Numerous personal connections existed between members of the two groups, the strongest perhaps involving the paleontologists, whose work kept them in close association with the Museum at all times. Neighbourhood was another factor; Wilson, for example, often had Harlan I. Smith from the next office dropping in to visit and air all his troubles. Wickenden collected birds for P. A. Taverner and mammals for R. M. Anderson when he worked on the prairies, and they taught him taxidermy. Many geologists who retained similar lively interest in other fields of natural history, often availed themselves of the expert knowledge of their fellows, and brought back from the field a good deal of new material for the Museum's collections.

Nevertheless, overcrowding was bound to make for tensions. Camsell in his report for 1923-24 observed that the Museum was crippled in displaying its materials by lack of space, and went on to say that "The housing of the staff of the Geological Survey makes for congestion in the Victoria Memorial Museum, and until other quarters are found for the Geological Survey, the development of a National Museum cannot be properly proceeded with."[19] Collins, on the other hand, felt that since the Survey and Museum belonged together, the proper solution would be to remove the extraneous elements—the National Art Gallery, the Dominion Fuel Board, even the administrative part of the Department of Mines. Then it would be possible to bring the Mineralogy Division back from its separate quarters and make a saving of $13,000 to $17,000 per annum. However the crowding continued for over thirty years more, until at last in 1959 the Museum was left in sole possession of the building.

Collins' control was in evidence almost immediately after McInnes' death. In 1924 he had hired W. F. Ferrier to undertake to assemble "the extremely valuable systematic mineral collection, which has been in process of accumulation for three-fourths of a century and includes over 25,000 specimens."[20] For efficiency and economy he planned to use the same type of cases for other purposes: "The type of exhibit case used for the systematic mineral collection will also be used for exhibits of fossils."[21] In the following years wooden and metal exhibition cases, and steel table cases were built and purchased, the 1930 acquisitions including "five upright cases for the Palaeontological hall, a large upright case for the Biological hall, a large wall case for a dinosaur mount, three table cases for geological and palaeontological exhibits, and . . . a case specially designed to carry sixteen, small, standard, insect cases."[22] In all these matters Collins made the decisions in accordance with his own ideas and tastes, over the strenuous objections of at least one division chief.

The editorial management of Museum publications also occasioned some difficulty. Its memoirs and bulletins were issued under the supervision of Nicolas, with major publications by Museum personnel appearing as memoirs of the Survey. The Survey's and department's control in at least one instance—the study by L. Bloomfield, *Sacred Stories of the Sweet Grass Cree*—amounted to censorship exercised by persons who presumably were not qualified to judge such matters.[23] The same control was even more strongly asserted on the question of providing funds to publish Museum manuscripts. These were cut back more severely than were geological subjects, especially during the depression, when little more than the annual reports and works that were guaranteed to pay their way, such as *The Birds of Canada* or *The Indians of Canada,* were published for the Museum.

The most obvious measure of the Museum's handicaps is indicated from its staff roster. The bright young men Brock hired before the First World War were not advanced to higher positions in the government service, nor were they joined by young, fresh scholars. They remained for thirty years a static unit, while age turned them from young men to mature scholars and disheartened older men. In fact, the staff actually was reduced as men left or died. The Anthropological Division, for example, included six specialists in 1920—H. I. Smith, E. Sapir, C. M. Barbeau, D. Jenness, F. W. Waugh, and W. J. Wintemberg. Waugh disappeared on September 20, 1924, while on a visit to or from Caughnawaga Indian Reserve near Montreal, and Sapir, the chief of the division, resigned in 1925 to join the staff of the University of Chicago, to be succeeded by Jenness. A single addition to the division, of J. D. Leechman, was made in 1925, but Smith and Wintemberg retired in 1936 and 1940, while Jenness left the Museum in 1942, so the division was practically reduced to Barbeau and Leechman. In the biological field the four specialists—Anderson, Taverner, C. Patch, and M. O. Malte, the chief botanist—were reduced to three when Malte died in 1933 and no replacement was sought for some years, the position being eventually filled by A. E Porsild (1936). The competent Museum specialists were trapped in a neglected, declining part of the public service with resulting frustration and lowered morale.

McInnes had entered into his directorship in high hopes that the establishment of the Museum as a division of the department "is the first step toward the creation for Canada of a National Museum that will worthily represent her great natural resources."[24] Collins, when he took charge, spoke grandly of the Museum as "an important feature in any scheme for the improvement and beautification of the capital city," and as "an expression of the interest of its country towards the cultural aspects of civilization."[25]

The reality was very different. The Museum was firmly under the personal control of Collins and always was starved for funds. As director, Collins divided the appropriation among the various divisions, giving what he chose for those of the Museum. Publication funds were inadequate throughout, and there was so little opportunity to publish that a great deal of work had to wait many years or be published elsewhere. Though extremely conscientious, strict and honest, Collins never allowed the Museum people to make their own decisions, and always interfered in the name of economy or in the light of his own opinions. Not until Alcock took charge in 1947 did the Museum begin to receive enough funds to catch up with its publications program, hire new staff, and advance beyond the iron confines within which it had been kept. Alcock found the survivors quite demoralized from thirty years of neglect; they had been starved so long for funds that they had given up trying to improve their situation.

The bond between the Survey and the Museum had a long history and strong justification, but it was unfortunate that it continued so long as it did. Collins recognized in 1925 that it would not be possible for both institutions to be administered by one head for much longer. Yet, knowing this, he proceeded to assume control over both for a decade. In fact, the separation ought to have taken place long before. The error, perhaps, lay in failing to appoint a separate director, with a separate budget, when the move to the Victoria Memorial Museum building was made, a man who would have given the Museum a chance to develop in its own way and not as an appendage to the Geological Survey. As it was, the Museum did not begin its truly separate existence until 1947; and even now, as it occupies the renovated building at the foot of Metcalfe Street and various other buildings in and around Ottawa, it has some distance to go to merit the title bestowed on it, in January 1927, in such mockery of the facts.

XVII-1 AVIATION OPENS THE NORTH
In 1934 travellers from Fort McMurray, Alta., to Great Bear Lake, N.W.T. could choose between 30 days by steamboat or 8 hours by air.

17

The Thirties – The Survey Downgraded

THE COMING OF THE GREAT DEPRESSION of the thirties drastically reduced the markets for minerals, the profitability of mining, and for some time, also, prospecting for new sources of minerals. In 1933, however, the United States government raised the price of gold from $20 to $35 an ounce, dramatically reversing this situation in the case of gold mining. Almost at once scores of new gold mines began to be brought into production in Canada, some in established mining districts, but many of them in distant parts of the Precambrian Shield and Cordillera, extending from northern Quebec to Yellowknife on Great Slave Lake. The Northwest Territories contained an additional centre of prospecting interest resulting from the discovery and development of an important silver-pitchblende deposit on Great Bear Lake into a major radium producer—a good whose value was great enough to be invulnerable to the depression. Mining suddenly became one of the few bright spots in the gloomy economic picture of Canada during the depression.

The development of many of these new mines was concrete illustration of the place aviation had assumed in the opening of Canada's northern frontiers. As early as 1920 the Imperial Oil Company had attempted to transport its men by aircraft to its oil-well drilling site at Norman Wells on the lower Mackenzie River. Airplanes played an important part in two mining rushes to Rouyn, Quebec, in 1924 and to Red Lake, Ontario, in 1925, by flying prospectors and equipment to these localities, enabling operations to proceed almost without pause through the winters. In 1925 a party of prospectors was flown to the Cassiar district of northwest British Columbia, where for three months prospectors were set down and picked up at point after point in the Dease and upper Liard basins. From 1927 aerial exploration companies crisscrossed the north, flying through distant parts of the Yukon, over the tundra west of Hudson Bay, and to Coronation Gulf and Victoria Island, placing company prospectors and geologists to seek out and investigate interesting mineral showings. By the thirties the buzz of aircraft resounded over all parts of the Canadian mainland. The country beyond the end-of-steel had been opened up to regular mining activity.

The coming of the depression forced the Survey to re-examine its programs to assist established mining industries with their ore supply problems and to aid searches for economic deposits of gold and other minerals by providing maps of the basic geology of promising northern districts. The more immediate concern, however, was to prevent the Survey's existing program from falling victim to the federal government's desperate economizing drive, let alone undertaking the expensive activity of helping extend mining development into new districts.

Camsell, conscious of the threat as well as the opportunity, pleaded for the expansion rather than the curtailment of the Department of Mines' operations:

> Canada today has well-established mining and metallurgical industries with a capital investment of nearly a billion dollars, giving direct employment to nearly a hundred thousand men. In a period of industrial depression there is all the more need for the greatest possible technical contribution toward lower costs of prospecting, operation and production, improved methods of treatment, higher recoveries, better products, more efficient utilization, and the development of Canadian resources to replace imported products. These are among the more important objects of national interest with which the department is actively concerned.[1]

His pleading, as the quotation shows, was mainly on behalf of the work of the Mines Branch, where he chose to highlight such achievements as those of the Ore Dressing and Metallurgical Division, about which he reported that of some fifty new gold milling plants that had entered production during 1934, "most of these plants are using ore treatment processes designed in the Ore Dressing and Metallurgical Laboratories at Ottawa."[2] In 1932-33 he had also singled out the Mines Branch's work in helping develop the successful radium treatment process used in the Port Hope refinery of Eldorado Gold Mines, Limited, as being "regarded by leading metallurgists as an outstanding contribution to metallurgical science."[3] In that same report he devoted almost the whole of a long paragraph to describing in detail the achievements of the Mines Branch's first quarter century, while dismissing the Survey with two simple sentences to the effect that it had gained recognition long before the department was created, and that "Its participation in the development of Canada's mineral resources is too well known to require comment."[4]

The Survey, despite such lukewarm support, attempted to make out the strongest case possible for a fair share of the government's purse. Thus Collins and M. E. Wilson prepared a short piece in the *Northern Miner* special edition of July 9, 1931, that described the Survey's contributions to the national economy. The lesson was hammered home that the Survey's role was not to discover mineral deposits, but to point out the most promising areas to prospectors: "The Survey geologists' place in the industrial scheme is to find out in which rock formations certain minerals can, and are likely to occur, to find out why and how they occur there, to find out where the various rock formations occur, and to tell this to the prospector in the form of maps and reports." The report went on to outline their work in helping the industry to locate supplies of various minerals, instancing its activity in the Rouyn district of Quebec of mapping the potential gold-bearing areas, so that "It is probably safe to say that a hundred million dollars' worth of metals have been found in the Rouyn field. The Geological Survey found none of this, yet the prospectors who did find it would probably be prompt to acknowledge the value of the assistance given them by the Survey."[5] The report concluded by reviewing contributions to other segments of society and government, ranging from discoveries of underground water resources to training the future leaders of the mining industry in its summer field parties.

The seriousness of the situation was indicated by a special meeting of the Logan Club, October 30, 1931, held to discuss how service to the public might be improved. Mainly stressed were publishing interim reports and maps to speed up reporting of promising prospecting areas, special bulletins emphasizing economic aspects, and starting a series, "Contributions to Economic Geology," similar to that of the United States Geological Survey. Another sort of suggestion was to open additional regional offices, on the model of the Vancouver office, in Calgary or Edmonton, Halifax or Saint John, and possibly in Winnipeg. Above all, contacts with the industry should be fostered; the director should devote the summers to getting acquainted with leaders of the industry and their problems, to explain Survey programs, or to make changes to attune its work more closely with the needs of the industry. Following the meeting some of the officers, A. H. Lang in particular, began collecting illustrations of the economic value of past surveys as ammunition for future efforts to gain favourable publicity and support.

The government, however, seemed unconvinced; every year the Minister of Mines reported a further reduction in the vote for explorations, surveys, and investigations:

> Mr. Gordon. Yes, the parties will be cut down this year. This again I recommended with a good deal of reluctance, but I think with the rearrangement in the field, the projects of the Geological Survey will not be seriously interfered with. The officials of the department have, I believe, rearranged the survey parties in such a way that the mining industry will be served, I admit not as well as it might otherwise have been, but for this year I think there is a justification for cutting down the survey parties.[6]

This attitude was not always approved by the mining industry or even by other members of Parliament. Another such explanation in the spring of 1934, evoked a protest from a private member:

> I suggest that this is the wrong place to practice economy. It is false economy, at a time like this, to cut down the amount spent on geological surveys. The object of these survey parties is to explore new areas particularly for the purpose of discovering any minerals that exist therein, and I believe that this offers at the present time some hope for employment in new industries. I think we should encourage the opening up of new mineral areas in Canada.[7]

The Survey, like every other part of the civil service, was pressed to reduce its expenditures in every possible way. Agents of the Department of Finance suggested that the field work should be canceled, and that purchases of new equipment or specimens for the Museum should be halted. In 1933, after field plans had been made under the already reduced budget, a last-minute order came out from the Minister of Finance to reduce "controllable expenditures" throughout the government service by $14 million. The upshot was that the field expenses were slashed from $90,500—with which it was planned to send from 32 to 35 parties to the field—to $40,406. As a result, there were only thirteen parties in the field that season. Of these, two had no assistants at all, while five field officers were limited to such investigations as they could

XVII-2 HORNE MINE AND SMELTER, ROUYN-NORANDA, QUEBEC

manage with allowances of $300-$400 apiece. Naturally, operations were curtailed in all but the most important areas.

Needless to say, the least 'practical' activities suffered the most so long as the austerity continued. Camsell reported that

> Field work in anthropology and biology, which in previous years has constituted the principal activity of the National Museum of Canada, had to be postponed altogether in 1931-32 owing to lack of funds . . . . Advantage has been taken by the Geological Survey and National Museum of the temporary reductions in field work to review the records of work undertaken in recent years, and to prepare for publication a number of reports that are mainly of scientific or technical importance.[8]

The same enforced concentration on indoor work prevailed for the next two years as well; not until 1934 did the Museum attempt "a small resumption of field work, which had been almost completely prohibited since 1930."[9] In 1935 the Museum divisions at last returned to something approaching their previous activities pattern, with a program of ten field projects—five anthropological, three biological, one mineralogical, and one paleontological.

The staff also suffered from the financial difficulties of the government. In 1932 a uniform 10 per cent reduction of all salaries was ordered, a freeze was put on new appointments, and retirements were accelerated by lowering the superannuation age to 65. Even the system of expenses advances that went back to 1884 was cut down. Instead of an advance of $200 on going to the field, the maximum was reduced to $50, and no item costing more than $5 was to be charged against it.

Routine retirements cost the Survey two veterans in the early years of the depression—C. O. Senécal, the long-time head of the Survey's mapping division, on August 23, 1931; and E. R. Faribault, the specialist on the gold fields of Nova Scotia, on December 31, 1932, after fifty years' service, probably the longest in the Survey's history. Three resignations of senior geologists in little over a year—J. F. Wright in June, 1934, and D. F. Kidd on July 20, 1935, both to enter private practice, and J. F. Walker on November 29, 1934, to become provincial mineralogist of British Columbia and deputy minister of mines two years later—created something of a crisis. Camsell appealed for a lifting of the freeze on new appointments, since the Survey was already operating with barely enough officers capable of directing field parties, thirty officers being "the minimum desirable to permit of the Survey having available specialists in different phases of geology and specialists on certain important geological areas." He closed by warning that "If the present policy of Government is continued much longer, we are bound when staff is depleted by retirements, deaths or resignations, to be reduced to a state of efficiency decidedly inferior to that which has won the Department the high esteem it now enjoys."[10] Whether or not his advice was heeded, two new appointments were made in 1935, of H. C. Horwood and A. W. Jolliffe, and a third on December 1, 1936—the day the new Department of Mines and Resources came into being—of H. M. A. Rice.

As an economy measure Collins had also to curtail work by outside specialists, and to discontinue such an activity as the peat investigations of A. Anrep (1932). T. T. Quirke, who had worked on the Huronian north of Lake Huron, was unable to secure work at the Survey after 1931. As the author of his obituary remarked, after Quirke's early death from a heart attack in 1947, "It is unfortunate for geology that so young and capable a man was forced to retire from a field of study in which he was so well prepared and to which he had already made such outstanding contributions."[11]

The depression and the difficulties in which the mining industry found itself caused the Survey to concentrate harder than ever on programs that might assist the industry and the economy immediately and directly. Thus, since there was little point in developing more base metal deposits, the Survey, like the industry, concentrated on helping find and develop deposits of gold, oil and gas, asbestos, or radium. Though the character of the work remained substantially unaltered, it was conducted in places where the results seemed likelier to be of immediate value, such as in the vicinity of mines, ore deposits under development, or districts that gave every indication under reconnaissance of having exceptional promise.

Work in the Precambrian Shield was largely confined to areas where mining was in progress or where new mines were springing into production. Tanton continued his studies around the Lakehead district of western Ontario where he was joined by A. W. Jolliffe, Kidd, and others; Collins still worked around Sudbury. In Quebec M. E. Wilson, after a struggle with Collins, undertook work in connection with the Noranda mine and helped the owners to find important new orebodies by working out the problem of the local structure. He afterwards worked on other mines of the Noranda district, while Lang and L. J. Weeks continued the earlier work of W. F. James and J. B. Mawdsley in the Quebec Gold Belt, to be followed by H. C. Gunning, J. W. Ambrose, G. W. H. Norman, and others. J. F. Wright and C. H. Stockwell worked separately and then together on a number of map-areas in northern Manitoba and Saskatchewan.

The spread of mining operations to the great lakes of the north in the early thirties caused the work to be extended into that part of northwestern Canada. The discovery that the silver ores of Great Bear Lake contained significant amounts of radium inaugurated production of an important mineral that until then could not be mined commercially in North America. Kidd was dispatched to this remote region in 1931, 1932, and again in 1934, to investigate the geology, particularly of the pitchblende occurrences. When important gold deposits were discovered and began to be developed after 1932 in the Yellowknife area of Great Slave Lake, Stockwell turned to work on the area. In 1932 he conducted one of the last traditional long-distance reconnaissances through country north of Yellowknife. The party travelled by canoe

XVII-3 H. C. GUNNING

XVII-4 J. F. WRIGHT

XVII-5 H. C. COOKE

XVII-6 F. J. ALCOCK

up Yellowknife River, crossed the height of land to Coppermine River, which they descended to Point Lake, and after exploring that region returned up the Coppermine, crossed to Barnston River and thence to the east arm of Great Slave Lake. This was truly an epochal trip, even though the party's path was greatly assisted by supplies the RCAF deposited for them at Winter Lake and Point Lake. While Stockwell returned to his work on the Precambrian of Manitoba in 1933, Kidd made a reconnaissance in 1934 of the route via Camsell River between Great Slave and Great Bear Lakes. In 1935 Jolliffe conducted a first study of the Yellowknife gold field.

In other regions the focus was on areas of current mining interest. In the Cordillera, H. S. Bostock, G. Hanson, W. E. Cockfield, Walker, C. E. Cairnes, Kerr and Gunning investigated lode and placer gold discoveries and other minerals in the Cariboo, Bridge River, Vancouver Island, Okanagan, Omineca and Portland Canal districts. In the Eastern Townships, H. C. Cooke, assisted by E. Poitevin, examined the important asbestos and chromite occurrences about Thetford, and Norman and F. J. Alcock studied salt and gold deposits in the Maritimes. In the Central Plains region G. S. Hume continued investigating the oil-bearing potentialities of the formations along the Alberta Foothills, assisted in 1933 by R. T. D. Wickenden; B. R. MacKay carried on his studies of the coal deposits in the Crowsnest Pass district of Alberta and British Columbia. In view of the needs of certain cities for water, and especially that of the farmers of the southern prairies who were faced by widespread, serious and sustained droughts, surficial studies were concentrated on the problem of groundwater resources. Wickenden studied the area near Regina in the years after 1930, and in 1935 and subsequent years this work was greatly expanded under the direction of MacKay.

The Survey in the depression years found itself faced by an unprecedented demand for its services by a desperate mining industry. Mineowners, forced to economize on development work, pleaded for very detailed maps of their areas to help them find new orebodies, or new ore shoots, in the deposits being worked. On the other hand, there was a constant stream of requests for information about the unmapped northern regions which were now accessible to airborne prospecting. Both types of requests would have been hard to meet even in normal times—the first, because the Survey could never hope to deal with more than a tiny fraction of the many mining areas across the country for which its help was requested; the second, because the policy of Collins since 1920 had largely abandoned reconnaissance operations and the mapping of remoter parts of Canada. In the depth of the depression, crippled by the government's austerity program, the Survey could do little more with its limited forces than meet the most pressing needs of the industry. Only a large-scale increase in money and manpower would permit the Survey to make a beginning on the systematic mapping of the vast areas north of the thin band of settlements across southern Canada, or do more detailed work in areas of economic or scientific importance.

The financial stringencies and the technological changes of the early thirties affected the other divisions of the Survey in various ways. By reducing the money available for field work the economic situation encouraged them to seek out, and stress, those aspects of their work that were especially practical, so as to justify their continuance through difficult times. The curtailment of funds for field work also encouraged these divisions—willingly or not—to concentrate on their office routine, as in the case of the Museum divisions. Others, for example the Palaeontological Division, had some of their personnel assigned to regular field mapping studies; in the Mineralogical Division Poitevin investigated areas of interest within the context of the economically-oriented field program. Still other divisions were affected by technological changes, such as the impact of air photography on the work of the Topographical Division which caused W. H. Boyd in 1930 and 1931 to convert more of his force of surveyors to doing the ground control work required in conjunction with the air photographs. One man, D. A. Nichols, was also trained as physiographer, to specialize in studying landforms.

For the Mineralogical Division, the onset of the depression brought a whole new flood of requests for information from complete amateurs, hopeful that they had stumbled on a fortune. The number of visitors to the office on Sparks Street increased, and somehow they seemed stranger than usual. Poitevin tells this story:

> Another time a woman came in. It was almost closing time. She said, "I've got something that's more than all diamonds of the British Crown." I said, "Fine, I'd like to see it." She opened the satchel. She had a group of quartz crystals, transparent, quite clear, nice crystal, and she said, "That's it." Well, I said, "Madame, these are quartz crystals, they are not diamonds. They're not common. Good crystals like that are not common in nature. I have some myself here. See, these are pure quartz." She said, "I went to Smith's Falls, and there's a woman that told me that." I said, "This woman, what does she do?" She said, "She has a ball and she looked at me, and she said, that I have in my purse something that will be worth more than all things." "Well," I said, "Surely you don't believe what this woman would say?" "Well," she says, "Well, little man, if Mackenzie King can go down there and believe her, why can't I?" I found out after that this was true. This Mackenzie King used to consult this lady, and it wasn't a secret. She went into hysterics. Fortunately we had a policeman downstairs. They had to take her downstairs and one man couldn't handle her. Finally one man opened the door, and on the corner of Bank and Sparks he called another policeman. I don't know what he did with her. A lot of them like that.[12]

XVII-7 W. H. BOYD
Chief, Topographical Division and Branch, 1908-40

Another of Poitevin's assignments during these years was most interesting and unusual. It was a study on behalf of the Ontario government, of mineral particles extracted from the lungs of miners who were known, or were suspected to have died of silicosis or other lung diseases. The study lasted for ten years and was related to Workmen's Compensation questions, and to the need for health regulations in the mining industry. Before long Poitevin was able to recognize the contaminating minerals in each lung, estimate their proportions, and from this identify the camp or the particular mine in which the victim had worked. It became possible to learn which minerals or combinations of minerals—even which mines—were especially dangerous, and the government could take appropriate action.

As part of his experiments, which he conducted after hours, Poitevin once asked for a completely uncontaminated lung for comparative purposes, and in due course received a sample from the lung of a one week-old baby. When he treated it by the usual procedure he found to his surprise and dismay that he was left with a very considerable mineral residue. The mystery was explained when examination revealed that the mineral was talc, evidently ingested as talcum powder. It would seem that the gross hazards of air pollution, indeed, follow us from the cradle to the grave. In 1935 the studies were extended to the asbestos mines of Quebec, this time also with the co-operation of the Quebec Department of Health.

After Ingall's retirement, Collins decided to broaden out the work of the Borings Division by adding to it Pleistocene geology which, though efficiently managed by W. A. Johnston, was something of an orphan in the organizational sense. In 1930 a new Division of Pleistocene Geology, Water Supply, and Borings was established, with Johnston at its head, assisted by Wickenden, D. C. Maddox, and F. J. Fraser. Its main activities, besides those of collecting and studying samples and records of well borings, and investigating water supply, included the study of Quaternary (Pleistocene and Recent) formations throughout Canada to assist agricultural surveys and other industrial purposes, placer deposits, peat bogs, and the like. Collins explained the importance of this type of work:

> With increasing density of population in southern Canada and corresponding increase in individual complexity the Pleistocene and Recent formations—those unconsolidated deposits of clay, sand, organic matter, etc., that overlie the glaciated bedrock formation of the country, which are popularly spoken of as soils—are assuming a growing importance. A large share of the materials required for our great program of road building comes from these deposits. They furnish much of the raw materials for manufacture of brick, tile, and other structural materials. They yield gold, platinum, and other metals from placer deposits, and fuel and other useful products from peat bogs. They are the basis of all agriculture, and are becoming more and more the subject of scientific survey and analyses from the agricultural standpoint. From these deposits is drawn much of the water supply needed for domestic and individual uses, and in places where population is dense, e.g., southwestern Ontario, or precipitation is scant, as in the southern parts of Alberta, Saskatchewan, and Manitoba, the amount and purity of the available supply are becoming of growing concern.[13]

XVII-8
EUGENE POITEVIN
Chief, Mineralogy Division, 1922-56

Under Johnston the emphasis changed to the Pleistocene, particularly to examining groundwater supplies of cities and rural areas in the prairies. Through such studies additional supplies of water had been discovered for Regina and Moose Jaw in 1929 and 1930. Now, in the dry years, the search for underground water supplies began with a vengeance, and soon the division had several thousand records of water wells. Maddox was mainly employed at this work, and Wickenden continued to specialize in fossil determinations from deep well samples. In 1935 the search for groundwater resources was greatly expanded throughout the southern prairie region.

The depression years brought great pressures on the Draughting and Reproducing Division, headed after 1931 by A. Dickison, and also on the editorial and publication side of the Survey, in the form of complaints about the slowness in making the findings of field surveys available to the public. In view of budgetary difficulties, men became uncertain when a memoir might be published and began increasingly to prepare elaborate preliminary reports, including maps and figures, to appear in the summary reports. Part A-1 of the *Summary Report for 1932* runs to 151 pages but comprises only two papers, both on the Cariboo gold deposits. These are certainly not preliminary papers and could not be processed very speedily. Nicolas grumbled that the summary reports were becoming fuller and less comprehensible, and suggested that he had better oversee all contributions. The lengthy reports also held back publication of papers with important information. It was a parallel situation—delays in waiting for the slowest member of a group—that had put an end to the Annual Reports 25 years earlier, and to the Reports of Progress 20 years before that.

XVII-9 W. A. JOHNSTON

XVII-10 ALEXANDER DICKISON
Chief Draftsman, 1931-45

A similar problem arose over maps which prospectors desired to use by the spring following the original field work. After a reconnaissance survey of the Waswanipi region of Quebec in 1931, Lang as an experiment compiled a map from various sources (including air photographs), had it reproduced photographically, and then coloured by a clerk. The published account in the Summary Report contained a footnote to the effect that hand-coloured photographic copies of the map could be obtained for a nominal charge. In the next year (1932) this was repeated following his work in the Palmarolle district of Quebec, and his report carried a similar footnote, as did one or two of the other reports.

Because of the pressure for the Survey to make its field results available at the earliest possible date, the Summary Reports were discontinued after 1933, and beginning in 1935 the policy was instituted of preparing and issuing preliminary maps and reports by mimeographing, blueprinting and other quick, cheap methods, shortly after the men returned from the field and in good time for use in the following summer's prospecting season. These Preliminary Series Papers, which lasted down to 1958, were numbered for reference purposes with a prefix to indicate the year of publication, as for example, *Preliminary Series Paper 37-6.* The papers included both reports and annotated maps accompanied only by marginal notes. Their unpretentious appearance made it clear these were true preliminary reports, and often they subsequently became the basis for more ambitious publications in the Memoir Series. Since every stage of processing and distribution was carried on inside the Survey by its own staff, the time spent in preparing and publishing the new papers and maps was greatly reduced. Henceforth the Memoirs and the Preliminary Series Papers became the chief repositories of the findings of the field work, research, and study of the geology of Canada.

* * *

THE EARLY THIRTIES MARKED the introduction of a significant change in the style of field operations through the application of aviation for transportation and reconnaissance purposes—a change that was to become so all-pervasive that geologists have been known to divide the history of the Survey into pre-airplane and airplane periods. Yet by the time the Survey began making any adequate use of the airplane its employment for these purposes had become almost commonplace, and a whole decade had elapsed since the vital technological breakthrough had occurred.

Why was the Survey so backward in putting to use such an obvious labour- and time-saving invention that was so beneficial for its field operations in many parts of the country? Since Collins was so concerned to advance the understanding of Canadian geology and improve the efficiency of its mapping program, why should he have failed for so long to act on the insurmountable evidence that revolutionary improvements could be secured by using aircraft on Survey operations? Collins did not have an entirely free hand to pursue his own wishes for the Survey because of its administrative subordination to the Department of Mines and its relatively limited financial resources even in the prosperous twenties. Still, Collins' reaction to the prospective new tool is abundant illustration of some of his major shortcomings as director—his innate conservatism and reluctance to recognize and cheerfully accept a new opportunity and role for the Survey; his obsession over money costs that inhibited him from using his own funds to inaugurate a desirable and important program; his inability to influence his superiors to institute a system that at least took account of the legitimate interests of the Survey; and perhaps worst of all, his failure to realize how, by tamely acquiescing to the existing program, he was fencing-in the Survey and failing to give it a dynamic, distinctive, progressive image. By not making an early and positive commitment to aviation and working out its implications for Survey activities, Collins kept the Survey on an unenterprising, even prosaic, course in a period of national desperation that was to cost it, and Collins, dearly.

The age of practical aviation had arrived in Canada by 1919, and Canadian industry, especially the mining industry, was quick to put it to use. Aviation rapidly proved the means *par excellence* for opening up isolated areas beyond the railways and roads of settled Canada. It was ideal, too, as a mapping tool, and the only practical means at the Survey's disposal to try to keep pace with the expansion of prospecting activities into vast areas that had never been adequately examined or mapped from a geological point of view. To maintain its position in relation to the industry, the Survey also had to extend its field of operations. Aviation had created the Survey's problem, and aviation alone could help solve it.

As a bare minimum, the Survey needed to consider using aircraft for transportation, to deliver parties of geologists and topographers to out-of-the-way places and lengthen their field season there. More important, however, were the possible values of aviation to assist the men with their activities. For instance, a great deal of topographical information might be secured from the air, that surveyors had to climb to the tops of mountains to obtain. Provided visibility was good, the airplane would permit very effective, rapid, accurate photographing, sketching, and mapping of lakes and other topographical features. Since topographical surveying was a serious drain on the Survey's resources, this use of aircraft represented a tremendous saving in time and effort.

Aviation held out an even greater promise of helping with geological investigations, as an interesting article published by E. L. Bruce early in 1922 indicated. That study was based on his experience the previous summer while doing some work for the Ontario Bureau of Mines. At Sioux Lookout the officer in charge of the Canadian Air Board station had flown him in a seaplane over the country to test whether geological reconnaissance could be conducted from the air, and what value seaplanes might have in this regard. As the flight was over territory Bruce had been surveying the previous summer, he was in an excellent position to evaluate the experience and make comparisons with traditional methods. In an hour's flight they reached Lake St. Joseph, a journey that had previously taken four days of travel. During the flight Bruce checked many of the lakes he had mapped; he noted he had made several errors in his mapping of major lakes and detected numberless small lakes, and some large ones, that he had missed altogether. He observed a possible canoe route that he might have taken had he known of its existence. As for geological information, he saw that a belt of granite he had examined at two separate places extended between those points and could now be mapped as a continuous belt. He also noted how this granite body extended across another six to eight miles of country he had been unable to reach. From the air, too, the glaciation was readily visible: "Kettle lakes with no outlets, hummocky hills and striking ridges of sand and gravel, which indicate the general south-westerly trend of the Patrician glacier." He summed up his experience:

> It seems evident that the use of sea-planes would relieve the geologist of a vast amount of work and would save a large part of his time. Areas of granite which are not important economically, and those areas of deep glacial cover where no rocks are exposed, could be eliminated. It would thus free him for the exploration of the

XVII-11 STOWING A PROSPECTOR'S BAGGAGE IN A VICKERS HS2L FLYING BOAT
Northern Quebec, the late 1920's

> areas of promising rocks, and so extend the field covered by each season's operations. It would, moreover, give him an idea of the most advantageous points for inland traverses and would show him the position of inland lakes which might be of use to him in fixing the boundaries of formations. It would not, of course, be economical to attach a sea-plane section to each survey party, because there will still remain a large part of the geological work that must be done by the old methods of canoe travel and traverses on foot; but it might be possible to so arrange field parties that several geological parties could each have the advantage of a sea-plane from a central station for a short time, and in this way the work of all would be facilitated.[14]

Collins was not oblivious to these uses of aircraft. In the winter of 1920-21 he applied to the Air Board for assistance in connection with the Survey's projected operations along Mackenzie River, including supplying food and mail for three geological and five topographical parties working north from Fort Simpson at intervals through the summer, and then bringing the parties back from the district around October 1. Early in 1922, perhaps reflecting Bruce's experience, he presented another list of possible work the Air Board might do on the Survey's behalf, among which was enumerated the following: "If flying operations for any other purpose are to be carried on in the vicinity of Lake St. John, Quebec, one of our geologist[s] would like to be given one or more observation flights of 50 to 100 miles for the purpose of observing from the air the distribution of the Palaeozoic formations in that region." He also suggested that the geologists working in the Mackenzie district might be taken on observation flights inland to obtain "birdseye views and photographs of the country, or even to land a party of two or three men inland, with supplies for a fortnight."[15] Thus Collins in the early twenties was sufficiently interested in the possible uses of aviation for Survey operations at least to desire to experiment with air transport and observations. Indeed, on March 13, 1923, he was appointed as Department of Mines representative on the permanent interdepartmental committee on government air operations.

But why did Collins not follow up this beginning? The Survey's experience of the first years was not at all satisfactory. Though Collins dutifully applied each year on behalf of the Survey as he was requested, he received no service in 1921 on behalf of the expedition to the Mackenzie valley, while the only result of his 1922 application was the transporting of R. C. McDonald and his party by air to the site of his topographical operation at Wuskwatim Lake in northern Manitoba, and "Unfortunately this was the only flying operation that could be carried out, on account of trouble with the machines later on in the season."[16] In 1923, after negotiations with the Department of National Defence (Air Service), which had taken over from the Air Board, it was agreed that the RCAF would do transportation work for two parties in Manitoba, one around Island Lake, the other 60 miles east of Victoria Beach, as well as "observation flights for officials of the survey working in that district, also oblique photography."[17] The Island Lake operation was apparently canceled, while the service on the other operation was quite inadequate, according to J. F. Wright:

> The Royal Canadian Air Force transported the parties, provisions, and mail from Victoria Beach to Long lake. Owing to bad weather conditions and other unfortunate and unforeseen events, the Air Force could not give as much assistance as had been hoped, but the writer wishes to express his indebtedness to Major D. B. Hobbs, Squadron Leader, for the assistance rendered and for his keen desire and efforts to do whatever was possible.[18]

The story in British Columbia was the same. V. Dolmage's party operating north of Vancouver at Chilko Lake in 1924 received none of the air support the RCAF had promised, because of the risks of flying in that mountainous country, and the less-than-perfect weather. The Survey faced obvious difficulties in obtaining service at this time. It did not seem to have any real claim on the facilities of the Air Board or RCAF, while the uncertainties of planes and weather meant that what service was provided was of limited real value.

Added to these was a problem of cost. The earlier offers had the Air Board absorbing the entire expense, but the RCAF's outlined proposal put a price tag of $3,000 on a budgeted 50 hours of flying time while asserting that the figure was merely for the record since "Provision has been made in our estimates for the total cost of these operations." The memorandum went on, however, to hint that "If arrangements can be made whereby the cost, or any portion of it, can be made from other appropriations on account of any saving made by doing the work from the air, the flying operations can be increased accordingly."[19] Later, the department was asked to contribute towards the cost of the air service about to be rendered to the Survey in 1923, and the Survey agreed to pay half the cost of what service it received, "on condition that the services requested are performed to our satisfaction and that we have a sufficient balance to make payment."[20] For this effort the Survey secured the limited assistance on which Wright had reported.

Indeed, as late as 1924 the Survey still was not entirely convinced of the value of aviation for its field work. Some of the doubts emerged in a cost-benefit study Young made in September 1924 in response to a draft of an article on the uses of air transport. He set the value of the air service against the value of an equivalent expenditure on additional geologists, less the possible inconvenience and expense of depending on the aircraft—notably having to adopt a rigid itinerary that might make it necessary to leave before the work was finished, or that might leave a party waiting around for the aircraft. Thus the unsatisfactory aspects of aviation at that time were stressed, while the benefits of air service apart from the saving in travel time were discounted. Young's conclusion was that the value of using aircraft depended on the merits of the particular case: "It would appear that the merit or demerit in the use of aircraft is not to be measured in terms of dollars nor of degrees of effectiveness, but according to the needs and desires of the geologist in charge and of his employer."[21]

This seems to have ended Collins' interest in using aircraft for the next half dozen years. In fact, in 1930 when Lang discussed the values of air transport with Collins he came away with the impression that the subject was a new one to the director.

By 1926, however, Collins had become aware of one benefit that the Survey might obtain from aircraft without having to spend its money for the privilege—air photographs, as supplied by the RCAF's aerial photography program. By now the techniques of aerial photography had been considerably improved, and so had the aircraft. Referring to the value of air photography for topographical work which had consumed "a large share of the energy of the Geological Survey" and had been performed "at the sacrifice of geological investigation" in the past, Collins foresaw how air photography would eventually greatly reduce the Survey's needs for traditional base maps. He went on, moreover, to report an important contribution to geological knowledge as well:

> During 1926, however, a far more important service was rendered by the vertical photography of an area of nearly 4,000 square miles in Rouyn mineral district, Quebec. The photographs, which were taken by the Royal Canadian Air Force, are being used jointly by the Geological Survey, Department of Mines, and the Topographical Survey Branch, Department of the Interior, for the preparation of maps richer in geographical detail than would be obtainable by ordinary ground survey methods. The photographs also permit of a more accurate delimitation of rocky, and of drift-covered, areas, a distinction of much value to prospectors. The brief experience that the Geological Survey has had with aerial photographs demonstrates that, in certain cases, geological information can be obtained that would be difficult and costly to secure by other methods, and it seems probable that, with increasing experience, geologists will benefit to a greater extent from these pictures.[22]

Once more dependence on other agencies of government kept the Survey from obtaining full benefit of this new development—dependence on the RCAF for taking photographs of areas of geological significance as requisitioned for by the Survey; and on the Topographical Survey Branch of the Department of the Interior for processing them and compiling the maps. The Survey had excellent topographical and photographic units of its own, but the government in 1925 had decided to consolidate all aerial photographic work in the hands of an Aerial Surveying Unit within the Topographical Survey Branch to prevent duplication of effort, maintain a centralized filing system, and control applications for work sent forward to the RCAF. Besides the difficulties in get-

XVII-12 PLANE OF WESTERN CANADA AIRWAYS EQUIPPED WITH SKIS FOR WINTER FLYING, LEAVING WINNIPEG FOR THE NORTH COUNTRY, WINTER 1929-30

ting desired work done by the Air Force, which had so many other calls on its services, there were delays and inadequacies in the subsequent processing and compilation work based on the photographs. W. H. Boyd, head of the Survey's Topographical Division, whose role was bypassed in this important new field of activity, complained on April 13, 1928, about the difficulties of using photographs prepared by the Topographical Survey for the purposes of the Geological Survey and Camsell proposed to the Department of the Interior that the Survey should do part of the work of compilation for its own needs:

> Difficulties which have arisen in embodying into our ... sheets the compilation from the aerial photographs which were prepared for us by the Topographical Survey, has compelled us to pay a good deal of attention to the matter .... These difficulties which were due to inaccuracies in the compilations supplied us and the exacting nature of our requirements are so serious that, except for the geological information on them, it will probably not pay us to use aerial photographs in future unless a high standard of photography is maintained and better results can be obtained from the photographs .... It does seem, however, that our needs would be better served by an arrangement whereby the Geological Survey would make its own compilations of geological and geographical information from the photographs and your department could give so much the more time to better photography. Compilations we would make under this plan would, of course, be available to your department in the same way that your compilations are to us.[23]

In 1929, Collins again complained about the air photographic service and declined to submit new projects for transmission to the RCAF on the grounds that the work performed to date was so unsatisfactory and tardy as to be of little or no use. He informed Camsell that in nearly every case where maps and reports depended on air photographs, they had been almost at a standstill for as much as three years. In view of the long unsatisfactory record he felt that "apparently it is the system which is defective."[24] The situation persisted for many years. As late as 1936 F.C.C. Lynch, Collins' successor, complained to Camsell in very similar tones about the air photography work being done for the Survey by the RCAF through the Topographical and Air Survey Bureau to the effect that the uncertainties prevented the intelligent planning of its own work. He proposed instead that the Survey should contract out its photographic needs to commercial operators, a course that soon began to be followed.

By the later twenties, when prospecting activities were supported by commercial aircraft in many parts of the Canadian north, the Survey was under obvious pressures to follow suit and use aircraft at least to fly its men to remote locations. Since the Survey still depended on the RCAF for service, it resorted again to this agency, for example, in 1929 when Collins submitted requests for observation flights over the Hudson Bay coast from Chesterfield to Rankin Inlet, and also to transport Weeks and his party to that district from Churchill in the spring of 1931. For 1931 he also proposed that two parties should be transported to the Great Bear Lake district, a topographical party under McDonald, and a geological party under Kidd. The job would include air photography as well as the transporting of men, food supplies, and equipment. Collins explained that a party depending on water transport would be limited to a four-week season in that locality; whereas, with air transport that could fly it 1,000 miles from railhead in a single day, an entire ten-week season of work could be completed.

Prior to 1931 the Survey does not appear to have had any greater success with its applications to the RCAF than in the early twenties. But by 1931 the situation had changed; the RCAF, facing threats of cutbacks in funds and activities, was now eager to demonstrate how useful it could be to other agencies of government. In 1931 and the next few years it placed its aircraft at the service of Survey parties working in locations handy to their own air operations. In 1931 and again in 1932 the geological and topographical parties were flown out to Great Bear Lake and were able to make much earlier starts on their season's work. In northern Quebec the RCAF also provided transportation and air reconnaissance service for Lang's large field operations in the Matagami and Waswanipi Lakes area in 1931 and 1932. In 1932 Gunning was supported in his work in the Zeballos gold camp on Vancouver Island by the RCAF squadron based at Jericho Beach near Vancouver. Again, in 1934 Norman was supplied with 18 hours of transportation time (enough to move his entire operation from railhead to the site of the work, 160 miles away) and with 10 hours of geological reconnaissance flying by RCAF planes doing air photography work in the Chibougamau Lake district.

Occasionally, too, and on an emergency basis, the Survey employed commercial aircraft for pay. In 1930 a Prospectors' Airways plane flew Lang and his party from their work in the Chibougamau district to the railhead at Senneterre on the understanding that if the Survey failed to approve the $300 fare the company would simply write it off. Collins authorized the expenditure which had saved the party an arduous canoe and portage journey, and he expressed interest in Lang's experience to the point of sending him to make a study of interpreting air photographs. No doubt geologists often were taken on air flights on an informal basis as part of the camaraderie and mutual assistance habitual to the time and place. Lang was carried on reconnaissance flights by Prospectors' Airways at the end of the 1931, 1932, and 1933 seasons, sometimes as far as Nemiskau, where the company was examining prospects of its own. He was

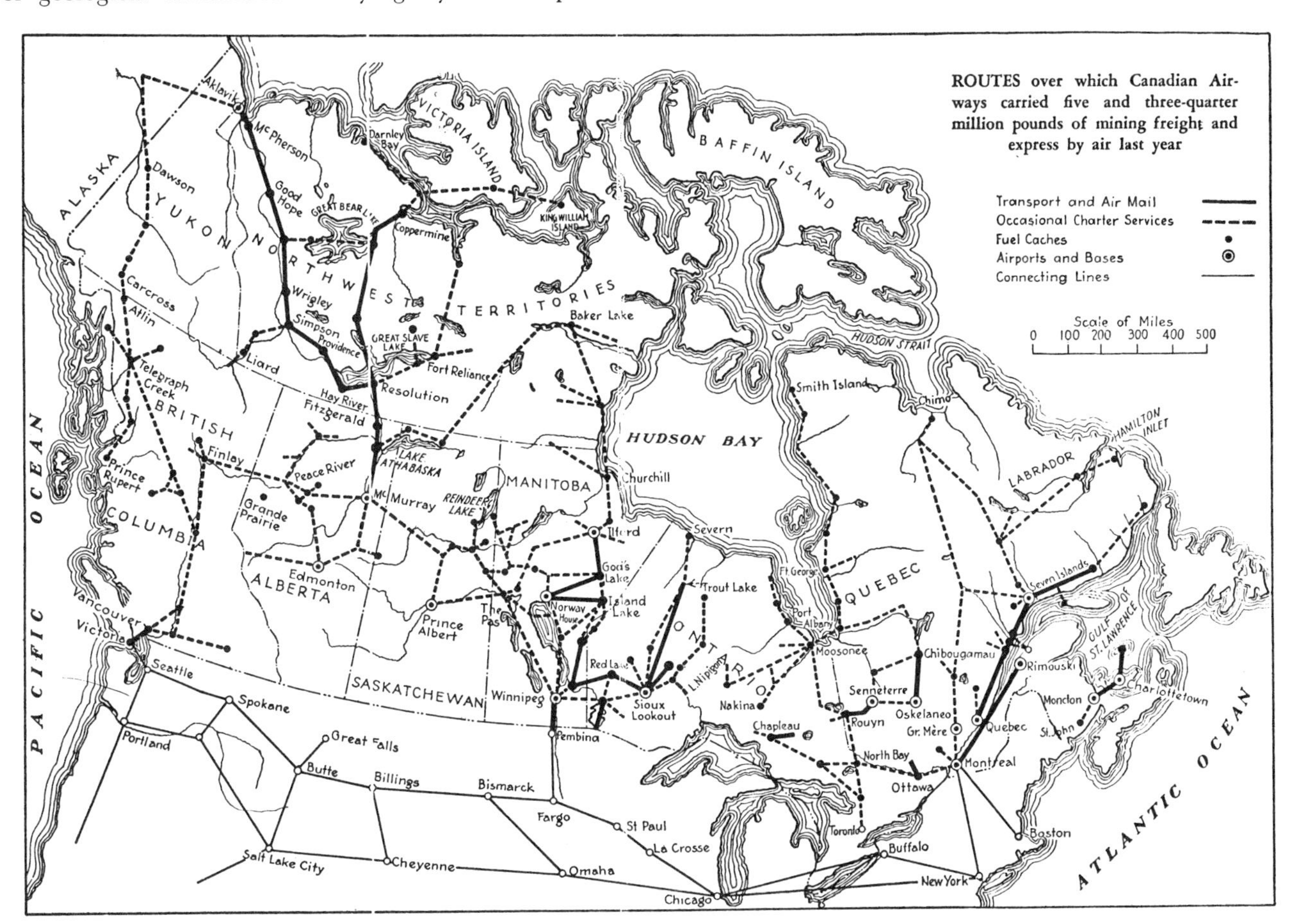

XVII-13 NORTHERN AIR SERVICES, 1934-35
With air service so widespread throughout the North, prospecting activity in the 1930's moved into unsurveyed territory, far outdistancing the Survey.

able to study areas of possible future operations and, in return, to give his hosts the benefit of his opinions.

Along with these few flights the geologists and the Survey were gaining greater experience in putting the air service to fuller use. Thus Collins was able to inform a correspondent in 1931 about the Survey's experience with using air photographs and what had been learned:

> Last year, for the first time, three of our geologists undertook to use aerial photographs in the field instead of base maps. One of those parties was mapping on the scale of 1″ to 1 mile in mountainous country with a local relief of about 4,000 feet and found the photographs satisfactory. A second was working on the same scale in comparatively flat country and found the photographs even more satisfactory. The third, working on a scale of 1″ to 800′ in a mining camp was unable to use the photographs as a base. On the whole, this plan is promising and will permit geological and topographical work to be carried on simultaneously, and then produced in map form correspondingly quickly, in contrast to the ordinary practice whereby the topographical base maps must be undertaken at least one year ahead of geological field work.[25]

The geologists had also become accustomed to using air photographs to indicate geographical features from which to plan camp sites and traverses to outcrops, and also to obtain an understanding of the general geology of the area, which could then be tested by ground observations.

One of the most advanced such operations for the period was conducted by Alcock in 1934 in his exploration of the Mudjatik and Haultain Rivers district of northern Saskatchewan. Thanks to special service provided by the Saskatchewan government which was far superior to what the Survey received from the RCAF, he was able to profit from air service in a way that his fellows were not. The survey had the frequent and regular use of the Saskatchewan government's Vedette amphibian aircraft, to which was added a set of oblique aerial photographs supplied to serve for the base map, plus a short-wave radio receiving set for the party's use. The pattern of the lakes and rivers on the photographs suggested to Alcock that the rocks were comprised of volcanics or sedimentaries intruded by oval-shaped granitic masses. The granites, being the harder rock, were less eroded than the volcanics or sedimentaries, whose positions were indicated by the drainage patterns. The possibility of mineral occurrences was confined to these softer formations, so the expedition concentrated on them. As much of the area as possible was covered in the course of a survey that combined aerial observations with inspections of rock outcrops. The photographs afforded a base map of the innumerable lakes and waterways which could not have been made in any other way. They facilitated planning the routes of ground travel by way of the larger lakes, the location of the main camps, and arranging the work of the sub-parties so as to examine all the outcrops and make traverses on foot to key places away from the water. Supplies were easily moved from point to point by the aircraft, and it was even possible for the geologist to have the plane set him down to examine a rock outcrop at a moment's notice whenever he saw something interesting from the air. The flying boat was very helpful for tracing geological boundaries, the contacts between different kinds of rock, and the continuity of formations. By combining air observations with ground examinations, large areas of country rock could be rapidly mapped in such operations, the effects of faults could be observed, and formations could be traced for long distances as they wound through and past the various lakes.

These experiences were mouth-watering tastes of what the Survey might do for itself if it had the money to hire and control its own air service, and if it were freed, or freed itself, from depending on the "authorized" channels of the RCAF and the Topographical and Air Services Bureau (the successor of the Topographical Surveys Branch of the former Department of the Interior). The failure to adopt modern methods and the Survey's persistence in traditional routines that were shown to be inefficient and wasteful of effort compared with the new methods were frustrating to the staff. More important, the Survey missed acquiring an activity and a role that could be dramatized before the aviation-conscious public and the politicians, and one that the deputy minister Camsell would have been bound to endorse and support. Instead, Collins tamely accepted the situation and apparently made no effort to force a change in policy that would have allowed the Survey to increase its services to the mining industry. Through Collins' slowness to recognize the importance and value of aviation, and his failure to resist the deadening effects of the system, the Survey for a whole decade derived little benefit from the airplane. As a result—cause or effect?—the Survey lost the chance to make an adequate response to the industry's requests for maps and reports on the northern Canadian districts opened up for prospecting and development by the airplane.

* * *

THE SURVEY'S INABILITY TO UTILIZE the technical capabilities of the aircraft for the more effective performance of its work was only one of the many difficulties in which it found itself in the early thirties. The Conservative government of R. B. Bennett, 1930 to 1935, drastically cut the appropriations and in order to provide for the staff the spending on field work had to be reduced time and again, until it reached disastrous depths. At the same time, the mining industry and provincial governments were pleading that the Survey's means and staff should be increased so as to accelerate the mapping of the more promising prospecting districts. The requests went largely unanswered; the Survey, despite its best endeavours, was able to do little to meet these many demands for its work. The favourable price of gold provided every incentive for companies to discover and develop new deposits that would now pay working, and the search for such deposits had been extended far into the north of the provinces and the territories beyond, to areas that had never been geologically mapped beyond the route traverses of early explorer-geologists, if that. Since a very high percentage of the country was completely unmapped, it was certain that many large mineral occurrences still awaited discovery and development. In the face of this immense challenge the Survey was hopelessly understaffed and under-financed. Only with a vigorous, costly program would the Survey be in a position to provide adequate help to an ailing mining industry, and be able to do its part to bolster the national economy.

Just as the Survey was experiencing such deprivation, so too were the geological and mining engineering professions. Many geologists were unemployed, while new graduates were unable to find positions, many students were abandoning their programs for lack of money to continue their studies. Besides the many human tragedies, there was danger that in the long run the mining industry and the national economy would be hopelessly crippled for years to come if so vital a part of the future professional, technical, and managerial mining classes was sharply diminished. The federal government, slowly and reluctantly drawn by the Roosevelt example into the unfamiliar activity of creating work for the casualties of the depression, had done little as yet to assist this group of students in the mining and geological sciences. In fact, the unfortunate curtailment of field work by the Survey only contributed to the situation. The expansion of the Survey's operations, needed in any event for its contribution to the progress of the industry, would provide some employment and give practical field experience to a class of persons important for the future of Canadian mining.

In the spring of 1935, the Dominion government, without much apparent forethought, took a dramatic, unprecedented, and totally unexpected step. After so many years of reducing the Survey's funds, it announced that one million dollars (a round figure obviously selected for its publicity and political value for a government facing a general election in the near future), would be allowed for special projects to be undertaken by the Survey. This was more than a tenfold increase in the funds that were allotted in the 1935 budget to field operations. The Minister of Mines, W. A. Gordon, who as a member of Parliament for a northern Ontario mining constituency had this program very much to heart, spoke on the subject in the House of Commons on April 17, 1935, taking as his theme the value of the proposed program in opening the north: "We have across the northern part of Canada from coast to coast geological conditions that would challenge the courage, the enthusiasm and the capacity for effort of the Canadian people for thousands of years."[26] The Supplementary Public Works Construction Act, which provided for a large number of activities

XVII-14 WESLEY A. GORDON, M.P.
Minister of Mines, 1930-35

and expenditures, carried an item under Schedule A, "Geological Surveys and investigations in the Northwest Territories and elsewhere in Canada . . . 1,000,000."[27]

This wording, as well as the minister's introductory address, seemed to imply that the effort would be concentrated on northern Canada. Yet as the Million Dollar Year developed, only one of its many projects actually worked in the Northwest Territories, and only one topographical and three geological parties in the Yukon Territory. However, a large part (though by no means all) of the activities carried on in the provinces were located in their more northerly districts. The bulk of the effort was undertaken in British Columbia, Ontario, and the Prairie Provinces.

From the beginning the program had two objectives which to some extent were in conflict: the creation of employment for a certain group of young people affected by the depression; and the furtherance of the Survey's work on behalf of the mining industry. The relief aspect, which was probably the more important objective from the government's point of view, was defended in a memorandum prepared later in 1935:

> As an unemployment relief scheme it had as its immediate objective the provision for the giving of practical experience, denied in the depression years, to a class of students and young college graduates upon whom hinges the future of Canada's mining industry, and the consequent economic expansion of the Dominion. The group in question was those who had taken academic courses in mining engineering and geology.[28]

On the other hand, Collins and Young were concerned above all to secure the maximum value for geological work and for the future well-being of the Survey from the unaccustomed windfall. One thing certain was that none of the million dollars would go to the Survey's depressed dependant, the National Museum. The appropriation was definitely restricted to expanding geological field activities, not for helping museums or other undertakings.

Meanwhile, as excitement and enthusiasm grew, the spring came and went; not until mid-June was the vote finally approved by Parliament and the Survey could at last begin making firm arrangements for the work. Finding people seemed no problem; around 4,000 applications were received for positions. The difficulty was to secure enough qualified persons to take charge of the many projects or to lead the sub-parties. As early as April 27 Camsell had written to the minister for authority to offer positions to 34 persons who were some of the "key geologists and topographers who have been with our field parties in the past and are to take full charge of parties this year."[29] Several of these men were already withdrawing their applications to accept other positions because of the delay, leaving the senior positions for possibly less capable persons. Certain projects also could not be begun until vertical air photographs of the areas were taken, so such work would have to be contracted out to air survey firms as early as possible.

Finding the leaders, crews, supplies, and equipment for the many additional field parties to be put in the field that summer was a major difficulty. Every available geologist capable of leading a field party was hired, every university was canvassed for students, the country was scoured for cooks, packers, canoemen. Horses, canoes, cars, tents, stoves, hammers, compasses—all the equipment geological or topographical parties needed, had to be purchased. Indeed, one of the important legacies of the Million Dollar Year was the great stock of supplies, instruments and equipment (including items as large as trucks) that devolved on the Survey; some continued to be used for two decades or more, reducing the need for future purchases for some time to come. Altogether 1,005 persons were taken on the payroll for the project, and no fewer than 188 field parties engaged in geological and topographical work in connection with the Million Dollar Year. The bulk of this manpower was credited to British Columbia, Ontario, and Quebec in connection with searches for ore minerals, principally gold, and asbestos. The large number of persons hired for work in Alberta and Saskatchewan were occupied with searches for water resources as well as for oil and natural gas.

Though 691 of the 1,005 persons were university students (of whom 658 were hired for field work), only a few had geological training. One officer found that his assistants included a third year mechanical engineering student, a second year commerce student, and a first year arts student, not one of whom knew the difference between a granite and a diorite. Since each Survey officer had to direct an average of five parties besides his own, they had to put the more responsible, experienced men at the head of the sub-parties. Graduate and senior undergraduate students in geology or mining engineering, who in the normal course of events could scarcely have expected to be appointed to a Survey party, thus found themselves leading field parties.

The senior geologists directed their own field parties and kept in as close contact as they could with the work being done by the detached sub-parties. This was not too difficult when fairly closely related operations were involved. But when a far-ranging operation was attempted, as in northern Manitoba or Saskatchewan, special arrangements entailing the use of aircraft and radio equipment had to be made. Alcock, who returned to northern Saskatchewan to work in the Lake Athabasca

XVII-15 AIR-SUPPORTED GSC GEOLOGICAL MAPPING PROGRAM, 1935
F. J. Alcock's parties working in the Lake Athabasca district, Northern Saskatchewan, in the Million Dollar Year, were regularly supplied by aircraft.

area, had charge of seven parties in addition to his own. The parties were comprised of students, some of whom had never studied geology, but all were headed by engineering geology students, mostly third or fourth year students at the University of Saskatchewan. He spent five days of each week with his own party, then on Saturdays and Sundays visited the other field parties in turn, using a plane that came up from Ile à la Crosse every weekend for the purpose. Contact among the parties was maintained through the use of radio, a small station being installed at Ile à la Crosse with which the parties communicated at 6:00 a.m. each morning to order further supplies or pass on information, which could then be dealt with by the weekly plane.

The largest operation of all was the search for supplies of groundwater or artesian water in the arid parts of the Prairie Provinces. MacKay, replacing the ailing W. A. Johnston, took charge of the work of about 35 different field parties, with an office staff of twenty. The work was more extensive than profound; small parties covered an estimated 100,000 square miles in southern Saskatchewan and southeastern Alberta, collecting all available data on the location of wells and the quality and extent of existing water supplies. Oil and gas searches also figured prominently, the area of investigations including almost all of southwestern Ontario, Manitoulin Island, and central Saskatchewan and east-central Alberta. In the Yellowknife district, the single reconnaissance program investigated a 10,000-square-mile area, and classified about 2,900 square miles of it as favourable territory for gold prospecting. In northern Ontario several projects were to map the geology between areas known to contain gold discoveries to see whether the same conditions extended across the intervening districts. In a number of operations the geological parties were accompanied by topographical surveying parties, organized in similar fashion under the supervision of the Topographical Division.

And so the Million Dollar Year came to an end in the autumn of 1935. What had it really achieved besides employing 1,005 university students and others? Beyond question it did much to fulfil the objective of creating work. Unemployed cooks, packers, and labourers found jobs, students earned funds to return to university and finish their education, and graduates could hold out a little longer until a place for them turned up in their chosen field. This applied especially to the geologists. The Survey was able to offer employment to many graduated and graduating geologists; some who were inducted into the Survey were subsequently added to the permanent staff, among them H. M. A. Rice, J. F. Henderson, and J. F. Caley, who began distinguished careers with the Survey, as well as J. W. Ambrose, E. J. Lees, A. W. Jolliffe, and R. C. McMurchy, who served for shorter periods. The introduction to field work caused some of the young men to adopt geology as a profession. The

XVII-16 B. R. MacKAY

Survey re-established stronger contact with the academic community, something that had been greatly weakened by the austerity of the recent years.

As for the work itself, the main results were in the form of information gathered by mapping the terrain, recording observations of the geology, and collecting data on wells. The information procured by the 35 field parties on the water resources survey in the prairies took years to collate. Indeed, the vast amount of data, organized according to municipal districts, was published as a series of 225 Water Supply Papers that by 1957 had grown to over 300 reports. Much work had been done to examine districts that seemed to have good prospects of sustaining mining industries; the findings either substantiated or ruled out such expectations, saving the Survey some fruitless searching in the future and enabling it to concentrate its efforts in more useful directions. At least one of the parties, that of J. R. Johnston in the Yukon, much later was credited with having reported the favourable showings that eventually resulted in the opening of a large base-metal mine at Anvil Creek. These were useful results, though (except for the water resources work) they were almost negligible in comparison with what the Survey might have achieved had the $1,000,000 been applied in more regular fashion to its on-going field program—say tripling or quadrupling the amount of field work over a three- or four-year period.

One important effect of the Million Dollar Year was the increased use of aircraft by the Survey, required by the pressing need to get base maps of some areas for work that summer, and to enable party chiefs to keep in contact with their many sub-parties. It coincided, too, with a federal government policy to favour the use of civilian air contractors as a way of helping the aviation industry over difficult times. The Million Dollar Year, in fact, was the first time the Survey ever had enough money to hire air service on a considerable scale on its own account. Some eleven of the chiefs directing geological parties employed aircraft in their work; one of them, A. W. Johnston, even had the full-time use of an airplane for moving his sub-parties about northernmost Manitoba. At least six commercial air firms provided transport service, while a seventh photographed and supplied prints in advance of field operations in New Brunswick and at Timiskaming. The Survey received a good deal of assistance from the aviation industry, and the companies, in turn, gained their share of the bounty that came the Survey's way that lucky year.

The Million Dollar Year seems to have finally broken the dam that held back the Survey from obtaining the full use of its natural ally, the airplane. In later years the Survey's appropriation was increased (including another Million Dollar Year in 1936 that again employed over 1,000 persons) to allow for contracting more air services. The advantages had been proved beyond dispute, even from the monetary standpoint, for in many regions air transport was cheaper than any other means, not to speak of the saving in time and comfort. Moreover, the use of aviation to view and photograph the geological structure of a region and help direct the work of the parties most effectively, gave results in mapping the geology and solving problems in all parts of Canada that were beyond calculating in dollars and cents terms.

The laborious topographical mapping burden could at last be reduced, and Survey parties could begin to concentrate on their proper role. As F. C. C. Lynch delightedly observed of the contracted air photographic work at Timiskaming, the photographs had been taken early in the season, and had become available to the geologists almost at once, enabling them to plot the geology directly on the photographs without waiting for a base map: "This co-ordination of the work enabled the Survey to complete the project in one season with a great saving of time."[30] The air photographs went directly to the Survey's Topographical Division, where the plotting could be done in line with the requirements of the Survey and according to their own schedule, without having to depend on the Topographical and Air Services Bureau. Lynch concluded that this experience had undoubtedly proved the merits of operating with commercial companies through contract. The aircraft were even more helpful in permitting the Survey to carry its work to more distant areas that had been too difficult to reach previously, and in speeding up the amount and effectiveness of the work that could be performed in a season. The airplane had revolutionized the conduct of field operations in undeveloped areas, turning the field work into a partnership between men and aircraft.

The report on the Million Dollar Year had this to say about the role of aviation:

> The range of the prospector's activities has been greatly extended by reason of the airplane, and, as a consequence, an urgent demand for topographic and geologic maps has arisen. The facilities . . . have been strained to cope with this demand, but thanks also to the use of the airplane in aerial surveying and the modern science of map-making, the prospector has had the satisfaction of obtaining maps of a number of areas that were formerly almost inaccessible.[31]

For the Survey, the use of aircraft achieved in the Million Dollar Year was something that would not be curtailed, but rather increased. In fact, the major result of the 1935 program was that never again would the Survey's work be reduced to the deplorably low levels of the early thirties. The bad years as regards field work were ended. Hereafter the Survey could proceed to bigger and better things in its field operations with a refreshed, revitalized field staff.

* * *

BY THE TIME THE SURVEY participated in the excitement of the Million Dollar Year, however, it had undergone one serious administrative reorganization, and was on the verge of participating in another. The first concerned mainly the position of W. H. Collins, the director. A brilliant geologist, honest, conscientious, and with a correct sense of scientific and economic priorities, he antagonized many of the staff by narrow vision, pettiness and a tendency to interfere to excess. His preoccupation with his own geological researches meant that the Survey's affairs, notably public relations, were neglected. Nor did he get along at all well with the deputy minister Camsell, who found it difficult to deal with the Survey.

Camsell's main difficulty with Collins arose over his feeling that Collins was not doing enough to make the work of the Survey relevant to the needs of the country—or what came to the same thing—that he was failing to make it appear relevant, as Logan and Brock had been careful to do. Camsell complained he had great difficulty making out a good case for government support of the Survey, with the result that the department and the Survey did not receive the support they deserved. The coming of the depression, with the accompanying heightened concern of government to eliminate every apparently superfluous expense, was an especially worrying time. Camsell's concern may be grasped from this memorandum he addressed to Collins on May 6, 1931, respecting the review of the Survey's field work for 1930 that Collins had sent him to incorporate into the book of memoranda for the minister to accompany the estimates:

> I have been looking this over and I am inclined to think if it is read by the Minister in its present form, he would perhaps draw the inference that the Survey was concerned only with doing field work and compiling information, because in the whole memorandum there is no comment on the particular reason for any one piece of field work, nor any information as to the value of the results of any piece of work nor any information to say to what extent the results of any piece of work have been made known to parties directly interested, or to the public. The memorandum should, I feel, be rewritten with a view to bringing out information as indicated above. It occurs to me, for example, that it would be of interest to tell of G. S. Hume making known through papers before the Geological Society of America and the Engineering Institute of Canada his conclusions with regard to Alberta oil fields from his field work up to and including that of 1930; that something be said of Johnston's work on water supply and of this information being made available to the cities interested and to the public; that something might be said of J. F. Wright's work, for I imagine that in the assignment given him he may have had opportunities of giving advice to quite a few claim owners; and that something be said of Cooke's work in the asbestos area and of the information probably passed along by him to asbestos property owners.
>
> It is desirable that you have revised memorandum on the field work of 1930 made available to me within, say, the next week.[32]

In the meantime other branches of the federal government concerned with the administration of natural resources were falling by the wayside, notably the many-sided Department of the Interior. Having lost most of its functions by the transfer to the western provinces of the control of their natural resources that Ottawa had exercised since 1870, the department was in the process of being disbanded. Many of its senior public servants were casting about for other positions in government. One such person was the chief of the National Development

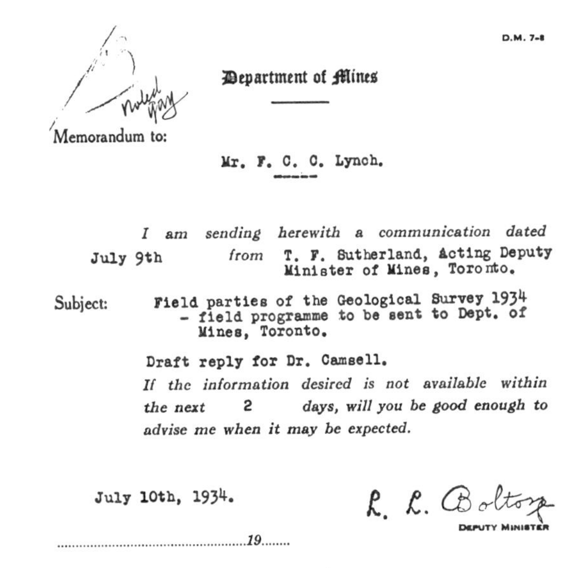
D.M. 7-8

Department of Mines

Memorandum to:

Mr. F. C. C. Lynch.

I am sending herewith a communication dated July 9th *from* T. F. Sutherland, Acting Deputy Minister of Mines, Toronto.

Subject: Field parties of the Geological Survey 1934 - field programme to be sent to Dept. of Mines, Toronto.

Draft reply for Dr. Camsell.

If the information desired is not available within the next 2 *days, will you be good enough to advise me when it may be expected.*

July 10th, 1934.

DEPUTY MINISTER

....19.......

XVII-17 SAMPLE MEMORANDUM, 1934, BOLTON TO LYNCH TO YOUNG

Bureau, the public relations unit of the Department of the Interior, F.C.C. Lynch. Camsell was very favourably impressed with Lynch, who had the reputation of being a man who got things done, did what he was told, and was well versed in public relations work. On April 1, 1933, in a surprising move, the organization and management of the Survey were completely altered by the creation in the Department of Mines of a Bureau of Economic Geology headed by Lynch which included little else than the Geological Survey, still under Collins' direction. A little later another thirteen staff members were transferred from the Department of the Interior to the Department of Mines.

Lynch was a career civil servant, born in Ottawa and educated at McGill University, who had entered the government service in 1906, became superintendent of the Railway and Swamp Lands Branch of the Department of the Interior, then in 1917 superintendent of the newly-formed Natural Resources Intelligence Service, which was renamed the National Development Bureau in 1930. Clearly the purpose of parachuting Lynch into the Department of Mines (besides providing for a displaced senior civil servant) was to put him in a position to shift the Survey's activities into more desirable directions.

As Camsell explained in a letter of May 29, 1934:

> It is a little difficult to reply to your letter of May 15th without appearing to be critical of the administration of the Geological Survey in the past. Hence my delay in replying.
>
> The new arrangement is that Dr. Collins retains the title and salary of Director of the Geological Survey, though the whole Branch will be under the administrative direction of Mr. F. C. C. Lynch, and for the purposes of the Civil Service Commission Mr. Lynch is given the title of Director of the Bureau of Economic Geology. In his administrative capacity Mr. Lynch will work through the Chiefs of Divisions, and tie them all in together into a properly coordinated unit. Under him, Dr. Collins directs the fundamental scientific geological work, Dr. G. A. Young the Economic Geology, Mr. W. H. Boyd the Topographic work, and Mr. A. Dickison the Draughting and Reproducing.
>
> The object of this re-arrangement is to develop a more fully coordinated machine in the Geological Survey, to speed up the production and issue of maps and reports, to give them a more practical character, and generally to provide a better and more efficient service to the industry.
>
> It may surprise many people that I have taken Mr. Lynch over from the Interior Department for this job, but I did it with a full knowledge and appreciation of his record and his capacity for organization and administration, and I am confident that the move will be to the advantage of the Geological Survey.[33]

Camsell's explanation in the *Annual Report* for 1934-35, was couched in more general terms:

> The Bureau of Economic Geology, established April 1, 1934, is a reorganization of the Geological Survey, effected to develop a more fully co-ordinated machine in the Geological Survey, to hasten the publication of maps and reports, as well as to give them a more practical character, and generally to intensify the services of the department to the industry.[34]

That is to say, the major divisions of the Survey had been detached from Collins, and their chiefs—Young, Boyd, and Dickison—were now immediately under Lynch. Collins retained only the oversight of the paleontological and mineralogical divisions and his acting directorship of the National Museum, and he shared control over the geologists with Young.

Lynch was not a geologist, nor did he hold a doctorate in a related science, so he was possibly ineligible to be appointed to the Survey under the 1907 act. Perhaps the new bureau was organized to get around this prohibition and make Lynch *de-facto* director. Because of his lack of training he could not manage the Survey without the assistance of the division heads, notably Young; nor could he effectively argue its needs before the deputy minister or others, still less guide its programs and policies. Instead, he was purely an administrator, devoting himself to carrying out the wishes and orders of his superior, which in the main meant producing a large volume of reports and maps quickly, completing as much work of immediate interest to the mining industry as possible, and publicizing the work of the bureau to the limit of his ability. It went without saying that he did not gain the confidence or respect of the geologists, who were bound to resent the manner of his coming to his position and the function he was performing. His former bureau had been notorious for the way it copied Survey work in its

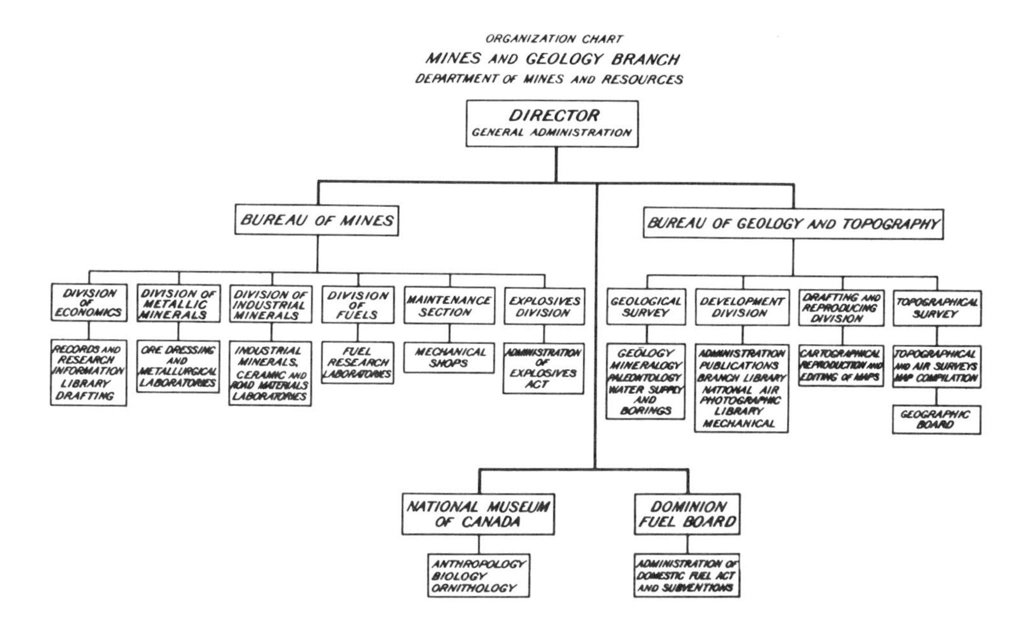

XVII-18

own publications without acknowledgment of either photographs or whole paragraphs of text. Perhaps, indeed, there was delicious irony in Lynch's being appointed to do officially what he had so long been doing surreptitiously.

The function of the Bureau was to promote and publicize the Survey's assistance to prospecting and mining through its mapping and exploration work, and this Lynch proceeded to do. In spite of the limited funds, in 1934, he placed 24 parties in the field and instituted the system of preliminary papers and maps to speed up the placing of the results of Survey work in the hands of the public. In 1935, notwithstanding the hectic pressures of the Million Dollar Year, he rushed the reports into publication, and his own report was full of new records—188 field parties, the busiest year ever for the Mineralogical Division, the large volume of work done by the British Columbia office, and the increased number of samples collected by the Pleistocene Geology, Water Supply, and Borings Division. The Survey's public relations now had the touch of the expert.

The situation, bad enough already from the standpoint of Survey morale, worsened in March 1936 when reports which alarmed the mining industry began to be heard of an impending sweeping reorganization of departments by the newly-re-elected Liberal government of W. L. Mackenzie King. These suggested that the departments of Mines, Interior, Indian Affairs, and Immigration were all to be merged, with obvious adverse effects on all the prospective ingredients of that merger. And so it proved; a new Department of Mines and Resources, presided over by T. A. Crerar as minister, was called into being. There was intense competition for positions in the new organization, and Camsell gained the prize over the other deputies. As the deputy minister of the unwieldy, new omnibus department, he was raised one level above his former position, which was now a mere small segment of his new responsibility. Under the reorganization, which went into effect on December 1, 1936, the minister and his deputy minister had five great branches under them that could have been—and most, indeed, until recently had been—complete departments of government.

Each of these branches—Mines and Geology; Lands, Parks, and Forests; Surveys and Engineering; Indian Affairs; and Immigration—was now under a chief who corresponded closely to a former deputy minister. The Mines and Geology Branch, which was virtually identical to the former Department of Mines, was headed by John McLeish, until recently the director of the Mines Branch. It comprised three main divisions: the Bureau of Mines (the former Mines Branch), under W. B. Timm; the Bureau of Geology and Topography (formerly Bureau of Economic Geology), which Lynch continued to head; and the National Museum of Canada, also controlled by Lynch. The Bureau of Geology and Topography was composed of four separate divisions—the Geological Survey, the Topographical Survey, the Draughting and Reproducing Division, and the Development Division—all of them made up of the former components of the Survey. In sum, then, the new Geological Survey division had lost all its associated scientific and service units and was left with only the four sections of Geology, Mineralogy, Paleontology, and Water Supply and Borings. Young, the manager of these sections and chief geologist, did not rank above Boyd, Dick-

XVII-19 JOHN McLEISH
Director, Mines and Geology Branch, 1936-41

ison, and Malcolm, who headed the other three divisions of the bureau. The Survey rated no higher on the administrative chart than did the Economic, Metallic Minerals, Industrial Minerals, or other divisions of the former Mines Branch, which had now become a full bureau. What had been a direct chain of command between Collins and Camsell barely three years earlier had become by late 1936 a four-step process, Young to Lynch to McLeish to Camsell. The Survey, with all its proud traditions, had plumbed its lowest administrative depths thus far.

Collins, the director of the Geological Survey, was the embarrassing, dispensable element in these changes, and his removal from the scene quickly followed. Despite the affront directed against him in the change of 1933, Collins continued with his scientific studies that made him one of the most eminent Precambrian specialists of his generation, as was acknowledged by his being elected president of the Geological Society of America in 1934, as his distinguished predecessor G. M. Dawson had been in 1900. Unfortunately, his health, precarious since a kidney operation in 1925, deteriorated in 1935 and 1936 to the point where he was largely bedridden. In 1936 he was very often too ill to come to the office and worked at home to finish his Precambrian researches and studies, which were published by the Royal Society of Canada, the last article posthumously:

> Twelve years ago Dr. Collins underwent a severe operation, and though fully aware that his life was hanging by a thread, his productivity continued unabated. A series of notable papers on the Sudbury district were among his very last contributions. The final months of his life, when he was already too weak to write, were utilized in dictating the manuscript of a work on the Canadian Shield that he was preparing in collaboration with some of his associates. Struck down at the peak of his powers, he has gone from us leaving an irremediable loss to geology and especially to Precambrian geology.[35]

On December 1, 1936, the date the departmental changes took place that lowered the Survey to the rank of a division, Collins was appointed to a new position of Chief Geologist Consultant, which was to have been a kind of freelance research post. At the same time he also relinquished control of the National Museum, which became the administrative responsibility of Lynch, with Wyatt Malcolm as its effective head under the title of assistant curator. Collins, however, was very far gone in his final illness, and on January 13, 1937, he died. He was

XVII-20 WYATT MALCOLM
Assistant Curator, National Museum, 1936-41

58, and had served 30 years with the Survey, 16 of them as director.

A kindly man, genuinely interested in the welfare of his staff, a first-class scientist with a world reputation, W. H. Collins' tragedy lay in his achieving the directorship in times that put demands on him which were outside his capabilities. His conservatism and obsession with economical management were not suited to the spirit of the go-ahead, expanding, rapidly changing, publicity-seeking twenties. At the same time his scholarly rectitude, his concern for sound, thorough, scientifically respectable work was equally out of step with the desperate mood of the early thirties that could be satisfied only by quick dramatic, dollars-and-cents, practical results. Above all, he was not suited to lead the kind of institution the Geological Survey had become during the twenties under Camsell's management of the Department of Mines. Directing a branch in a department of government with programs and goals of its own, required an order of talents—patience, suppleness, manoeuverability—that Collins simply did not possess. Hence the bitter, personal tragedy of his last years, and the diminished status of the Survey, which would persist until the imposing but unwieldy Department of Mines and Resources had run its course.

XVIII-1 THE SCIENTIFIC STAFF OF THE GEOLOGICAL SURVEY DIVISION, FEBRUARY 1943
Front row, seated (left to right): C. E. Cairnes, A. E. Wilson, F. H. McLearn, G. Hanson, G. A. Young, E. Poitevin, F. J. Alcock, B. R. MacKay, H. C. Cooke, W. A. Bell. *Middle row, standing:* H. S. Bostock, J. Spivak, J. D. Bateman, H. M. A. Rice, E. D. Kindle, R. J. C. Fabry, H. V. Ellsworth, J. R. Marshall, L. J. Weeks, H. H. Beach, J. F. Henderson, G. Shaw, A. F. Buckham, D. C. Maddox. *Back row:* A. H. Lang, J. F. Caley, C. H. Stockwell, G. W. H. Norman, J. S. Stewart, R. T. D. Wickenden, C. O. Hage, C. S. Lord, J. E. Armstrong. *Insets:* G. S. Hume, J. W. Ambrose, W. E. Cockfield, A. W. Jolliffe, M. E. Wilson, T. L. Tanton.

18

The Forties– Out of the Shadows and into the Light

WHAT DID IT MEAN for the Geological Survey, the senior scientific agency of the Canadian government with almost a century of matchless tradition, and a one-time department in itself, to be reduced to a fourth level of magnitude in the civil service hierarchy: to a division—of a bureau—of a branch—of a department of state? It would seem to be a simple case of rapid demotion and subordination. But was this really so? The Survey had not materially changed as far as its activities were concerned; it still had the same duties to map and study the geology of Canada, and to collect records, specimens of rocks, minerals and other materials for the National Museum. Its staff and personnel continued as before, in about the same numbers as before. Even its relationships with the other parts of its former organization were little altered. The Topographical Survey, Development Division, and Draughting and Reproducing Division had been created to meet the special requirements of the Survey, and as yet had no purpose beyond serving the needs of their former, and still only, customer. All four divisions, as well as the National Museum, were still neighbours in the same building. Their staffs met socially, and on business, while taking field notes to be made up in maps, checking references in the library, or securing a copy of a report or a map from the Development Division. Young and Boyd conferred on the field work plans and co-ordinated the mapping operations of their two groups as before. The staff of the Development Division looked after the equipment, instruments, and furnishings of the building, the equipping of field parties, the photographic section, the library, the distribution of publications on behalf of both surveys, and the Museum. This was little different from its previous functions as part of the Geological Survey. The Draughting Division, as before, prepared the maps for both surveys, plus maps and illustrations for Museum publications. The Topographical Survey continued to prepare the areal base maps for future geological work; as late as 1940 all thirteen of its parties were set to work in the foothills region of Alberta "to provide base maps for the extension of geological studies of the oil potentialities."[1]

In fact, what work did the Bureau of Geology and Topography do that was different from that of the former Geological Survey? The first report of that bureau, for 1936-37, listed its duties as being "to promote, by geological and related work, the discovery and development of the mineral resources of Canada; to contribute to the knowledge of the geology and geography of Canada; and to disseminate such knowledge by the issue of reports and maps, or by other means."[2] Was that not simply a restatement of the activities of the Geological Survey at their narrowest? Was the Bureau of Geology and Topography, then, nothing more than the Geological Survey under another name, simply renamed and placed under a new, more effective chain of command—the businesslike, accommodating Lynch rather than the introverted, stand-offish Collins and Young?

Comment on the reduced status of the Survey was aired in November 1938, in a paper by M. Y. Williams on "The Geological Survey and Mining Development" presented before the annual western meeting of the Canadian Institute of Mining and Metallurgy in Vancouver. Williams pointed out that the Survey had not only lost its direct contact with the deputy minister, but also its control of subsidiary services, that the directing geologist had been left with little authority, his requisitions and requests having to go to the hands of three non-geological officers before reaching the deputy minister, if indeed they ever did reach him. "From its time-honoured position, especially in western Canada, the name 'Geological Survey' has all but disappeared from government literature, and does not even appear on its stationery or on maps of its own making."[3] The chief geologist, Williams charged, could no longer direct his draftsmen or topographers but had to proceed through a cumbersome, interlocking, delay-creating system. The many channels of authority also meant delays in determining the Survey's appropriations, then in securing clearances to hire the summer assistants, so that the Survey was increasingly unable to secure the proper type of students for its field arrangements, which affected its chances of recruiting the best future staff. Williams went on to urge that the Survey should be returned to its position as an independent unit under a directing geologist responsible directly to the deputy minister; that auxiliary staff of topographers, draftsmen, photographers, etc., should be under the immediate control of the directing geologist; and that the Survey should secure a separate budget sufficient to its needs.

Young was given the assignment of rebutting this attack, with which he had every personal sympathy. The reply, as published in the *Northern Miner* of February 6, 1939, resorted to such quibbles as the claim that since McLeish had once been an employee of the Survey (from 1897 to 1907), the Survey had over it as director of the primary Mines and Geology Branch a person attuned to its needs. Young asserted that the efficiency of the administration was at least as great as it had been before, that students and university staff were being employed on field parties in the summers, that appropriations and programs were discussed in conferences of the deputy minister, the director of the branch, the chief of the bureau, with the chief geologist, and that the Survey controlled a separate budget, as before.

Many of Williams' criticisms, of course, were really directed against targets other than the Survey. The government, the Treasury Board, and the Civil Service Commission, for example, were mainly responsible for the lack of means to expand the staff and the work program or to relieve the situation within the Survey, of which Williams had said, "Several of the remaining staff are in ill health and all are discouraged beyond measure. From its position of administrative independence, the Geological Survey feels itself reduced to subserviency to bureaucratic and political forces."[4]

While a large part of Williams' attack was over the Survey's failure to do as much work in British Columbia as that province felt it needed and deserved, in Ontario the provincial Department of Mines was concerned about the Survey doing too much work in areas of current or potential mining activity. The appearance of the name, Bureau of Economic Geology in 1934, coupled with sending T. L. Tanton to work in the Atikokan district, gave rise to an exchange of agitated letters from provincial officials and eventually to a conference in Ottawa on November 22, 1934, to discuss the relationship between the two agencies. Camsell was at pains to explain that "no changes in relationships with the provinces were contemplated because of the new title," and that "the new title was a result of an effort of the Dominion Government to get more practical results from its expenditures in times when economy is necessary." Lynch also remarked that the Survey "had no intention of trying to compete with the Ontario Department in economic work," and "that the title, "Bureau of Economic Geology" has resulted in a rather unfortunate situation for the Geological Survey."

The discussion was noteworthy for the frank manner in which Camsell and Young pleaded the Survey's need to do a certain amount of economic work in Ontario as a question of its survival. Young remarked that "if the Geological Survey restricted its activities to fundamental geology, Government support would be lost," while Camsell revealed that, "in connection with the Survey field programmes, he has each year been asked by his Minister with regards to each item—"Is it economic?" The situation is that Mr. Gordon [the minister] can get support for work that will help prospectors, operators and financiers, but not for fundamental work only." Camsell went on to suggest that if economic work in a province was halted, the Survey would have to withdraw entirely from geological work there, and "that if the Geological Survey withdrew from work in Ontario, withdrawal in Quebec and elsewhere would inevitably follow, i.e., the Geological Survey would pass out of existence. In consequence if the provinces believe in the existence of the Geological Survey, it is in their ultimate interest that the Survey be allowed work which will enable it to get appropriations. The Survey, he stated, has been yielding ground to Ontario for twenty-five years and now has its back to the wall."[5]

The Ontario acting deputy minister, T. F. Sutherland, in subsequent correspondence described the Survey's situation in the following terms:

> The position taken by this Department was that while we welcomed the assistance of the Geological Survey in determining the general geology of the Province, this Department must necessarily continue to devote its attention to those areas in which valuable minerals are found or which appear to be favourable for their occurrence. This we regard as our peculiar field.
>
> We desire to maintain the cordial relations which have hitherto existed between this Department and yours, particularly with the

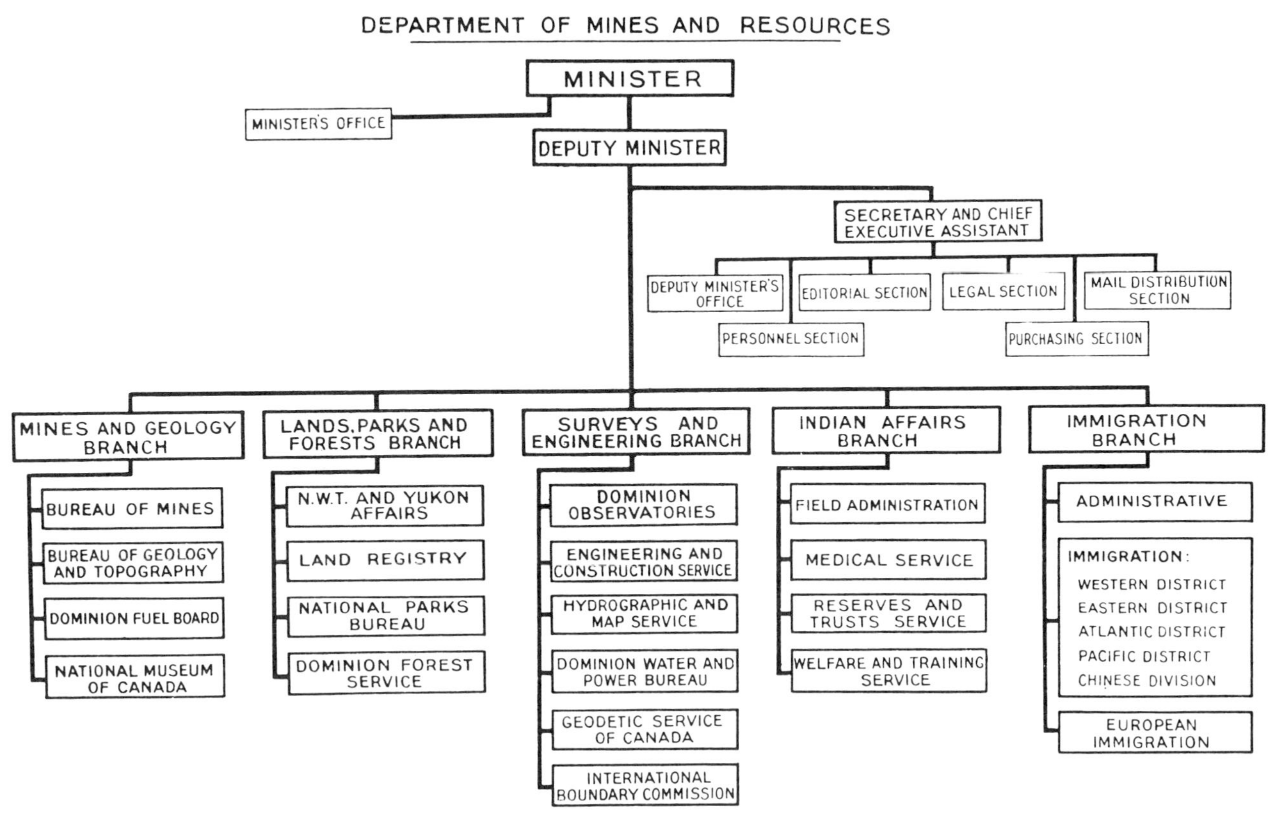

XVIII-2 ORGANIZATIONAL CHART, 1937

> Geological Survey whose work in some respects resembles our own. The recent formation by your Department of a Bureau of Economic Geology raised some apprehension in our minds as to whether a change in the Survey's policy was contemplated which might possibly lead to an invasion of what we deemed to be our particular sphere of work. We noted and appreciated the reasons given for the creation and naming of this Bureau, and we understand that no change in policy is proposed, and that probably the Bureau's title will be amended so as to remove any possible misunderstandings of its objectives. The difficulties attendant upon procuring the necessary appropriations for carrying on the Survey's work were explained, and we understood that such difficulties arose out of the pressure on the Survey to produce results of an economic character.[6]

In fact, in the 1936 reorganization the bureau received a revised title from which the offending word "Economic" was expunged—the Bureau of Geology and Topography.

It is probable that during the years from 1934 till after 1945 the Survey was more closely geared to meeting immediate short-range needs of the mining industry than during any period of its history. With the economy of Canada recovering from the depression, the industry was struggling to get back to normal footing. To reduce expenses in the face of low prices, mine operators had exploited the richest parts of their ore reserves, and cut down on exploration for new orebodies. On the other hand, the gold mining industry was still expanding rapidly, and studies of promising areas were much in demand. It was also necessary to keep up with the vast expansion of airborne mining prospecting activity. Thus the Survey had to spend the last years of the thirties providing immediate aid to the mining industry, and the field work was largely concentrated on areas that were known or considered likely to contain mineral deposits.

Hard on the heels of the depression came the Second World War, and Canada before long was called on to meet hitherto unimagined demands for manpower and war material. Vast quantities of metal were consumed for guns, aircraft, ships, automobiles, and munitions; the air training programs required great quantities of structural materials, and, incidentally, made frequent calls on the Survey for advice as to soil and water conditions. The output of mines was increased again and again, even though the companies found themselves hampered by shortages of labour and equipment. Exploration work fell badly in arrears; minerals were extracted without any consideration to finding reserves in their place. By the end of the war many major mines were severely depleted and the companies desperately needed to discover and develop new ore supplies.

The war had other effects. Besides the pressures on such minerals as nickel, copper-lead-zinc, asbestos, oil and gas, the economy was suddenly greatly in need of tin, antimony, mercury, tungsten, chromite, manganese, fluorspar, mica, quartz crystals, and other minerals, supplies of which Canada had formerly secured from foreign sources. Since the officers of the Survey knew as much about the geologies of these strategic minerals as anyone in the country, and Survey reports recorded the locations of many of the occurrences, the Survey was assigned the task of actually searching for mineral deposits. Geologists and field men, whatever their assignments, were instructed to keep an especially watchful eye for possible deposits of strategic minerals. So complete was the emphasis on wartime needs that the report of activities for 1943-44 began with the simple sentence: "Work was devoted entirely to the examination of and search for potential sources of oil, gas, and strategic minerals, and to other projects closely associated with the war effort."[7]

In the Cordillera region, H. M. A. Rice spent two seasons looking for antimony, mercury, tin, tungsten, and chromite, and at the same time supervised the exploratory diamond drilling of a base metal deposit in Banff National Park. J. E. Armstrong continued the work on the mercury deposits that he and J. G. Gray had discovered in the Pinchi Lake district just before the war, tracing an important mineralized fault zone for over 40 miles. Because of the great need for mercury, these deposits were rushed into production until by 1945 they were the largest source of mercury in the British Commonwealth. The need for chromite sent Armstrong, A. F. Buckham, W. E. Cockfield, and Rice searching for ultrabasic rocks that might contain deposits of chromite worth developing, while C. S. Lord supervised a diamond drilling project exploring for tin-bearing veins in a mine in the Takla Lake district. In the Northwest Territories a search was made among the pegmatite dykes for tungsten and other urgently needed rare-element minerals in the country north of Yellowknife.

In eastern Canada, M. E. Wilson worked for a number of years off and on in the Marmora-Madoc district of central Ontario in search of strategic minerals, including fluorspar. The search for chromite led H. C. Cooke back to the Eastern Townships on a notable re-survey of the copper-bearing zone. In the Maritimes Cooke hunted for tungsten in New Brunswick; Tanton visited the region in 1942 to inspect the iron reserves. Alcock examined the Saint John district for possible occurrences of manganese and other rare metals and directed diamond drilling operations in 1942 on both manganese and oil-shale prospects. Tanton's work in and around Steeprock Lake should also be mentioned. Under the stimulus of the war emergency, the iron ore deposit was undergoing dramatic development: the lake was drained to get to the iron bed beneath it, entailing one of the largest such operations in history.

Lynch summarized the wartime search for strategic minerals in his first report after the war as follows:

> Of particular interest, however, were the investigations in relation to the supply of strategic minerals. Although ordinarily the Bureau does not search for mineral deposits, several of its geologists during the war spent most of their field seasons in a search for strategic minerals, and these and other geologists were placed in charge of exploratory and development work on various properties. The work on strategic minerals included: investigation of the chromite resources of the Eastern Townships of Quebec, including detailed geological studies, active prospecting, and supervision of exploratory work and of mining; investigation of the mercury deposits in the Pinchi Lake area, British Columbia; investigation of and prospecting for tungsten deposits, which work contributed to a substantial production of tungsten ore from various properties; and investigation, prospecting for, and exploration of tin-bearing deposits in Northwest Territories, Yukon, and southeastern Manitoba. Although no production resulted from the last-mentioned investigation numerous discoveries were made, and geological conditions in the areas concerned warrant considerable optimism for future prospecting.[8]

XVIII-3 W. E. COCKFIELD

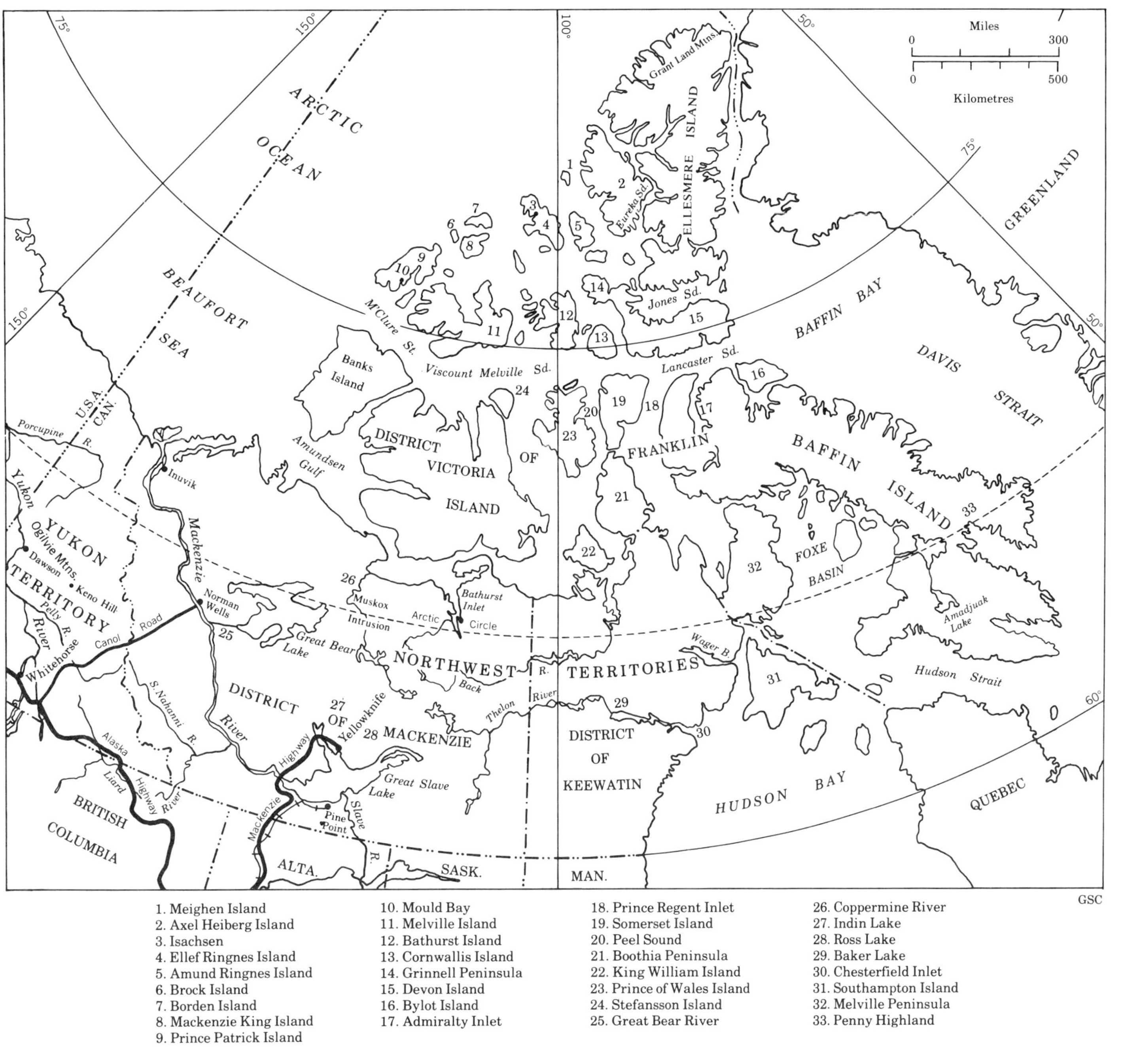

XVIII-4 NORTHERN CANADA AFTER 1940

Less attention was paid to locating deposits of strategic minerals after 1943, for it seemed fairly certain that any newly-discovered mineable deposits could not be brought into production in time to have much effect on the course of the war.

During the war years the Survey participated in a notable endeavour to develop new energy sources in Canada to meet the rising national requirements. Petroleum was essential to the mechanized war effort Canada was putting forth, and prior to 1939 about 90 per cent of

Canada's requirements were imported. Hence much effort was devoted to examining favourable structural conditions along the foothills of Alberta, and in view of the opening of a northwest front in Alaska, along the Alaska Highway and in the Mackenzie district. The great wartime need for other fuels also led to new emphasis being paid to the stratigraphy and structure of coal measures in the Maritimes and in western Canada, and to attempts to develop the Athabasca oil sands in Alberta.

Playing important parts in these wartime drives were two of the Survey's leading petroleum specialists. G. S. Hume, as technical adviser to the Oil Controller, supervised (and perhaps inspired) many of these activities. J. S. Stewart, who returned to the Survey after a 20-year absence, was assigned to the American contractors operating the Canol project. In this capacity he oversaw their drilling activities along the Mackenzie valley and their development of the Norman Wells oil field, and served as a liaison between the Americans and the Canadian authorities. In 1941 a great series of explorations for oil-bearing structures was mounted along the eastern margin of the entire Cordillera from the International Boundary to Liard River, and it was continued on into the postwar era. B. R. MacKay, C. O. Hage and H. H. Beach examined the structure east of the Rockies south of 52 degrees in Alberta in 1941, and no fewer than seven geological parties were moved into northeastern British Columbia near Hudson Hope in 1942 to investigate the geology, stratigraphy, and structures of that district. Nine parties were occupied in this same region of Alberta and British Columbia between the 52nd and 57th parallels tracing coal, gas, and oil prospects in 1943, as were seven parties in 1944 and four in 1945. Participating in these explorations were A. H. Lang, J. F. Henderson, E. J. W. Irish, as well as a number of geologists who did not remain long with the Survey. F. H. McLearn conducted paleontological studies in the Peace River district in conjunction with these efforts. W. A. Bell made a paleobotanical collection to help work out the succession between the Upper Cretaceous and the Paleocene formations. Interest in the oil-bearing formations did not diminish with the end of the war, for the major oil companies immediately increased their own prospecting activity, and the great Leduc oil discovery ensued in 1947. Since then the Survey's operations in connection with oil searches in the Plains region have never looked back.

The construction of the Alaska Highway, Northwest Staging Route and Canol oil pipeline projects attracted great interest to the resources of this hitherto unexplored region and to the role they could play in the postwar growth of Canada after the return to a peacetime footing. The Survey, as part of an integrated series of studies known as the North Pacific Planning Project conducted a noteworthy series of geological reconnaissances in the vicinity of these enterprises. In addition to the investigations in 1943 and 1944 of the oil potential along the Alaska Highway by Hage and M. Y. Williams, C. S. Lord examined the metal-bearing possibilities of the section between Watson Lake and Teslin Lake, a district where small deposits of gold, silver, lead, zinc, tungsten, and tin had been found. E. D. Kindle made a similar traverse in 1944 and 1945 along the Canol Road across the Tintina valley and near Pelly and Ross Rivers, a region of Paleozoic and older sedimentary rocks and some mineralization associated with granite intrusions; he returned to the area in years to come.

The Survey's interest in the Northwest Territories, aroused in the early thirties by discoveries in the Great Bear Lake, then in the Great Slave Lake regions, continued through the war years. From 1936 the Survey was active in the greater Yellowknife area. A. W. Jolliffe spent most of a decade and Henderson six seasons in this region, Y. O. Fortier three seasons in the Ross Lake district (1941-43), besides which R. E. Folinsbee, M. Feniak, and Lord also worked in the area. Investigations were extended to the Great Bear Lake district and the uranium-radium mine at Echo Bay by Jolliffe in 1936 and 1944, and along Camsell River in 1946.

Apart from the diversion to the hunt for strategic minerals in 1942 and 1943 and the intensified work on possible oil-bearing formations, most of the Survey's field operations followed fairly regular lines, with the same personnel working in the same districts. In the Cordillera Bostock worked in the Yukon, Armstrong in the northern interior, Rice and Cairnes in the southern interior, and Buckham on the historic Nanaimo coal-area. In the western part of the Precambrian Shield the 'regulars' were C. H. Stockwell in the Flin Flon region of northern Manitoba and Tanton in northwestern Ontario; the Lake Athabasca district was examined by Cooke, F. J. Alcock, Jolliffe, G. M. Furnival, and notably, by A. M. Christie in 1945 and 1946. Future officers involved in these surveys included R. A. C. Brown and M. J. Frarey, who assisted Stockwell in 1939 and 1946 respectively, and J. M. Harrison, who worked with J. D. Bateman in east-central Manitoba in 1943.

Before the war there had also been much interest in the district from Chibougamau to Rouyn—the gold belt of Quebec—with G. W. H. Norman and M. E. Wilson working most seasons there, and J. W. Ambrose, H. C. Gunning, G. Shaw, and E. D. Kindle less frequently. From 1943 onward much of the work in northern Quebec consisted of remapping the region in greater detail and in the light of more up-to-date information on the character of the volcanic and intruded rocks, and by other techniques, as in 1944 when "Information from

XVIII-5 ATLANTIC CANADA AFTER 1930

outcrops and drill holes was supplemented in places by that from magnetometric surveys conducted by the field party."[9] The St. Lawrence region was studied by J. F. Caley, who investigated geology and groundwater resources of the Toronto, Hamilton, Brantford, London areas, as well as oil and gas structures in the whole of southwestern Ontario as far north as the Bruce peninsula and Manitoulin Island. Alice E. Wilson continued her work on the Paleozoic of the Ottawa valley and the Cornwall district. In the Maritimes W. A. Bell made several studies of salt occurrences in the Malagash district and in Newfoundland besides his regular work on the coal deposits; and L. J. Weeks studied the metallic mineral deposits of the Cobequid district, Londonderry iron deposits, and a small lead-zinc operation on Cape Breton; Cooke worked two or three seasons on the gold deposits. Alcock was the main worker in New Brunswick, at first on the northern and central parts of New Brunswick, but from 1940 onward mainly about Saint John.

* * *

THE FIELD WORK DURING MOST of the Survey's first decade as part of the Department of Mines and Resources continued in spite of severe limitations of staff and funds. The Survey's total expenditure declined by about 25 per cent from $339,750 in 1938-39 to $249,690 in 1940-41, and remained at that level until the last year of the war. The expenditure on salaries also declined slightly, though not so drastically as the outlay on explorations, which moreover had benefited in the late thirties from the large supplementary grants given expressly to increase the amount of field work the Survey could perform. The table illustrates the comparable declines in the expenditures of the Topographical Survey and the Development Division, the even more drastic fall in the pittance allowed the luckless National Museum, but no decline in the expenditure on map-making until 1943. The Bureau of Mines investigations, in contrast, held more or less steady throughout the war at about 50 per cent above the total Geological Survey expenditures.

The shortages of money and then, in the war years, of available student manpower, created difficulties in obtaining summertime field staff and in planning. This last was made infinitely more difficult by the cumbersome hiring procedure of the time, particularly of having to wait on Treasury Board approval for every item, and even then without assurance that the amounts might not still be reduced or increased. Each year about January Lynch would call on Young, Boyd, and Malcolm (for the Museum) to submit their estimates of the student assistants, labourers, and cooks required for the field parties during the coming summer. In 1936, after the resulting list of 437 positions had been submitted by Lynch, the Treasury Board cut back the allotment by about 25 per cent, and Hanson (acting for Young) had to hurry to reduce the Survey's field operations proportionately, first cutting down the quotas for the various provinces, then reducing each field party's allotment by the appropriate percentage. When all this was done, the Treasury Board on May 19, 1936, presumably as a result of a governmental decision to repeat the Million Dollar Year experience of 1935, greatly increased the authorization, so that in the end the Bureau of Economic Geology found itself employing no fewer than 1,082 persons, including

Annual Expenditures of the Geological Survey and Certain Related Parts of the Department of Mines in the Second World War Years (in dollars)

Fiscal Year	Bureau of Geology and Topography	Topographical Survey	Draughting and Map Reproducing Division	Administration (Development Division)	Geological Survey Expenditures (total)	Geological Survey Salaries	Geological Survey Explorations	Bureau of Mines Investigations	National Museum
1938-39	$804,286	$210,797	$116,753	$136,985	$339,750	$172,251	$160,782	$407,464	$67,364
1939-40	719,114	205,256	115,771	132,148	265,938	164,450	81,128	407,111	70,252
1940-41	640,647	168,177	112,361	110,417	249,690	161,405	68,859	375,977	55,889
1941-42	652,713	173,580	120,441	99,529	259,162	164,263	77,496	383,351	45,663
1942-43	632,148	165,658	106,900	102,138	257,481	164,724	71,715	391,494	45,875
1943-44	614,424	168,624	82,870	105,755	257,174	162,585	67,777	390,769	44,631
1944-45	712,551	222,426	79,033	108,271	302,819	176,717	99,033	384,657	44,501

30 assistant geologists, 31 topographic engineers and junior topographers, 465 student assistants, and 460 cooks and labourers.

Such a system—or rather, lack of system—needless to say, created great uncertainties in each field season's work assignments and held up final arrangements until the last minute. Because there was so much unemployment, however, there did not seem to be much difficulty in finding the required personnel. Thus extra money was provided in the Supplementary Estimates in June 1938; but even at that late date the Survey was able to find five additional assistant geologists, 25 labourers, and 15 cooks to expand its field program.

The coming of the war, however, dried up the supply of summer staff, and made it necessary to increase the pay to get satisfactory help. In 1943 the expedient of renaming the student assistants "Wartime Student Assistants" was adopted, apparently to create a new category that could be paid higher wages than those fixed for summer assistants. The number of student assistants employed by the Survey was 135 in each of 1943 and 1944, then fell successively to 110 in 1945, 80 in 1946, and 70 in 1947, after which there was a return to the earlier levels as field operations were expanded.

As before, most of the Survey's future officers came from the ranks of the student assistants hired for summer work. The career of J. E. Armstrong, who began his field work in 1934 as a student assistant under H. S. Bostock in the Yukon, well illustrates this process. In his second summer under Bostock he operated a party largely on his own. Then in the next six seasons, in which period he also completed his Ph.D., he led his own field parties in north-central British Columbia except for one year in Manitoba. From being an assistant geologist on the temporary staff, he became employed on a year-round basis, and finally when the Treasury Board began approving an increase in the Survey's permanent establishment, he was promoted to assistant geologist on April 1, 1942, associate geologist on October 1, 1943, and geologist in 1947. Two future directors began in the same fashion during this period, J. M. Harrison by serving as a student assistant in 1935, 1938, 1939, 1940, and Y. O. Fortier in the summers of 1940 to 1942.

Since the Survey was prevented by the Civil Service regulations from hiring new permanent staff in the depression years, many future officers—like Armstrong—joined the Survey as temporary officers, even though they were employed on a 12-month basis. This was very reminiscent of the situation some 50 years previously, when Selwyn hired a considerable part of staff in this fashion, paying them out of the allotment for field explorations. In 1937, such "temporary" officers included J. S. Stewart, holding the rank of full geologist, and 13 assistant geologists. The number of temporary officers employed in this way increased year by year until 1942, when finally the Survey's establishment was expanded and some of the newcomers could be absorbed into the permanent staff. In the meantime, many were enrolled in the temporary staff under a strange variety of designations—Wartime Associate Geologist, Wartime Technologist, and Junior Wartime Technologist.

In the meantime, because of the regulations there were almost no promotions within the ranks of the permanent staff. It was even difficult to replace such retiring officers as C. S. Evans, F. A. Kerr, and L. S. Russell who left in 1937, H. C. Gunning and E. M. Kindle in 1938, W. A. Johnston in 1939, Beach in 1943, and J. W. Ambrose and Hage in 1945. Once men reached the top of their grades, there were no increases in salary either. Young

XVIII-6 GEORGE ALBERT YOUNG
Chief Geologist, 1924-43
In charge of Survey, 1936-43

XVIII-7 CAMERON BAY (PORT RADIUM) SETTLEMENT, GREAT BEAR LAKE, 1936

pointed out that such men were being unjustly held back and could be tempted into resigning to seek alternate employment. He also protested against the slowness of the whole system of promotion in terms of the training, experience, and responsibilities of the men. He felt the rank of assistant geologist should only be a probationary position for men who had completed seven years academic training or equivalent, and that promotion to associate geologist should follow automatically once such probation was completed. He also urged the establishment of three more associate geologist positions.[10] Lynch passed along the request to McLeish, with his support, but McLeish reported there was almost no hope of an increase in the Survey's establishment, commenting bitterly that "The Geological Survey has for many years served more or less as a training ground for geologists for industry and it would appear as if that situation cannot be changed at the present time."[11] The situation did not begin to improve until the Treasury Board had authorized the increase in the permanent establishment, and then in 1943 allowed promotions to go forward.

This easing of the restrictions on staff promotions was the first faint light that heralded the dawn of a better day for the Survey. It coincided also with the retirement in 1943 of the chief geologist, G. A. Young, who had headed the directorless Survey since 1936. Before that, as chief geologist under Collins he had been in effective control of the operations since 1924, and indeed, his office experience went back all the way to 1913 when Brock had appointed him technical editor. Thus in one way or another, Young played a very considerable part in Survey affairs for an entire generation—as fate would have it, possibly the most troubled and disappointing 30-year period in its history. Born in Montreal in 1878, Young, like his uncle A. P. Low, was a McGill graduate, obtaining a B.A.Sc. in 1898, his M.S. in 1900, and the doctorate at Yale in 1904. As a youth of 18 he had assisted Low on some of his difficult explorations of Labrador, before he joined the Survey as a geologist in 1904. Perhaps the most learned man of his time in Canadian geology, Young completed Brock's term as president of the Royal Society of Canada in 1935, then served in his own right in 1936.

Young's greatest contribution to Canadian geology was his work as scientific editor of the Survey's reports and maps. His vast knowledge enabled him to keep out many errors that might otherwise have crept in. He also did important service in training the men in effective expression and argument. He would edit two or three papers by each man most thoroughly until he was satisfied with the man's training and judgment, after which he might be content to examine his work more cursorily while he concentrated on other members of the staff who needed further guidance. As far as presentation was concerned, he was usually in the right; even if he sometimes criticized a man until he thought of resigning, Young usually could convince the author of his literary and logical lapses by going very carefully over the report with him.

The main complaints against Young's editing were over his judgment as to what should, or should not, be published. He argued with authors over how far they were entitled to proceed beyond their actual observations in making assumptions or inferences. More often it was over his refusal to allow authors to present novel theories or speculations of their own, which sometimes reduced the work's value, and occasionally prevented a man from being able to claim credit for discovering or enunciating a new idea. However, Young had to consider the Survey publications' reputation for strict accuracy and reliability; one inaccurate or faulty speculation could have done far more harm than any number of theories that were subsequently proved correct. This was simply the unending argument over publications that had gone on since Logan's day.

In his editing, as with everything else he did, Young was completely conscientious. His high opinion of the value of the Survey's work was reflected in a determination that any work done under his direction would have to live up to its highest standards. He would go to infinite trouble to help younger men with their problems to enable them to do more effective work for the Survey. When the Second World War came, he fought hard to keep his junior staff from enlisting or being drafted, considering they were far too important to the nation to be risked in the fortunes of war. However, he was extremely impatient with any one who, in his opinion, did not give his all on behalf of the Survey, who seemed lazy or 'unproductive.' He was thoroughly devoted to the Survey's ideals, and its staff, and he tried his best, with little success, to improve their lot as regards promotions and salaries. Though he was often critical of the men's shortcomings, he was actually proud of them and would not let anyone else interfere with them if he could help it.

After Young became chief geologist he probably had greater direct contact with the field work and field programs than any other person, and this control became more complete after 1934, owing to Lynch's need to rely on someone with that knowledge. Young, however, lacked the authority and responsibility that would have enabled him to manage or revise the program the way he felt was required. At the same time, he kept control of everything in his own hands and would not delegate responsibilities to others. He would not let the men make suggestions as to programs, or hold meetings with them to discuss aspects of the work; mainly, he dealt with them by sending around notes with his instructions. Glorying in hardship, he expected his men to be tough enough to abide by the rules and could see no reason for relaxing them. By not allowing departures from the regulations he got a name, like Collins, of being stingy, where he was only trying to meet the stringent requirements of the Treasury Board. Young, in fact, had a very difficult role to play as chief geologist, both under Collins and on his own. The increasing bureaucratization of the Department of Mines and its successor meant that the head of the Survey had less and less freedom to make decisions on field plans, appointments, or even on the type of work to be done.

With the Second World War came more restrictions and controls, and interferences with Survey plans in the interests of the Metals Controller and the Oil Controller. At the same time, the changed status of the former client divisions made them less amenable to Survey control and made administration even more difficult. Small wonder that Young felt very disappointed and embittered over the diminished status of the Survey that was so great a part of his life.

Under all these influences—especially the financial limitations and the deputy minister's views regarding the sort of program the Survey should follow—Young carried the Survey in the direction of economic work. In his draft reply to T. F. Sutherland about the work in Ontario in 1934, after defending the Survey's need to override provincial boundaries in its programs, Young stressed the economic orientation of the Survey's work.

> The Geological Survey if it is to continue to exist must do work that has a clearly apparent practical value . . . . It is my belief that the work performed anywhere by the Geological Survey should be geological work but always with an economic purpose behind it. At times problems must be investigated that may appear to be purely scientific but their investigation is not undertaken merely because of what might be termed scientific curiosity but because the solution of such problems is essential in the study of mineral resources. It is not, however, a proper purpose for the Geological Survey, to evaluate a mine nor to plan the development of a mine. The ultimate objective is to classify mineral deposits and determine their mode of origin.[12]

Young insisted that the work of the staff should be of the highest scientific standards and be permanent contributions to scientific knowledge. He was a careful force, for example, in the discussions leading to the reclassification of the Precambrian into Archean and Proterozoic eons, urging that the work go no further than could be completely supported at the time. As he wrote of these negotiations,

> I told Mr. Dresser, that while favouring the division of Precambrian time into two eras, I did not favour the rest of the time calendar because, in my opinion, it ignored the main difficulties confronting geologists and was not a time calendar but was a correlation table. The correlations advocated are not held by everyone, in the past firmly held correlations have proved erroneous, in the future some at least of the correlations now held will be discarded. A time table founded on correlations cannot be permanent. It is my opinion that the time table should be based on facts free from all assumptions. The terms employed should be pre-

cisely defined and should be based on facts alone. I think such a time table can be constructed but the construction should proceed step by step and each step should be very carefully considered.[13]

Rather like Collins, however, he was prejudiced against paleontological and sedimentary geology, as well as conservative about adopting the new tools and methods. Regular, traditional field mapping was his idea of the proper work the Survey should be doing.

Young, even more than Collins, was hampered by his personality from being an effective director of Survey affairs, however well he might have discharged his subordinate roles as technical editor or overseer of field programs. He was forever involved in arguments and would drop everything to debate a geological point. He made bitter enemies or ardent admirers, and left few people unimpressed one way or the other. He had very strong convictions and prejudices, tended to jump to conclusions about people without giving those in his bad books a chance to explain themselves, and seemed never to forget all too easily acquired dislikes. Cantankerous and negative, bothered by ulcers and by an inferiority complex, he gave the impression that the way to get along with him was to argue back at him, and that to gain his support for anything it was necessary to demand the opposite. He seemed to feel that anything worthwhile needed to be disputed, and he questioned, pried, and challenged every request to make sure it was well thought out and seriously meant. Many men who did not like this kind of cross-examination simply gave up rather than fight back, left the room, and sometimes quit the Survey as well. Young, who had only wanted an explanation, felt quite frustrated by this. His combative nature, his inability to compromise, and his constant fights over trifles, built up such strong prejudices against him that he could not persuade others when important points were involved.

He was even less skillful at managing the public or political relations of the Survey. Outspoken and tactless, he seemed incapable of avoiding contention, or smoothing over differences and working out compromises. He was so argumentative that he could not work with others. As early as the Geological Survey dinner of December 5, 1912, this was such common knowledge as to be made the basis of the limerick:

YOUNG—George Albert is still very Young,
And he works like a son of a gun,
He oft disagrees
With all committees,
And puts them all on the bum.

He could not help the Survey with its public relations, or lead it in new directions to enable it to play a meaningful role in a rapidly-changing world. This was left for his successors, Hanson, Hume, Bell, Harrison, and Fortier. But Young was not destined to see the Survey, in which he believed so strongly, regain its former status and go on to new triumphs. Four years after his retirement, on June 7, 1947, at his little farm near Brantford, Ontario, the stormy George Albert Young passed away quietly in his sleep.

* * *

IMMEDIATELY AFTER THE WAR a succession of spectacular mining achievements took place, to stimulate investment and touch off renewed prospecting activity. In the nickel industry there was the discovery and development of new deposits at Lynn Lake and Thompson in northern Manitoba, while gold mines were opened up in northern Quebec, northwestern Ontario, eastern Manitoba, the Northwest Territories and in the Cordillera of British Columbia. Base metal mines were developed all the way from Gaspé to the Yukon, and an asbestos mine was opened up in the Cassiar district, which had been made accessible by the completion of the Alaska Highway. In iron ore production, a field in which Canada was very much an importer, the impressive wartime operation at Steeprock Lake in western Ontario was followed by the beginnings of large-scale iron mining in Labrador utilizing deposits Low had reported as early as 1893. Here development reflected a favourable market situation: the need of American steel companies for an alternative source of supply to the Lake Superior district, and their reaching the conclusion that the quantity and grade of ore warranted the expenses of building a long-distance railway and the necessary concentrating and beneficiating plants.

These were less important than two very major mining developments of the period. The first of these was the discovery at Leduc in 1947 of a rich oil field, a discovery that suddenly unlocked the door to finding a dozen new oil and gas fields at great depth in Alberta and east as far as Manitoba. For a country that had been so dependent on imported supplies, the idea that the western provinces would become self-sufficient was exciting; but within two years it became evident that that region's oil and gas resources could transform Canada into a net exporter. Then followed the uranium rush, restrained by the Atomic Energy of Canada monopoly until 1948 when a policy of purchasing requirements from private companies was adopted. Spurred on by the very high prices the United States atomic energy authority was offering for uranium concentrates, prospectors swarmed over all areas of Canada that showed indications of radioactive minerals and staked claims that created many millionaires within a few weeks. Mines were brought into production on Lake Athabasca, but the largest uranium industry and the large new model town of Elliot Lake grew up in the district north of Lake Huron, where geologists from Murray to Collins had devoted many seasons to working out the structure.

At the same time the heavy pace of public and private construction created large demands for structural materials such as cement, bricks, gypsum, and asbestos. The coal and gold mining industries alone seemed in some trouble, the former because its markets were decreasing with the widespread turning to oil and gas for fuel and transportation, the latter because the once-generous fixed price of $35 per ounce had become inadequate in the face of rapidly rising costs. The government saved the gold mining industry by the Emergency Gold Mining Assistance Act which paid large enough subsidies to mines to keep many of them in profitable operation. In the years immediately after the war the mineral production of Canada soared by leaps and bounds. From an average annual value of about $250 million in the twenties, production reached levels of $624 million in 1947, $810 million in 1948, and passed the one billion dollar level in 1950, making minerals a major primary product in the Canadian economy.

All this meant a continuing need for up-to-date evaluations of the potentialities of the districts in which prospectors were interested, and also detailed work to assist the most effective development of Canada's mineral resources. Mainly this was in the form of traditional field mapping on the 1 inch = 4 mile or 1 inch = 1 mile scales, with the emphasis on solving structural problems. Air transport now was commonly used to carry parties to their bases of operations, but only occasionally did the Survey have to commission aircraft to carry out particular photographic projects. Instead, the men were able to use the air photographs in the National Air Photographic Library, as Bostock did in preparing a physiographic study of the Cordilleran region north of 55 degrees on the basis of more than 20,000 air photographs. Increasingly though, the Survey in the postwar years began to employ aircraft on aeromagnetic surveys, to the tune of $89,914 in 1949.

The annual reports indicate how wide the range of the work was in these years. That for 1947-48, for example, outlined the following activities:

> Much of the field work was again devoted to mapping-in areas of economic interest, including potential sources of gold, strategic minerals, base metals, oil, and natural gas. Further investigations were made of coalfields in Eastern and Western Canada, and of Pleistocene geology and ground-water supply in parts of Ontario and the Prairie Provinces. Standard geological mapping, on scales of either 1 or 4 miles to an inch, was conducted in 31 areas in various parts of Canada. Detail mapping was continued in the Yellowknife greenstone belt at Yellowknife, Northwest Territories, and in Beauchastel and Dasserat townships of western Quebec. Similar mapping was begun in Ossian township of eastern Ontario. Subsurface stratigraphic studies were made in the oil and gas fields of Alberta and western Saskatchewan. Other stratigraphic investigations, supplemented by palaeontological collections, were conducted in Alberta and British Columbia. Assistance was given when possible to industrial and engineering projects. Field work included a reconnaissance survey by aircraft over a large area of the Arctic regions, and an inspection of the iron ore deposits of Ontario, Quebec, and Labrador.[14]

The report represented operations by 57 field parties, almost double the lowest figure during the war years, and a considerable advance over the 41 parties in 1946. There had been talk of tripling the number of field parties after the war, and the totals continued to increase to 72 parties in 1949—including seven in Newfoundland, now Canada's tenth province—and to 89 in 1950, when the objective was reached.

Field operations, which received the attentions of the chief geologist George Hanson, consisted mainly of mapping the metal-producing Cordillera, Precambrian Shield, and Appalachian regions, or more detailed investigations associated with occurrences of gold, base metals, iron, uranium and rare-element minerals, and asbestos. In view of the opening up of large parts of northwestern Canada, the Survey put forth a considerable effort in the territories, eight or nine parties being assigned each year to the Northwest Territories alone. In fact, in 1948 at the height of the Yellowknife gold boom, a second branch office was opened in the government building at Yellowknife under a resident officer. The first appointee, M. Feniak, an assistant geologist who had worked half a dozen seasons in the region, was drowned in June 1949 while trying to recover a canoe that had gone adrift. His successors in the Yellowknife office were W. B. McQuarrie, A. B. Irwin, J. C. McGlynn, W. R. A. Baragar, and R. I. Thorpe. The major field work was J. F. Henderson's detailed study of the Yellowknife greenstone belt, besides which R. E. Folinsbee spent several seasons mapping in the Lac de Gras area, Tremblay in the Giauque Lake district, and Fortier around Indin Lake. In the Yukon Bostock examined a different area

XVIII-8 THE YELLOWKNIFE TOWNSITE IN 1947
With the old town in the background, on the peninsula and islands of Yellowknife Bay.

each year, while E. D. Kindle conducted a long-range survey of the Dezadeash area (completed in 1950). Besides these, J. O. Wheeler began a study of the Whitehorse area (completed 1951) and K. C. McTaggart worked in the Keno Hill district. Of special importance was the beginning of Y. O. Fortier's work in the Arctic islands. In 1947, he flew with the Dominion Observatory's magnetic survey party in the Arctic islands and mainland regions, observing the bedrock geology, Pleistocene glaciation, land emergence, and collecting fossils. He returned in 1949 to conduct a reconnaissance survey of southern Baffin Island.

As in the past, several parties were distributed each year among different regions of the Cordillera—among them R. L. Christie and L. L. Price in the Bennett and McDame region in the north of British Columbia; in the central interior, Armstrong, J. W. Hoadley, E. F. Roots, and S. Duffell (Whitesail Lake) and W. H. Tipper (Nechaco); in the southeast, Rice, Duffell, and others; and Hoadley in the Zeballos district of north Vancouver Island. The Vancouver office, under Cockfield, assisted for a time by Lang to 1940, was another important contribution to the mining economy of British Columbia.

In the Precambrian Shield Christie studied the uranium district north of Lake Athabasca, M. J. Frarey and N. R. Gadd worked in the Churchill River basin, and J. M. Harrison, Frarey and others in the Flin Flon district. In northern Quebec it was Stockwell, M. E. Wilson (who retired in 1948), Norman, and Tremblay; the Labrador Trough was examined by Tanton in 1946 and by Harrison in 1949. Work continued on a small scale by H. V. Ellsworth and others in the Precambrian of Ontario, particularly in connection with the search for rare-element minerals, while in the Eastern Townships Cooke completed in 1948 the careful, systematic mapping project he had begun in 1943, his last field work before he retired in November 1949. In the Maritimes, Weeks worked several seasons on a lead-zinc area in Cape Breton Island and in 1949 supervised the entry of the Survey into Newfoundland. In 1946 Alcock also completed his last field work on the Bay of Fundy islands, while in 1949 D. G. Crosby began a two-year study of the Wolfville district, containing the largest barite deposit in Canada.

The work of geological mapping was facilitated by the use of aircraft to supply transportation service to out-of-the-way places, and even more by aerial photographs, whose value for geological purposes increased as geologists gained more experience in their use and as interpretation techniques improved. But even perfectly-made air photographs had their limitations; only investigations by field parties seemed capable of ensuring a proper interpretation of the geology of all types of country.

An attempt to combine the flexibility of air travel with the accuracy of ground observations along lines followed in part by Alcock and others five or six years previously, was made in 1941 by G. Shaw, who mapped a 30,000-square-mile area in a single season on a 1 inch = 8 mile scale. The district included a little-known belt of Archean greenstones and metamorphic rocks along Eastmain River. From a low-flying airplane Shaw traced out the formations, and was landed from time to time to confirm his interpretation of the rock types, collect samples for study, and measure outcrops. Traverses were also made of places of particular interest, or where air observations did not provide sufficient information. Though the reconnaissance was only a superficial one, it was sufficient to lead Shaw to conclude that none of the area showed much promise, saving many seasons of useless regular work. However a fixed-wing aircraft could only be used in suitable terrain, and could not be employed in many other places. When helicopters were perfected a few years later, the technique immediately became the most efficient method of conducting field surveys, especially in difficult terrain.

Meanwhile, in governmental, industrial, and university laboratories, researches were proceeding into developing effective geophysical methods to assist with detecting and tracing buried unconformities (including orebodies) from the ground or air. Since the Survey's brief experimentation with the method in 1928 and 1929 much work had been done to discover new techniques and to refine them for practical use. As late as 1944, however, many difficulties remained to be overcome, as may be seen from a report by A. A. Brant on several years of tests (1937-44) of various techniques by the Ontario Department of Mines. Eight different methods were discussed, some of which were rejected as being unworkable in muskeg country, or lacking sufficient penetrating power, or too awkward to apply, or still too undeveloped for practical use. Brant concluded that magnetometric methods, which attempted to detect different rock types by minute differences in their magnetic qualities, held the greatest promise, though even there much more experimentation and developmental work were required.

By this time the Survey too was becoming interested in the subject and in 1946 a thorough study of current geophysical methods was made by Shaw. An airborne magnetometer of the United States Geological Survey was tested in September 1946. It was used in an experiment to map magnetically some 3,000 square miles in the Ottawa district in 1947 and then a 1 inch = ½ mile map was prepared. In 1948 the Survey committed itself to the method to the point of establishing a Geophysics Section under Shaw as chief and with a staff of twelve. In the first year a ground magnetic and geological reconnaissance was conducted of the area surveyed from the air in 1947, as well as further airborne magnetometer surveys of a 15,100-square-mile area, a small part of which was carried on by, and at the expense of, the Ontario Department of Mines.

In the following year, 1949, the staff was increased to 29, and three separate units were set up to deal with field work, compiling and plotting results, and research and technical services. An area larger than in 1948 was surveyed from the air and nine 1-mile map-sheets were prepared for publication. The director's report commented that "The results are expected to provide significant and helpful information in the search for iron and other minerals, and the operations of the Geological Survey were designed to provide basic information for research in this field."[15] A new element of the Survey of the future had been rapidly brought into being to do a very specific kind of work.

Also impressive was a new stress that began to be placed on the many facets of sedimentary geology in the immediate postwar years. Activities ranged from studies of the Pleistocene and Recent geology of Fraser River valley by Armstrong; to E. B. Owen's and E. I. K. Pollitt's investigations of the Pleistocene and groundwater resources of Prince Edward Island; to Y. O. Fortier's preliminary examination of glaciation and land emergence in the Arctic islands. They included an increasing number of studies in connection with such great engineering projects as the Columbia and Kootenay dams, on which the specialists in the geology of the region (Rice, Cockfield) collaborated with the engineering geologists Owen and E. Hall. Groundwater studies, grown important since 1935 because of the drought on the western prairies, were continued after the war in collaboration with the Dominion Water and Power Bureau and the Saskatchewan government's agent, Professor F. H. Edmunds. R. T. D. Wickenden, B. A. Latour, and A. M. Stalker of the Survey were associated with this work.

But above all, the continuing searches for oil and gas put great pressure on the paleontological work of the Survey, so often neglected in the past. Efforts were made to define the Paleozoic successions more carefully than ever before, and to apply the improved knowledge of fossil types to the practical problem of identifying potential oil-bearing horizons. Among the investigations were V. J. Okulitch's attempt to work out the boundary between Precambrian and Paleozoic at the base of the succession in the Selkirk Mountains of British Columbia; L. D. Burling's study of the Cambrian fossils on both sides of the Rockies; W. A. Bell's paleobotanical studies of the succession of coal-bearing sections in southern British Columbia, and the succession between Cretaceous and Tertiary; J. A. Jeletzky's work on the Mesozoic in the Vancouver area; and D. J. McLaren's on Devonian fossils.

Increasingly, the paleogeography of the prairie region was studied as a clue to possible oil and gas occurrences. Geologists were becoming aware that the character of sedimentary beds depended very much on the type of seas in which they were formed, and that those seas also supplied the favourable conditions for the production and concentration of deposits of oil and gas. Hence studies relating fossils to particular sea conditions such as depth, temperature, salinity, turbidity, and currents were begun. Specialists in sedimentary beds of certain ages were expected to know not only where the fossils belonged in the succession, but also the conditions under which they had flourished. For example, it was now recognized that coral reefs were highly suggestive of the possible presence of oil and gas. One geologist, having spent the summer examining Paleozoic reefs in the Rockies, in the true spirit of Hutton went to Florida the following winter to study coral reefs in the making.

A long series of stratigraphic studies was made of the foothills region from the International Boundary to the Peace River district, and drill core results of Leduc and other centres of intensive drilling for oil and gas were studied; in southern Ontario there was similar study of the stratigraphic successions important to the oil and gas industry. In central Canada the groundwater resources and the Pleistocene were examined in the Lake Simcoe district, and south and east to the St. Lawrence. Year by year the number of parties assigned to these tasks increased. By 1950 a total of 27 field parties (of the 89) were working on these fields: paleontology and sedimentary stratigraphy, Pleistocene, groundwater, oil, gas and coal geology, engineering geology, and the like.

A most welcome but embarrassingly large number of fossils, drill cores and well records poured from these operations, from oil companies, from provincial agencies, and from individuals, which quite overwhelmed the office staff of the paleontology section. Since a large fraction of these fossils originated from districts recently opened by the developments of the Second World War, a great deal of new material was received that required original classification and close study. The section was conscious of the oil companies' need, but with the numbers of specimens continuously on the rise, it became increasingly difficult to keep up with the work. New officers were added to the staff in 1948, outside specialists were employed for such tasks as indexing the Survey's holdings in particular periods, and collections were sometimes turned over to interested institutions for study and report.

There was heavy pressure, too, on the Water Supply and Borings Section. By contributing large numbers of samples, cores, and records, government agencies and oil companies added greatly to the tasks of cataloguing the material and making it available for users, who frequently were the contributing companies themselves. In 1944-45, for example "representatives of thirty American and Canadian oil companies spent as much as several months in the Geological Survey offices examining well samples, compiling logs, and studying available reports, maps, and other data pertaining to areas from which they contemplate exploratory surveys. Several of these geologists represented major American oil companies who have recently entered Canada in the search for oil."[16] The numbers of samples increased by leaps and bounds with the growth of oil exploration activity in Alberta. By 1946 the collection exceeded one million samples, and was increasing at a rate of about 100,000 samples per annum.

Both the Palaeontological, and the Water Supply and Borings Sections thus received fuller recognition than they had previously enjoyed. Indeed, the Survey received its first director from the paleontological field in this period—W. A. Bell, who became director under the reorganization of 1949. Bell gave up the headship of the section, which he had held since the retirement of E. M. Kindle in 1938, and the division (as it then became) came under the gentle, scholarly Nova Scotian McLearn, whose worth had so long gone unrecognized by Collins and Young. After the retirement of Alice Wilson and the transfer of the vertebrate fossil collector C. M. Sternberg to the Museum staff in 1948, McLearn and J. F. Caley were left as the veterans of the section around whom a whole new large division was built.

The field of sedimentary geology produced a third specialized section, the Coal Section, in the immediate postwar years to deal with coal resources. Despite the current economic difficulties the industry was facing—in fact, to a degree, because of those difficulties—research continued at various coal districts, notably at Nanaimo, B.C., in the Minto coal basin of New Brunswick, and especially in Nova Scotia where P. Hacquebard and others worked at the various coalfields. The establishment of the section arose from a 1946 request for assistance from the Royal Commission on Coal to the Water Supply and Borings Section. The result was the employment of B. R. MacKay, who had done much work on the coalfields of Nova Scotia and western Canada, to investigate the structures of the major coalfields across Canada, and the mining practices of companies. MacKay furnished the commission with memoranda, maps, statistical data, and eventually, with a monumental report. To continue this work, the Survey in 1947 established the Coal Section under MacKay, who recruited additional staff to do stratigraphic and microscopic studies, and to help with investigations of various coal deposits and mines. The work was partly financed and serviced by the Nova Scotia Department of Mines and the Nova Scotia Research Foundation, and it operated out of facilities in Sydney, N.S.

A similar trend towards new methods and specialties was

XVIII-9 E. M. KINDLE
Chief, Paleontology Division, 1918-38

XVIII-10 ALICE E. WILSON

XVIII-11 F. H. McLEARN
Chief, Paleontology Division, 1949-52

even more evident in the laboratory side of the Survey, heretofore the province of the Mineralogical Section. With the Second World War, the section set about helping Canada overcome her strategic mineral problems. New sources and supplies of rare, important, or strategic minerals, as vanadium, tungsten, chromium, and tantalum, were sought out. Searches were made for Canadian supplies of high-quality quartz crystals required in electronic equipment. The section received one unusual request to develop a suitable optical glass for the manufacture of glass eyes, the regular European sources no longer being available. The section even tested the sand ballast of free-flying incendiary balloons launched by Japan against the west coast of America, and from this clue helped identify the areas where those missiles were prepared.

In Ellsworth, the section possessed a foothold in the newly-important field of radioactive minerals, which had suddenly burst into terrifying prominence in the final days of the Second World War. In 1947 the section tested ores and mill products by Geiger counter (Ellsworth as early as the thirties had developed a portable one) and spectroscopic methods, did detailed research on ores from known occurrences, and made plans to expand the work in view of the expected opening of prospecting for radioactive minerals to private enterprise. In 1948 the chief of the Geological Survey was made the agent of the national Atomic Energy Control Board, and a radioactivity group was organized to deal with expected heavy demands that would arise from the public. A Geiger-Mueller counter laboratory and the necessary crushing and grinding machinery were installed and during the summer of 1948 samples began arriving in considerable numbers to be tested for radioactivity.

A related area of interest was X-ray and spectrographic work, which was begun by the Mineralogical Section in a small way in 1944, though it did not become fully operational until 1948 when modern X-ray diffraction equipment was obtained and a specialist was added to the staff to operate the equipment.

At the same time these new sections and activities were being developed, some of the Survey's former associates were moving farther away. The Topographical Division, for example, whose work in 1940 was given over entirely to meeting the requirements of the Geological Survey, during the war undertook special tasks for the RCAF: it made detailed maps of coastal defence artillery sites in Newfoundland, helped locate the important Goose Bay landing field, advised on aspects of the North West Staging Route, and assisted the RCAF prepare its part of an air map of the northern part of North America in conjunction with the United States Air Force. Soon after the end of the war the Topographical Survey was transferred to a different branch of the department and a long association had passed into history. Somewhat the same happened to the Survey's other associate, the Draughting and Reproducing Division, whose main task was to prepare maps from the records compiled from field work by the two surveys. In 1942-43 of its 63 maps,

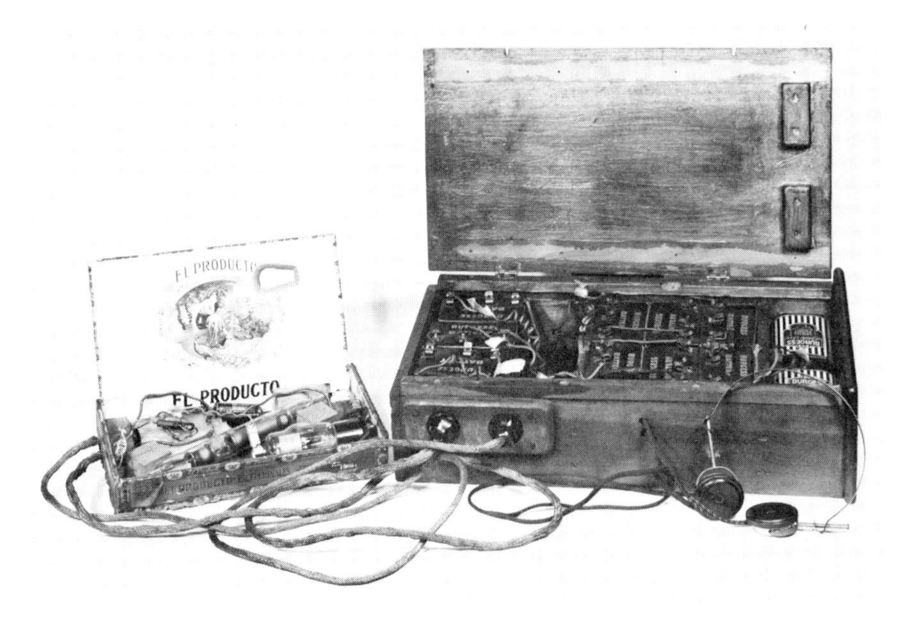

XVIII-12 H. V. ELLSWORTH'S EARLY GEIGER-MULLER COUNTER
Built in 1934 or 1935 with the assistance of A. W. Jolliffe and J. T. Wilson, then student assistants at the GSC.

25 were of geology and 38 of topography, but in the following year it produced 55 topographical maps against only 4 geological maps. This division also did wider work in the war years, lending two senior draftsmen to the Department of National Defence for important secret work, undertaking a duty with the Hydrographic and Mapping Service, and generally, performing "other draughting and related work necessary for staff, war departments, and public use."[17]

A change, too, was at long last in the making for the Survey's hapless *alter ego,* the National Museum. While the survivors of the original staff—Anderson, Barbeau, Taverner, and C. M. Sternberg—approached retirement age and expenditures were cut almost to the vanishing point, the Museum suddenly began to grow in the public consciousness, no doubt because of the strong upsurge of national patriotism. Clearly part-time direction by Collins, Bolton, Malcolm or Lynch would no longer suffice. Early in 1947, therefore, the Museum received as chief curator and as its second full-time director, the veteran geologist F. J. Alcock. Alcock had long been interested in the Museum, perhaps because of his original training as a paleontologist, and he gave his complete attention and considerable administrative talents to his new assignment. Though the connection with the Survey was formally retained by appointing C. E. Cairnes, Poitevin, and W. A. Bell as curators of the appropriate divisions, the staffs were completely separated now by transferring the Survey's work in vertebrate paleontology to the Museum. Using the larger grants that became available to him, Alcock recruited new staff, published some of the manuscripts held back since the onset of the depression, and developed the Museum, at long last, into an independent institution capable of standing on its own feet.

* * *

THE END OF THE WAR returned the work of the Survey to something approaching a near-normal basis, for almost the first time since the twenties. Since that nearly-forgotten time one emergency had succeeded another until the constant pursuit of short-range, immediately-useful programs became an entrenched policy. Now a return to longer-term programs and studies of a basic scientific nature appeared as a radical change. Much standard geological mapping of areas of interest on 1-mile and 4-mile scales needed to be done and completion of the geological map of Canada now lay within the realm of possibility. There was a need, also, for the Survey to undertake a wider measure of basic research as a contribution towards the progress of the mining industry.

Many large mining organizations had company geologists to work out in detail the controlling structures of their properties; the best help the Survey could provide was to develop better insight into the pertinent general principles by undertaking fundamental researches into such basic processes as ore deposition.

To enter on its new career, the Survey had the means at hand for more effective field work thanks to airplanes and aerial photography, while important new scientific tools were coming into use for field and laboratory work. A crowd of new graduates, mainly from the Canadian universities, was becoming available to replace the retiring veterans and to mount the expanded operations that the excellent financial position of the peacetime government at last permitted the Survey to contemplate.

The Survey's annual expenditure rose by 50 per cent in 1946-47 from the low wartime level of barely $300,000, and in just five years had almost quadrupled to $1,196,404 by 1949-50—at last approaching the level of the Million Dollar Year. This good fortune was not peculiar to the Survey: other agencies, for example, the National Museum and the Mines Branch, received similar increases, and the Topographical Survey and Draughting and Map Reproduction apparently made even larger gains. In any event, the Survey had far more money to carry on its operations, and to that extent had more freedom to initiate new activities that fulfilled current needs or to seize new opportunities for meaningful service.

Such was the situation that faced the Survey's head, the chief geologist George Hanson, who had assisted Young for almost a decade before his retirement in 1943. Entering into the administration as assistant chief geologist at the same time was Cairnes, who had been attending to the technical editing and other duties from about 1939. Late in 1947 H. M. A. Rice took over some of the editorial and office tasks under Cairnes.

An immediate problem was to attract enough well-qualified personnel to staff the expanded, more highly specialized operations of the new era. Mining activity was booming everywhere in Canada and other countries; university employment was also increasing. The Survey, therefore, would have to offer salaries that were competitive with these alternative employments, and here it came up against the restrictions of the Treasury Board. At least twice in these years, delegations of men went to the deputy minister to seek more adequate salaries, and when their requests were denied, several men resigned. Those who left to take other employment included Ambrose, Bateman and Hage (1945), Jolliffe and J. Tuzo Wilson (1946), and Buckham (1948). In the same period J. S. Stewart and M. Feniak died, Alcock assumed the position of curator of the National Museum (1947), and Maddox, M. E. Wilson, Alice Wilson, and Cooke were superannuated. Some of them saw their pensions cruelly devalued by the inflation of the next twenty years and were forced to pass their declining years in genteel poverty—small reward for their many years of talented, unselfish public service.

Salaries for the staff were improved in the autumn of 1945 when a 10 per cent increase was awarded throughout the Survey. Furthermore, a new category of senior geologist was opened for men who held major adminis-

XVIII-13 C. E. CAIRNES
Assistant Chief Geologist, 1943-53

XVIII-14 GEORGE SHERWOOD HUME
Chief, Bureau of Geology and Topography, 1947-50;
Director General of Scientific Services, 1950-56

trative posts and had made outstanding contributions to science. These included the assistant chief geologist, Cairnes; the section heads Bell and MacKay; Cockfield, in charge of the Vancouver office; and G. S. Hume, who besides his distinguished wartime service as assistant to the oil controller, needed the promotion to equal the higher salary he had been receiving since 1941 and—hint of coming changes—because "It is virtually certain that Dr. Hume will be appointed Regional Geologist, Plains and Foothills, when the Geological Survey is reorganized on an enlarged basis."[18] Later appointments to this rank included Tanton (1947), Poitevin and Cooke (1948), and Bostock (1950). At the same time Hanson was given a special classification as chief geologist.

During the war years a variety of classifications had grown up, with titles like assistant geologist; wartime assistant geologist (the "wartime" designation applied only to some of the men hired during the war and was dropped officially on November 16, 1945); junior wartime technologist; technical officer, reconstruction. Each was divided in grades analogous to those of the geologists—senior, geologist, associate, assistant. Finally there was the new category of junior geologist, established in November 1945 for young men with the necessary academic qualifications but with insufficient field experience to enable them to qualify at once as assistant geologists.

To clear away this labyrinth and help the Survey to compete with other employers for the available professional and scientific staff, a reclassification establishing a new system of grades and salaries was instituted in 1947. Five equal grades, each with from four to six salary steps, were established for geologists, technical officers, or engineers in the Survey's employ, ranging from the lowest, Geologist (etc.) Grade I, that corresponded to the former rank of junior geologist; Geologist Grade II (corresponding to the former assistant geologist); Geologist Grade III (corresponding to associate geologist); Geologist Grade IV and Grade V; and finally, the class of Senior Geologist as before. Thanks largely to this new system, by the end of the postwar period the Survey had acquired a considerably larger staff of qualified new personnel, under a salary and promotion system that allowed for a widened range of technical grades better suited to the diversified organization the Survey was once more in the process of becoming.

For the Survey was again being drawn into a series of administrative changes that would alter its status and open before it the opportunity to evolve new roles in keeping with the scientific and governmental climate of the postwar period. Beginning in 1947, these changes reinstated the Survey as a primary branch in a government department in an even shorter time than had been required for it to tumble to the third level of the departmental hierarchy. The first step occurred on March 24, 1947, when Lynch retired, and was succeeded as chief of the Bureau of Geology and Topography by the senior geologist Hume, who thus became Hanson's immediate superior. Hume's appointment presumably was a tribute to his important work as technical adviser to the Oil Controller in the Second World War, for which he received the O.B.E. in 1946.

Hume was a most unusual geologist in that he was a man with a "mission"—to develop the petroleum industry of Canada—and made no show of trying to maintain an objective or scientific outlook such as most other geologists profess, although sometimes they become so emotionally attached to areas, theories, or heroes as to lose most of their scientific objectivity on those subjects. Geologists who worked with Hume sometimes felt that his faith hampered his approach to geological problems, made him opinionated, rigid, stubborn, and occasionally hard to deal with, and that it led him to push ahead obstinately when any reasonable chance of success had passed. Along with this, though, went abounding energy, great vigour, and enthusiasm for any assignment he un-

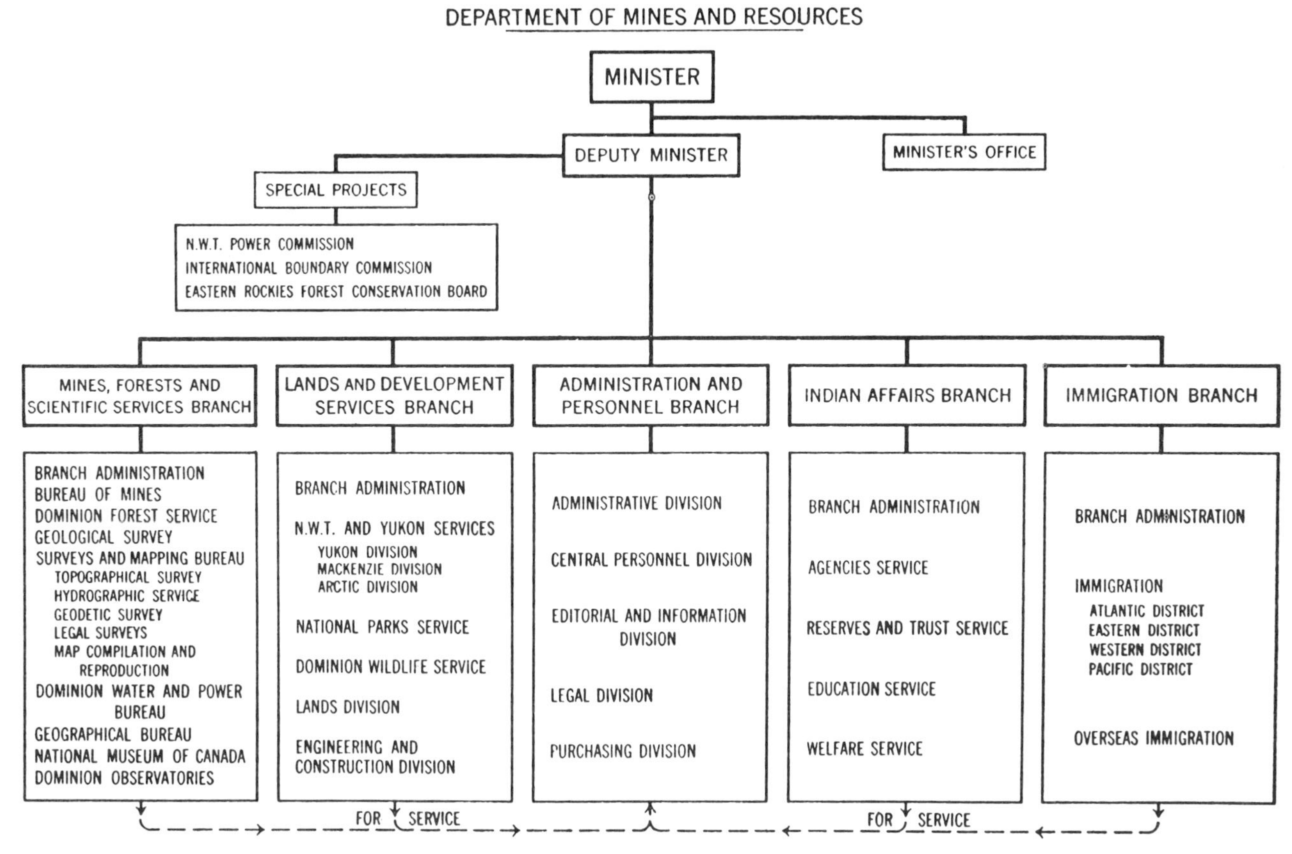

XVIII-15 ORGANIZATIONAL CHART, 1948

dertook. He was a good organizer and administrator, with demonstrated ability in dealing with governments, industry, politicians, and civil servants. As a field manager, he was very helpful and sympathetic to his assistants, giving them the benefit of his advice yet encouraging them to test their own ideas. He impressed his colleagues with his high principles, humanity, kindliness, interest in the growth and development of younger men, and his quiet sense of humour.

Later that year, on November 1, 1947, there came a thorough reorganization of the branches of the Department of Mines and Resources. Three of its major, primary bureaus, including the Mines and Geology Branch, were abolished and their component parts were rearranged into two new branches–Lands and Development Services Branch; and the Mines, Forests and Scientific Services Branch (to which the Geological Survey and its former dependencies belonged). The purpose was to bring together in the last-named branch "all the basic research activities and all the survey and mapping responsibilities of the Department."[19]

When the Bureau of Geology and Topography was abolished, its divisions were brought directly under the new Mines, Forests and Scientific Services Branch with the status of bureaus. The Geological Survey and the National Museum became two such bureaus in the eight-segment branch until 1950, when they parted administrative company altogether, the Museum passing to the newly-created Department of Resources and Development. The Survey's former Topographical Survey at this time was assigned to a Surveys and Mapping Bureau where it acquired new associates (the Canadian Hydrographic Service, the Geodetic Survey of Canada, and the Legal Surveys Division) and a host of new functions, which included conducting surveys for ground mapping; air control mapping for use in the development of mining, agricultural, and other industries; charting harbours; hydrographic work; and surveying airfields and the Alaska Highway. After 1947 the Topographical Survey no longer had any direct or personal connection with the Geological Survey. The National Air Photographic Library, which was part of the defunct Development Division, also passed to the Surveys and Mapping Bureau.

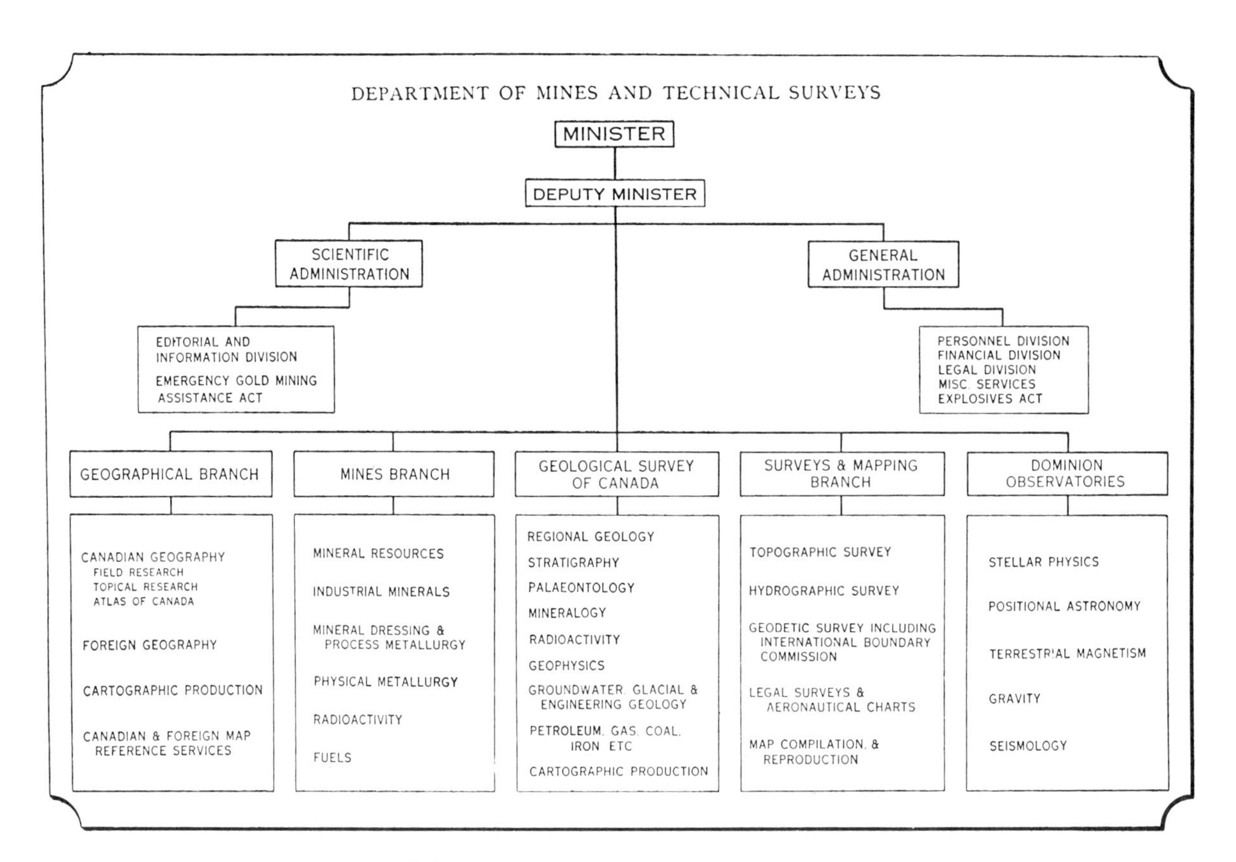

XVIII-16 ORGANIZATIONAL CHART, 1950

Offsetting the departure of the Topographical Survey was the return to the Survey of at least part of the Draughting and Reproducing Division. In its new locale in the Surveys and Mapping Bureau, the Topographical Survey began employing the services of a new "Map Compilation and Reproduction Division" which "was formed from parts of the former Hydrographic and Map Service and the draughting office of the former Bureau of Geology and Topography."[20] The remainder of that "draughting office" was returned to the Survey under the name of the Geological Mapping Division (or in 1950, the Geological Cartography Division), becoming a service unit within the Survey. In similar fashion the Survey in and after 1949 repossessed the library and photographic sections from the former Development Division to become once again a reasonably integrated, self-contained organization, that was generating new scientific units even while its old connections with the National Museum and the Topographical Survey were being dissolved.

With the abolition of the Bureau of Geology and Topography, Hume's position as its chief also disappeared, and for a time Hume became chief of the new bureau, the Geological Survey. While Hanson who was still chief geologist, continued the day-to-day direction of Survey operations, Hume managed the Survey's relations with the department and other authorities. In May-June 1949, the Civil Service Commission, after considering 13 names for the position, recommended the appointment of W. A. Bell as chief of the Geological Survey bureau. The appointment was posted June 20, 1949, but could not take place until Hume had vacated the position, which did not occur until January 23, 1950, in connection with another reorganization, this time at the departmental level. The ungainly Department of Mines and Resources (1936-50) was finally broken up into three departments, among them the Department of Mines and Technical Surveys, which was quite reminiscent of the old Department of Mines. It was comprised of five branches: the Mines Branch and the Geological Survey of Canada—the two components of the old Department of Mines—plus the Surveys and Mapping Branch, Dominion Observatories, and the Geographical Branch. The official explanation recited that

> The Department was established in view of the growing feeling, particularly among those interested in mining, that the importance of the mineral industry and of the Government's relations

with the industry was such that there might well be a Minister of the Crown who would devote his full attention to the fields of mines and mining.

. . . as now constituted the Department of Mines and Technical Surveys is an integrated organization whose primary function is to provide technological assistance in the development of Canada's mineral resources through studies, investigations, and research in the fields of geology, mineral dressing, and metallurgy, and of topographic, geodetic, and other surveys.[21]

The act, which came into force on January 20, 1950, listed the functions of the new department in seven subsectlons of article 6 reciting the minister's powers. That list of functions contained no distinction between those that were expressly the responsibility of either the Mines Branch or the Survey; it would be the task of the ministers, J. J. McCann and his successor George Prudham, as well as their deputy ministers, to keep the lines clear in some of the more doubtful, marginal areas.

On January 23, 1950, three days after thc new department began its operations, a new position was created for Hume as acting deputy minister and Director-General of Scientific Services; and at last Bell's appointment as director of the Geological Survey, which had hung fire for seven months, went into effect. Bell was, to all appearances, a reluctant candidate for the position, and his appointment came as a complete surprise to the Survey staff, which had assumed Hanson would obtain the position by virtue of seniority, though Hanson's uncertain health was possibly a factor. But Bell was a very hard worker, universally respected as a scientist and a man, and perhaps it was felt he would give a more even balance to all sides of the Survey's work, especially at a time when the oil and gas industries were assuming primary importance.

Thus, after slightly less than 16 years (April 1, 1934-January 20, 1950),—from the establishment of the Bureau of Economic Geology to the formation of the Department of Mines and Technical Surveys—the Survey at last recovered its status as a departmental branch headed by a director, and its primary units were elevated to divisional rank. Moreover, it was part of a department devoted largely to serving the mining industry. The times—as they had not been since before 1914—were once again opportune for the Survey to step forward and play a large, active role in shaping the prosperity of Canada. Mining was booming in every section of the country, and new media of communications were making it feasible to extend the range of activity into the remotest corners of the land. After two decades of disappointments, difficulties, and crises, the Survey at last had the status, the financial means, and a young staff of specialists to enter upon a new era of great service to the mining industry and the nation.

PART FOUR

The Survey Today 1950–1972

XIX-1 THE GEOLOGICAL SURVEY OF CANADA BUILDING
601 Booth Street, Ottawa

19

Since 1950–A General View

THE TWO DECADES SINCE 1950 have probably been the most productive in the Survey's history. The mining industry expanded rapidly with the development, in response to world demand, of vast resources of fossil fuels in Western Canada, and of the iron ore deposits in Labrador, as well as numerous other deposits of metallic minerals in the Precambrian Shield, Cordilleran and Appalachian regions. In developing these deposits, and in the accompanying search for others, the industry has found increasing need for maps, reports, and geological information. The public mood, which was reflected in the policies of governments of the period, had grown strongly sympathetic to the cause of scientific research; accordingly, there was greater readiness than ever before to allow the Survey a free hand to undertake scientific researches into general geological problems. Indeed, strong pressure arose in mining and university circles after the War for the Survey to assume a role of leadership in this field—to develop facilities and staff for advanced scientific studies; and to play a part in co-ordinating similar activities in the universities and elsewhere in Canada. Above all, in the new age of "big government" fiscal means were no longer a problem. Funds were made available to permit it to undertake far more ambitious programs of field and laboratory work in line with the opportunities and needs of the time.

Other developments opened new doorways for the Survey to advance to higher levels of effectiveness. The remarkable improvements in transportation and communications made it possible to operate successfully in hitherto inaccessible parts of Canada. Great scientific and technological advances in geology and the other physical sciences provided the means for mastering many previously insoluble problems, then with these answers in hand to review past work and move on to investigate new problems. Finally, the Survey had at last been placed in a department that was primarily concerned with its needs and those of related agencies, and was restored to a position of some influence within the federal bureaucracy. As a member of the Department of Mines and Technical Surveys formed in 1950, the Survey was in a better position to face the challenges and the opportunities of the new era.

Under these circumstances there was an unprecedented burgeoning of the Survey, and the result was unparalleled accomplishment. To begin with, there was an incredible increase in the rate of primary mapping of Canada's geology. By 1950, after more than a century of steady work, the Survey had still not succeeded in mapping in a reconnaissance way more than one-quarter of the bedrock geology of Canada. In the two decades after 1950, two and one-half times as great an area was mapped as in the whole of the previous century. Between 1950 and 1970 some two-thirds of the 3,852,000-square-mile land and lake area of Canada was mapped to at least a 1 inch = 8 mile scale, leaving only a few scattered, isolated areas to be completed to this level. In addition, using other methods, good progress was made in mapping the geology of the bedrock and surficial deposits beneath the bays and continental shelves adjoining Canada, which add a further 1,381,000-square-mile area to the Survey's responsibilities. The rise of such new subsciences as geophysics, geochemistry, tectonics or metallogenics made new techniques and approaches available to help clear up many of the problems of the mining industry and develop new scientific theories. Rock forma-

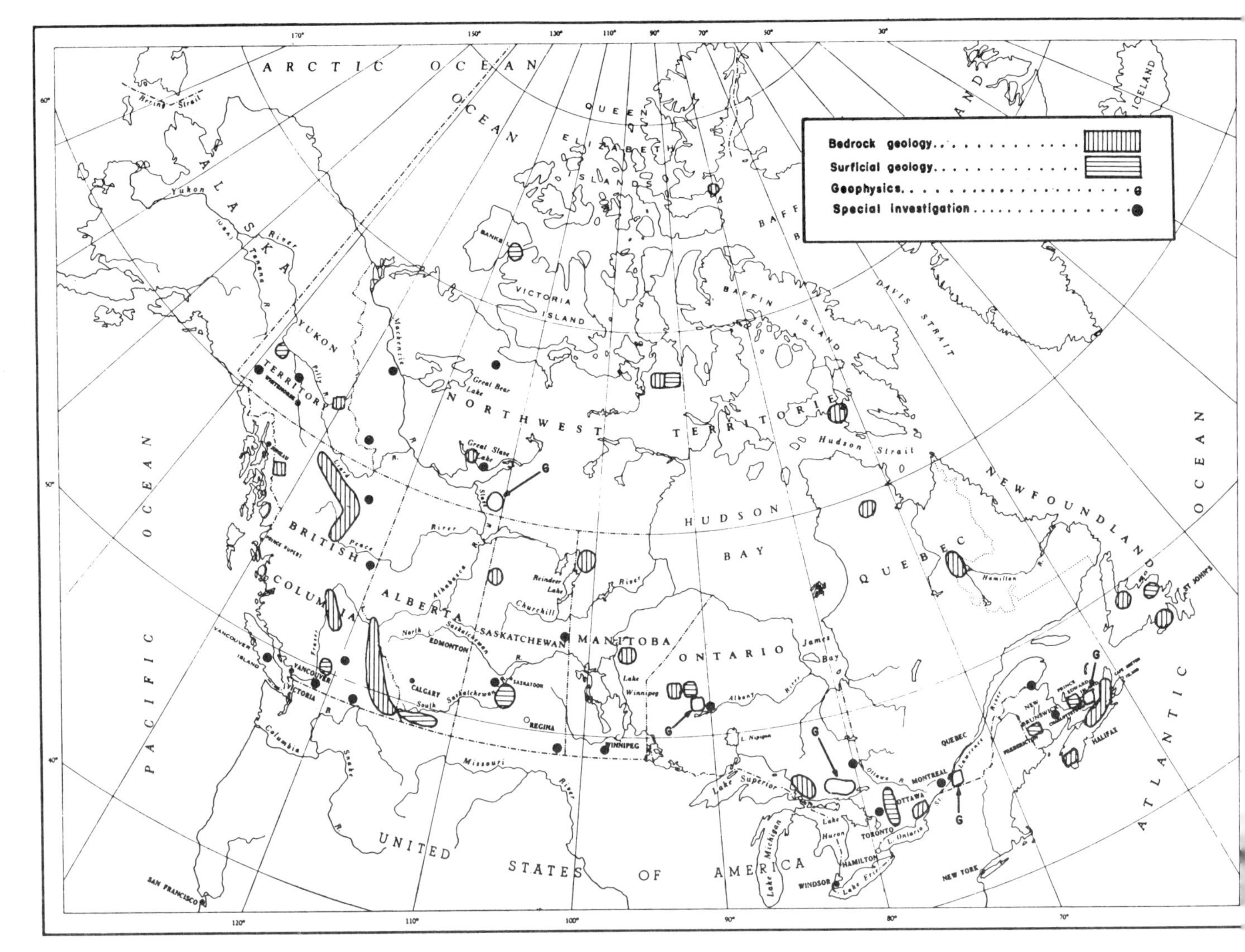

XIX-2 THE FIELD WORK PROGRAM, 1960

tions at depth could be investigated, including the physical processes that have determined the major geological features of the earth's crust; at the other end of the timescale, the surficial unconsolidated sediments and the contemporary geomorphological processes relating to everyday life and work could be studied much more effectively. The vastly improved understanding of the major characteristics of Canadian geology was reflected in a large, impressive, and important series of geological reports and maps of all, or parts of Canada, which are summarized in the latest 838-page edition (1970) of *Geology and Economic Minerals of Canada,* edited by R. J. W. Douglas.

The number and sophistication of the specialized laboratory services increased as operations in the field became more and more dependent on laboratory analysis and studies. The facilities of an excellent new headquarters building made it possible to expand the laboratory and research activities. New tools contributed to the more accurate understanding of the geology of Canada, and a host of advances in scientific concepts ensued. By again becoming a major force in the study of earth science in Canada, the Survey in these years helped considerably to elevate the levels of geological thought and enquiry. There was a remarkable increase in the numbers of major papers by individuals, and of the published symposia or collected papers on particular themes. Scientists from around the world came to visit and train in the Survey's new facilities, while the expert knowledge of Survey personnel was diffused abroad through their publications, participation on international commissions and panels,

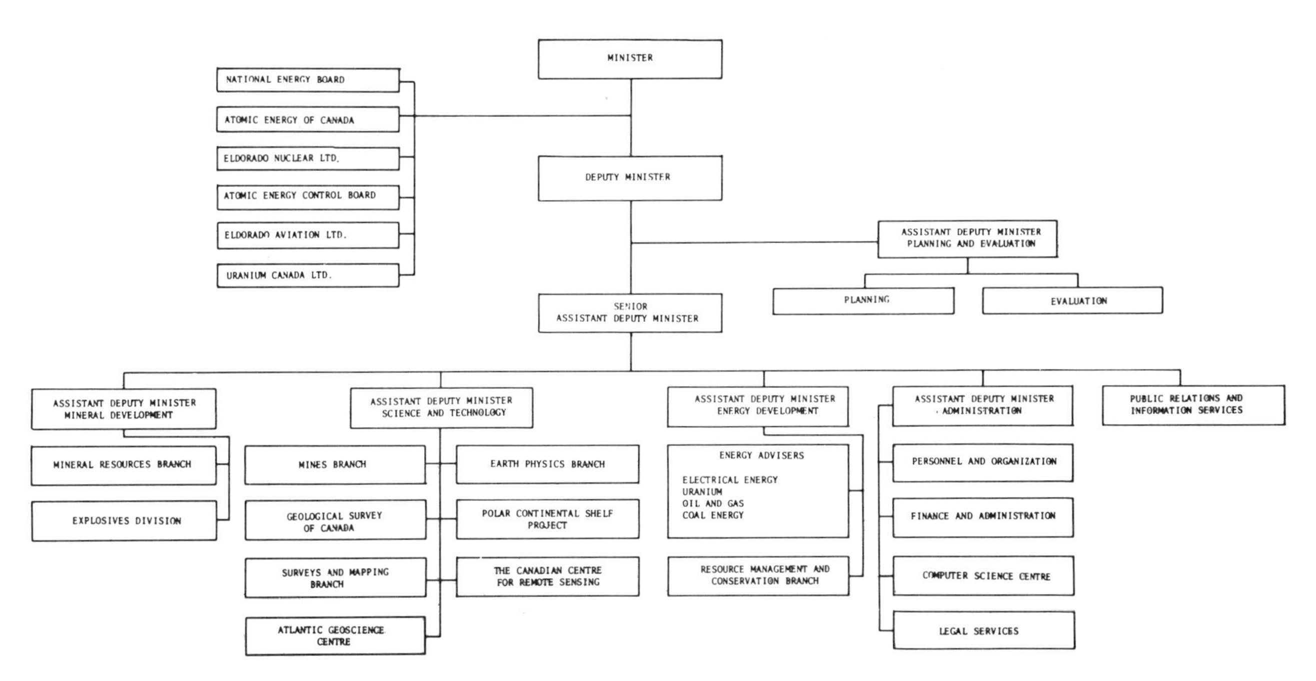

XIX-3 ORGANIZATIONAL PLAN OF DEPARTMENT OF ENERGY, MINES AND RESOURCES, 1972

and the assignment of staff members on temporary duty in foreign lands.

Seldom, if ever, has the Survey experienced such a sustained period of growth and advance; the undernourished institution of the thirties and forties was transformed almost beyond recognition. A large part of the explanation lies in the highly favourable circumstances in which the Survey found itself after 1950. But the success and the recognition that have come to it are also due in some degree to its continuing efforts to carry out its functions in the light of the country's needs, and to mobilize its strength in pursuit of these goals. The shaping of its policies and programs has been the work of those who were primarily responsible for its management: the ministers, deputy ministers, directors, chief geologists, and division heads. It might be proper, therefore, to begin by examining the executive structure in recent times.

Some difficulty arises in describing the precise roles of the series of ministers who have headed the departments of which the Survey was a part, since records of cabinet and ministerial decisions are not available for study and examination. In any event, the ministers presided over the affairs of large departments of which the Survey was only a small part. Even in the Department of Mines and Technical Surveys, which had been established expressly for the purpose of expanding federal government activities in the earth sciences and related fields, the Geological Survey was only one of five primary branches. A minister who might have been inclined to give the affairs of the department his full attention had little cause to devote much time to the day-to-day affairs of a long-established, stable branch like the Survey, which traditionally was capable of managing its own operations efficiently. With the formation of the Department of Energy, Mines and Resources in 1966, the Survey's relative position was reduced further, since the new department received additional functions of administering mineral exploration in areas under federal jurisdiction, making national water resource inventories, and studying and co-ordinating federal policies in the field of energy development. The principle of the new department's structure and functioning was that "all functions concerned with a given field or resource were combined under a single management. . . . Thus an integrated approach is being sought for the planning of both research programs and policy development."[1] Accordingly, the Survey, along with four of its five correlative branches, was once more reduced to being a part of a Mines and Geosciences Group within the department, flanked by parallel groups concerned with mineral development, water resources, and energy development, each headed by an assistant deputy minister. The more recent (1970) establishment of two new areas of government, the Department of the Environment, and the Minister of State for Science and Technology, may bring further reallocations of functions and of activities.

The Mines and Technical Surveys portfolio, perhaps because of its scientific orientation, tended to be held by ministers who were not inclined by temperament to make strong impacts on the branches and agencies they were administering. No fewer than seven political heads ruled the department in its 16-year lifetime, and most were transitory figures who did not play important roles in relation to it. George Prudham, the second minister and by far the longest-term holder of the portfolio (1950-57), might be singled out for giving his undivided attention to the administration of the department; his tenure coincided with the Survey's greatest advances. Of the remaining six ministers only Paul Comtois, who was minister from 1957 to 1961, held the position long enough to make any impact on the departmental administration.

This phase of the Survey's history came to a close with the formation in 1966 of the Department of Energy, Mines and Resources, with its much greater emphasis on policy planning in the vital natural resources field. The first minister, J.-L. Pepin, was concerned with organizing the water and energy aspects of his department, but his two successors, J. J. Greene and D. S. Macdonald, have concentrated more on resource management, planning, and conservation. The Survey was under strong constraints to accommodate its programs to the larger plans of the department, and to adjust its activities to these new governmental imperatives. By 1972, after several touchy transitional years, the process seemed largely completed. The Survey had embarked on a new role of assisting in resource policy making, and seemed once more to achieve a point of equilibrium relative to its administrative surroundings.

The successive ministerial heads have been responsible for carrying through Cabinet and Parliament the Survey's requests for funds, staff, and facilities; also overseeing how they were translated into new buildings and equipment, increases in staff establishments, and higher annual appropriations for operations. They bear the responsibility, too, for making and implementing major policy decisions affecting the programs, direction, and style of Survey activities. These decisions occur in the form of pieces of legislation that determine in a fundamental way the functions of the Survey and its relationships with other government agencies. Other directives that bring about a redirection of approaches and priorities come to the Survey as the result of the findings of various commissions, councils, task forces and committees, such as the Science Secretariat of the Cabinet.[2] Above all, from Parliament, filtered through the Treasury Board and retrospectively analyzed by the Auditor General, comes the budget that translates policies into practice. These various sorts of actions enable the Survey to continue its work in accordance with stated or indicated goals, and to meet its permanent responsibilities of studying and making known the geology of Canada in all its aspects.

If the "proof of the pudding is in the eating," then it must be concluded that the Survey was well supported by its political superiors at least till the later 1960's, and that they were most appreciative of its performance and value. The annual operations expenditures went up almost steadily after 1950, when they stood at $1,196,000 (for 1949-50), to $4,133,000 in 1960, to about $15,000,000 in 1971-72. At the same time, the full-time staff increased four times over, from about 150 to 565 persons.

As for improved facilities, there is the tangible evidence of the eight-story building completed in 1959 on Booth Street to house the scattered staffs and laboratories, and the beautiful regional headquarters built in Calgary in 1967 as a centre for its work in the sedimentary basins of the western interior and the Arctic islands. These buildings house many highly-sophisticated laboratories equipped for conducting complicated chemical analyses, petrological studies, age determinations, and for testing geophysical apparatus and methods; as well as largely self-contained ancillary units for meeting the drafting, cartographic, photographic, library, editorial and publication distribution needs of the Survey and its staff. Field work has been revolutionized through using new types of land and water vehicles and a variety of aircraft; costly and highly-sensitive equipment; even mobile field laboratories. These aids increase the efficiency of the Survey and enable special kinds of services to be undertaken.

As a branch of a government department the Survey comes under the purview of the senior civil servants of the department, to a degree under the chief deputy ministers (Marc Boyer, 1950-62, W. E. van Steenburgh, 1962-66, C. M. Isbister, 1966-70, and J. Austin, since 1970), but more especially under their immediate subordinates, the assistant deputy ministers to whom oversight of the Survey was committed. George S. Hume (1950-56) and W. E. van Steenburgh (1956-63) held the position of Director-General of Scientific Services. Afterwards the Survey was supervised by J. M. Harrison (1964-71), first as Assistant Deputy Minister (Research) and then as Assistant Deputy Minister (Science and Technology); and since 1971, by C. H. Smith in the last position. Hume, Harrison, and Smith all came to the administration from prominent positions within the Survey—Hume following a distinguished career as the leading specialist in the stratigraphy of the Plains region, Harrison as its twelfth director from 1956 to 1964, and Smith after periods as chief of the Petrological Sciences and Crustal

Geology Divisions. Under these men, the Survey could look for knowledgeable, considerate attention to its needs and interests.

Hume played a key role in the advancement of the Survey in the years after 1949, perhaps thanks to his long-standing connections with the important, rapidly-growing petroleum industry of western Canada and his own wartime record that provided him with an access to the powerful C. D. Howe, the Minister of Defence Production, who held the main reins of administrative control in the governments of the time. A determined man, Hume may be seen as taking a hand in such developments as the setting up of the Calgary office as a permanent centre of research and operations in Western Plains geology in 1950; the selection of the Survey to manage the uranium mining program on the government's behalf; the appointment in 1950 of a paleontologist as director of the Survey (the first in its history); and the decision in 1953 to provide the Survey with a new headquarters as part of the governmental building program for the Department of Mines and Technical Surveys. Hume stayed six years in this post, serving as acting deputy minister from time to time, notably during the illnesses or absences of Boyer. He retired from the government service in September 1956, a few months before the normal retiring age, to accept a position in the Western Canadian oil industry, where he served until his death in 1965.

Harrison did not have the opportunity to play as innovative a role as Hume. As assistant deputy minister after 1964, and later as senior assistant deputy minister, he kept a watchful eye on the Survey, helping to hold it on an even course in the reorganization attendant on the creation of the Department of Energy, Mines and Resources, and the subsequent shifts in governmental priorities. Since these entailed some transfers of units and personnel, he helped determine the Survey's position in relation to the new branches, notably those concerned with marine and groundwater resources. In January 1973 Harrison was seconded to UNESCO in Paris. C. H. Smith, the current Assistant Deputy Minister for Science and Technology, continues this supervision.

Mainly, however, the Survey has been managed by its administrative chief, the director, assisted by the chief geologist and the heads of the various divisions and sections. Here, also, the period was one of considerable changes, with four directors within a space of 15 years—a rate of turnover exceeded only in the period after 1894, when there were five directors within 13 years. Walter A. Bell, the director from 1950 to 1953, was a highly gifted, devoted paleobotanist, who secured his appointment on the basis of his excellent work in his chosen field and his seniority. Shy, elderly, and somewhat deaf, he was not a good public speaker or chairman, but for all that was a successful director. He was thoroughly conscientious, handled Survey business expeditiously, and was not afraid to make decisions or push for their implementation. He was fair, considerate, and gentlemanly in his treatment of others, and a competent judge and manager of the men and their work. The major policy decisions affecting the Survey in these years, as already indicated, were probably made by Hume, who had the temperament, and some talent, for dealing with politicians and top-level bureaucrats. The day-to-day operation of the Survey, which was still mainly a field mapping organization, continued in the hands of the chief geologist, George Hanson, and of the assistant chief geologist and technical editor, the extremely experienced, able C. E. Cairnes.

Bell had really not sought the directorship and was perhaps as surprised as anyone by his appointment, which took him away from his beloved fossil plants, trays of which always loaded the table behind his office desk waiting to be examined in any spare moments. He succeeded in advancing his retirement slightly to devote

XIX-4 WALTER ANDREW BELL
Director, 1950-53

himself completely to his special studies in stratigraphic paleobotany, for which there was great need. Emulating W. H. Collins, Bell applied for and received an appointment on July 4, 1953, to the long-unfilled position of Chief Geologist Consultant. That post had been established in 1936 to provide for Collins after the formation of the Department of Mines and Resources, but Collins held it for only two months before he died. Bell filled the position for just six months, and after January 4, 1954, which was Bell's normal superannuation date, it was abolished. Following his retirement Bell continued to be employed on a series of annual contracts, then on a number of part-time project contracts till shortly before his death in 1969. During these years Bell completed valuable work on fossil plants of the Cretaceous in western Canada, put the Survey's fossil plants collection in order, reported on specimens brought in from the field, and trained new specialists.

Bell's years as director saw the beginning of all the major advances that were to characterize the Survey over the next two decades. Most of these Bell supported to the full, though he did oppose the establishment of the Calgary office into a centre for reference, study, and research on behalf of the oil, gas, and coal mining industries. As director, Bell gave long-overdue attention to the stratigraphic and paleontological aspects of Canadian geology, an area that had been neglected by his predecessors at least as far back as Brock's day. The result was a considerable expansion of staff for paleontological laboratory and other work. Field operations were extended to the limit of procurable staff, especially in keeping with the needs of the oil and gas industries in the Western Plains region and the uranium mining industry in the Precambrian Shield. Geophysical aeromagnetic mapping activity was greatly expanded, and the field work even included the first of the large airborne reconnaissance mapping operations that were to become characteristic of the next two decades. Furthermore, Bell oversaw the advance into the field of special studies and a noteworthy expansion of the laboratory operations, notably in connection with the work in the radioactive minerals field. In sum, W. A. Bell presided over the early stages of a revolution in Survey activities, perhaps without having any clear idea where it was leading, as indeed no one had.

Bell's successor, George Hanson, was of the same generation as Bell, being a graduate in geology of the University of Manitoba and a member of field parties in the years before 1914. His academic and professional career, like those of Bell and Hume, was interrupted by a period of war service (in his case, as a pilot in the Royal Flying Corps), followed after the war by postgraduate study at Harvard and Massachusetts Institute of Technology. He worked for the Survey in the 1920 field season, then was appointed to a position of assistant geologist in the same competition as Hume, who secured the other of the two advertised positions. Hanson later worked in central and northern British Columbia, but after 1935 became increasingly involved in the office, helping G. A. Young with the field assignments, estimates, programs, and checking maps and reports preparatory to publication. When Young retired in 1943, Hanson was promoted chief geologist in his place, the highest position in the directorless Survey.

XIX-15 GEORGE HANSON
Chief Geologist, 1943-53; Director, 1953-56

Hanson might have gone further, but two heart attacks, in 1944 and in the winter of 1946-47, perhaps prevented him from being considered seriously for the position of chief of the Bureau of Geology and Topography in succession to Lynch; it was awarded instead to Hume early in 1947. Again in 1949 Hanson was passed over in the competition to succeed Hume; as chief geologist he was now subordinate both to Bell as director, and to Hume as Director-General of Scientific Services. Hanson worked loyally for Bell, and when Bell was arranging his own retirement, he made certain that Hanson would be considered for the directorship even though Hanson was visiting in Australia at the time. A cablegram was sent to

Hanson advising him of the impending vacancy and inviting him to apply. He received the appointment, and assumed the directorship on October 26, 1953; he was director for a little over three years, until his superannuation at the age of 65 on November 8, 1956. The post of chief geologist vacated by Hanson was filled late in 1954 by Clifford S. Lord; while H. M. A. Rice assumed the technical editorship of the publications, on the superannuation of Cairnes in 1953.

Lord's active tenure of the position of chief geologist continued till late in 1972 when ill health unexpectedly forced him into early retirement after a period of service comparable with Young's 19 years. A highly-respected field geologist interested in the Survey's traditional role of pointing the way to potentially mineral-bearing formations, Lord helped keep this important aspect from being overshadowed at the height of the "research boom." His concern that the field work be conducted as effectively and efficiently as possible led him to espouse wholeheartedly the fullest use of aircraft for transport and mapping in field operations. As other features of Survey work have expanded and been organized into strong, specialized divisions, the chief geologist's position has gradually evolved into that of planning and co-ordinating the scientific activities of the Survey. A hard-working and dedicated officer, Lord deserves much credit for the smooth operation of the Survey over the past two decades.

The director Hanson was a naturally gifted person to whom everything seemed to come so easily that in his earlier years he gave the impression of doing a minimum of work while enjoying life to the full, especially athletic sports. This light-hearted attitude did not sit well with more sombre geologists, who tend to view their work with utter seriousness and something approaching missionary fervour. In later years Hanson's health, of course, made it imperative for him to avoid exertion and emotional upset. Perhaps this accounts for his attitude of cool detachment, his frequently commented-on failure to exert himself to fight staff members' battles with the administration. On the other hand, he was informal, completely approachable, and was disposed to let the men suggest their own field programs; he gave them as much freedom of choice as he could. Years before, he himself had tried to do this for one season in Young's absence, only to have Young throw out his entire program. Having ceased his own geological work for many years before he became director, Hanson had no continuing personal geological interests. After he retired, he devoted his remaining years to travel, golf, and relaxation in a more agreeable climate than that of Ottawa, to which he returned periodically on social visits.

Hanson's policy of leaving the men to indicate their own programs allowed many of them to undertake the kinds of work dearest to their true interests, inspired them to their best efforts, and produced first-class results. Hanson was without prejudices for or against any aspect of the work as some of his predecessors had been; he realized that all types of work had to be allowed to progress and not be checked in the interests of some other activity. He had a preference perhaps, for the field mapping program; in any event the Hanson years saw the airborne mapping operations increased to the point where as much as 10 per cent of the area of Canada was being mapped in a single season. At the same time, however, he also assigned parties to revise and update previously mapped areas, and allowed the laboratories to expand by bringing in specialists in new field and laboratory studies. Work was also proceeding in these years on the designing of the new Survey headquarters, so that by the time Hanson retired the plans had been completed and construction was about to begin.

Bell and Hanson were elderly men approaching retirement age, and neither, perhaps, was at his peak of effectiveness when he achieved the directorship. Both men

XIX-6 CLIFFORD SYMINGTON LORD
Chief Geologist, 1954-73

XIX-7 JAMES MERRITT HARRISON
Director, 1956-64; Assistant Deputy Minister, 1964-72

had also experienced and been marked by the trials of the long, gloomy, disheartening years under Collins and Young. As their successor, the choice fell on a much younger man whose experience derived from the Survey's happier years. James M. Harrison, who was only 41 when he became director, was a science graduate of the University of Manitoba who proceeded to M.A. and Ph.D. work in geology at Queen's University. After several seasons as a summer assistant, he was appointed to the Survey in 1943 just prior to Young's retirement. His field work was entirely in the Precambrian Shield, mainly in northern Manitoba before 1949 and afterwards in the iron ranges of Labrador, which were just coming under development. He showed outstanding ability in his handling of such problems as the magmatic differentiation of anorthositic bodies, or the formation of granites by metamorphism, and he produced excellent structural studies on the Flin Flon area and Labrador Trough. His talents were quickly recognized, and he rose rapidly within the Survey; by 1947 he was a Geologist Grade IV, and he became chief of the Precambrian Division when it was organized in March 1955.

Harrison was a product of a more modern age, who as a student of E. L. Bruce at Queen's University, had been brought in contact with the latest geological ideas. Where his predecessors were men of the First World War era, Harrison was fully attuned to the remarkable new scientific and technological developments and realized their implications for the future role of the Survey. Under him the Survey wholeheartedly adopted the most modern technology for use in aerial reconnaissance and readily applied such new methods as geochemistry, biogeochemistry, or the improvement of airborne geophysical instrumentation. Study was begun of the marine geology of Canadian coastal areas and continental shelves, which add around 40% per cent more to the total area of Canada for the Survey to investigate and study. Harrison gave every encouragement to the development of new subsciences and their fuller exploitation, including the latest advances in areas as age determination of the rocks, geochemical studies, and the application of data processing to geology. Although the drive to complete the reconnaissance geological mapping of Canada continued in full course, an increasing proportion of the field work was directed to detailed field studies of key areas. The work also began to include such new types of study as the interpretation of metallogenic and physiographic regions; fundamental researches into geological processes; and efforts at explaining the geology of Canada in the context of the history of the earth. This growth was made possible by the move in 1959 from the old, overcrowded National Museum building, to the roomy, modern quarters on Booth Street—a step long in the making, but of which the Harrison regime was the immediate beneficiary. Before long, also, planning was begun on the new two-million-dollar headquarters in Calgary for the Survey's Western Plains operation.

A good organizer and manager of public relations, Harrison led the Survey into new scientific fields, and encouraged its officers to undertake many new responsibilities of a scientific nature. He arranged for the Survey to participate actively in international scientific activities. Putting the new facilities to good use, he had the Survey host many scientific conferences and other projects. Co-operative enterprises, in which the Survey took part along with provincial departments, universities, and private companies, were eagerly accepted, winning friends and enhancing the Survey's reputation in those quarters. An expanded program of scholarships and exchanges of scientists had a similar effect with the international academic and scientific communities.

Harrison was also an able diplomatist in dealing with the all-powerful Treasury Board and with the members of both Liberal and Conservative governments when in power. He quickly sensed opportunities for gaining public support for certain programs or activities, and seized them to further the interests of the Survey. For example, when the Conservative government announced its "Roads to Resources" program, he gained a leading role for the Survey in the ambitious ten-year federal-provincial multispecialty exploratory program designed to map northern districts for potential mineral deposits. Under his direction the annual budget of the Survey increased from $2.4 million to $7 million between 1956 and 1964, and the permanent staff practically doubled, to 420, including many outstanding specialists attracted by the opportunities for pursuing their desired research projects.

The Survey also benefited immensely in these years from the wide measure of public and government support, the acclaim of the mining industry and the scientific community, and the almost unrelieved prosperous times apart from the economic recession from 1961 to 1963. By exploiting these favourable conditions the Survey enjoyed one of the most successful periods in its entire history, comparable, perhaps, with the Brock years (1907–14) of the pre-World War I boom.

Because of his talents, Harrison was promoted up the administrative ladder to the position of assistant deputy minister, from which he could oversee the interests of the Survey, while the management passed to his contemporary and colleague, Yves O. Fortier, who became director in June, 1964. Fortier, who was born and educated in Quebec City, graduated from Laval University and subsequently did postgraduate studies at Queen's, McGill, and Stanford Universities, receiving his Ph.D. from Stanford in 1946. He worked as a summertime assistant in the Ross Lake district of the Northwest Territories in the early forties, and like Harrison was appointed to the full-time staff in 1943. He worked in the Precambrian of the Great Bear Lake-Indin Lake region, and increasingly in the Arctic islands, which he first investigated in 1947 on a reconnaissance in connection with a magnetic survey by the Dominion Observatory.

Fortier's work in the Arctic culminated in 1955 in the management of Operation Franklin, the most ambitious of all the airborne reconnaissance surveys. Through this operation and subsequent reports and addresses, Fortier kindled the interest of oil companies and the government in the potentialities of that remote region as a major petroleum and natural gas producer. His contributions were recognized by his being awarded the Massey Medal of the Royal Canadian Geographical Society in 1964. Fortier replaced Harrison as chief of the Precambrian Division, became chief of the newly-formed Economic Geology Division in 1960, and from there advanced to director.

The boom period continued for the first years of the Fortier regime, with 1966 to 1968 being highly productive years as a number of major projects reached fruition. After that, in spite of many important achievements, adverse economic and other conditions tended to hamper the Survey to some extent. The number of staff, for example, which had followed marked increases from 1943 onward, leveled off after 1964, with an increase over the next seven years of about 10 per cent, from around 420 to around 460 persons. Promotion was slow, the intake of fresh talent was reduced, and valued staff members left to take up more attractive positions in university teaching and research, often working part-time for the Survey on contract. For the first time in many years, notwithstanding the rise in the annual budgets from $7 million to $15 million between 1964 and 1972, a measure of financial stringency began to be felt. The budget increases, unfortunately, were largely eaten up by inflation and by higher fixed expenses, obliging the Survey to curtail or downgrade certain of its projects. For these and other reasons, it was difficult after 1968 to sustain the momentum of the previous 15 years. Indeed, the rapid growth of the Survey in those years was largely responsible for some of the problems of the post-1968 period.

The difficulties of the Survey were further increased at this time by numerous, sudden, quite drastic changes arising from governmental programs of administrative reorganization and reform; also from fundamental changes in public attitudes which those policies partly reflected. The sorting out of concerns and activities in the earth-sciences field consequent on the creation of the Department of Energy, Mines and Resources came to a head in the years 1968–72, entailing much planning and restructuring of the Survey. Changes in the operations of the public service—such as the introduction of collective bargaining and of steps to increase bilingualism—had marked effects on the style of day-to-day operations, and caused some disquietude and unsettlement.

The drive for greater economy and efficiency in government touched off by the report (1963) of the Royal Commission on Government Organization (the Glassco Commission) has been reflected by steadily-tightening Treasury Board controls over Survey spending and programs, and by increasing administrative preoccupations with cost-benefit analyses and "mission-oriented" programs that challenge the continuance of certain units and activities. As with earlier periods of financial or

XIX-8 PROGRAM AND OBJECTIVES CONFERENCE HELD IN OTTAWA, JUNE 29–JULY 3, 1970
Front row, left to right—R. G. Blackadar (Chief Scientific Editor), E. Hall (Technical Executive Assistant), S. C. Robinson (Chief, Economic Geology and Geochemistry Division), and S. Duffell (Acting Chief, Crustal Geology Division). *Middle row*—G. D. Hobson (Acting Chief, Exploration Geophysics Division), and D. J. McLaren (Director, Institute of Sedimentary and Petroleum Geology). *Back row*—J. G. Fyles (Chief, Quaternary Research and Geomorphology Division), Y. O. Fortier (Director), J. O. Wheeler (Chief, Cordilleran and Pacific Margin Section, Vancouver), P. Harker (Chief, Geological Information Processing Division), and C. S. Lord (Chief Geologist). G. M. Wright (Planning Office, Scientific Services) took the photograph.

other tensions, more efforts tended to be diverted into areas that appeared to hold high current economic or political validity. Even the autonomy of the Survey seemed imperiled by the tighter external controls and the overriding departmental administration.

After 1968 the Survey was also affected by, and responded to, the vigorous public debates on such fundamental policy questions as the devising and implementing of an effective national science policy; the replacement of the age-old emphasis on material growth with one that paid more consideration to the quality of life; and the efforts to develop resource-use policies appropriate to a whole-environment approach. The prolonged debate on national science policy reflected growing and widespread disillusionment at the past lack of economic and scientific planning by Canadian governments, and questioning by the public of the ideology of technological progress and material growth. It was felt that modern science should find new purpose in solving problems of the physical environment and meeting the problems of everyday living. Widespread concern was also expressed over environmental pollution. So vivid and emotional a subject was readily exploitable by the media, and soon government was faced with a very formidable pressure group too powerful to ignore. Survey planning had to keep all these new considerations in mind.

As a result, the Fortier years were among the most delicate and difficult in the history of the Survey. Organizational change after change was required in the effort to align operations with shifting administrative priorities. Responding to the government's stress on planning functions and its apparent de-emphasis of the data gathering and research that were traditional staples of Survey activity, a new direction began to assume greater significance in Survey operations—that of applying its knowledge and experience to advise federal governments on appropriate measures in mineral resource development and environment protection. Under the thoughtful, concerned guidance of Y. O. Fortier, the Survey by 1972 seemed to round a corner. Following some years of uncertain adjustment, the Survey appeared to attain a stability of function and status appropriate to its present-day situation within the Canadian government administration.

* * *

A FUNDAMENTAL POLICY QUESTION that has profoundly changed the Survey's structure relates to its role in a nation of widely-separated, disparate provinces and regions. The constitutional questions of the respective responsibilities of the federal and provincial governments in the mining, surveying, and geological research fields have never been resolved, but are always in a state of flux. Since the 1950's these have arisen with renewed vigour. Provincial challenges of federal authority in other areas–of which there have been a goodly number–induce in some provinces a certain sensitivity about Survey operations in their territories. The heightened importance of the mining industry in the economic life and politics of many provinces naturally increases the desire of provincial governments to accommodate this powerful clientele and win its favour. Moreover, the ability of provincial government agencies and the universities to do geological mapping and research improved markedly after 1950. These agencies sometimes became fearful lest the Survey impinge on, and curtail, their spheres of operations, and pressed for curbs on its freedom of action. Prosperity, or the reverse, has dramatically altered prevailing attitudes towards the Survey's working in a province, as have changes of government or of personnel occupying important political or administrative positions.

Faced with such a variety of situations, the Survey has worked out many special arrangements, usually along the lines of those reached earlier with Ontario, which retained for it such roles within the provinces as standard-scale mapping of geological features and studying geological problems of a national scope, while the provincial authorities undertook detail mapping and special studies in connection with their mining districts or mining industries. Thus the Survey recently decided to discontinue one-mile mapping in provinces that were able to do it themselves. Since no such difficulties exist in connection with work in areas which are the primary responsibility of the Dominion government, such as the northern territories or national parks, the tendency is to pay rather disproportionate attention to those regions. Specially-negotiated federal-provincial geological programs offer a different approach that has been adopted in recent years and is likely to receive increasing use.

Another reaction to the regional problem was to give the operations of the Survey a more acceptable regional and local orientation, as by decentralizing them according to the differing needs of Canada's regions. Almost from the beginning, the Survey was regarded as a national institution that afforded equal benefits to all parts of the country; but because it is centred in Ottawa (or earlier, in Montreal) many have felt that its benefits tended to be largely restricted to that locality, and that other regions were not so well served. Nowadays the argument takes a different direction: which is the better way to study the geology of a particular geological province–by means of operations based in Ottawa and using the facilities available there, or from a suitably equipped centre in the region concerned? It has been contended that the geological features of the flat-lying sedimentary rocks of the Prairie region could be studied more effectively and more conveniently from a headquarters in that region, making it unnecessary for scientists to travel to and from Ottawa to conduct their work. Stationing staff and facilities in the region would also help provincial agencies, universities, and private industries with their problems; hopefully, too, this would enlist the co-operation of these groups, making it more likely that the geological features and problems could be studied successfully. Having the specialists, samples, records, data, and literature of use to the private industries available locally would be another benefit. The Survey staff would also be given a greater sense of purpose through the closer contact with the industry and the public, and the Survey's reputation in that region would be improved.

Against decentralization, however, were the arguments on behalf of maintaining the consolidated, unified Survey–that it could provide much more specialized staff, laboratory and other facilities than would be possible by a regional office; that it could avail itself of the experience of personnel who covered the entire spectrum of geological specialties and periods, and who could supply relevant information on all of Canada's regions; that the materials and the experience derived from working in all regions were needed to help solve the geological problems of individual parts of Canada; and that the Survey's role of advising the federal government on policy questions required that its best, most complete, staff be based in Ottawa. The same question is being faced by other branches of the federal government; the Survey has studied it long and carefully.

Experience with regional offices goes back to before 1918; offices were set up in Vancouver and Calgary in the inter-war period, and after 1945, in Whitehorse and Yellowknife. They were mainly to aid local prospecting and mining interests with information and advice, and to serve as a base for Survey field operations and for contacts with local operators. The two offices in the Territories also helped with the administration of the regional mineral resources on behalf of the federal government. A Coal Research Centre was also established in Sydney, Nova Scotia, in the 1940's as a joint venture with the Nova Scotia government, but in 1959 it was transferred to Ottawa.

What was desired by those advocating regional offices in centres such as Vancouver or Calgary, however, was the equipping of these offices so that they could become the primary centres for study of the Cordilleran, Western Plains, and Arctic islands regions. In Calgary this hope has been realized. From 1950 onward a team of paleontologists, sedimentary geologists, and other specialists was stationed there in a downtown building that also became a repository for important drilling records submitted voluntarily by companies operating in the provinces, and supplied automatically by those active in areas where the natural resources were still under federal jurisdiction. The office also shared information with the Alberta government's regulatory authority administering petroleum and natural gas resources, and with the Alberta Research Council. To help the Calgary staff, well records, drill cores and samples, and fossils dating from the geological periods from which the oil and gas were derived, were transferred from Ottawa to Calgary.

The office, accordingly, gradually evolved into an Institute of Sedimentary and Petroleum Geology, which since 1967 has been located in a specially-built headquarters adjoining the campus of the University of Calgary to afford the best possible liaison with the scientists and facilities of the university. Besides scientific personnel, the institute acquired the full complement of support services to enable it to operate as a virtually self-contained unit. By the end of 1966, its staff of 40 included 24 scientists, five technicians, plus several clerical workers, draftsmen, and a librarian. By 1970 the staff had grown to 78, including three senior research scientists; its scientific sections include Arctic Islands, Structural Geology, Paleozoic, Mesozoic, Western Paleontology, and Petroleum Geology. Its Publications Section comprised a scientific editor, draftsmen, and distribution staff, under the general supervision of the chief scientific editor in Ottawa. Laboratory facilities included lapidary, photographic, macro- and micropaleontology, palynology and sedimentary petrology units, while the library numbered 50,000 volumes. The fossils in which the Calgary scientists are particularly interested (apart from those in the National Type Collection), as well as drill cores originating from the region, were transferred to Calgary to augment the collections already there. Today the main bulk of the Survey's paleontological facilities and staff are centred in the institute, which comprises over one-fifth of the whole and is almost a Survey in miniature except for the restricted area and periods with which it deals.

The facilities remaining in Ottawa included an Eastern Paleontology Section, a small group concerned with paleoecology, the National Type Collection, and the fossil samples collected from eastern Canada. The large collection of eastern Canadian well cores was later divided between the Survey's unit at the Bedford Institute of Oceanography at Dartmouth, N.S., and the Ontario Department of Mines. Then, with the transfer of the Eastern Petroleum Geology Section to the Bedford establishment, almost all resources for paleontological work have been shifted to two regional centres.

The establishment of the Institute of Sedimentary and Petroleum Geology increases the effectiveness of the Survey's paleontological work by consolidating it under the immediate supervision of its director, who could act in conjunction with local groups and according to locally-conceived but Ottawa-approved needs and priorities.

XIX-10 GROUP OF GEOLOGISTS, DECEMBER 1957
Front row, seated, left to right—J. M. Harrison (Director), Miss Gertrude E. Derry (Chief Geologist's Secretary), J. F. Caley (Chief, Fuels and Stratigraphic Geology Division), and H. S. Bostock. *Back row, standing*—H. M. A. Rice (Scientific Editor), C. H. Stockwell (Chief, Precambrian Division), L. J. Weeks (Chief, Post-Precambrian Division), and J. F. Wright.

The smaller size of the institute and its location away from the civil service atmosphere of Ottawa perhaps enhances the sense of camaraderie and improves staff morale. Against this, however, must be set the weakening of the ability of the Ottawa-based parts of the Survey (as well as the section stationed in Vancouver) to deal with the Phanerozoic (Paleozoic, Mesozoic and Cenozoic periods) rocks important in Cordilleran, Appalachian, and St. Lawrence Lowlands regional geology. In the opinion of those who have observed or dealt with the situation, the advantages of the 'division of labour' far outweigh the disadvantages, which improved communications have largely overcome. The problem of liaison is met by telephone and telex, quick visits of scientists back and forth on commercial planes, and the stationing of units on temporary duty from one establishment at the other. Thus the institute maintains a unit of four paleontologists in Ottawa, while officers of the Ottawa-based divisions are assigned to Calgary to help the institute with certain aspects of the regional geology.

Devolution of the Survey has not proceeded as yet beyond the fields of sedimentary basin and fossil fuel geology, and the somewhat analogous field of marine geology, which was centred in the Bedford Institute of Oceanography for reasons of convenience. A few years ago it seemed likely that Vancouver might become a similar autonomous centre for specialized studies of Cordilleran geology. At present the Survey maintains a Cordilleran and Pacific Margins Subdivision in Vancouver that includes a dozen geologists, a technician, and clerks. The subdivision does not possess a full complement of scientific, technical, administrative and support staffs and facilities as in Calgary; it is still dependent on Ottawa for those services. The development of a specialized institute like that in Calgary, however, is not beyond the realm of possibility. The Calgary experiment was undertaken with misgivings and fears for the continuing efficiency and integrity of the Survey. These have not been realized to date, but the problem remains whether such a process could be extended further, and if so, in what directions and to what limits? What would be the effect on the organization, for example, if every major geological province of Canada was furnished with a comprehensive and unique regional centre?

Survey activities and units were also partitioned and rearranged to correspond with the periodic reassignment of responsibilities among various federal scientific agencies after 1968. Within the Department of Mines and Technical Surveys arrangements had been worked out with the other branches, but in the later 1960's the governmental decision to set up a Water Group within the new Department of Energy, Mines and Resources occasioned some functional redefinition. The staff concerned with groundwater studies—a long-standing activity of the Survey—was formally joined to a newly-formed Water Research Branch. The Survey continues to work in the area of engineering geology, the smaller part of the former Engineering and Groundwater Geology Section. As a sort of exchange, however, the Survey acquired new staff and functions in the geomorphological field from the Geographical Branch.

More ambiguous has been the history of the Survey's activity in marine geology. A Marine Geology Unit was established in 1959, in response to the greatly-increased interest in submarine geology and the recently-acquired ability to study this formerly inaccessible terrain. Its personnel were later stationed at two newly-established outside centres: the Bedford Institute of Oceanography and the Canada Centre for Inland Waters at Burlington, Ontario, which the federal government had set up for multipurpose researches in these subjects. With later additions the contingent at Bedford has grown to considerable size as the Atlantic Geoscience Centre, a division of the Survey since 1972.

External organizational changes are few in comparison with the many periodic reassignments of staff and functions that have occurred since 1950 in response to changing emphases between field and laboratory activities, and between various types of work. The major field activity—the investigation, mapping, and reporting on the bedrock geology of Canada, which contains most of the metallic mineral deposits—is the present responsibility of the Division of Regional and Economic Geology, formed in 1972 with J. O. Wheeler as chief. This division is the successor of a Crustal Geology Division (previously styled "Regional Geology"), which existed during most of the years between 1950 and 1967. For a time in the later fifties, two divisions (of the total of six) dealt with this important aspect of the work: the Precambrian and Post-Precambrian Divisions. Perhaps the existence of the two divisions reflected the relative importance in which areal mapping and field studies were held in those years. At present, the division also embraces the Survey's interests in mineral deposits geology and economic geology, which were formerly a part of the Economic Geology Division (1960-67). The intermingling of field, laboratory, and inventory activities in recent years is readily illustrated by comparing the earlier divisions, which only had sections concerned with field work and related office study, with the present division, which includes laboratory support units to help in solving and interpreting geological problems, as well as computer and clerical facilities to aid with mineral resource appraisal work. The sections of the division, accordingly, include three that specialize in major geological regions: Canadian Shield Subdivision (W. F. Fahrig), Cordilleran and Pacific Margins Subdivision (H. Gabrielse), and Appalachian Section (W. H. Poole); Correlation and Standards Subdivision (J. E. Reesor) containing the division's laboratory facilities organized as Paleontology, Geochronology, Paleomagnetism, and Petrology Sections; and Economic Geology Subdivision (G. B. Leech), comprising the Mineral Deposits Geology Section, and the Geomathematics and the Mineral Data Bank Units.

A second division concerned primarily with field work and areal mapping, which reflects the sharply increasing importance of work in the Pleistocene and Recent, is the Terrain Sciences (formerly Quaternary Research and Geomorphology) Division under J. G. Fyles. This division, which studies the unconsolidated deposits, principally glacial in origin, was only formed in 1967. It is the successor, in the main, of an earlier Pleistocene and Engineering Geology Division set up by Hanson and Bell, who were so determined to enter this important field that they sent a number of staff members to various universities, mainly outside Canada, to become specialists in the Pleistocene. The division was downgraded during a reorganization in 1955, however, and for the next dozen years was merely a section in the Post-Precambrian (to 1960) and Economic Geology (1960-67) divisions. About 1966, however, a number of important developments changed the role and prospects of the section. The removal of groundwater work left a small remnant of the former Engineering and Groundwater Geology Section to be accommodated within the Survey's organizational structure. At about the same time, when the geographical programs of the Department of Energy, Mines and Resources were redistributed, the Survey acquired part of the staff of the Physical Geography Section of the former Geographical Branch, whose geomorphological and related work fitted very well into the Pleistocene Section. The present division was formed from those three elements, with the addition of the Radiocarbon Dating Laboratory, which performs a very important function for the division by its age determinations of Pleistocene and Recent materials. Since then other laboratories have been added, as well as workers in paleoecology and palynology. As a result, the nucleus was created for a sizable division concerned with activities that were becoming increasingly important from scientific or policy standpoints. The division is organized in the Regional and Stratigraphic Projects Section (B. G. Craig), Engineering Geology and Geodynamics Section (J. S. Scott), and Paleontology and Geochronology Section (W. Blake, Jr.), plus a Scientific Services Section that includes a photo-interpreter, geomathematician, and other specialists. In addition to its ambitious program of mapping the surficial deposits of the whole of Canada in parallel with the Regional Geology Division's work on bedrock geology, the Terrain Sciences Division conducts special field studies of largely unconsolidated Quaternary deposits and of other aspects of the present-day environment.

Research on the sedimentary rocks of the Phanerozoic periods, which is principally the concern of stratigraphers and paleontologists, and of petroleum and coal geologists, has given rise to another of the major divisions into which the Survey is divided. In 1950 this was the work of a Fuel Resources Division that included studies of coal and well-borings, and the offices and laboratory operations at Sydney and Calgary; and by a second, traditional Palaeontology Division (renamed Stratigraphic Palaeontology Division under Hanson) that carried on the activity launched by Billings a century before. The two were combined in the reorganization of March 1955, into the Division of Fuels and Stratigraphic Geology, with J. F. Caley as chief. Twelve years later, in 1967, the Calgary operation, under D. J. McLaren, the former head of the Palaeontology Section, was raised to divisional status under the title of Institute of Sedimentary and Petroleum Geology (ISPG), and took over most of the functions of the former Fuels and Stratigraphic Geology Division. The two sections that remained in Ottawa—Eastern Palaeontology and Coal Research—were elevated into a Biostratigraphy Division for a year until Caley's retirement, when they were added to the Crustal Geology Division. Since then, the Coal Research Section (P. Hacquebard) has also been transferred to the authority of the Calgary-based division and is to be relocated there in 1973. The institute is mainly responsible for investigations of the rocks of the Western Plains and Arctic regions lying between the Precambrian Shield and Cordillera, including the Rocky Mountains. It is presently organized in four major technical and scientific subdivisions: Regional Geology (with Arctic Islands, Northern Mainland, and Southern Mainland Sections, and the sedimentary laboratory) under D. F. Stott; Energy (with Geology of Petroleum, Clay Mineralogy and Chemical Laboratories, Coal Program, and Basin Analysis Program) under R. G. McCrossan; Paleontology (Macropaleontology and Micropaleontology Sections, Ottawa Section, and laboratories), under B. S. Norford; Administrative and Technical Services (A. W. Brusso); and Geological Information (E. J. W. Irish).

A second regional division, the recently-established Atlantic Geoscience Centre, is modeled on the ISPG in that it conducts a specialized activity and possesses field, laboratory, and research capabilities that make it almost self-sufficient. Situated in the Bedford Institute of Oceanography at Dartmouth, N.S., the centre studies the surficial and bedrock geology of the continental shelves off Canada's east coasts, appraises petroleum and natural gas potentialities of eastern and central Canada and the Atlantic continental shelves, and advises on technical and policy aspects of the exploration and exploitation of the seabed mineral resources. Its facilities include sedimentology and palynology laboratories for analysis of bottom sediments; a paleontology staff; a geophysics staff and instrument shop; a geochemistry laboratory for analysis of sediments and bedrock sample cuttings; and other service and support units, for a total complement of 65 staff.

Like the institute at Calgary, the centre evolved out of the former Fuels and Stratigraphy Division. A Marine Geology Unit (B. R. Pelletier) was established in that division in 1959 to study bottom sediments off the northern and eastern coasts, and later for logistic reasons moved to the Bedford Institute. Shortly afterwards, in view of the rapid growth of interest in submarine bedrock geology, a Marine Geophysics Unit (B. D. Loncarevic) was formed to work in this field from the base at Dartmouth. Finally in 1971 the Survey transferred its Eastern Petroleum Geology Section, under B. V. Sanford (then part of the Crustal Geology Division) there. On January 1, 1972, the three groups were combined into the Survey's seventh division with Loncarevic as director and the following scientific sections: Marine Geology (Pelletier), Marine Geophysics (D. I. Ross), and Eastern Petroleum (Sanford).

XIX-11 J. F. CALEY

XIX-12 BEDFORD INSTITUTE OF OCEANOGRAPHY, DARTMOUTH, NOVA SCOTIA
The Atlantic Geoscience Centre, one of the seven present divisions, is housed in these facilities.

By comparison with the foregoing divisions, the geophysical work of the Survey followed a steady, quite uneventful administrative history. Included briefly by W. A. Bell (1950-51) in a Radioactivity and Geophysics Division, its operations were sometimes listed in the annual reports of the department under those of Regional Geology, at other times separately as "Geophysics." In 1955, however, the operations in this field were formally constituted as one of the six technical divisions, under L. W. Morley, and so it continued thereafter until 1971, apart from a change of title to Exploration Geophysics Division in 1967. In 1971 the division acquired the much-transferred Geochemistry Section (E. M. Cameron) and assumed its present name, Resource Geophysics and Geochemistry Division, as a part of a new emphasis on assisting the discovery and appraisal of resources "in support of the inventory of the geological framework of Canada."[3] Changes inside the division reflected the increasing diversification of the work, which includes several types of airborne, marine, and ground mapping by geophysical methods of areas favourable for the occurrence of mineral deposits; also developing new types of instruments, field techniques, and interpretational methodology. As these activities were developed new sections were added, until the division, now headed by A. G. Darnley, includes Geochemistry, Electrical Methods, Magnetic Methods, Remote Sensing, Seismic Methods, Data Reduction, Experimental Airborne Surveys and Contract Surveys Sections. Rock magnetism, which studies the magnetic properties of rocks as a clue to their time and place of origination, was for many years an important section of the division; but since it is more closely identified with the work of the Regional and Economic Geology Division, it has been transferred there.

These divisions are chiefly concerned with field operations, and the function of their laboratory units is to help them achieve those objectives. Integrating laboratory and field mapping sections in the same divisions is a fairly recent development of the 1960's, one that reflects the changing character of geological work. Divisions in earlier times tended to group similar kinds of activity, that is, field mapping and research sections; and the laboratory sections were often assigned to separate divisions. Units that belonged to neither, such as mineral deposits geology or geomathematics, were usually lumped with the laboratory divisions. In 1951 there were two such divisions—the historic Mineralogy Division under E. Poitevin, who was succeeded by S. C. Robinson in 1957; and a newly-formed Radioactivity Resources Division, under H. V. Ellsworth, who was succeeded by A. H. Lang in 1952. In the reorganization of 1955, the latter division was renamed Mineral Deposits Division, which was considered more appropriate because of the diminished interest in discovering additional uranium deposits. Since most of the laboratory facilities of the former Radioactivity Resources Division had been transferred to the Mineralogy Division, during the late fifties the Mineralogy Division was the laboratory division *par excellence,* with the Analytical Chemistry, Mineralogy, Isotope Geology, Petrology, and Geochemistry Sections collected under the same roof. The Mineral Deposits Division, on the other hand, for some time was mainly a data-gathering, investigating and publishing division concerned with aiding the mining industry by preparing reports on the various metals.

Both divisions were reorganized in 1960. The Mineralogy Division, renamed Petrological Sciences Division, retained four of its sections but lost the Geochemistry Section to the Economic Geology Division. That division, the successor to the Mineral Deposits Division, took over its predecessor's functions as a Geology of Mineral Deposits Section; and also the Pleistocene Geology, and the Engineering and Groundwater Geology Sections from the Post-Precambrian Division (which at the same time was combined with the Precambrian Division to form the single Regional Geology Division). Thus the two divisions were now "balanced" at four sections apiece, though the Economic Geology Division seemed a rather artificial grouping of little-related units. A Data Processing Unit was added to the Petrological Sciences Division in 1965.

In 1967 the two divisions were again merged into a Geochemistry, Mineralogy, and Economic Geology Division, under S. C. Robinson, but only five of their eight sections were carried over. The Pleistocene and Engineering Geology Sections now became part of the new Quaternary Research and Geomorphology Division, while the Petrology and Isotope Geology Sections were transferred to the newly-formed Crustal Geology Division. Thus the Geochemistry, Mineralogy, and Eco-

XIX-13 A. G. DARNLEY

XIX-14 S. C. ROBINSON

nomic Geology Division included the following group of somewhat unrelated sections: Geology of Mineral Deposits, and Geochemistry (from Economic Geology); Mineralogy, and Analytical Chemistry (from Petrological Sciences); plus Geomathematics and Data Processing, the former Data Processing Unit of the same division. The division, still comprising the same sections, was renamed the Economic Geology and Geochemistry Division in 1969 but was abolished in the reorganization of 1971, when its sections were distributed among three different divisions. The present Central Laboratories and Technical Services Division (J. A. Maxwell), consisting mainly of the Analytical Chemistry and Mineralogy Sections, as well as the instrument development shop, is the lineal descendant of the original laboratory arms of the Survey.

XIX-15 PETER HARKER

The six scientific divisions account for only a part of the total Survey staff, for a large and growing part of its labour force is comprised of ancillary and support staff. Many serve as the clerical, administrative and labouring staffs of the scientific divisions. The majority, however, are members of the branch headquarters (such as branch executives, financial services, office services, branch registry, equipment office, stationery and supplies stores, etc.). Finally, there is the recently-formed (1970) Geological Information Processing Division (P. Harker), which includes editorial, cartographic and publications distributing units, the library and map collection staffs, the secretarial services, and the photographic section, comprising production and library units.

The preceding descriptions show how the work of the Survey has become organized over the period since 1950. The coming chapters will examine the activities of the various divisions and sections in more detail. The steady purpose behind the frequent administrative changes of 1955, 1960, 1967, and 1971-72 was to group together in the divisions those sections that had common interests, similar operational goals, or that were major users of one another's facilities. Most divisions, as the outlined events showed, were established, reorganized, or abolished in response to changing emphases of the general programs. The trend since 1950 has been away from organizing divisions according to the activities they pursue, in favour of divisions comprised of different units grouped together to work towards a common objective or mission. This can be seen in the present Regional and Economic Geology Division, which includes field mapping and research, laboratory, and economically-oriented sections. The combining of the economic geology with the other units has enabled the currently important resource-estimating function to be conducted more efficiently than before by integrating a wide spectrum of activities ranging from geomathematics to paleontology within the same division. The Institute of Sedimentary and Petroleum Geology and the Atlantic Geoscience Centre are even more mission oriented in organization. Sometimes, however, more traditional reasons were apparent in the administrative arrangements that led to the formation of new divisions: the need to group together sections in which a division chief had a personal interest or special training; the equalization of the responsibilities and work loads of the several division chiefs; or provision of a suitable position for a senior scientist. Some inconveniences arose in connection with these regroupings, but the advantages hopefully outweighed the drawbacks. The periodic reorganizations of the Survey's divisions, so vital to its effective functioning, proceeded partly through experimentation, with constant reassessments in the light of current means and future needs.

* * *

Since 1950 the staff of the Survey has been greatly increased, its work has grown far more specialized, its internal structure and relationships with other parts of the federal administration have become more formalized, and its professional and service staffs have become increasingly differentiated. How have these changes affected the general style of work of the Survey?

The number of staff members in each category has grown rapidly since 1950. The evolving earth sciences now require a wide variety of specialists to secure the full results of field operations and to study general problems effectively. The relative importance of laboratory and technical staffs has also grown. Aircraft and automated data-recording instrumentation have so markedly improved the ability to gather data that a single major airborne operation can yield enough results for years of laboratory analysis and office study. The changing nature of the scientific work, too, creates a need for additional service personnel to work in laboratories, compile and record data, prepare photographs, maps and diagrams, and attend to supplies, equipment, and to staff needs.

The high degree of specialization of the present-day earth sciences has led to the Survey becoming subdivided into ever-smaller, more specialized internal units. Although such specialization assists integrated multidisciplinary studies, such as preparing resource inventories, it does have the drawback that there may not be sufficient staff available to meet a major special program. For example, the urgent need in 1971 to investigate possible routes for a Mackenzie valley pipeline required the Survey to call on much temporary help from outside, which was in line with the current government policy of contracting out scientific requirements where possible. With a widely diversified staff of specialists, the management, in turn, must arrange proportioned, multifaceted programs that will keep all facilities suitably employed.

The appointment of so many specialists has raised problems similar to those that followed other sudden expansions of staff, such as at the end of the Selwyn era, or especially, the early 1930's. The scale of the problem, the highly specialized character of the work, and the rapidly changing scientific and technological situations in which the Survey has to operate, however, make the current situation unique. As the field staff grows older and less able to continue working in the field, or as field and laboratories specialties become relatively redundant in terms of Survey needs, the Survey faces a considerable problem of finding alternative useful employment for its specialized personnel, either inside the organization or elsewhere.

While opportunities for such employment in other parts of government remain regretfully scarce, the Survey itself can now offer more scope for continuing service. To help specialists stay abreast of the latest advances throughout their careers, the officers—especially in the Harrison years—were enabled to attend conferences and occasionally take study leaves. More opportunities were provided for a man to continue a useful career at the Survey when physical decline eventually reduced his effectiveness for field work. The days of the near-obsession with field mapping have passed—in this latest period it became accepted that experienced senior staff could be employed full-time at office research projects. Promotion and salary policies made it possible for staff members with strong research proclivities and important projects to avoid much involvement in administrative duties. Among those who were accorded this opportunity were the former division chiefs R. J. W. Douglas, A. H. Lang and C. H. Stockwell, and senior scientists such as H. R. Belyea, R. W. Boyle, H. W. Frebold, J. A. Jeletzky, V. K. Prest, R. Thorsteinsson, and E. T. Tozer.

However, the Public Service Commission's regulations do not always operate in the best interests of the Survey, for "Rules established to apply to the great bulk of the public service are not necessarily good for the few employers of scientists"[4] in government. The commission's control over hiring, for example, forces the Survey into a cumbersome, inflexible, inefficient routine. For many of the years after 1950, at a time when attractive alternative employment was available, the Survey had difficulty in getting its pick of the best summer assistants because of the ineffectual system for temporary appointments. As a result the Survey could not make full use of this most desirable means of securing, assessing and training highest-quality potential field officers. Instead, it often had to turn for its future officers to Ph.D. students trained in specialized research at the universities, men who with all their talents often lacked the several seasons of practical field training that had made their predecessors so effective in Survey work. These men were usually recruited under the system for longer term appointments, which offers a one-year probationary appointment followed by permanent tenure. This system also creates difficulties, because it does not give enough flexibility to meet sudden work emergencies, or to replace redundant or insufficiently productive staff. Instead, the Survey has been forced to rely more on hiring desired specialists on a temporary or contract basis.

The public service regulations largely determine the working conditions and the quality of staff relations. In 1968 Deputy Minister C. M. Isbister felt that the job classification system encouraged "personnel to relate themselves to an artificial system controlled from outside the Department, rather than to the tasks that have to be

XIX-16 GROUP OF DIVISION CHIEFS AND ADMINISTRATIVE OFFICERS
Ottawa, September 20, 1972. *Front row, left to right*—B. G. Craig (Assistant Chief, Terrain Sciences Division), B. D. Loncarevic (Director, Atlantic Geoscience Centre), Y. O. Fortier (Director), J. O. Wheeler (Chief, Regional and Economic Geology Division), E. I. K. Pollitt (Chief Administrative Officer), S. C. Robinson (Chief, Task Force on Special Projects). *Back row*—J. G. Fyles (Chief, Terrain Sciences Division), J. A. Maxwell (Chief, Central Laboratories and Technical Services Division), D. J. McLaren (Director, Institute of Sedimentary and Petroleum Geology), J. R. Hickson (Head, Financial Services Unit), A. G. Darnley (Chief, Resource Geophysics and Geochemistry Division), G. M. Wright (Task Force on Special Projects), T. S. Hillis (Administrative Officer, Atlantic Geoscience Centre), and P. Harker (Chief, Geological Information Processing Division).

done, and to the colleagues with whom they work."[5] On the same occasion he described the recently-introduced collective bargaining as "tending to introduce a rivalry between groups as to which can drive the best bargain, and to make group members more intent on preserving the rights they have won, than in bringing the projects they are involved in to a successful conclusion."[6]

The trend of the regulations towards standardizing grades and procedures throughout the public service has also done something to undermine the team morale that was a large part of the Survey's strength in times past. Service staff enter the federal public service and join the Geological Survey as positions become available there. The system is such, however, that they can readily transfer to more attractive and remunerative positions in larger, more powerful sectors of the government service. Thus the regulations designed to improve transferability work to encourage a sizable turnover among the service staff, particularly of the best, more ambitious workers. Besides having adverse effects on the quality of the work, valuable training and experience are lost, especially where there is need to recall the whereabouts of records or the routines of former years. The degree of control that the management of an agency like the Survey can exercise over the nontechnical support staff is reduced because this large part of the staff looks beyond the Survey for their future advancement, and follows work standards and practices that are determined outside the Survey for public employees as a whole.

Scientific and technical staff are also appointed through the Public Service Commission's competitions, but they apply there expressly for employment at the Survey in their specialty. In any event, their specialized training leaves them little likelihood of finding suitable positions elsewhere in the federal service; they almost certainly continue as employees of the Survey as long as they remain in government employment. They have a powerful incentive, therefore, to strive to achieve professional excellence and distinction in their work at the Survey. Indeed, salary increases and promotions in the professional groups today are based on the annual performance evaluations of each scientist relative to his peers rather than seniority. Furthermore, professional recognition is their best recommendation if they aspire to positions in the universities and mining or research organizations, which hold the best chances for alternative employment. The professional and scientific staffs are thus Survey members first in terms of their career commitments, and public servants second. Support staff are encouraged by the present system to belong first to the public service, though the loyalty and devotion of many of them to the Survey—especially those who have worked there for many years—is beyond challenge.

The growing number and complexity of public service regulations and procedures contribute to an increasing volume of administrative work that is making serious inroads into the productive output of many officers. An ever-widening segment of the professional staff has been drawn into the preparation of internal reports, position papers, the documenting of scientific progress, financial planning, or preparing estimates. The supply and financial procedures intended to facilitate oversight of spending by the branch, department, minister, Treasury Board, Auditor General, and various task forces adds to the weight of time-consuming administrative work. Government policies delegating financial responsibility upon the divisions has given them greater autonomy, but has also added to the administrative burdens of the division chiefs. Detailed estimates have to be prepared at every level, then be evaluated against division, branch, and department policies and requirements, and scrutinized further by the Treasury Board which imposes its own perspectives and priorities, usually strongly coloured by economic and statistical biases. The annual airing of all parts of the Survey's programs before the various levels of the department and the Treasury Board is time-consuming, frustrating, and often risky. Out of concern to present its programs in meaningful terms before the departmental management or Treasury Board, the Survey recently developed its own cost-benefit and performance-rating techniques, and has striven to work out appropriate methods of presenting its position properly before these powerful higher authorities.

Thus the Survey advances into the decade of the 1970's—its fourteenth decade—as a much larger, more complex institution than it was in 1950. It has also become heavily weighed down by procedures and administrative methods imposed on it by virtue of its greater size, complexity, and not least, by its membership in the increasingly bureaucratized public service of Canada. As an organization inheriting a proud tradition of service to Canada and to science, the Survey's present task is to shape its remarkably diversified staff, spread across the nation, into an efficient entity capable of coping with present and future challenges—challenges of working in remote, difficult districts; surmounting disruptive tendencies from within, and hazards of governmental experimentation from outside; keeping its place at the forefront of a rapidly advancing, expanding group of related earth sciences; and above all, as always, staying effective, alert and relevant to the needs of Canada.

XX-1 YVES OSCAR FORTIER
Director, 1964-73

20

Rounding Out the Geological Map of Canada

FIELD MAPPING IS A FUNDAMENTAL ACTIVITY of the Geological Survey of Canada, one that has engaged most of the attentions of its staff from 1843 onward. Through field work the Survey's geologists produced data on the nature, location, and distribution of rock formations and successions that are recorded in the published reports and maps, which are its end product and *raison d'être*. Although the term "surveying" implies mere measuring, the geologist faces a far greater task. From his observations in the field he must identify, interpret, and synthesize the features that make up the geological record as he proceeds, and be able to correlate them from one place to another. Notwithstanding the invaluable help he may secure from laboratory analysis and fossil determinations, the successful solving of geological problems on which correct mapping depends, rests in large measure on the care, experience, and reasoning power of the geologist in the field. The geological map of Canada, the result of 130 years' work, reflects millions of such reasoned decisions.

In the early years, field work was usually conducted in unmapped country, and topographical work was required to make the base map on which to plot the geology. By 1945, however, geologists generally did not investigate an area until published or manuscript topographical maps, or air photographs, were available to them. By then, too, the National Museum, Department of Agriculture, Mines Branch, Dominion Observatory, or other agencies handled most of the non-geological functions of the earlier all-purpose surveys, leaving geologists to concentrate on the geology alone. Many of the field operations in the nineteenth century were also limited to reconnaissances, recording what could be observed along the routes traversed, usually navigable waterways. Such maps indicated no more than the types of rocks encountered. The geologists were compelled to guess as best they could the distribution of the rocks beyond the outcrops they had observed. Proper mapping of regular map-areas necessitated field parties crossing and recrossing the area several times in each direction, or tracing particular beds or structures across the area. Accuracy and detail depended on the time and care available, and the frequency of observations made. The usual scale was 1 inch = 4 miles, the map-areas being of the order of 40 miles by 60 miles. Logan initiated this areal mapping program as early as 1863, and Selwyn continued it with the objective that eventually the whole of Canada would be covered. Other directors followed suit, with the result that by 1950 about one million square miles, a little over one-quarter of the land mass of Canada, had been mapped. Today, however, the Survey is on the threshold of realizing this long-term objective—of completing a primary geological map of Canada's 3,851,000 square miles.

From the beginning it was recognized that such primary mapping was only a preliminary phase that would enable the Survey to select areas for more intensive study wherever economic or scientific reasons warranted. In any event, neither the public nor the government would have accepted for a moment the Survey's neglecting districts where mining development was in progress, or in prospect, in order to pursue a general mapping program. Concurrent with that program, other parties were engaged in remapping certain areas in greater detail to describe complicated structures or mineralized districts, using scales of 1 inch = 1 mile, or less. Since there are still many subjects complex and important enough to justify careful, detailed study, field mapping will obviously occupy a large part of the staff for many years to come.

Besides, it should not be forgotten that the program of mapping the bedrock geology to standard 4-mile, or smaller, scales, is not the only mapping for which the Survey is responsible. Mastering the geology for purposes of producing an accurate general map requires collaboration of specialists from a number of subsciences and special fields, who help with the geological mapping through their specialized knowledge and insights, and also collect fossils and samples for their own special studies. Paleontologists, for example, found innumerable key areas for special work in the 50 per cent of Canada that was being mapped for the first time in the years between 1950 and 1970, and were of immense help with the correct mapping of the geology of those areas. However, they had no independent mapping program. Other specialties acting in conjunction with the large-scale areal mapping projects of these years secure data from very extensive parts of the country that are important enough in their own right to warrant being represented on local maps and as part of a national map series. This applies especially to surficial geology, the study of the unconsolidated materials that mantle the bedrock over most parts of Canada, which is not usually shown on the regular geological maps. Information on the nature and distribution of surficial deposits is increasingly helpful as the subject becomes better understood. These data are capable of being represented on maps that have many useful purposes such as in land-use planning, construction, or locating supplies of industrial minerals. Until recently Survey officers (apart from a few specialists) paid only incidental attention to the subject. At present, however, the work is in the hands of the large Terrain Sciences (formerly Quaternary Research) Division, which conducts investigations in almost every part of Canada and has a mapping program that may one day cover the whole country.

The same applies to several other major subsciences. The appearance of sophisticated laboratory techniques for studying physical characteristics, ages, and histories of rocks—coupled with pressures from the international scientific community—makes it desirable and feasible to map such special features of Canada's geology as tectonic structures, metamorphism, and especially the environments favourable to the occurrence of mineral deposits. Such metallogenic maps have immense practical value as a guide to government for inventories and administrative purposes, and to industry in its search for exploitable mineral deposits. As the structural geology of Canada and the principles governing the emplacement of mineral-bearing belts become better understood, it becomes feasible to prepare metallogenic maps of the entire country to illustrate the presence or absence of conditions favourable to the occurrence of particular minerals in certain areas.

Moreover, the geological map of Canada is constantly in need of updating and revision in view of the tremendous expansion of knowledge and research over the whole range of the earth sciences. Geological data are constantly being reappraised and revised in the light of new findings in the individual subsciences, or of broad general concepts of earth history. Hence a constant need exists to restudy areas that were investigated in the past, and to remap them in the light of recent findings. Consequently, the completion of the map of the bedrock geology of Canada, important as it is in marking the achievement of an effort of 130 years duration, will only be a milepost along the Survey's continuing way of systematically studying, investigating, describing and explaining the geology of Canada. Rather than an ending, the completion of the basic geological map is the foundation of new programs to investigate the geology of Canada along comprehensive, national lines.

The shape of this new course was discerned as early as 1969 by a Precambrian geologist who observed as follows:

> Coverage of the unmapped terrain will allow more of the resources of the [Precambrian] Subdivision to be available for the study of geological problems previously recognized, for closer examination of areas recommended as a result of the initial surveys, and for revision and remapping of other areas considered to be inadequately covered by modern standards. These future problem studies and remapping projects in the Canadian Shield will require long-term, continuing programs in order to attain an adequate level of knowledge and understanding and thus a meaningful geological foundation. These goals are essential for future mineral development throughout the Canadian Shield, particularly in the Northwest Territories where geological knowledge is still rudimentary over large regions. To facilitate future planning for the mainland Northwest Territories, all four-mile N.T.S. [National Topographical Series] blocks have been systematically appraised geologically and assigned priorities for future investigation.[1]

Today, in fact, much field work has progressed to specialized studies that include a wide variety of types of operation that were scarcely dreamed of even a few years ago. The number of field parties spending a minimum of four weeks in the field now ranges around 100 each year, though many projects also entail some time in the field. In 1969, for example, of 489 active "projects," 221 were reported as having a "field component." (The term "project" has only recently entered the Survey vocabulary as a way of identifying and tracing a given operation even though it may extend over several years and may involve changes of personnel as well.) A glance at the field work undertaken in 1969 furnishes a good illustration of how far this process had progressed by then. Areal mapping of bedrock geology involved only 42 of the parties, fewer than surficial geology which accounted for 45 projects. Geophysical projects accounted for 27 operations, mineral deposits studies for 20, paleontological studies for 15, and mineralogy for 2. In addition, 17 investigations were classed as "topical studies" that included such subjects as the study of volcanic rocks, and of structural mechanics and tectonics. The diversity of the operations and the degree of specialization may be seen by the activities being undertaken by some of the parties. Thus the geochemical operations included field studies of gold deposits, of biogeochemical indicators, and of trace element studies of streams and sediments, glacial till and micro-deposits. At the same time, other parties studied the geology of deposits of economic minerals, including iron, manganese, nickel, platinum metals, silver, gold, lead, zinc, uranium, thorium, beryllium, tin, the lithopile metals, and of coal and clays.

The ability to undertake such a wide variety of field activities while simultaneously mapping over half of Canada for the first time indicates the unprecedented pace at which areal mapping could be conducted with the new inventions, using only a fraction of the Survey's scientific personnel and effort. The Second World War had led to remarkable advances in travel and communications: in aerial navigation and photography, radar scanning and electronic measurement, in new types of aircraft ranging from the Canso flying boat to the helicopter, and in such assorted practical field aids as walkie-talkies, jeeps, motorized toboggans, rubber dinghies, weather-resistant fabrics, insect repellents and dehydrated foods. Work in northern districts, where much of the areal mapping effort was concentrated, was improved by the construction of the Alaska Highway and other roads, airfields, radio and weather stations, plus the continuing presence of military personnel and facilities. Since the War the tide of new invention and the process of opening the North have continued, with further benefits for the field work of the Survey.

Improvements in communications and technology that change the pattern of field operations are an old story. For 90 years field work had depended mainly on canoes, pack teams and backpacking, although in time the construction of railways and roads improved access to, and travel within, some of the areas being mapped. Photography, gradually introduced from the 1860's onward, also afforded a better means of recording certain data, while the introduction of the automobile and motorboat around the time of the First World War accelerated the pace of the work. The first major change, however, was the use of the airplane in field work, which did not begin to any extent at the Survey before 1935. Aircraft carried survey parties and supplies to their field areas, and sometimes within them; while photographic and aeromagnetic equipment mounted in a plane converted the aircraft into a rapidly-moving surveying instrument in its own right. The continuing refinement of such methods, coupled with advances in interpretation techniques and data analysis, greatly accelerated the process of mapping the geology of Canada.

The major breakthrough after the War that speeded up the completion of the basic geological map was the development of a technique using helicopters in conjunction with fixed-wing aircraft. These last were still used for the transport of men and supplies and for some air observations, but the helicopter added many further benefits. Helicopters had served since 1948 in control work for topographical surveys. The Survey in 1952 used them experimentally in an ambitious project, Operation Keewatin, in which it was proposed to map a large tract of Precambrian rocks in a 57,000-square-mile area west of Hudson Bay. This was almost ideal terrain for such a survey; the surface was of low relief, unforested, and covered with an abundance of irregularly-shaped lakes and rock outcrops.

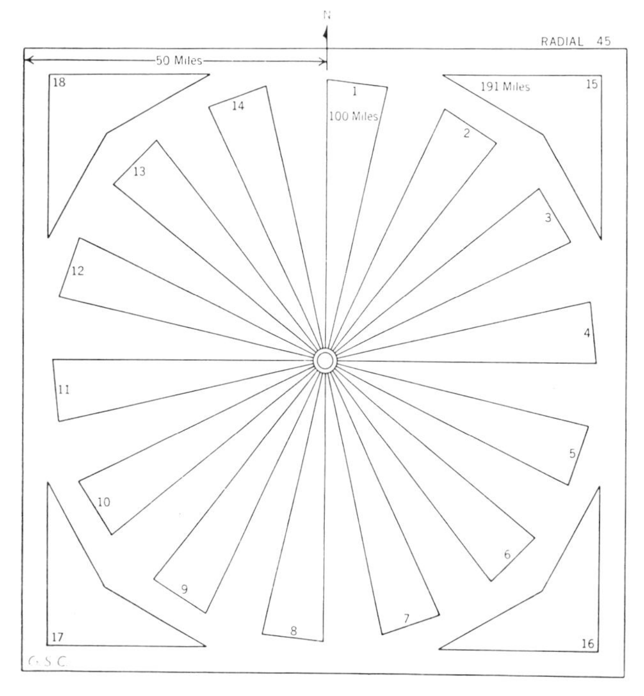

XX-2 THE FLYING PATTERN ADOPTED FOR OPERATION KEEWATIN

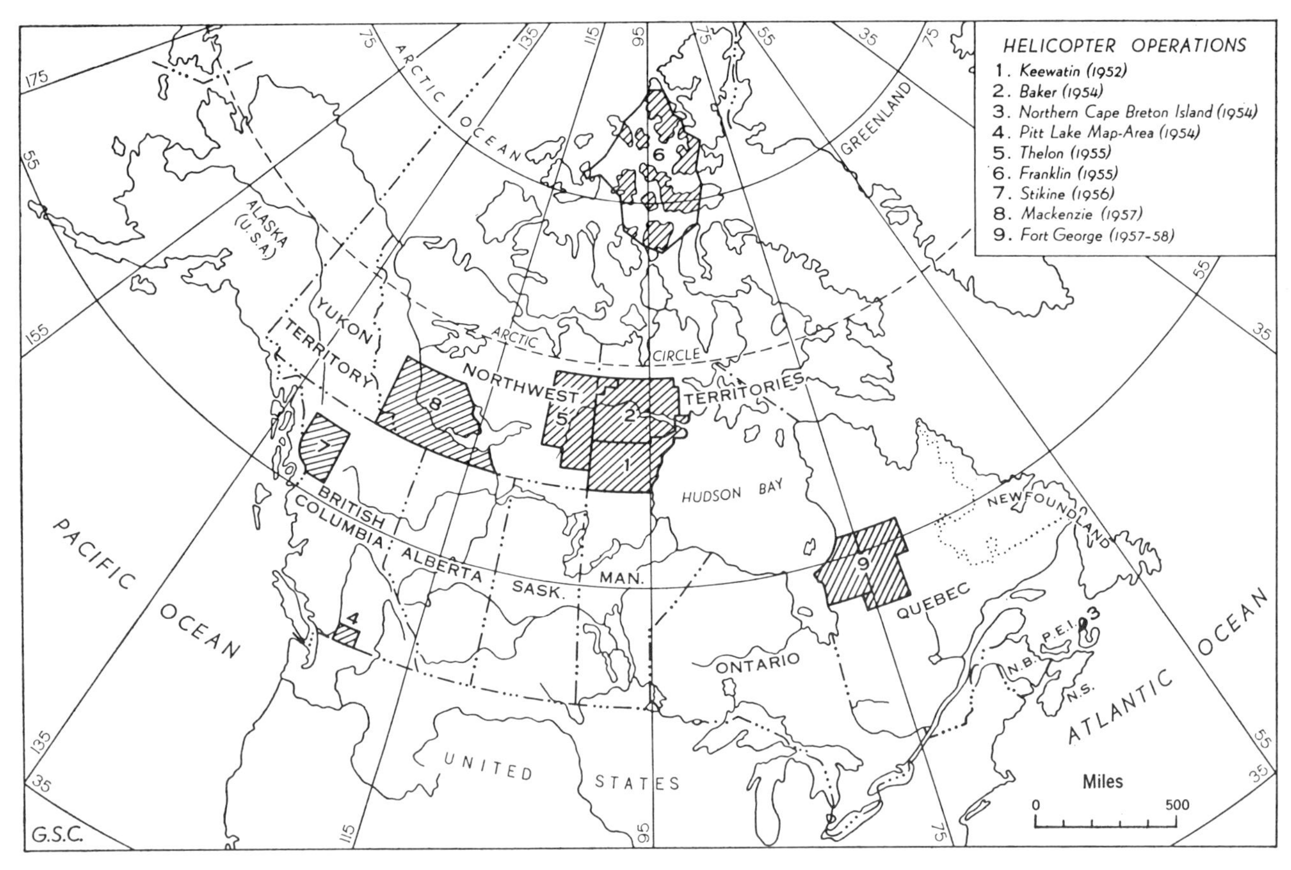

XX-3 THE MAJOR HELICOPTER OPERATIONS OF THE 1950's

Float-equipped light helicopters were sent out from a central base to fly 100-mile traverses, each one consisting of a triangular course of 40 to 50 miles outward, turning and proceeding 10 miles laterally, and then returning to base. Fourteen such traverses completed the survey of a circular area of 50 miles radius centred on the base camp, after which the operation was resumed from another centre, 100 miles distant from the first. Generally the helicopters were flown 100 to 300 feet above the ground, the speed of flight and altitude being varied to suit the needs of the geologist. The machine could be brought down to within 25 feet of the surface for close viewing, and could be landed at will to permit the geologist to examine outcrops and collect rock samples. During the flights the geologists' observations were plotted directly on a base map or on air photographs, and at the base camp the results from the traverses were compiled on a 1 inch = 8 mile map, the most practicable scale for such operations. Though the resulting maps were not as detailed as the 1 inch = 4 mile maps, they were acceptable for reconnaissance, and detailed enough to outline the areas most deserving of further, more concentrated study.

Fixed-wing aircraft were more economical for transporting men and supplies for the expedition to the base camps, eight or ten of which were occupied each season. Supplies, foodstuffs, and aircraft fuel caches were laid down in advance of the operation in predetermined locations, using fixed-wing aircraft where it was not feasible to use boats or trucks. By this means about 50,000 square miles—the equivalent of the field work by 20 or 25 conventional parties—could be mapped in a single season; with only a small fraction of trained geological (or other) staff, and producing results far more quickly and more cheaply on a per-square mile basis than by the older methods.

After the first successful experiment, the Survey began using helicopters and aircraft on one or more projects each year, varying the types of machines used, the flight patterns followed, and the sort of terrain investigated. In the next few years operations varied from a few hundred square miles to over one hundred thousand square miles. They were conducted in such diverse environments as forested, muskeg, rocky or mountainous regions, including several in the Arctic tundra. Seven large and two small operations conducted between 1952 and 1958 employed fewer than 60 geologists (less than nine a season), but nearly 500,000 square miles were geologically mapped, much of it being among the most inaccessible and difficult territory in Canada.

These operations indicated that while the helicopter was the ideal instrument for geological reconnaissance in certain areas, its usefulness at that time varied with the type of area being studied. It was not so effective in drift-covered or forested country where it was more difficult to follow trends than in the open Precambrian Shield country where the technique worked especially well. It was also harder to observe outcrops and to land geologists where they could make useful observations. In such areas it proved most practicable to place parties for periods of from three to ten days at a series of subcamps with radio communication with their base camp. Thus the helicopter was used mainly for local transportation.

There were certain drawbacks to using helicopters in mountainous terrain, arising from the constant need to protect the safety of the vehicle and its occupants, especially against very changeable weather conditions. Sudden clouds and storms, dangerous downdrafts, and gusty winds often made flying impossible, and forced curtailment of operations regardless of charter costs and disruption of a work schedule that depended on the availability of the machines. Furthermore, the earliest machines could not operate at sufficiently high altitudes to convey parties to the tops of mountains. After one operation (1956) which mapped 25,540 square miles of the Stikine district of northern British Columbia, it was concluded that in at least some areas it was more economic to conduct the bulk of the operation by pack-horse parties, supplemented by helicopters to help with filling in the gaps in areas too difficult of access by ground parties.

Greatly improved machines became available soon afterward, and helicopters began to be used more extensively on surveys in the Cordillera where appropriate. With these, geologists could be transported to formerly inaccessible places or to convenient elevated locations from which they could descend during their investigations, sparing them the difficult uphill climb. On some operations a helicopter was used to transport as many as five separate sub-parties from the base camp, usually to the tops of ridges which the men could then follow back down to the base camp. It might also convey the geologist in charge to each subcamp to study the progress of the work and make any necessary adjustments to the program. Sometimes a number of independent parties

XX-4 SURVEYING IN MOUNTAINOUS TERRAIN USING THE HELICOPTER, 1968

XX-5 A BALLOON-TIRED AIRCRAFT USED ON ARCTIC SURVEY WORK

operating in the same general vicinity in such country shared a helicopter for local transport and cleaning up gaps in the work of the ground parties. By reducing the amount of tiring backpacking and making it easier to collect and bring out samples from difficult locations, helicopters considerably decreased the time and effort required to map a given area.

Another area where helicopters were used in reconnaissance surveying was the high Arctic. Operation Franklin, conducted in 1955, was the most ambitious of all the large-scale operations, during which parts of Somerset, Prince of Wales, Melville, Cornwallis, Bathurst, Devon, Ellesmere, Axel Heiberg, Amund Ringnes, Ellef Ringnes and lesser islands—including some 100,000 square miles of land in a 200,000-square-mile zone—were mapped in a single season. The 28-man party headed by Y. O. Fortier consisted of 11 geologists, two or three of whom were field co-ordinators and the rest with the field teams, 10 geological assistants, a radio engineer and two cooks, plus four pilots and aircraft maintenance engineers. Two long-range Sikorsky S-55 helicopters were used to fly parties from four main bases 200 miles apart, and also served for observation purposes (for which the region was ideal) and transported field parties to various locations. The work was performed most effectively, but the cost of chartering and operating these aircraft was high. Since the land was open, and level in many places, it was decided to conduct such surveys by more economical fixed-wing aircraft, which needed only to be equipped with suitable landing gear (mainly low-pressure outsize tires) to permit them to land readily on the innumerable gravel beaches or ice floes. Piper Super-Cubs, equipped in this fashion, were used thereafter on several expeditions to the high Arctic, even though improvements in helicopter designs and efficiency have also made them more useful for this type of work.

Thanks to such large-scale operations, a few of which were mounted each year and covered up to 30,000 square miles or more of territory apiece, the mapping of the hitherto neglected parts of Canada proceeded at a rapid pace. The Yukon and Northwest Territories, large parts of which were still unmapped, were almost entirely mapped by these operations. Approximately 1,500,000 square miles of territory were covered in some 30 large operations, and by numerous smaller field surveys using helicopters or Piper Super-Cubs that were grandiloquently called "Operations" as well. The Precambrian of the mainland was mapped in Operations Keewatin (1952), Baker (1954), Thelon (1955), Coppermine (1959), Back River (1960), Bathurst (1962), and Wager (1964); the Mackenzie valley in Operations Mackenzie (1957) and Norman (1967-69); and the northern part of the Cordillera by Operations Pelly (1958-60), Ogilvie (1961), Porcupine (1962, 1970), Liard (1963-65), Keno (1964), Nahanni (1965), Yumack (1965), Selwyn (1965-67), and Snag (1968). The Arctic islands after Operation Franklin (1955) were the locale of Operations Eureka (1961-62), Prince of Wales (1962), Admiralty (1963), Amadjuak (1965), Grant Land (1965-66), Grinnell (1967), Bylot (1967-68), Southampton (1969), Penny

Highlands (1970), and Peel Sound (1970). Large tracts of the Precambrian and Hudson Bay Lowland of Ontario, Quebec, and Labrador were accounted for in similar large-scale airborne reconnaissances: Fort George (1957-59), Leaf River (1961-63), Northwest River (1965-66), Winisk (1967), Torngat (1967-69), as well as others associated with the Federal-Ontario "Road to Resources" program in northwestern Ontario, 1959-62. The work in the mountainous Cordilleran region was speeded up by Operations Stikine (1956), Bow-Athabasca (1965-66), Smoky (1969-70), Finlay (1970), and Stewart (1970).

Even though only three or four large operations, and perhaps a dozen regular parties, employed aircraft during each year's surveys, the total outlay on field operations increased considerably, though not necessarily in terms of dollars per square mile mapped. The cost of hiring the aircraft became the largest single expense in those operations in which they were used. In 1969, two-thirds of the operating budgets (a total of $424,057 of $650,300) of nine major aircraft-using projects were assigned to air services: Operations Norman and Torngat used one fixed-wing plane and two helicopters apiece, and seven other surveys used one helicopter apiece. The costs of individual operations varied from $317,000 for Operation Franklin or $168,000 for Operation Norman, to as little as $8,250 for Operation Skeena River, which was conducted by a single party and used aircraft sparingly. The three extensive operations that mapped most of the Barren Lands west of Hudson Bay (Keewatin, 1952; Baker, 1954; and Thelon, 1955) cost $486,000. But the large area mapped, a total of 185,000 square miles, reduced the average cost to only $2.63 per square mile, which is within the range of the cost of mounting the 45 to 60 field parties that would have been required to map the area by conventional methods. Moreover, the helicopter had many nonmonetary benefits. It extended the field season, permitted more flexibility in revising plans unexpectedly, and made the most effective use of highly-trained personnel. Geological features could be observed from a much broader perspective than by traditional methods, while at the same time direct contact with the geology was maintained. Sending the same members of staff on a single large operation, or on successive operations in adjoining territories, ensured uniform treatment over a very large area, leading to more reliable results and making future checking much easier. Small wonder, then, that C. S. Lord concluded his study (1958) of a number of early helicopter operations with the statement that "the helicopter technique, in its various modifications and in appropriate combinations with a host of other aids, is making possible professionally acceptable reconnaissance geological maps much more rapidly, and more economically in terms of cost per unit mapped, than any other field method known to the Geological Survey."[2]

The giant strides made possible by the new technique naturally enough gave rise to optimistic hopes that the reconnaissance mapping of Canada would be rapidly completed. In 1955 G. Hanson proclaimed that "Top priority was given to reconnaissance mapping and to the development of more rapid reconnaissance methods, the object being to provide, as soon as practicable, geological maps of all land areas of Canada at scales of 1 inch to 4, 8 or about 16 miles."[3] J. M. Harrison, in his first report in the following year, also referred to the objective of "completing the preliminary reconnaissance mapping of Canada at the earliest practicable date and of developing more rapid, economical and efficient field techniques for all major regions of the country."[4] In his study Lord also referred to the striking change that had occurred as a result of the use of the new surveying method:

> In any event, the impact of the helicopter on the rate of reconnaissance geological mapping in Canada has been truly spectacular. In the period 1842 to 1951 the Geological Survey mapped somewhat more than a million square miles. From 1952 to 1958 field work was completed within nearly half a million square miles by seven major helicopter Operations—in addition to much other field work accomplished by some 530 party-years' effort. Thus, since 1952, the Survey's staff mapped about half as much of Canada as in the previous 110 years—due, in large measure, to the helicopter. Clearly, this achievement represents the first major break-through in the Survey's century of effort to complete the initial or reconnaissance phase of the geological mapping of Canada. Less than a decade ago it looked as though a century or two would be required to complete this phase. Now there is reason to expect that it will be nearly completed in a decade or two.[5]

After eight seasons of such operations, when the preliminary field mapping was almost 60 per cent completed, the goal was proclaimed "to complete the preliminary mapping of the entire country by 1970."[6] As of 1963, about 65 per cent of the country was covered by published maps (the rate of publication actually lagged somewhat behind the pace of gathering the data in the field), with around 4 per cent being added each year. The western half of the Canadian Shield was completed in 1964, the Yukon Territory was almost completed in 1965; the Hudson Bay Lowland, 130,000 square miles, was mapped in a single season (1967) in Operation Winisk. By 1971 the reconnaissance-scale mapping was virtually completed; only a few areas then remained to be surveyed—mainly in northern Quebec, Melville Peninsula, and the most inaccessible parts of British Columbia and the Yukon. Completion of this longest of all Survey projects is expected in the mid-seventies.

* * *

A MAJOR USE OF AIRCRAFT was to obtain useful data about the area being overflown during the course of flight. Trained geologists immediately grasped the benefits of examining the country from the air, and aerial photography was perfected to give elevated views of the ground permanent form. From the mid-twenties such photographs were the basis of topographical mapping, and they were also extremely useful records of the geology when supplemented by ground observations. Just as aircraft carried photographing equipment to provide a record of the earth's surface so they might also be equipped with instruments capable of recording still other features of Canadian geology. Collecting such information was the function of the Geophysics Division (now Resource Geophysics and Geochemistry Division), whose activities in the fifties were chiefly geared to mapping Canada from the air.

The Geophysics Division in these years also undertook airborne reconnaissance surveying operations, and then interpreted the data to learn about the bedrock geology. The division's aircraft, equipped with magnetometers, were used to record changes in the magnetization of the rocks over which the planes were flying. Rather like field parties covering a map-area by traverses, these aircraft flew parallel courses back and forth across the area, the courses being half a mile apart, and the altitude 1,000 feet. Thus every two miles of flight yielded data on the rocks within one square mile of territory. Such aeromagnetic operations, which were flown in the early years by L. W. Morley, F. P. DuVernet, or other Survey personnel, mapped from 20,000 to 40,000 square miles a year. Compared with the amount of work involved in mapping such an area by regular field methods, the effort seemed miniscule. The greater part of the work, by far, lay in making efficient use of the data, and developing methods to make them more meaningful.

The resulting data, properly recorded on base maps, could reveal the locations of major faults and structures, or buried intrusive rocks, on the basis of the different reactions they registered on the magnetometer; such data provided valuable leads to geologists and prospectors about where to look for features of possible interest. Aeromagnetic maps are valued by the mineral industry because certain variations in the magnetic patterns (anomalies) give clues to the occurrence of various kinds of mineral deposits. A striking practical illustration of the utility of this technique occurred early in the Survey's geophysical operation. A joint aeromagnetic survey, financed largely by the Ontario Department of Mines but directed by the Survey, published a map-sheet in May 1950 recording an exceptionally strong magnetic anomaly near Marmora in eastern Ontario. Private interests investigated the area more closely and discovered a magnetite deposit in Precambrian rocks that was completely masked by a cover of Paleozoic sediments. By 1955 the Bethlehem Steel Corporation was developing this iron deposit into a large iron-ore open-pit mine with concentrating and pelletization plant that exported 7 million tons of iron pellets in its first 15 years. It was a convincing, tangible example of the value of the aeromagnetic method, and a great boost for the Geophysics Division. What was needed was to establish the role of geophysical surveying in the context of the Survey's total activities, and to refine and extend the techniques for the detection and interpretation of the geology of Canada.

Private mining and surveying companies were already actively engaged in this work, especially in western Canada, employing geophysical methods in their searches for deposits of petroleum and natural gas. Was it desirable

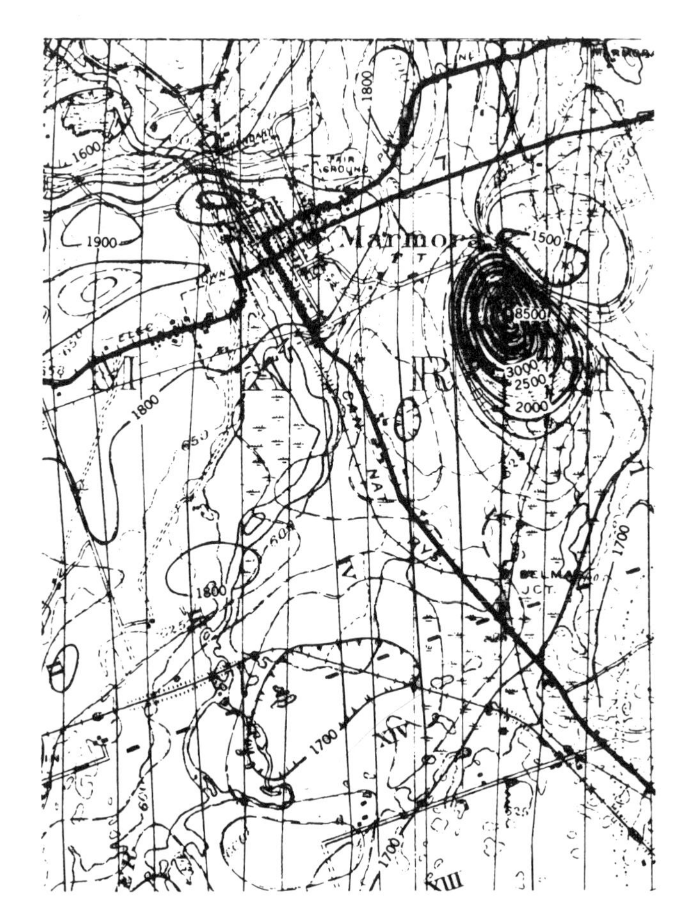

XX-6 THE AEROMAGNETIC ANOMALY NEAR MARMORA, ONTARIO
The vertical lines on the blueprint aeromagnetic map of 1950 represent the survey flight pattern, while the contours and numbers indicate the position and intensity of the magnetic mineral concentration.

XX-7 L. W. MORLEY

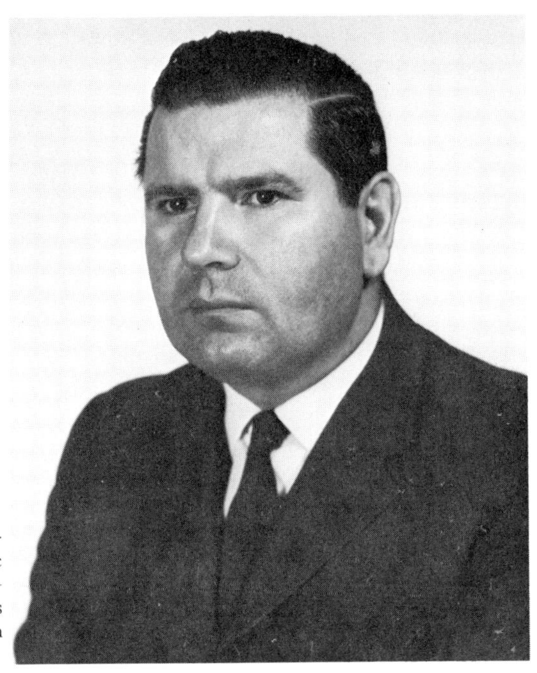

XX-8 P. J. HOOD
A leader in making improved electromagnetic geophysical instrumentation and methods available to industry in Canada and abroad.

for the Survey to compete with such enterprises, or for that matter, to perform services *gratis* that could possibly yield immense fortunes to companies without much effort on their part? Morley took the position that his division, a government agency, should confine itself to activities that were in the general interest. Thus geophysical operations should be carried on at the reconnaissance level to provide base maps from which mining companies could select favourable areas in which to concentrate further detailed geophysical and ground studies. The division should devote itself to mapping entire map-areas according to a general plan, and not concern itself with areas that were presently being investigated by private organizations. Instead it should co-operate with such companies by securing any useful non-secret data from them, and even be prepared to pay for the data where necessary. It should employ equipment designed to secure the maximum amount of geological information, rather than instruments designed for specifically economic purposes, such as searching out given kinds of mineral deposits. Finally, Morley envisaged an important role for his division in helping the aerial surveying and mining industries through researching, developing, and testing royalty-free instruments and techniques.

To be of maximum utility to the Survey and to industry, however, geophysical surveying had to be made into a fully effective geological tool. Numerous practical problems had to be overcome. Flying operations techniques had to be developed for surveying areas of irregular terrain that could not be flown according to the pre-set patterns. In the Cordillera a technique was developed whereby a light plane followed the contour lines, and the data it secured were automatically transmitted to a larger aircraft flying overhead which also recorded the position of the transmitting aircraft at the same instant. Equipment sturdy and compact enough to be carried in certain types of planes was also developed, as well as layouts inside the aircraft that did not interfere with the perfect functioning of the equipment. Another type of activity consisted of correlating the geophysical data recorded from the aircraft with the geology as mapped on the ground, in order to learn how to interpret similar data obtained in other geophysical surveys. For several years a bedrock geologist, A. S. MacLaren, worked with the division investigating the geology of an area in the Eastern Townships, and helped to calibrate the results recorded from the air against the actual geology. Through such efforts it was learned, for example, that certain kinds of associated rocks gave similar reactions and could not be differentiated by the existing geophysical techniques; and that certain conditions, like subsurface water in the overlying surficial deposits, produced anomalous reactions because of their greater conductivity. It became evident, therefore, that methods must be developed that would tune out unwanted signals and overcome certain distortions. This was solved partly by the kind of mathematical treatment applied to the data in the compiling process, that could smooth out the results to minimize accidental elements and produce configurations in which the effects of the underlying formations were sharpened and intensified.

Mainly, however, the attack proceeded through the development of special geophysical methods and tools to meet particular situations. One such method was the electromagnetic technique in which the aircraft was equipped with a transmitting device that emits an electric pulsed current, and also with a recording device that registers the "echoes" sent back from the rocks below. It was found that by varying the intensity, frequency, and the wave patterns of the emitted pulse, special sorts of signals could be produced to overcome the local surficial reaction and secure results from the underlying rocks whose character and position could thereby be interpreted much more successfully than before. On the other hand, an INPUT (Induced Pulse Transient) electromagnetic system was found to be especially sensitive to features of surficial deposits such as clay, till, gravel, and buried watercourses (aquifers), and so could be used very effectively to delineate these phenomena. Other types of equipment were designed or purchased, such as airborne scintillation equipment, which measures the levels of emitted radioactivity, to record responses from deposits containing radioactive minerals. Using such a method in 1969-70 "a cross country reconnaissance from Ottawa to Yellowknife with the high sensitivity gamma-ray spectrometer . . . identified all known uraniferous localities en route and indicated the presence of several other localities of probably comparable importance."[7] Another recently applied method is an airborne infrared scanning instrument that may be used to detect polluting effluents, thermal wastes, underwater obstructions and currents in bodies of water. Such experimentation goes forward unceasingly. Attempts have also been made to detect traces in the atmosphere of gaseous emissions of radon, or of hydrocarbons from petroleum and natural gas formations, as clues to the character of the rocks over which the aircraft is flying. Such methods are similar to some that are now being used in earth-circling or interplanetary satellites.

While these many activities have been taking place, the division has continued steadily in its original purpose of the geophysical mapping of Canada from the air. The project of a complete aeromagnetic map of Canada was enunciated in 1960, and by 1970 the work was approaching the halfway point. The problem is not so much the flying operation as processing the wealth of the data obtained. In the early days when only about 45,000 line-miles a year were being flown, a five-year backlog in compilation had developed within just six years of the start of operations. The need to perform complicated mathematical operations with the data further slowed the pace of compilation. Hence by the mid-fifties it became necessary to automate the recording and transcribing equipment and to introduce computer techniques to perform the mathematical adjustments. Thus the division was early to point the way for the other divisions into the modern-day realms of automated equipment and computer technology.

With steady improvement in methods and results, the division's maps became increasingly useful to geologists and the mining industries, and arrangements were made for aeromagnetic surveys to precede geological field work wherever possible. Even in areas already mapped, aeromagnetic surveys frequently produced further information about the formations that answered some questions and pointed out features deserving of further intensive study. Above all, the geophysical methods gave the Survey a means of learning something about the character and structure of formations hidden at depth or buried beneath the lakes and continental shelf waters of Canada. In 1958 surveys were begun over the bays, gulfs, and continental shelves to map and evaluate the thickness of the Carboniferous rocks underlying the Gulf of St. Lawrence and the Paleozoic and older sedimentary formations that lie beneath much of Hudson Bay. Such work has inspired, and been inspired by, recent activities of oil companies along the coasts and continental shelves of Canada, from the Grand Banks and Gulf of St. Lawrence to Hudson Bay and the Arctic islands. On land aeromagnetic surveys have figured in the discovery of base metal deposits in New Brunswick and at Matagami in northwestern Quebec, and have also indicated other iron ore deposits, aquifers, sedimentary basins, uranium-bearing deposits, and the like.

The rate of airborne geophysical mapping of Canada has been greatly accelerated in the years since 1962 as the result of an $18,000,000 federal-provincial cost-sharing program to map most of the Canadian Shield and adjoining areas by aeromagnetic methods with the objective of accelerating exploration for potential orebodies that would lead to future mines. By 1962 the total area flown by the Survey and some provinces amounted to 543,000 square miles; seven years later, under this grant program about 1,300,000 square miles were covered. The Survey only oversees such operations; the actual work in connection with the program (which is expected to continue into 1974) is done almost entirely by contractors, who not only collect but also compile the data. Survey policy since 1962, in fact, has been to have such routine surveying done by contract rather than hiring its own aircraft for the purpose. The staff, instead, now concentrates mainly on developing new techniques and on the experimental testing of new types of equipment (sometimes for operation on ships as well as planes); and its flying operations are now limited to individual mis-

XX-9 GENERALIZED FIELD LAYOUT ILLUSTRATING THE REFLECTION SEISMIC TECHNIQUE
The diagram, prepared by Seismic Service Supply (1958) Limited, was reproduced in *Mining and Groundwater Geophysics* (E.G. 26), a Survey publication that presented the results of an important international conference held at Niagara Falls, Ontario, in 1967.

sions of more purely scientific character. An example is the work being done since 1964, in conjunction with the National Aeronautical Establishment (of the National Research Council) and the Royal Canadian Air Force, to explore the bottoms of Foxe Basin in northern Hudson Bay, the Labrador Sea, and the Atlantic Ocean. This last is part of the investigations of the Mid-Atlantic Ridge in the latitude of Canada, contributing to the important work on the phenomena of ocean-floor spreading going on in several countries.

Not every sort of geophysical activity involves the use of aircraft; the seismic work, for example, is conducted on the ground or from ships. Shock waves are transmitted into the earth and the echoes that are recorded indicate the depths and attitudes of the various strata beneath the surface. This method is used by the Survey to trace aquifers and other superficial features, to determine thicknesses of sedimentary sequences, to detect potential oil-bearing structures (the main commercial use of the method), or other formations; also to increase the knowledge of the tectonic framework of the continent, and of the deep-seated structures of the earth.

* * *

SINCE 1950 THE GEOLOGICAL WORK of the Survey has expanded to take in the whole landmass of Canada, and much of the continental shelves as well. Its quest for knowledge has developed into a many-sided endeavour, employing specialists in a wide variety of fields. The work is conducted by one-man parties and by teams of scientists brought together in tightly-organized air-supported operations. Taken together, these field activities average a hundred or more per annum over the past 20 years, much more than a study such as this can hope to examine with any degree of comprehensiveness. Only a small fraction of the entire group can be referred to in this and following chapters.

A review of the progress of field operations since 1950, however, should at least examine activities in areas where the Survey had rarely, if ever, ventured before. The entrance of Newfoundland into Confederation in 1949 as Canada's tenth province extended the work 500 miles farther eastward into the Atlantic, and into a large and important part of the Quebec-Labrador peninsula. Two other areas—the Arctic Archipelago north of the mainland of Canada, and the adjoining offshore areas—were also investigated. In both, technological changes made productive research possible for the first time, and both areas were attracting the attention of the mineral industry in their unceasing search for new resources to exploit.

The extension of operations into Newfoundland and Labrador was provided for in the Terms of Union with Canada. Here the Survey was not entering on a new field that depended on the development of new techniques, nor one that was unknown geologically. It was no stranger in Newfoundland in the years after Alexander Murray's retirement in 1883. A. O. Hayes had published an outstanding memoir (Memoir 78) on the Wabana iron ore deposits, D. B. Dowling had examined the coal beds at St. George's Bay in 1920 on behalf of the Newfoundland government, and W. A. Bell published a report on its fossil plants. Earlier, of course, A. P. Low had traversed a large part of Labrador in his epic exploratory survey of 1893-94.

As for the state of geological knowledge, there was the work of Newfoundland's own Survey, done by Murray's assistant, James P. Howley, down to 1909; by H. S. Baker from 1926 to 1930; and after 1933, by A. K. Snelgrove and C. K. Howse. In the thirties and forties as many as six to ten parties a season worked in the field for the Newfoundland Survey. The results of these studies were published as *Bulletins* by the Survey, and as papers in professional journals. Other work had been done by large mining corporations interested in developing mineral concessions, and the island provided subjects for numerous theses and academic studies, principally by Princeton University students. The geological outlines of Canada's newest province were therefore reasonably well known.

The Survey began in 1949 by continuing two projects of the former Newfoundland Survey with the same men, and starting several other projects. L. J. Weeks came over from Nova Scotia to inspect the work of the field parties. Areal mapping was the main activity, but work also was quickly begun on special subjects including mineral deposits, paleontology, aeromagnetic surveying, surficial geology, and as early as 1952-53, a petrological study of ultrabasic rocks. A regular program of 4-mile mapping was initiated, and in 1954 the study of a geological cross-section across the island between the 48th and 49th parallels was begun as part of the general investigations of the Appalachian belt. Map-areas were studied by T. O. H. Patrick, G. C. Riley, S. E. Jenness, D. M. Baird, F. D. Anderson, J. W. Gillis, R. K. Stevens and H. Williams. The completion of Williams' map-area in 1967 ended the 4-mile mapping program throughout central Newfoundland.

As elsewhere, detailed studies were also made of problem areas and of areas of economic importance, among them stratigraphic and structural studies by E. R. W. Neale, W. A. Nash and M. J. Kennedy; and mineral deposits studies by E. R. Rose, W. D. McCartney and D. F. Sangster. Surficial work was left mainly to E. P. Henderson, who made a study of the important Avalon Peninsula; while several officers were involved in paleontological work, notably R. D. Hutchinson, and L. M. Cumming, who worked in Newfoundland for a number of years after 1954. During and after 1966 Cumming restudied Richardson's classical work of a hundred years earlier on the Paleozoic strata of the west coast. In the meantime work on the Northern Peninsula was also raising questions about structural relations with the Grenville of the adjoining mainland. Accordingly, Operation Strait of Belle Isle directed by Cumming, who was assisted by two other specialists, H. H. Bostock (Precambrian) and D. R. Grant (Pleistocene), was launched in 1969-70 to work out these relationships and, in the process, complete the geological reconnaissance of the province.

Newfoundland's entry into Confederation coincided with the major mining development in Labrador, the beginning of the mining of the enormous iron ore deposits first pointed out by Low. In 1949 J. M. Harrison began a detailed study of the Labrador Trough, which was continued by his assistant, J. E. Howell. Unfortunately, Howell drowned in 1954 in one of two canoe accidents

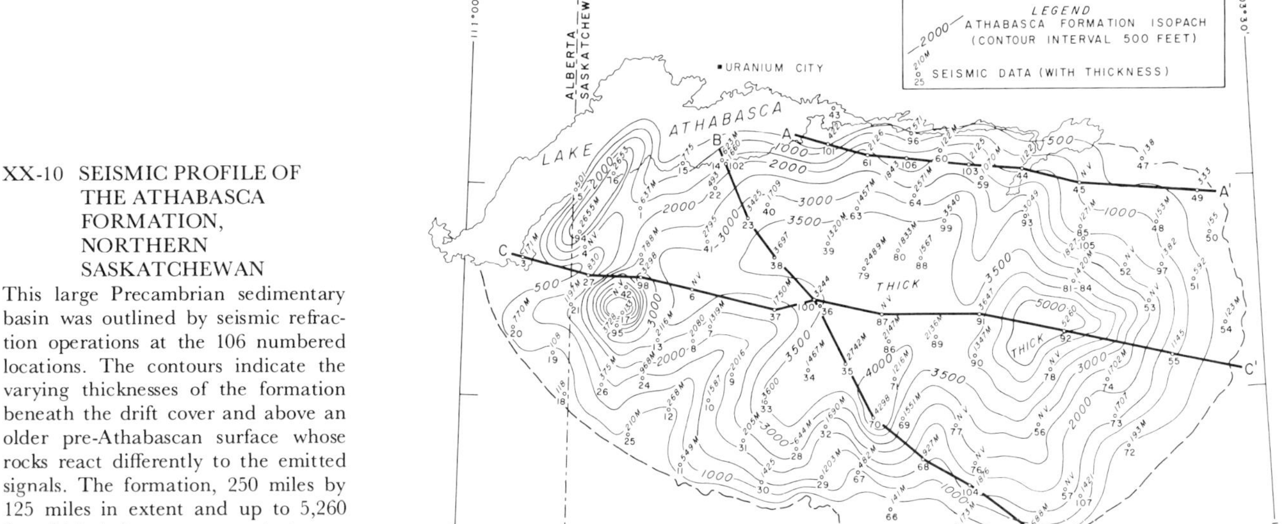

XX-10 SEISMIC PROFILE OF THE ATHABASCA FORMATION, NORTHERN SASKATCHEWAN
This large Precambrian sedimentary basin was outlined by seismic refraction operations at the 106 numbered locations. The contours indicate the varying thicknesses of the formation beneath the drift cover and above an older pre-Athabascan surface whose rocks react differently to the emitted signals. The formation, 250 miles by 125 miles in extent and up to 5,260 feet thick, is known to contain important uraniferous concentrations. The thick lines A-A1 etc., represent three sections through the formation.

in the area that summer, which also cost the lives of three more summer assistants. With Harrison as director the work remained suspended until 1960, when W. F. Fahrig completed the 1 inch=500 feet map of a strip across the iron-formation zone of the Labrador Trough. Since then the Survey's activity in the mining district concentrated mainly on the deposits themselves. For example, G. A. Gross examined them for his monumental general study of iron ores in Canada.

This detailed work was accompanied by regular field mapping of the region, which was directed first at areas known to contain iron deposits or other evidences of mineralization. A notable group of Precambrian geologists worked on the iron deposits: Fahrig, M. J. Frarey, K. E. Eade, S. M. Roscoe, W. R. A. Baragar, I. M. Stevenson, S. Duffell, H. R. Wynne-Edwards, and J. A. Donaldson. Other investigations included early reconnaissances up the coast of Labrador by A. M. Christie; and, in the sixties, studies of the Grenville Front by G. M. Wright and others; detailed studies of intrusives by R. F. Emslie, of diabase dyke swarms by Fahrig, and of possible meteorite craters by K. L. Currie. Two large-scale airborne reconnaissance programs were mounted to expedite the mapping of the region, Operation Northwest River in 1966, headed by I. M. Stevenson, and Operation Torngat, 1967, by F. C. Taylor. Surficial work, which E. P. Henderson began in 1953, was continued much later with helicopter-type operations in 1969 and 1970 by R. J. Fulton and D. A. Hodgson.

In the Arctic islands massed, large-scale airborne operations covered the enormous areas quickly, though single-party operations were also conducted simultaneously. Even a small operation, however, required air support to reach the field of study, and often helicopters for transport within the map-area. The Survey's experience with the Arctic Archipelago before 1950 had been very limited. Robert Bell had mapped part of the south coast of Baffin Island in 1897, and Low had examined the geology of places along the eastern margin of the Arctic Archipelago from Baffin to Ellesmere Islands on the *Neptune* expedition of 1903-04. Weeks also had made some useful observations during his unlucky sojourn on Baffin Island in 1926-27.

The Survey's first modern work in the region was by Y. O. Fortier in 1947, who accompanied a Dominion Observatory party investigating the North Magnetic Pole. The geology of parts of Victoria, Prince of Wales, Somerset, and King William Islands and Boothia Peninsula was viewed from the air and was supplemented at a few points by ground observations. These were far too limited to give more than very general impressions of the geology of the regions seen. In 1949 Fortier returned to begin areal mapping on south Baffin Island; the mapping was continued from 1950 to 1964 by W. L. Davison, G. C. Riley, and especially R. Blackadar. Precambrian terrane along the eastern margin of the archipelago on eastern Devon and southeastern Ellesmere Islands was examined by R. L. Christie in 1960-61

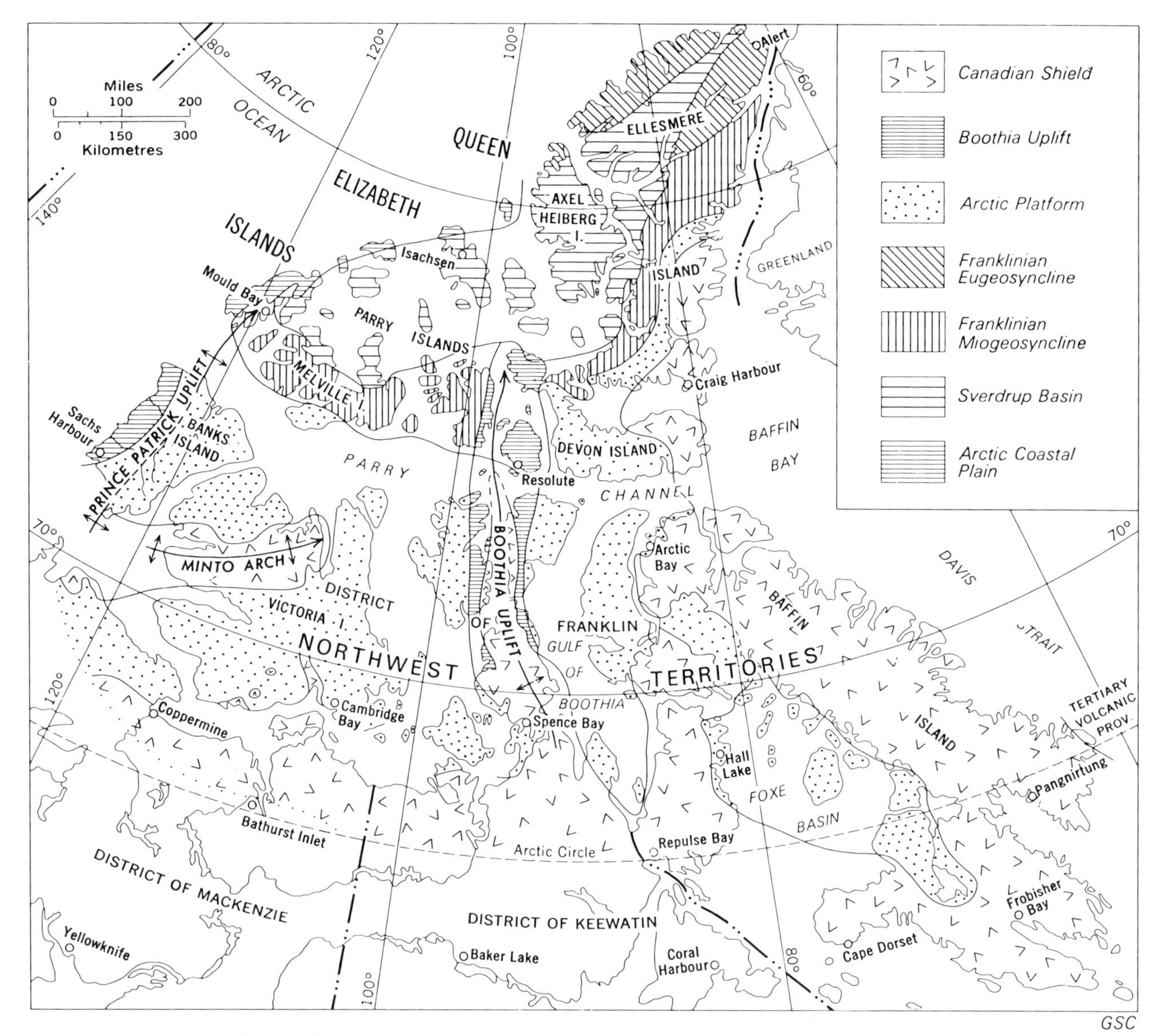

XX-11 GEOLOGICAL PROVINCES OF THE ARCTIC ARCHIPELAGO
As published in the fifth edition (1970) of *Geology and Economic Minerals of Canada* (E.G.1).

and 1968-69. The Precambrian spine of Boothia Peninsula and the sedimentary formations of adjoining areas were investigated in 1962 in Operation Prince of Wales, directed by Blackadar. He also headed the first two of the four great airborne operations that accelerated the mapping of Baffin Island: Admiralty (1963) and Amadjuak (1965). The other two, Bylot (1967-68) and Penny Highlands (1970), were directed by G. D. Jackson.

The younger formations of the western parts of the archipelago received much more attention. Fortier and R. Thorsteinsson began in 1950 by examining Cornwallis Island and making observation flights from there to northern Ellesmere Island. In the same year V. K. Prest was able to observe the geology and collect rock samples and fossils from the east coast of Ellesmere Island. Thorsteinsson returned to Cornwallis in 1951, 1952, and 1953, and reported the presence of petroliferous horizons among the Paleozoic sedimentary rocks. In 1953 W. W. Heywood, who was studying Ellef Ringnes Island, investigated several circular piercement domes. These structures, favourable to the accumulation of oil and gas, had previously been photographed from the air. The growing interest in the sedimentary formations resulted in Operation Franklin, which was carried out in 1955 by a geological party headed by Fortier and including four paleontologists. The party found the region, a circle of islands mainly in the Parry and Sverdrup groups, of much geologic and potential economic interest; and the reconnaissances of 1955 were followed by a series of more detailed investigations. Thorsteinsson and E. T. Tozer, paleontologists with the Franklin operation, examined the Ellesmere Island-Axel Heiberg Island region in 1956 and 1957; the Brock, Borden, Mackenzie King, and Prince Patrick Islands of the western margin in 1958; and they headed the 110,000-square-mile mapping operation of Banks, Victoria, and Stefansson Islands in 1959. They returned with H. P. Trettin and J. W. Kerr, to the Ellesmere-Axel Heiberg area in 1961 and 1962 to work out the succession in the important Eureka Sound district. As a result of this work, as early as 1960 Thorsteinsson and Tozer published an excellent account of the structural history of the archipelago *(Paper 60-7)*. It divided the region into seven structural provinces, described two great sedimentary basins—the early Paleozoic Franklinian geosyncline, and the later Paleozoic-Tertiary Sverdrup basin on the north—as well as the effects of major tectonic events, the trends of faults, and special structures like the piercement domes.

From 1963 Kerr and Trettin were actively engaged in stratigraphic studies, which Thorsteinsson also continued. They examined Cornwallis, Bathurst, central Ellesmere Islands, Grinnell Peninsula of Devon Island, and in 1964, the asphaltic sediments of Melville Island that had begun to attract the attention of oil companies. Some of these studies were air-supported operations, for example Grant Land, 1965-66 (R. L. Christie), which examined the central part of Ellesmere Island opposite Greenland. A feature of that project was the co-operation of a geologist from Greenland, supplied through the island government, to help correlate the geologies of the two large islands. Trettin returned in 1967 and 1970 to investigate some problems revealed by the Grant Land operation. He also worked far to the south in 1968, investigating the petroliferous prospects of northern Foxe Basin and its islands southwest of Baffin Island, and in 1969 he assisted with an investigation of Southampton Island which W. W. Heywood was conducting. Completing the picture, the region immediately west of Baffin Island—

XX-12 A PIERCEMENT DOME NEAR ISACHSEN, ELLEF RINGNES ISLAND, N.W.T.
Such structures are of interest to petroleum geologists. This gypsum and anhydrite dome, a rim of which is photographed from an altitude of about 1,000 feet, is about 5 miles by 7 miles in extent and between 200 and 300 feet high.

the locale in 1962 of Operation Prince of Wales—was examined further in 1970 by Christie, Kerr, and Thorsteinsson in Operation Peel Sound.

Thanks largely to the airborne operations, substantial headway had been made in less than two decades, not only with the reconnaissance mapping of this remotest region of Canada, but also with more detailed mapping and analysis of the sedimentary basins and structures so important to the great petroleum search that began in the early sixties, largely in response to the findings of Operation Franklin and its successors. With improving knowledge of the regional geology it became possible to recognize many previously identified formation types and to reassign other units on the basis of new evidence. The growing knowledge of the paleontology of the region, which was pushed forward through special efforts to collect microfossils and fossil plant spores, resulted in an improved ability to correlate strata throughout the region with North American and European successions.

While the study of the stratigraphy of the Paleozoic and Mesozoic rocks was of immediate concern, the most recent, or Quaternary, sediments are of equally great scientific interest and practical importance. The islands offer many highly interesting subjects for study. They include insights into the movements of the great continental ice sheets; the uplift and subsidence of land masses in isostatic movements; recent and present glaciers and their effects on the physiography of the region; how soils are formed as a result of glacial processes and what characteristics they possess; and the effect of permafrost on rock decomposition and soil formation. Here may be observed in plain view actual events and processes that elsewhere have had to be deduced from their aftereffects. The understanding obtained of such phenomena soon came to have immense practical value with the opening up of the North, when the impact of men on the Arctic environment became of prime concern.

Before 1959, however, only scattered observations of the Quaternary had been made by regular mapping parties; it was not until then that studies of the surficial geology began. In 1959, 1960, and 1961, J. G. Fyles travelled extensively through the region, notably on the mapping operation of 1959 of Victoria, Stefansson, and Banks Is-

XX-13 A FRAGILE ARCTIC ENVIRONMENT
This barren plain extending in all directions from the Isachsen Piercement Dome is composed of saturated clay derived from the underlying shales that thaws out every summer to a depth of a few inches. The tracks of the geologist's ankle-deep walk through the gumbo may remain imprinted on the terrain for many years.

XX-14 J. G. FYLES

lands. Banks Island was found to be an excellent subject for study, so Fyles returned there in 1960 to find evidences of three, possibly four, different glaciations. In 1961 he studied the fiord lands between Ellesmere and Axel Heiberg Islands, making about 250 landings from a Piper Super-Cub aircraft. Among materials transported and deposited in that remote locality by the great ice sheets he found buried soils, peat, and "an abundance of wood (diameter up to 10 inches) as well as buried soils and peat; in a few places the wood has been gnawed by beaver."[8] Fyles returned to the Arctic islands on other reconnaissances and investigations of particular areas in 1964, 1966, and 1968. B. G. Craig also examined the Pleistocene geology of the western part of Baffin Island and adjacent islands in the Operations Prince of Wales and Admiralty of 1962 and 1963, the main features of his interest being the elevated marine levels, dated by shells collected and tested by radiocarbon methods. W. Blake, Jr., also worked in the region, on Bathurst Island (1963, 1964), Operation Amadjuak (1965), and in southern Ellesmere Island (1967, 1968, and 1970).

The transfer of scientists from the Geographical Branch brought an increase in personnel available for Arctic islands work, as well as new capabilities in geomorphological and soil studies. In this way the Survey in 1968 inherited two ongoing projects in Baffin Island, a study of the effects of an ice-dammed lake, Generator Lake, and of an outwash plain. The work on Banks Island was also expanded; soil studies were made of the Beaufort Plain on the northwestern part of the island, and a beginning was made to subdivide the thick Beaufort Formation into a series of units that might be correlated with similar types of sediments elsewhere.

During its Arctic operations the Survey, like other organizations working in the Far North, received and gave a good deal of help, in the tradition of mutual assistance common to isolated locations. At times its staff obtained food, lodgings, the use of radio communications and air transport, and the loan of vehicles and other equipment. There was also much sharing of scientific information and some joint undertakings, with Survey staff being seconded to various operations. The defence bases and weather stations, Arctic Institute of North America projects, and those of the Canadian government's Defence Research Board and Polar Continental Shelf Project, were some of the agencies with which the Survey was drawn into contact in its Arctic work.

The last-mentioned project, also an enterprise of the Department of Mines and Technical Surveys, was to play an important part in the Survey's gradual involvement after 1958 with submarine geology. Such work is of two widely-differing general sorts. One kind is mainly concerned with the unconsolidated materials that cover the sea-bottom and indicate processes of sedimentation and the transformation of sediments into rock strata. Here the work consists mainly of collecting samples of the unconsolidated materials through dredging, and examining them from every conceivable point of view—chemical, mineralogical, mechanical, biological, age of deposition, and statistical configuration. The other is the investigation and mapping of the bedrock, which is mainly a task for geophysical methods of detection, supported by securing rock samples from key locations and formations.

The first sort was the work of a Marine Geology Unit, established in 1959 under B. R. Pelletier, which worked very closely from 1960 to 1964 with the Polar Continental Shelf Project based at Isachsen and Mould Bay. The unit studied the continental shelf in the area of the Sverdrup Islands and other western islands of the archipelago, outlining the submarine topography. The effects of riverine deposits and the patterns of the deposited material were investigated, climatic and other environments, past and present, were examined from evidence afforded from the sediments secured from the sea floor. At the same time, work was begun in 1961 in other areas; notably off Gaspé, on the Scotian Shelf off Nova Scotia, and in Hudson Bay. As it developed further, the program became more closely associated with the newly-formed Bedford Institute of Oceanography at Dartmouth, Nova Scotia, to which the unit and its laboratories moved in 1963. After 1964 its reports were no longer being carried in Survey publications. But under a later reorganization (1972) the work, still based on Dartmouth, has returned to the purview of the Survey as part of its Atlantic Geoscience Centre.

The other branch of the work was started in 1958 with an aeromagnetic survey of the Gulf of St. Lawrence to help correlate the geology of Newfoundland with that of the mainland. Surveys began in 1959, using seaborne magnetometer equipment, with a sensing head towed about 400 feet behind the ship. Using such equipment two ships of the Hydrographic Service traveling some 15,000 line-miles investigated an area of ocean floor south and southeast of the Nova Scotia mainland. Beginning in 1964 under the direction of G. D. Hobson, a series of refraction seismic studies was made of the submarine geology of the Gulf of St. Lawrence and Hudson Bay. The signal was produced usually by the electrical detonation of small explosive charges placed on the bottom in predetermined locations; but for ecological reasons, in more recent work the discharging of an air gun has been used. The echoes indicated the configuration of the bottom and of the buried geological structures. By means of dredging equipment and bottom corers, supplemented at times by submarine photography, material on which to make detailed investigations could then be secured from key locations. Such work, carried on over several years, has greatly improved the understanding of the geology underlying Hudson Bay and the Gulf of St. Lawrence.

Because aeromagnetic surveys can cover large areas very quickly, they continue to be used for studying submarine geological features. After the National Aeronautical Establishment's North Star aircraft's flights in 1964 over parts of Foxe Basin, Baffin Bay, and the Labrador Sea, the same aircraft flew some 7,000 line miles in 1965 gathering data for profiles of a part of central Hudson Bay. A similar reconnaissance was made in 1965 of Flemish Cap Bank, about 350 miles east of St. John's; while beginning in 1966, flights were made each year far out into the Atlantic to trace the extension of the Mid-Atlantic Ridge northward to Iceland, its branches west of Greenland, and the transverse fracture zones. In 1967 and 1968 geologically-significant seabed rocks were recovered and studied from Flemish Cap, San Pablo Seamount (about 400 miles south-southeast of Halifax), and from the Mid-Atlantic Ridge itself.

XX-15 G. D. HOBSON

Meanwhile this aspect of the work was extended to other parts of Canada's offshore areas. In 1966 at the University of British Columbia, a program of submarine studies had been initiated, using both seismic profiling coupled with recovered samples, and collecting samples of bottom sediments from inshore waters. The Survey began supporting this work financially in 1968, and it secured data from sedimentary material recovered off west coast inlets and in Queen Charlotte Sound, plus seismic profiles of the geological structure of the sea bottom westward from Vancouver Island to the edge of the continental shelf. Moreover, shallow seismic reflection profiling was undertaken in 1970 by Hobson in Beaufort Sea in the area between 69° and 72°N and 128° and 141°W to trace the features of the shallow, drowned coastal margin as a help to the Institute of Sedimentary and Petroleum Geology's study of the sedimentary succession of the region.

Thus the field work of the Survey has proceeded over the past two decades, developing and making use of swift, effective general reconnaissances of bedrock and surficial geology, and the aeromagnetic mapping of the land and marine areas of Canada. It has also advanced into using

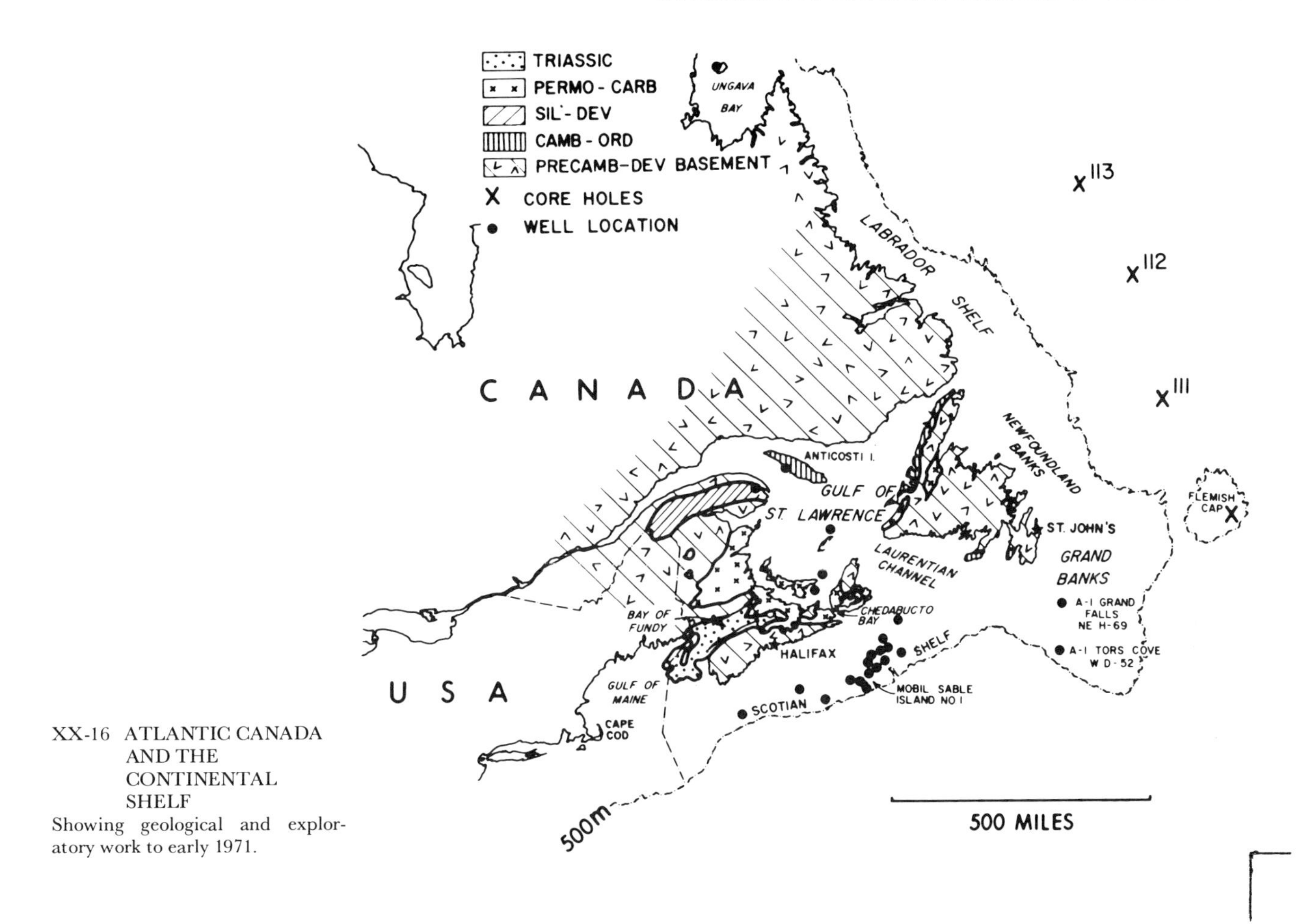

XX-16 ATLANTIC CANADA AND THE CONTINENTAL SHELF
Showing geological and exploratory work to early 1971.

more specialized techniques and undertaking studies that employ a variety of approaches in dealing with some of the many aspects of Canadian geology. Thanks to the aircraft, experienced, trained men can be carried to the remotest corners of Canada to contribute their knowledge and wisdom to solving geological problems; while the new geophysical tools allow men to probe through the thickness of the rocks on the land and to study those at the bottom of the oceans. And all the time the interpretation of the geology of Canada moves forward, advancing from describing to analyzing, and progressing from the particular to the general, from the details to the fuller understanding of geological processes in the context of the totality of earth history.

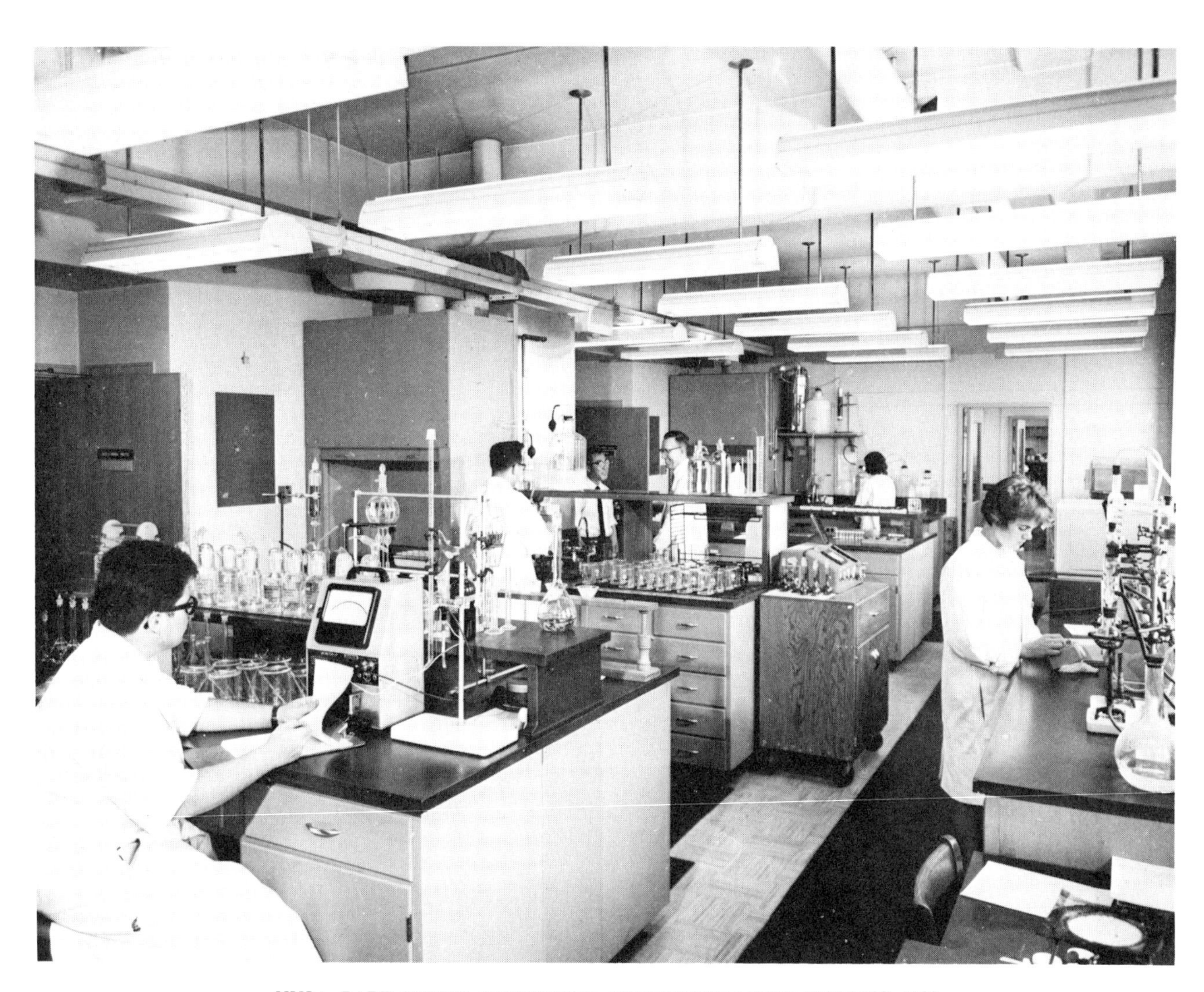

XXI-1 PART OF THE ANALYTICAL CHEMISTRY LABORATORIES, 1963

21

In 601 Booth Street–The Laboratory and Supporting Services

PERHAPS THE MOST SIGNIFICANT major development of the past generation is the emergence into greater prominence of the research and laboratory sides of the Survey. The laboratory had its beginnings in the early years when Logan at the first opportunity added a chemist, and then a paleontologist, to his staff. The role of these specialists, and of the petrologist who was added later, was essentially an auxiliary one of supporting the field operations. As Dawson asserted in 1894, "The operations of the Geological Survey in the field, constitute the basis of the entire work of the department." [1]With but a few illustrious exceptions, such as Hunt, Adams, Barlow, Ellsworth, the personnel of the laboratories seldom extended their work beyond meeting the demands pressed on them from the field geologists or the public.

Since 1950, however, the role of the laboratory units has noticeably altered, to include initiating and carrying through research projects of their own. J. M. Harrison commented in 1963 that while the "primary function" of the laboratories was "to provide specialized information on specific problems that arise from field studies, in recent years a small but increasing number of the Survey staff has been undertaking special laboratory investigations of problems pertaining to fundamental geological processes, and the development of techniques and equipment with which to tackle these problems."[2] Authorship credits on some recent papers published by the Survey demonstrate a new concept of partnership between field and laboratory scientists. Moreover, every year brings many articles by members of the Survey devoted to discussing and applying new theories and principles based largely, even wholly, on laboratory investigations and researches.

Why should so fundamental a change have come about at this time? A large part of the explanation obviously lies in the enhanced public recognition of research science. The war years had demonstrated what wonders scientific research could perform, given sufficient encouragement and the aid of industrial technology to immediately transform findings into concrete realities. The accomplishments of these years ranged from designing better vehicles and weapons or finding substitutes for scarce materials, to making revolutionary breakthroughs in the radar and atomic energy fields. Convinced by these achievements, the public came to feel that research science also held the key to limitless material progress for mankind through developing new products and discovering more efficient methods of operation. A flood of practical inventions like television or the computer, and especially the dramatic achievements since 1957 in space travel, drove the lesson home. "Pure" research, training research scientists, and developing research facilities and technologies, became accepted as entirely "practical" objectives in their own right, as things that "paid" and should be encouraged. The federal government, responding to the public mood, was far more willing to assist the research activities of a scientific agency like the Survey than in earlier times. The establishment of the Department of Mines and Technical Surveys in 1950 was tangible recognition of the heightened importance accorded the federal earth science agencies, and an indication of the government's willingness to support their expansion into new activities in line with national needs. The result was more generous budgets for staff, equipment, and field operations, and more freedom to undertake projects of "pure" research, the economic benefits of which might be reaped in some future time.

Moreover, the war years produced many developments which could now be used by the Survey to help solve present problems and furnish bridgeheads from which to advance further. The more efficient field operations necessitated the parallel expansion of the laboratory facilities and staff to cope with the increased volume of results from the field. There were great advances in equipment and technology in the shape of the application of new alloys, plastics, and other materials, in X-ray, radar, and other technologies, and in such new fields as isotopic chemistry and geochemistry. These developments elevated the work of the laboratories to previously unimaginable levels of complexity and precision. The tide of invention has continued to bring such further improvements as transistorized circuitry, computer technology, and for paperwork, the Xerox machine and other duplicating devices. Though considerable research has been needed to adapt some of them to the purposes of the Survey, such inventions hold forth the promise of more rapid, efficient operation.

These and similar improvements reflected new levels of specialized technology, and opened such fields as geochemistry, palynology, geomathematics, aeromagnetic survey compilation, or electromagnetic methods, for specialized study. To cope successfully with increasingly difficult problems and to undertake important new types of work, it became necessary for the Survey to recruit men with specialist training, and provide an impressive array of sophisticated equipment and facilities. Fortunately, an increasing number of such specialists—some of them the products of new postgraduate programs at Canadian universities—were now becoming available. To recruit and retain such personnel the Survey had to keep abreast of the standards of scientific achievement in the universities and industry so that it could offer them appropriate facilities and the opportunity to work in their chosen areas of study. Hence it became desirable to adopt a dynamic approach towards research that would permit the specialized manpower and equipment to fulfil their potentials.

These trends, surprisingly, were foreseen at the very start of the period. In its first annual report, dated March 15, 1951, the National Advisory Committee on Research in the Geological Sciences (NACRGS, a group of 18 scientists appointed by an order in council in 1949, only two of whom were officers of the Survey) made a strong plea for more trained personnel, more modern equipment, "and other facilities essential to the carrying out of fundamental or basic research in the Geological Sciences in a manner commensurate with the vital part mineral resources play in the economy of Canada."[3] The presentation centred directly on the more advanced research role that was envisaged for the laboratories:

XXI-2 J. F. HENDERSON
Secretary to the NACRGS, 1951-71

> Although all geological investigations involve some scientific research, most of these conducted by the Geological Survey and Provincial Departments of Mines have been based on routine field work and have resulted in the publication of geological maps and reports, the latter chiefly of a descriptive character, in the compilation of which the only common laboratory tool has been the petrographic and metallographic microscope. This work has been well done and has contributed greatly to the discovery and development of mineral resources in Canada, but it is the opinion of this Committee that more complete and exhaustive laboratory studies are needed to supplement this work if we are to meet the increasing demands of industry and defence.
>
> During the past twenty years, new laboratory techniques have been developed which may now be applied to geological problems. These include the use of X-ray diffraction apparatus, spectrographic equipment, the mass-spectrometer, equipment for differential thermal analyses and for the detection of radioactive minerals, the electron microscope for study of super-fine materials, and apparatus for the synthesis of minerals at high temperatures and high pressures in bombs made of new alloys arising from wartime investigations.[4]

The report went on to suggest the upgrading of the facilities of the Geological Survey which "should have first class equipment and personnel available for this type of research."[5] Among these it listed the establishment or expansion of nine distinct laboratory facilities—chemical, microchemical, thermal, spectrographic, mass spectrographic, X-ray diffraction, radioactivity and geophysical laboratories, and a laboratory for studying sedimentary rocks—plus a machine shop. It rated the existing facilities as being adequate in only the spectrographic and radioactivity categories.

That the Survey was not adjudged a failure in all nine categories was a result of the recent (since 1948), long overdue, stirring of modernity in the work of the Mineralogical Section. That section was at a low ebb for years, largely because of lack of support by the management, who concentrated on field work and mapping, and tended to look on laboratory analysis as a "frill." Located in the old Survey Building on Sussex and George Streets hard by Byward Market, the section was "out of sight and out of mind." Even its work was performed largely on behalf of the general public, with little relationship to the operations of the Survey as a whole, except, perhaps, for its rock and mineral sample collecting in connection with the ever-popular sets made up for presentation or sale. E. Poitevin and H. V. Ellsworth had struggled so long for trifling necessities like glassware and reagents that they had quite given up hope of ever securing modern laboratory instruments. The section had acquired an emission spectrograph before World War II, but without an effective operator it was not a very satisfactory tool. In the meantime, a whole new breed of laboratory instruments was becoming available. X-ray diffraction equipment that was capable of doing in 30 minutes work that required two weeks by traditional methods of chemical analysis, for example, had come into standard use for quick, accurate mineral identifications. Queen's University had had such an installation as early as 1931.

During the war years the Survey had organized many important searches for mineable deposits of scarce or strategic minerals. Notable among these were secret investigations in 1944 connected with the allied governments' atomic energy program, by A. W. Jolliffe, J. D. Bateman and others, of radioactive mineral occurrences in the Great Bear Lake and Lake Athabasca districts. As a result the Survey was selected in 1946 as the geological arm of the newly-formed Atomic Energy Control Board of Canada. In the next year when the government decided to meet the needs of the United States and Canadian atomic energy programs by throwing the door open to private enterprise, the Survey was appointed to manage this aspect of the program. It supervised all uranium prospecting activities in Canada, carried out field researches, tested all mineral samples in the laboratory, and published such highly useful guides for prospectors as *Notes on Prospecting for Uranium in Canada* (1949) and *Canadian Deposits of Uranium and Thorium* (1951). Ellsworth was given a separate position so he could concentrate on analysis and research on uranium-bearing mineral samples; a basement room at the National Museum served as his laboratory. To assist his work the Survey in the autumn of 1948 purchased a modern X-ray spectroscope for X-ray diffraction analysis and hired S. C. Robinson to operate it.

An enormous quantity of important and intricate work fell to the Survey once the uranium rush got under way. To cope with it G. S. Hume, the Director-General of Scientific Services, formed a specialized Radioactivity Section. Ellsworth, Robinson, and H. R. Steacy attended to the laboratory testing and research, while A. H. Lang looked after the field aspects—geological investigations and inspections of operations—and the office activities, of maintaining records, inventories, and preparing publications.

In the next few years the work of analyzing radioactive materials was greatly expanded. Partly because of Ellsworth's conscientious insistence on examining samples that were quite hopeless from the commercial point of view, no fewer than 6,820 prospectors' samples were analyzed for radioactivity content in 1949 alone. Staff officers regularly visited uranium deposits to test new equipment in the field and to collect samples for detailed study. The urgency of the work led to the acquisition of additional staff and specialized equipment, some of which was designed and built for the Survey by its own

XXI-3 A. H. LANG

XXI-4 R. K. WANLESS

XXI-5 J. A. MAXWELL

staff or that of the Mines Branch. By 1951, when a Radioactivity Resources Division was separated off from the Mineralogy Division, the new division included an X-ray diffraction laboratory for identifying minerals by their diffraction patterns, which are as distinctive as fingerprints. For purposes of reference a comprehensive collection of standard X-ray powder diffraction patterns of minerals in the National Mineral Collection was also begun. There was also a Radiometric Laboratory (H. R. Steacy) for determining the uranium and thorium contents of samples that were now being sent forward in large numbers because of the heightened prospecting activity.

Two more units were added in 1951: X-ray fluorescence equipment, believed to be the first in Canada, for rapid qualitative and quantitative analysis of heavy minerals (including the radioactive minerals uranium and thorium); and an electronics laboratory to build, repair, and redesign all sorts of equipment. Important later additions were an optical emission spectrographic laboratory in 1952 with W. H. Champ in charge, and a laboratory for age determination and isotopic studies in 1953-54. The heart of this last was a mass spectrometer built at McMaster University in Hamilton, Ontario, and installed by a graduate physicist of that university, R. K. Wanless, who stayed with the Survey to operate the instrument and its successors, and to develop the applications of the method to the geological studies. High-temperature furnace and vacuum systems were constructed for conversion to SO_2, and to extract, purify, and measure argon for age determinations by the potassium-argon (K-Ar) method.

In the meantime W. A. Bell had followed up the recommendations of the NACRGS by recruiting a number of laboratory specialists in 1952 and 1953 to inaugurate and expand modern laboratory activities. Among this group was J. A. Maxwell, a geologist-analytical chemist who had taken his major training in analysis and laboratory operation in the world-famous rock analysis laboratories of the University of Minnesota. He immediately set to work cleaning and thoroughly reorganizing the rock and mineral analysis facilities to suit the most efficient modern methods. These laboratories were the real "workhorses" of the Survey, not only providing a very large service to the public but also meeting the rigorous demands of the new-style field men conducting the large-scale operations of the time. Bell reported in 1954 that "the new chemical laboratory was put into operation to meet the growing demand by the Geological Survey for analytical work required for research projects."[6] By 1957 the organization was expanded to include a rapid-methods system which "is only slightly less accurate than the best established method, but four times as rapid."[7] At the same time the new optical emission spectrograph was developed into a major tool for trace-element determinations, and the X-ray fluorescence unit was updated to achieve greater sensitivity. Thus the analytical laboratories were transformed into an efficient, reliable, and smoothly operating organization that was on its way to achieving international recognition as one of the major installations of its kind.

At first the rock and mineral analysis laboratories also handled the thousands of samples required for geochemical studies. But the quantities became so great, and the materials so different (soils, stream sediments, water samples) from the rocks for which the system was designed to work efficiently, that in 1955-56 it was decided to start a geochemical laboratory especially for the requirements of the Geochemistry Section. By 1957 this small laboratory, the work of a chemist Mrs. M. A. Gilbert, and largely based on that at Imperial College, London, was in operation "for rapid semi-quantitative determination of specific elements in samples of stream sediments and soils. The high productivity and consistent accuracy of the laboratory have made possible precise geochemical studies, on a statistical basis, which can not be achieved by other techniques."[8]

There was also a thoroughgoing revision and modernization of the mineralogical laboratories, which was associated with R. J. Traill. To prepare the materials for their various purposes, new units were set up for crushing, sampling, and mineral separation, which "markedly facilitated the preparation of samples in adequate amount for rock and mineral analysis, isotopic analysis, age determination, etc., thus removing the previous bottleneck in this field."[9] To aid mineralogical studies new pieces of equipment were added, such as a high-temperature X-ray powder diffraction camera (1955).

The uranium boom subsided, the industry became stabilized around a dozen or more producing mines, the prospecting rush dwindled to insignificance–and the staff and equipment of the strengthened Radioactivity Resources Division were ready for wider work related to the vigorous mining activity currently under way in most of the country. Under the reorganization of 1955 the division, headed by Lang since the death of Ellsworth in October 1952, embarked on a program of studying and reporting on the whole range of mineral occurrences in Canada, under the new title of Mineral Deposits Division. The division proceeded to collect data on mineral occurrences and mining, and gradually built up a staff of specialists in the geology of mineral deposits. A series of reports, the Economic Geology Papers, was prepared on the economic geology of various minerals; an updated version (the third edition) of *Prospecting in Canada* and other works were issued, as well as the first Canadian metallogenic maps. The laboratory facilities of the former Radioactivity Resources Division, except for the radiometric laboratory, were eventually consolidated by 1958 in the new, enlarged Mineralogy Division, headed by S. C. Robinson.

All this time the Survey remained housed chiefly at the overcrowded, unsafe National Museum Building, which it had shared with the museum and the National Art Gallery since the end of the Great War. Such physical conditions hampered the progress of the laboratory activities, and may even have prevented the Survey from attracting qualified personnel. Yet in the National Museum Building many delicate instruments overcame their growing pains, to give way to much sounder pieces of equipment in a new headquarters. However, the staff was too scattered about the city to benefit from, or help in, the upgrading of the work. The Mineralogy Division, for example, was divided between the basement of the National Museum, and the old Survey building on Sussex Street where Poitevin held sway. In all, the Survey was distributed among nine separate buildings, the geophysics staff alone being scattered in five locations. The National Advisory Committee report of March 15, 1951, had taken cognizance of these deficiencies by recommending "that plans be prepared and steps be taken for the construction at an early date of a permanent building that will adequately care for the needs of the Geological Survey of Canada and the Research Laboratories recommended herewith."[10]

The unsatisfactory nature of the quarters in which the Survey had to work was underscored in March-April 1953 when a House of Commons special committee (chairman, G. J. McIlraith) investigated government operations in the atomic energy field. After hearing testimony from W. A. Bell and A. H. Lang, as well as from leading officers of the Mines Branch, the committee visited the laboratories of both branches, accompanied by the press, which devoted a good deal of unfavourable comment to the adverse conditions under which the government scientists were working. No doubt this event gave added force to the drive under way since 1951 for a new Survey building. At any rate, in June 1953 the staff was quietly canvassed to suggest projects and activities that might be conducted in a larger building so as to strengthen the case for a new headquarters.

Before the close of the fiscal year the government had also reached a decision. The annual report of the Department of Mines and Technical Surveys for 1953-54 announced plans to erect a complex of buildings along the southern end of Booth Street, where some of the laboratories of the Mines Branch had been established 20 years previously. First, the Mines Branch would be provided with a new building to accommodate its chemistry and radioactivity laboratories and administrative staff. Then would follow a new building for the Geological Survey—the first to be built expressly to its requirements in its entire history. The plans were under active consideration by 1955, tenders were called in July 1956, and the building was completed by Thomas Fuller Construction Company of Ottawa for occupation in 1959, to be formally opened by the then-minister Paul Comtois in May 1960. Foreknowledge of the move allowed plans to be made from 1954 onward for the various sections and units to expand their operations, and to develop staff and equipment with that end in view.

The move to the handsome eight-story building gave impetus to the expansion of the laboratory side of the Survey. As the director reported (1957):

> With the anticipated completion of its new $6.3 million building in 1959, the Geological Survey is also looking to the much needed expansion of its research effort to enable it to fulfil properly its task of carrying on essential supporting research. Of immediate importance is the expansion of certain phases of its laboratory research and other services to realize the best possible results from its extensive program of reconnaissance mapping, and the long overdue expansion of its work in ground-water surveys and surficial and engineering geology.[11]

Disruptions were inevitable, regardless of the advance planning. Despite the best endeavours, a large piece of equipment, the 3.4-metre Wadsworth mount grating emission spectrograph of the "grand piano" style acquired in 1952, was dropped to the ground while being hoisted by crane into position on the seventh floor. It suffered no more damage than a cracked grating support and for another 20 years continued to provide a large part of the emission spectrographic analyses. It was out of commission for four months being recalibrated, however, which for a time created a great backlog of samples.

A regrettable, but unavoidable, casualty of the move was the taking leave of longtime friends and associates of the museum staff. A tangible sign of this separation was the partitioning of the priceless library that had been over a century in the collecting, as well as the photographic collection. Dividing the library presented difficulties, because so many works were important for both the Survey and the National Museum. It was decided that the purely biological and anthropological materials should remain in the picturesque but highly impractical round room of the museum, while the Survey made off with the geographic and general scientific publications and most government publications, in addition to those expressly concerned with the geological and physical sciences. The satisfactory settlement from the Survey's point of view was credited "to the tact, patience and perseverance of Dr. Peter Harker, the Survey's representative."[12] Thus the Survey library, under Mrs. N. I. Kummermann, said farewell to the "crowded, old fashioned and inadequate quarters"[13] and began quickly repairing any breaches—as regards quantity if not quality—from amongst the torrent of new publications pouring off the presses of the world. The photographic collection was divided along similar lines, though sorting out the successor collections was delayed by the failure to obtain a photographic librarian until 1965.

The new building made it possible for the Survey to undertake new laboratory activities, such as a radiocarbon dating laboratory and a biogeochemistry establishment comprising a roof-top (on the eighth floor) greenhouse and a radiochemical laboratory. The coal research laboratories were brought in from Sydney, N.S., and the palynology laboratory found its own space. The Seismic Methods Section, which was only formed late in 1958, moved its staff and equipment directly into the new building at the end of the field season of 1959. Several sections were able to rearrange their operations on a more practical

XXI-6 MRS. N. I. KUMMERMANN

XXI-7 LOGAN HALL
Situated to the right of the entrance lobby, the hall offers displays of interesting rocks, minerals, fossils, maps and reports representative of the work of the Survey, as well as the portrait, field equipment, paintings, notebooks and medals of the Survey's founder.

basis; thus in age measurement work the opportunity of the move was used to construct new, more effective argon extraction lines with up-to-date equipment. The design of the building also permitted many practical improvements: better exhaust and cleaning facilities to reduce contamination; better protection against fires or explosions; better shielding of sensitive equipment from outside interference, and of humans from the hazardous effects of certain processes and pieces of equipment. Many collections stored for many years, or located apart from the sections with which they were concerned, now received proper and adequate display space for reference and study.

An added dividend from the move was improved morale in some of the formerly divided divisions. L. W. Morley, the chief of the Geophysics Division, which at last was able (for only a short time as it proved) to bring together its dispersed segments to a single floor of the new building, felt the better communications and the extra facilities would certainly raise the morale of the staff and result in more effective work. There was the hope, too, of greater co-operativeness among the different divisions, as the chief of another division reported:

> . . . there has been a distinct trend to cooperation of two or more specialists in many of the projects. This is well illustrated by the cooperation of field geologists, petrographers, mineralogists and physicists on the Survey's age determination programme. It is also apparent in the number of disciplines combining to provide data for the petrological and geochemical studies. The increase in co-operation with field officers and the more frequent visits by them to the laboratories since the move to the new building, has been a source of much satisfaction to this Division.
>
> The move during the summer to the new Geological Survey Building was particularly significant to this Division for three reasons. Most obvious of them is the incomparable improvement in laboratory facilities. Secondly, the Division was housed together for the first time which has greatly facilitated inter-laboratory co-operation. Finally, the Division is brought into much closer touch with the field aspects of the Survey's work, services for which constitute much of our work.[14]

* * *

BY THE TIME THE DUST had settled from the move and reorganization, the Booth Street building housed no fewer than 28 different laboratory units—the modern-day successors of the three historic specialties, paleontology, chemistry, and mineralogy-petrology. Moreover, some of these occupied whole suites of rooms. Numerous changes have occurred since 1959, with units being transferred or disbanded, and others being established or expanded. In essence, however, the laboratories, technical services units, and collections still occupy about six of the ten floors.

First are various units such as the lapidary, ore polishing, petrology, mineralogy, analytical chemistry, geochronology, geochemistry, and mineralographic laboratories that treat inorganic rock minerals and samples. Some of the laboratories in the main prepare samples for detailed studies elsewhere. Thus the rock and mineral preparation laboratories crush, grind, and screen rocks, divide the pulverized samples for testing purposes, separate heavy minerals, and prepare concentrates. In the lapidary laboratory thin sections are prepared for microscopic study, and rock and mineral specimens are cut and polished for study and display. A somewhat comparable work on samples of ore minerals is done by the ore polishing laboratory.

Some of the units also study the properties of the samples. In the sedimentology laboratories sedimentary material is analyzed by mechanical means such as sieve or pipette. Heavy mineral separations are made, and samples are measured and analyzed for size, shape, roundness, and other properties of the particles. The Terrain Sciences Division has two of these laboratories in Ottawa, and a third in Calgary. The petrology laboratory studies the physical characteristics of rocks by determining the specific gravity and mineral composition by optical methods.

A parallel operation, the Mineralogy Section (R. J. Traill), studies minerals, investigating them in rock samples according to physical properties, especially crystalline structure, which by the method of X-ray diffraction reveals highly characteristic patterns based on their atomic framework that provide positive identification in all but a few rare instances. Quantitative X-ray fluorescence analysis is available for rapid identification of rock and mineral samples, plus electron-beam scanning of small areas in thin sections. After the move to Booth Street two diffractometers were added, as well as two electron probe microanalyzers for studying individual mineral grains in thin sections. A third such instrument, a laser microprobe with attached direct reading spectrometer (the first in Canada) can volatilize a selected part of the sample which the spectrometer immediately analyzes. Modern automation includes printout facilities, and a mini-computer and magnetic tape unit for programming and for data processing.

The Mineralogy Section also continues early functions of the Survey, identifying and reporting on rock and mineral samples brought in by the public. Though the number reached a peak of 11,050 samples in 1953-54 during the uranium boom, it has since fallen off markedly to less than 1,000 samples a year. This is not the case, however, with the sets (boxed, named collections of typical rocks and mineral samples), of which as many as 10,000 may be sold in a single year, entailing the procurement of over 20 tons of a wide variety of rocks and minerals for making into specimens. The increase in sales reflects the greater interest in and knowledge of geology and rock collecting, which has been served by the preparation and publication of a number of guidebooks. Another historic activity, begun by Logan, is the National Mineral Collection in H. R. Steacy's charge, amounting to some 10,000 specimens of about 1,650 species from Canada and abroad. Steacy also has custody of the National Meteorite Collection in which a total of some 350 falls and finds are represented.

Complementing the work of the Mineralogy Section and comprising with it the larger part of the Central Labora-

XXI-8 AN ELECTRON PROBE MICROANALYZER OF THE MINERALOGY SECTION

tories and Technical Services Division, is the Analytical Chemistry Section (Sydney Abbey), the modern-day successor to the first of the Survey's laboratories. With the great expansion of field operations in the fifties, and of special studies in the sixties, the section has been faced both with large numbers of samples for relatively simple analyses, and with many more requests for special, highly-accurate studies of complex and unusual samples. To cope with this situation, the section adopted a streaming process, with rapid chemical analysis of samples mostly from reconnaissance and regional studies, for which analysis totals of 98.5-101.5 per cent accuracy were considered adequate. A step taken to improve the performance included the acquisition in 1962 of a multichannel X-ray fluorescence spectrometer that yielded more satisfactory results than the rapid chemical methods for the elements that are the principal constituents of rocks.

However, the highly accurate analysis of samples required for special research was still done by a modified classical analysis scheme that produced analysis totals within the range 99.5-100.5 per cent. Finally, there was provision for analyzing samples that could not be analyzed completely by either system because of their special nature, their small sample size, or because they included certain elements or combinations of elements for which no special technique had been developed.

The section works continually to improve and refine its techniques to increase the speed and accuracy of the results, or to accommodate itself to changing demand patterns for its services. New equipment is constantly being considered for these purposes. Atomic absorption spectroscopy has provided rapid methods of analysis with an accuracy equal to, or in some cases better than, those of the classical analysis scheme for a number of elements. Later experimentation brought improvements in the determinations of other elements.

A comparable but more limited role is that of the metallogenic or mineralographic laboratory (C. R. McLeod), which investigates ores and related rocks in connection with the studies of the Geology of Mineral Deposits Section. It makes mineral separations, routine sample preparations, microscopic examinations and photomicrographs, and conducts optical identifications of opaque mineral species of economic interest.

A whole new field of study based on the discovery and perfecting of methods for determining the ages of rocks has developed in the past 20 years, which is tremendously helpful in throwing new light on the geological evolution of Canada. Such determinations are the main activity of the mass spectrometric laboratories of the Isotope and Nuclear Geology (now Geochronology) Section (R. K. Wanless). The method grew out of the work on radioactive decay of uranium minerals by Ellsworth and others, during which it was discovered that uranium decays into lead (U-238 to Pb-206) at a fixed, constant rate, whose half-life (the time in which half the U-238 in a sample will be changed to Pb-206) is 4,507 million years. A similar decay process of U-235 to Pb-207 is also the basis for other age determinations. By measuring the present ratios of uranium and lead in a given sample, it is possible to calculate when the decay process started. A tool was provided for dating even the earliest Precambrian formations, and the geologist had gained a means for obtaining an accurate picture of the geological history of this earliest, longest segment of earth time. Such determinations were made with great difficulty by the early X-ray spectrographs of the Radioactivity Resources Division, but the development by 1954 (not by the Survey) of a practical mass spectrometer was followed by the setting up of a laboratory especially for the purpose, which made the task of age determinations much easier. Moreover, the Survey was able to utilize newly-developed methods that could be applied to minerals other than strongly radioactive ones containing uranium or thorium. The same procedures, it was learned, could be applied to potassium-base minerals such as the micas that are far more widely distributed in nature. These, too, are faintly radioactive, with some of the potassium having a half-life of 142 million years as it decays into argon (K-40 to Ar-40). Much testing was necessary before a technique was developed whose results were comparable with those secured from the uranium-lead analyses, but the Survey by 1959 was taking K-Ar age measurements. By 1961 still another method based on the radioactive decay of rubidium-87 to strontium-87 was also being used for dating purposes.

With all these methods, however, a serious problem was soon encountered: the determinations in the laboratory did not always accord with the unmistakable evidence in the field. It was discovered that if a rock had been exposed to temperatures above a certain critical point, its daughter products (such as argon in the K-Ar process) were not retained, and only the products of the subsequent decay process were left to be measured. Hence the results of the laboratory measurements would not refer to the time when the mineral was first formed, but perhaps only to the latest of a number of geological events in which the rock was involved—or even to an intermediate stage in which only part of the rock was metamorphosed, yielding a reading that was a composite of two or more stages. As the situation was described in 1964,

> Since the inception of the potassium-argon age-determination in 1959, more than 1,000 K-Ar age measurements have been com-

> pleted. These are used to unravel tectonic history and also to identify periods of intrusion and metamorphism in the Appalachian, Precambrian Shield and Cordilleran regions. In areas subjected to more than one period of metamorphism, this technique identifies only the most recent geological event. To penetrate this metamorphic "age barrier," increasing attention is now being given to the application of rubidium-strontium whole-rock isochron studies. Under certain conditions, these can indicate the initial or primary age of the rocks. Measurements based on them have been completed . . . .[15]

However, despite the difficulties associated with using each method, having four to choose among gave the Survey a much better chance of achieving a more accurate and useful understanding of the geological history of Canada.

To make age determinations by the several methods, and conduct analyses of minerals containing stable isotopes, the facilities of the laboratory had to be increased steadily; new equipment designed and assembled by the section was installed. The first mass-spectrometer of 1954 was followed by six others, designed to handle the several kinds of analyses—along with appropriate extraction lines to supply material in the desired form for each spectrometer. The preeminence of the work in the age determination field was shown by the Survey's being selected to develop a national system for recording all Canadian isotopic age data and to serve as the Canadian centre for such data.

Some work was also done in connection with stable isotopes. Tests carried out in 1956-57 of sulphur isotopes in the crude petroleums drawn from various Alberta oil wells indicated that deposits could be distinguished and identified by the differing patterns of their sulphur and sulphur isotopes, and this sort of analysis permitted the correlating of deposits separated by distance and even those with physical and chemical differences. (Such telltale identifications are also used to trace the origin of polluting discharges from ships at sea.) Similar studies of sulphur isotope distribution in the rocks at Yellowknife, and of lead isotopes in metallic mineral deposits in southern British Columbia, have been helpful in studying the genesis of particular orebodies and tracing their distribution. The lead isotopes study revealed that lead from all parts of the great Sullivan mine at Kimberley was of one type, while most of the smaller lead deposits in the region were isotopically different. The conclusion followed that the lead was introduced at different times from more than one source, which "narrows down the search for additional deposits to those structures and rocks in the favourable range of age."[16] Unfortunately, this promising line of study has tended to take a back seat to the age determination work, the change of focus being reflected by renaming the unit the Geochronology Section and transferring it to the Crustal, now Regional and Economic Geology, Division, where it serves to help with the understanding of rock relationships.

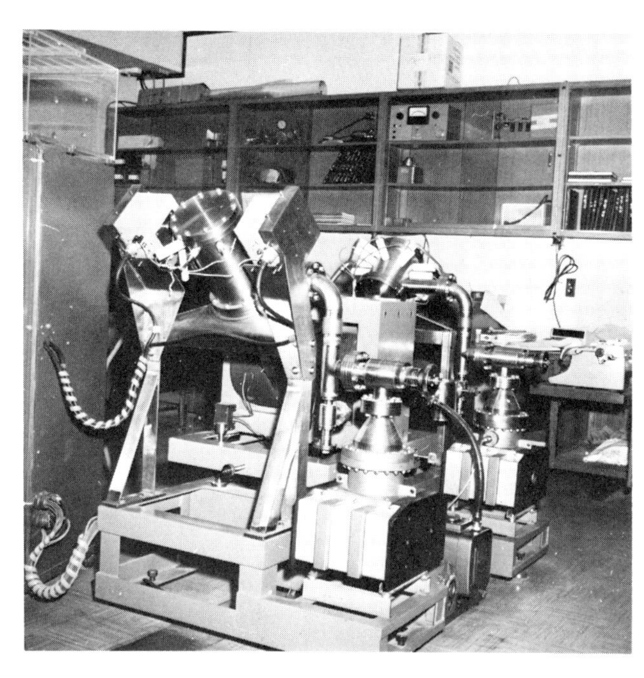

XXI-9 SOLID SOURCE MASS-SPECTROMETER FOR LEAD-URANIUM-THORIUM ISOTOPE ANALYSIS

Also concerned with age relationships, and also recently (1971) transferred to the Regional and Economic Geology Division, is the Paleomagnetism Section (W. F. Fahrig). The original presence of this section (then named the Rock Magnetism Section; head, A. Larochelle) in the Geophysics Division, indeed, was rather accidental, being an outgrowth of geophysical evidence that magnetization of rocks varies enormously in intensity and direction even in a given locality. Investigations showed that the magnetization results from the way in which certain minerals, either during their original crystallization or their re-deposition as sedimentary particles, orient themselves in the direction of the magnetic field of the earth, like miniscule compass needles. As the material hardened, this paleomagnetic orientation became imprinted on the rocks, giving a permanent record of the direction of the earth's magnetic field at that time. To study fossil magnetism it is necessary to examine rocks whose field positions are very carefully measured when the samples are taken. Once the pattern of pole wandering is known and a time-scale has been worked out, the paleomagnetic lines can help to date the rocks, furnishing a supplementary method for age determination. Even more, however, paleomagnetism furnishes additional evidence on continental drift. It was found that rocks formed in the same period on different continents do not always show an orientation to the same pole, indicating the possibility that the positions of the continents relative to one another have changed over time. This is the principal use made at present of the findings of rock magnetism.

Large numbers of carefully-collected rock samples are examined in the section's laboratory; for example, an analysis of 533 samples collected from Newfoundland and adjoining parts of Labrador in 1963 for a time was claimed to indicate that the northern peninsula of Newfoundland had shifted 30 degrees in a counter-clockwise direction relative to New Brunswick and other Atlantic areas. To help with investigating these phenomena, a considerable amount of special equipment, in which A. Larochelle introduced a good many improvements in the direction of higher sensitivity, was developed and put into service. Instruments used in the section include spinner-type and astatic magnetometers, some automated to produce printed results for data processing; an instrument to measure the Curie point of particular samples (the critical temperature at which the record of rock magnetization is destroyed and replaced by a new pattern dating from that time); and an electromagnetic "cleaner" unit that removes unstable components of magnetization and enables the researcher to determine the fundamental paleomagnetism. Appropriate portable equipment has also been designed to permit studies to be carried on in the field.

Another newly-recognized subscience, geochemistry (E. M. Cameron), now part of the Resource Geophysics and Geochemistry Division, has a series of laboratories to support its research programs. Chemical methods are employed to detect mineral occurrences, especially by means of the traces these impart to surrounding rocks, soils, sediments, waters, vegetation, marine organisms, and the atmosphere in the form of dispersion "trains" or "haloes." Each year the section mounts field investigations ranging from studies of the dispersion patterns of ore molecules in the containing walls of country rock in a mine, to tracing gold occurrences back along a watershed from the minute amounts in sediments and waters of the streams. The materials studied in the laboratory and in the field can give rise to case studies of purely local phenomena, or to geochemical maps of entire regions.

The main laboratory activity consists of analyzing field samples, with emphasis on detecting and making quantitative measurements of minute amounts of "trace elements." The section's suite of laboratories has expanded enormously since their inception in 1955 in the basement of the National Museum. A 22-channel direct-reading quantigraph was secured in 1961 to provide rapid routine trace-element analysis of the large numbers of highly diversified samples that a single study can produce. In 1970 it was replaced by a 50-channel quantimeter controlled by a small computer that governs the calibration of the spectrometer, the analysis of the samples, and the printing out of the completed results.

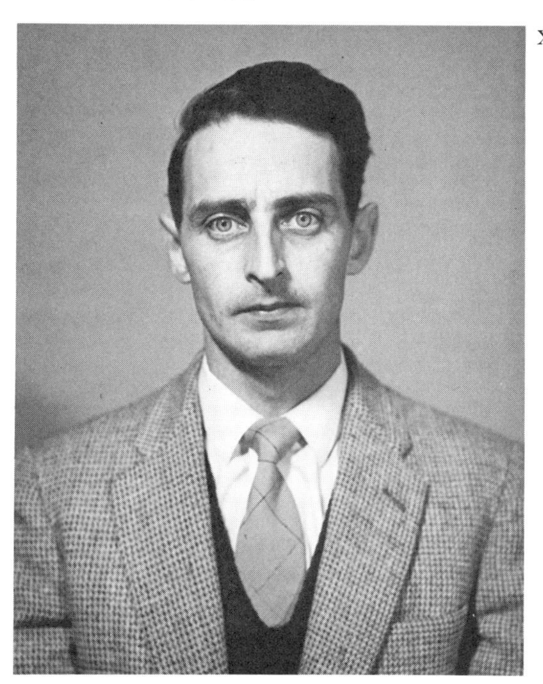

XXI-10 A. LAROCHELLE

XXI-11 E. M. CAMERON

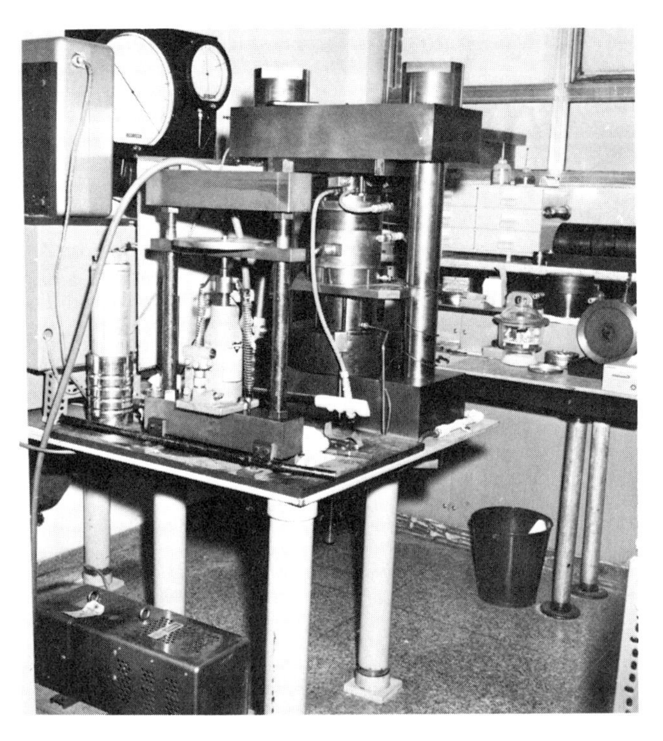

XXI-12 APPARATUS FOR HIGH-PRESSURE REACTION STUDIES ON ROCKS AND MINERALS

The Geochemistry Section also develops methods to facilitate field operations and help geochemical prospecting in general; these methods include colorimetric and radiometric procedures, and the use of various types of portable analytical and other equipment. An investigation of the feasibility of detecting petroleum and natural gas occurrences through minute traces of hydrocarbon in the overlying soil led to the establishment of a gas analysis laboratory. To help with investigations of gaseous emanations of mercury a gas chromatograph-mass spectrometer was recently acquired. An effort to locate uranium deposits by measuring the radon content of soils, gases, and waters inspired the designing of portable equipment for this purpose. The long-known relationship between growing plants and the presence of particular elements was also studied, beginning in 1962; a biogeochemical garden was established in a greenhouse on the eighth-floor roof of the Survey building to observe, by means of radioactive tracers, how the plants absorbed particular minerals from the soil and to detect the effects of selective removal of those minerals. After the completion of this study, field observations were made in the vicinity of known orebodies.

The section has also launched a program of analyzing samples in the field by building and operating mobile laboratories. The first one was used in 1960 for work in northern New Brunswick, where colorimetric and other tests were made to detect traces of heavy metals in samples as they were collected, and the course of future sampling was determined on the basis of the findings. By 1968 there were four such mobile geochemical laboratories, two fitted with optical emission spectrographs, and a third with an atomic absorption spectrophotometer. A mobile laboratory was also used in the biogeochemistry work to investigate the traces of metals in vegetation around Timmins, Ontario, with a view to developing methods of recognizing clues to hidden mineral deposits.

In keeping with a suggestion of the NACRGS, the Survey equipped one of the laboratories in its present building for experimental studies duplicating conditions occurring in nature as a help to understanding how geological phenomena operate. The major piece of equipment for creating artificial environments is a pressure vessel (or bomb) capable of producing up to 50 kilobars (50,000 atmospheres) of pressure and temperatures up to 2000° Centigrade. Most of the studies attempted so far have been undertaken by the Petrology and Geochemistry Sections into certain problems of anorthositic and related rocks. One line of inquiry, reminiscent of the bold theorizing of the versatile Sterry Hunt, examined the behaviour of super-critical water in association with silicates and other rock minerals.

* * *

THE LABORATORIES AND SECTIONS that deal with fossiliferous and organic materials lack the variety of those described earlier. The paleontological side of the Survey has now been largely relocated in Calgary, and the facilities and collections in Ottawa are far less than they were a decade or so ago. The original collection of well cores has become the responsibility of the Ontario government, and a large part of the fossil material has been moved to Calgary.

Paleontological preparation facilities remain in Ottawa, however. These are available to clean, cut, grind, polish, etch with acid, or make thin sections, casts, or moulds of the familiar macrofossils; preparatory work on the tiny microfossils is done in a separate laboratory. Here rocks containing microfossils are disintegrated in carefully controlled chemical baths, and the fossils freed thereby are separated from the residue and sent forward for study. This relatively recent branch of paleontology was initiated in the late 1920's and has only become generally used since 1945. Being tiny, the fossils are likelier to survive intact, and they are extremely numerous and varied, so their diagnostic potential is very promising. As knowledge of their types and successions improves, they afford an increasingly precise means of dating Paleozoic strata in particular. Investigations in other laboratories in the even newer field of palynology, "the study of the spores and pollen of plants to determine the age of the sediments in which they are found"[17] complete the paleontological studies in this very active field. Even after nearly two centuries of fossil study around the world, each year's collecting by the Survey adds scores of new fossil types to be catalogued, described in publications, and added to the National Type Collection of invertebrate and plant fossils. Moreover, the earlier collections, stored in the sub-basement after having been checked, studied and reported upon, are constantly referred to for comparative purposes, and, on occasion, for restudy in the light of new findings.

In the National Type Collection (T. E. Bolton, curator) are stored the differentiated and described types of invertebrate and paleobotanical fossils, ranging from the Precambrian to the Pleistocene, over 30,000 specimens in all. Most originated from collections made in the field by Survey officers from Logan and Billings onward, but a significant part has also been donated by outside investigators. The collection, while by no means exhaustive of the recorded types from all of Canada, is certainly the largest in the country and is constantly being referred to by geologists from all over the world. It was largely put in order by Alice E. Wilson, and since about 1910 it has been systematically maintained by checking through the annual collections brought in from the field, and by rechecking the specimens in the light of new discoveries and interpretations. To assist researchers, four volumes of catalogues have been published to date. A few specimens are always out on loan to students, while photographs and casts are provided in response to requests. The entire collection is also available to be consulted by authorized researchers.

The facilities at the Institute of Sedimentary and Petroleum Geology, similar to those in Ottawa, handle all fossiliferous material derived from the Western Plains and the Arctic islands that is collected or received by the Survey. The institute has its own lapidary, sedimentary petrology, clay mineralogy and chemistry, organic geochemistry, macropaleontology, micropaleontology, and palynology laboratories, drill core and sample collections, as well as collections of fossils of the different periods made by, and studied by, members of the institute.

Still housed at Ottawa pending completion of an extension to the institute's building to which it will shortly move, is the coal research laboratory (P. Hacquebard), first formed at Sydney, Nova Scotia, in the 1940's. Besides the preparation rooms for coal samples and thin sections, a special palynology laboratory is required. Plant spores, which are almost indestructible, are widely distributed throughout all coalbeds and in other sedimentary beds of nonmarine origin. They are valuable for correlating the formations, and may give clues both to beds of comparable ages and to the type and character of the coal that may be discovered.

The section also investigates the qualities and uses of coals. Petrographic studies examine coke-making qualities, and detailed microscopic investigations of screened coals may reveal the best ways of preparing them for metallurgical operations. Column sample studies indicate the variations of the composition of the coal, and coal rank within beds. Other studies investigate coals in their environments to learn more about working properties and potential mining hazards. These activities, carried on in conjunction with the Mines Branch and the provincial mining agencies, are an important aid to an industry that has gone through difficult times before its recent return to importance.

Also concerned with organic remains are two of the three major laboratories of the Terrain Sciences Division (J. G. Fyles). These are the Pleistocene palynology laboratory (R. J. Mott), which is concerned with making use of fossil pollen and plant spores for stratigraphic correlation work, age studies, and studies of climatic changes of unconsolidated surficial sediments; and the radiocarbon dating laboratory (J. A. Lowdon). The radiocarbon dating laboratory works out ages of sedimentary materials on principles similar to those used in the radioactive isotope measurements of the Geochronology Section, of which the radiocarbon dating laboratory was actually a part till 1963. The method consists of checking the ratios of the carbon isotopes C-12 and C-14 in organic matter, and from the changing isotopic ratios determining the ages of the organic substances in which the carbon occurs. The two carbons are found regularly in nature, the C-14 being formed from atmospheric nitrogen (N-14) by the constant bombardment of cosmic rays. All plants (and animals through consuming the plants) acquire a given amount of C-14, along with C-12, as part of the normal life process, the ratio of the two carbons being

XXI-13 THE RADIOCARBON DATING LABORATORY OF THE TERRAIN SCIENCES DIVISION

fixed, but after the organism dies, the C-14 slowly decays back into nitrogen. Hence the amount of C-14 remaining in a sample can indicate its age and that of the deposit from which it came.

In the laboratory, which is in the sub-basement and heavily shielded against cosmic rays and other outside interference, samples are converted to pure CO_2 gas, and the amount of C-14 present is determined by counting the beta particles emitted from a given volume of the gas. Since the half-life of C-14 is only 5,600 years, the amount of C-14 remaining in samples over a certain age becomes too attenuated for accurate measurement; the limit depends on the sensitivity of the counting equipment. With a 2-litre proportional counter the limit was about 35,000 years, but in 1962 a 5-litre counter sensitive enough to extend the dating limit to 54,000 years was installed. Efforts to reduce background interference continue, and also to work out the precise C-12/C-14 ratios over past centuries, which are known to have varied through natural causes long before the last two centuries when mankind began radically altering the situation. To help with this last problem, precisely datable material like tree rings, a succession of which has been worked out over many thousands of years, is studied. Radiocarbon dating is an essential tool for work on the Recent period, providing data on changing shorelines, rates of sedimentation, advances and retreats of glacial ice and similar events over the past 50,000 years. The laboratory also helps other government agencies occasionally with land-use questions, or dating archeological sites, and even the ages of works of art for art historians.

The foregoing laboratories all handle, treat, identify, measure and study rocks, fossils, and sediments for scientific purposes. Superficially, there may seem to be some duplication at the rather low levels of sample and slide preparation, microscopic and spectroscopic examinations, and chemical analysis. But usually there are important differences in the types of material treated, the methods used, the purposes for which the results are derived, or the form in which they are required, that explain the seeming duplication. Sometimes, too, a second laboratory has been set up because the volume required of a certain kind of work makes it more efficient to establish a new laboratory specializing in that activity. The setting up of a separate geochemistry laboratory is an example. Convenience is another consideration: close co-operation is best achieved when the laboratory and its chief customer are associated in the same administrative unit or division. Therefore, each division tries to be as self-sufficient as possible in its laboratory requirements, with due regard for economy and efficiency. The present plan of attaching to divisions those laboratories that serve their operations—as for example, linking petrology, geochronology, and paleomagnetism to the Regional and Economic Geology Division—was in line with this aim.

Laboratory economy and efficiency have been investigated from time to time. It is concluded that work of equal quality could not be obtained more cheaply outside. Besides, the in-house operations provide speedier service, and the units for which the work is being done can always know exactly how the samples were treated, what the results were based on, and how reliable they might be. The laboratories also afford training for Survey staff, and the opportunity to test and develop new equipment.

Laboratories do not exist in a vacuum; they must make worthwhile contributions to sections, divisions, and to the Survey as a whole. When they do not, usually because the function for which they were established has been dropped or taken away, they are transferred or abolished. On the other hand, when an existing or new Survey function requires laboratory support, this duty is grafted on an existing laboratory, or assigned to a new laboratory unit. Thus laboratories have grown up to help the Survey discharge certain functions; and units within laboratories have been established to help sections extend into new directions. The histories of two recent laboratories, paleomagnetism and radiocarbon dating, illustrate how such units begin and evolve into separate existences. Both grew out of the current activities of existing sections and divisions, in which they acquired staff, equipment, and experience, and developed distinctive roles. Eventually they were transferred to different divisions where they were needed, becoming useful tools rather than scientific novelties.

* * *

ANOTHER LABORATORY ACTIVITY, which deals mainly with field data and instrumentation and only incidentally with rocks, is that conducted by the present Resource Geophysics and Geochemistry Division. Though aeromagnetic surveying was initially the *raison d'être* of the division, by 1957 its work was listed as including: "aeromagnetic interpretation, palaeomagnetism and magnetic properties of rocks, magnetic resonance as a means of detection and nondestructive analysis of minerals and rocks, seismic reflection and shallow refraction, a pulsed method of electromagnetic prospecting, and radio frequency field strength measurements as a means of mapping near surface conductivity."[18] A later statement (1964) added to these, "mathematical methods of interpreting geophysical data, photogeologic research, and research in methods of remote airborne sensing."[19] With the contracting out of aeromagnetic surveying and the transfer of paleomagnetism to the Regional and Economic Geology Division, geophysical laboratory work has been mainly of two kinds: compiling, interpreting, and processing data provided from airborne, seaborne and ground surveys; and developing and testing instruments for field and public use.

For aeromagnetic surveying the laboratory worked at improving the magnetometers and developing a sturdier, less cumbersome instrument, until by 1969 a lightweight, high-resolution, digital-recording aeromagnetic surveying system had been developed, "the design and fabrication of . . . which is probably the most advanced system of its kind in the world today."[20] This was part of a regular plan to develop instrumentation not merely for Survey use, but also to improve geophysical surveying methods by the industry: "The capabilities of geophysical prospecting methods lie, to a large extent, in the instruments. Therefore every effort should be made to keep abreast of developments in geophysical instrumentation and to try to devise new methods. The only way to do this is be active both in theoretical studies and in the construction of experimental instruments."[21] An electronics laboratory was started in 1953 for this purpose, and in due course an Instrumentation and Electronics (now Electrical Methods) Section (L. S. Collett) was established. In 1960 this section provided two modern free-precession sea magnetometers (patented to the Survey) for installation on ships and a third for a base ashore. It also developed a lightweight airborne magnetometer and built two spectrometers for magnetic resonance research. In 1961 work proceeded on devising a direct-reading magnetometer, and in 1963 on a portable three-channel gamma-ray field spectrometer. From 1964 much attention was devoted to modifying equipment so that the data would be recorded on digital magnetic tape for computer storage and retrieval.

This division was not unique in designing new equipment, though its efforts were perhaps more extensive than those of other divisions; an Instrument Development Section (P. Sawatzky) was even provided for the purpose. Several other laboratory units, especially in the 1950's, designed their own instruments and operational layouts—notably spectrometers and fluorescence equipment—when suitable commercial equipment was lacking. In fact, it was claimed in 1963 that "all mass

XXI-14 L. S. COLLETT

spectrometric, high vacuum, and ancillary apparatus has been designed and assembled by members of the [present-day Geochronology] section."[22] The Geochemistry Section also designed and built its own analysis equipment for direct use in field studies, as well as the mobile laboratories now in wide use. For projects of this sort, the Survey has a very capable electronic services and instrument development officer, J. W. Jones, and maintains an instrument development shop (G. A. Meilleur) to provide "technical services available to all laboratory units in the building in the design, fabrication, assembly, modification, and repair of scientific apparatus."[23] In the early sixties this shop, besides repairing the whole range of regular equipment used by the Survey, worked on developing new X-ray and optical spectrographic equipment; astatic air and sea magnetometers; rock conductivity measuring apparatus; a magnetic field washer; gas and solid source mass spectrometers; accessories for the microprobe analyzer; the 5-litre radiocarbon counter; high vacuum equipment; sediment samplers; apparatus for flow studies; and a portable diamond drill.

Along with the Geophysics Division's efforts to design new equipment went the development, testing, and perfecting of practicable user techniques. Much experimentation was needed with airborne electromagnetic systems to find the most effective frequencies and signals, to improve the reception and recording of the data, and to develop filters that reject interference. One system, tested in 1969, was proclaimed as being "the most advanced system of its kind in the world today, and . . . a significant contribution towards maintaining Canada's leadership in airborne geophysical surveying techniques."[24] The Seismic Methods Section also tested much equipment in the field in the course of its hammer-seismic studies.

Another objective of the laboratory work of the division was to devise ways of making the data more meaningful and usable. In 1963 a computer was programmed to make the mathematical adjustments in connection with the mapping of second derivative values of aeromagnetic data. In the following year an airborne magnetometer was designed that supplied its data on digital magnetic tape. To deal more effectively with problems of this type, the division from 1965 to 1971 maintained a Theoretical Geophysics Section (B. K. Bhattacharyya) of which the announced function was "to devise new computer-oriented methods of interpreting geophysical data and to carry out interpretations using existing methods . . . also to do research on and to establish a system for the storage and retrieval of geophysical data compatible with the electronically-stored geological data."[25]

As many of the foregoing activities indicate, the Survey has increasingly had recourse to mathematical and computer technologies to deal with its mathematical and data problems. By the late fifties a number of divisions and units were showing an interest, but the Geophysics Division apparently was the first to move. In 1957 that division made "a study of the usefulness of treating aeromagnetic data analytically through the use of an electronic computer service obtained on contract. By this means, it constructed a second derivative map of the Eastern Townships area, which bears a better correlation with mapped geology than does a total field map."[26] In May 1959, after attending a course in programming and operating a computer, a staff member applied it to magnetometer readings, "doing in three minutes calculations that would have taken several days by hand."[27] Another area in which the computer was used for making mathematical calculations in these early years was in connection with the K-Ar age determination work of the Isotope and Nuclear Geology Section.

Mainly, however, the staffs of a number of units looked to the computer to help store the masses of data being compiled for various purposes, such as all the geochemical data contained in Survey publications (1957); the data on mineral deposits and mining; or data on minerals and X-ray diffraction. The institute in Calgary was anxious to co-ordinate the data on rock formations supplied for thousands of oil and gas wells drilled since 1949. With chemical analyses of over 28,000 samples and 175,000 individual determinations completed in 1964, Y. O. Fortier in his first report as director predicted that "the problem of handling, storage and retrieval, and the future processing of such large volumes of data is leading to the introduction of electronic data-processing methods."[28] Work was begun on an electronic file to contain all the analytical data the Survey possessed on geological subjects, while a data-processing unit (K. R. Dawson) was set up to design forms for the use of every section that would be of comparable format and content, and capable of being processed directly for computer. Data records were obtained of emission, spectrographic, chemical and age determinations, and of the photographic collection, while a start was made on indexing all data contained in the Survey's publications. Many highly-sophisticated automatic-reading pieces of apparatus were reprogrammed to record and directly print data in a form suitable for computer storage.

By the end of the sixties as many as 464,000 data cards from such operations were punched in a single year. Conversion of the data to conform to such uniform principles as the Universal Transverse Mercator (U.T.M.) grid for geographical co-ordinates also proceeded steadily. The data are transcribed on magnetic tape kept in the tape library of the Department of Energy, Mines and Resources' Computer Science Division (now Computer Science Centre), while two other copies are stored elsewhere for safety.

At the same time the Survey was playing an important part in developing a Canada-wide system of geological data-processing. In 1965 S. C. Robinson chaired an *ad hoc* committee of the NACRGS that studied and reported on the storage and retrieval of geological data in Canada. Its interim report (August 1966) recommended setting up a national system and suggested a format that the files should follow. The Survey became responsible for the maintenance and operation of a national index to geological data in Canada. Its own records were indexed according to these decisions as a pilot study for the national index under B. A. McGee. The National Index was launched under C. F. Burk, Jr., in 1969, with six original member agencies, among them the mines departments of Ontario, Saskatchewan, and Newfoundland.

So far, however, most of the effort has gone into collecting masses of data rather than putting them to use. The most ambitious effort to apply the data is by the Geomathematics Section (F. P. Agterberg), which has applied mathematical techniques to such geological problems as frequency distributions of trace elements in rock samples, trend analyses of ore deposits, and probability studies of regional mineral potential derived from analyses of regional geological data. The assessment of geological information by statistical methods and the interpretation of scientific results in quantitative form will probably become increasingly used in view of the Survey's growing involvement with governmental planning and with making mineral resource inventories.

Numerous service and administrative units and groups that work with, or for, the Survey occupy a considerable part of the new headquarters. Some are seconded from other parts of the public service, including several members of the department's administration branch who are responsible for the personnel records and computer services of the Survey, representatives of the Public Service Commission's official languages bureau, of the Department of National Health and Welfare medical services, and of the Department of Public Works for building maintenance. There is also the headquarters staff of the branch, including the director and chief geologist and their advisory and technical staff, personnel seconded to such associated organizations as the International Geological Congress, the NACRGS, or the Canadian Centre for Geoscience Data; as well as the secretarial and clerical staff of the branch registry office, and of the financial, administrative, secretarial, supplies and equipment services.

XXI-15 MRS. D. M. SUTHERLAND

XXI-16 PART OF THE SURVEY'S LIBRARY COLLECTION
Most of the Survey's 130-year output of printed works is shelved on the six large sections immediately behind the globe. The entire collection of books, periodicals, microfilms and documents occupies over two miles of shelving.

Apart from these are the technical personnel who do not deal with geological materials, but with the findings and publications arising from the field and research activities of the Survey. These units, formed into the Geological Information Processing Division, include such sectors as the library, geological cartography, photographic, scientific editing, publications, and data processing units and activities. All play their part in making the results of the Survey's work known to the outside world, and most also offer important help to the Survey staff with their research and study programs. This last applies especially to the library and map collection (Mrs. D. M. Sutherland), without which it would be hard to conceive of any worthwhile scientific or scholarly work being done. Notwithstanding the 1959 partition, the library possesses over 120,000 books and periodicals, and is probably the largest and best Canadian collection on geology and related fields, widely consulted by industry, government, and students. The inexorable pressure of over 15,000 acquisitions every year has caused it to overflow the large original space and fill several additional rooms, even though valuable but marginally-useful materials have been banished to the Public Archives and elsewhere, while the map collection has been relegated to a basement room in the company of that of the Surveys and Mapping Branch.

When the geological cartography unit (C. E. McNeil) moved to the new building it was able to gather up its scattered staff for a time, and add new personnel and equipment. A special camera that was built into the building can make negatives up to 40 by 48 inches, has a range from 1:7 enlargement to 5:1 reduction, automatic adjustments, and push-button controls. Geologists receive copies of suitable topographical base maps adjusted to the desired scale on which to draw and colour the map units, legends are set up on a proofing press, and the final maps are photographed to furnish the colour separation negatives, with printing done either by the Surveys and Mapping Branch, or through Information Canada. Aeromagnetic maps, for which the flight lines and contours have been plotted on topographical base maps, are photographed and printed as blue-line maps on the blueprinting machine.

A much smaller but important unit is the Photographic Section. Its work has become greatly diversified; some of its newer tasks include processing aeromagnetic and radioactive emission film, photographing microfossils, and preparing and developing colour film. The negatives of the photographic collection were not brought over from the National Museum till 1963, but since then they have been put in good order, with a cataloguing system to provide quick and ready reference to desired photographs.

The editorial staff has always consisted of experienced field scientists, and in recent years has grown to require a permanent complement of four, headed by the chief scientific editor, R. G. Blackadar. He is assisted by a scientific editor responsible for the French language requirements of the branch. Each year the number and size of the publications have increased, the list published in the year 1971-72 including 6 memoirs, 22 bulletins, 1 economic geology report, 31 open file reports, 59 preliminary series papers (the major category of productions), 1 catalogue, and 31 multicoloured and 19 preliminary separate maps besides. Also made available for distribution that same year were re-issues of previously out-of-print items of the various series, including no fewer than 508 aeromagnetic maps. The scale of the operation by the distribution office can be seen from the fact that 105,510 copies of reports and 160,991 maps were distributed in 1971-72. A student, looking over the publications of the last ten years, cannot fail to be impressed by the evidence they present of the high standards of accomplishment that have been attained in the years since the move to Booth Street.

XXI-17 C. E. McNEIL

The remarkable achievements of the Survey in the last 20 years owe much to the expansion of the laboratories, which truly came into their own in this period and completely altered the style of the work. As late as 1950 the only reasonably adequate laboratories were those that provided routine chemical and mineralogical analysis, or that related to the preparation of rock and mineral sets for sale or display. Sydney Abbey, who entered the Survey in 1952 as a youthful industrial chemist, recalled from his first visit to the Sussex Street premises that the laboratory looked as though time had stood still for half a century, and that no one had tried to keep the place clean. When he went to work as Ellsworth's assistant his first workroom in the basement of the Victoria Memorial Museum was poorly equipped with a sink and a fume hood, both of which were out of order for months at a time. Work benches lining three walls were completely filled with rock specimens, leaving no room for chemical apparatus. All the laboratory furnishings had a home-made look about them, and the only sign of progress was a single ancient spectrograph. Morale was low, the staff was largely untrained, and even the great Ellsworth, whose work bridged the old and new technologies, was hopelessly out of touch with modern developments.

Since then the transformation has been as from night to day. The present-day laboratories, mostly at Booth Street but also at Calgary and Dartmouth, are highly specialized. Scientists trained in the latest methods operate the most modern equipment. Wherever possible, the equipment is automated, with recording devices and computer linkages to process the data and store them for future uses. One of many evidences of the Survey laboratories' high standards of achievement was a well-deserved tribute from the National Aeronautical and Space Administration (NASA) of the United States, which selected their proposals for chemical, mineralogical, geochronological, electrical conductivity, and magnetic studies for inclusion in the program of studies of the lunar material returned by the Apollo Manned Lunar Missions.

The Survey's laboratory facilities were not assembled at a single stage, although the move to Booth Street fur-

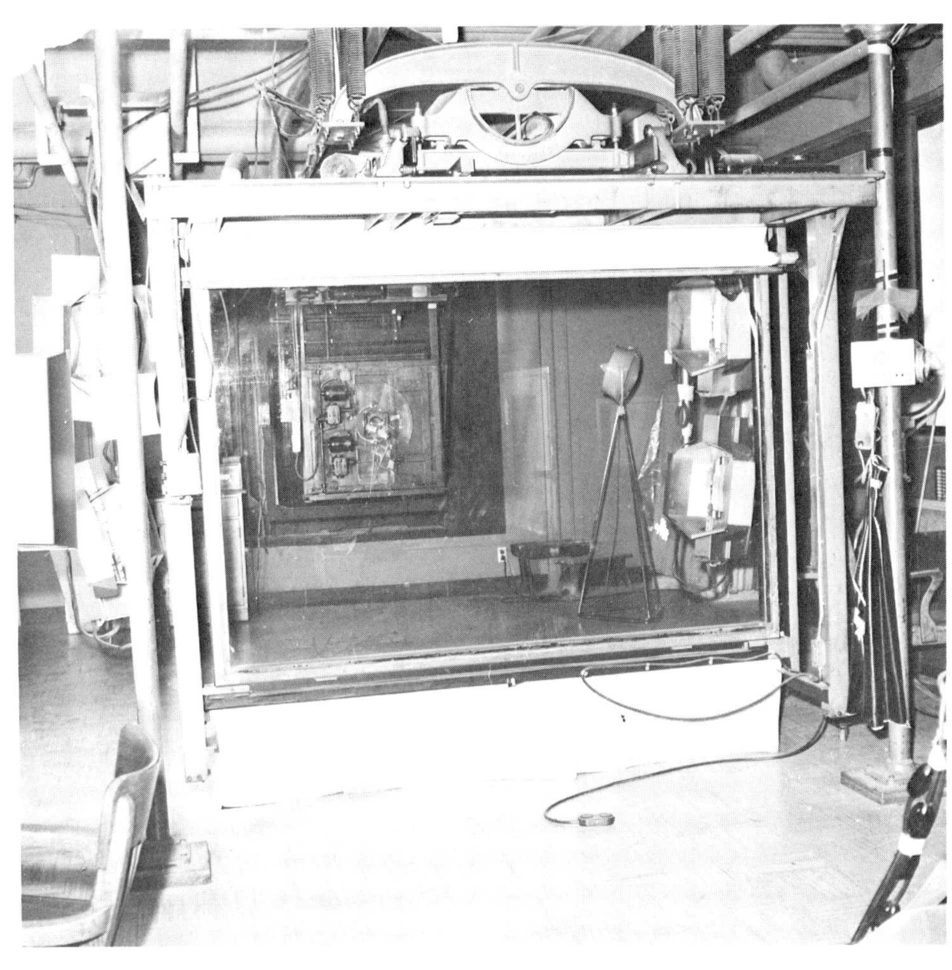

XXI-18 THE PHOTOMECHANICAL UNIT'S MAPPING CAMERA

nished a matchless opportunity to reorganize and to acquire new facilities and equipment. The drive to improve the laboratory operations was a continuing process, directed by specialists like Robinson, Maxwell, Traill, Wanless, Reesor, and Larochelle, all of whom joined the Survey between 1948 and 1953. These men took the existing facilities and transformed them. Often the Survey's laboratories were the first in Canada to use a new type of equipment. More than copying existing installations, they constantly tested new methods and machines to see whether they could be adapted to the Survey's needs. Moreover, they often designed and built their own equipment, thanks to the machine shop and the design staff. The Geochronology Section, for example, designed and assembled no fewer than seven mass spectrometers for its own requirements in the years 1955-72. Equipment was constantly modernized and updated; in recent years, by substituting transistorized circuitry for electronic tubes, and connecting the equipment with computers for purposes of operations control, recording and storing data, and performing complicated mathematical calculations.

The main function of the laboratories, as always, was to assist the work of the units devoted to studying and mapping the geology of Canada. But with their increased sophistication, the laboratory sections were able to supply the field-based operations with better results than ever before. The speeded-up methods of analysis, for example, made it possible to provide as many as 100,000 determinations for a single project. Men like the geochemist R. W. Boyle were given the data that helped them reach more accurate conclusions and make important discoveries. The laboratories also furnished the new tools that helped answer a host of previously insoluble questions. Behind C. H. Stockwell's interpretation of the tectonic history of the Canadian Shield and his improved perception of Precambrian time, for example, lay the revolutionary developments in geochronological instrumentation and methods. Similarly, isotopic studies gave deeper insights into the principles of metallogenesis that significantly advanced the study of ore deposits geology. Improved geophysical instrumentation made it possible to examine the geology of the submerged parts of the earth's surface and discover the phenomena of seafloor spreading, which along with paleomagnetic evidence brought greater credibility to the theory of continental drift. These are only a few of the ways Canadian geology reached new heights of accomplishment thanks to the availability of sophisticated laboratory technology. Since the accuracy of present-day research-based findings depends so much on laboratory determinations, the Survey is constantly striving to extend the range of uses, the sensitivity, and the discriminatory power of its various laboratory units and their equipment. Thus the laboratory sections, like those concerned with field investigations, work towards the Survey's objective of becoming better able to serve the needs of the science and the nation.

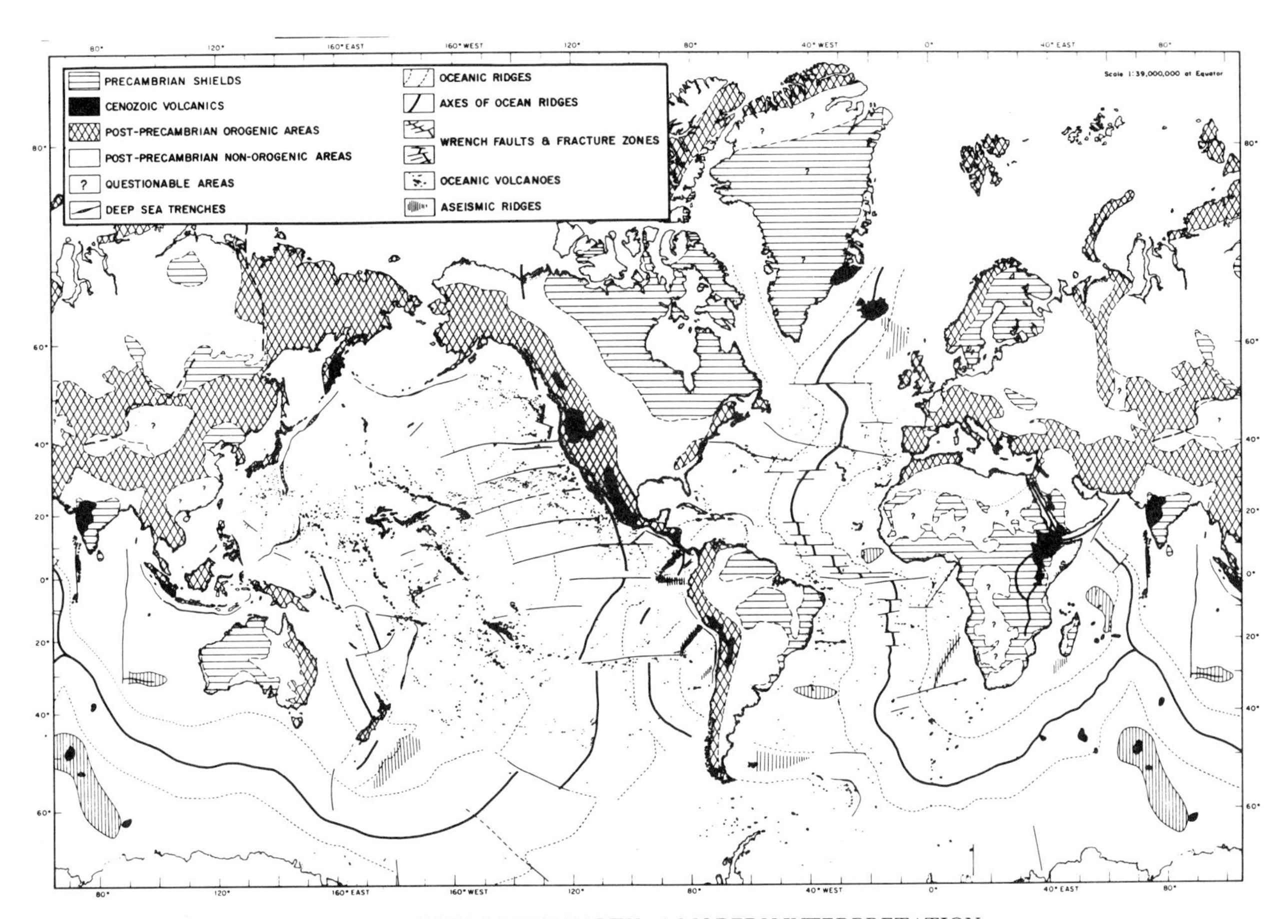

XXII-1 MAJOR GEOLOGICAL FEATURES OF THE EARTH—A MODERN INTERPRETATION
This map, based on work in the early 1960's of B. C. Heezen, M. Ewing, R. L. Fisher, H. H. Hess, and H. W. Menard, demonstrates the developing view of the role of tectonics in shaping the present-day world. It was compiled for a contribution on "Geothermal Studies of Continental Margins and Island Arcs" by W. H. K. Lee, S. Uyeda, and P. I. Taylor to one of the symposia of the International Upper Mantle Committee held in Ottawa in 1965 and published by the Survey as Paper No. 66-15. Compilations such as this led to current theories of plate tectonics.

22

Research Provides New Answers

MOST OF THE SURVEY'S FIRST CENTURY was spent in gathering increasingly complete and accurate data on the geology of Canada, and interpreting that information in the light of current scientific knowledge. After the Second World War, however, the Survey was able to add to these traditional functions a new activity—carrying out general or basic scientific research on its own account as a means of better understanding the fundamental forces that determine the geology and control the utilization of Canada's mineral resources. The time was opportune; an immense array of technical aids and scientific discoveries had just become available for studying general problems that were especially relevant for Canadian geology. The scientific community of Canada, whose prestige and influence had never stood higher, strongly stressed the vital importance of such basic studies, and through the Royal Society of Canada pressed Canadian scientists and scientific agencies to engage in them. In 1949 the National Advisory Committee on Research in the Geological Sciences (NACRGS) was established by order in council to advise government on science policy in these areas. The value placed on general and fundamental research was shown by the formation of this committee, which was a prime agency for the next years in mobilizing private industry, the universities, and government agencies to add such studies to their current activities.

The first annual report of the NACRGS, on March 15, 1951, contained a long list of problems the solutions to which were considered important for the economic and scientific progress of Canada. They had been compiled by six subcommittees appointed to deal with the following fields: physical methods as applied to geological problems; metallic mineral deposits; structural geology; petrology, mineralogy, and chemistry; Pleistocene, glacial, water supply, engineering geology, and geomorphology; nonmetallic deposits, industrial minerals, coal and oil; and stratigraphy and paleontology. Something of the diversity and range of the suggestions may be envisioned from the list on metallic mineral deposits:

—An independent investigation to evaluate the methods recently developed for determining temperature and pressure of vein minerals at time of their formation
—Support of current biogeochemical research at the University of British Columbia
—Establishment of Geological Field stations
—Investigation of use of the refraction and reflection seismograph in the mapping of crystalline rocks
—Statistical studies of spectrographic analyses of glacial drift
—Studies of the correlation of geophysical and geological surveys for the purpose of better interpretation
—A study of the problems relative to replacement in the deposition of ore
—Further detailed studies east and west of Steeprock Lake, Ont., with particular reference to iron ore
—Mathematical studies of frequency curves of assay values with particular reference to gold deposits
—A statistical study of past and producing gold mines to arrive to the importance of mineral associations, types of deposits, associated rocks, structures, etc.
—Spectrographic studies of ore and gangue minerals as well as associated intrusives and wall rocks in order to correlate them
—A study of the Quadeville pegmatite mass, Renfrew County, Ontario
—Delineation of metallogenic provinces in Canada.[1]

Of course the Survey was not the sole organization in Canada interested in, or capable of, investigating these and related subjects. Other agencies—provincial government branches, private industry, and especially the universities—felt the same needs, and responded according to their abilities to the call of the NACRGS for research into important geological problems. But the Survey was unquestionably in a good position at this time to undertake scientific studies in the earth sciences because of the specialized staff and laboratory equipment it was acquiring, and because of its long experience with all kinds of geological work and with the full range of the Canadian terrain. Many of the subjects suggested for investigation were entirely in keeping with the broad functions of the Survey of collecting geological information, and developing techniques and concepts to assist the search for mineral deposits and the evaluation of potential mineral resources. It was reasonable, perhaps even inevitable, that given the necessary means and personnel, the Survey should undertake work in most of the directions indicated by the subcommittees of the NACRGS. That so little similar research was being conducted in Canada made the Survey's action all the more desirable—and welcome.

In fact, in addition to its own special research projects the Survey made notable efforts to promote and assist geological researches and general studies by these various agencies. Its officers participated in, supported, and encouraged such work by many professional and scientific societies, for example, the Geological Association of Canada, which was founded in 1947 to draw together the varied segments of the profession across Canada. In addition, they helped to organize and co-ordinate research in the earth sciences by serving as members and directors of relevant associate committees of the National Research Council, such as geology and geodesy, and of its subcommittees in special fields such as soil mechanics or hydrology.

They were also major supporters of and participants in workshops, conferences, and symposia organized under the auspices of the NACRGS; in the important series of volumes on geological subjects published periodically by the Royal Society of Canada; in the symposia and conferences of professional groups, such as the Alberta Society of Petroleum Geologists; and in the nation-wide Canadian Institute of Mining and Metallurgy (CIMM), whose co-operation was enlisted for such work. That society of practical mining men was inspired to organize a geology division and to sponsor studies of general problems of significance to mining and exploration geology. Largely through the efforts of M. E. Wilson and A. H. Lang, a pioneering volume, *The Structural Geology of Canadian Ore Deposits,* appeared in 1948. It proved to be the first of a useful continuing series, the latest of which, *Geochemical Exploration: Proceedings* (1971) had R. W. Boyle as a technical editor.

The Survey has also encouraged and supported the Canadian universities' progression into the field of scientific researches and specialized studies in the earth sciences. Survey personnel frequently gave lectures on their current research at virtually every university in the country, and some offered courses in their fields of specialization. They acted as advisors and visitors to university departments, advised graduate students, and served on thesis boards. Specialists were loaned to universities to set up new programs and laboratories that facilitated the entry of those institutions into additional fields of research.

A number of Survey officers also joined university staffs after 1945 to take up teaching, research, and administrative appointments. With so many universities and earth science departments being inaugurated or expanded into specialized graduate programs, these men made valuable contributions in raising the new institutions to high levels of scientific competence. Moreover, the presence of former officers in the departments facilitated the interchange of ideas, talents and facilities between the university and the Survey, helping in some measure to co-ordinate their programs of research, and generally making each party more understanding of the problems of the other. The Survey also supported certain specific research programs at given universities, or by agreement took over other research programs pioneered by them.

The Survey managed the annual distribution of NACRGS funds instituted in 1951 as grants in aid to stimulate and support geological research conducted in the universities, which research was receiving almost negligible aid at the time. The NACRGS reviewed the applications from university faculty members and recommended to the Survey to whom to award the grants. Year after year as the applications grew in number and amount the Survey requested and received additional sums, so that the totals climbed from $10,000 to $270,000 in 1970-71 (by which time the grant was being contributed directly by the Survey), which supported 100 projects at some 25 universities. These furnished small but very useful and much appreciated assistance to many important projects.

They were soon overshadowed by a new grants program initiated by the National Research Council that was eight or ten times as large in sum as those distributed by the Survey, and by grants from several other federal agencies. Altogether, between 1958 and 1969, federal government spending in support of earth science studies at Canadian institutions of higher learning increased thirtyfold, compared with a fourfold increase in faculty but a mere doubling of the graduate student enrolment. Many departments, moreover, received considerable additional aid from their respective provincial governments and some from industry as well. As a result, there was a great expansion of specialized staff, and sophisticated research facilities in the earth sciences were acquired by several universities, which proceeded to embark on ambitious research programs. Complaints began to be heard about the lack of co-ordination, the duplication and misdirection of much of these efforts. The study *Earth Sciences Serving the Nation* (1971) commented on these criticisms:

> Most earth scientists outside the universities think that the earth science research training in our academic institutions has swung too far toward esoteric research and laboratory investigations. The opinion is widespread that the universities are not meeting adequately the nation's needs for specialists in several fields of earth sciences . . . and are producing too many Ph.D. graduates in other fields . . . .
>
> That the universities should be free to undertake basic research of their own choice is sound in principle, but it has resulted in concentrating research and producing most Ph.D. graduates in fields chosen by professors. In addition to the development of science for its own sake, universities should be encouraged to direct a greater part of their research activities towards fields in which the employment opportunities are greatest.[2]

That report recommended that, in future awards, more weight should be given to the projects being proposed as well as to the reputations of the professorial applicants; also that some part of the grants should be assigned to more immediately practical projects. The Survey began considering the desirability of awarding its grants in such a way as to coax university research into areas that are potentially helpful to the orderly development of Canada's resources, and more in tune with the Survey's own research priorities.

The remarkable expansion of research activities at two dozen independent universities occasioned certain tensions with Survey programs. Survey scientists who are deeply committed to long-term major research projects into important fundamental problems, feel that progress with these is hampered by a lack of planning, direction, or continuity of effort under the present organization. They consider that university faculty and graduate students who receive research grants should forward the solution of problems of concern to the Survey. University staff, on the other hand, feel their freedom of inquiry should not be interfered with in any way, that the universities are the proper place for carrying out "pure" or "basic" research projects, and that the Survey should defer to them by withdrawing from areas of research in which universities are both interested and capable of carrying out the work. Occasional cases of competition over the conduct of specific studies or over the control of certain materials and bodies of data arose to trouble the long-standing good relations between the Survey and the universities. Such questions can only be resolved by mutual concessions by the interests concerned, unless or until the continuing debate over a national science policy eventually produces a drastic realignment of the financial support for future research in the earth sciences in Canada.

The debate over national scientific priorities and objectives was first raised a decade ago by the Glassco Commission, when it emphasized the need for "wise and appropriate national science policies,"[3] called for continuing government study and oversight, and suggested setting up certain administrative bodies expressly to perform these functions. In line with the commission's recommendations, a Science Secretariat was set up in the Privy Council Office to advise the cabinet on science policy matters; a Science Council of Canada (analogous to the Economic Council of Canada) composed of prominent scientists was appointed to investigate and report on specific or general questions of science policy; and eventually, a Ministry of State for Science and Technology was created to take charge of the Science Secretariat. Every facet of the subject apparently was examined during the 1960's, especially in the profusion of special studies reports issued by the Science Council (no less than 41 in the five years from 1967 to 1971); in an excellent study of 1969 prepared by the Organization of Economic Co-operation (OECD), a United Nations agency based in Paris; and in the voluminous proceedings and reports of a Special Senate Committee on Science Policy, chaired by Maurice Lamontagne.

These reports emphasized that Canada's total expenditures on R. & D. (Research and Development) were far below those of other technically advanced countries, and called for a substantial proportionate increase in such spending to around 2.5 per cent of the Gross National Product by 1980, double the 1967 rate. Investigation also revealed that a disproportionate amount of effort (in the opinion of the Lamontagne Committee) was being devoted to basic or fundamental research, rather than to applied research or development that offered better prospects of good economic returns; also, not surprisingly, that the research done at the universities was principally in the "basic" category, which accounted for 70 per cent of its total effort between 1962 and 1970. The major role of the federal government—amounting to 60 per cent of the total Canadian spending on R. & D. in 1971—was another finding that inspired comments that the federal government's share of future spending on research should be lowered while that of Canadian industry should be greatly increased.

It was further claimed that some federal government agencies did not exercise proper concern for practical goals in their research programs. The government's scientific effort was criticized as being too widely diffused, sometimes overlapping and inefficient, and too slow to become translated into practical development:

> Numerous suggestions have been made that basic research of interest to government missions should be done in universities or in industry. The argument is that in universities basic research serves the function of education and that in industry it indirectly raises the level of scientific capability and hence productivity in the industry, with the possibility of foreign sales of expertise gained. In government, it often remains more isolated and produces fewer indirect benefits. Contracting work out may also make it easier to initiate and . . . terminate projects.[4]

The Survey has complied with these suggestions (as well as earlier ones based on economic considerations) by contracting out certain work in data collecting and instrument development. But implementation has proved more difficult in the area of laboratory research where it is felt that most of its research programs should be kept under the Survey's control. Performance or quality control cannot be assured so well when parts of projects are delegated to others. Delays that could have serious effects on important programs cannot be so easily prevented when control is not entirely in the Survey's hands. Besides, in many projects, field and laboratory research are too intertwined for either to be treated in isolation; useful concepts are best secured from the field and laboratory data by handling both concurrently. The Survey's staff members also need to be able to view their projects in their entirety, and to be free to pursue any aspects of them as required.

The main study to examine the national effort in the field of the earth sciences was sponsored by the Science Council of Canada, its Special Study No. 13 (1971), *Earth Sciences Serving the Nation.* This was the work of a Solid-Earth Sciences Study Group, chairman R. A. Blais, in which C. H. Smith participated as project officer and Y. O. Fortier as a committee member. The study formed the basis of the council's recommendations, which were presented in its Special Report No. 7. It advised consolidation and closer oversight of earth science research by several government agencies, and proposed a continuing interdepartmental committee on the geosciences to appraise and co-ordinate their various earth science interests, suggest new programs of action, and conduct special studies. The report advocated the concentration of effort into certain key areas for more effective results from the national point of view. It called for the co-operation of governmental, university, and industrial staffs and facilities on work in such fields as geotechnical, metallogenic, geochemical, and geophysical research to increase the effectiveness of prospecting for mineral deposits; on developing new insights into the origin and accumulation of petroleum; and on paleostructure and continental drift studies as they relate to Canada. Specifically recommended were:

> A comprehensive and multidisciplinary program of research into the origin and evolution of the Canadian Shield should be undertaken during the next decade, with particular reference to geodynamics of proto-continents, and Precambrian sedimentation, volcanism, plutonism, metamorphism, and orogenesis. This program should be under the general direction of the Geological Survey of Canada, and entail the active co-operation of provincial agencies, industry and universities . . . .
>
> Fundamental knowledge on the nature and history of the major elements of Canada's geological provinces and their correlation with similar elements in contiguous regions of the world should remain a prime research objective of the government agencies and university departments engaged in earth science activities . . . .
>
> The following fields are exceptionally suited for Canadian research (in alphabetical order): geomagnetism, geotectonics, glaciology (including snow and ice), mineral deposits geology, mining exploration geophysics, muskeg studies, permafrost studies, Precambrian research, and Quaternary geology.[5]

To assure co-operation among governmental, industrial, and university groups for integrated research efforts, the Solid-Earth Sciences Study Group advised setting up regional centres where challenging problems of concern to the region could be attacked in a planned, logical manner by the combined efforts of such groups. Such centres should aim at becoming "Centres of Excellence," capable of making significant contributions to given sorts of general problems. Indeed, the Survey already had such a centre to serve as a prototype or model—the Institute of Sedimentary and Petroleum Geology, situated adjacent to the campus of the University of Calgary, and operating in close touch with the oil and gas industries of the Western Plains and Arctic islands regions.

In the meantime, as discussion of a national science policy continues, the Survey proceeds to carry out its field and laboratory studies, to assist university research programs in financial and other ways, and to co-operate in professional and scientific research undertakings at the regional, national, and international levels. As for its own research, financial constraints and the closer oversight of the Treasury Board have tended to stress immediately practical activities. In recent years, accordingly, basic researches have held a somewhat lessened priority. Programs that received much attention earlier in the 1950's or sixties, such as studies of ultrabasic rocks and granites, or research into Precambrian life-forms, tend to languish, while the hydrothermal laboratory goes unused. Nevertheless, the Survey's basic research effort achieved such success in these years and acquired such a momentum—in terms of facilities and personnel, in commitments made, projects in progress, and in a high reputation won in the world scientific community—that the Survey will certainly continue to figure importantly in the never-ending drive to master the general principles that shape the geology of Canada.

* * *

SINCE 1950 THE SURVEY has undertaken many special studies in an effort to understand some of the problems most often encountered in the field. Through co-operation among specialists, the findings, approaches, and insights from a number of subsciences have been brought together to provide more accurate interpretations of individual phenomena and of universal forces. These studies were conducted in parallel with, and grew out of, the mapping of a large amount of previously unstudied territory. They have helped answer some of the many questions, and have assisted the Survey to achieve a more accurate, comprehensive and useful account of Canadian geology.

Some general studies have consisted of examining in great detail characteristic field structures and relationships in the light of the latest scientific knowledge. Others have included highly sophisticated laboratory studies of rocks and other materials; and in a few, the natural processes themselves were simulated under controlled conditions to afford better insights into how they operate in nature. The work has ranged from detailed studies of rock relations and other phenomena in a single mine, to investigating aspects of recurring phenomena, to studies of massive episodes such as the advances and retreats of ice sheets, to the preparation of maps showing the tectonic structure or geochronology of the country, and to relating Canadian geology as a whole to the grand panorama of planetary evolution.

Such works of synthesis, collating and analyzing results derived from all regions and settings, have only become possible as a result of many successful local investigations conducted by Survey geologists and others over a period of more than a century. Most of the investigations, however, have been carried out over the past 25 years, when scientific advances and large specially-trained staffs made these operations possible. By such a process of accumulating field data, supplemented by radiocarbon dating of organic remains, an increasingly accurate picture is emerging of the glacial events of the past 100,000 years that affect over 97 per cent of the Canadian landmass. Studies are proceeding into the succession of four major glaciations and three warmer interglacial intervals, the results of which may be found in V.K. Prest's chapter on "The Quaternary Geology of Canada" in the latest edition of *Geology and Economic Minerals of Canada*, and his fine map of the stages in the retreat of the Wisconsin ice-sheet over the last 18,000 years.

XXII-2 STROMATOLITES FROM THE PRECAMBRIAN
Believed to be the remains of former life, stromatolites have proved of considerable value in correlating Precambrian rocks. This example is from the Epworth Group, Takiyuak Lake, N.W.T.

Similar trends may be seen even in a subscience like paleontology, which has been intensively investigated for nearly two centuries, since the dawn of geological study in Europe. Paleontological work, in the main, follows the traditional tasks of describing and identifying fossils, and studying fossil assemblages in strata to establish relative ages as an aid to mapping. With so much of Canada coming under active exploration and development since the Second World War, collecting and identifying the fossils received from newly-mapped districts has become a formidable task. Two almost completely new subspecialties—microfossils and palynology—were also developed practically from their foundations in this period, and their fossils correlated with the classical fossil (macrofossil, or megafossil) successions and with the geology of every part of Canada.

After barely 15 years of such research, microfossils have proved uniquely helpful in identifying and correlating potentially oil-bearing horizons. Fossil spores are more effective than megafossils in the study of coal-bearing deposits. Specialists have been appointed to study these new subsciences as well as the older one of paleobotany, and to concentrate on particular geological periods and regions. A full range of faunal zones, each of which developed over an average period of 1.8 million years (m.y.), has been worked out that takes account of regional variations in the fossil record and spans most of the interval since the earliest Cambrian. Workers in the Phanerozoic rocks have thereby been furnished with an extremely useful, accurate geological time-scale. By the end of the sixties local or regional zonations had been erected in many areas that were more highly refined than those accepted internationally: "In [Canada] . . . over vast areas, rigorous stratigraphic and biochronologic observations date back hardly more than 20 years. It is a remarkable vindication of the methods of biochronology that such accurate time-scales may be erected for most periods."[6]

Investigations were also extended back in time to pre-Cambrian life forms, such as stromatolites that occur at rare intervals in the Proterozoic—to serve possibly as an eventual dating tool and as an aid to understanding certain geological problems. Another profitable field of study beginning to receive some attention is paleoecology—the study of the environments in which the organisms lived and in which the sediments burying them were formed.

In an outline in 1968 of the Survey's work in this field D. J. McLaren, then director of the Institute of Sedimentary and Petroleum Geology and now director of the Geological Survey, listed the accomplishments since the Second World War as including erecting accurate time-scales for most periods; extending rigorous stratigraphic and biochronological investigations into virtually unknown territory in western, northwestern, and in Arctic Canada; and restudying earlier areas with greater refinement and in line with the newest concepts. Among specific achievements McLaren stressed the work on Precambrian fossils; the extension of Cambrian trilobite zonation to new areas; accurate zonation of bottom-dwelling organisms in the Ordovician and Silurian of the Cordilleran region; extending graptolite zonation into the Lower Devonian; coral-brachiopod zonation of the Middle and Upper Devonian in the west and the Arctic, and correlating them with standard sequences; the discovery in the Cordillera and Arctic of the most complete sequence of Marine Triassic faunas in the world; the subdivision of the Lower Triassic into four new stages of worldwide significance; the extension of European ammonite zonation to the Jurassic of the west and Arctic; and the discovery of important major sequences of late Paleozoic rocks in the Arctic islands and Yukon Territory. Looking ahead, McLaren saw continuing activity along similar lines: in describing the many, diverse faunas to be collected in future explorations, further refinement of zonation for all areas and periods, and increasing attention to the fields of micropaleontology, palynology, and paleoecology.[7]

An increasingly important scientific concern of the Survey, both in the field and in the laboratories, is studies of rocks and structures that bear significantly on mineral deposition. Such, for example, are the investigations of intrusive rocks (ultrabasic or ultramafic rocks, granites, pegmatites) that may occur in any age and almost any region except the flat-lying strata of the Plains; or studies of reef and dome structures in the deposits of sediments of the oil- and gas-producing periods of the Phanerozoic. By the late fifties the Survey had laboratory sections for the highly demanding, specialized tasks that these careful studies require. Important new fields like geophysics, geochemistry, gravity studies, isotopic age determination, and geomathematics were providing new approaches and techniques that were immensely helpful in solving complex problems. The widespread explorations of large parts of the country had begun to indicate geological patterns that opened the way to new avenues of study. Above all, a large, well-trained staff recruited mainly in the decade following the war, was reaching by the late fifties a level of experience and competence capable of dealing with major fundamental problems.

One such field of generalized study was the intrusive rocks, innumerable examples of which are found in different settings across Canada. Hunt and Logan had studied anorthositic intrusions north of the St. Lawrence, as had Adams, Ferrier, and Barlow, who worked also on

XXII-3 FOUR GSC PALEONTOLOGISTS WHO HAVE ACHIEVED WORLD RECOGNITION
Upper left, H. W. Frebold (Jurassic); *upper right,* J. A. Jeletzky (Cretaceous); *lower left,* E. T. Tozer (Triassic); *lower right,* R. Thorsteinsson (Paleozoic biostratigraphy of Arctic Canada).

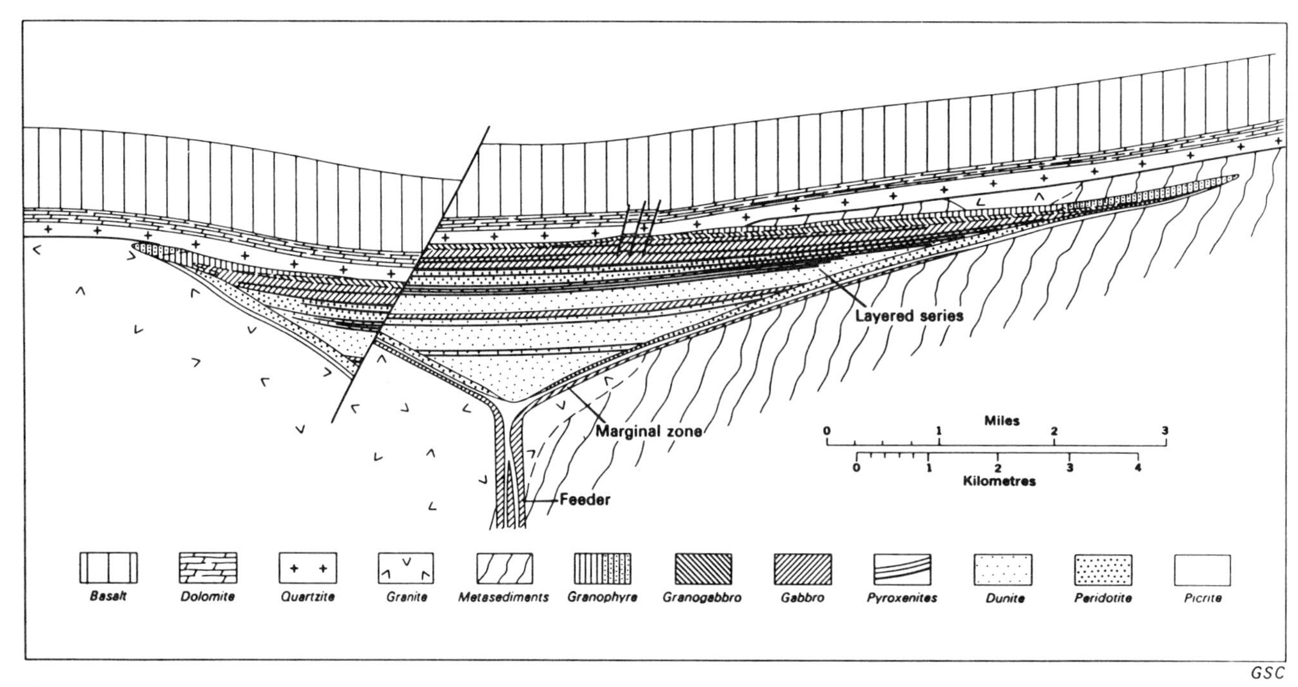

XXII-4 CROSS-SECTION OF THE MUSKOX INTRUSION

the magmatic mass at Sudbury, and on various batholiths and other structures along the Frontenac Axis of Ontario. The Monteregian Hills and the intrusions in the Eastern Townships were studied by LeRoy, Dresser, and others. Studies, mainly by K. R. Dawson, were continued during the fifties on the Anstruther batholith near Bancroft, Ontario, and the Preissac-Lacorne batholith near Rouyn, Quebec, while E. R. Rose studied the structural relations of the Morin, St. Urbain, and other anorthosites with particular reference to iron-titanium occurrences. Petrologists paid special attention to the ultramafic, or ultrabasic, intrusions because of their association with metallic minerals and asbestos.

In 1957 C. H. Smith, who had studied ultrabasic rocks in Newfoundland in the early fifties, began "a systematic study of the ultrabasic rocks of Canada to establish the reasons why deposits of chromite, asbestos, magnesia, nickel, etc., are usually associated with them."[8] Occurrences were investigated in Gaspé, at several localities in the Cordillera from the International Boundary nearly to the Arctic Circle west of Dawson, and in the Coppermine River district of the Precambrian Shield. In 1959 it was decided to study four widely-separated major plutons, each representing a different kind of magmatic material: the Bay of Islands complex in Newfoundland, the Mount Albert pluton in Gaspé, the Tulameen complex in British Columbia, and the Muskox Intrusion in the bend of Coppermine River in the Northwest Territories. Such detailed and comparative studies, it was hoped, would improve the understanding of such phenomena as the sequence and conditions of the differentiation of the several contained minerals, the forms each assumed, and their relationships to the masses of which they formed part; also of secondary factors that modified the mineralogy and geochemistry of the intrusions—metamorphism, absorption of materials with which they came in contact, the role of water in dissolving and distributing metallic components, or processes of alteration such as those that caused asbestos to form from serpentine. Such investigations involved careful measurements in the field, as well as detailed petrological, mineralogical, and chemical studies by specialists in several laboratories. Following such studies, mainly by D. C. Findlay, the Tulameen complex was found to be a combination of two separate "affinitive" magmas with some differing constituents, which were also affected by additional alkaline intrusions.

Most work was devoted to the investigation of the Muskox Intrusion, first discovered in 1956. This body was mapped geologically and geophysically between 1959 and 1962, and was investigated through extensive drilling in 1963 as part of Canada's contribution to the International Upper Mantle Program. The data and materials derived from this work have been studied in detail by T. N. Irvine, and the site was the subject of an excursion of the International Geological Congress in 1972. The

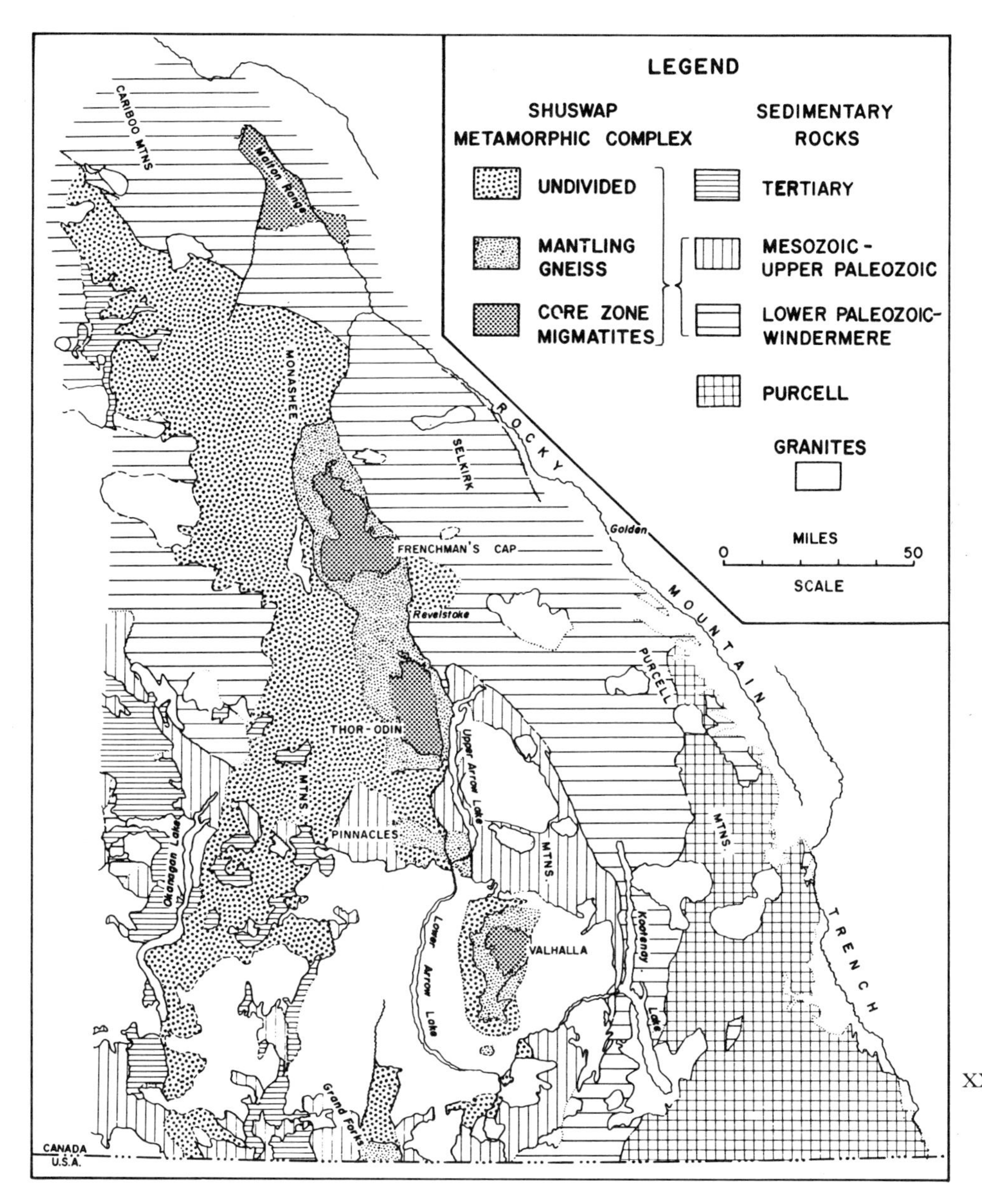

XXII-5 THE SHUSWAP METAMORPHIC COMPLEX, SOUTHEASTERN BRITISH COLUMBIA

geological mapping traced the exposed part of the body, which is 74 miles long with a maximum width of about seven miles, but later aeromagnetic and gravity investigations indicated that it has an additional subsurface length of at least 75 miles and broadens as it extends to the north. The selection of the occurrence for study under the International Upper Mantle Program led to the drilling of three holes to a maximum depth of 4,000 feet into and through the mass, with subsequent physical, chemical, and geophysical studies of the holes and cores. These and the later studies indicate that the intrusion was emplaced about 1,200 m.y. ago, in Proterozoic time, between a basement complex and the overlying Hornby sandstone, altering some of the rocks along the roof contact but leaving most of them unaltered, while the body of magma solidified into layers of basic and ultrabasic rocks. The Muskox Intrusion is an almost perfect case study in the emplacement of igneous rock, in that the basaltic magma cooled, solidified, and settled in a crucible-like reservoir with very little subsequent disturbance, by processes that can be duplicated in the laboratory; much has been learned about the general behaviour of such magmatic materials through the Muskox studies. Investigations have been made more recently on other ultrabasic bodies, near Gordon Lake, N.W.T., and Jennings River in northwestern British Columbia.

An equally ambitious study was begun in the same period on granitic rocks and their characteristic relations to their enclosing country rocks. Granitic rocks are among the most common of all igneous rocks, emplaced in all ages, occurring in almost every tectonic province, and extending over an estimated 1,400,000 square miles, or 40 per cent of the Canadian landmass. To learn as much as possible about the characteristics, behaviour, and relations of such rocks a special project, the granites of Canada, was begun by J. E. Reesor in 1957. After investigating several possible areas, Reesor chose the Valhalla, Thor-Odin and Pinnacles complexes in the Revelstoke district, B.C., for detailed study as a well-exposed area of deep-seated, closely interrelated granites, granitic gneisses, and mixtures of leuco-granitic material and highly metamorphosed sedimentary rocks. This combination is characteristic of granitic rocks in many parts of the world. The study included a number of granitic and gneissic domal complexes that were members of a single, well-defined metamorphic belt occurring in the core zone of the eastern Cordillera fold belt in the Monashee and Selkirk Mountains. The geologists had an excellent opportunity to observe how the individual bodies were related to a wide variety of settings as well as to one another. They hoped to learn more about the sequence and duration of the cycle of intrusion, and the relationship of the granitic emplacement and metamorphism to the larger processes of mountain-building. Detailed field mapping of the bodies and their enclosing rocks was required to reveal more about their origin and field relations. With a number of co-workers and students, Reesor investigated the Valhalla complex and other batholiths and domes forming part of the same general metamorphic, tectonic, and intrusive event.

XXII-6 J. E. REESOR

Extensive laboratory work was needed for the related mineralogical, chemical, X-ray, spectrographic and other tests of the samples brought in from the field. Indeed, the number of required analyses overtaxed the available facilities, and the project for a time had to be restricted to the pace of the laboratory determinations. Comparable studies have been made on granites in other parts of the country, on pegmatites (by R. Kretz), on swarms of diabase dykes (by W. F. Fahrig), on micas (by J. Y. H. Rimsaite), and on alkaline rocks. Despite their difficulty and expense, such projects can add to the understanding of the rocks and structures, and of the mineral deposits associated with them. Indeed, the Survey from time to time in a small way also investigated certain natural geological processes experimentally in the laboratory.

Somewhat related to these studies are other investigations of true and suspected meteoric craters, which are first discovered, as a rule, from studies of air photographs, or by field geologists in the course of areal mapping. The craters, even in the usual absence of any surviving meteorite material, are usually characterized by the generally circular shape of the craters and surrounded by rocks displaying the effects of shock metamorphism and fracturing. These physical features are not so easy to identify as one might think, because similar structures often result from magmatic intrusion from below. Indeed, there is a hypothesis that the two actions may be related—that sometimes the impact of a meteorite may have so weakened and fractured the crust as to have created a favourable locale for the intrusion of magmatic material. The shape and position of the ovate, saucer-shaped, magmatic nickel-copper intrusion near Sudbury, Ontario, has been seen by some as arising from the previous cracking of the crust there by a large meteorite. Because of the scientific and possible economic importance of such structures, they have received intensive study in recent years, notably by K. L. Currie.

Another category of fundamental studies deals with ore generation, a subject of enormous economic significance because of its value to the exploration geologist in improving prospecting and mining techniques. An example of a most effective and highly influential study of a particular kind of ore deposit is S. M. Roscoe's *Huronian*

Rocks and Uraniferous Conglomerates in the Canadian Shield (GSC Paper 68-40), which incorporates the results of studies in the Elliot Lake area since 1954.

Roscoe found that the uranium deposits originated as concentrations of uraninite in placer beds; therefore, their occurrence can be readily delimited within relatively narrow boundaries in space and time:

> Deposition of Huronian uraniferous conglomerates was controlled by the disposition of Huronian lava fields and by tectonic movements during sedimentation as well as by local and sourceward topographic features. . . . Uraninite and many other heavy minerals . . . are detrital. . . . The highest grade ores represent headward and medial concentration of the heaviest detrital grains within more extensive placers.[9]

He concluded also that the deposits were formed in very early time "when the earth's atmosphere contained no free oxygen," so that "similar important deposits may not be found in younger rocks."[10] He listed several key criteria that could be used to identify comparable deposits, chiefly, that the rocks should be of Aphebian age or older (2,000 m.y.) and should include "thick continental sandstone formations evidently derived from a weathered granitic source."[11] Such studies drew attention to drainage patterns and other paleogeographic aspects of Precambrian time, and yielded insights that were directly applicable to the investigation of similar uranium-bearing rocks in the Athabasca sandstone region of northern Saskatchewan. The theory that the lack of oxygen in the primitive Precambrian atmosphere permitted sulphide minerals to be preserved as detritus in placer deposits has also received wider application, as in its use to explain the sedimentary iron beds of the Labrador Trough and other deposits. Here again an element of selectivity, helpful to the prospector, is suggested by indicating that searches for sulphide minerals formed at the ancient land-surface should be directed to rocks of certain ages. These and similar principles governing the occurrence of certain rocks and minerals have been studied increasingly in line with the new approach that sees mineral deposition as an incident in the interaction between components of intruded and host rocks leading to the establishment of a stable equilibrium. R. W. Boyle carried out studies, which gained world acceptance, into the formation at Yellowknife of gold-bearing quartz from enclosing metamorphosed basalts and andesites, and on characteristics and locations of the resulting ore shoots. Also near Yellowknife, R. Kretz studied the "mobility of rock-forming elements under metamorphic conditions,"[12] especially the associations arising from the intrusion of pegmatites. At Goldfields (now Uranium

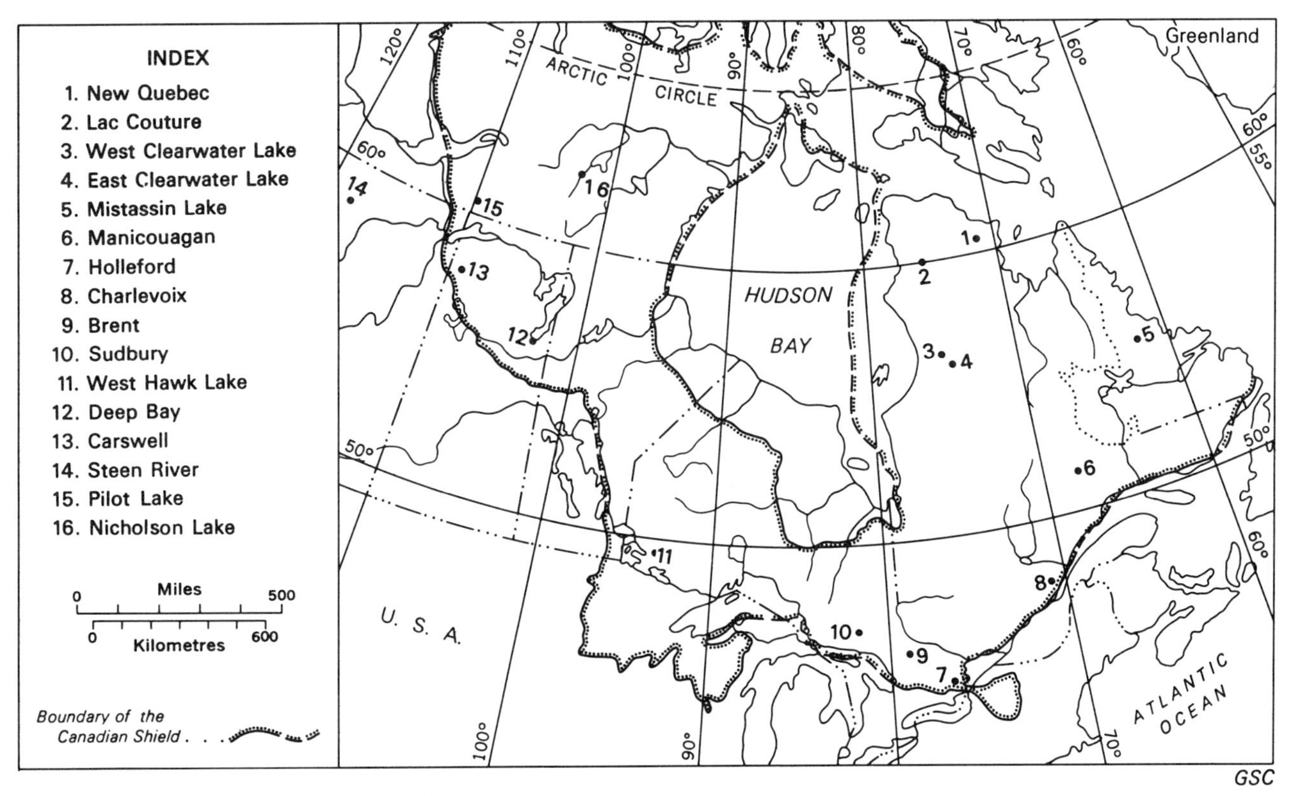

XXII-7 SOME SHOCK METAMORPHIC STRUCTURES IN THE CANADIAN SHIELD

City), Saskatchewan, K. R. Dawson examined the altered wall-rocks around known uranium deposits to determine whether characteristic types of alteration could be used as an aid in finding further deposits. Boyle carried out other outstanding field and laboratory researches on the geochemistry and origin of silver-lead-zinc deposits in the Yukon, base metals in New Brunswick, barite and base metals in Nova Scotia, and silver deposits of Cobalt, Ontario. He has won international recognition as an original thinker and an important contributor to the furtherance of geochemical studies.

In 1960 the opportunity arose to make a comprehensive study of ore deposition and its relations to its geological environment, a project suggested in 1957 by the NACRGS. The Coronation Mine, a new mine being developed on the Saskatchewan side of the Flin Flon district by the Hudson Bay Mining and Smelting Company was made available for scientific studies by university and government geologists under the general supervision of the Survey, with D. R. E. Whitmore as co-ordinator. It thus became possible to study the deposit from every point of view while it was being mined (1960-64), to work out the geological history of the entire deposit and its surroundings. The 26 studies included investigations of the geology, geochemistry, and petrology of the orebody and its setting, the tectonic history of the area, the structure of intruded rocks, and the effects of mining operations. After this "anatomical dissection" of an ore deposit, a similar study was begun on a copper deposit at Whales Back Pond, Newfoundland, the Survey being once again made responsible for the general oversight of the project. Structural geology and tectonophysics form a relatively new but increasingly important type of local and regional study. G. H. Eisbacher and H. U. Bielenstein are investigating the physical stresses associated with particular structures, the "remanent components of former tectonic stresses," such as the gradual release of stress in an Elliot Lake mine by the "slow regional arching over the last 1,000 m.y."[13] Similar studies have been undertaken of the tectonic frameworks of certain mountain-building events in Nova Scotia and British Columbia.

Another approach to understanding the processes of ore deposition and localization is the analysis of large numbers of phenomena with the object of discovering their general or common elements. Latterly the tools of statistical analysis and of computer technology have been applied to this task, although such studies continue to depend mainly on patient, careful field investigations of large numbers of deposits of a given metal. These analyses go back to the early years of the present century; and the work of T. L. Tanton, who carried on until 1954, provided a link with that period. Modern studies deal with relations between minerals and host rocks: conditions of deposition; how the components of the deposits were transported and emplaced, and perhaps subsequently metamorphosed, concentrated, or altered by physical or chemical agencies; the tectonic settings and the structural controls that localized or influenced the depositional processes. Such studies go far beyond making inventories of the resources, although this remains an important objective. They examine the deposits "genetically, relating them to one another and to the geological framework in which they are found, seeking knowledge that may be applied to improve the techniques of mineral exploration."[14] Since 1945, studies have been made of many elements: silver, titanium, nickel, niobium, tungsten, molybdenum, lithium, beryllium, copper, lead, zinc, and especially of the economically-important uranium and iron; also of coal, which is the special duty of the Coal Research Section under P. A. Hacquebard. G. A. Gross's massive three-volume report, *Geology of Iron Deposits in Canada* (Econ. Geol. Series, No. 22), which was published in 1965 and 1968 and may eventually reach five volumes, is the outstanding example of studies of specific mineral commodities.

Such studies have recently moved into investigations of the metallogeny of geological regions, which will ultimately cover the whole of Canada. As early as 1957, maps were prepared of the country-wide distribution of known occurrences of the elements uranium, beryllium, and niobium classified according to the geological features of typical deposits, as an aid to prospectors and others. These maps also provide a first step to working out the features of metallogenic provinces. Since 1966 a comprehensive metallogenic map of Canada has been in preparation, under the direction of G. B. Leech.

The great advances in metallogenic work are indicated by the chapters on the economic minerals of the Canadian Shield and the other major regions in the 1970 edition of *Geology and Economic Minerals of Canada.* They are demonstrated also in the chart "Metallogenic Study, Canadian Appalachians," prepared in 1965 by W. D. McCartney as an early attempt at correlating the numerous mineral occurrences with the tectonic setting of this structurally disturbed region of Canada.

Some of the major scientific advances since the Second World War have occurred in the field of historical geology, in working out a more accurate time-scale extending back into early Precambrian time, and in developing and applying a more satisfactory, comprehensive interpretation of earth history. This is an achievement in which many nations and individuals have shared, but the Canadian Survey has played an important part in determining the correct sequence of geological events

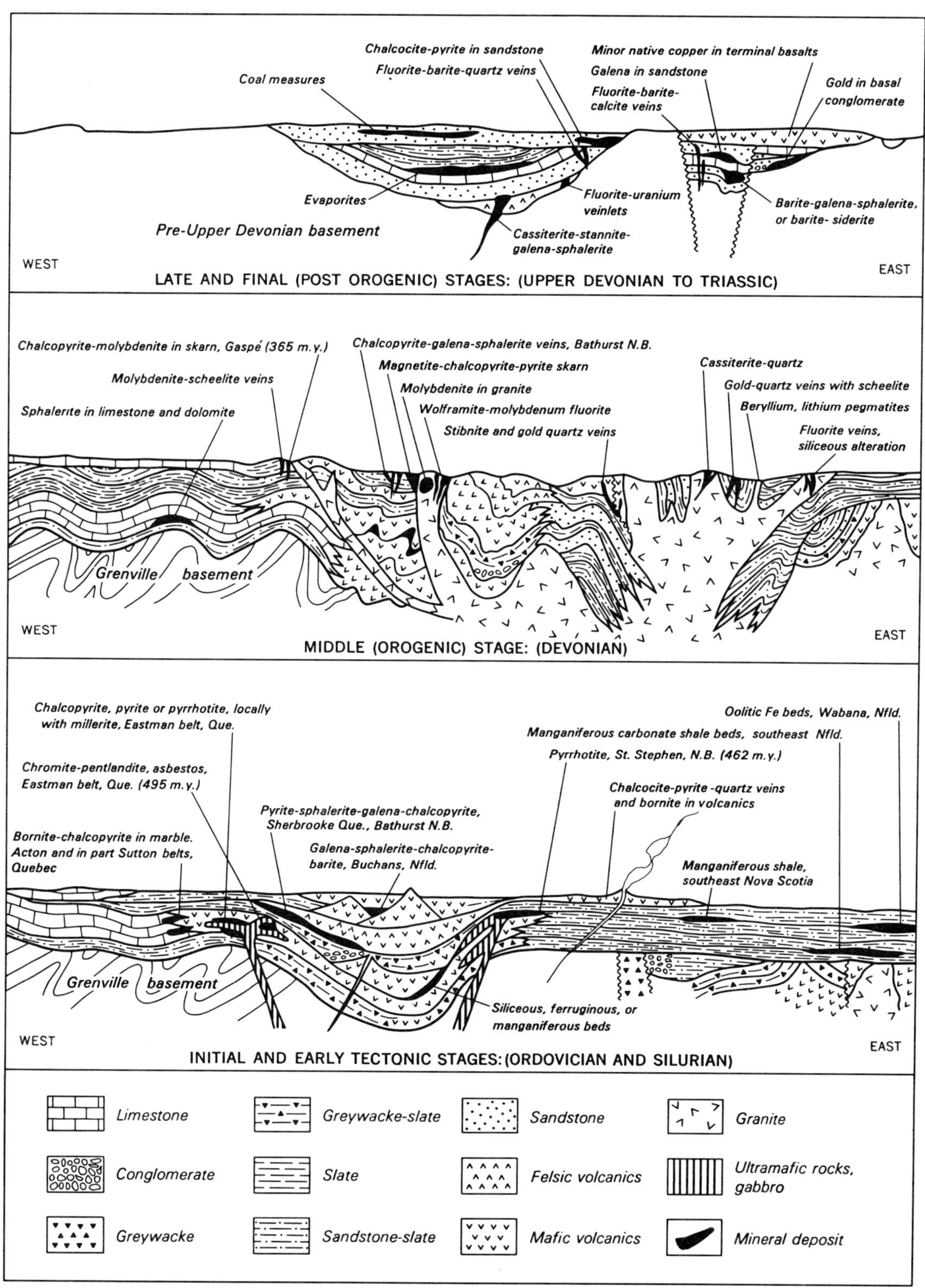

XXII-8 MODEL OF THE METALLOGENESIS OF CANADA'S APPALACHIAN REGION

XXII-9 C. H. STOCKWELL

XXII-10 R. J. W. DOUGLAS
Editor, *Geology and Economic Minerals of Canada*, fifth edition.

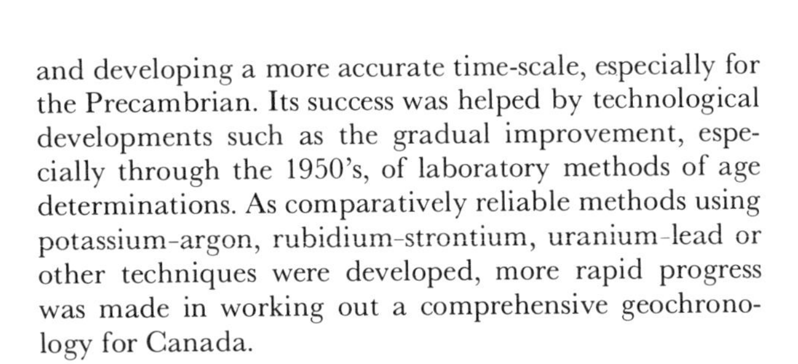

and developing a more accurate time-scale, especially for the Precambrian. Its success was helped by technological developments such as the gradual improvement, especially through the 1950's, of laboratory methods of age determinations. As comparatively reliable methods using potassium-argon, rubidium-strontium, uranium-lead or other techniques were developed, more rapid progress was made in working out a comprehensive geochronology for Canada.

The leader in this field was C. H. Stockwell, who had investigated many different parts of the Precambrian during his long career in the Survey. As a result of his broad first-hand knowledge of field relations of the rocks of the Canadian Shield, he was ideally fitted for checking laboratory age determinations against field conditions to ensure that they were consistent with the observable relationships. In his later years, Stockwell has devoted himself to collecting specially-selected samples from a wide variety of districts for testing and dating, as well as to investigating which laboratory methods furnish the most reliable results. He gradually established probable ages for the various regions and orogenic events of the Canadian Shield, and from them he derived a revised scheme of provinces, or regions, of the Precambrian.

The subdivision of the Precambrian into geological provinces had been attempted by other geologists from Logan onward, including Collins, Young, and M. E. Wilson, but they had based their conclusions mainly on stratigraphic and structural evidence, which was all that was available to them. By the 1960's, however, geologists could apply age determinations, and also the results of seismic and gravity measurements, to identify differences in crustal thicknesses that might mark the boundaries between blocks or plates of differing geological qualities. Stockwell used a wide combination of methods to secure results that became more accurate and more representative of the region as a whole. In 1961 he published a first report, *Structural Provinces, Orogenies, and Time-Classification of Rocks of the Canadian Precambrian Shield;* others of note include a Fourth Report in 1964, a supplementary article in 1968, and the important tectonic map of Canada in 1969.

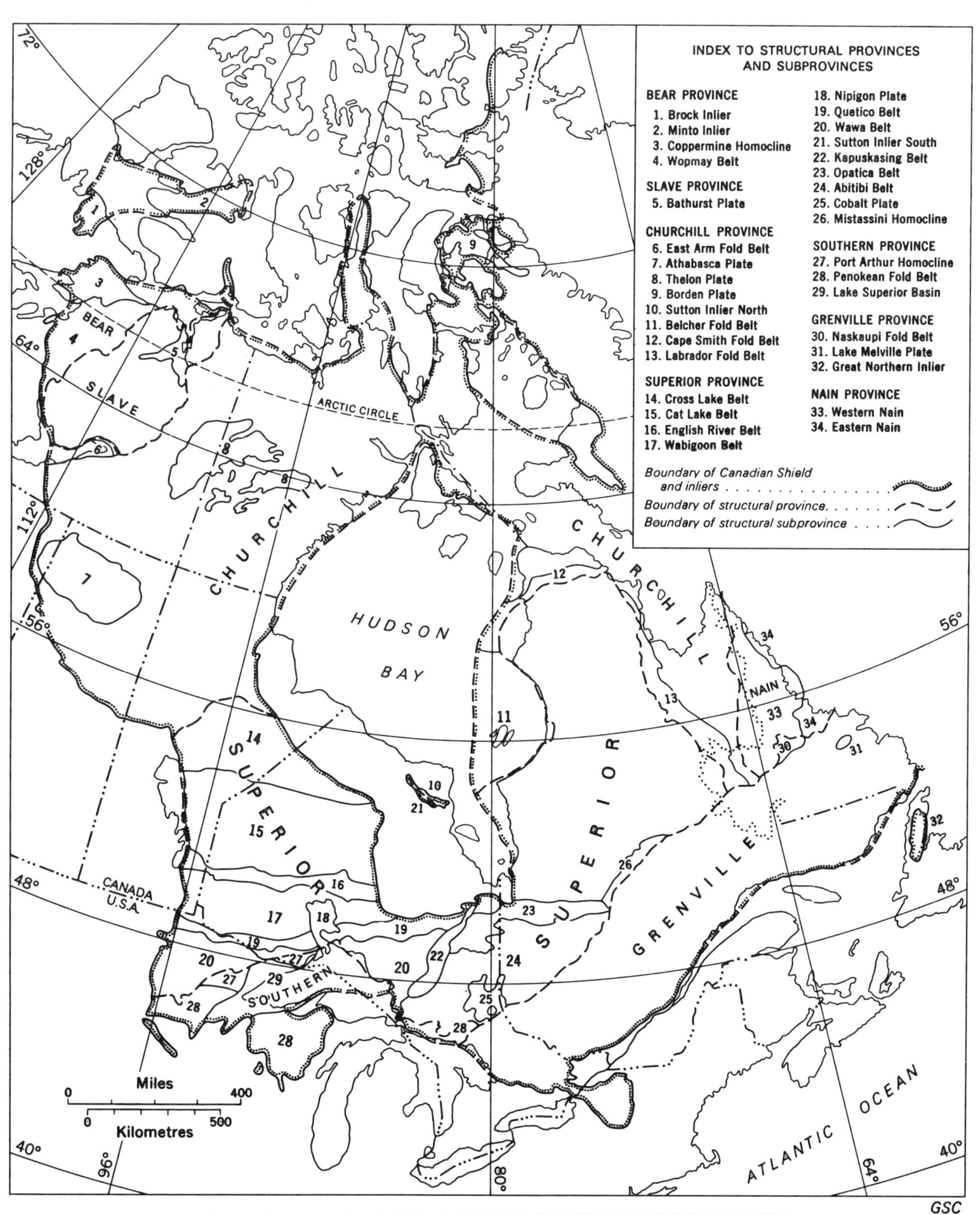

XXII-11 PROVINCES AND SUB-PROVINCES OF THE CANADIAN SHIELD (1970)

Stockwell in 1961 divided the Canadian Shield into three great provinces—Superior, Churchill, and Grenville—as earlier geologists had done on structural grounds. His division, based mainly on age determination results by the potassium-argon method, was generally according to the time of the major structural deformation, the introduction of granitic intrusive material, and the widespread metamorphism associated with the three main orogenic episodes, which occurred about 2,500 m.y. ago (Kenoran), 1,700 m.y. ago (Hudsonian), and 950 m.y. ago (Grenville). By 1964, based on the results from many other samples analyzed by more reliable methods, he raised the number of structural provinces to six, adding the Bear and Slave provinces (named from the great lakes of the Northwest Territories included in them) and the Nain province underlying the northern half of Labrador. At the same time, parts of these provinces were differentiated into 23 subprovinces. The Nain province was dated around 1,370 m.y., reflecting the effects of a hitherto-unrecognized orogenic event called Elsonian. The work of identifying areal units of the Canadian Shield has continued. The 1970 edition of *Geology and Economic Minerals of Canada* divides the Precambrian Shield into seven structural provinces (the seventh, the Southern province, is mainly in the United States and includes only a small part of Canada around Lake Superior) and 34 structural subprovinces, variously described as belts, plates, inliers, and homoclines.

The main thrust of Stockwell's work that puts him in a direct line with previous Survey officers was his development of an improved time-scale and a suggested new nomenclature for the Proterozoic. In his 1964 report he proposed replacement of the terms Lower, Middle, and Upper Proterozoic by Aphebian, Helikian, and Hadrynian, with the further subdivision of the Helikian into two sub-eras, Paleohelikian and Neohelikian. Archean was retained, undivided and undifferentiated, for the time being:

> The Archaean contains at least one unconformity over older granite but has not yet been formally subdivided because of difficulties in obtaining any clearly pre-Kenoran isotopic ages. Consequently, correlation of the units has not been possible and there is no point in making formal local sub-division until such time as correlation, which is the prime object of time-stratigraphic classification, becomes practical. Time-honoured methods of Precambrian correlation by means of lithological similarity, comparison of sequences, and relation to unconformities and intrusions are still useful where isotopic age methods fail.[15]

The same situation continued in 1968: "Profound unconformities are known within the Archean rocks but, because of the strong overprinting of the Kenoran orogeny it has not yet been possible to date them nor to subdivide the Archean on a time-stratigraphic basis."[16]

Stockwell's latest scheme (1968-69), with revisions in 1973, presented the following sequence of eras; each bounded by a major orogenic phase:

Eon	Era	Sub-Era	Orogeny	Mean Date (K-Ar ages)
Proterozoic	Hadrynian		(none)	
	Helikian	Neohelikian	Grenvillian	1,000-1,070 m.y.
		Paleohelikian	Elsonian	ca.1,400 m.y.
	Aphebian		Hudsonian	1,750-1,850 m.y.
Archean			Kenoran	2,540-2,690 m.y.[17]

No orogeny marked the close of the Hadrynian, the latest Precambrian era, which was followed directly by the Cambrian. Because he took the endings of orogenic periods as the boundaries for the eras, no hiatuses were left between them, though the dates for the eras remained rather uncertain because the orogenies extended over long periods of time. Age determination methods still had to allow for a margin of error, making it necessary to rely on mean, rather than absolute, results from as many samples as possible.

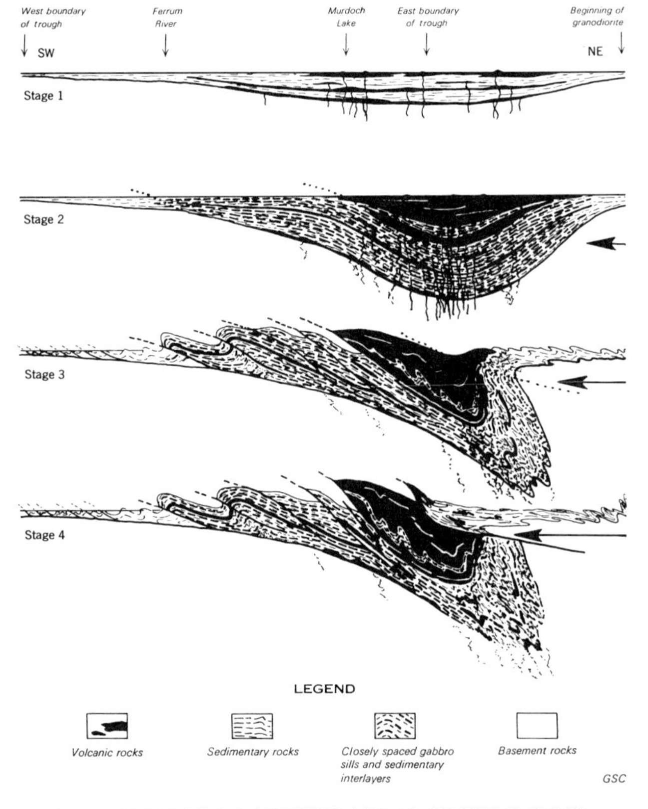

XXII-12 EVOLUTION OF THE LABRADOR FOLD BELT
A geosynclinal interpretation, by W. R. A. Baragar.

Since the structural provinces are mainly based on areas that were affected in a major way by the same orogenic movement, the provinces can also be approximately dated from the time of the last extensive orogeny that affected each of them, as follows:

Archean (Kenoran)–Superior and Slave provinces, Eastern Nain subprovince.
Aphebian (Hudsonian)–Churchill, Bear and Southern provinces.
Paleohelikian (Elsonian)–Western Nain subprovince.
Neohelikian (Grenvillian)–Grenville province.

Approximate absolute ages are now assigned to the historic groupings, or series, that the earlier field geologists had distinguished mainly from careful stratigraphic studies. For example, Stockwell (1968) assigned the following series to the Archean: the Keewatin (typical age, 2,550 m.y.) and Coutchiching (2,495 m.y.) of the southwestern district of the Superior province, and Abitibi (2,630 m.y.) and Timiskaming (2,475 m.y.) of the Timiskaming district of the province. Huronian, Animikie, and classical Grenville strata were assigned to the Aphebian; Lower and Middle Keweenawan to the Helikian; and Upper Keweenawan to the Hadrynian.

This progress in working out the ages and historical sequence of the Precambrian contributed to, and was aided by, advances in structural geology. As R. J. W. Douglas reported in 1970, the application of the principles of tectonics "has only in the last decade emerged . . . to its current position of being a vital, dynamic and essential part of the earth sciences."[18] Under the new view, the Precambrian Shield was seen as a collection of large, relatively stable blocks (cratons), each having a distinctive crustal thickness that sets it off from the bordering terrains. The interspaces between the blocks were much altered by the Kenoran, Hudsonian, or Grenvillian orogenies (hence the name of orogens for these zones), which were accompanied by volcanic activity, magmatic intrusions, faulting and folding, and horizontal and vertical displacements. These activities also affected the cratons, but to a smaller degree, by depositing materials on them, subjecting them to heat and chemical action that altered some of their parts, or imposing mechanical strains that gave rise to faulting and folding. This interpretation obviously has implications for prospecting and mining operations.

In the Tectonic Map of Canada, published in 1969 as Map 1251A, the work of identifying and dating orogenic episodes was extended into the Paleozoic and later periods, as follows:

Eon	Era	Orogeny	Mean Age (m.y.)	Region of Activity
Cenozoic	Quaternary			
	Tertiary	Late Laramide	50	Cordilleran, Arctic
Mesozoic		Early Laramide	80	Cordilleran
	Cretaceous	Late Columbian	100	Cordilleran
		Mid Columbian	110	Cordilleran
	Jurassic	Early Columbian	140	Cordilleran
		Nassian	165	Cordilleran
		Inklinian	190	Cordilleran
	Triassic	Tahltanian	210	Cordilleran
Paleozoic	Permian			
	Pennsylvanian (Carboniferous)	Melvillian	290	Arctic
		Late Appalachian	310	Atlantic, Arctic
	Mississippian (Carboniferous)	Early Appalachian	320	Atlantic, Arctic
		Ellesmerian	340	Arctic
	Devonian	Acadian	360	Atlantic, Arctic
	Silurian			
	Ordovician	Taconian	440	Atlantic
	Cambrian			
Proterozoic	Hadrynian	East Kootenay	730	Cordilleran

It may be noted that these disturbances, which took place after the close of the Precambrian, were associated with three major geological regions. The Cordilleran region experienced most of the later orogenic movements, especially in the Mesozoic and Cenozoic. Nearly all the tectonic activity in the Atlantic and Arctic regions, on the other hand, occurred during the Paleozoic. The Canadian Shield was now a stable area (craton), so the Phanerozoic strata deposited upon it in the St. Lawrence Lowland, Hudson Bay Lowland, and Interior Plains regions were never disturbed and remain flat lying. The western margin of the Plains region, however, was affected by the Late Laramide Orogeny, when some 150 miles of flat-lying sedimentary strata were crumpled and telescoped by the uplifting of the Rocky Mountains.

* * *

THE GRADUAL GROWTH OF KNOWLEDGE of the orogenic and tectonic events of Canadian geological history based on age determination and other work has in the last few years merged with another world current of scientific thought that already is having a profound effect on future studies. This new movement, to understand the processes of geological evolution through the ages, has since 1969 served to unite almost the entire geological spectrum along lines that would have delighted a Hutton or Lyell. Yet at first glance, the new view, a revival of a somewhat similar theory advanced half a century or more before, would seem to be the wildest of imaginings. How else can one describe a concept that the seemingly stable, permanent, unyielding earth's crust is actually in constant motion in both vertical and horizontal directions, that the amplitude of the horizontal movement can be measured in thousands of miles; also that continental masses split apart, drift over great distances, and afterwards may regroup as new continents?

The greatest tangible discovery of recent years, which arose from the rapid progress of geological mapping around the world and of improved dating of major geological periods and orogenic movements, was the finding and mapping of the Mid-Atlantic Ridge by many agencies from several countries, in which the Survey participated. Investigation since 1958 seems to indicate that magmatic material from the Upper Mantle or asthenosphere (mainly basalt) is being pushed up along a rift at the crest of the ridge, moving the walls outward and spreading basaltic matter along the adjoining ocean floor. Studies of changes in the direction of magnetization of the rocks in successive belts on either side of the rift show that the process has been long and continuous, causing the crust to spread out from the rift, mirror-image fashion, at rates of but a few centimetres a year, but large enough in total to account for the entire width of the Atlantic Ocean. The regular increases in the ages of many islands in proportion to their distances from the mid-oceanic ridge are cited as pointing to the same conclusion. So does the geology of Iceland, which stands squarely on the Mid-Atlantic Ridge. Much of the centre of the island is a wide *graben* lined with young basaltic matter from the rift.

Mid-ocean rift spreading is considered to account for the notion of continental drift, explaining the parallelism between the opposite coasts of the Atlantic. Similar rifts, sometimes interconnected, have been found in virtually every ocean and sea and in many major bays and gulfs, and are regarded as being responsible for the separation of similar landmasses on either side. Such rifts cross the Indian Ocean, run along the eastern margin of the Pacific, and extend from the Jordan valley via the Red Sea down East Africa. Also part of the East Pacific rift traverses the adjoining continental mass in western Mexico, California, and again in southern Yukon and Alaska.

What causes the phenomenon of mid-ocean rift spreading, and how does the mechanism operate? Much research and thought have gone into this question over the past decade. Scientists have related the profound crustal movements to chemical and physical forces that alter the volumes of the mantle rocks, causing mantle material to well up to the surface and spread out there. The great Russian scientist V. V. Beloussov in 1963 attributed the formation of the primary deep faults to the uneven expansion of the interior of the earth under the influence of radioactive heating.[19] Some scholars have concluded that the earth was expanding, but more recent workers (including the Canadian university professor J. Tuzo Wilson) maintain that the size of the earth remains constant, and that matter added to the lithosphere (crust) is offset by other matter returned to the Upper Mantle.

This exchange, which is also a heat transfer mechanism, is visualized as being effected by an immensely slow-moving convection current that carries matter upward to the crust through the rifts, causing those to spread apart, and draws other crustal matter back into the asthenosphere. The matter extruded at the mid-oceanic ridges forms large crustal flows, or plates, that are pushed or drawn outward by the currents. When such plates collide with larger continental masses, various results may occur—the crumpling of the continental margin into ridges, the depositing of matter from the oceanic plate on the continental margin, or the overriding of the coast by part of the plate. Most often such plates are downfolded beneath the edge of the greater continental block into the magma of the asthenosphere. The downfolding produces a trench along the edges of the contact between the two blocks, which receives sediments subsequently eroded from the continent. Light, more volatile material from the depressed part of the downfolded plate can merge with the continental margin, thickening the continental block and causing it to be uplifted. This frequently gives rise to volcanic activity and other disturbances along the margin of the continent. As it loses its lighter elements, the depressed plate descends deeper into the magma, reaching, according to seismic evidence, depths of as much as 700 kilometres. The Survey has contributed to the working out of this hypothesis by staff participation in a 1964 conference of the Royal Society of Canada on continental drift; and especially by sponsoring, organizing, participating in, and publishing the findings of three important special symposia convened by the International Upper Mantle Committee of the International Unions of Geological Sciences and of Geodesy and Geophysics in Ottawa in September 1965. These symposia brought together leading world authorities to discuss "The World Rift System," "Continental Margins and Island Arcs," and "Drilling for Scientific Purposes."

The margins of Canada have been relatively stable long enough for erosive and other forces to modify the results of plate tectonic activity, but the almost contemporary results of such movements can be studied elsewhere in the world. Along the southwestern coast of the United States during the past 30 m.y. a group of plates moving in a northwesterly direction parallel to the San Andreas Fault have opened the Gulf of California and have regrouped as part of the state of California. During that time individual plates are believed to have moved as much as 650 miles, and to have completely changed positions relative to one another. The process is still under way, accounting for the earthquakes and upheavals of that part of California. A contemporary subduction zone (or "sink area") of downward-moving plates is located along the western margin of the Pacific Ocean beneath geologically-active island arcs extending from the Philippines to the Aleutians.

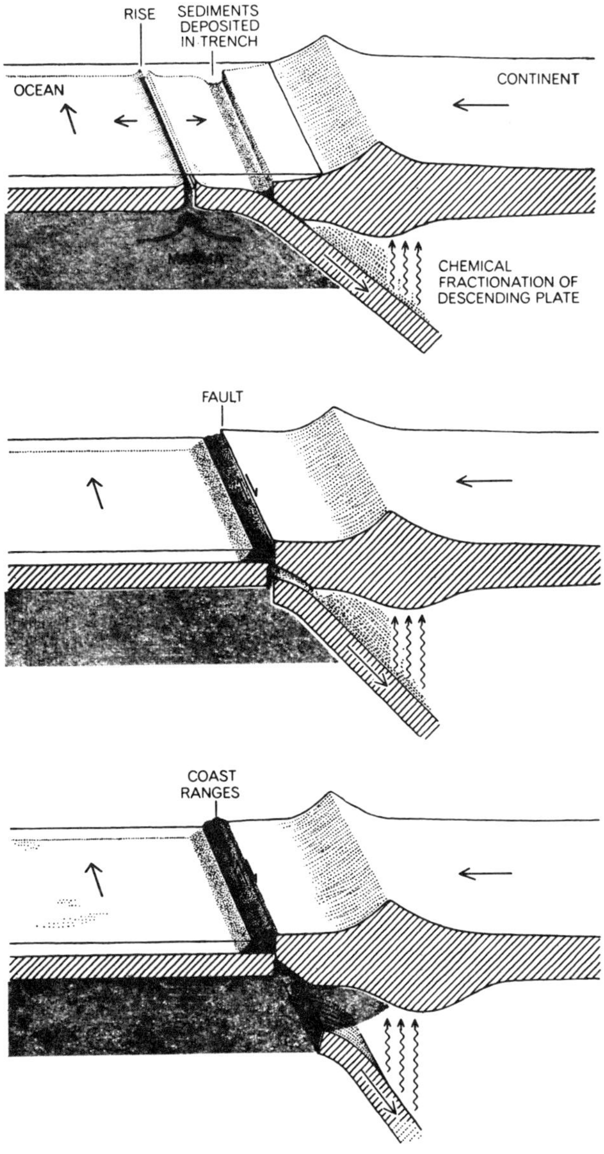

XXII-13 A PLATE TECTONIC MODEL
These diagrams, from Don L. Anderson's article, "The San Andreas Fault" in the November 1971 issue of *Scientific American,* illustrate the successive interactions between an outward-spreading plate of oceanic crust and the coastal margin of a larger continental block.

In such fashion, it is contended, during a period of 210 m.y. a single great landmass divided and spread out to take up the configurations of the present continents. At the beginning of the Mesozoic a rift began to develop that became the Atlantic Ocean; in Tertiary time, about 60 m.y. ago, a new rift was started that became the Indian Ocean, separating Africa, Australasia, Antarctica, and Asia from one another. If such enormous changes could occur in a comparatively short and recent (geologically speaking) interval of earth history, how many earlier movements may have happened in the preceding two or three billion years since the oldest rocks of the continental blocks began forming early in Precambrian time, at least 3,000 m.y. ago? The answer may be found within the rocks of the Canadian Shield, when we can read them.

Survey geologists are engaged in working out new interpretations of the geological evolution of the more recently affected parts of Canada, primarily the Atlantic and Cordilleran regions, in the light of these new concepts. The plate tectonic approach provides the most satisfying and complete explanation for the evolution of these coastal areas, one that goes a long way towards clearing up many unanswered questions. As stated in a recent paper by J. W. H. Monger, J. G. Souther, and H. Gabrielse, on the "Evolution of the Canadian Cordillera: A Plate Tectonic Model,"

> Generally the paleogeography and tectonic evolution of the North American Cordillera have been discussed within the context of geosynclinal theory .... However, the Cordillera remained a unique type of geosyncline by comparison with the classical geosynclines of the Appalachians and Alps. It is unique in the sense of its long history that includes several major orogenies each of which adds to the final picture. Because of this, geosynclinal models of the Cordillera resembled a rather confused listing of events with no unifying thread running through them .... Not until the advent of the concepts of crustal plate tectonics was any unified picture possible.[20]

The authors' detailed explanation of the evolution of the Cordilleran region is too complicated to outline here, involving as it does five distinct zones (Insular Belt, Coast Plutonic Complex, Intermontane Belt, Omineca Crystalline Belt, and Rocky Mountain Belt) that have been affected over time by recurring cycles of mid-oceanic spreading ridges, outward-moving oceanic crustal plates,

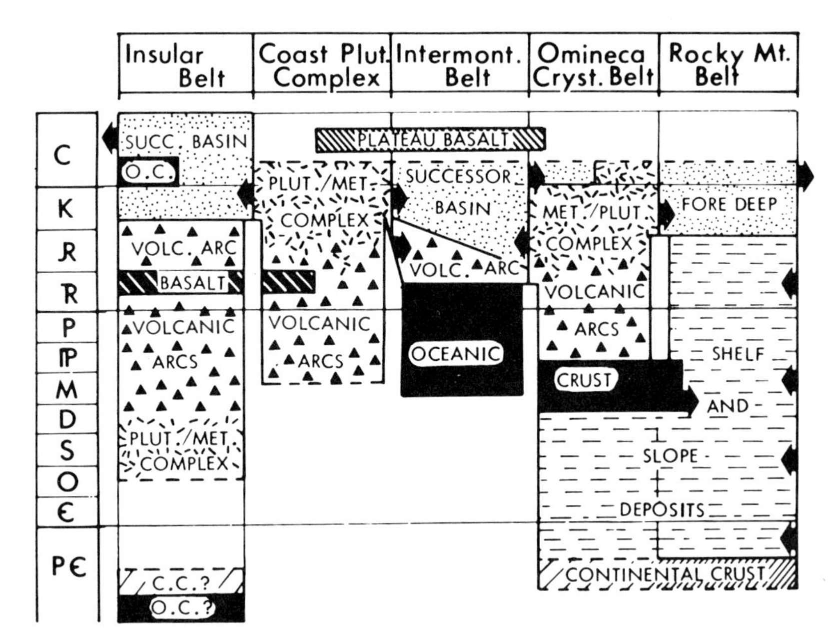

XXII-14 THE TECTONIC HISTORY OF THE CANADIAN CORDILLERA
This interpretation of the space and time distribution of lithological assemblages within the various belts of the Canadian Cordillera was prepared as a tabular summary of the article "Evolution of the Canadian Cordillera: A Plate Tectonic Model," published in the Summer 1972 issue of *American Journal of Science*. The time scale at the left extends from Precambrian to Cenozoic, with Carboniferous separated into Mississippian and Pennsylvanian.

and subduction movements. These have generated ridge-like structures, oceanic-type rocks, areas of tectonic thickening, island arcs, widespread magmatic intrusions, metamorphic zones, as well as extensive and frequently-repeated structural effects such as overthrusting, block faulting, and gravity slides. The abstract for the paper provides a short summary:

> The Canadian Cordillera is presumed to have been initiated by an episode of rifting in the mid Proterozoic. Subsequently a miogeoclinal wedge developed until late Devonian time. During late Paleozoic and later time complex interaction between the continental margin and a Pacific plate resulted in large scale overthrusting of oceanic crust in the early Mississippian, development of island arcs and a deep metamorphic root zone during the early Mesozoic and finally rebound of the continental margin accompanied by block faulting and emplacement of granitic rocks during the late Mesozoic and Cenozoic.[21]

The results for the five zones are shown diagrammatically.

Somewhat less complicated is the geological evolution of the Atlantic region of Canada as interpreted in terms of the plate tectonic concepts by W. H. Poole. In late Precambrian time a continental crust was fractured along what is now the northwestern edge of the Appalachians, and the two continental blocks were driven apart by a spreading mid-ocean ridge. Both parts continued to evolve separately through erosion, developing sedimentary aprons along their margins and adjoining ocean floors, as well as distinctive faunas that reached their most marked divergence during the Cambrian period.

Later, in Cambro-Ordovician time, a subduction zone developed along the North American edge of the break, and an island arc formed above it. Basaltic matter and other intruded material being generated at the mid-ocean ridge in the centre of the proto-Atlantic Ocean was carried to this coast by an oceanic plate, which on reaching the subduction zone plastered basalts, ultramafic and gabbroic rocks, and deep-sea cherts against the island arc before being drawn downward into the asthenosphere. During this period of disturbance (the Taconian Orogeny), the volcanic, ultramafic belt of the Eastern Townships was formed and received all or most of its deformation, while uplifted areas spread clastic sediments over parts of the original continental platform for a considerable distance inland. The orogeny also produced gravity slide masses (allochthons or *klippen*) that extend discontinuously from the Taconic region of New York to the northern tip of Newfoundland. The leading edge of these masses, first traced around Quebec City and in the Eastern Townships by Logan, bears his name—Logan's Line.

In the Devonian the two original continental masses again approached one another, driving out the proto-Atlantic Ocean and compressing together the sedimentary and volcanic rocks that had formed along their facing margins. Almost the whole area to Logan's Line was deformed during this Acadian Orogeny, becoming highly metamorphosed and intruded by granite batholiths. The entire region was uplifted and the land was eroded. A later fracture (the Cabot Fault zone) developed from the Bay of Fundy to White Bay, Newfoundland, and Carboniferous continental sediments, together with a short interval of marine carbonates and evaporites, were deposited within depressed fault-bounded basins.

Finally, in the Triassic, when the new and present Atlantic Ocean cycle began, the fracture line along which the present mid-Atlantic Ridge developed split the eastern continental block, and left its westernmost part joined to the westward-moving North American block. The rocks of present-day southern New Brunswick, all of Nova Scotia, plus eastern Newfoundland, were attached to the American continent, and the hitherto distinctive faunas of the region (Atlantic Realm) were brought into line with those of the rest of the continent (Pacific Realm). Thus the tectonic evolution of the Atlantic region of Canada was completed more than 200 m.y. ago to give the district its basic present-day structure.

The plate tectonic interpretation gives the geologist new insight in his task of indicating the best locations to search for workable mineral deposits. As its great Canadian advocate, J. Tuzo Wilson, has observed,

> The replacement of ideas about a relatively stable earth with fixed continents by ideas about a highly mobile one with moving continents will profoundly affect many ideas about the origin, sources and distribution of ore-bodies and petroleum deposits .... A greater value lies in developing a proper framework and mechanism to aid in ore prediction and herein the new ideas have untold value for the future.[22]

The concept of plate tectonics and continental drift supplies fresh knowledge on how geological regions evolved, what elements have gone into them, and from what source they are derived. It may indicate regions now separated by thousands of miles that were linked to one another in the past, and features which attract interest in one area may now be sought out in the other. Yet it is no substitute for local field studies and fundamental investigations of the sorts indicated earlier, even though it may change their focus and direction.

The recent rise of the plate tectonic interpretation marks a most significant advance in the work of the Geological Survey of Canada. For more than a century data were accumulated mainly through field studies and mapping, small bits at a time, for limited areas, rather like seemingly unrelated pieces of a large, complicated, highly deceptive jigsaw puzzle. Today, when most pieces of the puzzle have received some study, the geologist has suddenly reached a position of being able to arrange them in order and recognize previously undetected broad patterns. These may show special features of metallogenic significance or the course of general geological development. Undoubtedly the plate tectonic approach, although it seems to meet most questions raised thus far, will need much revision in essentials as well as details as a result of future findings. Nevertheless, as they proceed with their regular studies, geologists in Canada, and elsewhere, will hereafter be able to throw light on the broader meanings of local and regional phenomena in the context of the ever-changing earth.

XXIII-1 THE ROCK MAP OF CANADA
As its contribution to the Canadian Pavilion at the 1967 World's Fair in Montreal (EXPO), the Survey prepared this large map of Canada (approximately 32 feet, or 10 metres in width) on which the areas corresponding to various major geological formations were represented by their characteristic rocks. The exhibit attracted much attention, with interest at recognizing rocks from one's home locality being a frequent reaction.

23

Looking Beyond Canada's Boundaries

As THE SCIENCE OF GEOLOGY becomes more and more involved with global concepts, it is not surprising that the Survey finds itself caught up to an unprecedented degree in international projects and organizations. The wide diffusion of rapidly-advancing scientific knowledge and technology makes it vital for specialists to be continually in touch with the latest developments in their sciences, not merely through the literature but through personal contacts and discussions with their counterparts in other lands. Because of the desirability of establishing common standards of geological reporting and mapping to facilitate the preparation of world maps, Survey officers participate in the bodies that work out these rules. Besides, decisions reached there may vitally affect the Survey's operations and its priorities, even in respect of its Canadian work. Moreover, international collaboration offers mankind the means of achieving mighty accomplishments—to study the beds of the oceans, to probe the earth's mantle, to scan the earth by orbiting satellites, and to investigate geology beyond our own planet. The desire to share in these and other dazzling opportunities encourages the drawing together of scientists of many nations, commonly transcending even political differences.

Such developments occurred at a most opportune time in the Survey's history. By the mid-1950's a generation of young specialists trained in the most advanced methods was coming to the fore at the Survey. These men were keenly aware of the dramatic explosion of scientific knowledge all around them, and were confident of their ability to hold their own in international company. They belonged to the wartime generation, and they felt the stirrings of national identity aroused by Canadian achievements in world affairs. They were also members of a generation that had seen the shining ideal of international co-operation take shape in the United Nations Organization and its affiliated agencies, and they felt the call to serve humanity. The large, bright, new building containing the products of the latest technology seemed to invite the Survey to parade its possessions and its accomplishments before the world of science. Above all, the direction of the Survey was in the hands of James M. Harrison, himself one of the new generation, who recognized the value of gaining for it a proper position in the hierarchy of scientific organizations. As director from 1956 to 1964, Harrison used the Survey's stable financial position, its new facilities, and its superb staff of specialists, to pursue a policy of *la gloire,* of accepting international involvements and responsibilities that would inspire the staff to still greater heights of accomplishment and, as in the days of Logan and Brock, would again bring distinction and honour to the Survey.

There were also strong practical and emotional considerations urging the Survey to share part of its knowledge and experience with other, less fortunate, countries. A powerful social and humanitarian concern had sprung up for the overcrowded, insecure, technologically backward peoples of the 'Third World.' It was recognized that the most practical way of overcoming their great need was to bring about the rapid and large-scale economic transformation of their lands—for which no better method exists than by discovering and developing mineral resources. The Survey has almost limitless opportunities to put its facilities and personnel at the service of these nations to show them how to make most effective use of their resources. Such assistance, properly applied, can make important tangible contributions towards a more prosperous, and hopefully, a more peaceful world.

The Survey was relatively slow to embark on external operations, particularly in sending its staff members on foreign service assignments as technical advisers. For many years it lacked the money, the manpower, and perhaps, the will to become involved in such work. Moreover, such activity was not encouraged by the government, the department, or even by the management of the Survey. Such inaction reflected Canadians' rather limited interest in technical assistance as opposed to trade. While fading and rising imperialist powers vied with each other to gain customers, friends, and allies through generous foreign aid programs, Canada kept a low profile.

Even after the Survey had begun to supply technical assistance, the scale of its operations was held back by financial restraints and reluctance to release key personnel for extended tours of duty abroad. Besides, any requests for assistance in the field of geology and mining that came to the federal government were funneled principally to the Mineral Resources Division of the Mines Branch (now the Mineral Resources Branch of the Department of Energy, Mines and Resources). The Survey was called on for help only when that division judged that its services were required. Thus the Survey's part was essentially a passive one of supporting the plans and efforts of other government agencies, and participating in programs that had been arranged without reference to it. Often the Survey was called in belatedly and with little warning to meet difficult situations. What was worse, there was generally no provision for the continuing association of the Survey with the projects in which it had become involved.

Because the Survey's contributions were irregular and lacked continuity, they were kept on an *ad hoc* basis, and no consistent approach or organization was developed for the work. The role of supplying specialized technical advice to foreign government clients on demand was further limited by shortages of staff, even though the budget eventually was slightly increased to allow for small-scale participation in these activities. All in all, the Survey was greatly hampered from rendering the prompt or effective service that might have been desirable; or from giving, and receiving, the benefits that could have resulted from an active, regular, positive technical assistance commitment.

Still, despite the difficulties outlined, the Survey's external aid reached considerable proportions by the mid-sixties. Projects that initially were undertaken mainly in connection with the Colombo Plan or United Nations special agencies, in later years were more commonly accepted at the behest of the federal government's own External Aid Office and its current successor, the Canadian International Development Agency (CIDA). The list of activities in 1960 and the years that followed shows a remarkable increase in the number and variety of functions in which the Survey was invited to share. In 1960 L. W. Morley spent a month in Bangkok at the headquarters of the UN's Economic Commission for Asia and the Far East at a UN seminar on aerial surveying methods and equipment, while R. T. D. Wickenden was loaned for a year's work on micropaleontology in India under the Colombo Plan. In 1961 Morley presented a course in aerogeophysics at Tokyo, on behalf of the same agency, which attracted 32 representatives from six Asian countries.

In 1962 A. F. Gregory spent three months in Nigeria under the auspices of the Colombo Plan, preparing recommendations for an aeromagnetic survey program that were afterwards adopted. A geological and economic investigation by G. A. Gross, the iron ore specialist, was also begun in 1962, to advise the government of British Guiana (now Guyana) on the possibility of establishing an iron and steel industry. In 1963, on behalf of the UN Special Fund, Y. O. Fortier evaluated a mineral exploration program submitted by Tunisia, while a more unconventional activity was the participation of the Quaternary geologist R. J. Fulton in a celebrated UNESCO archeological program in conjunction with the Aswan High Dam project in the United Arab Republic. In 1963, too, C. S. Lord began the first of a remarkable series of visits to Southeast Asia and Africa to inspect and help arrange the detailed follow-up studies arising from a Ford Foundation economic study of part of Malaysia.

The year 1964 saw the number of operations further increased and the involvement of yet another important part of the Survey, the Geochemistry Section. R. H. C. Holman conducted a geochemical survey program in Venezuela useful as a prototype for future work, the emphasis being on setting up field laboratories in test locations, working out procedures, and training local geologists, chemists, and technicians to operate geochemical surveys in their own country. In the same year, the frequently-consulted Geophysics Division was represented abroad by Gregory in Nigeria and Kenya, and by A. S. MacLaren in Madagascar; also by L. S. Collett, who attended a meeting in Addis Ababa of the Economic Commission for Africa as a special consultant on a plan to set up regional training centres in aerial surveying methods. Gross, again on behalf of a UN agency, spent two months studying and reporting on iron and manganese deposits in Ceylon and India. Lord and a mining geologist from Kirkland Lake, G. H. Charlewood, visited India, Uganda, and Kenya at the request of the External Aid Office, to help organize mineral exploration programs. They recommended areas to be investigated and the methods, staffing, and equipment that would be required. Equally varied kinds of service were supplied in subsequent years to several emerging nations of Africa, Asia, and the Commonwealth.

The frequent requests for assistance with geophysical surveys were undoubtedly a reflection of the favourable reputations of the Survey and of the Canadian firms specializing in that field. In 1966 and 1967 Morley was appointed by the UN Economic Commission for Africa to a three-man committee that inspected proposed regional centres at which local personnel were to be trained in photogrammetry and aerogeophysics. Moreover, the Geophysics Division, taking advantage of an important and widely-attended conference on mining and groundwater geophysics being held in 1967 at Niagara Falls, mounted a two-week seminar at Minden, Ontario. The program, conducted by the staff of the division, trained scientists from developing countries to use the most modern equipment and familiarized them with the latest techniques. This single program did much to enhance the credit of the Canadian industry and the Survey in a number of countries. In the following years, A. Becker, L. P. Tremblay, A. S. MacLaren, E. E. Ready, and A. Larochelle visited Niger, the Cameroons, Ethiopia, and Upper Volta to help the governments of those countries with their geophysical surveying programs and inspect contractors' work. In 1970 another group made a seismic study in Guyana and worked out the details for an aeromagnetic survey, for which they drafted the tender and then supervised the contractor's performance in executing the survey.

The Survey was often consulted for advice on reconnaissance and detailed surveys to identify areas likely to contain exploitable mineral resources. In this, it was represented abroad by such men as D. R. E. Whitmore (1965), K. E. Eade (1970), L. P. Tremblay (1970), and especially by the chief geologist, C. S. Lord. Lord's work was principally in Asia—in Malaysia (1963), Malaysia and Thailand (1967, 1968), India (1964, 1968), and Burma (1971)—interspersed with visits in 1964 and 1966 to Kenya, Uganda, and Tanzania. His advice was sought on a wide range of activities from preliminary reconnaissances to detailed studies of limited areas for deposits of economic minerals. The assignments tested his long experience in the field and in directing and evaluating every type of field operation. The visits to Malaysia and Thailand were in connection with the mineral resource assessment aspect of a land classification program in western Malaysia, and included follow-up examinations of the most encouraging districts. A helicopter survey was recommended for a mountainous, little-known but promising area; for other districts, field mapping at various scales, supplemented by geochemical and geophysical surveys.

In Burma it was a question of deciding which one of more than a dozen proposals Canada should sponsor and assist. The decision reached was to try to locate ore reserves for a mining plant and townsite in the Mandalay district that had been built without adequate advance planning. A two-phase exploration program was recommended: first a qualified geologist should work out the controls that govern the occurrence of base metals in the locality; then there should follow geochemical and ground geophysical surveys of the favourable areas indicated by that regional study, supplemented by diamond drilling. The visits to East Africa also were concerned with extending airborne geophysical surveys to new areas and planning the follow-up stages of more concentrated work. In both regions the work of other Canadian geological and mining personnel serving under other programs of external aid was also inspected. Lord's visits to India called for a different sort of expertise. Lord and Charlewood were to help decide the most effective way to implement a $10,000,000 Canadian government grant for promoting the exploitation of the mineral resources of India. The major drilling program originally contemplated was eventually exchanged for one that concentrated on ground and airborne geophysical surveys; but because of political difficulties neither program was carried out.

Another aspect of external aid can be observed in the activities of G. A. Gross. His work in Guyana (1962) and India-Ceylon (1964) was comparable with Lord's in bringing the knowledge and experience of an unbiased expert to assist governments with their mineral development programs. However, most of his overseas duties were on the level of scientific and technical exchanges with specialists in his own field. Working chiefly for international organizations (notably the UN Technical Assistance Fund), he went from Guyana to Brazil, Peru, and Venezuela where he examined iron ore deposits and mining operations. Later, when he travelled to the U.S.S.R. in 1966, he inspected ten research institutes and two major iron mining regions (Krivoy Rog and Novosibirsk). He also visited iron ore research institutes and industries in Germany and France during 1967. The knowledge gathered from these visits, along with his experience of Canadian iron deposits, was valuable for his work as a member of an eight-man panel updating the United Nations' *Survey of World Iron Ore Resources.* Gross wrote three chapters and took part in the concluding conference at Geneva (1967).

XXIII-2 G. A. GROSS

Such visits were a two-way exchange; while Gross gave his hosts the benefit of his experience and reflections, his own understanding was broadened, as well as his perspectives on the world situation. The Canadian terrain does not offer many opportunities for observing the effects on local geology and mineral deposition of desert, humid or tropical weathering conditions, however much such climatic conditions may have contributed in the past to shaping Canada's present geology. Gross' visits to South America, Asia, the U.S.S.R., and western Europe gave him new insight into comparable Canadian deposits, experience with technical processes that Canadian industry could copy or avoid, and ideas as to the competitiveness of different grades of Canadian iron ores with those from other parts of the world. They increased his effectiveness as a scientist, as a specialist in economic aspects of iron mining, and as an expert in handling inquiries from industry or visiting geologists. He also made important contacts with foreign scientists, industrialists, and government officials that heightened the Survey's reputation abroad and aided the progress of the iron mining industry in Canada.

Other Survey personnel have had similar experiences. The eminent geochemist R. W. Boyle, and P. A. Hacquebard, the coal specialist, were both chosen, as was Gross, by the National Research Council to visit Russia as part of its scientific exchange program with the Soviet Academy of Science in the 1960's. Three more Survey scientists—J. W. Kerr, J. E. Reesor, and E. T. Tozer—were similarly honoured during the decade. Through contacts with different geological environments and with specialists in the same subject-areas whose experience differed from their own, they were able to make their own work more effective. So well was this recognized that regular exchanges of staff were instituted with the geological surveys of other lands, notably Australia and Sweden, while other officers were given extended leaves to accept assignments to various Commonwealth or United Nations agencies.

The urgent need of the underdeveloped countries to put their mineral resources to good use placed much stress on scientific and technical upgrading of local personnel. From his wide experience, Lord concluded that a major purpose in all Canadian programs should be the training of local staffs, and that more attention should be devoted to it. The government agencies in the countries being aided should be assisted in ways that would strengthen their authority, and that would build up the confidence of their men as well as raise their proficiency in the performance of their duties. He recommended lending Canadian geologists for limited terms to speed up essential mapping and publication work until the national surveys were in a position to meet the needs of their own lands. Specialists in fields such as geochemistry, engineering geology, or groundwater geology should train the local scientists and familiarize them with the instruments, procedures, and techniques best suited to the local conditions. In some situations special types of skilled tradesmen, such as drilling crews and foremen, could usefully be sent out to help train local technicians, mechanics, and workmen.

The administration of the programs was a source of further difficulties. Official proposals from foreign governments needed to be processed more quickly and decisively by the responsible agencies in Ottawa so the applicant countries could make definite plans as early as possible. For its part, the Survey should remain in contact with the programs with which it was associated until they were completed. Administration of the programs should permit constant oversight and appraisal by Ottawa, yet be flexible enough to meet any changing conditions in the country where the work was being done. It was also found that Canadian staffs abroad sometimes lived and worked under unsatisfactory conditions because of inadequate local arrangements. Distinctively Canadian-sponsored and Canadian-administered programs should be chosen wherever possible, as being less likely to cause friction in the field and more likely to serve Canadian interests best.

Most of these views, founded on experience in the field, were echoed in the Science Council of Canada's special study, *Earth Sciences Serving the Nation:*

> A realistic target for the magnitude of the earth science aid to developing countries is $30 million for 1975 [i.e., an almost fivefold increase over 1968-69]. . . .
>
> Natural resource development, with an indigenous training component, should figure pre-eminently in Canada's external aid programs. To ensure the availability of suitable experts for resource development work abroad, CIDA should define its earth science manpower requirements for a five-year period to allow for the orderly training and recruiting of Canadians for Canada's bilateral and multilateral assistance programs.
>
> In the resource field, Canadian technical assistance to developing countries should be part of a concerted effort, based on an integrated approach to natural resource development, and with good priority setting. The individual programs should be multistage and sustained. Their performance should be assessed through mid-project reviews and post-project appraisals.
>
> The Canadian International Development Agency should solicit more actively the co-operation of other government agencies for establishing the policies and objectives, developing the planning and supervising the execution of earth science programs of foreign aid. . . .
>
> The Department of Energy, Mines and Resources should establish an Overseas Branch, with a permanent cadre of earth science specialists having the scientific or technological competence, proper attitudes, personal qualities and interest demanded for foreign work. This Branch should be financially supported by the Canadian International Development Agency.[1]

The difficulties blocking an effective foreign role did not stem solely from the Survey's situation or its position in the Canadian bureaucracy. Experience showed that the local administrative and political conditions sometimes presented further impediments. In some countries, technical competence to assess and develop mineral resources was lacking, while domestic political considerations would not permit the employment of capable foreigners. In others, ideological reasons prevented the use of foreign capital or experienced commercial mining concerns. As a result, almost nothing effective could be done, and any advice given was bound to have limited value. A situation more often encountered was an understandable desire to reserve managerial, administrative, and employment opportunities as far as possible to indigenous workers and civil servants. Accepting this situation, Survey officers saw their role as helping the assisted country to become as self-sufficient as possible. They worked closely with the local authorities and stressed the training of local staffs.

While some training could be given locally, in conjunction with the aid programs, a certain number of scientists and technicians needed longer periods of more advanced instruction and experience at a university or modern laboratory. Accordingly, the Survey became actively involved in training foreign students at its facilities in Ottawa and Calgary. From the early fifties, while the Survey was still located in the overcrowded National Museum building, a succession of trainees began arriving. They came from Asia, the Middle East, West Africa, and the Americas, and were sponsored by the Colombo Plan, the various UN agencies, or later, CIDA. They sought training in almost every sort of facility, from X-ray spectroscopy, rapid methods chemical analyses, testing, operating and servicing geophysical equipment and interpreting data, to working in the Cartographic Section or the radio workshop. Many trainees who were attending universities in Canada or the United States under various aid programs complemented their studies with practical laboratory or field experience at the Survey. Senior scientists and teachers from other lands

XXIII-3 R. W. BOYLE

XXIII-4 AT THE MINDEN SEMINAR, 1967
G. D. Hobson of the Seismic Section explains the making of a chart record of a hammer seismograph operation to S. Samaniego of the Philippines (*centre*) and Richard Belletty of Guyana (*right*).

improved and updated their knowledge, while visiting officials of foreign surveys examined the Survey's organization and procedures as guides for their own efforts.

Once again, the lack of a well-planned, continuing program reduced the effectiveness of the results and increased the strain on Survey staff and facilities. Much thought had to be devoted to the applications, especially those that were too ambitious or too vague. The applicants' previous training and ability to benefit from the proposed programs had to be carefully evaluated, as well as how they might be fitted into the Survey's activities. When the facilities were not available for a particular type of training, arrangements had to be made to direct the candidates to suitable private companies that were prepared to accept them.

Despite extreme care in the selection process, some applicants failed to make full use of the opportunities, or required an inordinate amount of attention from a fully-occupied staff. Moreover, as with foreign students generally, a rather large proportion failed to return home after their training ended, depriving their homelands of their much-needed skills. Hence the desirability of giving as much training at home as possible was stressed, with Canadian aid being directed at strengthening the training facilities in the countries concerned, and concentrating the training in Canada on selected key specialists and senior administrative and instructional staff. In accepting a number of foreign personnel each year for training, the Survey was doing its best to help overcome the great shortage of trained scientific and technical staff in the underdeveloped countries.

In a somewhat different category are the holders of National Research Council postdoctoral fellowships, whom the Survey accepted in numbers of one to five each year from 1956 onward. This program, established at the instigation of the NACRGS, entailed selecting highly qualified young scientists, mainly from foreign universities or government agencies, to work at the Survey for one or two years on mutually-agreeable projects of advanced research that would equip them as specialists in their fields. This form of aid was by no means a one-sided affair. The postdoctoral fellows sometimes introduced new methods and approaches developed at their home institutions, and they often provided valued assistance to their supervisors' projects. Some made useful contributions to the laboratory phases of the Muskox, Coronation Mine or granites projects, to studies of fossil plant spores, coal petrology, ore deposits, theoretical geophysics, and to various field projects.

The foreign activities, limited though they are and strewn with difficulties, indicate a new role for the Survey, one that adds a new dimension to its original functions. In an age when scientific training and industrial technology have become important diplomatic currency, external aid and training are fulfilling a semi-diplomatic national service. The participation of the Survey's staff in these programs strengthens Canada's image abroad, for its men have made a good impression in providing sound, disinterested advice, being able to work under trying physical conditions, and displaying resourcefulness and versatility coupled with sympathetic understanding of the needs of the countries and the objectives of the aid programs. The expanding role of the Survey on the world scientific stage as the operative agent of the national government in this and other special fields, adds a new rationale to the existing practical, scientific, and legal arguments for maintaining a strong, progressive Geological Survey of Canada.

* * *

THROUGH THE YEARS the Survey has adapted discoveries made by international science to the Canadian situation, has tested and revised them, and sometimes has developed new approaches and principles to be applied and checked, in their turn, by the world at large. For the Survey has been more than a borrower. Over its history it has produced its fair share of improvements, refinements, and new interpretations. Individual staff members in every generation, in almost every decade, have ranked among the foremost earth scientists of their time; while the Survey's collective annual contributions have constituted important additions to world science.

Much of the interchange of ideas and information—even more today than in the past—takes place through scientific societies and organizations, so it is appropriate to examine the Survey's part in those bodies. In the nineteenth century, when there were scarcely any such societies in Canada, and few anywhere, the staff kept abreast of current developments through the scholarly journals and by their memberships in international (chiefly British and American) geological, geographical, and natural history societies. As the Survey became more completely committed to geological work, staff participation became concentrated in general or omnibus geological societies, as well as in the increasing number of specialized societies devoted to such major subsciences as paleontology or mineralogy. With the increase in the numbers of earth scientists since the Second World War and the proliferation of special fields, present Survey staff are in a position to be informed and inspired by a bewildering variety of general and specialized societies.

The Survey's strongest identification with a foreign society is probably with the Geological Society of America (GSA), founded in 1889, which attracted generations of Canadian geologists because of the opportunity to discuss the mutual geological problems of related regions (Appalachians, Precambrian Shield, and Cordillera) or fields of specialization, such as paleontology, the Pleistocene, or mineral deposits geology. Canada held its share of annual meetings, Canadian scientists were regularly appointed to the council, and the presidency was held by G. M. Dawson in 1900, F. D. Adams in 1917, and W. H. Collins in 1934. Such links survive, as witnessed by the succession of G. S. Hume to the presidency in 1956 (and R. F. Leggett of the National Research Council in 1966, the only one of the five Canadians to that time who had never worked for the Survey). Occasional annual meetings are still held in Canada (Ottawa 1947, Toronto 1953), and perhaps most significant of all, the coveted Penrose Medal was awarded in 1950 to M. E. Wilson. The second Penrose award to a Canadian, in 1968, also went to a Wilson—unrelated to the first—the one-time Survey officer, J. Tuzo Wilson.

Gradually this traditional relationship has changed, partly because a largely Canadian-educated generation of geologists has now emerged in the Survey, and partly because of the formation in 1947 of the Geological Association of Canada and more specialized bodies such as the Alberta Society of Petroleum Geologists. Survey staff, however, meet frequently with neighbouring societies across the border that are concerned with similar geological problems. Members of the Appalachian Section, for example, meet periodically with geologists of the United States Geological Survey's Northeast Region, the northeastern section of the GSA, or the New England Intercollegiate Conference. They attend one another's meetings, go on tours, and take turns at serving as hosts. Members of the Institute of Sedimentary and Petroleum Geology have a similar exchange with the American Association of Petroleum Geologists, or regional groups as the Billings (Montana) Geological Society.

Members also take part in the work of many other international scientific associations. The multidisciplinary Pacific Science Association, which includes scientists and scholars from all the countries bordering on the Pacific Ocean, is a good example. The Survey, in the interest of improving its understanding of the forces shaping the geology of the Pacific region of Canada, sends representatives who present papers at the various congresses. Survey scientists have participated in, and addressed, conferences and congresses of a wide variety of international scientific associations in fields of their specialization. Examples include the Commonwealth Mining Congress, World Petroleum Congress, Carboniferous Congress, International Jurassic Colloquium, International Sedimentological Congress, International Union for Quaternary Research (INQUA), and British Commonwealth Geological Liaison Office symposia.

The Survey also has developed links with governmental agencies in other countries, which are based on their mutual interest in exchanging information and ideas, working co-operatively for common objectives, providing training and experience to their staffs, and on occasion, perhaps, helping each other with their administrative problems. In its technical assistance work, the Survey has often had dealings with surveying or mining agencies of other countries. In 1958, for example, Harrison, accompanied by a soil specialist, A. Leahey, visited Jamaica under the auspices of the Colombo Plan to report on the activities, requirements, and plans of the island's geological survey in connection with a proposed External Aid Office assistance program.

Strong ties of training, interest, and outlook frequently bring the national agencies of Canada and the United States together. In earlier generations such co-operation was exemplified in the discussions and conferences between the directors or groups of specialists on questions of nomenclature, such as the conferences that led to the working out of agreements on the Appalachian or Precambrian successions. Joint committees of the two surveys are often formed to deal with specific questions. The Survey has been represented on the North American Commission on Stratigraphic Nomenclature, which has representatives from federal, state, and learned institutions in Canada, Mexico, and the United States. T. E. Bolton and P. Harker have both acted as chairman of the commission.

An unusual example of collaboration was the designation of the Survey as one of the agencies from six foreign

XXIII-5 IN DISTANT LANDS
A. Larochelle standing in front of the Maison du Préfect, with M. Goukoye Karimon (*centre*), Préfect of Agades, Republic of Niger. Following a mineral prospecting program conducted under the United Nations Special Funds, the Canadian government was requested in March 1968 to provide geophysical information on the areas previously studied and on some adjoining tracts. Larochelle and L. P. Tremblay went to Niger in April-May 1969 under CIDA auspices to make the necessary on-the-spot investigations and work out a feasible program of activities. The geophysical and photogeological program they suggested was approved and was successfully carried out.

countries entrusted to study lunar material brought back by the Apollo missions of the National Aeronautical and Space Administration of the United States. As early as 1966 six members of the Geophysics and Petrological Sciences Divisions were appointed as principal investigators to undertake analyses and experiments on the first group of Apollo XI samples recovered from the moon. When the material (several crystalline rock pieces and some fines) was received, J. A. Maxwell was relieved of his administrative duties with the Analytical Chemistry Section for several months in 1969-1970 to conduct his tests and to co-ordinate the various studies made by different Survey laboratories. The program included chemical, petrological, and mineralogical analyses, isotopic age determinations, and studies of the electrical conductivity, magnetic susceptibility, and remanent magnetism. The highlights of the investigation were the discovery of a new mineral, pyroxferroite, and an unusual spinel group mineral containing titanium and chromium. The chemical analyses of the standard samples were reported as placing the Survey among the ten foremost rock analysis laboratories in the world. Maxwell continued as a principal investigator with the Apollo program, directing the analyses of samples from the later Apollo missions. L. W. Morley also represented the Survey at NASA headquarters in Houston in briefings to work out possible geological photographing and data recording activities from satellites. Later he transferred from the Survey to direct the newly-established Canada Centre for Remote Sensing.

Since 1960 the national surveys of both Canada and the United States have been caught up in a new kind of international scientific co-operation that reflects their common allegiance to a wider cause. They are involved at present in preparing an ongoing series of world maps intended to consolidate the knowledge of the geology of the earth in a number of fields. On the surface, a national contribution for a world map may seem relatively simple to prepare, entailing nothing more than selecting the data from existing Canadian maps, and compiling it on the prescribed scale. Each world map project, however, requires that every national contribution should conform to a specific basic approach and standard of nomenclature established by the sponsoring agency, usually a commission of the International Geological Congress (IGC) or of the International Union of Geological Sciences (IUGS). As may be imagined, this can create serious difficulties. National surveys have adopted mapping conventions and standards that are unique in certain respects; therefore, their standards will never coincide precisely with those demanded for the world map. Thus the project may have to be remapped from existing data or new field work may be needed to fill in unsurveyed areas or detail omitted in the present maps. Thanks largely to the Survey's long, sustained, technically advanced, geological work, Canada has a good background for participating in these projects of international scientific collaboration, notwithstanding the vast size of the country and the division of effort, in some matters, between federal and provincial agencies. This country was in a position to prepare its contribution to the geological and tectonic maps of the world with comparative ease. But in the case of certain suggested projects, the data presently available for Canada may be wholly inadequate. Then gigantic research and data collection efforts might be necessary. Because the area of Canada is so great, collecting and preparing the data for each world map requires as much effort from Canadian scientists as may be demanded from the much wealthier, more populous United States, or from the whole of Europe.

It is very important, therefore, that Canadian interests should be carefully considered when world map subjects, standards, and schedules are established. Unfortunately, such matters are invariably decided by international committees that are dominated by powerful, populous, and wealthy countries, in which Canada's voice is in no way commensurate with its assured burdens and responsibilities. Furthermore, these international maps are prepared on the basis of continents, with an appointed subcommittee to co-ordinate and direct the work for each. The North American part of these map projects, in which Canada is a major contributor, is dominated by United States personnel, outlooks, and interests. So, Canada finds itself with a subordinate, almost colonial, role in these important projects of international science. Yet how can its co-operation be withheld from a project that has been launched with full international agreement, and which, like all such proposals, is worth being undertaken in its own right?

The Survey had worked for many years on the map of the general geology under the auspices of the IGC, but as late as 1948 the North American maps were reported as being delayed for many reasons, including discrepancies between Canadian and United States interpretations of the Precambrian. In succeeding years much additional data was collected, new methods of rock analysis became available, and new concepts in geochronology and geodynamics revolutionized the understanding of the tectonic history of Canada. Under these favourable conditions the general geology 1:5,000,000 map was completed during the decade of the 1960's, almost as part of the regular Survey work. It was compiled in the early years by A. H. Lang, and afterwards by R. J. W. Douglas in connection with the latest edition of *Geology and Economic Minerals of Canada.*

XXIII-6 WHEN THE MOON ROCKS WENT ON DISPLAY
Above: On October 18, 1969, crowds lined up in front of the Survey building to inspect the moon rocks on display in Logan Hall, and also to see the dignitaries, who were given an advance showing and explanation of the lunar material.
Below: A group of visitors to the display. *Left to right:* Y. O. Fortier (Director), J. M. Harrison (Assistant Deputy Minister, Science and Technology), H. R. H. Prince Philip, Duke of Edinburgh, and J. J. Greene, Minister of Energy, Mines and Resources. *In the background:* C. M. Isbister, Deputy Minister, Department of Energy, Mines and Resources.

The Survey has also been involved with three other world maps: the tectonic and metallogenic maps of the world set in motion by the Mexico meeting (1956) of the International Geological Congress, and the metamorphic map approved in Paris by the Commission for the Geological Map of the World. The tectonic map was comparatively straightforward, since it fitted easily into the current drive by the Survey to expand and update past work through applying the new age determination techniques. A preliminary tectonic map of the Canadian Shield was published in 1965; and the finished Canadian map, the work of a committee headed by C. H. Stockwell and drawn from the Survey, the universities, and provincial mines branches, was published in 1968. It was incorporated in the map for North America that appeared the following year.

The Survey's experience with the two other maps was very different, however. Progress on the metallogenic map was slowed by intrinsic difficulties—the enormous quantities of geological, tectonic, metamorphic, and other data that had to be obtained and analyzed for every mineral deposit and its setting, compounded by failure of the international committee to agree on standards, then later, by a change of approach instituted after the project had already begun. The metamorphic map presents equally great difficulties, this time of a practical sort, in working out the metamorphic history of the vast areas of much-altered, ancient Precambrian rocks of the Canadian Shield. Because of these difficulties, as well as the financial stringency and diversion of manpower to more pressing problems, progress on the metallogenic and metamorphic maps has been much slower than on the two previous projects.

In the meantime, new international enterprises continue to be trotted forward by interested parties, for instance, a neo-tectonic map (earthquake belts, rising and falling coasts, etc.) and a hydrogeological map. Even under the best of conditions the burdens of being a contributor to world mapping projects imposes an onerous responsibility on Canada, whose geography endows it with scientific responsibilities on the scale of the great powers. It is to be hoped that the demands of such worthy projects, or of any others that may follow, will not be permitted to impose undue pressures on the Survey's limited means and staff, or to interfere with its other priorities.

* * *

THE PROGRESS OF THE WORLD MAPPING PROGRAM strikingly demonstrated the improvements in international co-operation in the earth sciences field in the sixties, a movement in which the Canadian Survey and its director Harrison played an important part. Indeed, the years of growth associated with the move to the Booth Street headquarters, and the higher scale of government spending elevated the Survey in the sixties into a considerable power in international scientific organization. Among other things, Survey personnel in these years contributed to the formation of the International Union of the Geological Sciences (IUGS) as the world geological organization, and to the bringing of the XXIVth meeting of the International Geological Congress to Canada.

The institution that has spoken for the world community of geological scientists since 1878 has been the International Geological Congress, in which the Survey had a long-standing interest. The founding committee included such associates of the Survey as T. Sterry Hunt, who was the president and the secretary, and James Hall, the director of the New York State Museum. Sir William Dawson was a member, while A. R. C. Selwyn was on the organizing committee for the first congress. The meetings of the IGC, which were usually held at four-year intervals, afforded opportunities for discussing the latest advances in scientific thought, establishing relations with fellow scientists in other countries on the basis of shared special interests, and for visiting and studying points of geological and mining interest in the host countries. There were occasions also to discuss such matters of common concern as definitions of geological periods, development of standard map legends and nomenclature, and organization of co-operative research projects on a variety of topics of general interest.

Between congresses, however, the organization dwindled almost to nothingness; a local arrangements committee had to take up the heavy burdens of organizing the forthcoming congress from the beginning. It soon became apparent something more permanent was needed to represent the community of earth sciences and to coordinate activities between congresses on a continuing basis. By the time of the Toronto meeting in 1913, a full-time secretary for the executive committee had recently been appointed, and consideration was being given to electing an eight-man permanent committee to direct the affairs of the organization between congresses. R. W. Brock chaired a committee on the constitution that met in London in 1921 to discuss the question of a more permanent organization. However, at the next IGC, held in Belgium in 1922, as a later committee reported, "it appears that the Council of that Session did not consider that the formation of a Union would usefully supplement the existing functions of the Congress."[2] There the matter rested for over 25 years.

In the meantime, under the aegis of the League of Nations, a useful prototype for international scientific co-operation was emerging. This was a series of fourteen great international 'unions' of scientific organizations concerned with the various disciplines, each being comprised of national member organizations. These unions were united, in turn, in the International Council of Scientific Unions (ICSU), formed in 1931 as the crowning arch of the organizational structure of international science. The principal constituent union of the ICSU where many geologists found a haven, was the International Union of Geodesy and Geophysics (IUGG).

After the Second World War the project of establishing an international union exclusively for geology was revived at the XVIIIth Congress (1948) meeting in London. There was now an added practical incentive. UNESCO, the United Nations agency which had assumed the role of promoting scientific co-operation and was actively associated with ICSU, had expressed the desire for "a permanent international geological executive to which UNESCO may turn for advice" on geological questions, including the awarding of funds "for the prosecution of scientific projects of international importance."[3] As a result, the proposal to establish an international union of geologists separate and distinct from the IGC but in close liaison with it, was approved and sent forward, along with an outline statute for the new organization, to the next congress scheduled to meet at Algiers in 1952. At Algiers, however, the statute was rejected, partly because the prepared draft was considered too elaborate and unwieldy, but mainly because of the opposition of Soviet and other Eastern European delegates to the proposed association with UNESCO. Therefore, the project had to mark time until the 'Cold War' era changed to one of 'thaw' or *détente*.

In the meantime, organizations and individual geologists found a substitute by associating themselves with the IUGG and taking part in some of its major activities. Chief among these was the highly successful, spectacular International Geophysical Year (1957-59), in which organizations from 66 countries were drawn into a vast series of related programs under the guidance of a special ICSU committee. The International Geophysical Year also demonstrated the need for geologists to be represented in an organization that would be devoted completely to their discipline and that would give top priority to their interests in future similar projects of international co-operation.

Action was therefore initiated at the next or XXIst IGC meeting of 1960 in Copenhagen. By then a new spirit was becoming evident in international relations, and Soviet delegates to the congress intimated that they might be able to respond favourably to a new proposal. Harrison took a leading part in the discussions, coming out strongly in support of the formation of the new union. As a result, he was named to a small organizing committee that investigated the project, and when it reported in favour of proceeding, he was appointed to the new committee charged with organizing the International Union of Geological Sciences. Harrison was one of the working group that met in Stockholm in December 1960 to organize the arrangements for the new union, which the ICSU ratified early in 1961. On March 9, 1961, when the International Union of Geological Sciences came into being, its first president was the Survey's twelfth director, J. M. Harrison.

In the next few years Harrison went on to become a member of the executive board or bureau of the International Council of Scientific Unions in 1962, and a senior vice-president in 1963. From 1966 to 1968 he served as its president, carrying Canada's name and the Survey's reputation to this topmost pinnacle of organized world science. In these years he was also associated with the ICSU's newer programs of international scientific operations, namely the International Indian Ocean Expedition, and the International Upper Mantle Project.

This last-named project became for a time a leading interest of the Survey. It was first proposed at the International Union of Geodesy and Geophysics meeting in 1960, was subsequently discussed at the next meeting at Berkeley, Calif., in August 1963, and at the IGC's XXIInd meeting at New Delhi in December 1964. Initially the programs included an ambitious plan to drill a deep borehole at a selected site where the earth's crust was believed to be thinnest, to penetrate to the Mohorovicic discontinuity (the Mohole project) and gain a more accurate idea of the character and dynamics of the deeper crust of the earth. The guidelines for the program were the responsibility of the International Upper Mantle Committee of the ICSU, of which V. V. Beloussov was the chairman and J. M. Harrison, the Canadian representative. Six working groups were set up to deal with various aspects of the program. However, the national responses were somewhat indifferent. The United States and some of the other countries scaled down their original plans (including abandonment of the Mohole project) after which a revised program went forward in earnest. C. H. Smith of the Survey probably was the main figure in the implementation of this program, and its ultimate success owed much to his efforts.

XXIII-7 C. H. SMITH

Smith served as deputy secretary-general of the International Upper Mantle Commission that was set up under Beloussov to oversee the implementation of the program. He was the chairman of the joint committee of the IUGS and IUGG that represented the interests of those scientific unions in the project. Finally, he was secretary of the Canadian Upper Mantle Project that was responsible, through ten subcommittees, for preparing and carrying out the program in this country. Moreover, he co-ordinated the Survey's efforts in the field, notably on the studies and drilling projects of the Muskox Intrusion. This remarkable little-disturbed layered ultrabasic dyke-like intrusion recently discovered near Coppermine, N.W.T., was valuable in discovering the nature of the magma derived from the earth's upper mantle. Smith also helped organize the important scholarly meetings of the Upper Mantle Committee hosted by the Survey between September 2 and 11, 1965, the three large symposia of which were attended by 180 delegates from 16 countries. He organized the symposium on aspects of drilling deep holes (he also arranged the program), the noted American authority B. Heezen being the organizer of the other two symposia on the world rift system, and on continental margins and island arcs. The Survey also undertook the onerous task of publishing the symposia, which appeared as three of the Survey's largest volumes, its *Papers* 66-13, 66-14, and 66-15, comprising some 1,200 pages of text. Although the studies associated with the Muskox Intrusion were the major Canadian activity, the full register of Canadian contributions to the program actually numbered 152 projects. Canada was the second largest contributor to the International Upper Mantle Project, her effort being surpassed only by that of the USSR.

Other officers of the Survey served on various committees and commissions of the IUGS during the sixties. S. C. Robinson, in his capacity as chairman of its Committee on Storage, Processing, and Retrieval of Geological Data, in 1963 visited a total of seven European countries from Britain to Czechoslovakia. Specialists frequently attended meetings of such associations as the Commission of Stratigraphy's Sous-Commission du Jurassique, or participated in special colloquia on granites or other subjects. In this way staff members worked for the general interests of Canadian science and the Survey, and also acted as ambassadors for Canada in the world of science.

These contacts helped bring the XXIVth International Geological Congress to Montreal in August 1972. The Survey has always been represented at the congresses, the numbers depending on both the willingness of those in charge to let staff attend, and on the distance of the locale from Canada and the ease of travel to it. Thus W. H. Collins was the only one to attend the XVth Congress in South Africa in 1929, whereas the chief geologist, G. Hanson, headed a delegation of six at the XVIIIth meeting in London in 1948. The much greater speed and convenience of air travel is an obvious factor behind the larger attendance at recent meetings.

The project of sponsoring a meeting of the IGC in Canada engaged the interest of the Survey from 1960, when Harrison made the suggestion at Copenhagen—until August 1968, when the official proposal was formally presented to the XXIIIrd Congress meeting in Prague. Authorization was obtained in 1967 from the Canadian government for an official invitation to go forward, following which a Canadian National Committee for Geology (formed from the executive of the NACRGS and the presidents of five major geological societies in Canada) became responsible for co-ordinating the arrangements, selecting the interim officers, and securing the support and sponsorship of interested agencies. In December 1967 R. E. Folinsbee of the University of Alberta was appointed president-elect and given the responsibility of leading the Canadian delegation to Prague and issuing the formal invitation. At that meeting it was also decided that Montreal would be the site of the meeting, and that J. E. Armstrong, in charge of the Vancouver office, would serve as secretary-general of the organizing committee.

A large delegation of 16 Survey officers, headed by the director Y. O. Fortier, proceeded to Prague in the summer of 1968 where, like other delegates and guests, they were treated to more than they had bargained for. The breakdown of local transport facilities and other services after the entry of foreign troops into Prague to restore order brought the meeting to a premature halt almost before the program had begun. However, on the morning of the invasion Fortier presented the invitation at an abbreviated council meeting, which Folinsbee was unable to reach in time from his hotel. The invitation was accepted, and Canada gained the honour of hosting the meeting of the XXIVth Congress in 1972. Then the Canadian contingent made a troubled, hasty exit, mostly by train, to Austria and the Federal German Republic.

XXIII-8 J. E. ARMSTRONG

Folinsbee and Armstrong were appointed president and secretary-general of the organizing and executive committees, with Mrs. P. Moyd of the Geophysics Division as organizing secretary. Later I. M. Stevenson of the Survey was made assistant secretary-general. For a period of more than four years the Survey furnished the headquarters, administrative and office services for the congress, until the business of the meetings was finally wound up in the summer of 1973. Fortier, Lord, Smith, and Tremblay were members of the 19-member executive committee of the National Organizing Committee; a further 17 Survey officers were included in the full 87-member National Organizing Committee that represented all the geoscience organizations and interests in Canada. C. H. Smith was appointed chairman of the technical program, while 11 of the 36 program conveners were Survey officers, plus another seven former staff.

From the headquarters, the staff mailed a first circular, dated October 31, 1969, to about 55,000 persons. A second follow-up circular, dated March 1971, was sent to more than 15,000 persons in 115 countries who had responded to the first circular. In each of these, and still more in the third circular (March 1972), the program was more firmly defined. That circular listed 17 major technical sessions and two symposia arranged by the IGC, while 16 additional symposia were organized by affiliated associations concerned with various aspects of the earth sciences. An attendance of 5,000 geoscientists was anticipated, and in the end 3,896 geoscientists and 1,480 relatives and friends were registered at Montreal, plus an additional 804 non-attending members. The attendance was much the largest for any IGC to date, and about nine times as great as at the previous Canadian meeting in 1913 at Toronto. Incidentally, three other international scientific organizations also met at Montreal and Ottawa in July-August, 1972—the 22nd International Geographical Congress (founded 1871), the 12th congress of the International Society of Photogrammetry, and the 6th congress of the International Cartographic Association—which, including the membership of the IGC, brought about 15,000 scientists to these cities. Because of the scale of the operations, the IGC needed to employ the services of professional firms for the physical staging of its meetings and exhibitions, and handling the many, varied travel arrangements.

Numerous excursions were considered before a group of 73 was selected. They were of bewildering diversity, traversing the country from coast to coast, and from the International Boundary to northern Ellesmere Island, 450 miles from the North Pole. At the other extreme were 27 local geological tours around Montreal, Quebec City, and Ottawa, some repeated several times. The Survey

XXIII-9 AT THE XXIVth IGC IN MONTREAL, 1972
Sir Kingsley Dunham (*left*), president of the IUGS, and D. S. Macdonald (*second from right*), Minister of Energy, Mines and Resources, are being received by the officers of the IGC executive committee: R. E. Folinsbee, president (*second from left*), J. E. Armstrong, secretary-general (*centre*), and P. E. Auger, executive vice-president.

was heavily involved with these excursions. The 56 members of the staff who acted as leaders, co-leaders, or guides comprised about one-quarter of the total personnel required. The 50- to 100-page guidebooks specially prepared for these tours contain a wealth of information on the geology of Canada, much of it hitherto unpublished.

The heart of the technical sessions at Montreal were the discussions of some 880 scientific papers, organized in 17 specialized sections. These were the survivors of the many hundreds of proposed papers and presentations submitted to the organizing committee, and referred to panels of specialists for evaluation according to the standards followed by leading scientific journals. Since the papers were to be printed in advance of the meetings, they had to be submitted in final form late in 1971. Abstracts (300 words) of the accepted papers were published and made available to the registrants, while the individual papers and reports (4,000 words or equivalent space) of each of the sections (up to as many as 74 papers in the case of tectonics) were published under the general editorship of J. E. Gill, of McGill University.

At Montreal the program for technical sessions allowed 30 minutes for the presentation and discussion of a paper, the day usually consisting of two sessions of five papers each. The papers in each section were grouped by themes. Those of Section 3 (Tectonics—conveners, R. J. W. Douglas and J. Tuzo Wilson) for example, dealt with the following topics: problems of the earth's interior (13 papers), plate tectonics and continental drift (13), great strike-slip faults (6), tectonic styles of continental blocks (19), evolution of geosynclines (6), temporal relationships between deformation and metamorphism (6), aspects of brittle deformation (6), and Quaternary deformation and neotectonics (5). Only a fraction of these papers were by Canadian authors, and since the accent was on general studies, most of those were produced by university scientists, with a sprinkling of papers by scientists employed by government agencies and by private industry. The staff of the Survey contributed 17 technical papers and co-authored 6 others; they also figured prominently among the conveners of the sections and in the selection groups for the papers, while the secretariat which the Survey provided was heavily involved with every aspect of the programs.

The papers, as might be expected, reflected the strong current interest in plate tectonics and continental drift. Besides those papers given in the tectonics section, the sections concerned with stratigraphy and sedimentation and with paleontology also devoted sessions to discussing evidence for or against the relative movement of continental blocks in the context of those specialties. Through being on the program these scholarly papers dealing with aspects of geological problems received wide exposure, and early study and evaluation by leading specialists, to ensure that the authors' conclusions were balanced and accurate. The sessions gave participants the opportunity to meet specialists in their own fields, or from other fields that had a bearing on their work.

The general symposia dealt with two aspects of the interdependence of mankind—the environmental challenge, in "Earth Sciences and the Quality of Life," the major symposium which was sponsored by the IUGS and the XXIVth IGC; and the sharing of scientific knowledge for the well-being of humanity, "Earth Sciences Aid to Developing Countries." The meetings also featured Georama '72, a large exhibition hall filled with many large displays put on by governments, corporations, and

XXIII-10 THE CANADIAN DISPLAY IN THE EXHIBITION HALL (GEORAMA) OF THE IGC

educational institutions; while ladies' and children's programs, special sightseeing tours, a fashion show, film festival, and cultural programs were available to the registrants' dependants.

Among the subcommissions of the IGC that reported in conjunction with the Montreal meeting was one on stratigraphic classification, set up in 1952 at the XIXth Congress in Algiers. Four reports (nos. 3 to 6), edited by H. D. Hedberg of Princeton University and concerned with working out standard principles for naming types of strata and kinds of lithological, biological, and chronological units, were issued in 1970, in line with the subcommission's belief "that real and lasting progress in stratigraphic classification, terminology and usage is achieved only as geologists in general become convinced of the validity and desirability of the principles, rules and terms proposed."[4]

The sessions of the XXIVth Congress came to an end on August 31, 1972, but the Survey's connection with the work of the IGC was far from completed. For another year, or more, winding up the financial accounts, scientific programs, publications (notably the General Proceedings/Comptes-Rendus Généraux), and correspondence arising from the congress, as well as assisting with preparations for the XXVth Congress in Australia, continued to occupy the secretary-general and organizing secretary, and their office staff at the Survey. Indeed, the problem of the relationship between the periodic congresses and the permanent IUGS grows more serious with each congress, as does the long-standing problem of working out organizational and financial structures to carry on between congresses and to assist future host committees with their local programs without impinging unduly on their autonomy or their responsibilities. Armstrong was appointed chairman of an international commission charged with arriving at possible solutions for these problems.

The scale, both physical and scientific, of the XXIVth Congress was such that a single major section at Montreal was comparable with the entire program of papers offered at the XIIth Congress. The excursions also far exceeded those of 1913 in number and in distances and area covered. In 1913 the excursions had to be confined to places accessible by rail, ship, or road; while, in 1972 the longer excursions were mainly by airplane, and the whole of Canada was open to examination. The style of the 1972 program contrasted completely with the leisurely, intimate, convivial meetings, social gatherings, and excursions of the Toronto meetings two generations before. Still, the quantity, the scope, and the sophistication of the papers presented at the XXIVth Congress would surely have impressed even the great minds of the XIIth IGC, who started geology along some of the paths now coming to fruition.

Altogether, the congress was a very fine achievement in significantly advancing the frontiers of the geological sciences, bringing together scientists from all over the world in the spirit of international co-operation, and showing off the accomplishment of Canadian scientists and technicians, as well as the wealth and variety of the geological features of the land. The new president of the IUGS, Philip H. Abelson, paid fitting tribute to the Canadians who had taken part in organizing the congress:

> The Canadian scientists who arranged the Montreal meetings had unusually great responsibilities which they discharged very well. Before and after the congress they provided about a hundred geological field trips throughout almost all of Canada. They edited and published about 5000 printed pages of scientific material. They arranged for the majority of the scientific talks and handled all the other logistics of a congress. Most important of all for the facilitating of useful human contacts, the Canadians were excellent hosts, and their guests were comfortable. The Montreal meeting ranked with the best ever and represented a new high in Canadian contributions to international science and human progress.[5]

The Survey's part in this result was both extensive and indispensable, even though it played a relatively smaller role in 1972 than it had for the XIIth IGC, which to a considerable degree was a Survey enterprise. The Survey and its men were intimately involved in every aspect of the XXIVth Congress, working to bring it to Canada, helping organize the programs, serving as guides, preparing contributions to guidebooks and papers for the sessions, and notably, maintaining the general office and secretariat. These contributions were entirely appropriate, since the congress served to carry the Survey along the paths of international scientific co-operation and the rapidly evolving earth sciences, and towards more complete understanding of the geology of Canada.

THE SURVEY'S ROLE in international science grew naturally out of more than a century's work and experience. Its operations bring it into contact with most of the broad range of earth science activities and groups in Canada—universities, private industry, and provincial mines branches. It is thus well equipped to represent the interests of nearly the whole of the Canadian earth science community before the rest of the world. Having made the entire country its field of study over such a long period, the Survey is better suited than any other agency to satisfy any requests for Canadian contributions to world mapping projects or other international scientific enterprises. Its long association with the great scientific organizations of the world and its competence in most fields of geological science make it the natural choice to direct the great projects of scientific co-operation in Canada, and to advance Canadians' interests in the supra-national scientific organizations.

Many of the countries with which Canada deals—particularly Communist lands and newly emerging nations—entrust activity in the earth sciences to public agencies, which find it easier and more natural to deal with the Survey, their Canadian equivalent, than with associations staffed by private citizens. In its external aid programs, too, the Survey's dealings are with mining or geology branches of the governments concerned. The Survey is fulfilling a variety of special functions on behalf of the federal government that no other organization in Canada, certainly no one provincial agency or a combination of several of them, could perform with equal effectiveness. The Survey gives the Canadian government a useful entry to important governmental agencies and academic-scientific circles in other countries that cannot otherwise be easily reached.

Its international record during the past dozen years was without precedent in the Survey's history, except possibly for participation in conferences and exhibitions that went back to Logan's day. It included many remarkable accomplishments, offset by some frustrations and disappointments. The Survey represented Canada in a variety of multinational mapping and research activities, and it was accorded representation on the bodies regulating such activities. Its rapport with agencies in other lands was excellent, while its staff enjoyed good relations with their foreign counterparts. The distinguished scientists and administrators who made a point of visiting the Survey, as well as the number of foreign students who trained at its facilities, reflected the international esteem in which the Survey was held. The Survey's technical assistance work was rather limited, but was highly regarded for its realistic, disinterested, and effective qualities. At the same time, its effectiveness—for some years at least—was hampered by the way in which such programs were administered. In the years since 1968, too, the greater financial stringency, as well as greater concentration on domestic activities more closely aligned with national economic and other objectives, lessened the earlier emphasis on research into general problems that enhances the Survey's reputation in international scientific circles.

Since the earth sciences have now become international in character and scope, the operations of the Survey can be adversely affected by a slackening of its external activities and associations, or by excessive preoccupation with narrow domestic concerns. Firm, frequent international contacts are needed to keep the institution at peak efficiency and to enhance the morale and well-being of a first-class professional staff. The Science Council's special study has indicated the growing importance of international earth science involvements from economic, technological, and scientific standpoints. Given the importance for Canada of an active international role in the earth science field, the national interest is bound to be well served by maintaining a large, meaningful international Survey presence.

XXIV-1 THE LOGAN BOULDER IN WINTER
At Survey headquarters, Ottawa

24

Logan's Legacy–Some Achievements of the Survey

How can the Survey's multifaceted 130-year experience be summarized and evaluated in the space afforded by a study such as this? How can one assess the activities of its men, so few in numbers, who extended their work from the vicinity of Montreal to the utmost limits of northern Canada; who followed a youthful science and helped it evolve to its present diversity, sophistication, and period of syntheses; who expanded an informal two-man operation into the highly-specialized, differentiated, co-ordinated, regionalized institution of today?

The operations of the Survey have been determined throughout by the legislative and administrative decisions of its political superiors, embodied mainly in the acts and orders in council establishing or renewing the Survey or the government departments of which it formed a part. In addition to such explicit actions, governments have also controlled it less directly but even more profoundly by the financial means and staff that they placed at its disposal, and by their attitudes towards its freedom of action. These included the presence or absence of political pressure on such matters as supply contracts, appointment of temporary or permanent staff, or the specific choices of field work to be undertaken. A second sort of pressure concerned the Survey's freedom of initiative–whether it should be allowed to select its own objectives, or be made to accept others desired by government. Finally, there are the changing constraints that bureaucratic methods of the public service impose on its operations.

In the course of its history, the Survey has experienced all these pressures many times over, since they are inherent in the relationship between democratically-elected governments and publicly-paid civil servants whose activities are those governments' responsibility. It may be said, however, that as government agencies go, the Survey has suffered surprisingly little interference over its long-term career. The small size of the institution, its autonomous position, the specialized scientific tasks it was performing, the public and international prestige of its directors Logan, Selwyn, Dawson, or of such officers as Bell, Tyrrell, McConnell, Macoun and Dowling, all helped protect the organization against the worst excesses of patronage during its first 60 years.

As the Survey came increasingly under governmental sway, it experienced the traditional forms of political patronage, to a greater degree in the period 1900-1940 perhaps, than either before or since. Even then the Survey remained relatively free of gross political interference because of its small size in terms of men and money, the down-to-earth management by men like Brock, Collins, or Young, and the professional nature of its work and its staff. Since 1940 the much larger scale of its operations and funding leaves greater scope for mismanagement; but against this there are the safeguards of the Public Service Commission's control over appointments, plus the elaborate apparatus of financial control by the Treasury Board and Auditor General. One may be excused from wondering, however, whether in the case of the Survey, the array of cures is not worse than the possible disease.

The story is otherwise as regards political control over policy, and administrative control over the character and quality of the operations. The early Survey to all intents and purposes was as independent of government as a Crown corporation and was almost entirely at liberty to pursue its own programs. Its status as a separate department of government from 1890 to 1907 also left it relatively free except for the meagre funds placed at its disposal. The ability to communicate the results of its work directly to the general public in addresses and newspaper reports made the agency and its staff highly visible, and secured the backing of an influential, well-disposed public following capable of deterring governments from laying violent hands on it.

The attempt of Camsell in the 1920's to adjust the work program to harmonize with the objectives of the Department of Mines was the first serious effort to impose a pattern of work from outside. Since then, apart from the period of the Department of Mines and Technical Surveys (1950-66), this form of control has increased, until the policy approaches of the present Department of Energy, Mines and Resources constitute perhaps the most thoroughgoing effort in the history of the Survey to bring its programs and priorities in line with departmental objectives. The expansion of scientific specialization and the increase in capital equipment also changed the character of the work after 1950, giving the Survey certain attributes of a factory-like industrial institution, with the accompanying controls inherent in such a situation. At the same time too, the increasing bureaucratization of the Canadian public service is having remarkable effects on its day-to-day functioning, contributing to the increase in the number of staff and to the over-all cost of operation.

Under each of its successive mandates the Survey's central function has remained the one which Logan inaugurated in 1842—to explore, map, study, assess, and publish as full, accurate information about the geology of Canada as possible, as a means of assisting the development of the country. The precise activities or roles of the Survey in any given time however, have deviated from this central theme in keeping with governmental policies or the ideas of its directors. Indeed, over most of its history the Survey performed much broader functions than those normally included in geological mapping and study. Logan quickly discovered that certain additional tasks—topographical surveying, astronomical observations, or operating a geological museum—were essential if the work of geological surveying was to be done effectively. Other functions were acquired at different times, usually at the government's request, under the impression that they could be conducted more economically and efficiently by Survey personnel in the course of their regular work in the field. In this fashion the agency, over time, began collecting natural history and ethnological specimens and records, and making assessments of the waters, forests, arable soils, and wildlife in addition to the potential mineral deposits of the lands they traversed. Some of these activities were thereby introduced into the federal government service, and were developed to the point of achieving separate existence as new agencies.

These more or less extraneous operations were not uniformly well performed, especially when they bore little relevance to the central function of mastering the secrets of Canada's geology, or when shortages of funds placed difficulties in the way of doing justice to that all-important responsibility. The Survey developed and maintained an excellent, proficient topographical mapping service which was only separated from it at the time of the Second World War. On the other hand, while the Survey did the best it could, according to its lights, to support and develop the National Museum, that best—especially under Collins—simply was not good enough. The National Museum of Canada, which the Survey had largely created from its collections from Logan onward and had firmly set on its feet under Brock's able direction, was enormously handicapped by its overlong dependence on the Geological Survey. But should the blame for this deplorable state of affairs not belong more properly to the federal governments of the time, who failed to give the Museum the financial means and administrative autonomy for it to develop in line with its own needs and those of the nation?

In addition to fostering the National Museum, the Survey undertook other activities that led to the creation of new branches of the public service, especially in the field of the earth sciences. Thus when its studies of the mining, processing, and marketing of minerals (the basis for a Survey section in the 1890's) proved inadequate to the aspirations of the mining industry of the time, a new institution, the Mines Branch, was established in 1907 to deal exclusively with these functions. In similar fashion, the Survey has been involved on the periphery of the transfers of function and administrative reorganizations attendant on the creation of new departments in the past few years. These, however, did not materially affect its over-all program or structure. Groundwater and submarine geology, geomorphology, and remote sensing were the main fields of transfer. The Survey's areas of jurisdiction became more precisely defined in relation to the Earth Physics and Mineral Resources Branches, and to certain agencies of other departments.

Yet another side of the constant redefining and reshaping of historic functions concerns its relationships with emerging regional, provincial, and private geological surveying and research institutions. From time to time the Survey's operations were challenged by certain provinces on economic and constitutional grounds. As a re-

XXIV-2 THE MASS SPECTROMETRY LABORATORY
The Survey's first mass spectrometer, built by R. K. Wanless in 1953 and used in geochronology and stable isotope work since then, occupies the centre of the photograph, with the second mass spectrometer beside it.

sult, notably since 1950, the Survey has moved away from such areas as detailed mapping and from the studies of individual mineral deposits. Considerable progress has also been made in establishing regional centres to co-operate with and serve local mining and oil interests, and governments. Where the scale and specialized character of a particular kind of work has made it feasible to do so, the Survey has established a relatively self-contained, specialized regional office or institute.

Another sort of challenge has arisen in recent years over basic research studies, a field the Survey entered in force after 1950 in line with new developments in the earth sciences and largely as a service to the nation. Having contributed in these years to the building up of powerful research facilities in the universities, other governmental agencies, and in private industry, the Survey is now seeking an acceptable relationship with the possessors of this recently-developed research capacity. In these and other ways, it is responding voluntarily to imperatives of the present Canadian situation, trying to anticipate the public need and adjust to it. This present change of front, like earlier ones, has necessitated appropriate internal changes in its programs, facilities, staff, and structure. In working to evolve a co-operative style, the Survey is providing a practical guide—as so many of its policy decisions of past years have done—for larger federal government responses to similar situations.

The most striking long-term result of the Survey's evolution has been the increasingly concentrated, specialized character it has assumed over time. For all its size, the present-day Geological Survey functions in a mere corner of the broad range of activities conducted by Logan, Selwyn, Dawson, Bell, or Brock. Indeed, the many different sorts of work to which those gifted men turned their hand now constitute the basis of numerous scientific and technical agencies of governments, and of the private sector as well. The professional, scientific, and administrative staff of today operate over a much larger spatial area than those great pioneers, however, one co-extensive with the landmass of Canada and the continental shelves besides. Moreover, as has been well said, where the mapping ability of the geologist of yesterday was "limited to the two dimensions of the surface," his successor deals with aspects of problems that formerly had to be ignored: "the enhanced ability to measure . . . has suddenly extended precision in geology from the two dimensions of the earth's surface to the third dimension of depth, and to the further dimension of geological time."[1] Today, consequently, results are achieved that the forerunners with all their talents could scarcely have envisaged, though their contributions made the work of the present-day Survey possible.

* * *

HOW DID THE SURVEY translate its responsibilities to the governments and people of Canada into actions? Though its tasks changed from time to time, they fell under the following six categories: exploring and studying the geology and other aspects of the land that is Canada; making its findings known to all manner of users and contributing to the raising up of a class of professionally-trained persons; sustaining and expanding the mining and other industries using the natural resources of Canada; advising and assisting Canadian governments with the management of the mineral and other natural resources of the country; relating the scientific effort of Canada with those of international organizations and governments; and contributing to the furtherance of the earth sciences. Over most of its existence the organization has acted in all these directions, though the emphasis has constantly shifted.

First and foremost, in its 130 years the Survey has traversed, gathered data, mapped, studied, and reported on the geological and other aspects of Canada's vast area, until, today—notwithstanding the limited number of personnel available for the work—the geology of Canada is about as well known as other parts of the world of comparable extent. Often men from the Survey were the first scientists of any sort to study an area, and they almost always presented the first reports on the surficial and bedrock geology. Over the first 70 years, when mapping and reporting of many aspects of the lands of Canada were the Survey's most noteworthy, most highly publicized activities, the officers made their mark with the professional world and general public by their explorations and geographical discoveries. This was the natural outcome of the need to investigate the geology of the primitive, sparsely-settled, unsurveyed or poorly surveyed, parts of Canada at a time when no comparable agency of government existed that was capable of performing this function satisfactorily. The emphasis on geographical work was given stronger force by the government's action in 1877 expressly empowering the Survey to investigate all manner of physical and human resources of the areas in which it operated.

In keeping with this mandate, the officers from the 1850's onward explored, and to some extent carried the evidence of Canada's authority, into the Precambrian Shield country of the upper Ottawa, inland from the North Shore of the Gulf of St. Lawrence, and north and west of Lakes Huron and Superior. After 1870 they began mapping and studying the limitless western prairies and the massed mountain ranges of the Cordilleran region. Next the geographical work advanced into the Middle North in the 1880's to the basins of Nelson and Churchill, Athabasca, Peace, and Yukon Rivers. With the 1890's, field work moved into the Arctic by extending into the basins of the Mackenzie, Hamilton (Churchill), Koksoak and Thelon, inland from the shores of Hudson Strait and Bay, and, a little later, into the eastern and central Arctic coasts and islands. These were mainly reconnaissance surveys, which gave only a general idea of the geology exposed at the surface, especially in the vicinity of waterways and coastlines. That their principal scientific contribution perhaps was to geographical knowledge was reflected in the honours the leading officers of that day received from the major geographical societies. Yet these broad-ranging surveys revealed enough about the geological framework of the country to indicate numerous features of interest for further, more systematic investigations.

Concurrent with the reconnaissance mapping, more detailed surveys of entire map-areas were undertaken in the settled, most accessible parts of Canada, especially those that were of present or prospective economic or scientific importance. The Survey also conducted special large-scale mapping studies of such mining districts as the Nova Scotia coalfields. Its work, which was almost wholly oriented to the field mapping program in those days, was always a blend of reconnaissance, areal, and special studies, with the directors determining the priorities and courses the work should follow. It was recognized that ideally the work should proceed ahead of mining development, since accurate geological maps were fundamental for any effective resource development, but the manpower for such a program was always lacking. What staff was available often had to be largely diverted to activities that would maintain its political and public support, leading to a scattering of effort and distortion of the long-term program. Directors, faced with these pressures, often neglected the reconnaissance work and studies of potentially-important areas, in favour of work in established regions that was more immediately helpful to the mining industry. Some, also, put undue stress on field studies but failed to invest enough effort in obtaining adequate supporting paleontological or laboratory data. The directors differed markedly in their talent for accurately gauging and deploying the manpower resources available to them in the light of the permanent objectives and the immediate pressures with which the Survey was confronted.

But the work of geological mapping went forward steadily at the hands of 2, 10, 30, then 100 and more field parties per season, until with the advent of air-supported operations it became possible to speed up the geological mapping of Canada to a hitherto-unimagined pace. Through the energy, endurance, experience, and ability of five generations of field geologists, the geological map-

XXIV-3 THE BADLANDS OF THE MILK RIVER DISTRICT, SOUTHERN ALBERTA
G. M. Dawson's photograph of 1881 was printed by lithography as the frontispiece to his *Preliminary Report on the Geology of the Bow and Belly Region, N.W. Territory* (Montreal, 1883). The weathered sandstone rocks are known as "hoodoos."

XXIV-4 SCOW BRIGADE AT THE HEAD OF GRAND RAPID, ATHABASCA RIVER
For 50 years Survey parties proceeded by this route to their work in the Mackenzie basin and beyond.

XXIV-5 PACKTRAIN TRANSPORT IN THE ROCKY MOUNTAINS
Until the days of the helicopter, packtrains such as this afforded the only practical means for Survey parties to travel and work in much of Canada's rugged terrain.

XXIV-6 LEAVING THE FIELD OF WORK, MODERN STYLE
R. G. Blackadar's party returns by S55 helicopter to its base at Resolute in 1955. The dogs travelled in the aircraft, while the sledge was lashed beneath the plane.

ping has been executed at a very high level of competence, with virtually the whole landmass of Canada mapped in a general way, while a great many problem areas and subjects have received more concentrated study besides. The task is far from over, for geological mapping must go on indefinitely in the light of changing needs and insights. Provincial surveys, earth science departments in universities, and private mining and surveying corporations have contributed to unraveling the geological map of Canada, especially in the last 20 years, but the long-term achievement is mainly the accomplishment of the Survey. Its performance in the central role implied in its name and its original mandate, "of causing a Geological Survey of Canada to be made," must be rated as outstanding.

The physical exertions of the officers cannot be minimized. There were the difficulties of travelling along the rivers and lakes of the Precambrian or Cordillera in canoes manned largely by Métis and Indian crews, steering through dangerous white water, and making long, backbreaking portages between the navigable stretches. In the mountains there were the supply packtrains and the scrambles over or around fallen timbers and through thick brush, clambering up steep mountainsides to collect samples and to sketch or photograph long perspectives. Work in the extensive forests and marshlands of the Middle North presented further difficulties for travelling and finding rock outcrops. Common to these operations were the discomforts of biting insects, accidents on the water, falls from heights, axe-cuts, or occasional encounters with dangerous moose or bears. For most men, too, there were the summer-long separations from wives and children. From such experiences came the data which the men spent the ensuing winters reducing to reports and maps. By means of such efforts the courses of many of Canada's northern waterways were traced, and the character of much territory was described.

Today the field work is far different, covering complete map-areas, not merely the waterways, and concentrating on the geology, which is examined at depth and often in detail by collecting large numbers of samples for intensive study. In the 20 years between 1950 and 1970 it has extended over most of Canada, with nearly three times as much territory mapped geologically as in the previous 110 years. To some extent the exploitation of new technologies developed since the Second World War have once again given the Survey a prominent role in the geographical exploration of Canada's newest frontiers, the northwesternmost Arctic islands and sea beds of the continental shelves. Yet this exploration gains little of the recognition the earlier program of reconnaissances did. No doubt the changed character of the work accounts for some of the difference between the two responses. The present-day geologist has at his disposal base maps previously prepared by air photography, motorboats and motor vehicles, concentrated foods, insect repellents, weather-resistant clothing and bedding, and two-way radio communication. Above all, he has aircraft for transport into the area of his operations and to aid in the field traverses. Thanks to these, and to such laboratory-based specialties as geophysics, geochemistry, or age determinations, infinitely more can be learned today about the geology of Canada than was possible in earlier times.

The Survey's operations in the field lead to another role, one that may be termed educational in its broadest sense. Data secured through study of Canada's physical features are analyzed, assessed, arranged and made known to a whole spectrum of users, the form of the presentation varying according to the purpose and the audience it is hoped to reach. This objective is served above all by the publications, amounting today to over 3,900 reports and 12,000 maps, the reports alone occupying some 100 shelf feet in the Survey's library. They are of a great variety, including annual reports, special reports on local and regional geology, memoirs, papers, geological and museum bulletins, economic geology reports, and miscellaneous publications. The publications have reached record levels in recent years, an estimated 28,000 printed pages for the years 1965-70 including 312 papers, 21 memoirs, 66 bulletins, 7 economic geology reports, 11 miscellaneous reports, as well as 2,647 maps of several kinds. Moreover, much of the publication is done outside the Survey; in the years 1965-70, for example, 613 papers (over 100 a year) were published by members of its staff in scientific journals, or volumes of collaboration prepared under the auspices of other institutions or organizations. The Survey's findings are also offered to the public in the form of reports in the press and other news media, and in addresses before general and professional audiences. Special users may look to a good deal of unpublished material as well—oral and written counsel, technical files, economic reports, photographic and library collections, the national collections of type fossils and minerals, borings and drilling records and samples.

This information has been directed mainly at the professional and scientific groups, whose technical knowledge in earlier times was not developed far beyond the range of the well-educated layman. Over time most of the publications have grown more technical and less helpful to general readers; but this merely parallels the specialization of the sciences and of the scientific audiences for whom the publications are intended. However, the Survey is also aware of a responsibility to the general reading public and does its best to educate it on the importance of geological issues and on opportunities in the fields of the earth sciences. Good examples are the long-

XXIV-7 TAKING A FIELD OBSERVATION, 1962
C. K. Bell photographing an outcrop of brecciated anorthosite, Nelson River Group, in Manitoba.

term activities of preparing and distributing rock and mineral sample collections for schools and individuals; or the publication of various handbooks, aimed at less scholarly audiences, on such general subjects as rock and mineral collecting, the geology of national parks and other areas of local interest, and prospecting. While the National Museum was still under the Survey, the educational responsibilities of that institution were also served by publishing several excellent, though all-too-few, works on birds, plants, wildlife, and the native peoples of Canada. Moreover, while the means for conducting a major popular publications program are lacking, the Survey gladly gives every possible assistance to would-be writers of such popular or educational works.

The Survey has also helped to educate generations of university students and other young professionals, thousands of whom were trained in the field, and more recently, in the laboratories too. Nearly every officer taught a few dozen students how to operate the instruments, how to get around in the bush or on the water, and to recognize rocks and geological formations encountered in the field. They also imparted to the students their passion for accuracy and a conscientious attitude towards the work that turned many an undecided assistant into a career geologist. One appreciative assistant, who later became a university professor in the United States, afterwards wrote about his experience:

> I was a green college student when I assisted [H. C.] Cooke in the summer of 1917 when available Canadian students were so scarce because of the war. Working in the Matachewan District of Northern Ontario (Mem. 115, 1919) I remember how Cooke did his best to teach me some geology so I could be of some help to him. He could stand for an hour on an outcrop of Pillow lava, working out the top and bottom of beds, while the mosquitos were in clouds around his head, and the black flies came into his ears, nose and under his eyelids. He calmly chewed his plug of tobacco and ignored the insects. I remember how disgusted he was when I first got into an area of basic serpentized rocks and my traverse went crazy because of magnetic disturbance. Cooke was a hard, no-nonsense worker who accomplished a great deal in the field under adverse conditions before the days of helicopters.[2]

Several thousand undergraduates and graduate students gained similar experience through one or more seasons on a Survey field party. In addition to their training, they earned much-needed income to help support them through their studies, and were encouraged to make their future contributions as prospectors and mine managers, as consultants, university teachers, or as officers in the Canadian Survey, or those of the provinces or of other lands.

The Survey assisted in raising the technical and scientific competence of the industry through more formal participation in the instruction process. Throughout its history members of its staff gave guest lectures or offered special courses at the universities. Over the years, former Survey personnel have also constituted a significant fraction of the teaching staffs in earth-science departments at Canadian universities. They have comprised an important part of many new departments started since 1900, and in acquiring them the university gained the nucleus of a fully-qualified, highly-respected department. Some credit is due the Survey for helping to shape the quality of university instruction in the earth sciences, and through this, the training of future scholars and professional men. The scores of university teachers and researchers who were able to impart something of their Survey experience to successive generations of university students at almost every Canadian university include such men as F. D. Adams, J. C. K. Laflamme, L. W. Bailey, J. J. O'Neill, R. W. Brock, M. Y. Williams, H. C. Gunning, J. B. Mawdsley, E. L. Bruce, J. T. Wilson, and E. R. W. Neale; or A. C. Lawson, R. A. Daly, T. T. Quirke, and R. C. Emmons, who worked at universities in the United States. These few names, from among the many, give some idea of the extent of the Survey's part in elevating the standards of training and practice in the earth sciences. The many contacts with the universities and with geologists in industry forged a strong bond of sentiment and a community of interests that powerfully strengthened the Survey in its dealings with governments and helped the officers in the performance of their duties.

A central purpose behind the establishment and continuance of the Survey—perhaps the main purpose as far as successive Canadian governments were concerned—was their expectation that it would facilitate the utilization of those resources of Canada that fell within its range of operations. For the most part, though never entirely so, the development of the mineral resources of Canada was the particular goal they had in view, and the expected beneficiary of the Survey's endeavours was the mining industry. "I think one ultimately judges the work, if you will, of the Geological Survey by looking at the Canadian mining industry,"[3] C. M. Isbister said in 1968. These industries have grown from almost nothing when Logan began his labours in 1842, to a present level of production valued at nearly $6,000 million per annum. They directly or indirectly employ a large part of the nation's labour force, account for much of Canada's material prosperity, and are a major factor in its foreign trade and international balance of payments. They are also the most far-reaching element in the nation's economy, being important to every province and territory. They have caused railways, highways, airfields and pipelines to be built, have accounted for most of the settlements scattered around the Canadian Shield and Cordilleran regions, and have added as many as one million persons to the population of the prairie provinces between 1947 and 1968 as a result of the oil and gas discoveries. A significant, though indeterminate, share of this development of Canada's mining economy unquestionably results from the long-range mapping and other services provided by the Geological Survey.

Logan was at pains to affirm the concern of his organization to assist the mining industry of Canada as early as 1854 when he appeared before the legislative committee appointed to investigate the Survey:

> My whole connection with Geology is of a practical character. I am by profession a Miner and a Metallurgist and for many years, was one of the active managing partners in an establishment in Wales, where we annually melted 60,000 tons of copper ore, and excavated 60,000 tons of coal. It was my constant occupation to superintend and direct the minutest details of every branch of the business. A due regard to my own interests forced me into the practice of Geology, and it was more particularly to the economic bearings of the Science that my attention was devoted.[4]

At this time—as the works of Logan and Hunt testify—the economic functions entailed assisting the mining industry at every level—exploring, mapping, and working out the geology and favourable geological environments for the occurrence of given minerals as an aid to prospectors; publishing reports, and preparing displays at exhibitions in Canada and abroad; and giving first-hand advice in the field on practical mining questions. Moreover, prior to the establishment of the Mines Branch, Hunt and his successors in the laboratory also assisted the development of the industry through studying, reporting, and advising on mining techniques and on new processes of concentrating, smelting, and refining ores. They experimented with new uses, collected statistics of mineral production, and helped to develop markets for the products.

After 1907 the Survey's role in respect of the mineral industries was limited to studying every aspect of Canadian geology, identifying specific localities where mineral occurrences were most likely to be found in economic quantities and concentrations, and conducting or assisting basic researches intended to help achieve this goal. With improved knowledge and experience of the geological framework of Canada, these localities can be identified with ever-increasing precision. Modern technology, new methods in geophysics, geochemistry and geochronology, and new concepts of tectonics are making it possible to point out localities favourable for mineral occurrence that in former times would never even have been suspected.

The prospector is thus assisted both in positive and negative ways—positively, by focusing his attention on locations in previously unexamined districts, where promising indications such as the extension of rocks or structures known to contain valuable mineral deposits give his efforts the greatest chances for success; and negatively, by ruling out areas where the prospects are poor or nonexistent. By providing accurate geological maps based on careful field study, the Survey facilitates the rapid expansion of mining whenever it is made possible by improved communications or mineral prices, or the discovery of an important relationship, such as the fact that one of the rock units carefully drawn on Collins' Blind River map of 1925 was uraniferous in quantities capable of economic exploitation. Another service to the mining industry, especially valuable in a time of changing technology, is the development of instruments and surveying techniques that aid in the discovery of various kinds of mineral deposits or other phenomena.

In all these ways the Survey has contributed to building up the gigantic Canadian mining industries of today, making them leaders not only in terms of scale, but in technical progressiveness and efficiency. As existing mineral deposits are depleted, the Survey continues to assist in the discovery of new mineral occurrences that can maintain the industry at its present advanced levels. How can the real contributions to the mining industry be properly evaluated? Certainly the industry could never have achieved its present size or effectiveness without the thousands of maps and reports on all manner of geological conditions from the iron ores of Bell Island, Nfld., to the mercury of Pinchi Lake, B.C., to outlining the 465,000-square-mile basin of unmetamorphosed sediments in structures favourable to the concentration of fossil fuels in the northernmost Arctic islands of Canada. A. H. Lang, who investigated over 100 such examples of economic services to work out some method of making a cost-benefit analysis, concluded, on a most conservative estimate and using only directly traceable cases, that the full cost of the Survey had been repaid many times over from these 100 cases alone. In fact, Lang calculated that the $96,000,000 spent on the Survey from its inception to 1968 amounted to no more than 1.8 per cent of the value of just the uranium ore discovered and developed as a result of the services of the Survey to that industry from 1918 onward. And this, of course, is only one of a half-dozen major mining industries, as well as dozens of lesser ones, in whose development it has participated in a fundamental way.

Though assisting the mining industry is a major part of the Survey's practical achievement, it has never constituted its entire economic function, even in those most economically-oriented periods of the two world wars or the depressions of the 1890's and 1930's. In the first 70 years noteworthy service was done in bringing the whole range of Canada's potential resources to the attention of settlers and developers in Canada and other countries. The Survey also helped discover adequate water supplies for agricultural uses and for urban living, investigated possible uneconomic or hazardous construction and habitation sites (ironically, however, not including the site of the Victoria Memorial Museum, which was its own home for many years). Today it continues to study and report on land-use, settlement, construction, and recreation problems arising from, or relating to, geological factors. It worked, and is working, on the characteristics of the permanently frozen ground that extends over a large part of Canada; on features of possible significance for tourist and recreational purposes such as the remains of dinosaurs or of mammoths; on aspects of structures like the Rocky Mountains or Niagara Falls; or on potential park sites. Moreover, in keeping with the heightened public interest in outdoor recreation and concern for protecting the environment against excessive abuse, the Survey has expanded its activities in such directions as detecting sources of pollution by geochemical methods, paying more attention to avoiding the damaging side-effects associated with the mining of certain types of deposits, and devoting greater effort to studies in engineering geology and to Quaternary geology generally.

* * *

THE GEOLOGICAL SURVEY, as a component part of the federal public service, has always had the privilege and duty of serving the governments and people of Canada in whatever way it is called to do, and with whatever means it has at its command. It has assisted governments in a whole range of tasks, from upholding Canadian sovereignty in the Yukon and Arctic islands, to advising on the location of military camps or public buildings. The findings arising from its chief work of tracing, mapping, describing, and assessing geological information have always been available to governments as well as industry, though it was industry rather than government that usu-

XXIV-8 DEFENDING CANADIAN SOVEREIGNTY
The Canadian government ship *Neptune*, commander A. P. Low, spent the winter of 1903–04 at Fullerton, N.W.T., adjacent to the whaling vessel *Era*, the sole United States ship wintering in Hudson Bay.

ally made the most use of those findings. In recent years this has been changing.

Today the federal government is becoming much more involved with managing the national economy, the natural resources, and the environment. As a result, the Survey has become more closely identified with furnishing mineral resource evaluations and inventories, which serve as a basis for formulating government policies appropriate to the public interest in such areas as mining resource or northern regional development, land-use and environmental conservation programs, and the promulgation of a national science policy. Indeed, as geological surveying and research capabilities are built up in other parts of the public and private sectors, the Survey's roles of advising the federal government, investigating special problems, or administering aspects of its programs tend to assume even more prominence.

Some of the Survey's greatest contributions are associated with its career as a long-lived institution within the Canadian government system. A great deal of administrative experience was secured merely by functioning under a broad range of conditions, and responding to innumerable challenges. As one of the first, if not *the* first, part of government especially formed to undertake a scientific activity, it fulfilled an important and unique function in the administrative history of Canada. It discovered and reported objectively on scientific data, and it was staffed by qualified persons, appointed not through the usual patronage channels but hand-picked by the director because of their special abilities. For 70 years, down to the First World War, it was probably the largest employer of scientific staff in the Canadian public service in Ottawa.

The work was scientific in being aimed at seeking complete truth from the evidence of the geological and natural history phenomena. The directors fought persistently to maintain this scientific character, while satisfying the government and public that the needs of the developing economy were not being neglected. They tried to give their men scientifically respectable tasks in which they could feel they were developing their skills and contributing in a positive way to the welfare of the country. By insisting on the highest possible standards for its own staff in order to obtain the talented scientists required for effective conduct of the work, they led the struggle in government to upgrade the academic qualifications of civil servants. From the 1870's a few men were being encouraged to do postgraduate studies in Germany or the United States. Early in the present century the Survey was looking to the Ph.D. or its equivalent as the desired professional standard; and was helping its best prospective young scientists to higher levels of specialization by supporting them while they were earning their doctorates at foremost universities in the United States. By the time of the First World War these high standards of professional training were being passed along to the Civil Service Commission, which could take them over and apply them for the scientific and professional staffs that were now springing up in various sectors of the public service. Having played a part from the 1880's onward in the struggle for adequate pay and recognition for scientific personnel, the staff also took a lead in the drive to organize a professional association within the civil service in order to defend this status.

The Survey has striven throughout its history to manage its operations with a minimum of outside control. The work did not readily lend itself to control by non-specialists, while success of individual tasks was often dependent on the officers being allowed wide discretion to manage their assignments. The freedom of action tested the mettle of the staff and enhanced its morale. To protect the highly-prized autonomy of the agency the directors tried to ensure that the work was conducted as economically and efficiently as possible, so no complaints could be raised on those grounds. In the nineteenth century this chiefly meant maintaining as many parties as they could at field mapping, meeting as many requests from the public as possible, and publishing promptly and voluminously. The directors were especially zealous to guard the reputation of the Survey and its men for industry, honesty, and integrity. High ethical standards, intensified by the scientific impartiality that underlay their work, characterized the actions of the staff in the performance of its duties. The strict orders of the first director, later re-enforced by legislation, prohibited speculation through using prior or special knowledge for gain, or acting in collusion with others. The Survey throughout its history has maintained this reputation of unbiased public service, and rarely has the honesty of even a single staff member been questioned.

By the beginning of this century, increasing specialization within the agency made it desirable for Brock to introduce a number of functional divisions that persisted until the 1930's when the Survey's activities were again reduced to a part of their pre-1914 range. Since 1945, however, as the geological work grew more specialized, the situation has been met by hiring a wide variety of earth-science specialists and working constantly at organizing them in the most efficient, practical manner. In more recent years the configuration of divisions and sections was frequently re-arranged to conform more satisfactorily to the changing functional and activities patterns. But, throughout, the effort has been to conduct the scientific and service operations with maximum efficiency, which meant also in as autonomous and self-contained a manner as possible. The Glassco Commission, which reported in 1963, expressed some concern about the perils of self-sufficiency being carried to extremes, but gave the Survey a clean bill of health on its current operations:

> At one time the Geological Survey was completely autonomous. There are signs that the proper desire to maintain an efficient and progressive organization may lead the Branch to become too self-contained. Some overlapping may be expected, which is not harmful unless it leads to competition for scarce personnel or to duplication of expensive facilities. There is a danger of this becoming significant in the Geological Survey Branch, which now maintains an independent library, separate map drafting facilities and increasingly elaborate chemical laboratories.
>
> The Survey is an efficient organization, well staffed, with high scientific standards and an excellent national and international reputation. It has expanded rapidly and will no doubt continue to do so, particularly in areas related to geological field work. This is proper, but care must be taken not to duplicate or overlap the work of existing government facilities.[5]

Despite its efforts to win immunity from interference by providing efficient, honest, objective, economical service, the Survey was unable to retain its wide autonomy into modern times. Gradually, in the process of bringing the Survey into line with the highly-structured, regimented, working-to-rule, paper-ridden bureaucracy of today, something of the special character of the agency and the morale of the professional and scientific staff has been eroded. As a government agency devoted to a certain range of research activities, the Survey joins with other scientific organizations in fearing that excessive controls can lead to the diversion of much effort away from its important and proper tasks. Perhaps its experience may yet help guide governmental administrative practice towards a less regimented, more efficient style for managing the operations of the specialized agencies.

On firmer ground, the Survey has certainly pointed the way to the entire federal civil service by its careful experimentation over many years with the benefits and drawbacks of decentralized operations. While such moves are normally made for political, economic, or administrative reasons, in the Survey's case they have been adopted mainly out of scientific considerations, that is, the ability to operate certain special activities more effectively, and the prospect of improved collaboration with local scientific and business groups.

A dominant quality of the institution was a strong sense of responsibility; in the best traditions of the public service, the Survey strove to carry out the instructions it received. When directors felt there was a conflict between these and what they regarded as the long-term national interest, they were prepared to use the legitimate channels available to a government agency to try to influence their administrative and political superiors in the desired directions. When this failed, they conformed to the in-

XXIV-9 PUTTING MODERN TECHNOLOGY TO WORK
With these two aircraft, the Short Skyvan CF-GSC (*front*) and the Queenair CF-WZG-X, the Survey tested various geophysical methods and apparatus, developing its airborne gamma-ray spectrometer capability to a point where it can be used to measure variations in soil moisture conditions over wide areas through minute changes in the concentrations of radioactive elements.

structions of their superiors. The most successful directors—Logan, Brock, and Harrison—were pre-eminently men who could adjust to changing conditions and use them to strengthen the organization and the importance and value of its contributions. At their best, the directors were keenly sensitive to the currents of changing times. By anticipating developments and making the necessary adjustments in advance, they minimized the disruptions that might otherwise have occurred. The way in which the general criticisms expressed by recent inquiries into government science policies and environmental pollution have been so largely anticipated, is in this long tradition. The flexibility of the Survey has been one of its strengths, and the principal explanation for its continued survival down to the present.

Important in the Survey's modern role as an agency and representative of government are its activities in the foreign aid and scientific fields. Here the branch and its officers, in response to requests to the Canadian government from those of other lands, put their knowledge and experience at the service of those governments to help them with their mineral resource and mining development problems. Scientific exchanges with other lands that also have a semi-official character are participation in the International Union of Geological Sciences or similar organizations, or in projects of international collaboration such as the Upper Mantle Program, the world maps of geological features, or the meetings in Canada of the International Geological Congress or other associations. With the approval and backing of the federal government the Survey represents Canada in the international scientific world, and at the same time helps to co-ordinate Canadian scientific activities towards the achievement of certain desired national and international programs.

Finally, in the performance of all these functions, the Survey has been mindful of, and concerned with, promoting the development of the sciences with which it is involved. Even in the most narrowly practical, materialistically-inclined periods it was not content with undertaking work that was merely of direct benefit to the economy; it always sought to make full use of the scientific talents of its staff to help extend the frontiers of scientific knowledge. Economic and scientific activities are by no means antithetical, as Logan pointed out in 1854:

> Thus economics leads to science, and science to economics. The physical structure of the area examined is of course especially attended to, as it is by means of it that the range or distribution of useful materials, both discovered and to be discovered, can be made intelligible . . . . I do not describe fossils, but use them. They are geological friends who direct me in the way to what is valuable . . . . Some tell of Coal; they are cosmopolites; while some give local intelligence of Gypsum, or Salt, or Building Stone, and so on. One of them whose family name is *Cythere,* but who is not yet specifically baptized, helped me last year to trace out upwards of fifty miles of hydraulic limestone.[6]

The solution of a general problem like the conditions surrounding the emplacement of ore deposits is both an event in the progression of the science of metallogenesis and an important discovery for the furtherance of the mining industry. Dawson, Brock, and Collins to a lesser degree, followed the principle that properly selected field subjects should be practical in economic terms as well as applying the scientific talents of personnel to the best advantage.

The men of the Survey have always been obliged to keep abreast of the latest scientific advances in order to do the best job possible. Recent findings, from newly-recognized fossil types to the latest version of the continental drift thesis, have been used to help gain more accurate, useful understanding of Canada's geology. As these discoveries were tested in the field they were often found to need revision, or sometimes they suggested new approaches that have general applicability. Hence the best scientists have been more than borrowers; they have improved the understanding of particular problems. Throughout the Survey's history the insights of such men as Logan and Hunt; G. M. Dawson, A. C. Lawson, and F. D. Adams; A. E. Barlow, Collins, and M. E. Wilson; H. V. Ellsworth, W. A. Bell, F. H. McLearn, H. W. Frebold, and C. H. Stockwell—as well as many of the present generation of specialists—have brought them well-deserved reputations as innovative scientists and as world authorities in their fields.

These original contributions did not come about entirely as a matter of course, through the normal functioning of the Survey and the reactions of a Logan or Dawson to puzzling phenomena encountered in the field. The most original advances required a considerable degree of faith from directors or efforts by the men themselves to try out new ideas and methods. Scarce resources had sometimes to be diverted from comparatively standard but politically-desirable work in order to allow a specialist to work at such a problem. In buoyant times like those of Brock, or especially of Harrison, it was relatively easy to find the means to secure the specialists and the necessary equipment to push forward along frontiers of scientific achievement that were important for Canadian geological progress. But this was more difficult to do in the hard years which were more typical of the Survey's history. Directors like Dawson, or the often-maligned Selwyn deserve more credit than they have received for persisting in the face of such difficulties to strive for world standards of excellence. Somewhat the reverse applies to the Collins-Young-Camsell era, when the management lacked the imagination to pursue this goal, and threw obstacles in the way of gifted men like Ellsworth and McLearn, who were forced to make their contributions to science largely on their own.

XXIV-10 DIGBY JOHNS McLAREN
Appointed director of the Geological Survey of Canada in 1973.

The Survey kept more or less abreast of the latest advances in science and technology over most of its history, though in some periods there were lapses in certain areas. While Precambrian studies seem always to have been maintained at the highest possible level, surficial geology was more often inadequately served than otherwise, and the paleontological work seems also to have fallen behind in the late nineteenth century. The W. A. Bell-Hanson-Harrison years faced a similar burden of overcoming another period of neglect in these fields, and especially of winning a position of stature in the laboratory sciences. In these years, as in the Brock years, the Survey not only caught up with past neglect, but also moved forward to reach the highest world standards in many areas.

After 1950, the Survey began to undertake special studies and researches into general principles, assisted by the new laboratory tools and by the findings of the developing subsciences. Important contributions in radioactivity and age determination work, in paleomagnetic studies, in studies of various rock types and structures, and in other areas, ensued. In this latest period of specialization many of its leading scientists have become recognized world authorities. They exchange visits with their fellow scholars around the world to assist in solving such worldwide geological problems as those of ultrabasic rocks and paleontological successions, or applying geochemical or geophysical air-supported survey technologies to the practical needs of other lands. The modern-day style and range of scientific research, as well as the more significant achievements over the past ten years, 1962-1972, are expressed in this selected list of outstanding contributions:

> It is impossible in a brief space to mention more than a few examples to illustrate the scope of the Survey's contributions to the geosciences. In 1963 L. W. Morley and A. Larochelle first attributed the magnetic reversals on the floor of the Pacific Ocean to ocean floor spreading. The work of E. T. Tozer in developing the systematic time scale of the Triassic has gained world-wide recognition. C. H. Smith and T. N. Irvine's studies of the structure and cyclic differentiation of the Muskox ultramafic intrusion gave new insight into the provenance and genetic processes of these bodies. J. G. Souther's studies of the Pliocene to Recent Mt. Edziza volcano, now exposed through great vertical range by erosion, has added to knowledge of evolution of volcanoes on the Pacific Rim. A major contribution to the subdivision of Precambrian time based on tectonic analysis and isotopic geochronology of the Canadian Shield was made by C. H. Stockwell. The structural analysis of the Foothills of the Rocky Mountains by R. J. W. Douglas, R. A. Price and others provided a geological basis for interpretation of geophysical anomalies as they apply to the search for fuels. The work of G. A. Gross on iron formations and G. A. Gross and R. H. Ridler on equivalent exhalative facies has significantly affected appraisals of world resources of iron and other metals. [R. W.] Boyle's recent work on the geochemistry of silver has already become a standard reference. The discovery and interpretation of an exceptional fossil vertebrate record by A. M. Stalker and C. S. Churcher has added significantly to our knowledge of Pleistocene time including possible evidence of the earliest presence of man in Canada. The work of B. D. Loncarevic and others on the Mid Atlantic Ridge has materially advanced our knowledge of the rate of continental drift. Recent development by R. J. Fulton and others of a new method of rapid reconnaissance mapping of surficial materials may well have world-wide applications.[7]

As for the over-all effort, the lists in the annual bibliographic volume of the NACRGS, *Current Researches in the Geological Sciences,* indicate that even in this present period, when research efforts are widely diffused throughout Canada, between 15 and 25 per cent of all titles of published works represent work that was performed by the staff of the Survey, or was done under its auspices. A most useful indication of the vast scientific advances achieved by the Survey over time can be found in the two great volumes of synthesis published almost a century apart: Logan's *The Geology of Canada* (1863), and the fifth edition of *Geology and Economic Minerals of Canada,* edited by R. J. W. Douglas (1970).

* * *

THE SURVEY'S CONTRIBUTIONS to Canada's progress and well-being are an important facet of the history of this country in the last 130 years. Much of Canada's achievement as a nation has revolved around the exploration and settlement of its territory and the development of its natural resources for the use of mankind. No agency of government can claim a better record in this phase of Canadian history than the Geological Survey. The one thousand or so scientific and professional personnel who toiled for the Survey over its 130 year history, for the most part with limited technical aids, mapped an enormous territory and secured a great quantity of useful knowledge of its geological structures, rocks, minerals, soils, and other resources, and of their value for the Canadian people. Canada's huge present-day mining industries to a considerable degree reflect their work, which serves a great variety of other economic and public functions as well. Finally, the Survey offers Canada and the world a fine example by its long, distinguished record of public service, and by the remarkably high quality of its staff over the entire period. Some of the ablest, most devoted public servants—in a country where the civil service has generally been one of the most honoured outlets for the energies and talents of its people—chose the Survey as the vehicle by which to pursue their lifetime careers of service. By their efforts they helped elevate the international reputation of Canadian science and of the Canadian public service while they went about their business of assisting the small colony of 1842 to become the mighty nation of today.

NOTES

GSC: Geological Survey of Canada
RSC: Royal Society of Canada
PAC: Public Archives of Canada

Chapter 1. FOUNDING THE GEOLOGICAL SURVEY OF CANADA

1. McGill University Archives, The Logan Papers, Box M15943, W. E. Logan to James Logan, 6 Sept. 1841.
2. H. B. Woodward, *The History of the Geological Society of London* (London: Burlington House, 1907), p. 14.
3. Sir Edward Bailey, *Geological Survey of Great Britain* (London: Thomas Murby, 1952), p. 26.
4. Charles Schuchert, "A century of geology—the progress of historical geology in North America," *American Journal of Science,* Fourth Series, Vol. 46, No. 271 (July 1918), p. 70.
5. John Warkentin, ed., *The Western Interior of Canada: a Record of Geographical Discovery, 1612-1917* (Toronto: McClelland and Stewart, 1964. The Carleton Library No. 15), p. 125.
6. Upper Canada, House of Assembly, *Journals,* 1831-32, p. 100 (18 Jan. 1832).
7. *Ibid.,* 1832-33, p. 52 (10 Dec. 1832).
8. *Ibid.,* 1836, p. 179 (7 Feb. 1836).
9. *Ibid.,* 1836, Part III, Appendix, pp. 3-4.
10. *Ibid.,* 1836-37, p. 58 (17 Nov. 1836).
11. *Ibid.,* p. 71 (14 Dec. 1836).
12. Canada (Province), Legislative Assembly, *Journals,* 1841, p. 559 (10 Sept. 1841).
13. Logan Papers, Box M15943A.
14. *Ibid.,* W. E. Logan to James Logan, 2 Oct. 1841.
15. Montreal *Gazette,* 16 May, 1842.
16. PAC, R.G.5, B.3, The Bagot Correspondence, Vol. 6, No. 35, pp. 134-135, Bagot to Stanley, 18 Feb. 1842.
17. B. J. Harrington, *Life of Sir William E. Logan, Kt., Ll.D., F.R.S., F.G.S.,&c.* (Montreal: Dawson Brothers, 1883), p. 128.
18. *Ibid.,* p. 132.
19. Logan Papers, Box M15943A, A. Murray to W. E. Logan, 2 Dec. 1841.
20. Montreal *Gazette,* 23 June 1842, quoted *in* Harrington, *Life of Sir William Logan,* p. 131.
21. Logan Papers, Box M15943, Anthony Murray to Hart Logan, 16 June 1842.
22. *Ibid.,* Box M15943A, A. Murray to W. E. Logan, 11 July 1842.

Chapter 2. THE SCIENCE AND THE MAN

1. Charles Lyell, *Principles of Geology* (London: John Murray, 1833, 3 vols.), III, p. 1.
2. *Ibid.,* p. 2.
3. *Ibid.,* p. 3.
4. *Ibid.,* p. 6.
5. T. G. Bonney, *Charles Lyell and Modern Geology* (London: Cassell, 1895), p. 219.
6. Sir Archibald Geikie, "Obituary of Sir Charles Lyell," *Nature,* Vol. 12 (26 Aug. 1875), p. 326.
7. Baron G. Cuvier, *Discours sur les Revolutions de la Surface du Globe* (1826), pp. 8-9, quoted *in* F. D. Adams, "Sir Charles Lyell, His Place in Geological Science . . . .", pamphlet reprinted from *Science,* Vol. 78, 2018, 1 Sept. 1933, pp. 177-183), p. 4.
8. Abraham Gesner, *Remarks on the Geology and Mineralogy of Nova Scotia* (Halifax: Gossip and Coade, 1836), pp. 121-122.
9. British Association for the Advancement of Science, Seventh Meeting, Liverpool, Sept. 1837, *Report,* Vol. 6 (1837), pp. 83-85.
10. C. G. Winder, "Logan and South Wales," *in* Geological Association of Canada, *Proceedings,* Vol. 16 (1965), p. 111.
11. *Ibid.,* p. 113.
12. Bailey, *Geological Survey of Great Britain,* p. 33.
13. Logan Papers, Box M15943A, W. E. Logan to James Logan, 6 Sept. 1841.
14. Geological Society of London, *Transactions,* Second Series, Vol. 6 (1841), pp. 491-497; summarized in *Proceedings,* Vol. 3 (1840), pp. 275-277.
15. *Ibid.,* p. 494; and *Proceedings,* Vol. 3 (1840), p. 275.
16. Logan Papers, Box M15943B, W. E. Logan to Dr. G. M. Douglas, 18 Mar. 1845.
17. GSC, Director's Letterbook, 1865-69, p. 106, W. E. Logan to Dr. James Stinson, 4 May 1866.
18. Harrington, *Life of Sir William Logan,* pp. 397-398.

Chapter 3. EARLY TRIALS AND TRIUMPHS

1. GSC, *Report of Progress,* 1843, p. 34.
2. *Ibid.,* p. 50.
3. Harrington, *Life of Sir William Logan,* p. 154, quoting Logan's journal entry of 25 July 1843.
4. GSC, *Report of Progress,* 1845-46, p. 40.
5. *Ibid.,* p. 42.
6. *Ibid.,* p. 73.
7. *Ibid.,* p. 116.
8. W. E. Logan to Henry De la Beche, 12 May 1845, cited *in* Harrington, *Life of Sir William Logan,* p. 233.
9. Canada (Province), Legislative Assembly, *Journals,* 1846, Appendix WW, "Return to an address . . . copies of a Report or Reports from E. S. de Rottermund, heretofore Chemist to the Provincial Geological Department, or to the Provincial Government

. . ., p. [8]. Item 8, de Rottermund to Provincial Secretary, 23 April 1846.
10. Harrington, *Life of Sir William Logan,* p. 136.
11. Canada (Province), *Report of the Select Committee on the Geological Survey, printed by Order of the Legislative Assembly* (Quebec: Lovell and Lamoureux, 1855), p. 36. (Also in Legislative Assembly, *Journals,* 1854-55, Appendix L.)
12. Quoted *in* Harrington, *Life of Sir William Logan,* p. 180.
13. *Ibid.,* pp. 180-181, W. E. Logan to A. Murray, 7 March 1844.
14. *Ibid.,* pp. 234, 236, W. E. Logan to Henry De la Beche, 12 May, 1845.
15. Logan Papers, Box M15943B, W. E. Logan to James Logan, 13 Feb. 1850.
16. *Ibid.*
17. Robert Bell, "Alexander Murray, F.G.S., F.R.S.C., C.M.G.," *Canadian Record of Science,* Vol. 5 (1892), p. 82.
18. GSC, *Report of Progress,* 1852-53, p. 8.
19. Montreal *Gazette,* 29 Sept. 1851.
20. Canada (Province), *Report of the Select Committee on the Geological Survey* (1855), p. 26.
21. *Ibid.,* p. iv.
22. Montreal *Gazette,* 23 Aug. 1855.
23. *Ibid.,* 23 June 1855.
24. Logan Papers, Box M15943C, Robert Ellice to W. E. Logan, 15 Nov. 1855.
25. Geological Society of London, *Proceedings,* Vol. 12 (1856), p. xxiv.
26. See Frontispiece. We thank the Royal Canadian Institute, Toronto, owners of the portrait, and the Geological Association of Canada for the loan of their colour plate.
27. Harrington, *Life of Sir William Logan,* p. 313.
28. *Ibid.,* pp. 319-320. See also C. G. Winder, "Where is Logan's silver fountain?" Geological Association of Canada, *Proceedings,* Vol. 18 (1967), pp. 115-118.

Chapter 4. GEOLOGY OF CANADA, 1863

1. GSC, *Report of Progress from its Commencement to 1863* (Montreal: Dawson Brothers, 1863), p. iv.
2. Robert Bell, *Sir William E. Logan and the Geological Survey of Canada* (Ottawa: n.p., n.d.), p. 5.
3. PAC, M.G.29, C.23, The Robert Bell Papers, Vol. 1, W. E. Logan to Robert Bell, 14 May 1868.
4. GSC, *Report of Progress,* 1858, p. 328.
5. *Ibid.,* 1853-54-55-56, p. 250.
6. *Ibid.,* 1857, p. 147.
7. *Ibid.,* 1853-54-55-56, p. 180.
8. Ontario Nickel Commission, 1917, *Report,* p. 29.
9. GSC, *Report of Progress,* 1853-54-55-56, p. 245.
10. *Ibid.,* 1858, p. 7.
11. GSC, *Report of Progress from its Commencement to 1863,* pp. 42-43.
12. W. E. Logan, "Remarks on the fauna of the Quebec Group of rocks, and the Primordial Zone of Canada, addressed to Mr. Joachim Barrande," *Canadian Journal,* New Series, Vol. 6 (1861), p. 44. See also *Report of Progress from its Commencement to 1863,* pp. vi-vii.
13. Wallace M. Cady, "Tectonic setting and mechanism of the Taconic slide," *American Journal of Science,* Vol. 266 (Summer 1968), pp. 563-564.
14. T. S. Hunt, "On some points in chemical geology," Geological Society of London, *Quarterly Journal,* Vol. 15 (1859), p. 493.
15. F. D. Adams, *Biographical Memoir of Thomas Sterry Hunt, 1826-1892* (Washington: National Academy of Sciences, 1932. Biographical Memoirs, Vol. 15—Seventh Memoir), p. 217.
16. T. S. Hunt, *A Notice of the Chemical Geology of Mr. D. Forbes* (Montreal, 1867), p. 3. (Pamphlet in GSC library.)
17. T. S. Hunt, "On the chemical and mineralogical relations of metamorphic rocks," *Canadian Naturalist and Geologist,* Vol. 8, No. 3 (June 1863), p. 195. Earlier versions were published by Hunt in 1858 and 1859 in *American Journal of Science* and Geological Society of London, *Quarterly Journal.*
18. P. Frazer, "Thomas Sterry Hunt, M.A., D.S.C., LL.D., F.R.S.," *American Geologist,* Vol. 11 (1893), p. 10.
19. GSC, *Report of Progress,* 1853-54-55-56, p. 480.
20. T. S. Hunt, "On the theory of igneous rocks and volcanos," *Canadian Naturalist and Geologist,* New Series, Vol. 3, No. 3 (June 1858), p. 201.
21. F. D. Adams, *Biographical Memoir of Hunt,* pp. 221-227.
22. Harrington, *Life of Sir William Logan,* p. 331.
23. *Ibid.,* p. 360.
24. *Canadian Naturalist and Geologist,* New Series, Vol. 1 (1864), pp. 66-67.
25. "President's address, Canadian Institute, 8 Jan. 1859," *Canadian Journal,* New Series, Vol. 4, No. 20 (March 1859), p. 93.
26. Quoted *in* Harrington, *Life of Sir William Logan,* pp. 351-352.

Chapter 5. FROM PROVINCIAL TO DOMINION SURVEY

1. Harrington, *Life of Sir William Logan,* pp. 358-359.
2. *American Journal of Science and Arts,* Second Series, Vol. 38, No. 113 (July-Nov. 1864), p. 232.
3. GSC, *Report of Progress from its Commencement to 1863,* p. 48.
4. GSC, *Report of Progress,* 1863 to 1866, p. 15.
5. *Ibid.*
6. British Association for the Advancement of Science, 34th Meeting, Bath, Sept. 1864, *Report,* p. lxxv.
7. William King and T. H. Rowney, "On the so-called "Eozoonal Rock"," Geological Society of London, *Quarterly Journal,* Vol. 22 (1866), p. 215. For the controversy, see also Charles F. O'Brien, *Sir William Dawson, A Life in Science and Religion* (Philadelphia: American Philosophical Society, 1971. Its Memoir No. 84), Chapter VI.
8. Geological Society of London, *Transactions,* Vol. 23 (Aug. 1867), p. 260.
9. Quoted *in* J. M. Harrison and E. Hall, "William Edmond Logan," Geological Association of Canada, *Proceedings*, Vol. 15 (1963), p. 39.
10. GSC, *Report of Progress,* 1863 to 1866, p. 93, footnote.
11. *Ibid.,* 1866 to 1869, p. 321.
12. *Ibid.,* pp. 472, 475.
13. *Ibid.,* 1863 to 1866, p. 223.
14. GSC, Director's Letterbook, 1865-69, p. 260, W. E. Logan to John Machar, 22 July 1867.
15. *Ibid.,* p. 305, Logan to Sir John Rose, 14 Feb. 1868.
16. *Ibid.,* p. 406, Logan to A. R. C. Selwyn, 4 Dec. 1868.
17. Canada, Order in Council, PC928a, 19 Nov. 1869.
18. *Ibid.,* PC359, 10 March 1868, enclosure, Logan to Sir John Rose, 29 Feb. 1868
19. GSC, Director's Letterbook, 1865-69, pp. 305-306, Logan to Rose, 14 Feb. 1868.

20. Canada, Order in Council, PC359, 10 March 1868, encl. Logan to Rose, 29 Feb. 1868.
21. McGill University Archives, Sir J. W. Dawson Papers, Logan to J. W. Dawson, 7 May 1868.
22. Canada, House of Commons, *Report of the Select Committee . . . to obtain Information as to Geological Surveys &c.&c.,* (Ottawa: Maclean, Roger, 1884), p. 93. (Also published in House of Commons, *Journals,* 1884, Appendix No. 8.)
23. GSC, Director's Letterbook, 1865-69, pp. 305-306, Logan to Rose, 14 Feb. 1868.
24. James Douglas, *A Memoir of Thomas Sterry Hunt, M.D., LL.D.* (Philadelphia: MacCalla, 1898), p. 23.
25. GSC, Director's Letterbook, 1865-69, p. 418, Logan to Rose, 30 Dec. 1868.
26. A. E. B.[arlow], "Dr. Alfred R. C. Selwyn," *Ottawa Naturalist,* Vol. 16 (Dec. 1902), p. 172.
27. Robert Bell Papers, Vol. 27, for Extract from Debate in Legislative Assembly of Victoria, 19 June 1867.
28. Canada, House of Commons, *Report of the Select Committee* (1884), p. 103.
29. James Douglas, "Biographical sketch of Thomas Sterry Hunt" (*Transactions* of the American Association of Mining Engineers, meeting of June 1892), p. 6.
30. Quoted *in* Harrington, *Life of Sir William Logan,* p. 387.
31. J. W. Bailey, *Loring Woart Bailey, the Story of a Man of Science* (Saint John, N.B.: J. & A. MacMillan, 1925), pp. 108-109. But see also GSC, Director's Letterbook No. 2, Logan to L. W. Bailey, 26 Dec. 1871.
32. Logan Papers, Box M15943G, Logan to James Richardson, 17 Aug. 1874.
33. Canada, House of Commons, *Report of the Select Committee* (1884), p. 104.
34. *Ibid.,* p. 96.
35. F. D. Adams, *Biographical Memoir of Hunt,* p. 217.

Chapter 6. THE SURVEY FROM SEA TO SEA

1. H. Y. Hind, *Narrative of the Canadian Red River Exploring Expedition of 1857 and of the Assiniboine and Saskatchewan Exploring Expedition of 1858* (London: Longman, Green, Longman and Roberts, 1860, 2 vols.), II, p. 253.
2. PAC, R.G.6, C.2, Secretary of State for the Provinces Letterbook, No. 6, E. A. Meredith to A. R. C. Selwyn, 9 June 1871.
3. GSC, *Report of Progress,* 1871-72, p. 17.
4. PAC, M.G.29, A.8, The Sandford Fleming Papers, Vol. 64, Letterbook "Pacific Railway matters, 1871-1872," Fleming to George Watt, Victoria, B.C., 24 June 1871.
5. GSC, *Report of Progress,* 1871-72, p. 33.
6. *Ibid.,* p. 72.
7. GSC, *Report of Progress,* 1875-76, p. 35.
8. *Ibid.,* p. 42.
9. *Ibid.,* p. 164.
10. G. M. Dawson, *Report on the Geology and Resources of the Region in the Vicinity of the Forty-Ninth Parallel from Lake of the Woods to the Rocky Mountains . . .* (Montreal: Dawson Brothers, 1875), p. iv.
11. GSC, *Report of Progress,* 1879-80, Part B: "Report of an exploration from Port Simpson on the Pacific Coast to Edmonton on the Saskatchewan, . . .," by G. M. Dawson, p. 1B.
12. GSC, *Report of Progress,* 1874-75, p. 26.
13. GSC, Director's Letterbook, No. 3, p. 5, Selwyn to E. A. Meredith, 19 Nov. 1873.
14. GSC, *Report of Progress,* 1874-75, p. 3; also GSC, Central Registry, Historical File 3-0-1 #3.
15. GSC, Director's Letterbook, No. 4, p. 106, Selwyn to L. W. Bailey, 7 Jan. 1881.
16. GSC, *Report of Progress,* 1874-75, pp. 2-3.
17. GSC, Director's Letterbook, No. 2, p. 144, Selwyn to E. A. Meredith, 4 May 1872.
18. *Ibid.,* No. 1, p. 316, Selwyn to Meredith, 3 Jan. 1871.
19. *Ibid.,* p. 315.
20. *Ibid.,* p. 286, Selwyn to H. Y. Hind, 9 Dec. 1870.
21. GSC, *Report of Progress,* 1877-78, p. 14A.
22. *Ibid.,* 1878-79, p. 5.
23. *Ibid.,* 1866-69, pp. 316-317.
24. GSC, Director's Letterbook, No. 2, p. 69, Selwyn to Joseph Howe, 6 Feb. 1872.
25. *Ibid.,* p. 133, Selwyn to Alexander Mackenzie, M.P., 25 April 1872.
26. By Imperial Orders in Council: 16 May 1871 (British Columbia), and 26 June 1873 (Prince Edward Island); by Canada, *Statutes:* 33 Vict. (1870). Chap. 3, Sect. 26(7), for Manitoba, and 12-13 Geo. VI (1949, First Session). Chap. 1, Sect. 31(h), for Newfoundland.
27. GSC, Director's Letterbook, 1865-69, pp. 306-307, Logan to Sir John Rose, 14 Feb. 1868.
28. Canada, Auditor General, *Annual Report,* 1883-84 (*Sessional Papers,* 1885, No. 5), citing reply of John Marshall to J. L. McDougall, 18 Sept. 1884.
29. The Reverend S. Massey, in Montreal *Gazette,* 12 Oct. 1876.
30. GSC, Director's Letterbook, No. 2, p. 447, memorandum by Selwyn, 28 April 1873.

Chapter 7. WASHING DIRTY LINEN

1. Sir J. W. Dawson, "Some points in which American science is indebted to Canada," Presidential Address, RSC, *Proceedings and Transactions,* First Series, Vol. 4 (1886), Sect. IV, p. 8.
2. GSC, Director's Letterbook, No. 1, p. 321, Selwyn to E. A. Meredith, 3 Jan. 1871.
3. Canada, Auditor General, *Annual Report,* 1886-87, p. 268, J. L. McDougall to Selwyn, 24 Nov. 1887.
4. *Ibid.,* 1883-84, p. 212, McDougall to Selwyn, 18 Dec. 1884.
5. *Ibid.,* 1886-87, p. 270, Selwyn to McDougall, 5 Dec. 1887.
6. *Ibid.,* 1882-83, p. 327, McDougall to Selwyn, 28 Nov. 1883.
7. Canada, House of Commons, *Debates,* 1884, p. 575.
8. Canada, House of Commons, *Report of the Select Committee* (1884), p. [3].
9. GSC, Director's Letterbook, No. 4, p. 454, Selwyn to A. C. Perry, 5 April 1884.
10. Canada, House of Commons, *Report of the Select Committee,* p. 73.
11. McGill University Archives, The Thomas Macfarlane Papers, Macfarlane to Robert Bell, 30 Nov. 1885—quoting R. N. Hall, M.P.
12. *Ibid.,* Macfarlane to Mrs. Macfarlane, 16 April 1884.
13. Canada, House of Commons, *Report of the Select Committee,* p. 6.
14. *Ibid.,* p. 162.
15. *Ibid.,* p. 59.
16. *Ibid.,* p. 52.
17. *Ibid.,* p. 44.
18. *Ibid.,* p. 87.

19. *Ibid.,* p. 98.
20. *Ibid.,* pp. 144, 145.
21. *Ibid.,* p. 145.
22. *Ibid.,* p. 8.
23. *Ibid.,* p. 7.
24. *Ibid.,* pp. 7-8.
25. *Ibid.,* pp. 5, 11.
26. *Ibid.,* pp. 10-11.
27. GSC, *Report of Progress,* 1882-83-84, p. 28.
28. GSC, *Annual Report,* New Series, Vol. 4 (1888-89), p. 5A (Summary Report for 1889).
29. *Ibid.,* Vol. 3 (1887-88), p. 2A.
30. *Ibid.,* Vol. 5 (1890-91), p. 8A.
31. Canada, *Statutes,* 53 Vict., Chap. 11, Sect. 4(d).
32. GSC, Central Registry, File 105-1-13, "Liaison & Co-operation: Logan Club," Vol. 1.
33. McGill University Archives, The G. M. Dawson Papers, Private Diary, pp. 1-2, Memorandum, 1895.
34. *Ibid.,* p. 1.
35. Robert Bell Papers, Vol. 7, "General Correspondence, 1892-1895," Bell to John Haggart, M.P. (copy), 5 Jan. 1895.
36. *Ibid.,* Bell to Sir Mackenzie Bowell (copy), 10 Jan. 1895.
37. G. M. Dawson Papers, Private Diary, pp. 6-7, Memorandum, 1895.
38. Geological Society of London, *Quarterly Journal,* Vol. 59 (1903), p. lxiii, "Obituary of A. R. C. Selwyn," by H. M. A. and H. W.
39. Canada, House of Commons, *Debates,* p. 3409 (S. J. Dawson, M.P. for Algoma).
40. GSC, Director's Letterbook, No. 1, p. 320, Selwyn to Meredith, 3 Jan. 1871.

Chapter 8. ROLLING BACK THE FRONTIERS OF CANADA

1. GSC, *Annual Report,* Vol. 9 (1896), p. 7A.
2. *Ibid.,* Vol. 14 (1901), p. 16A.
3. *Ibid.,* Vol. 15 (1902-03), pp. 5A-6A (Summary Report for 1902).
4. G. M. Dawson, "Geographical work in Canada in 1893," *Geographical Journal,* Vol. 3, No. 3 (March 1894), p. 206.
5. GSC, *Summary Report,* 1905, p. 1.
6. *Ibid.*
7. GSC, *Annual Report,* Vol. 14 (1901), p. 17A.
8. *Ibid.*
9. *Ibid.*
10. GSC, *Summary Report,* 1905, p. 3.
11. *Ibid.,* p. 16.
12. *Ibid.*
13. GSC, *Annual Report,* Vol. 9 (1896), p. 88A.
14. *Ibid.,* Vol. 14 (1901), p. 269A.
15. *Ibid.,* Vol. 9 (1896), p. 71A.
16. GSC, Director's Letterbook, No. 34, p. 128, Bell to O. O'Sullivan, 6 May 1905.
17. G. M. Dawson, "On some of the larger unexplored regions of Canada," *Ottawa Naturalist,* Vol. 4 (1890-91), pp. 39-40.
18. GSC, *Annual Report,* Vol. 14 (1901), p. 14A.
19. *Ibid.,* p. 12A.
20. GSC, *Report of Progress,* 1882-83-84, pp. 5-6.
21. GSC, *Annual Report,* Vol. 16 (1904), p. xxxi.
22. *Geographical Journal,* Vol. 27, No. 5 (May 1906), p. 500.
23. GSC, *Report of Progress,* 1882-83-84, Part C, p. 5C.
24. G. M. Dawson, "Geological record of the Rocky Mountains Region in Canada," Presidential Address, Geological Society of America, *Bulletin,* Vol. 12 (1901), p. 84.
25. GSC, *Annual Report,* Vol. 4 (1888-89), Part D, p. 7D.
26. *Ibid.,* pp. 31D-32D.
27. GSC, *Annual Report,* Vol. 7 (1894), Part C, "Report on an exploration of the Finlay and Omineca Rivers," by R. G. McConnell, p. 6C.
28. *Ibid.,* Vol. 6 (1893), p. 4A.
29. G. M. Dawson, "On the later physiographic geology of the Rocky Mountain Region in Canada, with special reference to changes in elevation and to the history of the glacial period," RSC, *Proceedings and Transactions,* First Series, Vol. 8 (1890), Sect. IV, pp. 3-74.
30. G. M. Dawson, "Physical geography and geology," *in* British Association for the Advancement of Science, Toronto Meeting, Aug. 1897, *Handbook of Canada* (Toronto: Local Committee of the Association, 1897), p. 47.
31. *Ibid.,* p. 48.
32. GSC, *Annual Report,* Vol. 13 (1900), Part FF, "Report on geological explorations in Athabaska, Saskatchewan and Keewatin Districts," by D. B. Dowling, pp. 15-16.
33. GSC, *Annual Report,* Vol. 12 (1899), p. 103A.
34. *Ibid.,* p. 105A.
35. *Ibid.,* Part C, "Report on the topography and geology of Great Bear Lake and of a chain of lakes and streams thence to Great Slave Lake," by J. Mackintosh Bell, p. 16C.
36. *Ibid.,* p. 17C.
37. *Ibid.,* p. 27C.
38. *Ibid.,* Vol. 1 (1885), pp. 9A, 10A.
39. GSC, Field Notebook, No. 2450, A. P. Low, 7 May 1894-30, June 1894, p. 7.
40. GSC, *Annual Report,* Vol. 8 (1895), Part L, "Report on explorations in the Labrador Peninsula, along the East Main, Koksoak, Hamilton, Manicuagan, and portions of other rivers, in 1892-93-94-95," by A. P. Low, p. 7L.
41. Fabien Caron, "Albert Peter Low et l'exploration du Québec-Labrador," *Cahiers de Géographie de Québec,* Year 9, No. 18 (April-Sept. 1965), p. 176.
42. GSC, *Annual Report,* Vol. 8 (1895), p. 284L.
43. GSC, *Annual Report,* Vol. 13 (1900), Part K, "Report on the geology of the basin of Nottaway River," by Robert Bell, p. 8K.
44. *Ibid.,* Vol. 1 (1885), Part DD, "Observations on the geology, zoology and botany of Hudson Strait and the Bay made in 1885," by Robert Bell, p. 18DD.
45. GSC, *Report of Progress,* 1882-83-84, Part DD, "Observations on the geology, mineralogy, zoology and botany of the Labrador Coast, Hudson Strait and Bay," by Robert Bell, p. 36DD.
46. A. P. Low, *Report on the Dominion Government Exploration to Hudson Bay and the Arctic Islands on Board the D.G.S. Neptune 1903-1904* [Also titled as "The Cruise of the Neptune, 1903-04"] (Ottawa: Government Printing Bureau, 1906), p. 48.
47. GSC, *Annual Report,* Vol. 16 (1904), p. 142A.
48. Low, *Cruise of the Neptune,* p. 46.
49. *Ibid.,* p. 181.
50. *Ibid.,* p. 182.

Chapter 9. STARTING THE SEARCH FOR ANSWERS

1. GSC, *Annual Report,* Vol. 15 (1902-03), p. 3A (Summary Report for 1902).
2. GSC, Central Technical Files, C-127, manuscript of draft for C. Camsell radio address, ca. 1935.

3. GSC, *Annual Report,* Vol. 7 (1894), pp. 3A-4A.
4. *Ibid.,* Vol. 1 (1885), p. 73A.
5. *Ibid.,* Vol. 4 (1888-89), p. 4A (Summary Report for 1889).
6. *Canadian Mining Review,* Vol. 20, No. 12 (Dec. 1901), p. 282.
7. GSC, *Annual Report,* Vol. 14 (1901), p. 23A.
8. GSC, *Report of Progress,* 1882-83-84, p. 5.
9. GSC, *Annual Report,* Vol. 1 (1885), p. 6A.
10. *Ibid.*
11. *Ibid.,* Vol. 2 (1886), p. 14A.
12. Quoted *in* G. A. Young, "The geological investigation of the Canadian Shield (Canadian portion) 1882 to 1932," RSC, *Proceedings and Transactions,* Third Series, Vol. 26 (May 1932), Sect. IV, p. 348.
13. Geological Society of America, *Proceedings, Annual Report for 1953,* "Memorial to Andrew Cooper Lawson (1861-1952)," by P. Byerly and G. D. Louderback, pp. 141-142.
14. GSC, *Annual Report,* Vol. 10 (1897), Part H, "Report on the geology of the area covered by the Seine River and Lake Shebandowan map-sheets, comprising portions of Rainy River and Thunder Bay Districts, Ontario," by W. McInnes, p. 13H.
15. *Ibid.,* Vol. 8 (1895), Part J, "Report on the geology of a portion of the Laurentian area lying north of the island of Montreal," by F. D. Adams, p. 3J.
16. *Ibid.,* Vol. 3 (1887-88), p. 85A (Summary Report for 1888).
17. See G. A. Young, "The geological investigation of the Canadian Shield . . .," p. 366.
18. GSC, *Annual Report,* Vol. 5 (1890-91), "On the nickel and copper deposits of Sudbury, Ontario," by A. E. Barlow, *in* Part S, "Division of Mineral Statistics and Mines, Annual Report for 1890," p. 126S.
19. *Handbook of Canada* (Toronto, Local Committee of the British Association for the Advancement of Science, 1897), p. 26.
20. G. A. Young, "The geological investigation of the Canadian Shield . . . (Second Part)," *in* RSC, *Proceedings and Transactions,* Third Series, Vol. 27 (1933), Sect. IV, pp. 73-74, based on A. P. Coleman, "Clastic Huronian Rocks of Western Ontario," Geological Society of America, *Bulletin,* Vol. 9 (1897), pp. 223-238.
21. GSC, *Annual Report,* Vol. 12 (1899), Part J, "Report on the geology of Argenteuil, Ottawa and Pontiac Counties, Province of Quebec, and portions of Carleton, Russell and Prescott Counties, Province of Ontario," by R. W. Ells, pp. 15J-16J.
22. GSC, *Annual Report,* Vol. 3 (1887-88), p. 22A (Summary Report for 1887).
23. *Ibid.,* Vol. 9 (1896), Part I, "Report on the geology of the French River sheet, Ontario," by Robert Bell, p. 7I.
24. *Ibid.,* Vol. 1 (1885), Part CC, "Report on the geology of the Lake of the Woods region with special reference to the Keewatin (Huronian?) belt of Archaean rocks," by A. C. Lawson, p. 14CC.
25. *Ibid.,* Vol. 6 (1893), p. 32A.
26. *Ibid.,* Vol. 8 (1895), p. 62A.
27. *Ibid.,* Vol. 10 (1897), Part I, p. 94I.
28. Ottawa *Journal,* 30 Dec. 1892.
29. GSC, *Summary Report,* 1906, p. 12.
30. *Ibid.,* 1907, pp. 4-5.
31. GSC, *Annual Report,* Vol. 5 (1890-91), Part S, "On the nickel and copper deposits of Sudbury, Ontario," by A. E. Barlow, pp. 127S, 128S.
32. F. D. Adams, "On the igneous origin of certain ore deposits," *Canadian Mining Review,* Vol. 13, No. 1 (Jan. 1894), p. 8.
33. GSC, *Annual Report,* Vol. 14 (1901), p. 8A.
34. Geological Society of America, *Bulletin,* Vol. 26, Part I (31 March 1915), "Memoir of Alfred Ernest Barlow," by F. D. Adams, p. 14.
35. GSC, *Annual Report,* Vol. 5 (1890-91), Part F, "Report on the Sudbury mining district," by Robert Bell, pp. 46F-51F.
36. A. E. Barlow, *Report on the Origin, Geological Relations and Composition of the Nickel and Copper Deposits of the Sudbury Mining District, Ontario, Canada* (Ottawa: S. E. Dawson, 1904), p. 6. [Also in GSC, *Annual Report,* Vol. 14 (1901), Part H.]
37. *Ibid.,* p. 124.
38. *Ibid.,* p. 131.
39. *Ibid.,* p. 124.
40. *Ibid.,* p. 131.
41. *Ibid.*
42. GSC, *Summary Report,* 1906, p. 23.

Chapter 10. A MATTER OF DIRECTION

1. B. J. H[arrington], "George Mercer Dawson," *American Geologist,* Vol. 28, No. 2 (Aug. 1901), p. 73, quoting W. J. McGee.
2. *Canadian Mining Review,* Vol. 20, No. 3 (March 1901), quoting R. W. Shannon.
3. B. J. H., "George Mercer Dawson," p. 76.
4. GSC, *Annual Report,* Vol. 8 (1895), p. 7A.
5. *Ibid.,* Vol. 9 (1896), p. 6A.
6. *Ibid.,* Vol. 12 (1899), p. 5A.
7. G. M. Dawson Papers, Diary for 11 Dec. 1896.
8. *Ibid.*
9. GSC, Scrapbook No. 1 (Ami's), item dated "1897" (i.e. April).
10. G. M. Dawson Papers, Diary for 1 June 1898.
11. GSC, Director's Letterbook, No. 26, p. 212, Dawson to A. P. Coleman, 2 Feb. 1900.
12. G. M. Dawson Papers, Diary for 9 Dec. 1896.
13. GSC, Director's Letterbook, No. 14, pp. 37-38, Memorandum by G. M. Dawson.
14. G. M. Dawson Papers, Diary for 18 June 1895.
15. *Ibid.,* 27 July 1896.
16. PAC, M.G.27 II, D.15, The Sifton Papers, Vol. 17, p. 10780, Memorandum of staff to G. M. Dawson, 29 Jan. 1897.
17. M. Edith Tyrrell, *I Was There, a Book of Reminiscences* (Toronto: Ryerson Press, 1938), pp. 65-66.
18. G. M. Dawson Papers, Diary for 29 Nov. 1897.
19. *Ibid.,* 7 May 1898.
20. PAC, R.G.32, C-2, Historical Personnel Files, No. 535, "Albert Peter Low," Low to G. M. Dawson (copy).
21. GSC, Director's Letterbook, No. 16, p. 542, Dawson to Oliver, 8 Sept. 1896.
22. GSC, *Annual Report,* Vol. 9 (1896) p. 7A.
23. GSC, Director's Letterbook, No. 16, p. 95, Dawson to J. W. Tyrrell, 29 April 1896.
24. GSC, *Annual Report,* Vol. 10 (1897), pp. 9A-10A.
25. *Ibid.,* Vol. 9 (1896), p. 7A.
26. *Ibid.,* Vol. 11 (1898), p. 9A.
27. RSC, *Proceedings and Transactions,* Second Series, Vol. 7 (1901), p. XXXIX.
28. *Geographical Journal,* Vol. 17, No. 4 (April 1901), p. 439.
29. From "Ode to 'Dr. George'," by Capt. Clive Phillips-Wolley, as reproduced *in* H. M. Ami, "The late George Mercer Dawson," *Ottawa Naturalist,* Vol. 15, No. 2 (May 1901), p. 52.
30. Robert Bell Papers, Vol. 8, "General Correspondence, 1896-1898," Bell to Clifford Sifton, 15 Jan. 1898; also in Sifton Papers, p. 25647.
31. Sifton Papers, Vol. 76, p. 57265, Bell to Sifton, 4 July 1900.
32. GSC, *Annual Report,* Vol. 14 (1901), p. 5A.
33. GSC, Director's Letterbook, No. 31, p. 116, Bell to Lieut.-Col. F. Bailey, 16 July 1902.

34. Canada, House of Commons, *Debates,* 1906, p. 5980 (22 June 1906).
35. GSC, *Annual Report,* Vol. 14 (1901), p. 18A.
36. Robert Bell Papers, Vol. 11, "General Correspondence, 1904-1905," Sifton to Bell, 14 March 1904.
37. Canada, House of Commons, *Debates,* 1897, col. 2429 (N. A. Belcourt, Ottawa); reported in *Ottawa Evening Citizen,* 18 May 1897.
38. Canada, House of Commons, *Debates,* 1901, col. 5681 (20 May 1901).
39. Robert Bell Papers, Vol. 29, "Subject Files No. 4, Geological Survey," memo from Senécal, undated, probably 1902.
40. *Ibid.,* undated memo from Mrs. J. Alexander.
41. *Ibid.,* undated "Report of a committee of the Geological Survey staff in regard to the library."
42. *Canadian Mining Review,* Vol. 23, No. 7 (July 1904), p. 138.
43. GSC, Director's Letterbook, No. 29, p. 486, Bell to R. W. Brock, 3 July 1901.
44. Sifton Papers, Vol. 116, p. 92199, letter of 1 March 1902.
45. PAC, M.G.26, G.1(a), The Laurier Papers, Vol. 360, p. 96154, R. A. Daly to Laurier, 3 April 1905.
46. Robert Bell Papers, Vol. 11, "General Correspondenee, 1904-1905," Bell to Frank Oliver, 26 May 1905.
47. GSC, Director's Letterbook, No. 30, p. 279, Bell to Ami, 16 Jan. 1902.
48. Laurier Papers, Vol. 401, pp. 106819-20, Ami to Laurier, 5 Feb. 1906.
49. Canada, House of Commons, *Debates,* 1903, col. 6816 (F. D. Monk, Jacques Cartier).
50. *Canadian Mining Review,* Vol. 24, No. 6 (June 1905), p. 124.
51. Robert Bell Papers, Vol. 21, "Robert Bell-Agnes S. Bell Correspondence, 1905-1911, 1915," Bell to Mrs. Agnes Bell, 7 March 1906.
52. Laurier Papers, pp. 108079-80.
53. Canada, House of Commons, *Debates,* 1906, col. 406 (20 March 1906).
54. *Ibid.,* col. 412.
55. Robert Bell Papers, Vol. 12, "General Correspondence, 1906-1907," Oliver to Bell.

Chapter 11. THE SURVEY AND CANADA'S CENTURY

1. GSC, *Annual Report,* Vol. 12 (1899), p. 22A.
2. GSC, *Summary Report,* 1905, p. 53.
3. GSC, *Annual Report,* Vol. 8 (1895), p. 19A.
4. GSC, *Summary Report,* 1905, p. 3; R. W. Brock, *Preliminary Report on the Rossland, B.C., Mining District* (Ottawa: Government Printing Bureau, 1906. GSC, Publication No. 939), pp. 7, 8.
5. GSC, *Summary Report,* 1907, p. 24.
6. Ontario, Bureau of Mines, *First Report* (1891), p. 4.
7. GSC, *Annual Report,* Vol. 16 (1904), p. 198A.
8. *Ibid.,* p. 199A.
9. *Ibid.*
10. GSC, *Summary Report,* 1906, pp. 7-8.
11. GSC, *Annual Report,* Vol. 10 (1897), Part H, "Report on the geology of the area covered by the Seine River and Lake Shebandowan map-sheets, comprising portions of Rainy River and Thunder Bay Districts, Ontario," by W. McInnes, p. 57H.
12. *Ibid.,* Vol. 8 (1895), p. 7A.
13. GSC, *Summary Report,* 1907, p. 72.
14. A. P. Low, *Geological Report on the Chibougamau Mining Region in the Northern Part of the Province of Quebec* (Ottawa: Government Printing Bureau, 1906. GSC, Publication No. 923), p. 5.
15. GSC, *Annual Report,* Vol. 14 (1901), p. 209A.
16. *Ibid.,* Vol. 15 (1902-03), p. 339A (Summary Report for 1902).
17. *Ibid.,* p. 362A (1902).
18. Mining Society of Nova Scotia, *Journal,* Vol. 8 (1903-04), p. 62.
19. GSC, *Annual Report,* Vol. 16 (1904), pp. 1A-2A.
20. GSC, *Summary Report,* 1909, p. 1.
21. "Dedication of a Memorial Tablet to Hugh Fletcher at the Nova Scotia Technical College, Halifax, N.S., May Twenty-first, 1926," p. 14 (In GSC library).
22. *Stratigraphy of Plains of Southern Alberta: Donaldson Bogart Dowling Memorial Symposium* (Tulsa: American Association of Petroleum Geologists, 1931), p. viii.
23. GSC, *Summary Report,* 1906, p. 66.
24. *Ibid.,* 1907, p. 32.
25. *Ibid.,* 1905, p. 57.
26. *Ibid.,* p. 59.
27. *Ibid.,* 1907, p. 23.
28. J. W. Spencer, *The Falls of Niagara; Their Evolution and Varying Relations to the Great Lakes; Characteristics of the Power and the Effects of its Diversion* (Ottawa: S. E. Dawson, 1907. GSC, Publication No. 970), p. xi.
29. James M. Macoun, *Report on the Peace River Region* (Ottawa: King's Printer, 1904. GSC, Publication No. 855), p. 35E.
30. GSC, *Summary Report,* 1906, p. 81.
31. GSC, *Annual Report,* Vol. 15 (1902-03), p. 14A (Summary Report for 1903).
32. *Ibid.,* Vol. 16 (1904), p. xxxiA.
33. GSC, *Summary Report,* 1906, p. 16.

Chapter 12. INTO THE DEPARTMENT OF MINES

1. Canada, House of Commons, *Report of the Select Committee* (1884), p. 7.
2. *Ibid.,* p. 8.
3. GSC, *Annual Report,* Vol. 9 (1896), pp. 7A-8A.
4. G. M. Dawson Papers, Diary for 11 Sept. 1896.
5. Sifton Papers, Vol. 21, p. 13182, E. Haanel to Sifton, 15 July 1897.
6. *Ibid.,* Vol. 101, p. 79403, Haanel to Sifton, 9 April 1901.
7. *Canadian Mining Review,* Vol. 19, No. 1 (Jan. 1900), p. 5.
8. *Ibid.,* Vol. 20, No. 3 (March 1901), p. 70.
9. Sifton Papers, Vol. 104, pp. 81600-01, Low to Sifton, 5 March 1901.
10. *Ibid.,* p. 79396, Haanel to Sifton, 4 March 1901.
11. Historical Personnel Files, Vol. 18, No. 56, "Robert Bell," copy of undated (1908-11?) memorandum from Brock to W. Templeman *re* Bell's claim for arrears of salary. Brock reported that Bell had never been officially appointed acting director, as no order in council was issued.
12. GSC, Director's Letterbooks, No. 36, p. 1.
13. Robert Bell Papers, Vol. 9, "General Correspondence, 1899-1901," Broad to Bell, 24 April 1901.
14. *Ibid.,* Memorandum of staff to Bell, 13 March 1901, printed copy. There are a number of typographical errors.
15. Canada, House of Commons, *Debates,* col. 5317.
16. Sifton Papers, Vol. 104, pp. 79405-06, Haanel to Sifton, 20 May 1901.
17. Canada, Department of the Interior, *Annual Report,* 1900-01, p. ii.

18. *Ibid.,* 1901-02, Part VI, pp. 3, 10.
19. *Canadian Mining Review,* Vol. 20, No. 11 (Nov. 1901), p. 263.
20. Sifton Papers, Vol. 116, pp. 92196-98, A. E. Barlow to Sifton, 1 March 1902.
21. Canadian Mining Institute, *Journal,* Vol. 5 (1902), p. 119.
22. *Ibid.,* p. 589.
23. *Ibid.,* p. 590.
24. *Ibid.,* p. 596.
25. *Ibid.,* p. 592.
26. *Ibid.,* p. 597.
27. *Canadian Mining Review,* Vol. 21, No. 4 (April 1902), pp. 73, 74.
28. Robert Bell Papers, Vol. 10, "General Correspondence, 1902-1903," Fletcher to Bell, 10 Nov. 1902.
29. Canada, House of Commons, *Debates,* 1903, col. 6816.
30. *Ibid.,* cols. 6818-19.
31. *Ibid.,* col. 13762.
32. *Ibid.,* cols. 13762-63.
33. *Ibid.,* col. 13763.
34. Robert Bell Papers, Vol. 11, "General Correspondence for 1904-1905," Bell to Sifton, 1 June 1904; also Sifton Papers, Vol. 154, p. 123529.
35. Canada, Department of the Interior, *Annual Report,* 1905-06, Part VIII: Superintendent of Mines, p. 17.
36. *Ibid.,* 1902-03, Part VIII, p. 8.
37. Canada, House of Commons, *Debates,* 1906, col. 409 (E. B. Osler, West Toronto, 20 March 1906).
38. Canada, Department of the Interior, *Annual Report,* 1904-05, Part VIII, p. 16.
39. Canada, House of Commons, Debates, 1906, col. 5941 (J. E. Armstrong, East Lambton, 22 June 1906).
40. *Ibid.,* 1904, cols. 7218-19 (21 July).
41. *Ibid.,* 1906, col. 5942 (22 June).
42. *Canadian Mining Review,* Vol. 25, No. 1 (Aug. 1905), pp. 5-7.
43. *Ibid.,* p. 7.
44. *Ibid.,* Vol. 25, No. 5 (Dec. 1905), p. 144.
45. Canadian Mining Institute, *Journal,* Vol. 9 (1906), p. 87.
46. *Ibid.,* p. 103.
47. *Ibid.,* pp. 105-06.
48. Section 7(a).
49. Sections 8, 7(a).
50. Section 12.
51. Sections 15, 14.
52. Sections 17(d), 17(c).

Chapter 13. THE SURVEY REORGANIZED

1. *Canadian Mining Journal,* Vol. 28, No. 11 (15 Aug. 1907), p. 340.
2. *Ibid.,* Vol. 28, No. 17 (15 Nov. 1907), p. 515.
3. *Ibid.,* Vol. 28, No. 19 (15 Dec. 1907), p. 577.
4. *Ibid.,* p. 582.
5. GSC, *Summary Report,* 1908, p. 10.
6. For example, see T. M. Daly (Minister of the Interior) to J. A. Ouimet (Minister of Public Works), 12 May 1893, in Department of Public Works, File No. 688-1-A, "Victoria Memorial Museum. Site and Construction," letter no. 141110. This file contains most of the information with respect to planning, negotiating for the Stewart property, tenders, etc., for the paragraphs that follow.
7. Letter in possession of Mr. T. L. Brock, Mount Royal, Quebec.
8. "Ottawa's 'Leaning Tower'," *Contract Record and Engineering Review,* Vol. 26 (Sept. 1912), p. 42.
9. Canada, *Statutes,* 6-7 Edw. VII (1907), Chap. 29, "An Act to create a Department of Mines," Sects. 7(e), 7(a), and 8.
10. GSC, *Summary Report,* 1910, p. 10.
11. *Ibid.,* p. 9.
12. *Ibid.*
13. *Ibid.,* 1912, p. 448.
14. *Ibid.,* 1913, p. 11.
15. The figures in Column 1 are based on the annual reports of the Auditor General; those of the other columns, on financial statements in the *Summary Reports.*
16. GSC, *Summary Report,* 1910, p. 2.
17. *Ibid.,* 1913, pp. 1-2.
18. *Ibid.,* 1908, p. 1.
19. *Ibid.*
20. *Ibid.,* 1912, p. 8.
21. *Ibid.,* p. 5.
22. *Ibid.,* 1914, p. 117.
23. *Ibid.,* 1910, p. 255.
24. *Ibid.,* 1908, p. 10.
25. *Ibid.,* p. 1.
26. *Ibid.,* 1912, p. 507.
27. *Ibid.,* 1913, p. 390.
28. *Ibid.,* 1912, p. 6.
29. *Ibid.,* 1913, p. 391.
30. *Ibid.,* 1907, p. 120.
31. *Ibid.,* 1912, p. 6.
32. Canadian Mining Institute, *Bulletin,* No. 93 (Jan. 1915), p. 31.
33. *Canadian Mining Journal,* Vol. 36, No. 6 (15 March 1915), p. 165.
34. GSC, *Summary Report,* 1908, p. 9.
35. *Ibid.,* 1910, p. 7.
36. National Museum of Canada, Anthropology Division, manuscript, item P/ B86/ mss. (In National Museum library.)
37. GSC, *Summary Report,* 1909, p. 9.
38. *Ibid.,* p. 8.
39. *Ibid.,* 1910, p. 7.
40. *Ibid.,* 1911, p. 191.
41. *Ibid.,* 1912, p. 448.
42. *Ibid.,* p. 443.

Chapter 14. THE SURVEY ON DISPLAY

1. GSC, *Summary Report,* 1906, pp. 3-4.
2. *Ibid.,* 1908, pp. 10-11.
3. *Ibid.,* 1910, p. 2.
4. *Ibid.,* 1906, p. 5.
5. *Ibid.,* 1911, p. 3.
6. Robert Bell Papers, Vol. 12, "General Correspondence, 1906-1907," Bell to Sifton, 24 April 1907 (copy). No original can be traced in the Sifton Papers.
7. *Ibid.,* William Templeman to Bell, 15 Nov. 1907.
8. Historical Personnel Files, Vol. 18, No. 56, "Robert Bell," undated item by Percy Selwyn, 1908.
9. GSC, *Summary Report,* 1909, p. 2.
10. *Canadian Mining Journal,* Vol. 32, No. 11 (1 June 1911), p. 332.
11. GSC, *Summary Report,* 1909, p. 285.
12. *Ibid.,* 1908, p. 2.
13. *Ibid.,* p. 17.

14. *Ibid.,* 1909, p. 37.
15. *Ibid.,* p. 285.
16. *Ibid.,* 1908, p. 2.
17. *Ibid.,* 1909, p. 5.
18. *Ibid.,* 1910, p. 3.
19. *Ibid.,* 1909, p. 6.
20. *Ibid.,* p. 22.
21. *Ibid.,* 1910, p. 113.
22. "Effusives from the GSC Dinner, December 5th, 1912," print, copy in GSC, Scrapbook No. 2 (Dowling's).
23. GSC, Publication No. 1035 (1909), later revised and re-issued as Memoir No. 53 (1914).
24. GSC, *Summary Report,* 1909, p. 30.
25. *Canadian Mining Journal,* Vol. 30, No. 23 (Dec. 1909), p. 708.
26. GSC, *Summary Report,* 1910, pp. 2-3.
27. *Ibid.,* p. 4.
28. *Ibid.,* 1908, p. 180.
29. *Ibid.,* p. 183.
30. *Ibid.,* p. 176.
31. *Ibid.,* 1910, p. 5.
32. *Ibid.,* p. 239.
33. *Ibid.,* 1911, p. 242.
34. A. C. Lawson, *The Archaean Geology of Rainy Lake Re-studied* (GSC, Memoir No. 40, 1913), p. 40.
35. Robert Bell Papers, Vol. 11, "General Correspondence, 1904-1905," Joseph Pope to Bell, 28 Nov. 1905.
36. GSC, *Summary Report,* 1906, p. 9.
37. *Ibid.,* p. 10.
38. PAC, M.G.28, I. 18, International Geological Congress, Vol. 1, "Minutes 1910-14," Minutes of Executive Council, 17th Meeting, 28 May 1913, p. 4.
39. *Ibid.,* p. 5.
40. GSC, *Summary Report,* 1912, p. 6.
41. W. McInnes, D. B. Dowling, and W. W. Leach, eds., *The Coal Resources of the World* (Toronto: Morang, 1913, 3 vols.) Vol. 1, p. x.
42. Paris, Académie des Sciences, *Comptes Rendus,* Vol. 157 (1913), p. 747.
43. B. Hobson, "The Twelfth International Geological Congress in Canada," *The Geological Magazine,* New Series, Vol. 10 (1913), p. 490.
44. Congrès Géologique International, *Compte-Rendu de la XIIe Session, Canada, 1913* (Ottawa: Imprimerie du Gouvernement, 1914), p. 1006.
45. A. C. Lawson, "A standard scale for the Pre-Cambrian rocks of North America," *in* Congrès Géologique International, *Compte-Rendu de la XIIe session,* pp. 349-70, and discussion, pp. 421-25.
46. B. Hobson, "The Twelfth International Geological Congress in Canada," p. 488.
47. Congrès Géologique International, *Compte-Rendu de la XIIe Session, Canada, 1913,* p. 157.
48. GSC, *Summary Report,* 1913, p. 113.
49. Brock to Borden, 19 Dec. 1913, enclosing his memorandum to Louis Coderre, Minister of Mines, of the same date, copy in possession of T. L. Brock, Mount Royal, Quebec.
50. *Ibid.*
51. Canada, Order in Council PC 3287/1913 and PC 25/1914.
52. Canadian Mining Institute, *Bulletin,* No. 23 (Jan. 1914), p. 20.
53. *Ibid.,* No. 25 (April 1914), p. 8; also *Transactions,* 1914, p. 7.
54. Geological Society of America, *Proceedings,* 1935, "Memoir of Reginald Walter Brock," by M. Y. Williams, p. 161.
55. Letter of 16 April 1901, in possession of T. L. Brock, Mount Royal, Quebec.
56. Historical Personnel Files, Vol. 171, No. 616, "R. G. McConnell," copy of Order in Council, 15 Aug. 1914.
57. Canadian Mining Institute, *Bulletin,* No. 30 (Oct. 1914), p. 9.
58. *Ibid.,* p. 18.

Chapter 15. THE SURVEY GOES TO WAR

1. GSC, *Summary Report,* 1914, p. 2.
2. *Ibid.,* 1916, pp. 1-2.
3. "The fine arts versus Styracosaurus Albertensis," *Canadian Mining Journal,* Vol. 41, No. 23 (11 June 1920), p. 472. See also letter by T. C. Denis, *Canadian Mining Journal,* Vol. 41, No. 24 (15 June 1920), p. 495.
4. GSC, *Summary Report,* 1916, p. 2.
5. *Ibid.,* 1917, Part A, p. 1A.
6. *Ibid.,* 1918, Part B, p. 1B.
7. Historical Personnel Files, Vol. 40, No. 141, "Charles Camsell," memo by R. G. McConnell, 24 April 1918, "Re: Opening of Branch Office of Geological Survey, at Vancouver, B.C."
8. GSC, *Summary Report,* 1919, Part C, p. 21C.
9. GSC, Memoir No. 116 (1919) *Investigations in the Gas and Oil Fields of Alberta, Saskatchewan and Manitoba,* by D. B. Dowling, S. E. Slipper, and F. H. McLearn.
10. GSC, *Summary Report,* 1919, Part E, p. 18E.
11. *Canadian Mining Journal,* Vol. 40, No. 46 (19 Nov. 1919), p. 863.
12. GSC, *Summary Report,* 1916, p. 57.
13. *Ibid.,* p. 58.
14. PAC, M.G.26,H, The Borden Papers, p. 130237, letter of 14 Feb. 1913—copy of an earlier letter from Brock to W. J. Roche, Minister of Interior, 4 Feb. 1913, in PAC, R.G. 42, Marine, Vol. 490, "Canadian Arctic Expedition," File 84-2-55.
15. "Canadian Arctic Expedition," File 84-2-55—true copy of PC 406/1913.
16. George M. Douglas, *Lands Forlorn, a Story of an Expedition to Hearne's Coppermine River* (New York: G. P. Putnam's Sons, 1914).
17. PAC, Marine, Vol. 463, File 84-2-1, Brock to Desbarats, covering letter, 28 May 1913.
18. *Ibid.,* p. 3, of enclosed instructions.
19. GSC, *Summary Report,* 1913, p. 9.
20. PAC, Marine, Vol. 490, File 84-2-55, letter of 5 May 1914, pp. 2-3.
21. GSC, *Summary Report,* 1916, p. 333.
22. *Canadian Mining Journal,* Vol. 39, No. 11 (19 March 1920), p. 210.
23. *Ibid.,* Vol. 38, No. 14 (15 July 1917), p. 283.
24. *Ibid.,* Vol. 39, No. 5 (1 March 1918), p. 69.
25. *Ibid.,* p. 68.
26. Canada, *Statutes,* 8-9 Edw. VII (1909), Chap. 27, Sect. 10.
27. Quoted in *Canadian Mining Journal,* Vol. 39, No. 11 (1 June 1918), p. 181.
28. *Ibid.,* Vol. 41, No. 5 (6 Feb. 1920), p. 98.
29. PAC, R.G.45, Geological Survey, Vol. 35, File 749, "Smithsonian Institution," McConnell to Walcott, 9 April 1920.
30. Quoted in *Canadian Mining Journal,* Vol. 41, No. 3 (23 Jan. 1920), p. 65.
31. Canadian Mining Institute, *Bulletin,* No. 95 (March 1920), p. 183.
32. *Canadian Mining Journal,* Vol. 41, No. 5 (6 Feb. 1921), p. 97.
33. *Ibid.*
34. *Ibid.,* Vol. 41, No. 6 (11 Feb. 1921), p. 103.
35. J. Swettenham and D. Keely, *Serving the State, a History of the Professional Institute of the Public Service of Canada, 1920-1970* (Ottawa: Le Droit, 1970), p. 8.
36. Canadian Mining Institute, *Bulletin,* No. 94 (Feb. 1920), p. 164.
37. Quoted in *Canadian Mining Journal,* Vol. 41, No. 10 (12 March 1920), p. 202.
38. *Ibid.,* No. 16 (23 April 1920), p. 321.
39. Historical Personnel Files, "R.G. McConnell," McConnell to Meighen, 31 May 1920 (copy).
40. Canada, Department of Mines, *Annual Report,* 1921-22, p. 4.

Chapter 16. THE TWENTIES—OPERATING IN A DECADE OF PROSPERITY

1. Department of Mines, *Annual Report,* 1923-24, p. 8.
2. RSC, *Proceedings and Transactions,* Third Series, Vol. 59 (1959), p. 80, "Charles Camsell," by H. S. Bostock.
3. Geological Society of America, *Proceedings,* 1937, "Memorial to William Henry Collins," by T. T. Quirke, p. 159.
4. RSC, *Proceedings and Transactions,* Third Series, Vol. 31 (1937), p. xii, "William Henry Collins."
5. "Memorial to William Henry Collins," by T. T. Quirke, p. 160.
6. G. A. Young to H. M. A. Rice (personal communication from H. M. A. Rice).
7. Department of Mines, *Annual Report,* 1922-23, p. 8.
8. *Ibid.* p. 1.
9. *Ibid.,* 1928-29, p. 14.
10. *Ibid.,* 1925-26, p. 13.
11. T. T. Quirke and W. H. Collins, *The Disappearance of the Huronian* (GSC, Memoir No. 160, 1930), p. 1.
12. *Ibid.,* p. 102.
13. GSC, *Summary Report,* 1924, Part C, "Gunflint iron-bearing formation," by J. E. Gill, p. 28C.
14. Department of Mines, *Annual Report,* 1922-23, p. 10.
15. *Ibid.,* 1921-22, p. 12.
16. *Ibid.,* 1930-31, p. 10.
17. *Ibid.,* 1927-28, p. 20.
18. W. H. Collins, *The National Museum of Canada* (GSC, Museum Bulletin No. 50, 1927), p. 55.
19. Department of Mines, *Annual Report,* 1923-24, p. 5.
20. *Ibid.,* 1924-25, p. 13.
21. *Ibid.*
22. Department of Mines, *Annual Report,* 1930-31, p. 29.
23. PAC, Geological Survey, Vol. 49, File 3112, Memorandum of conference of Camsell, Collins, Jenness, Young, Nicolas and Bolton, 29 Jan. 1929, on publication of anthropological reports; and subsequent report of Bolton, 9 Jan. 1930.
24. Department of Mines, *Annual Report,* 1920-21, p. 17.
25. *Ibid.,* 1927-28, p. 33.

Chapter 17. THE THIRTIES—THE SURVEY DOWNGRADED

1. Department of Mines, *Annual Report,* 1930-31, p. 1.
2. *Ibid.,* 1934-35, p. 1.
3. *Ibid.,* 1932-33, p. 2.
4. *Ibid.,* p. 1.
5. W. H. Collins and M. E. Wilson, "Geological Survey serves Canadian public in an important way," *Northern Miner,* 9 July 1931 (annual special edition), p. 31.
6. Canada, House of Commons, *Debates,* 1932, p. 1714 (2 April 1932).
7. *Ibid.,* 1934, p. 1191 (5 March 1934).
8. Department of Mines, *Annual Report,* 1931-32, pp. 1-2.
9. *Ibid.,* 1934-35, p. 22.
10. PAC, Geological Survey, Vol. 49, File 3097E, vol. 2, Camsell to W. A. Gordon, Minister of Mines, 9 July 1934.
11. Geological Society of America, *Proceedings,* 1947, "Memorial to Terence Thomas Quirke," by C. A. Chapman, p. 169.
12. Interview with Eugène Poitevin.
13. Department of Mines, *Annual Report,* 1930-31, p. 10.
14. Canadian Institute of Mining and Metallurgy, *Bulletin,* No. 118 (Feb. 1922), pp. 228-229. For a copy of an earlier draft, see R.G.45, Geological Survey, Vol. 22, File 176, vol. 1.
15. PAC, Geological Survey, Vol. 22, File 176, vol. 1, Copy of memorandum, Collins to Camsell, 9 Dec. 1921, "Proposals for assistance from Air Board in Geological Survey field operations, 1922."
16. Department of Mines, *Annual Report,* 1922-23, p. 12.
17. PAC, Geological Survey, Vol. 22, File 176, vol. 1, G. J. Desbarats to Camsell, 26 April 1923, "Flying operations for Geological Survey."
18. GSC, *Summary Report,* 1923, Part B, "Geology and mineral prospects of the northern part of Beresford Lake map-area, Southeast Manitoba," by J. F. Wright, p. 87B.
19. PAC, Geological Survey, Vol. 22, File 176, vol. 1, Desbarats to Camsell, 26 April 1923, "Flying operations for Geological Survey."
20. *Ibid.,* Camsell to Desbarats, 29 May 1923.
21. *Ibid.,* Young to Bolton, 2 Sept. 1924, commenting on a paper by a Mr. Mackenzie on "The use of aircraft in mineral exploration and development."
22. Department of Mines, *Annual Report,* 1926-27, p. 13.
23. PAC, Geological Survey, Vol. 22, File 176, vol. 2, Camsell to W. W. Cory, 11 May 1928, relaying a copy of W. H. Boyd to Camsell, dated 13 April 1928.
24. *Ibid.,* Collins to Camsell, 12 Nov. 1929, "Memorandum re aerial photography, 1930."
25. *Ibid.,* Vol. 23, File 176, vol. 3, Collins to Donald Gill, 18 Dec. 1931 (copy).
26. Canada, House of Commons, *Debates,* 1935, p. 2826.
27. Canada, *Statutes,* 25-26 Geo. V (1935), Chap. 34, Schedule A, item 2.
28. PAC, Geological Survey, Vol. 29, File 401A2, Camsell to W. M. Dickson, 29 Nov. 1935, enclosing "Memorandum re the Department of Mines million dollar geological field programme as an unemployment relief scheme" (secret).
29. *Ibid.,* Camsell to W. A. Gordon, 27 April 1935, Memorandum "Re: $1,000,000 Geological Survey project."
30. PAC, Geological Survey, Vol. 22, File 176, vol. 1, Lynch to Camsell, 26 March 1936.
31. Department of Mines, *Annual Report,* 1935-36, p. 10.
32. PAC, Geological Survey, Vol. 29, File 401A, vol. 3, Camsell to Collins, 6 May 1931.
33. *Ibid.,* File 401B, vol. 2, Camsell to E. J. Carlyle (secretary-treasurer, Canadian Institute of Mining and Metallurgy, Montreal), 29 May 1934.
34. Department of Mines, *Annual Report,* 1934-35, p. 1.
35. *American Journal of Science,* Fifth Series, Vol. 33, No. 197 (Jan.-June 1937), p. 398.

Chapter 18. THE FORTIES — OUT OF THE SHADOWS AND INTO THE LIGHT

1. Canada, Department of Mines and Resources, *Annual Report,* 1940-41, p. 12.
2. *Ibid.,* 1936-37, p. 16.
3. Canadian Institute of Mining and Metallurgy, *Bulletin* No. 320 (Dec. 1938), p. 614.
4. *Ibid.,* p. 615.
5. GSC, Central Registry, File 269A, "Memorandum re conference on the fields of geological work of the Dominion Department of Mines and the Ontario Department of Mines, Nov. 22, 1934." (copy).

6. *Ibid.,* T. F. Sutherland to Camsell, 28 Nov. 1934.
7. Department of Mines and Resources, *Annual Report,* 1943-44, p. 15.
8. *Ibid.,* 1945-46, p. 18.
9. *Ibid.,* 1944-45, p. 17.
10. GSC, Central Registry, File 270F "Establishment and Reclassification File," Young to Lynch, 8 Nov. 1937.
11. *Ibid.,* McLeish to Young, 29 March 1938.
12. GSC, Central Registry, File 269A, Young to T. F. Sutherland, 26 July 1934 (copy).
13. *Ibid.,* Young to H. C. Rickaby, 29 Jan. 1935.
14. Department of Mines and Resources, *Annual Report,* 1947-48, p. 71.
15. Canada, Department of Mines and Technical Surveys, *Annual Report,* 1949-50, p. 49.
16. Department of Mines and Resources, *Annual Report,* 1944-45, p. 19.
17. *Ibid.,* 1942-43, p. 33.
18. GSC, Central Registry, File 270E "Field staff and establishment," memorandum dated 7 Dec. 1945.
19. Department of Mines and Resources, *Annual Report,* 1947-48, p. 7.
20. *Ibid.,* p. 121.
21. Department of Mines and Technical Surveys, *Annual Report,* 1949-50. p. 7.

Chapter 19. SINCE 1950—A GENERAL VIEW

1. Canada, Senate, Special Committee on Science Policy, *Proceedings,* No. 16 (Wed., 11 Dec. 1968), p. 2408 (item 84).
2. For examples of this *genre,* see Canada, Royal Commission on Government Organization (the Glassco Commission), *Report,* Vol. 4 (Ottawa: Queen's Printer, 1963); and Science Council of Canada, Special Study No. 13, *Earth Sciences Serving the Nation* (Ottawa: Information Canada, 1971).
3. S. C. Robinson, *The Geological Survey of Canada: Into the Seventies—the Fourteenth Decade* (GSC, Miscellaneous Report, No. 18, 1972), p. 10.
4. Canada, Senate, Special Committee on Science Policy, *Proceedings,* No. 16, p. 2348 (C. M. Isbister).
5. *Ibid.,* p. 2398.
6. *Ibid.*

Chapter 20. ROUNDING OUT THE GEOLOGICAL MAP OF CANADA

1. GSC, *Annual Report,* 1969-70, p. 5.
2. GSC, Bulletin No. 54 (1959) *Helicopter Operations of the Geological Survey of Canada,* 1959, p. 6.
3. Department of Mines and Technical Surveys, *Annual Report,* 1954-55, p. 55.
4. *Ibid.,* 1956-57, p. 29.
5. GSC, Bulletin No. 54, 1959, p. 6.
6. Department of Mines and Technical Surveys, *Annual Report,* 1960, p. 25.
7. GSC, *Annual Report,* 1969-70, p. 109.
8. S. E. Jenness, compiler, *Field Work, 1961* (GSC, Information Circular No. 5, Feb. 1962), p. 4.

Chapter 21. IN 601 BOOTH STREET—THE LABORATORY AND SUPPORTING SERVICES

1. GSC, *Annual Report,* Vol. 7 (1894), p. 3A.
2. GSC, *Annual Report,* 1963, p. 1.
3. National Advisory Committee on Research in the Geological Sciences (NACRGS), *Report,* 1950-51, p. 16.
4. *Ibid.,* p. 17.
5. *Ibid.,* p. 18.
6. Department of Mines and Technical Surveys, *Annual Report,* 1954-55, p. 67.
7. *Ibid.,* 1957, p. 42.
8. *Ibid.*
9. *Ibid.*
10. NACRGS, *Report,* 1950-51, p. 21.
11. Department of Mines and Technical Surveys, *Annual Report,* 1957, p. 24.
12. GSC, *Annual Report,* 1959, p. 129.
13. *Ibid.*
14. *Ibid.,* p. 48.
15. Department of Mines and Technical Surveys, *Annual Report,* 1964, p. 29.
16. *Ibid.,* 1957, p. 40.
17. *Ibid.,* 1956-57, p. 41.
18. *Ibid.,* 1957, p. 39.
19. GSC, *Annual Report,* 1964, p. 96.
20. *Ibid.,* 1969-70, pp. 104-105.
21. L. W. Morley, "The Geophysics Division of the Geological Survey of Canada," *Canadian Mining and Metallurgical Bulletin,* Vol. 56, No. 613 (May 1963), p. 359.
22. P. Harker, compiler, *Geological Survey of Canada, Laboratory Facilities and Technical Services* (Ottawa: Department of Mines and Technical Surveys, 1963), p. 38.
23. *Ibid.,* p. 6.
24. GSC, *Annual Report,* 1964, p. 93.
25. *Ibid.,* 1965, p. 127.
26. Department of Mines and Technical Surveys, *Annual Report,* 1957, p. 39.
27. GSC, *Annual Report,* 1959, p. 100.
28. Department of Mines and Technical Surveys, *Annual Report,* 1964, p. 24.

Chapter 22. RESEARCH PROVIDES NEW ANSWERS

1. NACRGS, *Report,* 1950-51, pp. 24-25.
2. Science Council of Canada, *Earth Sciences Serving the Nation* (Ottawa: Information Canada, The Council, Special Study No. 13), pp. 160-61.
3. Canada, Royal Commission on Government Organization (the Glassco Commission), *Report,* 1963, Vol. 4, p. 217.
4. Science Council of Canada, *Basic Research,* by P. Kruus (Ottawa: Information Canada, Dec. 1971. The Council, Special Study

No. 21), p. 38.
5. Science Council of Canada, *Earth Sciences Serving the Nation,* pp. 132, 135.
6. R. J. W. Douglas, ed. *Geology and Economic Minerals of Canada* (GSC, Economic Geology Series No. 1, 5th edition, 1970), p. 593.
7. D. J. McLaren, "Palaeontology in Canada," *Journal of Palaeontology,* Vol. 42, No. 6 (November 1968), pp. 1334-35.
8. Department of Mines and Technical Surveys, *Annual Report,* 1957, p. 34.
9. S. M. Roscoe, *Huronian Rocks and Uraniferous Conglomerates in the Canadian Shield* (GSC, Paper No. 68-40, 1968), p. ix.
10. *Ibid.,* p. 2.
11. *Ibid.,* p. 167.
12. C. S. Lord and S. E. Jenness, compilers, *Field Work, 1960* (GSC, Information Circular No. 4, March 1961), p. 6.
13. *Abstracts of Publications in Scientific Journals by Officers of the Geological Survey of Canada, April 1969 to March 1970* (GSC, Paper 70-4, 1970), p. 8.
14. Department of Mines and Technical Surveys, *Annual Report,* 1964, p. 28.
15. C. H. Stockwell, *Fourth Report on Structural Provinces, Orogenies, and Time-classification of Rocks of the Canadian Precambrian Shield* (GSC, Paper 64-17, Part II), p. 9.
16. C. H. Stockwell, "Geochronology of stratified rocks of the Canadian Shield," *Canadian Journal of Earth Sciences,* Vol. 5, No. 3, Part II (June 1968), p. 696.
17. C. H. Stockwell, *Revised Precambrian Time Scale for the Canadian Shield* (GSC, Paper 72-52, 1973), p. 2.
18. R. J. W. Douglas, "Tectonics and Geotectonics," *in* C. H. Smith, ed., *Background Papers on the Earth Sciences in Canada* (GSC, Paper 69-56, 1970), p. 96.
19. V. V. Beloussov, "Development of the Earth and Tectonogenesis," *Journal of Geophysical Research,* Vol. 65, No. 12 (Dec. 1960), pp. 4127-4146.
20. From a preliminary draft of the paper, "Plate tectonic evolution of the Canadian Cordillera," prepared for presentation at the 1971 meeting of the Geological Society of America at Washington, D.C., later published in *American Journal of Science,* Vol. 272, No. 7 (Summer 1972), pp. 577-602.
21. Geological Society of America, *1971 Annual Meetings, Abstracts with Programs,* Vol. 3, No. 7 (Oct. 1971), p. 714.
22. J. Tuzo Wilson, "Tectonophysics," *in* C. H. Smith, ed., *Background Papers,* p. 100.

Chapter 23. LOOKING BEYOND CANADA'S BOUNDARIES

1. Science Council of Canada, *Earth Sciences Serving the Nation,* pp. 16-17.
2. International Geological Congress, XVIIIth, London, 1948, *General Proceedings* (London: The Congress, 1950), Part I, p. 147.
3. *Ibid.*
4. H. D. Hedberg, ed., *Preliminary Report on Stratotypes* (Montreal: I.G.C. XXIVth, International Subcommission on Stratigraphic Classification, Report No. 4, 1970), p. 4.
5. *Science,* Vol. 177, No. 4053 (15 Sept. 1972), p. 947.

Chapter 24. LOGAN'S LEGACY: SOME ACHIEVEMENTS OF THE SURVEY

1. H. R. Wynne-Edwards and J. F. Henderson, "Trends in geological research in Canada," *in* E. R. W. Neale, ed., *The Earth Sciences in Canada, a Centennial Appraisal and Forecast* (Toronto: University of Toronto Press, 1968, RSC, Special Publication No. 11), p. 25.
2. Copy of letter of D. Jerome Fischer, Visiting Professor, State University of Arizona, Tempe, Ariz., to Y. O. Fortier, 12 March 1969.
3. Canada, Senate, Special Committee on Science Policy, *Proceedings,* No. 16 (Wed., 11 Dec. 1968), p. 2350.
4. Canada (Province), *Report of the Select Committee on the Geological Survey* (Quebec: Lovell and Lamoureux, 1855), p. 39.
5. Canada, Royal Commission on Government Organization, *Report,* 1963, Vol. 4, p. 246.
6. Canada (Province), *Report. . . . Geological Survey,* p. 39.
7. S. C. Robinson, *The Geological Survey of Canada: Into the Seventies—The Fourteenth Decade* (GSC, Miscellaneous Report, No. 18, 1972), p. 20.

ESSAY ON SOURCES

The complete sum of source materials for the present study is probably more than could be thoroughly uncovered and mastered in a working lifetime; indeed, one would be hard put to keep up even with the annual increment. I am only too conscious of having merely scratched the surface, though preparing this study has given me some ideas about the sorts of material available, as well as the uses to which they might be put. This essay will first discuss the published records, beginning with the Survey's publications, going on to those of other government agencies, and then to outside sources. The second major division will discuss the manuscript materials, generally following the same sequence. Each category will examine records pertaining to individual members of staff as well as those relating to the Survey as an agency.

Published Sources

GSC PUBLICATIONS

These, the natural starting point for almost every aspect of the Survey's history, are grouped in several series, but include many other works that defy compartmentalization. The Survey, having early recognized a need for referring to its own findings, has prepared a number of guides and indexes to its major publications and maps. The latest publications guide (1961) is A. G. Johnston compiler, *Index of Publications of the Geological Survey of Canada 1845-1958,* which has been supplemented by later volumes carrying the list beyond 1958. A chronological list, keyed to geographical areas, and with a separate author index, this work is in need of updating and of some revision. The data from the publications have been thoroughly indexed in a series of *General Index* volumes covering the years 1863-1884 (D. B. Dowling, 1900), 1885-1906 (F. J. Nicolas, 1908), 1905-1916 (Nicolas, 1923), 1910-1926 (Nicolas, 1932), 1926-1950 (W. E. Cockfield, E. Hall, and J. F. Wright, 1962), and 1950-1959 (Wright, 1965). Nicolas was also responsible for two index volumes of paleontological data covering from 1847 to 1926. Of primary value for historical purposes are publications that chronicle the annual activities in field, laboratory, office, administration and personnel. There were several such series during the 130-year period since 1842, though no annual report was published by the Survey itself for the years between 1920 and 1958. For the first 42 years, to 1884, the activities were presented in the series titled *Report of Progress . . . ,* which began with the director's report, proceeded to individual reports of field results, then closed with the reports of the specialists (paleontologist, chemist-mineralogist). The *Reports* varied greatly in size, becoming rather voluminous as the staff increased. The period was covered in 28 volumes, counting the large *Report of Progress from its Commencement to 1863* (1863). Most are for one year though two are for four years apiece, two more cover three-year intervals, and no report at all was issued for the years 1858-1862 while the great 1863 report was being prepared. Despite its title, that work does not review the annual operations since 1843, not even for the missing years. Instead it consolidates the scientific results of the 20-year period and presents them as a compendium of what had been learned about the geology of Canada by 1863.

A second series was launched in 1885, the *Annual Reports (New Series),* of which 16 volumes, spanning the years 1884-1904, were published. These massive tomes, usually bound in governmental orange-yellow cloth, sometimes cover two years, though in two instances second volumes were required to present the reports of a given year, or years. The first section of each *Annual Report,* Part A, was the director's "Summary Report" of the year's activities (as was a Part AA if a summary report for a second year was included), which incorporated excerpts from geologists' interim reports, letters, etc. This was followed by a varying number of final reports prepared by the reconnaissance geologists and by the regional and subject specialists. A newcomer was that of the Section of Mines, which presented mining statistics and findings of examinations of mining operations. Since the individual reports were published when a volume was ready, each volume included the work from several years, and the *Annual Reports* were later and later coming before the public, until that for 1904 did not appear until 1907. Most of the Parts, some of which were of considerable length, were also published as separate volumes, as were the envelopes of large folded coloured maps belonging to each *Annual Report,* which also were sold as individual maps.

The *Annual Report* series ended with the volume for 1904, to be replaced by a paperbound annual *Summary Report,* which in essence continued the Part A of the previous series. The results of the field work were supposed to be presented in other publications, but many interim reports by individual geologists came to be included in the *Summary Reports,* so these quickly increased in bulk until that for 1912 contained over 40 separate reports and made a volume of more than 400 pages. By 1917 the annual *Summary Report* was being issued in Part form, with Part A devoted to the director's report and Parts B to F to reports of field activities across Canada by regions from west (B) to east (E). After 1920 the director's reports appeared in the annual departmental reports, so the *Summary Reports* thereafter dealt exclusively with the regional field work results. In 1933 they were discontinued.

For many years after 1920 the only summary report of Survey activities was the director's report in the annual departmental report. But during the 1950's these gradually became so brief, general and lacking in specific detail as to have little value for Survey purposes. Moreover, important events that did not belong with the scientific reports were going unrecorded. Accordingly, in 1958 a mimeographed *Annual Report* was prepared that outlined work done, listed staff, publications and administrative changes, and the like. These were designed purely for internal purposes—to keep employees informed about what went on, and to give administrators something to which they could turn for information on current and recent activities. The submissions were unedited and merely tentative; they were not official reports, and the volumes are not for public distribution. Thanks to the efforts of the History Committee, I received access to these reports, which were indispensable for many aspects of recent Survey history. Since 1962 there is also an annual report—the *Report of Activities*—that is designed for the public who may be interested in knowing what the current projects are, and how they are progressing.

The several types of annual reports by no means exhaust the list of publications that are pertinent to the history of the Survey or of its staff. The Survey exists to perform a certain kind of scientific work on the public's behalf, the results of which are required to be communicated to the public for its use. Before 1904 these were mainly incorporated in the *Reports of Progress* and *Annual Reports*. There one finds virtually all the classical reports of the early "greats," plus others too numerous to mention. With the replacement of the *Annual Report* by the *Summary Report* some new vehicle for publishing the matured results of the work in the field and laboratory was needed. For a number of years the Survey simply published the individual reports, then in 1910 a new series, the *Memoirs,* was inaugurated to present the results of research activities. At first they dealt with every aspect of the work, including topographical, anthropological and biological topics as well as geological ones, but since 1922 (or after *Memoir* No. 126) the series was confined to the geological spectrum of subjects. The *Memoirs* continue to be issued from time to time; by 1973 the number reached 375. Many volumes have won deserved fame for their authors and have become classics in their fields, while the series as a whole became the major source of scientific information produced by Survey personnel over the next fifty or sixty years.

Not long afterwards a second series was inaugurated (1913)—the *Museum Bulletins*—which were designed to publicize researches into anthropological, biological, mineralogical and paleontological subjects as well as general articles on geological questions. These became primarily the vehicle for publishing the work by staff associated with the Museum, so when the National Museum went its separate administrative way after 1945, it took this series along. The Survey then started another in its place, the *Geological Survey Bulletins,* or *Bulletins* for short, to publish researches in paleontological, mineralogical, structural geology, geochemical or other special areas. The series now (1973) numbers 222 volumes.

The expansion of prospecting activity in the Depression years of the 1930's, coupled with the disappearance of the *Summary Reports* in 1933, created a great demand for the Survey to make the results of each season's field work available as speedily as possible in preliminary form. After experimenting for a year or two with issuing preliminary maps (some hand-coloured) of districts recently examined, the Survey in 1935 launched a *Papers Series.* The items were mostly hurriedly-prepared maps of recently-completed surveys of favourable prospecting areas in the Canadian Shield or the Cordillera, and they often took the form of blueprinted maps with marginal notations packed in a brown paper envelope open at the top. They were numbered according to year of issue, as 35-1, 35-2, etc. The number varied from 4 to 42, the average being about 25 per annum for the period 1935 to 1958. Also beginning in 1935 and paralleling the *Papers* was an extended series of *Water Supply Papers* that grew out of the groundwater resource surveys of the Prairie Provinces carried out under the impetus of the Million Dollar Year. This series, extended throughout southern Canada since 1945, listed 317 items by 1958.

The *Papers* began to be published in regular book form in 1953, and during the 1960's they have become the principal vehicle for offering the finished scientific work of Survey staff to the public. Their numbers varied from year to year, but as many as 71 *Papers* have appeared in a single year (1970), and the 20-year total (1953-1972) stands as high as 830 volumes. In the main they present the results of field and laboratory studies by permanent or temporary staff, but in recent years they have also come to include many works of collaboration, such as symposia on a wide variety of topics held under Survey auspices, the most celebrated probably being the three volumes issued in connection with the International Upper Mantle Program. Other items were published on behalf of organizations in which the Survey was interested, such as the NACRGS, for which it has published studies emanating from its projects, conferences, subcommittees, as well as its annual reports and annual catalogues of current research activities. Another recent adoption was the publication, as a *Paper,* of the background papers prepared for the Solid-Earth Science Project of the Science Council of Canada.

Another series of publications—two series in fact—involve studies about the geologies of various minerals. The first of these, a short-lived series of 15 *Mineral Resources Bulletins,* appeared between 1903 and 1907 to meet the needs of the mining industry and uphold the Survey's claims in that direction. Also appearing in the interval before the commencement of the *Memoirs* was the first edition of *Geology and Economic Minerals of Canada* by G. A. Young (1909), which was to become a regular feature of the Survey's publication record. The *Economic Geology Series* proper began with a volume on *Talc Deposits of Canada* (1926), and since 1950 has published several large, important works, notably updated editions of *Geology and Economic Minerals of Canada* (E.G.1) which appeared in 1970 in its fifth edition, a completely rewritten one; and *Prospecting in Canada* (E.G.7) which reached its fourth edition in 1970. Like the *Papers,* this series recently acquired an important collaborative work, the 722-page symposium edited by L. W. Morley (E.G.26, 1970) on *Mining and Groundwater Geophysics, 1967,* containing the results of an international conference held at Niagara Falls, Ont., which the Survey had co-sponsored.

These more than 2,000 volumes still do not exhaust the official publications. From the earliest years numerous separate items have been issued for various purposes that do not fit the existing formats. Early examples include Logan's report on the Lake Superior mining district (1847), Hunt and Michel's report on the gold region of Canada (1866), or Logan and Hunt's summary of Canadian geology for the Paris Exposition of 1855. Paleontology reports accounted for 11 separate volumes during Logan's directorship, plus another 30 by 1904. One or two catalogues of the Survey's museum were also issued in the early years. After 1900 a few volumes were issued as separates apparently in order to reach a wider audience, a case in point being A. P. Low's celebrated *Report on the Dominion Government Expedition to Hudson's Bay and the Arctic Islands on Board the D.G.S. "Neptune" 1903-1904* (1906). The hiatus between the *Annual Reports* and the *Memoirs* (1905-1910) also left some 32 reports of research to be issued as separates. Though they include several important items (such as A. E. Barlow's reports on the Sudbury and Timiskaming districts) they are elusive items to find, whether in libraries or in private hands. Other miscellaneous items issued over the years have included the several indexes; four volumes (so far) by T. E. Bolton cataloguing the fossils in the Survey's National Type Collection; and the important *Coal Resources of Canada* by B. R. MacKay (1947), based on his work for a federal royal commission.

Since 1960 an effort has been made to organize such individual items into a *Miscellaneous Series*—surely a contradiction in terms!—that now numbers 19 items. The group includes several guidebooks on the geology of certain National Parks and of the National Capital District (mostly by D. M. Baird), regional guides to rock and mineral collecting in Canada, an editorial guide for preparation of maps and reports, two descriptions (1963, 1972) of the laboratories, scientific and technical services present at the Survey building, and two volumes of photographs.

The Survey has also published important co-operative works based mainly on the work of its staff members outside of the series already described, such as certain reports of the NACRGS or the Canadian Centre for Geoscience Data, before these were added to the *Papers Series.* In partnership with the Department of the Naval Service (and its successor department) it helped prepare and publish the findings of the Canadian Arctic Expedition of 1913-1918, most of which were the work of its employees in the Survey or the Museum. Useful reviews of the geology of Canada were prepared for a series of international meetings in Canada, notably the *Handbooks of Canada* published in connection with the meetings of the British Association for the Advancement of Science in Montreal (Selwyn and Dawson, 1884) and Toronto (Dawson, 1897). These offer very concise, convenient summaries of the state of geological knowledge of the country at those dates. This aspect of the Survey's publication activity reached a peak in 1913 with the visit of the International Geological Congress (the Twelfth) to Canada. Its staff prepared the large key study, *Coal Resources of the World*, the *Compte Rendu* of the meetings, and most of the *Guide Books* (11 of 13 parts, the Ontario Bureau of Mines being responsible for the other two). By contrast, the publication role in connection with the recent (1972) IGC meeting in Canada was limited to helping with selecting and editing papers for the various programs, and to contributing papers as individuals.

Such papers are a reminder that throughout its history a considerable part of the scientific output has taken the form of presentations and articles to outside organizations—as addresses and papers at meetings, conferences, and symposia convoked by learned societies, regional groups, professional and specialist societies, and journals. These include, from the early years, such periodicals as *American Journal of Science, Science*, Geological Society of London, *Journal, Geographical Journal, Canadian Journal of Science*, or Royal Society of Canada, *Transactions*. Today the range extends from the *Northern Miner* and *Canadian Mining Journal*, to Geological Society of America, *Bulletin*, to *Geophysics, Palaeontology, Analytica Chimica Acta, Geoderma, Canadian Journal of Earth Science*, and *Geos;* plus the whole gamut of collected special volumes published by any number of sponsoring associations. Such publications can be traced through the many catalogues of scientific publications by referring to authors, geographical or geological regions, or subjects. A useful source for geological work about Canada is a series of *Bulletins* of the United States Geological Survey, beginning with W. H. Darton, compiler, *Catalogue and Index of Contributions to North American Geology, 1732-1891* (Washington, 1896) and for recent years comprising an annual more than 1,000 page volume. "Canada" is a major subject in these, with numerous subheadings. Similar catalogues of the Geological Society of America can be consulted, also those of great libraries, or the card catalogues of good research libraries. To deal expressly with the Canadian scene the NACRGS recently started an annual volume listing and summarizing *Current Research in the Geological Sciences in Canada*, while the Survey now prepares an annual Paper, *Abstracts of Publications in Scientific Journals by Officers of the Geological Survey of Canada.*

The foregoing series and individual publications present the work of the Survey and of its staff to the public. These thousands of items vary in their value for historical purposes according to the goal of the researcher. They afford tangible evidence of the work the Survey was conducting at any given time, and they comprise significant milestones in the lives of their authors. They shed light on conditions in, and knowledge of, the areas about which they are written; and they are enduring illustrations and commentaries on the technology and the state of scientific knowledge at the time of their publication.

Falling in much the same category as the Survey's early publications as historical sources are the records of the official geological surveying activities in the colonies and regions that were later joined with Canada. The legislatures of two future provinces sponsored such operations as early as 1839: Newfoundland, which subsidized J. B. Jukes' work that appeared in London in 1842 as *Excursion in and about Newfoundland, during the Years 1839 and 1840*, 2 volumes, and again in 1843 as: *General Report of the Geological Survey of Newfoundland . . . during the Years 1839 and 1840*. New Brunswick commissioned Abraham Gesner, published his reports between 1839 and 1842, and followed these with a culminating *Report on the Geological Survey of the Province of New Brunswick . . . by Abraham Gesner* (Saint John, 1843). Twenty years later New Brunswick sponsored and published the results of two separate surveys: *Observations on the Geology of Southern New Brunswick, by L. W. Bailey*, and *A Preliminary Report on the Geology of New Brunswick . . . by H. Y. Hind* (both Fredericton, 1865). For Nova Scotia annual reports of mining operations were published by the Chief Gold Commissioner from 1862 onward, as well as a report, *Mines and Minerals, Report upon the Mines and Minerals of the Province for 1863* (Halifax, 1864), by S. P. Fairbanks of the Crown Lands Office.

No province-sponsored geological surveys seem to have been conducted in British Columbia before its entry in Confederation, but parts of the future Prairie Provinces were explored and described by several private travellers and by two official expeditions sponsored by the British and Canada governments in the late 1850's while the territory was still under the weakening control of the Hudson's Bay Company. The final report for the British expedition is United Kingdom, Parliamentary Paper, *Exploration—British North America: The Journals, Detailed Reports, and Observations Relative to the Exploration, by Captain Palliser, of that Portion of British North America. . . between the Western Shore of Lake Superior and the Pacific Ocean During the Years 1857, 1858, 1859, 1860* (London, 1863). A convenient reprint, with very much besides, is Irene M. Spry, editor, *The Papers of the Palliser Expedition, 1857-1860* (Toronto, 1968. The Champlain Society, Vol. 44). For the Canadian expedition, see Province of Canada, Legislative Assembly, *Report on the Exploration of the Country between Lake Superior and the Red River Settlement, and between the Latter Place and the Assiniboine and Saskatchewan*, by S. J. Dawson (Toronto, 1859).

In Newfoundland a continuing government-sponsored geological survey that published regular *Annual Reports* was begun by Alexander Murray in 1864. The reports were reprinted in two consolidations, the first for the years 1864 to 1880 inclusive: *Geological Survey of Newfoundland, Alexander Murray, Director, James P. Howley, Assistant* (London, 1881); and the second, *Reports of Geological Survey of Newfoundland by Alex Murray and James P. Howley from 1881 to 1909* (St. John's, 1918). Reports were published intermittently afterwards, notably in the late 1920's, but not until 1934 were two series of *Bulletins* and *Information Circulars* started under the Commission Government that have continued since Newfoundland joined Confederation in 1949.

THE GSC IN THE PUBLICATIONS OF OTHER GOVERNMENTAL AGENCIES

The chief official publication dealing with the history of the Survey—apart from its own works—is the annual report of the department to which it has belonged, for while the Survey has been part of a department of the federal government its director has supplied a report for that publication. Hence for the years from 1880 to 1890 the *Annual Reports* of the Department of the Interior contained a section on the activities of the Survey that was a somewhat watered-down version of the director's Summary Report in the Survey's own annual publication. This ceased when the Survey became an autonomous government department in 1890, and Part A of the *Annual Report (New Series)* or the later *Summary Report* was the ultimate report on the Survey's activities. This also held true under the Department of Mines, since it published no annual departmental report until the old regime of McConnell-McInnes-Haanel had been succeeded by the new one of Camsell-Collins-McLeish. From 1920-21 the director or chief geologist has presented his deputy minister annually with a submission summarizing the field and other activities of the year. The amount of space given to the Survey in the various annual reports has depended on the Survey's relative importance within the department to which it belonged, and since the mid-1950's, on the tendency throughout government to reduce annual departmental reports to little more than abbreviated public relations exercises. The directors' annual reports have grown shorter and briefer with each of the past five decades, as the Survey's annual reports have transferred, along with the branch, from the Department of Mines (to 1935-36), to Mines and Resources (1936-37 to 1949-50), Mines and Technical Surveys (1950-51 to 1965-66), and since 1966, Energy, Mines and Resources. Their value for historical purposes has decreased accordingly.

Publications of other government departments also contain records of the Survey's operations, some fairly regularly, others only intermittently or occasionally. The prime example of the former are the annual reports of the *Auditor-General* or of *Estimates* and *Public Accounts*, since all changes in staff, spending and organizational structure are bound to be reflected in these. The amount of detail on Survey activity and the manner in which that information is presented, varies from year to year so the value of these records fluctuates a good deal in different periods. The *Auditor-General* reports for the period 1880-1936 were especially helpful, for they gave such complete details of the staff and the operations that it was possible to work out dates of hiring and retirement or even of absences and illnesses of staff; also names of individual members of field parties, and minor expenditures in the field or at headquarters. After 1936, however, the records were not so complete, and the Survey's altered administrative position in the large Department of Mines and Resources made it more difficult to isolate its operations from other segments of the department, greatly inhibiting the usefulness of this kind of record. Present-day financial records are prepared in such a fashion as to render them almost entirely worthless as a research source.

The Survey turns up occasionally in reports of the Department of the Interior, Naval Service, National Defence, Public Works, Railways and Canals, or Indian Affairs and Northern Development, or Environment. Activities of its members were reported in the publications of more ephemeral agencies, such as the Canadian Pacific Railway surveys of the 1870's, or of the Commission of Conservation in the 1910-1920 period, and sometimes in those of the provinces as well. A good case in point is Ontario, Royal Commission on the Mineral Resources of Ontario, *Report,* 1890, which contained a long interview with Selwyn. Provincial government departments concerned with land surveying and natural resource management also sometimes refer to the Survey and its personnel. The published *Debates* of the federal House of Commons and Senate yield occasional questions about, and comments on, Survey operations. Only rarely did Survey affairs receive serious attention, notably in the 1901-07 period, when its position in relation to government plans for aiding the mining industry came under discussion. Officers have also given recorded testimony before many Parliamentary committees and commissions investigating mining activities, northern development, transportation, immigration and colonization, civil service conditions, and similar matters. For this last there are the *Proceedings* and *Reports* of the House of Commons Special Committees on the Civil Service of Canada in 1892, 1908, 1923, and 1961. In view of the disastrous effect of the classification of 1919 on staff morale, the Civil Service Commission, *Classification of the Civil Service of Canada* in 1919, 1923 and 1927, as well as *Report of Transmission to Accompany the Classification of the Civil Service of Canada, by Arthur Young & Co.* are worth consulting. For the latest phase there is also the Heeney Report—Civil Service Commission, *Personnel Administration in the Civil Service, A Review of Civil Service Legislation* (Ottawa, 1958).

The Survey is also concerned with the more or less continuous discussions during the past decade of government-science questions. The Glassco Commission, for example, investigated and reported on Survey operations; see, Canada, Royal Commission on Government Organization, *Report* (Ottawa, 1963). The testimony and the three-volume final report of Canada, Senate, Special Committee on Science Policy, *Proceedings* and *Report* (Ottawa, 1968-73) also have relevance. Even more so does some of the material published by such government-appointed bodies as the Science Council of Canada, especially its Special Study No. 13, *Earth Sciences Serving the Nation* (Ottawa, 1971), and its own Special Report No. 13 on the same subject.

In a special category among government publications, because they relate directly to the Survey, are the rare official inquiries into its operations, two of which were most important as exposés of current situations and goals, and as guides to its future programs. For that of 1854 under the Province of Canada, see "Report of the Select Committee on the Geological Survey. . .," (*Journals,* 1854-55, Appendix L), which offered an insight into Logan's ideas, style of managing, and talent for dealing with people. Even more important was the inquiry of 1884 by the federal House of Commons, featured in "Report of the Select Committee. . . to Obtain Information as to the Geological Survey, &c., &c.," (House of Commons, *Journals,* 1884, Appendix No. 8) because of the contemporary views of the Survey expressed therein, and the dramatic, valuable glimpses into the personalities of the contemporary and former staff of the Survey. The briefer, more limited enquiry into the de Rottermond affair, "Return to an Address. . . E.S. de Rottermund, heretofore Chemist in the Provincial Geological Department, or to the Provincial Government. . .," (*Journals,* 1846, Appendix WW) has some value for that particular early period of Survey history.

Finally, the several statutes of the Parliaments of the Province of Canada and of the Dominion relating to the Survey hold vital importance, for they embody the legislation that shaped its structure and operations. These are awkward to seek out, but a very useful collection has set those before 1926 down in one handy place—W. H. Collins, "The National Museum of Canada," published as the Survey's *Museum Bulletin* No. 50 in 1927.

CONTEMPORARY NON-GOVERNMENT PUBLISHED SOURCES

News reports are an obvious source of contemporary information, but one from which usable information is obtained with difficulty and has to be treated with some caution after it has been secured. In the early years, while the Survey was exploring the natural resources and other aspects of little-known parts of Canada, the reports and speeches by its staff were written up in considerable detail. But later, especially after 1914, its activities have received relatively little coverage. Ottawa newspapers are more likely to reflect day-to-day developments at the Survey, or the major and minor episodes in the lives of its staff, but only a few scattered references are made to these. Newspapers concerned with districts in which operations are proceeding may also report on these and on the personnel involved. Few newspapers are indexed, so it is best to note the dates and places of significant or recurring events, or other "leads" that are likely to be reported, then look through the newspaper files (nowadays usually microfilmed) for specific reports on them. Some newspapers have good archives with well-organized subject files, which scholars may be permitted to consult. A researcher should rejoice when he encounters a well-ordered newspaper archives or turns up a scrapbook, pile of clippings, or other memorabilia collected by a former Survey member, especially when the items give details of place and date—which is not always the case. Two scrapbooks preserved at the Survey, those of D. B. Dowling and H. M. Ami, offered very useful bits of information and a good deal of local colour for the periods of their coverage, roughly 1885-1920.

The activities of the Survey and its staff are more likely to be reported in specialized news media, such as those aimed at the mining industry. These include the monthly *Canadian Mining Journal* (since 1906), which goes back to the *Canadian Mining Review,* founded in 1882; and the *Northern Miner,* a weekly newspaper, published since 1915. A number of regional journals or newspapers are also devoted to mining prospects and activities in their districts, such as *British Columbia Miner* (1928), a monthly which became *The Miner* in 1931, and after 1945 *The Western Miner;* a similar monthly *Pre-Cambrian* (1930), which became *Mining in Canada* in 1959 but has been merged with the *Canadian Mining Journal.* The western Canadian petroleum industry is served by such periodicals as *Canadian Oil and Gas,* a monthly, since 1940, and by the weeklies *Oilweek* (from 1948) and *Oil in Canada* (from 1949). Any of these, plus the publications of professional associations such as the Canadian Institute of Mining and Metallurgy established in 1896 (which also publishes a monthly magazine) may at least repay glancing through their annual indexes for items relating to significant persons, subjects or events. The same holds true for appropriate Canadian or foreign economic or scientific journals, also those in related fields such as geography, natural history, surveying, or Arctic science.

Such periodicals are also helpful as sources of biographical information on Survey personnel, for scientific publications print valuable memorial articles on deceased scientists customarily prepared by their colleagues or friends. These can be found in such periodicals as Royal Society of Canada, *Proceedings and Transactions*, Geological Society of America, *Proceedings*, Geological Society of London, *Journal*, or in the pages of *American Journal of Science, Nature, Science, Geographical Journal, Canadian Record of Science, Ottawa Field Naturalist, Canadian Mining Journal, Canadian Geographical Journal*, and *Arctic*. Such pieces vary greatly in informational quality, as well as the perceptiveness and candour of the writer, though they tend to be more accurate than ordinary news reports. They often suffer, however, from the natural sentiment of *nil nisi bonum*. Even Ami felt constrained to make his obituary of his arch-enemy Robert Bell a dispassionate one, though he found the task "not an easy one."

WORKS ON THE GSC AND ITS STAFF

Histories of the Survey are few, and generally not very satisfactory in facing up to the complexities of the subject. The fullest of these, though it is still a very brief work, is F. J. Alcock, *A Century in the History of the Geological Survey of Canada* (National Museum of Canada, Special Contribution No. 47-1, 1947). This concentrates on the picturesque side of exploration and travel, and on the biographical details of the early directors and exploring geologists, without really coming to grips with the Survey as an organization, or looking into the administrative, scientific, economic, or sociological dimensions of its history. In some ways the earlier work by W. H. Collins, *The National Museum of Canada* (see above) is a more satisfying account, reflecting its author's painstaking and profound but opinionated character. Despite the name, the booklet dealt mainly with the Survey, for Collins saw the Museum as an outgrowth of its work and began his account with the establishment and growth of the Survey, the only concession to the given subject being a slight emphasis on the museum activities. Indeed, when he published a second booklet, *The Geological Survey of Canada* (n.p., n.d.) shortly afterwards he simply borrowed the relevant material from the earlier pamphlet, deleting only the valuable collection of statutes relating to the Survey, for which readers were referred back to the other work.

Shorter, article-length treatments of the Survey include two concise, authoritative descriptions, "The Geological Survey of Canada," by J. M. Harrison, in the *Canada Year Book, 1960* (Ottawa, 1960), and S. C. Robinson, *The Geological Survey of Canada: Into the Seventies – the Fourteenth Decade* (Miscellaneous Publication No. 18), prepared as background information for the delegates to the XXIVth International Geological Congress. A number of popular articles stress the exploring, adventure side of the work in early times, two examples being Ralph Purser, "Canada's Geological Survey," in *Canadian Geographical Journal*, Vol. 13 (1936), and Robert Collins, "Rock Finders," in *Imperial Oil Review*, Vol. 56 (1972).

As some of the foregoing secondary works indicate, the biographical side of the Survey's early history tends to receive the most attention, no doubt in part because of the availability of books on some of the earliest figures of Survey history. Thus Logan is commemorated by B. J. Harrington's fine *The Life of Sir William E. Logan, Kt., Ll.D., F.R.S., F.G.S., etc.* (Montreal, 1883). For the remarkable G. M. Dawson there still is only Lois Winslow-Spragge's gentle, affectionate family portrait, *The Life of George Mercer Dawson, 1849-1901* (Montreal, 1962), but nothing substantive exists on his scientific and intellectual attainments. The elder Dawson, however, is represented by his autobiography, *Fifty Years of Work in Canada* (London, 1901) and by C. E. O'Brien's recent *Sir William Dawson, A Life in Science and Religion* (Philadelphia, 1971). T. C. Weston also left his own monument in the delightful *Reminiscences Among the Rocks* (Toronto, 1899), while L. W. Bailey was commemorated with the more solid *Loring Woart Bailey, the Story of a Man of Science* (Saint John, N.B., 1925), by J. W. Bailey. For the redoubtable A. C. Lawson there is his own early *Outside of Beaten Paths* (University of California Chronicle, Jan. 1926), and the somewhat curious, recent F. E. Vaughan, *Andrew C. Lawson, Scientist, Teacher, Philosopher* (Glendale, Calif., 1970). J. B. Tyrrell, who proved he was a successful man of business as well as of science, inspired a book by W. J. Loudon, *A Canadian Geologist* (Toronto, 1930) in his lifetime, and since his passing, a more objective study by W. E. Eagan, "Joseph Burr Tyrrell, 1858-1957" (Ph. D. dissertation, University of Western Ontario, 1971).

On the other hand, the careers of other early officers of the Survey seem curiously neglected. Hunt, for example, whose remarkable personality and intellectual career merits a full-scale biography to do him justice, is memorialized only by several short works, notably by his friends James Douglas, *A Memoir of Thomas Sterry Hunt, M.D.* (Philadelphia, 1898), and F. D. Adams, *Biographical Memoir of Thomas Sterry Hunt* (National Academy of Science, Washington, Biographical Memoirs). This also applies to Robert Bell, who is presently served only by a pamphlet by Charles Hallock, *One of Canada's Great Explorers* (Washington, 1901), the obituary article of H. M. Ami in the Geological Society of America, *Bulletin*, Vol. 38 (1927), and a few modern-day popular articles, such as J. D. Leechman, "The Father of Place Names in Canada," *The Beaver*, Autumn, 1954. For Alexander Murray there is only Robert Bell's article, "Alexander Murray, F.G.S., F.R.S.C.,C.M.G.," in *Canadian Record of Science*, Vol. 5 (1892). As for Selwyn, who served longest as director (25 years) and in possibly the most interesting period of the Survey's history, there are only routine memorials by A. E. Barlow, and most notably by Ami, "A Brief Biographical Sketch of the Life and Work of the Late Dr. A. R. C. Selwyn, Director of the Geological Survey of Canada from 1869 to 1894," *American Geologist*, Vol. 31 (1903), plus a short piece by E. J. Dunn on his hectic Australian career, "The Founders of the Geological Survey of Victoria, Alfred Richard Cecil Selwyn," in Victoria, Geological Survey, *Bulletin* No. 23 (1910).

Fewer twentieth century figures have reached the position of being made subjects of full-scale biographies, unless, of course, they undertake this task for themselves. One may cite Charles Camsell, *Son of the North* (Toronto, 1954), James Mackintosh Bell, *Far Places* (Toronto, 1931), and the controversial Vilhjalmur Stefansson, whose many writings were essentially exercises in autobiography even before he sealed his long career, as it were, with the outstanding formal autobiography, *Discovery* (New York, 1964). Stefansson's career, especially for the years when he was most closely associated with Canada and the Survey, has been studied and written about by D. M. LeBourdais in *Stefansson, Ambassador of the North* (Montreal, 1963), and most thoroughly in the recent doctoral dissertation, R. J. Diubaldo, "The Canadian Career of Vilhjalmur Stefansson," (University of Western Ontario, 1972). A greater likelihood nowadays seems to lie in being memorialized by a biographical sketch in a *festschrift*, a volume of essays prepared in the scientist's honour. D. B. Dowling and Diamond Jenness—to give two instances—have inspired such volumes.

Biographical sketches of Survey officers seem to follow two tendencies. One, the traditional one aimed at the general public and written for the mass-circulation magazines or weekend supplements of the newspapers, concentrates mainly on the half a dozen best-known persons and seldom goes beyond the published sources for their material. The other, a recent development, reflects the emergence of scholarly interest in the history of the sciences and in the lives and work of distinguished pioneers and innovators. While there are still only a few students of the history of geology in Canada (as opposed to "historical geology" which is the study of sedimentary successions, mainly on the basis of paleontological records) their number is growing. The Geological Association of Canada, for example, has launched a series, "History of Canadian Geologists" that has generated two groups of papers thus far in the *Proceedings*, Vol. 23 (1971) and Vol. 24, No. 2 (Jan. 1972). C. G. Winder, who has made the study of Logan's life an avocation, is a good example of this trend. He has investigated and published three articles on previously neglected facets of Logan's career and background, plus an informative biography in the new *Dictionary of Canadian Biography*, Vol. 10 (1972). Indeed, as this latest ambitious venture in Canadian biography continues, with individual pieces of up to 5,000 words and more for major figures, it should inspire a great many studies and re-evaluations of the careers of leading Survey scientists.

WORKS ON RELATED SUBJECTS

A glance at the published literature on a few important subjects that are indirectly related to the history of the Survey is also in order. Works on the history of geology examined in connection with this study were, for the early period, F. D. Adams, *The Birth and Development of the Geological Sciences* (Baltimore, 1938); K. A. Von Zittel, *History of Geology and Palaeontology to the End of the Nineteenth Century* (English edition, London, 1901); Sir Archibald Geikie, *The Founders of Geology* (Baltimore, 1901, The G. H. Williams Memorial Lectures); and the biographies by Sir Edward Bailey on *James Hutton—the Founder of Modern Geology* (New York, 1967), and *Charles Lyell* (Garden City, N.Y., 1962). For the United States there is the comprehensive, useful G. P. Merrill, *The First One Hundred Years of American Geology* (New Haven, 1924), which deals mainly with the work of men and organizations but examines them against the background of changing theories and styles of work. Helpful for the more purely scientific developments was the collection of articles by C. Schuchert, H. E. Gregory, J. Barrell, and other specialists in *American Journal of Science*, Series IV, Vol. 46, No. 271 (July 1918), commemorating the centenary of the birth of that journal in 1818. The fiftieth anniversary of Johns Hopkins University generated another group of papers, *Fifty Years' Progress in Geology, 1876-1926* (Baltimore, 1927, Johns Hopkins University Studies in Geology No. 8), and the Geological Society of America brought out the book *Geology, 1888-1938, Fiftieth Anniversary Volume* (1941) as well as several historical articles in the March 1939 issue (Vol. 50, Part I) of the *Bulletin*. These works concentrated heavily on developments in the United States, perhaps slighting European science, but not neglecting Canadian achievements. A considerable strength of Merrill's book, in fact, was the space it devoted to British North American and Canadian developments.

In contrast, the history of geology in Canada is represented by nothing more than two short essays in volumes devoted to the broader spectrum of Canadian science: H. M. Tory, editor, *A History of Science in Canada* (Toronto, 1939), in which F. D. Adams prepared the article "The History of Geology in Canada;" and W. S. Wallace, editor, *The Royal Canadian Institute Volume, 1849-1949* (Toronto, 1949), to which F. J. Alcock contributed an article of ten pages. The Royal Society of Canada, inspired by the centennial of Confederation, devoted a volume, partly historical, to *The Earth Sciences in Canada, a Centennial Appraisal and Forecast* (Toronto, 1968, Special Publication No. 11), edited by E. R. W. Neale. These few examples illustrate the limited interest that Canadian scholars, historians or scientists, have given to the evolution of scientific ideas in Canada.

Another area of research that intersects with the history of the Geological Survey of Canada is the record of other geological organizations and societies, both because of their relationships with the Survey, and because they afford some comparisons with Canada's Survey. Thus for the British scene there are Sir J. S. Flett, *The First Hundred Years of the Geological Society of Great Britain, 1835-1935* (London, 1937), and Sir Edward Bailey, *Geological Survey of Great Britain* (London, 1952) as well as his interesting footnote, "A Hundred Years of Geology," in *Advancement of Science*, Vol. 9, No. 33 (1952); also H. B. Woodward, *The History of the Geological Society of London* (London, 1907). For the United States, G. P. Merrill's *The First One Hundred Years of American Geology*, mentioned above, provides a useful review of institutional developments, from state and national surveys to universities, and of the noteworthy journals and publications. Several works have appeared on the United States Geological Survey, mainly by the early leaders, but including a most interesting modern book by T. G. Manning, *Government in Science, the U.S. Geological Survey, 1867-1894* (Lexington, Ky., 1967) that focuses on the personalities and issues that dominated the troubled early years of that institution. The competitive, aggressive, dynamic, empire-builders of that day could hardly have presented a more complete contrast with the Canadian Survey in the same years, notwithstanding the brief turmoil that arose in 1884, which probably owed some of its force to the contemporary gales blowing across the border.

Canada is better represented in the field of administrative history than in that of intellectual history; at least there is a growing number of competent histories of several government agencies, some of which have important relationships with the Survey. Thus D. S. Thomson, *Men and Meridians, the History of Surveying and Mapping in Canada*, 3 vols. (Ottawa, 1967-69), and the shorter C. C. J. Bond, *Surveyors of Canada, 1867-1967* (Ottawa, 1966) both give considerable space to the topographical surveying role of the Survey, which they see as a major ingredient in the progress of that science. Thomson devotes entire chapters to the Survey, while Bond strikingly indicates the Survey's role with a map, "Major Exploratory Surveys 1868-1918" (pages 43-44), in which the Survey is shown as doing the lion's share of the work of exploration and reconnaissance. Another departmental history that contains scattered specialized references to aspects of Survey history (e.g., the Canadian Arctic Expedition) is T. E. Appleton, *Usque Ad Mare: A History of the Canadian Coast Guard and Marine Services* (Ottawa, 1968). Other works that hold implications for the Survey include: M. Thistle, *The Inner Ring: An Early History of the National Research Council of Canada* (Toronto, 1966), D. J. Goodspeed, *DRB-A History of the Defence Research Board of Canada* (Ottawa, 1958), Wilfrid Eggleston, *The Queen's Choice, A Story of Canada's Capital* (Ottawa, 1961), and H. B. Hachey, *History of the Fisheries Research Board of Canada* (Ottawa, 1965, Fisheries Research Board MSS Report Series). Noteworthy among the provincial administrative histories are two on Ontario agencies concerned with natural resources, Merrill Denison, *The People's Power; The History of Ontario Hydro* (Toronto, 1960), and the important R. S. Lambert and Paul Pross, *Renewing Nature's Wealth, A Centennial History* (Toronto, 1967), a history of forestry administration since the earliest times.

The vast expansion of Governmental operations in the last thirty years and the growth of political science studies at Canadian universities have combined to inspire the appearance of many works on aspects of public administration in Canada. Following in the steps of such pioneer works as R. MacGregor Dawson, *The Civil Service of Canada* (London, 1929) and Taylor Cole, *The Canadian Bureaucracy, A Study of Canadian Civil Servants and Other Public Employees, 1937-1947* (Durham, N.C., 1949), there appeared the excellent J. E. Hodgetts, *Pioneer Public Service: An Administrative History of the United Canadas, 1841-1867* (Toronto, 1955), his *The Canadian Public Service, A Physiology of Government 1867-1970* (Toronto, 1973), and Taylor Cole, *The Canadian Bureaucracy and Federalism: 1947-1965* (Denver, Colo., 1966). On the subject of financial control, there are Norman Ward, *The Public Purse, A Study in Canadian Democracy* (Ottawa, 1962), and W. L. White and J. C. Strick, *Policy, Politics and the Treasury Board in Canadian Government* (Don Mills, Ont., 1970). On staff relations and organizations there are S. J. Frankel, *Staff Relations in the Civil Service: The Canadian Experience* (Montreal, 1962); J. E. Hodgetts, W. McCloskey, R. Whittaker and V. S. Wilson, *The Biography of an Institution: The Civil Service Commission of Canada, 1908-1967* (Montreal, 1972), and J. Swettenham and D. Kealy, *Serving the State, A History of the Professional Institute of the Public Service of Canada, 1920-1970* (Ottawa, 1970). The number and variety of the periodical literature in this field (including the quarterlies *Canadian Public Administration* and *Canadian Journal of Political Science*) may be seen in the recent *Canadian Public Administration Bibliography* (Toronto, 1972) compiled by W. E. Grasham and G. Julien, and in such collections as W. S. K. Kernaghan, ed., *Bureaucracy in Canadian Government* (Toronto, 1969), and W. S. K. Kernaghan and A. M. Willms, eds., *Public Administration in Canada: Selected Readings* (Toronto, 1971).

Unpublished (Manuscript) Sources

OFFICIAL RECORDS

Almost every sort of published record previously listed is supported by, and often supplemented by, a variety of unpublished materials, which can be consulted in connection with a specific topic, or even to secure a single piece of desired information. Manuscript materials, like the published ones, vary greatly in their usefulness to the historical researcher, both intrinsically, and in the light of the purposes of his investigation.

A number of valuable manuscript sources exist for the history of the Survey, primarily in the form of directors' letterbooks and organized files of correspondence. For the important 40-year period down to 1906 there is available an important succession of directors' letterbooks beginning with one from Logan's final years for 1866-1869, and carrying on in 36 large volumes through the administrations of Selwyn, Dawson, and Bell. These volumes, now in the Public Archives of Canada, are the nucleus of their Record Group (R.G.) 45. They contain letterpress copies of the outgoing correspondence of the directors or their assistants, usually in their own handwriting. These were preserved because they were the chief filing system of their day, and were needed for reference purposes. Copies of the letters are placed in the order in which they were written, but each letterbook contained an index section or notebook in which the page numbers were listed against the correspondents' surnames (and occasionally against important subjects). These indexes made it possible to trace the sequence of a given correspondence, making the letterbooks a sort of filing system. The great expansion of correspondence and correspondents, the advent of the typewriter, and the spread of the modern filing system for outgoing and incoming correspondence, spelled *finis* to this most helpful kind of record. The final letterbook or two of the series indicated the coming trend, being filled with many pasted-in carbon copies of typed letters, occasionally together with the incoming letters they were answering. Without the indexes, or when they fail to produce a desired lead, one faces a time-consuming, often frustrating exercise of going through a volume page by page, for many of the letters are merely routine. Occasionally, however, a very valuable, revealing and helpful letter will suddenly repay the effort. Other officers besides the directors also kept letterbooks. One by Ells survives from the late 1890's, and no doubt others do, or did, exist.

Not many Survey records apart from the letterbooks remain from this early period. Some can be found in, or may be reconstructed from, the several bodies of private correspondence dating from those years. A few items, such as the early records of the Logan Club or others dealing with organizational matters, remain at the Survey's Branch Registry, which normally transfers files prior to 1950 to the Public Archives. Virtually no papers survive for the period 1906 to 1920 either, a serious loss since the letterbooks had passed out of existence by this time. One suspects that these records may have fallen victim to an overzealous paper salvaging operation in the period of the First World War, though the Survey's frequent moves and a reported fire in W MacInnes' office may have played some part in this loss.

After 1920, fortunately, the researcher returns to firmer ground, for R.G.45 also includes a large group of files dating from this time onward to 1949, which seem to owe something to the businesslike methods of the Bureau of Geology and Topography régime. The records for the latest period, from 1950 onward, are mainly in the Branch Registry, where they can be traced with the assistance of a filing key. The files in a room in the Survey library known as the Central Technical Files also includes some material of interest for an administrative or scientific history, notably records and reports relating to external aid activities, from which an informative and interesting book could be prepared. Many other recent files and collections that are valuable for historical purposes remain in institute, division, section and unit offices, or in individuals' filing cabinets. It is to be hoped that these will not be destroyed without prior examination, and that historical material will eventually make its way to R.G.45 at the Public Archives—as the Chief Geologist's Files are in the process of doing—to ensure that useful, potentially valuable records do not become lost.

A small but important official record source for the history of the Survey is a series of documents in the Public Archives known as the Historical Personnel collection. These files, turned over by the Civil Service (now Public Service) Commission, relate to the central figures of the Survey from Bell onward, including such men as Bell, Low, Brock, McConnell, Camsell, Collins, Young, Hume, and W. A. Bell. They deal with these men's employment, promotion or superannuation records in the main, but copies (or originals) of significant documents relating to the men and their careers are often included. The papers are restricted, however, to persons who possess written permission to examine an individual's record from the department or agency with which the man served. Through the good offices of Y. O. Fortier and former Dominion Archivist W. Kaye Lamb, I was able to examine and make notes on a number of these files.

The Public Archives and other repositories contain other official records that are worth examination in connection with various historical topics. Those used for the present work included at the Public Archives the collected Orders in Council, often with appended letters and other supporting documentation; records of such defunct agencies as the Air Board and its successors; Naval Service (for the large, important collection of Canadian Arctic Expedition papers); the papers of the Twelfth International Geological Congress; the papers of the Prime Ministers, especially Sir John Macdonald, Sir Wilfrid Laurier, Sir Robert Borden, and Arthur Meighen, and of such cabinet ministers as Edgar Dewdney and Clifford Sifton; of the deputy ministers R. A. Meredith and Charles Camsell; the Sandford Fleming Papers, especially for the "Pacific Railway Matters" letterbook, etc. In Ottawa, further useful materials were found in the National Museum library and at the headquarters of the Department of Public Works, and similar records undoubtedly remain in other departments and branches there. Outside of Ottawa, the papers of David Mills, M.P. and Minister of the Interior at the time of the Survey act of 1877, are held in the University of Western Ontario library, while those of his chief, Alexander Mackenzie, are at Queen's University and at the Public Archives. I uncovered some traces of records of value for a variety of purposes in the mining branches of several provinces or in their archives; in the geology departments of Canadian and United States universities; at the offices of national and international scientific or professional societies, organizations, or of periodicals and publishers. Prime examples of this genre would be the Smithsonian Institution and the United States Geological Survey in Washington.

The original records of the scientific work of the Survey—in contrast with the administrative and personnel records which have survived only in part—are generally extremely well preserved. This applies especially to the thousands of field notebooks written by Survey officers from Logan onward, in which they recorded their travels, observations, measurements, and tentative conclusions in the course of their work. These have been carefully preserved, and nearly all of them survive, the pre-1900 notebooks being in R.G.45 at the Public Archives, and the post-1900 ones still at the Branch Registry of the Survey. Field notebooks usually contain matter-of-fact descriptions, sketches, lists, observations, calculations representing the daily record of what was seen, noted, or felt while the work was in progress. Usually these were afterwards worked up and incorporated in the published reports. The originals are still consulted, however, by field officers studying the areas with which they deal, by paleontologists or mineralogists wishing to compare notes with earlier workers, by geographers interested in more precise details of the observations than those offered in the published reports, or by historians looking for local colour or for some insights into the diarist's personality. Occasionally notebooks contain telling personal touches in the form of philosophical thoughts, poems, drawings, drafts of letters, and comments on contemporaries. A finding list helps in locating desired notebooks by author.

PERSONAL RECORDS

The foregoing records, while mainly relating to aspects of the administrative history of the Survey, are also records of the careers of its personnel, whether directors or ordinary members of the staff. They may, accordingly, be consulted while investigating the life and work of a given individual. Equally, many of the personal papers that were consulted for this study (see below) explain, fill in gaps, and extend the scope of available knowledge about the history of the Survey as a whole.

The historian is fortunate, all things considered, in the personal papers that have survived from the first half of the Survey's history, even though the lack of success in unearthing any body of Selwyn papers was a considerable disappointment for a time. In the end this did not prove too serious, since so much of Selwyn's correspondence and ideas survive in the numerous letterbooks, notebooks, and journals. An extensive collection of Logan Papers was preserved at the McCord Museum, which recently transferred it to the McGill University Archives. Other material relating to Logan is found at the Rare Book Room of McGill University Library and at the Baldwin Room of the Metropolitan Toronto Library; the Public Archives in Ottawa also possesses a few of his early account books of Survey expenditures. G. M. Dawson's very informative private diaries and his correspondence are also at the McGill University Archives, along with the extensive Sir William Dawson Papers, which include many of the son's letters. The large Robert Bell Papers collection at the Public Archives is another highly revealing, interesting source for Survey history. Other pertinent collections from the nineteenth century include the F. D. Adams, Thomas Macfarlane, T. Sterry Hunt, and Montreal Natural History Society Papers at the various libraries at McGill University; the J. B. Tyrrell Papers at the Rare Book Room of the University of Toronto Library; and those of L. W. Bailey at the University of New Brunswick Library. The Archives of the City of Montreal were helpful for details of the Logan Farm and the Survey's buildings during the Montreal phase of its history. In Montreal, also, certain treasured important family papers of the two Dawsons in the possession of Mrs. Lois Winslow-Spragge (née Harrington) and of R. W. Brock held by his son Mr. T. L. Brock (the only ones uncovered outside of official collections) were made available. I wish to thank both Mrs. Winslow-Spragge and Mr. Brock for their kindness and also for their hospitality. Many other rich holdings of private papers undoubtedly remain in the hands of the several hundred persons who worked for the Survey or had dealings with it. It is hoped that the present owners of these papers will recognize their value to scholarship and will take steps to see that they are properly preserved in public archives. There they would be protected against loss or deterioration, sorted and arranged for useful study, and made available for consultation by properly-accredited scholars in convenient surroundings. They would be brought to the attention of potential users by being listed in archival inventories and comprehensive catalogues such as future editions of the *Union List of Manuscripts in Canadian Repositories*, issued by the Public Archives of Canada in 1968.

ILLUSTRATIONS

Chapter 1

Figure I-1, p. 2–GSC
I-2, p. 5–GSC
I-3, p. 7–Metropolitan Toronto Library Board
I-4, p. 8–GSC 202318*
I-5, p. 9–Public Archives of Canada C 2614
I-6, p. 10–GSC 202419
I-7, p. 11–GSC 202375
I-8, p. 12–GSC 202397
I-9, p. 13–GSC 202404
I-10, p. 14–GSC 202398
I-11, p. 14–GSC 20237
I-12, p. 14–GSC 202399
I-13, p. 15–GSC 202387
I-14, p. 15–GSC 100798
I-15, p. 16–GSC 202376
I-16, p. 19–Metropolitan Toronto Library Board
I-17, p. 21–Metropolitan Toronto Library Board

*Geological Survey of Canada negative number

Chapter 2

Figure II-1, p. 22–GSC 154571
II-2, p. 25–GSC 202373
II-3, p. 25–GSC 202367
II-4, p. 25–GSC 202374
II-5, p. 25–GSC 202453
II-6, p. 26–GSC 202368
II-7, p. 26–GSC 202370
II-8, p. 26–GSC 202369
II-9, p. 29–GSC 202450
II-10, p. 29–GSC 202372
II-11, p. 32–GSC
II-12, p. 33–GSC
II-13, p. 35–GSC
II-14, p. 35–GSC
II-15, p. 37–GSC 202337
II-16, p. 38–GSC 55957

Chapter 3

Figure III-1, p. 40–GSC 69409
III-2, p. 42–GSC
III-3, p. 45–Public Archives of Canada
III-4, p. 46–Notman Photographic Archives, McCord Museum of McGill University, No. 24701-I
III-5, p. 47–GSC 69324
III-6, p. 49–Public Archives of Canada
III-7, p. 51–GSC
III-8, p. 52–GSC 200121
III-9, p. 52–GSC 81367
III-10, p. 56–GSC 202429
III-11, p. 59–McGill University Archives, Accession No. 1207
III-12, p. 60–GSC 91410

Chapter 4

Figure IV-1, p. 62–GSC
IV-2, p. 64–Public Archives of Canada C 2831
IV-3, p. 65–GSC 69323
IV-4, p. 65–GSC 77284
IV-5, p. 65–GSC 77281
IV-6, p. 66, left–GSC 77280
right–Public Archives of Canada C 7606
IV-7, p. 67–GSC 202434
IV-8, p. 70–Public Archives of Canada PA 38059
IV-9, p. 71–GSC 123853
IV-10, p. 71–GSC 77282
IV-11, p. 72, upper–GSC 202432
centre–GSC
lower–GSC 202433
IV-12, p. 73–GSC 202430
IV-13, p. 74, left–GSC 200446-F
right–GSC 200446-A
IV-14, p. 78–Notman Photographic Archives, McCord Museum of McGill University, No. 4408-I

Chapter 5

Figure V-I, p. 82–GSC 68772
V-2, p. 84–Public Archives of Canada PA 25471
V-3, p. 85–Public Archives of Canada PA 13008
V-4, p. 86–Public Archives of Canada PA 12855
V-5, p. 87, upper–GSC 200446-C
lower–GSC 200446-D
V-6, p. 89–Notman Photographic Archives, McCord Museum of McGill University, No. 3330-I
V-7, p. 90–Notman Photographic Archives, No. 28898-I
V-8, p. 92–GSC 202316
V-9, p. 94–GSC
V-10, p. 95, left–GSC 202391
centre–GSC 202379
right–GSC 202380
V-11, p. 96–Public Archives of Canada PA 25959
V-12, p. 97–GSC 202425
V-13, p. 99–Public Archives of Canada
V-14, p. 100–GSC 202377
V-15, p. 101–GSC 202381

Chapter 6

Figure VI-1, p. 104, upper–GSC 202410
lower–Public Archives of Canada PA 38051
VI-2, p. 106–GSC
VI-3, p. 107–GSC 201993
VI-4, p. 109–GSC 202421
VI-5, p. 110–GSC 201779
VI-6, p. 111–Public Archives of Canada C 4206
VI-7, p. 111–GSC 251
VI-8, p. 113–Public Archives of Canada C 61614
VI-9, p. 115–Public Archives of Canada C 18100

VI-10, p. 115—GSC 402
VI-11, p. 117—T.C. Weston, No. 293, 1879
VI-12, p. 119—GSC
VI-13, p. 120—GSC 68776
VI-14, p. 121—GSC 202412
VI-15, p. 123—Public Archives of Canada PA 25486

Chapter 7

Figure VII-1, p. 130—Public Archives of Canada PA 8438
VII-2, p. 132—GSC 201735
VII-3, p. 133—National Museums of Canada 31874
VII-4, p. 135—GSC 201736
VII-5, p. 137—GSC 202378
VII-6, p. 137—GSC 202253
VII-7, p. 139—GSC 200096
VII-8, p. 139—GSC 201778
VII-9, p. 141—GSC 202400
VI-10, p. 141—GSC 202403
VII-11, p. 141—GSC 202402
VII-12, p. 143—GSC 68774
VII-13, p. 144—GSC 201746
VII-14, p. 145—GSC 201739
VII-15, p. 146—GSC 81450
VII-16, p. 147—GSC 97342
VII-17, p. 148—Public Archives of Canada PA 33706
VII-18, p. 148—Public Archives of Canada PA 25581
VII-19, p. 148—Public Archives of Canada PA 25639
VII-20, p. 149—GSC 201481

Chapter 8

Figure VIII-1, p. 150—Public Archives of Canada PA 38226
VIII-2, p. 150—Public Archives of Canada PA 38279
VIII-3, p. 153—GSC
VIII-4, p. 156—GSC 202411
VIII-5, p. 157—GSC 386
VIII-6, p. 158—GSC
VIII-7, p. 159—Public Archives of Canada PA 38009
VIII-8, p. 160—Public Archives of Canada PA 37984
VIII-9, p. 161—GSC
VIII-10, p. 163—Public Archives of Canada PA 38193
VIII-11, p. 164—Public Archives of Canada PA 38145
VIII-12, p. 167—GSC
VIII-13, p. 169—Public Archives of Canada GSC 1904
VIII-14, p. 172—GSC
VIII-15, p. 173—A.P. Low, No. 2795, 1903
VIII-16, p. 174—Public Archives of Canada PA 38265

Chapter 9

Figure IX-1, p. 176—Public Archives of Canada GSC 2132
IX-2, p. 178—GSC
IX-3, p. 179—GSC 202390
IX-4, p. 180—GSC 81451
IX-5, p. 181—GSC 82062
IX-6, p. 181—GSC 202428
IX-7, p. 182—GSC 81516
IX-8, p. 182—GSC 202389
IX-9, p. 183—Public Archives of Canada PA 39835
IX-10, p. 184—GSC 201995-A
IX-11, p. 185—GSC 202420
IX-12, p. 188—GSC 201646
IX-13, p. 190—GSC 202385
IX-14, p. 191—GSC 202413
IX-15, p. 195—GSC 384
IX-16, p. 196—GSC 202388

Chapter 10

Figure X-1, p. 198—GSC 68773
X-2, p. 200—GSC 202431
X-3, p. 201—GSC
X-4, p. 203—GSC 201735
X-5, p. 205—GSC 202313
X-6, p. 205—GSC 201740
X-7, p. 207—GSC 201753
X-8, p. 210—GSC 68775-B107
X-9, p. 211—GSC 201736-A
X-10, p. 212—GSC 89436
X-11, p. 213, top left—Public Archives of Canada, Map Division
top right—Public Archives of Canada Mss. C 57374
bottom left—Public Archives of Canada Mss. C 57371
bottom right—Public Archives of Canada Mss. C 57372
X-12, p. 216—Public Archives of Canada Mss. C 57375
X-13, p. 217—GSC 202383
X-14, p. 218—GSC 73888

Chapter 11

Figure XI-1, p. 220—Public Archives of Canada GSC 3086
XI-2, p. 222—GSC
XI-3, p. 223—Canada Dept. of the Interior, *The Yukon Territory* (Ottawa, 1916), p. 10
XI-4, p. 225—Public Archives of Canada PA 39971
XI-5, p. 226—GSC
XI-6, p. 227—GSC 25982
XI-7, p. 229—Public Archives of Canada PA 39792
XI-8, p. 231—Public Archives of Canada PA 39851
XI-9, p. 233—GSC 201742
XI-10, p. 235—GSC 1616
XI-11, p. 237—Public Archives of Canada C 22991
XI-12, p. 239—Public Archives of Canada PA 39798

Chapter 12

Figure XII-1, p. 240—Public Archives of Canada PA 25940
XII-2, p. 242—GSC 14278
XII-3, p. 244—GSC 83752
XII-4, p. 247—GSC 202407
XII-5, p. 248—Public Archives of Canada, National Map Collection, C 57211
XII-6, p. 250—Public Archives of Canada PA 25998
XII-7, p. 251—GSC 202406
XII-8, p. 252—GSC
XII-9, p. 256—Public Archives of Canada PA 25976
XII-10, p. 258—GSC 72071-C30

Chapter 13

Figure XIII-1, p. 262–GSC 201772
XIII-2, p. 265–GSC 202418
XIII-3, p. 267–Public Archives of Canada PA 42281
XIII-4, p. 267–National Museums of Canada 18806
XIII-5, p. 271–GSC 202250
XIII-6, p. 272–GSC 109384
XIII-7, p. 273–National Museums of Canada 80792
XIII-8, p. 273–GSC 201744
XIII-9, p. 274–GSC 47109
XIII-10, p. 274–GSC 202386
XIII-11, p. 275–Public Archives of Canada C 57974
XIII-12, p. 277–National Museums of Canada 29062
XIII-13, p. 279–National Museums of Canada 85901
XIII-14, p. 279–National Museums of Canada 99477
XIII-15, p. 280–National Museums of Canada J4840
XIII-16, p. 281–National Museums of Canada 76087
XIII-17, p. 282–National Museums of Canada 100028
XIII-18, p. 282–GSC 202423

Chapter 14

Figure XIV-1, p. 284–GSC 202451
XIV-2, p. 286–GSC 113473
XIV-3, p. 288–GSC
XIV-4, p. 291–Public Archives of Canada GSC 5485
XIV-5, p. 292–GSC 202417
XIV-6, p. 293–GSC 78622
XIV-7, p. 294–GSC 74492
XIV-8, p. 295–GSC 202384
XIV-9, p. 295–GSC 57720
XIV-10, p. 298–GSC 202394
XIV-11, p. 299–GSC 201995
XIV-12, p. 303–GSC 26719
XIV-13, p. 305–GSC 202458
XIV-14, p. 307–GSC 68776

Chapter 15

Figure XV-1, p. 308–National Museums of Canada 81449
XV-2, p. 311–Public Archives of Canada PA 22433
XV-3, p. 312–National Museums of Canada 89197
XV-4, p. 313–GSC 68777
XV-5, p. 313–GSC 68778
XV-6, p. 317–GSC 201745
XV-7, p. 318–GSC 202424
XV-8, p. 319–GSC 68779
XV-9, p. 320–GSC
XV-10, p. 321–GSC 202257
XV-11, p. 322–Public Archives of Canada C 57953
XV-12, p. 323–National Museums of Canada 51548
XV-13, p. 324–National Museums of Canada J6611
XV-14, p. 324–National Museums of Canada 67220
XV-15, p. 326, left–National Museums of Canada 51623
right–National Museums of Canada 51624
XV-16, p. 327, left–GSC 202416
right–GSC 202415
XV-17, p. 329–National Museums of Canada 37080
XV-18, p. 330–GSC 47110
XV-19, p. 331–GSC 201484
XV-20, p. 333–GSC 201743

Chapter 16

Figure XVI-1, p. 334–GSC 91823
XVI-2, p. 336–GSC 202466
XVI-3, p. 337–GSC 72122
XVI-4, p. 338–GSC
XVI-5, p. 339, top left–GSC 202255
top centre–GSC 202252
top right–GSC 201734
bottom left–GSC 74800
bottom centre–GSC 91389
bottom right–GSC 201649
XVI-6, p. 342–GSC 202392
XVI-7, p. 343–GSC
XVI-8, p. 344–GSC
XVI-9, p. 347–GSC 202457
XVI-10, p. 349–GSC 74799
XVI-11, p. 351–GSC 99601
XVI-12, p. 352–National Museums of Canada 67938
XVI-13, p. 353–GSC 200128
XVI-14, p. 354–GSC 202405
XVI-15, p. 355, top left–GSC 202395
top right–National Museums of Canada 83906
lower right–National Museums of Canada 96951
centre left–National Museums of Canada 54584
lower left–National Museums of Canada 53312

Chapter 17

Figure XVII-1, p. 358–*Canadian Geographical Journal,* Vol. 10, No. 5. (May 1935), p. 241
XVII-2, p. 361–GSC 202452
XVII-3, p. 362–GSC 81519
XVII-4, p. 362–GSC 202254
XVII-5, p. 363–GSC 72070
XVII-6, p. 363–GSC 201773
XVII-7, p. 364–GSC 82061
XVII-8, p. 365–GSC 201777
XVII-9, p. 366–GSC 68780
XVII-10, p. 366–GSC 92981
XVII-11, p. 368–GSC 202456
XVII-12, p. 370–GSC 202455
XVII-13, p. 371–*Engineering and Mining Journal,* Vol. 136, No. 11, p. 548
XVII-14, p. 373–GSC 73395
XVII-15, p. 375–National Museums of Canada 79197
XVII-16, p. 376–GSC 201650
XVII-17, p. 378–GSC
XVII-18, p. 379–GSC
XVII-19, p. 380–GSC 88258
XVII-20, p. 380–GSC 69242

Chapter 18

Figure XVIII-1, p. 382–GSC 94252
XVIII-2, p. 385–GSC 202408
XVIII-3, p. 386–GSC 107007
XVIII-4, p. 387–GSC
XVIII-5, p. 389–GSC
XVIII-6, p. 391–GSC 200244
XVIII-7, p. 392–GSC 202454
XVIII-8, p. 396–GSC 202393
XVIII-9, p. 399–GSC 202256
XVIII-10, p. 399–GSC 112040

XVIII-11, p. 399–GSC 112055
XVIII-12, p. 400–GSC 105697
XVIII-13, p. 401–GSC 106869
XVIII-14, p. 402–GSC 94797
XVIII-15, p. 403–GSC 202409
XVIII-16, p. 404–GSC 202422

Chapter 19

Figure XIX-1, p. 408–GSC 112907-A
XIX-2, p. 410–GSC
XIX-3, p. 411–Dept. of Energy, Mines and Resources
XIX-4, p. 413–GSC 201114
XIX-5, p. 414–GSC 92956
XIX-6, p. 415–GSC 201112
XIX-7, p. 416–GSC 200127
XIX-8, p. 418–GSC
XIX-9, p. 420–GSC 201752
XIX-10, p. 421–GSC
XIX-11, p. 423–GSC 200233
XIX-12, p. 424–GSC 202034
XIX-13, p. 425–GSC 201598
XIX-14, p. 425–GSC 201776
XIX-15, p. 426–GSC 201722
XIX-16, p. 428–GSC 202081

Chapter 20

Figure XX-1, p. 430–GSC 200161
XX-2, p. 433–GSC
XX-3, p. 434–GSC
XX-4, p. 435–GSC 118966
XX-5, p. 436–GSC 8-4-63
XX-6, p. 438–GSC 202427
XX-7, p. 439–GSC 201209-A
XX-8, p. 439–GSC 201517
XX-9, p. 441–GSC
XX-10, p. 443–GSC
XX-11, p. 444–GSC
XX-12, p. 445–GSC 1-1-53
XX-13, p. 446–GSC 202331
XX-14, p. 447–GSC 201227
XX-15, p. 448–GSC 200234
XX-16, p. 449–GSC

Chapter 21

Figure XXI-1, p. 450–GSC 112891-A
XXI-2, p. 452–GSC 99706
XXI-3, p. 453–GSC 99707
XXI-4, p. 454–GSC
XXI-5, p. 454–GSC 200840
XXI-6, p. 456–GSC 200114
XXI-7, p. 457–GSC 201193-B
XXI-8, p. 458–GSC 202032-N
XXI-9, p. 460–GSC 202032-G
XXI-10, p. 461–GSC 111173
XXI-11, p. 461–GSC 200929
XXI-12, p. 462–GSC 202032-K
XXI-13, p. 464–GSC 113470
XXI-14, p. 466–GSC 201749
XXI-15, p. 467–GSC 201754
XXI-16, p. 468–GSC 202032-W
XXI-17, p. 469–GSC 202295
XXI-18, p. 470–GSC 202032-Z

Chapter 22

Figure XXII-1, p. 472–GSC
XXII-2, p. 477–GSC 124123
XXII-3, p. 479, upper left–GSC 108184
upper right–GSC 202057
lower left–GSC 200838
lower right–GSC 111315
XXII-4, p. 480–GSC
XXII-5, p. 481–GSC
XXII-6, p. 482–GSC 201203
XXII-7, p. 483–GSC
XXII-8, p. 485–GSC
XXII-9, p. 486–GSC 201755
XXII-10, p. 486–GSC
XXII-11, p. 487–GSC
XXII-12, p. 488–GSC
XXII-13, p. 491–GSC 202467
XXII-14, p. 492–GSC 202426

Chapter 23

Figure XXIII-1, p. 494–GSC
XXIII-2, p. 498–GSC
XXIII-3, p. 499-GSC 201600
XXIII-4, p. 500–National Film Board
XXIII-5, p. 502–GSC 202323-D
XXIII-6, p. 504, above–GSC
below–GSC
XXIII-7, p. 507–GSC 201688
XXIII-8, p. 508–GSC 201525
XXIII-9, p. 509–GSC
XXIII-10, p. 510–GSC

Chapter 24

Figure XXIV-1, p. 512–GSC 201590
XXIV-2, p. 515–GSC 112891
XXIV-3, p. 517–GSC 363
XXIV-4, p. 517–GSC 14532
XXIV-5, p. 517–GSC 202364
XXIV-6, p. 517–GSC 202056
XXIV-7, p. 519–GSC 117963
XXIV-8, p. 522–Public Archives of Canada PA 38272
XVIV-9, p. 524–GSC 202036-A
XXIV-10, p. 525–GSC

Appendix I

THE ADMINISTRATIVE CHAIN OF COMMAND, 1842-1972

POLITICAL CHIEFS	DEPUTY MINISTERS	DIRECTORS
	UNDER THE PROVINCE OF CANADA	
CIVIL SECRETARY TO THE GOVERNOR-GENERAL		
Rawson W. Rawson, 1842-44	Christopher Dunkin, 1842-47 (for Canada East)	William Edmond Logan (later Sir William), 1842-69
	James Hopkirk, 1842-46 (for Canada West)	
PROVINCIAL SECRETARIES		
Dominic Daly, 1841-48	Etienne Parent, 1847-67 (for Canada East)	
	Edmund A. Meredith, 1847-67 (for Canada West)	
R. B. Sullivan, 1848		
James Leslie, 1848-51		
A. N. Morin, 1851-53		
P. J. O. Chauveau, 1853-55		
G. E. Cartier, 1855-56		
T. L. Terrill, 1856-57		
T. J. J. Loranger, 1857-58		
Oliver Mowat, 1858		
Charles Alleyn, 1858-62		
A. A. Dorion, 1862-63		
J. O. Bureau, 1863		
A. J. Fergusson Blair, 1863-64		
John Simpson, 1864		
William McDougall, 1864-67		
	UNDER THE DOMINION OF CANADA	
SECRETARIES OF STATE FOR THE PROVINCES		
Adams G. Archibald, 1867-68	E. A. Meredith, 1867-78	
Joseph Howe, 1869-73		Alfred Richard Cecil Selwyn, 1869-95
Thomas N. Gibbs, 1873		
MINISTERS OF THE INTERIOR		
Alexander Campbell, 1873		
David Laird, 1873-76		
David Mills, 1876-78		
Sir J. A. Macdonald, 1878-83	John S. Dennis, 1878-81	
	Lindsay Russell, 1882-83	
Sir D. L. Macpherson, 1883-85	Alexander M. Burgess, 1883-97	
Thomas White, 1885-88		
Edgar Dewdney, 1888-92		
T. M. Daly, 1892-96		
H. J. Macdonald, 1896		
Clifford Sifton, 1896-1905		George Mercer Dawson, 1895-1901
	James A. Smart, 1897-1904	
		Robert Bell, 1901-06 (Acting Director)
Frank Oliver, 1905-(11)	William W. Cory, 1905-(31)	
		Albert Peter Low, 1906-07

POLITICAL CHIEFS	DEPUTY MINISTERS			DIRECTORS	
MINISTERS OF MINES William Templeman, 1907-11	Albert P. Low, 1907-14			Reginald Walter Brock, 1907-14	
W. B. Nantel, 1911-12 Robert Rogers, 1912 W. J. Roche, 1912-13 Louis Coderre, 1913-15	Reginald W. Brock, 1914 Richard G. McConnell, 1914-20			William McInnes, 1914-20	
P. E. Blondin, 1915-17 E. L. Patenaude, 1917 Arthur Meighen, 1917 Martin Burrell, 1917-19 Arthur Meighen, 1919-20 Sir J. A. Lougheed, 1920-21	Charles Camsell, 1920-46			William Henry Collins, 1920-36	
Charles Stewart, 1921-26, 1926-30					CHIEF GEOLOGISTS George A. Young, 1924-43
W. A. Gordon, 1930-35			BUREAU OF ECONOMIC GEOLOGY Francis C. C. Lynch, 1934-36		
T. A. Crerar, 1935-36					
MINISTERS OF MINES AND RESOURCES T. A. Crerar, 1936-45		MINES AND GEOLOGY BRANCH John McLeish, 1936-41 William B. Timm, 1941-50	BUREAU OF GEOLOGY AND TOPOGRAPHY Francis C. C. Lynch, 1936-47		
					George Hanson, 1943-53
J. A. Glen, 1945-48	Hugh L. Keenleyside, 1947-50		George S. Hume, 1947-50		
J. A. MacKinnon, 1948-49					
MINISTERS OF MINES AND TECHNICAL SURVEYS J. J. McCann, 1950 George Prudham, 1950-57	Marc Boyer, 1950-62	DIRECTORS GENERAL OF SCIENTIFIC SERVICES George S. Hume, 1950-56		Walter Andrew Bell, 1950-53 George Hanson, 1953-56	Clifford S. Lord 1954-73
		Wm. E. van Steenburgh, 1956-63		James Merritt Harrison, 1956-64	
Paul Comtois, 1957-61 Jacques Flynn, 1961-63 Paul Martineau, 1963	Wm. E. van Steenburgh, 1963-66				
W. M. Benedickson, 1963-65					
J. W. MacNaught, 1965 (July-November) Jean-Luc Pepin, 1965-66		ASSISTANT DEPUTY MINISTER (RESEARCH) James M. Harrison, 1964-66		Yves Oscar Fortier, 1964-73	
MINISTERS OF ENERGY, MINES AND RESOURCES Jean-Luc Pepin, 1966-68 J. J. Greene, 1968-70 D. S. Macdonald, 1970-	Claude M. Isbister, 1966-70 Jacob Austin, 1970-	ASSISTANT DEPUTY MINISTERS (SCIENCE AND TECHNOLOGY) James M. Harrison, 1966-71 Charles H. Smith, 1971-			

Appendix II

A SELECTED LIST OF FORMER EMPLOYEES OF THE GEOLOGICAL SURVEY OF CANADA

NOTE—This list is compiled from a variety of sources, not all of which could be thoroughly checked, while the choice of entries was sometimes governed by the availability of reliable information. Despite these limitations, however, it is felt that the information possesses enough interest and value to justify publication.

Dates of employment, especially of temporary employment, were a particular problem because the records have not always survived. Many persons served for varying periods as part-time and full-time temporary staff before being appointed to the permanent staff, which is indicated by the letter "P" in the Period of Service column. "T" denotes temporary status, usually of a part-time or intermittent nature. "C" refers to specialists employed as consultants, notably to study paleontological collections.

Name	Profession/Trade	Period of Service	Remarks
Achard, R. A.	Pleistocene Geologist	1966-70T	
Adams, F. D.	Petrographer	1876-89P, 1890-1902T	McGill University
Aitken, G. G.	Draftsman	1906-12P	
Alcock, F. J.	Geologist	1911-21T, 1921-47P	Curator, National Museum of Canada, 1947-56
Alexander, Mrs. J.	Librarian	1889-1912P	Survey Librarian 1908-12
Alexander, S. G.	Draftsman	1910-55P	Supt. of Cartography 1948-55
Allan, J. A.	Geologist	1916-18T	Univ. of Alberta
Ambrose, J. W.	Geologist	1934-36T, 1936-45P	Queen's University
Ami, H. M. I.	Paleontologist	1882-1911P	
Anderson, D. T.	Geophysicist	1962-65	Univ. of Manitoba
Anderson, R. M.	Zoologist	1913-45P	Chief, Biological Div.
Anrep, Aleph	Peat Specialist	1919-32T	
Antevs, E. V.	Glaciologist	1923-29T	
Arnold, W. W.	Map Engraver	1921-40P	
Aumento, Fabrizio	Submarine Geologist	1965-69P	Dalhousie University
Backman, O. L.	Geologist	1931-33T	
Baer, A. J.	Structural Geologist	1962-71P	Univ. of Ottawa
Bailey, L. W.	Geologist	1868-1904T	Univ. of New Brunswick
Bain, G. W.	Geologist	1921-24T	Amherst College
Baird, D. M.	Geologist	1958-60T	Director, National Museum of Science & Technology, 1966-
Bancroft, J. A.	Geologist	1907T	McGill University
Bancroft, M. F.	Geologist	1913-26T	
Bannerman, H. M.	Geologist	1926-30T	
Barbeau, C. M.	Cultural Anthropologist	1911-48P	
Barlow, A. E.	Petrographer	1879-1907P	
Barlow, Robert	Draftsman	1856-80P	Chief Draftsman, 1856-80
Barlow, Scott	Draftsman	1856-94	Chief Draftsman, 1880-94
Barrowman, G. D.	Instrument Custodian	1911-47P	Mechanical Supt.
Barry, Miss M. H.	Library Assistant	1882-1934P	
Bartlett, G. A.	Marine Geologist	1962-67P	Queen's University
Bartlett, R.	Topographer	1910-20P, 1921-36T	
Bateman, A. M.	Geologist	1912T	Yale University
Bateman, J. D.	Geologist	1940-45P	
Beach, H. H.	Petroleum Geologist	1933-40T, 1940-44P	
Bell, J. J.	Editorial Assistant	1914-21P	
Bell, J. M.	Exploration Geologist	1900-01T	
Bell, Robert	Geologist	1857-69T, 1869-1908P	Acting Director 1901-06
Bell, R. T.	Geologist	1966-69P	Brock University
Bell, W. A.	Paleobotanist	1911-20T, 1920-54P, 1954-67T	Chief, Paleont. Div., 1938-50, Director 1950-53

Name	Profession/Trade	Period of Service	Remarks
Bentley, W. K.	Photographer	1908-after 1937P	
Beuchat, Henri	Anthropologist	1913-14P	Died, Wrangel Island
Bhattacharyya, B. K.	Mathem. Geophysicist	1961-71P	Canada Centre for Remote Sensing
Billings, Elkanah	Paleontologist	1856-76P	
Black, R. F.	Paleomagnetist	1957-65P	Public Service Commission of Canada
Blanchard, J. L.	Radio Technician	1959-67P	Finance and Admin. Branch, Dept. of Energy, Mines & Res.
Bloomfield, Leonard	Ethnologist	1925T	University of Chicago
Bolton, L. L.	Administrator	1916-47P	Asst. Deputy Minister, Dept. of Mines 1925-36, Exec. Assistant 1936-47
Bostock, H. S.	Geologist	1926-65P	
Bowen, N. L.	Geologist	1910-12T	Carnegie Institute, Washington
Bowman, Amos	Geologist	1876-89P	
Boyd, W. H.	Topographer	1900-36P	Chief, Topog. Div., 1908-36, Topog. Survey Branch 1936-40
Brady, W. B.	Cordilleran Geologist	1952-59P	
Braidwood, A. W.	Draftsman	1913-22P	
Brandon, L. V.	Engineering Geologist	1958-65P	
Broad, Wallace	Geologist	1882-83P	
Broadbent, R. L.	Rock and Mineral Collector	1882-1911P	
Brock, R. W.	Geologist	1887-97T, 1897-1902P, 1902-06T, 1906-14P, 1920T	Director, 1908-14, Deputy Minister of Mines, 1914
Broom, Gordon	Chemist	1869-72P	
Brophy, L. L.	Statistician	1891-97P	
Brown, I. C.	Engineering Geologist	1946-66P	To Water Research Branch
Brown, Miss M. C.	Photographer	1914-36P	
Brown, R. A. C.	Geologist	1939, 1943, 1946T	
Bruce, E. L.	Geologist	1912-15T, 1915-20P	Queen's University
Brumell, H. P. H.	Statistician	1882-84T, 1884-95P	
Brunton, Sir Stopford	Geologist	1914, 1921T	
Buckham, A. F.	Geologist	1939-48P	
Buffam, B. S. W.	Geologist	1925-26T	
Burling, L. D.	Paleontologist	1913-20P, 1946-49T	
Butterworth, J. V.	Topographer	1922-26T, 1926-36P	To Topographical Surveys Branch
Byers, A. R.	Geologist	1932-34T	Univ. of Saskatchewan
Cairnes, C. E.	Geologist	1922-53P	Technical Editor 1943-53
Cairnes, D. D.	Geologist	1905-17P	
Caley, J. F.	Paleontologist	1935-36T, 1936-69P	Chief, Fuels & Stratig. Geology Div. 1955-69
Calhoun, Miss M.	Librarian	1909-18P	Survey Librarian 1912-18
Cameron, A. E.	Geologist	1916-18T	
Camsell, Charles	Geologist	1899-1904T, 1904-20P	Deputy Minister, 1920-46
Carr, J. J.	Draftsman	1910-47P	
Carr, P. A.	Groundwater Geologist	1959-66P	To Water Research Branch
Chalmers, Robert	Surficial Geologist	1882-84T, 1884-1908P	
Chamberlain, J. A.	Geologist	1959-68P	
Chapman, R. H.	Topographer	1909-11P	Major, U.S. Army
Charron, J. E.	Groundwater Geologist	1959-66P	To Water Research Branch
Chipman, K. G.	Topographer	1908-10T, 1910-36P	To Topographical Surveys Branch
Christie, A. M.	Geologist	1945T, 1946-54P	
Church, M.	Geographer	1962-68T	University of British Columbia
Clapp, C. H.	Geologist	1908-13T	Montana School of Mines
Clark, A. F.	Draftsman	1911-44P	
Clark, K. A.	Mining Geologist	1916-20T	Research Council of Alberta
Clark, T. H.	Paleontologist	1927-31T	McGill University
Clarke, C. H. D.	Naturalist	1936T	Ontario Dept. of Lands
Clarke, G. G.	Photographer	1911-41P	Chief Photographer, 1920-41
Clarke, L. Y.	Photographer	1915-26P	
Cochrane, A. S.	Topographer	1877-87T, 1887-94P	
Cockfield, W. E.	Geologist	1912-20T, 1920-55P	Head, Vancouver Office, 1929-55

Name	Profession/Trade	Period of Service	Remarks
Cole, A. A.	Statistician	1891-95T, 1895-98P	Ontario Dept. of Mines
Coleman, A. P.	Geologist	1916T	University of Toronto
Collins, W. H.	Geologist	1905-37P	Director, 1920-36
Connor, M.F.	Metallurgist	1903-07P, 1918-26P	To Mines Branch 1907-18
Cooke, F. J.	Photographer	1950-72P	
Cooke, H. C.	Geologist	1905-21T, 1921-49P	
Cope, E. D	Vertebrate Paleontologist	1884-97C	American Museum of Natural History
Coste, Eugene	Mining Geologist	1883-89P	
Cox, J. R.	Topographer	1910-14T, 1914-20P	To Mines Branch
Creighton, J. G. A.	Explorer	1890T	
Crickmay, C. H.	Geologist	1919-43T	
Crickmay, G. W.	Stratig. Paleontologist	1929-30T	
Crombie, G. P.	Geologist	1943-44T, 1948-49T	
Cruickshank, J. M.	Field Assistant	1890-1922T	
Daly, R. A.	Geologist	1901-06P, 1911-13T	Harvard University and Mass. Inst. of Technology
Daly, W. P.	Administrator	1947-58P	
Dart, Miss J. D.	Stratig. Paleontologist	1923T	
Daughtry, G. S.	Cartographer	1927-31T, 1931-69P	Supt. of Cartography, 1959-69
David, P. P.	Aeolian Geologist	1967-68T	Université de Montréal
Davis, N. F. G.	Geologist	1929, 1935, 1936T	
Dawson, G. M.	Geologist	1875-1901P	Director, 1895-1901
Dawson, Sir J. W.	Paleobotanist	1857-97C	Principal, McGill University
DeLury, J. S.	Geologist	1923-24T	University of Manitoba
Denis, T. C.	Mining Geologist	1898-1907P	Province of Quebec Geologist
Dickison, Alexander	Draftsman	1905-45P	Chief Draftsman, 1931-45
Dobbs, W. S.	Exploring Geologist	1905T	
Dolmage, Victor	Geologist	1918-20T, 1920-29P	Head, Vancouver Office, 1922-29
Donaldson, J. A.	Precambrian Geologist	1959-60T, 1960-68P	Carleton University
Douglas, G. V.	Geologist	1946-47T	Dalhousie University
Douglas, J. A. V.	Meteorite Specialist	1963-70P	
Dowling, D. B.	Topographer; Geologist	1885-95T, 1895-1925P	
Downie, D. L.	Geologist	1935-38T	
Dreimanis, Aleksis	Quaternary Geologist	1966-70T	Univ. of Western Ontario
Dresser, J. A.	Geologist	1898-1907T, 1909-11P	
Drysdale, C. W.	Geologist	1908-13T, 1913-17P	Drowned while on field work
DuBois, P.M.	Paleomagnetist	1957-59P	
D'Urban, W. S. M.	Botanist	1858T	
DuVernet, F. P.	Topographer; Geophysicist	1935-36T, 1947-56P	
Dyer, W. S.	Geologist	1919-27T	Provincial Geologist, Ontario
Eakins, P. R.	Structural Geologist	1961T	McGill University
Eaton, D. I. V.	Topographer	1890-96T	
Elliott, E. C.	Photographer	1935-66P	Head, Photographic Section, 1953-66
Ells, R. H.	Field Assistant	1899-1903T	
Ells, R. W.	Geologist	1872-1911P	
Ells, S. C.	Geologist	1908, 1935T	To Mines Branch, Ottawa
Ellsworth, H. V.	Mineralogist	1912-20T, 1920-52P	Chief, Radioactivity Resources Division, 1948-52
Elson, J. A.	Surficial Geologist	1946-56T	McGill University
Emmons, R. C.	Petrologist and Geologist	1920-28T	University of Wisconsin
Emslie, J. B.	Photographer	1961-71P	Head, Photographic Section, 1966-71
Erdman, O. A.	Petroleum Geologist	1943-46T	
Esdale, D. A.	Carpenter	1888-1934P	
Esdale, Matthew	Carpenter	1888-94P	
Eskola, Pentti	Precambrian Geologist	1922T	
Evans, C.S.	Geologist	1925-28T, 1928-37P	
Eve, A. S.	Geophysicist	1929-30T	
Fabry, R. J. C.	Mineralogist	1926-62P	
Fairbairn, H. W.	Petrologist	1931, 1942T	Mass. Inst. of Technology

Name	Profession/Trade	Period of Service	Remarks
Falconer, F.S.	Topographer	1911-13T, 1913-20P	
Faribault, E. R.	Geologist	1882-84T, 1884-1932P	
Feniak, Michael	Geologist	1943-44T, 1945-49P	Head, Yellowknife Office, drowned
Ferguson, S. A.	Geologist	1944-46T	Ontario Dept. of Mines
Ferrier, W. F.	Petrographer	1889-98P, 1924-28T	
Findlay, D. C.	Geologist	1962-70P	Head, Whitehorse Office, 1966-70
Flaherty, G. F.	Geologist	1930-36T	
Fletcher, Hugh	Geologist	1872-75T, 1875-1909P	
Foerste, A. F.	Paleontologist	1911-12T	
Folinsbee, R. E.	Geologist	1941-46T	University of Alberta
Foord, A. S.	Museum Specialist	1878-83P	
Forsey, Mrs. F. B.	Librarian	1913-41P	Survey Librarian, 1918-41
Fortescue, J. A. C.	Biogeochemist	1964-67P	Brock University
Fortin, J. O.	Draftsman	1909-after 36P	
Franklin, J. M.	Mineral Deposits Geologist	1967-69T	Lakehead University
Fraser, F. J.	Borings Engineer	1927-54P	
Fratta, M.	Uranium Geologist	1967-69T	University of Ottawa
Frebold, H. W. L. A. H.	Paleontologist	1949-68P, 1968- T	
Freeland, E. E.	Topographer	1911-14T, 1914-24P	
Freeman, B. C.	Precambrian Geologist	1929-38T	Ohio State University
Freeman, C. H.	Topographer	1915-18T, 1919-23P	To Mines Branch
Freeze, R. A.	Groundwater Geologist	1961-66P	To Water Research Branch
Frèreault, Paul	Draftsman	1901-06P	
Fritz, Miss M. A.	Paleontologist	1924-36T&C	Royal Ontario Museum, Toronto
Fry, W. L.	Paleobotanist	1953-57P	University of California, Berkeley
Furnival, G. M.	Economic Geologist	1931-32T, 1939-42P	Department of Mines, Manitoba
Gaucher, E. H. S.	Paleomagnetist	1962-66P	SOQUEM, Quebec Government
Gilbert, Mrs. M. A.	Geochemist	1957-61P	
Gilchrist, Lachlan	Geophysicist	1929-30T	University of Toronto
Gill, J. E.	Geologist	1924T	McGill University
Gilliland, J. A.	Groundwater Geologist	1965-66P	To Water Research Branch
Gillis, J. W.	Precambrian Geologist	1962-67P	
Giroux, N.J.	Geologist	1883-88T, 1888-96P	
Gleeson, C. F.	Geochemist	1961-65P	
Goldenweiser, A. A.	Social Anthropologist	1911-15T	Columbia University
Goldthwait, J. W.	Geomorphologist	1908-16, 1925-26T	Dartmouth College
Goodwin, A. M.	Geologist	1965-69P	University of Toronto
Goranson, E. A.	Economic Geologist	1930-31T	
Goranson, R. W.	Geologist, Petrologist	1921-25T	
Graham, R. P. D.	Geologist	1908T	McGill University
Grant, J. C. B.	Physical Anthropologist	1927-30, 1937T	University of Manitoba
Grant, J. L.	Assistant Librarian	1882-88P	
Gray, J. D.	Geologist	1926-27T	
Gray, J. G.	Geologist	1936-38T	
Green, L. H.	Geologist	1953-68P	Head, Whitehorse Office, 1965-66
Gregory, A. F.	Geophysicist	1958-66P	Carleton University
Guernsey, T. D.	Geologist	1920-27T	
Gunning, H. C.	Geologist	1927-28T, 1928-38P	University of British Columbia
Gussow, W. C.	Draftsman and Geologist	1927-39T	
Gwillim, J. C.	Geologist	1891-99T, 1899-1901P	Queen's University
Hage, C. O.	Geologist	1933-37T, 1937-45P	
Hale, A. E.	Cartographer	1928-35T, 1939-59P	Supt. of Cartography, 1955-59
Hall, James	Paleontologist	1855-65C	New York State Museum
Halstead, E. C.	Groundwater Geologist	1946-66P	To Water Research Branch
Hanson, George	Geologist	1913-21T, 1921-56P	Chief Geologist, 1943-53; Director, 1953-56
Harper, Francis	Zoologist	1914T	Cornell University
Harrington, B. J.	Petrographer	1872-79T	McGill University
Harrison, J. M.	Geologist	1943-44T, 1944-64P	Director, 1956-64, Asst. Deputy Minister 1964-71
Hartley, Edward	Geologist	1868-70P	

Name	Profession/Trade	Period of Service	Remarks
Harvie, Robert	Geologist	1905-11T, 1911-24P	
Haultain, A. G.	Topographer	1911-13T, 1913-36P	To Topographical Surveys Branch
Havard, Mrs. C. J.	Stratig. Paleontologist	1966-72P	
Hawkes, E. W.	Anthropologist	1914-15T	
Haycock, E. W.	Geologist	1902-05T	
Hayes, A. O.	Geologist	1907-14T, 1914-21P, 1925-26T	
Helmstaedt, Herwart	Structural Geologist	1969-72P	McGill University
Henderson, J. F.	Geologist	1933-36T, 1936-71P	Secretary, NACRGS, 1951-71
Herr, R. L.	Petroleum Geologist	1959-65P	To Inland Waters Branch
Herring, Samuel	Taxidermist	1883-1919P	
Hills, L. V.	Palynologist	1967-70T	University of Calgary
Hill-Tout, Charles	Archeologist	1899-1900T	
Hoadley, J. W.	Geologist	1945-49T, 1949-56P	
Hoffmann, G. C.	Chemist	1872-1907P	Chief, Chem. & Mineralogy Div., 1879-1907
Hofmann, H. J.	Precambrian Fossils	1966-69P	Université de Montréal
Holman, R. H. C.	Geochemist	1957-65P	
Honeyman, David	Geologist	1868T	
Horn, D. R.	Marine Geologist	1963-64P	
Horwood, H. S.	Geologist	1931-35T, 1935-36P	Provincial Geologist, Ontario
How, Henry	Geologist	1868T	King's College, Windsor, N.S.
Howard, W. V.	Geologist	1915-26T	
Howell, J. E.	Geologist	1952-54T, 1954P	Drowned while on field work
Howells, W. C.	Petroleum Geologist	1938T	
Hueber, F. M.	Paleobotanist	1961-62P	To Smithsonian Institution
Hume, G. S.	Petroleum Geologist	1915-21T, 1921-47P	Chief, Bur. of Geol. & Topog., 1947-50, Dir.-Gen. of Scient. Services, 1950-56
Hunt, T. Sterry	Chemist, Mineralogist	1848-72P	
Hunter, A. F.	Surficial Geologist	1902-05T	
Hurst, M. E.	Geologist	1919-22T, 1923-24P	Provincial Geologist, Ontario
Hutton, W. S.	Photographer	1914—after 1936P	
Hyde, J. E.	Stratigraphic Paleontologist	1911-14T	Western Reserve University
Ingall, E. D.	Mining Engineer	1883-1928P	Chief, Min. Statistics Div. 1889-1907, Borings Div. 1907-28
Irwin, A. B.	Geologist	1950-54P	Head, Yellowknife Office 1950-53 Department of Indian Affairs and Northern Development
Jacob, F. D.	Secretary	1890-93P	Director's Secretary
James, W. F.	Geologist	1919-23T, 1923-28P	
Jenness, Diamond	Anthropologist	1913-17T, 1920-42P	Chief, Anthropological Div., 1925-31, 1936-42
Jenness, S. E.	Economic Geologist; Scientific Editor	1954-66P	To National Research Council of Canada
Joanes, Arthur	Draftsman	1914-48P	Supt. of Cartography 1945-48
Johansen, Frits	Biologist	1916-18P	
Johnston, A. G.	Geological Records Compiler	1954-71P	To Mineral Resources Branch
Johnston, A. W.	Geologist	1935-37T	
Johnston, J. F. E.	Draftsman; Explorer	1887-1904T, 1905-10P	
Johnston, J. R.	Geologist	1935-47T	
Johnston, R. A. A.	Mineralogist	1887-1922P	Chief, Mineralogy Div. 1907-22
Johnston, W. A.	Surficial Geologist	1901-03T, 1905-39P	
Jolliffe, A. W.	Geologist	1933-35T, 1935-46P	Queen's University
Jones, R. H. B.	Geologist	1922-28T	
Jones, T. Rupert	Paleontologist	1858C	
Jost, A. S.	Draftsman	1909-21P	
Kalliokoski, J. O. K.	Geochemist	1949-53P, 1953-59T	Princeton University
Karrow, P. F.	Pleistocene Geologist	1955-70T	University of Waterloo
Keele, Joseph	Topographer; Geologist	1898-1916, 1921-23P	To Mines Branch 1916-21
Keevil, N. B.	Geophysicist	1935-36T	University of Toronto
Keith, M. L.	Geochemist	1938T	Penn State University
Kelley, D. G.	Geologist	1955-69P	To Dept. of National Health

Name	Profession/Trade	Period of Service	Remarks
Kennedy, M. J.	Structural Geologist	1966-68T	Memorial University of Newfoundland
Kenrick, E. B.	Chemist	1885-87P	
Kerr, F. A.	Geologist	1920-24T, 1924-37P	
Keys, D. A.	Geophysicist	1929-30T	University of Toronto
Kidd, D. F.	Geologist	1930-35P	
Kindle, C. H.	Paleontologist	1928-30T	City College of New York
Kindle, E. D.	Economic Geologist	1933-36T, 1936-71P	
Kindle, E. M.	Paleontologist	1910-12T, 1912-38P	Chief, Paleontology Div. 1919-38
King, C. F.	Museum Assistant	1902-06P	
Kirk, S. R.	Geologist	1928-31T	
Knowles, Sir F. H. S., Baronet	Physical Anthropologist	1914-19P	
Knox, J. K.	Geologist	1912-16T	
Kramm, M. E.	Geologist	1908-11T	
Kranck, Miss K. M.	Marine Geologist	1962-70T	To Bedford Institute
Kranck, S. H.	Precambrian Geologist	1959-60P	
Kretz, Ralph	Petrologist	1958-61P	University of Ottawa
Kummermann, Mrs. N. I.	Librarian	1930-66P	Survey Librarian, 1941-66
Lacroix, C. J. A. A.	Mineral Sets Preparator	1927-65P	
Laflamme, J. C. K.	Geologist	1883-1901T	Université Laval
Laing, H. M.	Zoologist	1921-30, 1935-39T	
Lambe, L. M.	Vertebrate Paleontologist	1885-1919P	Chief, Paleontol. Div. 1912-1919
Landes, R. W.	Stratigraphic Paleontologist	1934-38T	
Landles, W.	Engraver; Photo Retoucher	1939-51P	
Lang, A. H.	Geologist	1929-30T, 1930-65P, 1965- T	Chief, Mineral Deposits Div. 1955-65
Lawson, A. C.	Geologist	1884-90P, 1911-13T	University of California, Berkeley
Lawson, Charles	Photographer	1941-53P	Head, Photographic Sect. 1941-53
Lawson, W. E.	Topographer	1909-11T, 1911-18P	Killed in France
Lawson, William	Topographer	1890-97T	
Leach, W. W.	Topographer; Geologist	1892-1903T, 1906-13P	
Lee, H. A.	Pleistocene Geologist	1950-69P	
Leechman, J. D.	Anthropologist	1923-55P	
Lees, E. J.	Geologist	1935-37T	
Lefebvre, J. S. H.	Artist	1899-1939P	
LeRoy, O. E.	Draftsman; Geologist	1900-03T, 1906-17P	Killed in France
Leslie, R. J.	Marine Geologist	1962-64P	
Leverett, Franklin	Glacial Geologist	1910-12T	U.S. Geological Survey
Liberty, B. A.	Stratigraphic Paleontologist	1952-66P	Brock University
Lissey, Allan	Hydrogeologist	1964-66P	Brock University
Locke, Donald	Metallurgist	1902-03P	
Logan, Sir W. E.	Geologist	1842-69P, 1870-74T	Founder and director, 1842-69
Lopatin, I. A.	Anthropologist	1929-30T	University of Washington
Low, A. P.	Geologist	1881-82T, 1882-1901P, 1902-07P	Director 1906-07, Deputy Minister 1907-13
Lowe, James	Field Assistant	1858-72T	
Lynch, F. C. C.	Administrator	1934-47P	Chief, Bureau of Economic Geology 1934-36. Bureau of Geology & Topography 1936-47
McArthur, A.	Explorer	1887T	
McCann, W. S.	Geologist	1912-13T, 1919-21P	
McCartney, W. D.	Metallogenic Geologist	1952-67P	Queen's University
McConnell, R. B.	Geologist	1930-31P	Overseas G. S., United Kingdom
McConnell, R. G.	Geologist	1879-1914P	Deputy Minister, 1914-20
Macdonald, J. A.	Topographer	1922-36P	To Topographical Surveys Branch
McDonald, R. C.	Topographer	1912-22T, 1922-36P	To Topographical Surveys Branch
McEvoy, James	Topographer; Geologist	1885-1901P	
McEwan, W. R.	Library Assistant	1885-90P	
Macfarlane, Thomas	Geologist	1865T	
McGee, J. J.	Clerk	1904-28P	
McGerrigle, H. W.	Appalachian Geologist	1928-29T	Chief Geologist, Quebec
McGregor, Adam	Draftsman	1911-47P	
McIlwraith, T. F.	Anthropologist	1922T	University of Toronto

Name	Profession/Trade	Period of Service	Remarks
McInnes, William	Geologist	1882-84T, 1884-1925P	Director 1914-20, Museum Director 1920-25
MacKay, B. R.	Topographer; Geologist	1907-13T, 1913-20, 1923-52P	Chief, Water Supply & Borings Div.
MacKenzie, J. D.	Geologist	1908-15T, 1920-22P	Head, Vancouver Office 1920-22
McKinnon, A. T.	Rock Collector	1889-1934P	
MacLaren, A. S.	Geologist	1946-51T, 1951-71P	
MacLaren, F. H.	Draftsman; Explorer	1904-11P	
McLarty, D. M. E.	Geologist	1935-36T	
MacLean, A. H.	Geologist	1913-18T	University of Toronto
McLean, S. C.	Topographer	1909-13T, 1913-36P	To Topographical Surveys Branch
McLearn, F. H.	Paleontologist	1913-20T, 1920-52P	Chief, Paleontology Div. 1950-52
McLeish, John	Mining Statistician	1897-1907P	To Mines Branch, 1907, Chief, Mines & Geology Br., 1936-41
McLeod, M. H.	Field Assistant	1888-1911T	
McMillan, John	Field Assistant	1882-89P	
McMullen, John	Field Assistant	1858T	
McMullen, R. M.	Marine Geologist	1965-67P	To Marine Sciences Branch
McMurchy, R. C.	Geologist	1935-39P	
McNaughtan, John	Topographer	1845-46T	
McNaughton, D. A.	Geologist	1935-37T	
McOuat, Walter	Geologist	1869-73T, 1873-75P	
Macoun, J. M.	Topographer; Botanist	1883-96T, 1897-1920P	Chief, Biological Div. 1913-20
Macoun, John	Botanist	1875T, 1882-1920P	Chief, Biological Div. 1910-13
MacQuarrie, W. E.	Geologist	1949-50T	Head, Yellowknife Office, 1949-50
MacVicar, J.	Geologist	1916-20T	
Maddox, D. C.	Geologist	1919-26T, 1926-47P	
Mailhiot, A.	Mining Engineer	1909-14T	Ecole Polytechnique, Montreal
Malcolm, Wyatt	Editor; Geol. Compiler	1908-09T, 1909-41P	Asst. Curator, National Museum 1936-41
Malloch, G. S.	Geologist	1906-07T, 1907-14P	Died on Canadian Arctic Expedition
Malte, M. O.	Botanist	1921-33P	Dominion Botanist
Marlowe, J. I.	Marine Geologist	1962-70P	Miami—Dade Junior College
Marshall, John	Accountant	1872-1920P	
Marshall, J. R.	Geologist	1905-21T, 1921-51P	
Mason, J. A.	Naturalist	1913T	
Massicotte, E. Z.	Folklorist	1920-21T	Archivist, Montreal Judicial Dist.
Matheson, H.	Topographer	1907-11P	
Matthew, G. F.	Paleontologist	1868-1901T	Canada Customs Service
Mawdsley, J. B.	Geologist	1921-25T, 1925-29P, 1930-46T	Univ. of Saskatchewan
Mechling, W. H.	Social Anthropologist	1912-13T	
Meyboom, Peter	Groundwater Geologist	1961-66P	To Water Research Branch
Michel, A.	Mining Engineer	1865-67T	
Mihailov, G.	Geochemistry Assistant	1960-67P	
Miles, A.	Artist	1914-33P	
Miller, A. H.	Geophysicist	1929-35T	
Miller, W. G.	Geologist	1891-95T	Chief Geologist, Ontario
Miller, W. H.	Topographer	1913-20T, 1920-36P	To Topographical Surveys Branch
Mirynech, Edward	Pleistocene Geologist	1958-64P	Brock University
Moore, E. S.	Geologist	1912T	University of Toronto
Moore, J. C. G.	Economic Geologist	1948-51P	Mount Allison University
Moore, J. M.	Precambrian Geologist	1966-71T	Carleton University
Morley, L. W.	Geophysicist	1957-69P	Chief, Geophysics Div. 1957-69
Mountjoy, E. W.	Stratig. Paleontologist	1958-63P, 1967-69T	McGill University
Murray, Alexander	Geologist	1842-64P	Director, Newfoundland G.S., 1864-83
Neale, E. R. W.	Geologist	1954-68P	Memorial University of Newfoundland
Nettell, A. J. C.	Metallurgist	1914T, 1918-38P	
Nichols, D. A.	Topographer	1908-13T, 1913-36P	To Topographical Surveys Branch
Nicolas, Frank	Editor	1904-08, 1920-33P	Editor-in-chief, Dept. of Mines, 1925-33
Nigrini, Andrew	Geochemist	1968-71T	
Norman, G. W. H.	Geologist	1927-29T, 1929-50P	
Northrop, S. A.	Stratig. Paleontologist	1928T	University of New Mexico

Name	Profession/Trade	Period of Service	Remarks
O'Farrell, Finbar	Draftsman	1904-10P	
O'Farrell, Michael	Caretaker	1842-89P	
Okulitch, V. J.	Geologist	1936-49T	Univ. of British Columbia
O'Neill, J. J.	Geologist	1911-13T, 1913-20P	McGill University
Ord, L. R.	Exploring Geologist	1876-81P	
Osann, Alfred	Petrographer	1899T	
Osborn, H. F.	Vertebrate Paleontologist	1902C	American Museum of Natural History, New York
O'Sullivan, Owen	Draftsman; Explorer	1901-10T	
Parks, W. A.	Geologist	1900-05T	University of Toronto
Parsons, M. L.	Groundwater & Engrg. Geologist	1965-66P	To Water Research Branch
Patch, Clyde	Herpetologist	1913-42P	
Pearce, G. W.	Paleomagnetist	1967-69P	
Pegrum, R. H.	Geologist	1922-27T	New York State University, Buffalo
Penhallow, D. P.	Paleontologist	1903-07C	McGill University
Perrin, V.	Draftsman	1901-05P	
Perry, A. B.	Library Assistant	1882T	Commissioner NWMP, 1900-22
Phemister, T. C.	Geologist	1928T	
Picher, R. H.	Road Material Specialist	1917T	
Poitevin, Eugene	Mineralogist	1910T, 1913-57P	Chief, Mineralogy Div. 1922-57
Poole, H. S.	Coal Geologist	1901-04T	
Porsild, A. E.	Botanist	1936-67P	
Price, R. A.	Structural Geologist	1952-58T, 1958-68P	Queen's University
Prud'homme, O. E.	Draftsman; Artist	1889-1933P	
Quinn, H. A.	Precambrian Geologist	1952-60T	
Quirke, T. T.	Geologist	1914-31T	University of Illinois
Radin, Paul	Social Anthropologist	1911-17T	
Rand, A. L.	Zoologist	1942-47P	Chief Curator, Field Museum, Chicago
Raup, H. M.	Botanist	1928-30T	
Raymond, P. E.	Paleontologist	1910-12P	Chief, Paleontology Div. 1910-12; Harvard University
Reinecke, Leopold	Geologist	1908-11T, 1911-20P	
Retty, J. A.	Geologist	1936T	
Rice, H. M. A.	Geologist	1934-36T, 1936-65P	
Richard, L. N.	Clerk; Surveyor	1882-1925P	
Richardson, James	Geologist	1846-55T, 1855-80P, 1880-83T	
Ries, Heinrich	Clay Specialist	1909-13T	Cornell University
Riley, G. C.	Appalachian Geologist	1950-60P	
Ritchie, J. C.	Palynologist	1967-68T	Trent University
Roach, R. A.	Precambrian Geologist	1957-59P	Keele University
Robb, Charles	Geologist	1868-75P	
Robert, J. A.	Draftsman	1881-89T, 1901-31P	
Robertson, W. A.	Paleomagnetist	1962-64T	To Earth Physics Branch
Robinson, C. W.	Field Assistant	1913-19T	
Rochon, J. A.	Osteological Preparator	1919-33P	
Roots, E. F.	Cordilleran Geologist	1948-58P	To Polar Continental Shelf Project
Roscoe, S. M.	Precambrian Geologist	1953-68P	
Rose, Bruce	Geologist	1911-14T, 1914-20P, 1924-36T	Queen's University
Ross, J. V.	Petrologist	1958-65T	Univ. of British Columbia
Rottermond, E. S. de	Chemist	1844-46P	
Rousseau, Jacques	Ethnologist	1929-30T	Director, Canadian Museum of Human History
Russel, H. Y.	Topographer and Explorer	1891-96P	
Russell, L. S.	Vertebrate Paleontologist	1930-31T, 1931-37P	Director, Natural History Br., National Museum of Canada 1956-63
Rutherford, R. L.	Geologist	1917-19, 1936T	University of Alberta
Salt, Miss L. A.	Photographer	1913-49P	
Salter, J. W.	Paleontologist	1856-59C	Great Britain G.S.
Sapir, Edward	Linguist	1910-25P	Chief, Anthropological Div. 1910-25, University of Chicago

Name	Profession/Trade	Period of Service	Remarks
Sasaki, Akira	Isotopic Geologist	1962-64T	University of Calgary
Satterly, Jack	Mineralogist	1926-30T	Ontario Dept. of Mines
Sauvelle, Marc	French Translator	1917-20P	
Schiller, E. A.	Geologist	1963-65P	Head, Yellowknife Office 1963-64
Schofield, S. J.	Geologist	1907-11T, 1911-20P, 1921T	University of British Columbia
Schuchert, Charles	Paleontologist	1912, 1920T	Yale University
Scudder, S. H.	Paleontologist	1895-1900C	U.S. Geological Survey, etc.
Selwyn, A. R. C.	Geologist	1869-95P	Director 1869-95
Selwyn, P. H.	Secretary	1893-1924P	
Senécal, C. O.	Draftsman	1890-1931P	Chief Draftsman, 1899-1931
Shaw, E. W.	Geologist	1935-36T	
Shaw, George	Geophysicist	1937-39T, 1939-52P	UN Technical Assistance Programs
Sheppard, A. C. T.	Topographer	1909-36P	To Topographical Surveys Branch
Shimer, H. S.	Paleontologist	1910, 1917T	Mass. Inst. of Technology
Siddeley, G.	Geochemist	1968-70P	
Simpson, H. E.	Groundwater Geologist	1929T	
Sinclair, G. W.	Paleontologist	1954-70P	
Skibo, D. N.	Theoretical Geophysicist	1967-68P	
Slipper, S. E.	Petroleum Engineer	1910-14T, 1914-17P, 1918T	
Smith, A. Y.	Geochemist	1959T, 1967 69P	
Smith, C. H.	Petrographer	1951-70P	Assistant Deputy Minister
Smith, H. I.	Archeologist	1911-37P	
Smith, Horace	Artist	1856-71P	
Smith, W. H. C.	Geologist	1884T, 1885-93P	
Smitheringale, W. G.	Economic Geologist	1955-58T	Memorial University of Newfoundland
Snow, W. E.	Geologist	1937-39T	
Soper, J. D.	Zoologist	1923-27T	
Sparks, Mrs. W.	Statistics Clerk	1898-1908P	To Mines Branch
Speck, F. G.	Anthropologist	1914T	
Spence, H. N.	Topographer	1919-22T, 1923-36P	To Topographical Surveys Branch
Spencer, J. W.	Geologist	1874, 1905-07T	Geological Survey of Georgia, etc.
Spivak, Joseph	Geologist	1942-44P	
Spreadborough, William	Botanical Collector	1889-1919T	
Sproule, J. C.	Geologist	1927-30, 1936-39T	
Stansfield, John	Geologist	1911-18T	
Stanton, M. S.	Geologist	1946-48P	
Stauffer, G. R.	Paleontologist	1910-12T	
Stefansson, Vilhjalmur	Anthropologist	1908-12T	Commander, Canadian Arctic Expedition, 1913-18
Sternberg, C. H.	Fossil Collector	1912-16P	
Sternberg, C. M.	Fossil Collector and Vertebrate Paleontol.	1912-14T, 1914-57P	To National Museum
Sternberg, G. F.	Fossil Collector	1913-18P	
Sternberg, Levi	Fossil Collector	1913-16T	To Royal Ontario Museum
Stewart, J. S.	Petroleum Geologist	1912-14T, 1914-21P, 1935-48T	
Stockwell, C. H.	Geologist	1922-27T, 1927-68P, 1968- T	Chief, Precambrian Division, 1955-58
Stopes, Miss M. C.	Paleobotanist	1911-12T	Women's Rights Advocate
Sullivan, Arthur	Draftsman	1935-53P	
Swanson, C. O.	Geologist	1921-22T	Michigan College of Mines
Symons, D. T. A.	Paleomagnetist	1966-70T	University of Windsor
Tanton, T. L.	Geologist	1914-20T, 1920-55P	
Tassonyi, E. J.	Subsurface Geologist	1961-66, 1970-71P	
Taverner, P. A.	Ornithologist	1911-42P	Head, Ornithological Division
Taylor, F. B.	Pleistocene Geologist	1908-12T	U.S. Geological Survey
Teit, J. A.	Social Anthropologist	1913-19T	
Terasmae, Jaan	Palynologist	1955-69P	Brock University
Thomson, J. Ellis	Geologist and Mineralogist	1920-26T	University of Toronto
Thorburn, John	Librarian	1882-1907P	Chairman, Civil Service Board of Canada
Thorpe, Edward	Photographer	1956-69P	
Thurber, J. B.	Geologist	1944-45T	
Tolman, Carl	Geologist	1922, 1926-36T	George Washington University

Name	Profession/Trade	Period of Service	Remarks
Topley, H. N.	Photographer	1894-1908T	Dept. of the Interior Photographer
Torrance, J. F.	Geologist	1883-84P	
Toth, A. M.	Groundwater Geologist	1959-63P	
Tremblay, J. J. L.	Groundwater Geologist	1959-63P	
Trueman, J. D.	Geologist	1908-12T	Drowned while on field work
Tufts, H. F.	Draftsman and Explorer	1906-08T	
Tuttle, J. W.	Map Engraver	1921-40P	
Twenhofel, W. H.	Paleontologist	1912T	University of Wisconsin
Tyrrell, J. B.	Geologist	1881-99P	
Tyrrell, J. W.	Topographer	1883-85, 1893T	
Uglow, W. L.	Geologist	1903-12, 1921-25T	University of British Columbia
Upham, Warren	Pleistocene Geologist	1888T	Minnesota Historical Society
Usher, J. L.	Geologist	1945-49T	Queen's University
Usik, Lily	Geobotanist	1967-68P	
Van Everdingen, R. D.	Groundwater Geologist	1962-66P	To Inland Waters Branch
Veitch, R.	Map Engraver	1921-46P	
Vennor, H. G.	Geologist; Ornithologist	1866-81P	
Vokes, F. M.	Mineral Geologist	1957-58P, 1972-73T	University of Oslo
Wait, F. G.	Chemist & Mineralogist	1890-1907P	To Mines Branch
Walcott, C. D.	Paleontologist	1910-12T	Secretary, Smithsonian Institution
Walker, J. F.	Geologist	1921-25T, 1925-34P	Deputy Minister of Mines, British Columbia
Walker, T. L.	Geologist	1890, 1905, 1912T	University of Toronto
Wallace, R. C.	Geochemist	1911-14T	Principal, Queen's University
Warren, H. V.	Geologist	1935T	Univ. of British Columbia
Warren, P. S.	Geologist	1923-30T	University of Alberta
Washington, R. A.	Biogeochemist	1959-65P	Atomic Energy of Canada
Watson, L. W.	Naturalist	1901-02T	
Waugh, F. W.	Social Anthropologist	1913-25P	
Webster, Arthur	Topographer	1868-82P, 1902T	
Weeks, L. J.	Geologist	1921-25T, 1925-64P	Chief, Post-Precambrian Division, 1955-64
Weir, J. D.	Geologist	1935-36T	
Wesemayer, H. H.	Geophysics Instruments	1958-61P	
Weston, T. C.	Fossil Collector	1857-94P	
White, James	Draftsman	1884-99P, 1899-1907T	Dominion Geographer, 1898-1909
Whiteaves, J. F.	Paleontologist	1874-76C, 1876-1909P	Chief, Paleontological Division
Whittaker, E. J.	Fossil Collector & Preparator	1913-24P	
Wickenden, R. T. D.	Paleontologist	1927-30T, 1930-66P	Head, Western Plains Office, 1950-65
Williams, Harold	Geologist	1958T, 1965-69P, 1969-70T	Memorial University of Newfoundland
Williams, L. A.	Draftsman	1930-35T, 1935-70P	
Williams, M. Y.	Stratigraphic Paleontologist	1909-13T, 1913-21P, 1921-43T	Univ. of British Columbia
Willimott, C. W.	Rock Collector	1871-1908P	
Wills, Miss N. I.—See Kummermann, Mrs.			
Wilson, Miss A. E.	Paleontologist	1909-11T, 1911-46P	
Wilson, A. W. G.	Mining Geologist	1901-08T	To Mines Branch
Wilson, J. Tuzo	Geologist	1936-40P, 1946T	University of Toronto
Wilson, M. E.	Geologist	1904-07T, 1907-47P	
Wilson, W. J.	Paleobotanist	1890T, 1891-1920P	
Winder, C. G.	Geologist	1950-54T	Univ. of Western Ontario
Wintemberg, W. J.	Archeologist	1913-41P	
Wright, F. E.	Geologist	1903T	
Wright, J. F.	Geologist	1919-34T, 1953-60P	
Wright, W. J.	Geologist	1909-20T, 1935T	New Brunswick Provincial Geologist
Wyder, J. E.	Surficial Geophysicist	1960-71P	
Wynne-Edwards, H. R.	Precambrian Geologist	1957-67T	Queen's University
Yorston, R. B.	Draftsman	1907-08T, 1908-36P	
Young, C. H.	Collector and Taxidermist	1907-37P	
Young, G. A.	Geologist	1896-1905T, 1905-43P	Chief Geologist, 1924-43

Appendix III

THE PERMANENT STAFF OF THE GEOLOGICAL SURVEY OF CANADA AS OF 31 DECEMBER 1972

AGC	Atlantic Geoscience Centre, Bedford, N.S.
CL & TS	Central Laboratories and Technical Services Division
G.Info.Proc.	Geological Information Processing Division
ISPG	Institute of Sedimentary and Petroleum Geology, Calgary
R & EG	Regional and Economic Geology Division
R & EG (Econ.Geol.)	Regional and Economic Geology Division, Economic Geology Subdivision
RG & G	Resource Geophysics and Geochemistry Division
Ter.Science	Terrain Sciences Division

Name	Trade/Profession	Division	Section/Unit	GSC Service from
Abbey, S.	Section Head; Chemist	CL & TS	Analytical Chemistry	1952
Abbinett, D. D.	Geophysical Map Compiler	RG & G	Digital Compilation	1965
Agterberg, F. P.	Unit Head; Geomathematician	R & EG(Econ.Geol.)	Geomathematics	1962
Ahrens, R. N.	Electronic Engineer	RG & G	Electrical Methods	1960
Aitken, J. D.	Stratigrapher	ISPG(Reg.Geol.)	Southern Mainland	1972
Ajram, Mrs. M. V.	Clerk	ISPG(Admin.)	Office Services	1972
Allan, R. J.	Surficial Geochemist	RG & G	Geochemistry	1969
Anderson, F. D.	Regional Geologist	R & EG	Appalachian	1948
Anderson, T. W.	Quaternary Palynologist	Ter.Science	Paleont. & Geochronology	1972
Ansell, H. G.	Mineralogist	CL & TS	National Collections	1972
Armstrong, J. E.	Administrator; Regional Geol.	Branch Hqrs.	IGC Office	1939
Arnold, J. G.	Administrative Officer	R & EG	Division Office	1949
Artichuk, G.	Administrative Officer	RG & G	Division Office	1968
Ascoli, P.	Micropaleontologist	AGC	Eastern Petrol. Geol.	1971
Atkinson, Mrs. C. M.	Clerk	Branch Hqrs.	Secretarial Services	1972
Babcock, L. W.	Unit Head; Draftsman	G.Info.Proc.	Cartography Unit B	1951
Baker, Mrs. D. K. B.	Secretary	ISPG(Energy)	Subdivision Office	1972
Balkwill, H. R.	Regional Geologist	ISPG(Reg.Geol.)	Arctic Islands	1970
Ball, N. L.	Geologist	ISPG(Energy)	Petroleum Geology	1966
Bamber, E. W.	Permo-Carboniferous Paleont.	ISPG(Paleont.)	Macropaleontology	1961
Banning, W. J.	Repository Clerk	ISPG(Energy)	Petroleum Geology	1962
Baragar, W. R. A.	Volcanologist	R & EG	Precambrian Shield	1957
Barbery, G. J.	Draftsman	G.Info.Proc.	Cartography Unit A	1951
Barefoot, R. R.	Laboratory Technician	ISPG(Energy)	Geochemistry	1967
Barnett, D. M.	Geomorphologist	Ter.Science	Reg. & Strat. Projects	1964
Barrett, D. L.	Geophysicist	AGC	Marine Geophysics	1963
Barss, M. S.	Palynologist	AGC	Eastern Petrol. Geol.	1949
Beaulne, J. M.	Lapidary Assistant	R & EG	Petrology	1969
Beauvais, Mrs. C. J.	Staffing and Pay Clerk	AGC	Exec. Assist., Personnel	1972
Beckstead, D. C.	Photographic Librarian	G.Info.Proc.	Photographic Service	1965
Bégin, Mrs. G. M.	Clerk	Branch Hqrs.	Branch Registry	1965
Belanger, Mrs. G. C.	Laboratory Technician	CL & TS	Analytical Chemistry	1967
Belanger, Miss M.	Clerk	Branch Hqrs.	Branch Registry	1967
Belanger, P. G.	Laboratory Technician	CL & TS	Spectrographic Labs	1972
Bell, C. K.	Regional Geologist	R & EG	Precambrian Shield	1954
Belyea, Miss H. R.	Biostratigrapher	ISPG(Reg.Geol.)	Northern Mainland	1945
Bencik, K. C.	Draftsman	G.Info.Proc.	Cartography Unit B	1959
Bender, A. G. P.	Laboratory Technician	CL & TS	Spectrographic Labs	1965
Benson, D. G.	Regional Geologist	R & EG	Appalachian	1960
Bigras, Mrs. B. D.	Secretary	Branch Hqrs.	Admin. Services	1969
Bik, M. J. J.	Geomorphologist	Ter.Science	Special Projects	1964
Birmingham, T. F.	Coal Petrographer	ISPG(Energy)	Coal Petrology(Ottawa)	1952

Name	Trade/Profession	Division	Section/Unit	GSC Service from
Birtch, Mrs. E. J.	Secretary	Branch Hqrs.	Director's Office	1972
Bisson, J. C.	Technician	R & EG	Geochronology	1967
Bisson, J. G.	Technician	Ter.Science	Engineering Geology	1965
Bjarnason, D. A.	Repository Storeman	ISPG(Energy)	Petroleum Geology	1968
Blackadar, R. G.	Chief Scientific Editor	G.Info.Proc.	Division Hqrs.	1953
Blake, W., Jr.	Section Head; Quaternary Geol.	Ter.Science	Paleont. & Geochronology	1962
Blouin, Mrs. L.	Secretary	Branch Hqrs.	Director's Office	1972
Blunden, Miss V. L.	Typist	ISPG(Admin.)	Admin. Services	1971
Blusson, S. W.	Regional Geologist	R & EG	Cord. & Pacific Margins	1964
Bolton, T. E.	Curator; Paleontologist	Branch Hqrs.	National Type Collection Secretary, NACRGS	1952
Bonardi, M. B.	Laboratory Technician	CL & TS	Mineralogical Lab	1968
Bostock. H. H.	Regional Geologist	R & EG	Precambrian Shield	1960
Botte, B. J.	Chief Fossil Preparator	R & EG	Eastern Paleontology	1965
Bouvier, J. L.	Laboratory Technician	CL & TS	Analytical Chemistry Labs	1961
Bower, Miss M. E.	Geophysicist	RG & G	Magnetic Methods	1950
Boyer, A.	Uranium Geologist	R & EG(Econ.Geol.)	Special Projects	1971
Boyle, R. W.	Geochemist	RG & G	Special Projects	1952
Brideaux, W. W.	Micropaleobotanist	ISPG(Paleont.)	Micropaleontology	1970
Bristow, Q.	Instrumentation Scientist	RG & G	Geochemistry	1970
Brooks, Mrs. M. H.	Section Head; Publications	ISPG(G.Info.Proc.)	Publications Distribution	1963
Brown, A. G.	Laboratory Technician	CL & TS	Mineral Separation	1968
Brown, D. A.	Laboratory Technician	CL & TS	Spectrographic Labs	1964
Brown, D. G.	Draftsman	G.Info.Proc.	Cartography Unit C	1955
Brown, L. C.	Technician	AGC	Scient. & Tech. Support	1966
Browne, J. A.	Draftsman	ISPG(G.Info.Proc.)	Cartography	1963
Brusso, A. W.	Administrative Officer	ISPG(Admin.)	Institute Office	1967
Brydges, G. E.	Plant Engineer	ISPG(Admin.)	Bldg. & Engrg. Services	1971
Buck, N. E.	Head, Photomechanical Unit	G.Info.Proc.	Cartography Photo Unit	1956
Buckley, D. E.	Geochemist	AGC	Environ. Marine Geol.	1960
Burk, C. F.	Information Specialist	Branch Hqrs.	Can. Centre for Geoscience Data	1960
Burke, R. D.	Laboratory Technician	R & EG(Econ.Geol.)	Mineral Deposits Geology	1964
Burns, Miss E. M.	Secretary	R & EG(Econ.Geol.)	Mineral Data Bank	1953
Burns, R. A.	Electronic Technician	RG & G	Seismic Methods	1970
Butler, E. B.	Clerk	G.Info.Proc.	Branch Library	1972
Butterfield, D. C.	Electronic Technician	RG & G	Electrical Methods	1967
Cade, Mrs. M. R.	Accounting Clerk	Ter.Science	Admin. & Financial Services	1972
Callahan, J. J.	Assistant Preparator	R & EG	Eastern Paleontology	1948
Cameron, A. R.	Coal Petrographer	ISPG(Energy)	Coal Petrology (Ottawa)	1957
Cameron, B. E. B.	Regional Geologist	R & EG	Cord. & Pacific Margins	1969
Cameron, E. M.	Section Head; Geochemist	RG & G	Geochemistry	1957
Campbell, F. H. A.	Regional Geologist	R & EG	Precambrian Shield	1972
Campbell, R. B.	Regional Geologist	R & EG	Cord. & Pacific Margins	1959
Cantin, Mrs. L. S.	Typist	Branch Hqrs.	Secretarial Services	1966
Carbone, S.	Paleontological Preparator	ISPG(Paleont.)	Paleontology Laboratory	1964
Carr, Mrs. E. J.	Secretary	Branch Hqrs.	Chief Geologist's Office	1971
Carrière, Mrs. J. J.	Scientific Support	R & EG	Mineral Deposits Geology	1971
Cascadden, Mrs. M. B. I.	Secretary	ISPG	Institute Office	1969
Chamney, T. P.	Mesozoic Micropaleontologist	ISPG(Paleont.)	Micropaleontology	1961
Champ, W. H.	Chemist	CL & TS	Spectrographic Labs	1946
Champagne, L. A.	Storeman	G.Info.Proc.	Cartography Office	1970
Chandler, F. W.	Regional Geologist	R & EG	Precambrian Shield	1972
Charbonneau, B. W.	Geologist	RG & G	Radiation Methods	1965
Charbonneau, J. E. R.	Laboratory Technician	CL & TS	Mineral Separation	1967
Charest, Mrs. J.	Library Clerk	AGC	Bedford Institute Library	1972
Charlebois, G. W.	Stores Keeper	Branch Hqrs.	Purchasing & Supplies	1948
Chrétien, Miss M. B.	Data Compiler	RG & G	Contract Surveys	1949
Christensen, Miss C. J.	Librarian	G.Info.Proc.	Branch Library	1968
Christie, K. W.	Paleomagnetism Support	R & EG	Paleomagnetism	1965
Christie, R. L.	Regional Geologist	ISPG(Reg.Geol.)	Arctic Islands	1954
Chung, C. J. F.	Geomathematician	R & EG(Econ.Geol.)	Geomathematic Support	1972
Church, K. A.	Laboratory Technician	CL & TS	Spectrographic Labs	1965
Clark, Mrs. C. J.	Secretary	AGC	Scient. Program Co-ordin. Office	1972
Clark, D. F.	Paleontologist	AGC	Eastern Petrol. Geology	1967

Name	Trade/Profession	Division	Section/Unit	GSC Service from
Clark, K. R. C.	Technician	R & EG	Paleomagnetism	1972
Clarke, C. R.	Accounts Clerk	G.Info.Proc.	Publications & Information	1959
Clattenburg, D.	Sedimentary Technician	AGC	Eastern Petrol. Geology	1967
Clyburne, Miss S. A.	Technician	AGC	Eastern Petrol. Geology	1972
Coady, V. F.	Instrument Operations Technician	AGC	Scient. & Tech. Support	1963
Collett, L. S.	Section Head; Geophysicist	RG & G	Electrical Methods	1953
Connors, Mrs. M. E. C.	Typist	Branch Hqrs.	Secretarial Services	1972
Cook, D. G.	Structural Geologist	ISPG(Energy)	Geology of Coal	1967
Cooke, F. J.	Section Head	G.Info.Proc.	Photographic Services	1951
Copeland, M. J.	Micropaleontologist	R & EG	Eastern Paleontology	1952
Corbett, T. J.	Head Marine Geophysics Technician	AGC	Scient. & Tech. Support	1969
Cormier, B. R.	Geological Technician	ISPG(Energy)	Geology of Coal	1972
Corriveau, J. P.	Draftsman	G.Info.Proc.	Cartography Unit B	1959
Coulthart, I. A.	Draftsman	G.Info.Proc.	Cartography Unit C	1964
Courtney, T. F.	Seismic Technician	AGC	Scient. & Tech. Support	1967
Courville, S.	Chemist	CL & TS	Analytical Chemistry Labs	1956
Craig, B. G.	Assist. Chief; Surficial Geologist	Ter.Science	Division Office	1950
Cranston, R. E.	Inorganic Geochemist	AGC	Environment Marine Geol.	1969
Cregheur, A. Y.	Machinist	CL & TS	Mech. Serv., Instrument Shop	1965
Crepin, G. J.	Draftsman	G.Info.Proc.	Cartography Unit C	1966
Crump, Miss J. M.	Typist	Branch Hqrs.	Secretarial Services	1972
Cumming, L. M.	Regional Geologist	R & EG	Appalachian	1952
Currie, K. L.	Alkaline Rock Studies	R & EG	Petrology	1960
Daley, L. A.	Draftsman	G.Info.Proc.	Cartography Unit A	1959
Darnley, A. G.	Division Head; Geophysicist	RG & G	Division Office	1966
Daugherty, R. F.	Supervising Draftsman	G.Info.Proc.	Cartography Unit B	1951
Davidson, A.	Granite Studies	R & EG	Precambrian Shield	1966
Davidson, D. O.	Stores Clerk	Branch Hqrs.	Purchasing & Supplies	1971
Davies, G. R.	Sedimentologist	ISPG(Reg.Geol.)	Arctic Islands	1971
Davison, W. L.	Regional Geologist	R & EG	Precambrian Shield	1952
Dawson, K. R.	Petrographer; Ore Deposits Geol.	R & EG(Econ.Geol.)	Mineral Deposits Geology	1946
Dean, W. T.	Section Head; Paleontologist	R & EG	Eastern Paleontology	1969
Debain, P. C. J.	Supervising Draftsman	G.Info.Proc.	Cartography Unit A	1950
Delabio, R. N.	Laboratory Technician	CL & TS	Mineralogical Lab	1960
Delay, B. G.	Plant Engineer	ISPG(Admin.)	Bldg. & Engrg. Services	1969
Demers, A. Y.	Lapidary Assistant	R & EG	Petrology	1962
Derouin, E. J.	Data Compiler	RG & G	Contract Surveys	1949
Derry, Miss G. E.	Secretary	Branch Hqrs.	Scientific Programs Office	1936
Deslauriers, Mrs. I. M.	Clerk	Branch Hqrs.	Accounts Office	1972
De-Vreeze, Miss T. A.	Scientific Editor	Ter.Science	Scientific Services	1972
Dicaire, A. J. L.	Electronic Technician	RG & G	Experimental Air Operations	1960
DiMillo, Mrs. M.	Secretary	G.Info.Proc.	Cartography Office	1967
Dnistransky, P. V.	Storeman	R & EG	Cord. & Pacific Margins	1972
Dodds, C. J.	Support Geologist	R & EG	Cord. & Pacific Margins	1972
Dods, S. D.	Computer Programmer	RG & G	Digital Compilation	1957
Donaldson, J. R.	Coal Petrographer	ISPG(Energy)	Coal Petrology (Ottawa)	1951
Douglas, R. J. W.	Geotectonicist; Editor	Branch Hqrs.	Special Projects	1947
Drapeau, Mrs. G. M.	Secretary	G.Info.Proc.	Publications & Information	1961
Duffell, S.	Assistant Chief; Regional Geologist	R & EG	Division Office	1945
Dumbrell, E. A.	Unit Head; Draftsman	G.Info.Proc.	Cartography Unit C	1949
Durham, C. C.	Laboratory Technician	RG & G	Geochemistry Labs	1959
Dyck, A. V.	Geophysicist	RG & G	Electrical Methods	1968
Dyck, W.	Geochemist	RG & G	Geochemistry	1959
Eade, K. E.	Regional Geologist	R & EG	Precambrian Shield	1955
Eckstrand, O. R.	Ore Deposits Geologist	R & EG(Econ.Geol)	Mineral Deposits Geology	1969
Eisbacher, G. H.	Structural Geologist	R & EG	Cord. & Pacific Margins	1968
Emslie, R. F.	Anorthosites Studies	R & EG	Petrology	1960
Enright, M. L.	Draftsman	G.Info.Proc.	Cartography Unit C	1964
Ermanovics, I. V.	Regional Geologist	R & EG	Precambrian Shield	1968
Eyre, W. H.	File Clerk	G.Info.Proc.	Central Technical Files	1969
Fabbri, A. G.	Geomathematician	R & EG(Econ.Geol.)	Geomathematics Unit	1969

Name	Trade/Profession	Division	Section/Unit	GSC Service from
Fader, G. B. J.	Marine Geologist	AGC	Regional Reconnaissance	1970
Fahrig, W. F.	Section Head; Precambrian Geologist	R & EG	Paleomagnetism	1950
Fairfield, R. D. J.	Draftsman	G.Info.Proc.	Cartography Unit A	1961
Federovich, Mrs. S.	Quaternary Palynology Lab. Tech.	Ter. Science	Paleon. & Geochronology	1971
Fiddes, Mrs. L. V.	Supervisor, Typing Pool	ISPG(Admin.)	Office Services	1967
Field, D. E.	Technician	Ter. Science	Engineering Geology	1949
Findlay, Mrs. C. E.	Editorial Assistant	ISPG(G.Info.Proc.)	Subdivision Office	1971
Finn, H. J.	Supervising Draftsman	G.Info.Proc.	Cartography Unit C	1945
Fitzgerald, R. A.	Geochemical Technician	AGC	Environmental Marine Geol.	1971
Flint, T. R.	Electronic Technician	RG & G	Experimental Air Operations	1964
Fortier, Y. O.	Director			1943
Foscolos, A. E.	Clay Mineralogist	ISPG(Energy)	Geochemistry	1967
Foster, J. H.	Paleomagnetist	R & EG	Paleomagnetism	1971
Fouchard, G. W.	Draftsman	G.Info.Proc.	Cartography Unit B	1966
Fournier, J. P.	Machinist	CL & TS	Mech. Serv. & Instrument Shop	1961
Frape, Miss F. E.	Micropaleont. Lab Technician	AGC	Environmental Marine Geol.	1970
Frarey, M. J.	Section Head; Precambrian Geologist	R & EG	Precambrian Shield	1947
Fraser, J. A.	Regional Geologist	R & EG	Precambrian Shield	1953
Frechette, J. P.	Electronic Technician	RG & G	Electrical Methods	1966
Freda, G. N. J.	Technician	R & EG	Paleomagnetism	1960
Frisch, T.	Metamorphic Geologist	R & EG	Precambrian Shield	1972
Frith, R. A.	Geochronologist	R & EG	Precambrian Shield	1972
Fritz, W. H.	Stratigraphic Paleontologist	R & EG	Eastern Paleontology	1964
Froese, E.	Metamorphic Rock Studies	R & EG	Petrology	1965
Frydecky, I. I.	Electronics Technician	R & EG	Cord. & Pacific Margins	1972
Fulton, R. J.	Quaternary Geologist	Ter.Science	Reg. & Strat. Projects	1963
Fyles, J. G.	Division Chief; Quaternary Geol.	Ter.Science	Division Hqrs.	1950
Gabrielse, H.	Section Head; Cordilleran Geologist	R & EG	Cord. & Pacific Margins	1953
Gadd, N. R.	Quaternary Geologist	Ter.Science	Reg. & Strat. Projects	1948
Gagne, R. M.	Seismic Technician	RG & G	Seismic Methods	1965
Gagnon, J. G. E.	Supervising Draftsman	G.Info.Proc.	Cartography Unit C	1949
Gange-Harris, C. F.	Laboratory Technician	ISPG(Energy)	Coal Petrology (Ottawa)	1971
Ganim, T. N.	File Clerk	Branch Hqrs.	Branch Registry	1968
Garner, Mrs. I. J.	Typist	ISPG(Admin.)	Office Services	1972
Garrett, R. G.	Geochemist	RG & G	Geochemistry	1967
Gaumont, G. A.	Electronic Technician	RG & G	Geochemistry	1971
Gauthier, G.	Chemical Technician	RG & G	Geochemistry	1965
Gauvreau, C.	Electronic Technician	RG & G	Electrical Methods	1957
Gibson, D. W.	Regional Geologist	ISPG(Reg.Geol.)	Southern Mainland	1965
Giles, Mrs. C. M. A.	Typist	ISPG(Admin.)	Office Services	1971
Godden, C. A. D.	Technician	AGC	Scient. & Tech. Support	1966
Going, Mrs. M. I.	Clerk	Branch Hqrs.	Financial Services	1965
Goodman, Mrs. N. D.	Secretary	RG & G	Division Office	1964
Gordon, T. M.	Metamorphic Rock Studies	R & EG	Petrology	1970
Gorveatt, M. E.	Technician	AGC	Scient. & Tech. Support	1967
Gougeon, Mrs. C. L.	Secretary	R & EG	Division Office	1961
Grant, A. C.	Section Head; Geophysicist	AGC	Regional Reconnaissance	1965
Grant, D. R.	Quaternary Geologist	Ter.Science	Reg. & Strat. Projects	1968
Grant, G. M.	Compiler-Draftsman	AGC	Eastern Petrol. Geology	1972
Grasty, R. L.	Physicist	RG & G	Radiation Methods	1968
Gravel, R. J.	Laboratory Technician	CL & TS	Mineralogical Labs	1967
Greener, Miss P. L.	Library Clerk	ISPG(G.Info.Proc.)	Institute Library	1965
Grenier, N. A.	Draftsman	G.Info.Proc.	Cartography Unit A	1965
Gross, G. A.	Iron & Manganese Geologist	R & EG(Econ.Geol.)	Special Projects	1956
Grushman, Mrs. V. E.	Laboratory Technician	CL & TS	Analytical Chemistry Labs	1967
Hacquebard, P. A.	Section Head; Coal Geologist	ISPG(Energy)	Coal Petrology (Ottawa)	1948
Haden, D. F.	Laboratory Technician	ISPG(Paleont.)	Paleontology Labs	1971
Haley, E. L.	Data Compiler	RG & G	Contract Surveys	1958
Hall, E.	Technical Executive Assist.	Branch Hqrs.	Director's Office	1946
Hardy, Mrs. I. A.	Regional Geologist	AGC	Eastern Petrol. Geology	1972
Harker, P.	Division Chief; Paleontologist	G.Info.Proc.	Division Office	1948
Harris, I. M.	Regional Geologist	AGC	Regional Reconnaissance	1971

Name	Trade/Profession	Division	Section/Unit	GSC Service from
Harrison, J. E.	Quaternary Geologist	Ter.Science	Engineering Geology	1972
Haworth, R. T.	Geophysicist	AGC	Regional Reconnaissance	1968
Heffler, D. E.	Geophysicist	AGC	Regional Reconnaissance	1971
Heginbottom, J. A.	Geomorphologist	Ter.Science	Engineering Geology	1968
Heinrich, A. G.	Laboratory Technician	ISPG(Energy)	Geochemistry	1966
Henderson, E. P.	Quaternary Geologist	Ter.Science	Reg. & Strat. Projects	1951
Henderson, J. B.	Precambrian Sedimentologist	R & EG	Precambrian Shield	1969
Heney, F. J.	Draftsman	G.Info.Proc.	Cartography Unit B	1957
Heyendal, H. A.	Draftsman	G.Info.Proc.	Cartography Unit A	1953
Heywood, W. W.	Regional Geologist	R & EG	Precambrian Shield	1952
Hickson, J. R.	Branch Financial Officer	Branch Hqrs.	Financial Services	1968
Hill, B. G.	Draftsman	G.Info.Proc.	Cartography Unit C	1965
Hill, R. S.	Draftsman	G.Info. Proc.	Cartography Unit B	1956
Hillis, T. S.	Executive Assistant	AGC	Director's Office	1972
Hobbs, J. D.	Computer Programmer	RG & G	Geochemistry	1967
Hobson, G. D.	Section Head; Geophysicist	RG & G	Seismic Methods	1958
Hodgson, D. A.	Geomorphologist	Ter.Science	Reg. & Strat. Projects	1966
Hoffman, P. F.	Precambrian Sedimentologist	R & EG	Precambrian Shield	1969
Holman, P. B.	Geophysical Instrum. Operator	RG & G	Radiation Methods	1971
Holroyd, M. T.	Section Head; Computer Geophys.	RG & G	Digital Compilation	1969
Hood, P. J.	Section Head; Geophysicist	RG & G	Magnetic Methods	1961
Hopkins, W. S.	Micropaleobotanist	ISPG(Paleont.)	Micropaleontology	1968
Hornbook, E. H. W.	Geochemist	RG & G	Geochemistry	1962
Horton, R. E.	Chemist	RG & G	Geochemistry Labs	1964
Houston, W. N.	Isotope Analyst	RG & G	Geochronology	1972
Howe, K. G.	Draftsman	G.Info.Proc.	Cartography Unit C	1951
Howie, R. D.	Stratigrapher	AGC	Eastern Petrol. Geol.	1952
Huard, F. E.	Clerk	G.Info.Proc.	Publications & Information	1972
Hughes, M. D.	Marine Geophysics Technician	AGC	Scient. & Tech. Support	1966
Hughes, O. L.	Quaternary Geologist	Ter.Science	Reg. & Strat. Projects	1953
Hunter, J. A. M.	Geophysicist	RG & G	Seismic Methods	1970
Huot, J. M. R.	Laboratory Technician	CL & TS	Mineral Separation	1967
Hutchinson, J. D.	Electronic Technician	AGC	Scient. & Tech. Support	1969
Hutchison, W. W.	Regional Geologist	R & EG	Cord. & Pacific Margins	1962
Irish, E. J. W.	Subdivision Head; Geol. Editor	ISPG(G.Info.Proc.)	Subdivision Office	1943
Irvine, J. A.	Coal Geologist	ISPG(Energy)	Geology of Coal	1972
Irvine, T. N.	Ultrabasic Rock Studies	R & EG	Petrology	1962
Isaacs, R. M. F.	Engineering Geologist	Ter.Science	Engineering Geology	1970
Jackson, A. E.	Geophysicist; Data Technician	AGC	Regional Reconnaissance	1972
Jackson, G. D.	Regional Geologist	R & EG	Precambrian Shield	1958
Jackson, L. A.	Administrative Officer	Ter.Science	Division Office	1968
Jambor, J. L.	Ore Deposits Geologist	R & EG(Econ.Geol.)	Mineral Deposits Geology	1960
Jamieson, A.	Office Manager	ISPG(Admin.)	Office Services	1967
Jansa, L. F.	Sedimentologist	AGC	Eastern Petrol. Geology	1971
Jeletzky, J. A.	Cretaceous Biostratigrapher	ISPG(Paleont.)	Paleontology (Ottawa)	1948
Johnston, B. L.	Marine Geophysics Technician	AGC	Scient. & Tech. Support	1969
Jonasson, I. R.	Geochemist; Chemist	RG & G	Geochemistry	1971
Jones, F. W.	Electronics Technician	CL & TS	Electronic Services	1960
Jones, Mrs. L. E.	Typist	Branch Hqrs.	Secretarial Services	1972
Jones, Mrs. M.	Librarian	ISPG(G.Info.Proc.)	Institute Library	1966
Josenhans, H. W.	Technician	AGC	Regional Reconnaissance	1970
Katsube, T. J.	Electrical Rock Prop. Geophys.	RG & G	Electrical Methods	1971
Keen, Miss C. E.	Geophysicist	AGC	Regional Reconnaissance	1970
Kelly, A. M.	Computer Scientist	R & EG(Econ.Geol.)	Geomathematics	1971
Kelly, R. G.	Technician	Ter.Science	Engineering Geology	1960
Kempt, J. W.	Photographer	G.Info.Proc.	Photographic Services	1956
Kerr, J. W.	Regional Geologist	ISPG(Reg.Geol.)	Arctic Islands	1961
Khan, M. R.	Laboratory Technician	ISPG(Energy)	Geochemistry	1971
King, J. A.	Draftsman	G.Info.Proc.	Cartography Unit A	1957
King, L. H.	Marine Geologist	AGC	Regional Reconnaissance	1954
Kirkham, R. V.	Ore Deposits Geologist	R & EG(Econ.Geol.)	Mineral Deposits Geology	1969
Klassen, R. W.	Quaternary Geologist	Ter.Science	Reg. & Strat. Projects	1965

Name	Trade/Profession	Division	Section/Unit	GSC Service from
Knapp, H. W. C.	Electronic Technician	RG & G	Experimental Air Operations	1954
Knappers, W. A. K.	Data Compiler	RG & G	Contract Surveys	1972
Koops, Miss A. H.	Typist	Branch Hqrs.	Secretarial Services	1971
Koops, Mrs. M. F.	Clerk	Branch Hqrs.	Financial Services	1967
Kornik, L. J.	Geologist	RG & G	Magnetic Methods	1967
Kovachic, Mrs. H.	Draftsman	G.Info.Proc.	Cartography Unit B	1960
Lachance, G. R.	Mineralogist	CL & TS	Mineralogical	1960
Lafrance, Mrs. C. L. M.	Typist	Branch Hqrs.	Secretarial Services	1972
Lagler, Mrs. E.	Switchboard Receptionist	ISPG(Admin.)	Office Services	1972
Lajoie, L. J.	Administrative Officer	Branch Hqrs.	Office Services	1945
Lambert, M. B.	Volcanologist	R & EG	Precambrian Shield	1972
Lane, Miss P. D.	Clerk	R & EG	Cord. & Pacific Margins	1969
Langlois, R. J. Y.	Programmer; Map Compiler	RG & G	Digital Compilation	1950
Lapp, J. H.	Administrative Officer	CL & TS	Division Office	1969
Larochelle, J. E. A	Section Head; Geophysicist	RG & G	Contract Surveys	1952
Larose, J. M.	Laboratory Technician	CL & TS	Mineral & Rock Set Preparation	1953
Latour, B. A.	Coal Deposits Evaluator	ISPG(Energy)	Geology of Coal	1945
Laurendeau, P. E.	Map Librarian	G.Info.Proc.	Branch Library	1965
Lavergne, P. J.	Laboratory Technician	RG & G	Geochemistry	1957
Lavigne, G. H.	Draftsman	G.Info.Proc.	Cartography Unit A	1949
Lawrence, D. E.	Quaternary Geologist	Ter.Science	Engineering Geology	1971
Lawrence, D. G.	Photographer	ISPG(G.Info.Proc.)	Photography	1972
Leader, R. E.	Supervising Draftsman	G.Info.Proc.	Cartography Unit B	1945
Leaming, S. F.	Info. Officer; Rock Collector	R & EG	Cord. & Pacific Margins	1960
Leech, G. B.	Subdivision Head; Econ. Geol.	R & EG(Econ.Geol.)	Subdivision Hqrs.	1949
Leonard, J. D.	Organic Geochemist	AGC	Environmental Marine Geol.	1971
Letang, E. G.	Clerk	G.Info.Proc.	Publications & Information	1954
Lewis, C. F. M.	Quaternary Geologist	Ter.Science	Reg. & Strat. Projects	1965
Little, H. W.	Uranium Geologist	R & EG(Econ.Geol.)	Special Projects	1947
Locke, D. R.	Electronic Research Technician	AGC	Scient. & Tech. Support	1965
Loncarevic, B. D.	Director, Marine Geophysicist	AGC	Institute Hqrs.	1971
Lord, C. S.	Chief Geologist	Branch Hqrs.	Chief Geologist's Office	1937
Loveridge, W. D.	Isotopic Analyst	R & EG	Geochronology	1959
Lowdon, J. A.	Radiocarbon Lab Supervisor	Ter.Science	Paleont. & Geochronology	1954
Lynch, J. J.	Chief Analyst	RG & G	Geochemistry	1955
McAllister, M. E.	Draftsman	RG & G	Division Office	1971
MacAulay, H. A.	Seismic Technician	RG & G	Seismic Methods	1959
McBryde, Miss C. M. L.	Clerk	Branch Hqrs.	Supplies Office	1972
McCracken, J. N.	Draftsman	G.Info.Proc.	Cartography Unit B	1959
McCrossan, R. G.	Subdivision Head; Petroleum Geol.	ISPG(Energy)	Subdivision Office	1967
McDonald, B. C.	Quaternary Geologist	Ter.Science	Engineering Geology	1966
McDonald, Miss H. M.	Laboratory Technician	ISPG(Paleont.)	Paleontology Labs	1971
McDowell, Mrs. J. K.	Secretary	RG & G	Division Office	1970
McEwan, W. O.	Laboratory Technician	ISPG(Reg.Geol.)	Sedimentology Labs	1955
McGee, B. A.	Information Specialist	Branch Hqrs.	Can. Centre for Geoscience Data	1967
McGlynn, J. C.	Regional Geologist	R & EG	Precambrian Shield	1950
McGrath, P. H.	Instrumentation Geophysicist	RG & G	Magnetic Methods	1967
McGregor, D. C.	Palynologist	R & EG	Eastern Petrol. Geology	1957
Machin, B. D.	Laboratory Technician	CL & TS	Mineral Separation	1962
MacIntyre, J. B.	Technician	AGC	Regional Reconnaissance	1970
McKenzie, N. M.	Draftsman	G.Info.Proc.	Cartography Unit B	1965
Mackenzie, R. J. G.	Platemarker	G.Info.Proc.	Cartography Photomech.	1960
Mackenzie, W. S.	Regional Geologist	ISPG(Reg.Geol.)	Northern Mainland	1965
McKinley, Mrs. S. M.	Keyboard Operator	G.Info.Proc.	Central Technical Files	1966
MacLachlan, L.	Section Head; Cartographer	ISPG(G.Info.Proc)	Cartography	1959
McLaren, D. J.	Institute Director; Paleontologist	ISPG		1948
MacLaurin, Mrs. A. I.	Laboratory Technician	RG & G	Geochemistry	1972
Maclean, B.	Marine Geologist	AGC	Regional Reconnaissance	1953
McLeod, C. R.	Laboratory Chief	R & EG(Econ.Geol.)	Mineral Deposits Geology	1955
MacMillan, W. G.	Palynology Technician	AGC	Eastern Petrol. Geology	1972
MacNab, R. F.	Geophysicist	AGC	Regional Reconnaissance	1970
McNeil, C. E. A.	Section Head; Cartographer	G.Info.Proc.	Cartography Office	1933

Name	Trade/Profession	Division	Section/Unit	GSC Service from
McNeill, D. G.	Halftone Camera Operator	G.Info.Proc.	Cartography Photomech.	1968
Macqueen, R. W.	Regional Geologist	ISPG(Reg.Geol.)	Southern Mainland	1965
Macrae, J. L.	Technician	R & EG	Geochronology	1966
Mahoney, Mrs. L. R.	Editorial Assistant	G.Info.Proc.	Division Office	1955
Maiden, F.	Publications Clerk	ISPG(Energy)	Geology of Petroleum	1970
Mainville, B.	Draftsman	G.Info.Proc.	Cartography Unit A	1951
Major, A. C.	Clerk; Storeman	G.Info.Proc.	Cartograph. Photomech.	1943
Maland, Mrs. D. M.	Registry Clerk	ISPG(Admin.)	Office Services	1971
Manchester, K. S.	Section Head; Geophysicist	AGC	Scient. & Tech. Support	1965
Marble, Mrs. A.	Secretary	R & EG	Cord. & Pacific Margins	1950
Matiisen, Miss T.	Reference Librarian	ISPG(G.Info.Proc.)	Institute Library	1971
Maxwell, J. A.	Division Chief; Chemist	CL & TS	Division Hqrs.	1953
Meijer-Drees, N. C.	Subsurface Stratigrapher	ISPG(Reg.Geol.)	Southern Mainland	1971
Meilleur, G. A.	Unit Head; Machinist	CL & TS	Mech. Serv. & Instrument Shop	1952
Miall, A. D.	Sedimentologist	ISPG(Reg.Geol.)	Arctic Islands	1972
Michie, R. D.	Laboratory Technician	ISPG(Paleont.)	Paleontology Labs	1970
Miller, Mrs. D. E.	Typist	Ter.Science	Admin. & Financial Services	1958
Minning, Miss G. V.	Geomorphologist	Ter.Science	Scientific Services	1967
Mizerovsky, Mrs. G. J.	Photo-Interpreter	Ter.Science	Scientific Services	1959
Monger, J. W. H.	Regional Geologist	R & EG	Cord. & Pacific Margins	1965
Moreau, V. L.	Stores Clerk	Branch Hqrs.	Purchasing & Supplies	1960
Morel, D. M.	Technician	Ter.Science	Engineering Geology	1970
Morency, Miss L. S.	Secretary	Ter.Science	Division Office	1962
Morgan, W. C.	Regional Geologist	R & EG	Precambrian Shield	1968
Mott, R. J.	Palynology Lab Supervisor	Ter.Science	Paleont. & Geochronology	1958
Moyd, Mrs. P.	Organizing Secretary	Branch Hqrs.	I.G.C. Office	1966
Muller, J. E.	Regional Geologist	R & EG	Cord. & Pacific Margins	1948
Mulligan, R.	Ore Deposits Geologist	R & EG(Econ.Geol.)	Mineral Deposits Geology	1950
Murphy, B. P.	Shipping Clerk	G.Info.Proc.	Publications & Information	1967
Murphy, Miss F. M.	Typist	Branch Hqrs.	Secretarial Services	1962
Myhr, D. W.	Subsurface Stratigrapher	ISPG(Reg.Geol.)	Northern Mainland	1972
Nash, Mrs. D. K.	Micropaleontol. Support	R & EG	Eastern Paleontology	1968
Nassichuk, W. W.	Biostratigrapher	ISPG(Paleont.)	Micropaleontology	1965
Nichol, H. S.	Draftsman	G.Info.Proc.	Cartography Unit C	1959
Nicol, Miss S. A.	Kardex Clerk	G.Info.Proc.	Central Technical Files	1972
Nielsen, J. A.	Marine Geophysics Technician	AGC	Scient. & Tech. Support	1971
Norford, B. S.	Subdivision Head; Paleontologist	ISPG(Paleont.)	Subdivision Office	1960
Norris, A. W.	Biostratigrapher	ISPG(Paleont.)	Micropaleontology	1955
Norris, D. K.	Structural & Coal Geologist	ISPG(Reg.Geol.)	Northern Mainland	1951
Northcott, Mrs. E. M.	Laboratory Technician	ISPG(Energy)	Geochemistry	1971
Nunn, E. P.	Unit Head; Draftsman	G.Info.Proc.	Cartography Unit A	1935
Ollerenshaw, N. C.	Regional Geologist	ISPG(Reg.Geol.)	Southern Mainland	1962
Olson, D. G.	Electronic Technician	RG & G	Experimental Air Operations	1971
Ortman, B. H. A.	Draftsman	ISPG(G.Info.Proc.)	Cartography	1967
Ortman, K. M.	Draftsman	ISPG(G.Info.Proc.)	Cartography	1971
Overton, A.	Geophysicist	RG & G	Seismic Methods	1960
Owen, E. B.	Engineering Geologist	Ter.Science	Engineering Geology	1946
Owens, E. H.	Marine Geologist	AGC	Environmental Marine Geol.	1971
Owens, K. H.	Geophysical Surveyor	RG & G	Magnetic Methods	1948
Papps, T. L.	Draftsman	G.Info.Proc.	Cartography Unit C	1959
Paris, J. C.	Laboratory Technician	CL & TS	Mineral Separation	1949
Parker, J.	Electronic Technician	RG & G	Radiation Methods	1967
Parke-Taylor, Miss A. H.	Secretary	CL & TS	Division Office	1970
Parnham, Mrs. S. J.	Typist	Branch Hqrs.	Secretarial Services	1972
Peatman, D.	Stores Clerk	ISPG(Admin.)	Office Services	1967
Pedder, A. E. H.	Devonian Paleontologist	ISPG(Paleont.)	Macropaleontology	1968
Pelchat, J. C.	Chemical Lab Technician	RG & G	Geochemistry Labs	1965
Pelletier, B. R.	Administrator; Sedimentologist	AGC	Scient. Program Co-ordin. Office	1957
Penley, Miss W. L.	Registry Clerk	ISPG(Admin.)	Office Services	1963
Penney, Miss E. G.	Secretary	AGC	Director's Office	1972
Perron, R. R.	Draftsman	G.Info.Proc.	Cartography Unit C	1966

Name	Trade/Profession	Division	Section/Unit	GSC Service from
Peterkin, G. M.	Supervisor, Bldg. & Engrg. Serv.	ISPG(Admin.)	Bldg. & Engrg. Services	1967
Pettis, D. F.	Financial Clerk	Branch Hqrs.	Financial Services	1964
Plant, A. G.	Mineralogist	CL & TS	Mineralogy	1968
Pollitt, E. I. K.	Administrative Officer	Branch Hqrs.	Admin. Services	1947
Poole, W. H.	Section Head; Regional Geologist	R & EG	Appalachian	1952
Potvin, Y. R.	Draftsman	G.Info.Proc.	Cartography Unit C	1963
Prasad, N.	Scientific Support	R & EG(Econ.Geol.)	Mineral Deposits Geology	1968
Pratt, J. A. Y.	Draftsman	G.Info.Proc.	Cartography Unit A	1971
Prest, V. K.	Glacial Geologist	Ter.Science	Special Projects	1950
Price, L. L.	Regional Geologist	ISPG(Reg.Geol.)	Southern Mainland	1949
Procter, R. M.	Research Manager	ISPG(Energy)	Petroleum Geology	1960
Pugh, D. C.	Regional Geologist	ISPG(Reg.Geol.)	Southern Mainland	1954
Quigg, F. B.	Technician	R & EG	Geochronology	1966
Racine, T. H.	Laboratory Technician	CL & TS	Mineral & Rock Set Preparation	1966
Raddatz, Miss M. A.	Draftsman	G.Info.Proc.	Cartography Unit B	1957
Rail, E. J.	Shipping Clerk	G.Info.Proc.	Publications & Information	1966
Rampton, V. N.	Quaternary Geologist	Ter.Science	Reg. & Strat. Projects	1969
Rashid, M. A.	Organic Geochemist	AGC	Environmental Marine Geol.	1966
Rawlings, Mrs. J.	Information Clerk	G.Info.Proc.	Publications & Information	1971
Ready, E. E.	Data Compiler	RG & G	Contract Surveys	1950
Rees, J. L.	Laboratory Technician	ISPG(Reg.Geol.)	Sedimentology Labs	1972
Reesor, J. E.	Section Head; Petrologist	R & EG	Petrology	1952
Reinhardt, E. W.	Metamorphic Geologist	R & EG	Precambrian Shield	1965
Reveler, D. A.	Data Compiler	RG & G	Contract Surveys	1952
Rhoades, W. A.	Publications Clerk	ISPG(G.Info.Proc.)	Publications Distribution	1972
Richard, Mrs. B. G.	Office Manager	Branch Hqrs.	Secretarial Services	1964
Richard, S. H.	Geomorphologist	Ter.Science	Reg. & Strat. Projects	1957
Richards, T. A.	Regional Geologist	R & EG	Cord. & Pacific Margins	1972
Richardson, K. A.	Section Head; Geophysicist	RG & G	Radiation Methods	1971
Ridler, R. H.	Volcanic Geologist	R & EG	Precambrian Shield	1970
Rimsaite, Miss J. H. Y.	Mineralogist	R & EG(Econ.Geol.)	Mineral Deposits Geology	1959
Robertson, I. M.	Technician	Ter.Science	Paleont. & Geochronology	1964
Robertson, K. R.	Senior Geochemical Technician	AGC	Environmental Marine Geol.	1966
Robertson, Mrs. W.	Secretary	G.Info.Proc.	Division Office	1972
Robillard, R. D.	Office Manager	Branch Hqrs.	Branch Registry	1969
Robinson, S. C.	Chief, Task Force on Special Projects	Branch Hqrs.	Special Projects	1948
Roddick, J. A.	Regional Geologist	R & EG	Cord. & Pacific Margins	1951
Rose, D. G.	Ore Deposits Geologist	R & EG(Econ.Geol.)	Mineral Data Bank	1972
Rose, E. R.	Ore Deposits Geologist	R & EG(Econ.Geol.)	Mineral Deposits Geology	1949
Rose, Mrs. I. A.	Accounts Clerk	ISPG(Admin.)	Office Services	1972
Ross, D. I.	Subdivision Head; Geophysicist	AGC	Regional Reconnaissance	1966
Roy, K. J.	Lithostratigrapher	ISPG(Reg.Geol.)	Arctic Islands	1971
Rozon, R. J. A.	Clerk	Branch Hqrs.	Equipment Office	1960
Ruddy, Miss L. L.	Laboratory Technician	ISPG(Paleont.)	Paleontology Labs	1970
Russell, Mrs. B. C.	Clerk	R & EG	Cord. & Pacific Margins	1967
Rutter, N. W.	Quaternary Geologist	Ter.Science	Reg. & Strat. Projects	1965
Saffin, R. E.	Draftsman	G.Info.Proc.	Cartography Unit C	1961
Saint-Onge, D. A.	Geomorphologist	Ter.Science	Reg. & Strat. Projects	1958
Saint Pierre, A. L. P.	Administrative Officer	G.Info.Proc.	Division Office	1971
Saint Pierre, J. M.	Draftsman	G.Info.Proc.	Cartography Unit B	1961
Salter, I. C. G.	Storeman	Branch Hqrs.	Purchasing & Supplies	1971
Sanford, B. V.	Subdivision Head; Regional Geol.	AGC	Eastern Petrol. Geology	1949
Sangster, D. F.	Ore Deposits Geologist	R & EG(Econ.Geol.)	Mineral Deposits Geology	1965
Santowski, K.	Technician	R & EG	Geochronology	1968
Sauvageau, J. A. R.	Draftsman	G.Info.Proc.	Cartography Unit A	1965
Sawatzky, P.	Electrical Engineer	RG & G	Experimental Air Operations	1957
Schafer, C. T.	Marine Geologist	AGC	Environmental Marine Geol.	1967
Schau, M. P.	Regional Geologist	R & EG	Precambrian Shield	1972
Schwarz, E. J.	Paleomagnetist	R & EG	Paleomagnetism	1965
Scott, J. S.	Section Head; Engineering Geologist	Ter.Science	Engineering Geology	1969
Scott, W. J.	Geophysicist	RG & G	Electrical Methods	1970

Name	Trade/Profession	Division	Section/Unit	GSC Service from
Seeman, D. A.	Engineering Support	R & EG	Cord. & Pacific Margins	1972
Segall, Mrs. C. D.	Data Processer	G.Info.Proc.	Central Technical Files	1966
Seguin, R. J. C.	Technician	R & EG	Geochronology	1957
Sen Gupta, J. G.	Chemist	CL & TS	Analytical Chemistry Labs	1962
Seymour, Miss L. J.	Laboratory Technician	CL & TS	Analytical Chemistry Labs	1964
Shearer, J. M.	Marine Geologist	Ter.Science	Reg. & Strat. Projects	1970
Shih, K. G.	Data Analyst	AGC	Scient. & Tech. Support	1969
Shilts, W. W.	Quaternary Geologist	Ter.Science	Engineering Geology	1970
Shimizu, K.	Geomathematician	Ter.Science	Scientific Services	1953
Shurban, Mrs. P. K.	Administrative Clerk	R & EG	Cord. & Pacific Margins	1969
Siewert, Mrs. S. M.	Draftsman	ISPG(G.Info.Proc.)	Cartography	1968
Sime, Miss D. A.	Publications Clerk	ISPG(G.Info.Proc.)	Publications & Distribution	1967
Simonds, Mrs. B. J.	Draftsman	G.Info.Proc.	Cartography Unit B	1960
Sinha, A. K.	Geophysicist	RG & G	Electrical Methods	1970
Skinner, R.	Regional Geologist	R & EG	Appalachian	1950
Skinner, R. G.	Quaternary Geologist	Ter.Science	Reg. & Strat. Projects	1970
Skuce, C. A.	Photographer	G.Info.Proc.	Photographic Services	1965
Slaney, V. R.	Photogeologist	RG & G	Radiation Methods	1966
Sliter, W. V.	Micropaleontologist	ISPG(Energy)	Micropaleontology	1972
Snowdon, L. R.	Chemist	ISPG(Energy)	Geochemistry	1969
Solowan, Miss P. A.	Secretary	AGC	Marine Geophysics	1972
Souther, J. G.	Volcanologist	R & EG	Card. & Pacific Margins	1956
Sparkes, R.	Technician	AGC	Scient. & Tech. Support	1969
Srivastava, S. P.	Geophysicist	AGC	Regional Reconnaissance	1965
Stafford, W. G.	Photographer	G.Info.Proc.	Photographic Services	1947
Stalker, A. M.	Quaternary Geologist	Ter.Science	Reg. & Strat. Projects	1950
Stauffer, W. J.	Electronic Technician	RG & G	Electrical Methods	1959
Steacy, H. R.	Mineralogist	CL & TS	National Collections	1946
Stenson, Mrs. A. P.	Mineralogist	CL & TS	Mineralogy	1952
Stevens, R. D.	Isotopic Geologist	R & EG	Geochronology	1958
Stevenson, I. M.	Research Manager	R & EG	Precambrian Shield	1953
Stewart, Miss T. G.	Cataloguer	G.Info.Proc.	Branch Library	1947
Stott, D. F.	Subdivision Head; Stratigrapher	ISPG(Reg.Geol.)	Subdivision Office	1957
Sullivan, R. W.	Chemist	R & EG	Geochronology	1963
Sutherland, Mrs. D. M.	Head Librarian	G.Info.Proc.	Branch Library	1966
Sweet, A. R.	Coal Palynologist	ISPG(Paleont.)	Micropaleontology	1971
Taylor, F. C.	Regional Geologist	R & EG	Precambrian Shield	1954
Taylor, G. C.	Regional Geologist	ISPG(Reg.Geol.)	Southern Mainland	1959
Tempelman-Kluit, D. J.	Regional Geologist	R & EG	Cord. & Pacific Margins	1966
Ter Haar Romeny, W. U.	Petrology Curator	R & EG	Division Office	1958
Thomson, H. A.	Draftsman	G.Info.Proc.	Cartography Unit A	1961
Thomson, J. W.	Draftsman	ISPG(G.Info.Proc.)	Cartography	1958
Thorpe, R. I.	Ore Deposits Geologist	R & EG(Econ.Geol.)	Mineral Deposits Geology	1965
Thorsteinsson, R.	Biostratigrapher	ISPG(Reg.Geol.)	Arctic Islands	1952
Tiffin, D. L.	Marine Geologist	R & EG	Cord. & Pacific Margins	1969
Tipper, H. W.	Regional Geologist	R & EG	Cord. & Pacific Margins	1946
Touchette, J. L. L.	Section Head; Clerk	G.Info.Proc.	Publications & Information	1952
Tozer, E. T.	Triassic Paleontologist	ISPG(Reg.Geol.)	Paleontology (Ottawa)	1952
Traill, R. J.	Section Head; Mineralogist	CL & TS	Mineralogy	1953
Trapnell, Miss M. J.	Secretary	AGC	Eastern Petrol. Geology	1972
Tremblay, L. P.	Regional Geologist	R & EG	Precambrian Shield	1947
Trettin, H. P.	Regional Geologist	ISPG(Reg.Geol.)	Arctic Islands	1961
Turay, Mrs. M. N. H.	Laboratory Assistant	R & EG	Petrology	1970
Turpin, J.	Laboratory Technician	CL & TS	Mineral & Rock Set Preparation	1954
Unger, J.	Storeman	ISPG(Admin.)	Office Services	1970
Uyeno, T. T.	Conodont Micropaleontologist	ISPG(Paleont.)	Micropaleontology	1960
van der Linden, W. J. M.	Regional Geologist	AGC	Regional Reconnaissance	1972
Vermette, W. P.	Draftsman	ISPG(G.Info.Proc.)	Cartography	1958
Vickers, G. J.	Laboratory Technician	CL & TS	Analytical Chemistry Labs	1968
Vilks, G.	Micropaleontologist	AGC	Environmental Marine Geol.	1962
Vilonyay, A. G.	Technician	Ter.Science	Engineering Geology	1967
Vincent, L. E.	French Editor	G.Info.Proc.	Division Office	1971

Name	Trade/Profession	Division	Section/Unit	GSC Service from
Wade, J. A.	Regional Geologist	AGC	Eastern Petrol. Geology	1972
Wagner, Miss F. J. E.	Stratigraphic Paleontologist	AGC	Environmental Marine Geol.	1959
Walker, B. A.	Machinist	CL & TS	Mech. Serv. & Instrument Shop	1972
Walker, D. A.	Marine Biologist	AGC	Environmental Marine Geol.	1968
Wallace, Mrs. M. D.	Draftsman	ISPG(G.Info.Proc.)	Cartography	1967
Walter, D. J.	Draftsman	ISPG(G.Info.Proc.)	Cartography	1953
Wanless, G. G.	Instrumentation Scientist	ISPG(Energy)	Geochemistry	1971
Wanless, R. K.	Section Head; Geochronologist	R & EG	Geochronology	1953
Wardle, L. J.	Photomechanical Platemaker	ISPG(G.Info.Proc.)	Cartography	1971
Washkurak, S.	Electronic Technician	RG & G	Magnetic Methods	1952
Watson, D. T. W.	Machinist	ISPG(Admin.)	Bldg. & Engrg. Services	1972
Watson, Mrs. N. F. J.	Laboratory Technician	CL & TS	Analytical Chemistry Labs	1959
Wheeler, J. O.	Division Chief; Cordilleran Geol.	R & EG	Division Office	1951
White, Miss J. I.	Photolab Technician	G.Info.Proc.	Photographic Services	1953
White, Miss M. E.	Typist	Branch Hqrs.	Secretarial Services	1972
Whitehead, A. E.	Lapidary Supervisor	R & EG	Petrology	1953
Whitmore, D. R. E.	Ore Deposits Geologist	R & EG(Econ.Geol.)	Mineral Data Bank	1958
Wilkinson, R. W.	Platemaker	G.Info.Proc.	Cartography Photomech.	1971
Williams, G. K.	Regional Geologist	ISPG(Reg.Geol.)	Southern Mainland	1972
Williams, G. L.	Palynologist	AGC	Eastern Petrol. Geology	1971
Williams, J. B. F.	Supervising Draftsman	G.Info.Proc.	Cartography Unit A	1952
Williams, Miss M. E.	Librarian	G.Info.Proc.	Branch Library	1963
Wilson, G. B.	Typesetter	G.Info.Proc.	Cartography Unit C	1949
Wilson, Mrs. L. D.	Technician	Ter.Science	Paleontol. & Geochronology	1969
Wright, G. M.	Assistant Chief Geologist	Branch Hqrs.	Chief Geologist's Office	1947
Wylie, W. G.	Platemaker	G.Info.Proc.	Cartography Photomech.	1967
Yeager, F. S.	Draftsman	G.Info.Proc.	Cartography Unit C	1959
Yee, Miss I. F.	Cataloguer	G.Info.Proc.	Branch Library	1970
Yelle, J. S.	Draftsman	G.Info.Proc.	Cartography Unit B	1962
Yorath, C. J.	Regional Geologist	ISPG(Reg.Geol.)	Northern Mainland	1967
Young, F. G.	Stratigrapher	ISPG(Reg.Geol.)	Northern Mainland	1969
Young, W. G.	Draftsman	G.Info.Proc.	Cartography Unit C	1966
Zieman, F. W.	Map Compiler	RG & G	Digital Compilation	1958

INDEX